AEROSOL MEASUREME

AEROSOL MEASUREMENT

EDITORS

Dale A. Lundgren
Environmental Engineering Sciences
University of Florida
Gainesville, Florida

Morton Lippmann
Institute of Environmental Medicine
New York University Medical Center
New York, New York

Franklin S. Harris, Jr.
Physics and Geophysical Sciences
Old Dominion University
Norfolk, Virginia

William E. Clark
Environmental Engineering Department
California Polytechnic State University
San Luis Obispo, California

William H. Marlow
Brookhaven National Laboratory
Associated Universities
Upton, New York

Michael D. Durham
Environmental Engineering Sciences
University of Florida
Gainesville, Florida

A University of Florida Book

University Presses of Florida
Gainesville / 1979

Library of Congress Cataloging in Publication Data

Aerosol Measurement Workshop, University of Florida, 1976.
Aerosol measurement.

"A University of Florida book."
Includes bibliographies and index.
1. Aerosols—Measurement—Addresses, essays, lectures.
2. Particle size determination—Addresses, essays, lectures. I. Lundgren, Dale A. II. Title.
TD884.5.A34 1976 628.5′3 78-15424
ISBN 0-8130-0603-1

The University Presses of Florida is the scholarly publishing agency for the State University System of Florida.

PRINTED BY STORTER PRINTING COMPANY, GAINESVILLE, FLORIDA

CONTENTS

Part II: Light Scattering Particle Counters

Part III: Electrical Aerosol Analyzer

Part IV: Condensation Nuclei Counter and Diffusion Battery

PREFACE

In this volume we present a complete text on the basic principles, mechanical operation, calibration, data gathering procedures, data interpretation, instrument limitations, and specific applications of aerosol measurement instrumentation. Information contained herein is not found in normal technical papers or manufacturers' literature. It should enable the engineer and scientist to better understand the application and use of aerosol measurement devices. All answers are not provided because knowledge is incomplete and differences of opinion exist among the leading experts. However, different viewpoints are presented by various writers. Manufacturers point out instrument advantages while users point out shortcomings and limitations. Several articles by users of these instruments present performance data unobtainable elsewhere. Summary articles, complete with extensive literature review lists, enable the reader to become better aware of the extensive published literature.

This book is a result of the interest, support, and hard work of many individuals actively involved in the field of aerosol measurement. The Aerosol Measurement Workshop held at the University of Florida, March 24–26, 1976, could not have been held nor this book published without them. I wish each person and his or her contribution could be individually acknowledged.

A group of about twenty persons, while attending a 1975 Gordon Research Conference on Atmospheric Aerosols, discussed, supported, suggested funding methods, and helped formulate the workshop which led to this book. Three of these individuals, Drs. Morton Lippmann, William Clark, and William Marlow, have assisted from the formulation of the workshop idea through the organization and review of three sections of this volume. A fourth section was organized and reviewed by Dr. Franklin Harris. All sections of this book were assembled, edited, and coordinated with the able assistance of Ms. Ann Root and Ms. Kathy Sheridan.

In addition to the more than fifty persons who authored or co-authored the articles contained herein, many others assisted and contributed through active workshop audience participation which resulted in the discussions, questions, and answers also found in this book.

I have relied heavily on my graduate students to assist me with everything from the initial workshop on through to the final publication of this volume. One student in particular, Dr. Michael Durham, was involved in all aspects of this task, including the book's indexing, during his four year ''apprenticeship,'' and his name deservedly appears as a co-editor.

Special thanks go to Al Breslin and Bob Beadle of the Energy Research and Development Administration, now the U.S. Department of Energy, who helped obtain funding; to Edwin Stokely and John Kane in the Office of Public Affairs, who administered the funding; to Parker Reist and Lawrence Kornreich from the Triangle Universities Consortium on Air Pollution (TUCAP) for providing funds from the Environmental Protection Agency (EPA); to Major Peter Daley for his assistance in obtaining funds from the U.S. Air Force Office of Scientific Research; and to Ronald Vance, who administered these funds which were used to support the Aerosol Measurement Workshop and to help in the preparation of this book. Thanks go to Richard Palmer, Division of Continuing Education, University of Florida, for his help in presenting the workshop and administering these grants.

Receipts from the sale of this royalty-free book will be used to pay all costs of publication, including typesetting, printing, and binding. Final publication was delayed, however, because all costs had to be covered in advance of sale. Therefore, special thanks are extended to Thermo-Systems, Inc. (now TSI), St. Paul, Minnesota, for providing the financial backing required to guarantee this book becoming a reality.

DALE A. LUNDGREN

INTRODUCTION

Fine particle measurement has been an important discipline to many areas of science and industry, and the topic of many excellent books. Articles on the techniques and significance of particle size distribution measurements date back to about 1900. Industries as diverse as paint pigment manufacturing and flour milling have been affected by particle size distribution since their beginnings.

Particle size distribution can be measured in a powder or an aerosol. Measurement in an aerosol differs from that in a powder in several ways and for several reasons. Powders consist of relatively large particles (normally 1–1000 μm diameter) with large quantities of material available for the particle size measurement (1 gram or greater quantities). Aerosols, however, involve much smaller particles covering a wider size range (0.001–100 μm diameter) with much smaller quantities of particulate mass available for an analysis (often 1 milligram or less).

Measurement of the particle size distribution of an aerosol by instrumental techniques is the topic of this book. It is intended as a reference text for those who use or anticipate using commercially available instrumentation to study, measure, monitor, or analyze particle size distribution or size fractionated samples of an aerosol. Four specific techniques are covered: inertial classification (including centrifuges, cyclones and inertial impactors), light-scattering particle counter (single-particle optical counters), electrical aerosol analyzer (electrical mobility analyzer), and condensation nuclei counter–diffusion battery (combined for size analysis). These four techniques are essentially the researcher's only methods for studying the particle size-dependent properties of an aerosol such as formation, condensation/evaporation, or coagulation. As such, they form the basis for discussion.

Sieving, sedimentation, elutriation, simple air classification, and some air permeability and surface area techniques used for size measurement of powders are not directly applicable to aerosol measurement and as such are excluded from this volume. These methods of size classification are discussed in books by R. D. Cadle (1955), C. Orr and J. M. DallaValle (1959), G. Herdan (1960), L. Silverman, C. E. Billings, and M. W. First (1971), T. T. Mercer (1973), R. Dennis (1976), and others (see attached bibliography).

Microscopy, both light and electron, is equally useful in the measurement of powders and aerosols and is very basic to the calibration of other measurement techniques and to the description of particle shape. No measurement instrument is more important, useful or versatile than the light microscope. Similar comments are applicable to the relatively complicated electron microscope and the scanning electron microscope (SEM). Although it is not discussed as a subject in this text, it is assumed that the reader has a basic knowledge of microscopy and understands its uses and limitations. Several of the books referred to above include discussions of microscopy and there are several specific texts on the subject. These works also treat the techniques and problems of aerosol sampling and sample preparation for viewing. (General aerosol literature sources are reviewed later.)

PART I: INERTIAL CLASSIFICATION

A particle size characteristic of primary importance in aerosol studies is the particle aerodynamic diameter—the diameter of a unit density spherical particle which has the same steady state velocity in a gravitational or centrifugal force field as the particle of interest. This diameter allows determination of residence time, potential inhalation hazard, and particle inertial properties such as losses in a pipe bend or impaction against an obstacle in a flow stream. Aerodynamic diameter is a measurement parameter describing how a particle will act, not how it will look (except for a non-volatile, solid, spherical particle of unit density). Because most particles of interest are of irregular shape or of unknown density, it is often difficult to relate this diameter to optical appearance. For example, during the collection of gas-borne particles onto a solid surface suitable for subsequent microscopic viewing, liquid particles may coalesce, volatile particles may evaporate, particles may agglomerate, or agglomerated particles may break up.

Inertial classification devices separate particles from a gas stream but do not, by themselves, measure aerosol concentration or size distribution. This means that gas-borne particles are fractionated, or removed from a gas stream, in an aerodynamic size-dependent manner for subsequent analysis of the known size particulate fraction. Size-fractionated samples containing many particles are obtained in this manner, with the analysis being performed on the gross sample and normally not on single particles.

Aerodynamic diameter classification devices include gravitational sedimentation chambers, air elutriators, aerosol centrifuges, cyclones, and inertial impactors. Sedimentation chambers and air elutriators are frequently used for size-fractionation of aerosolized powders or dusts but not often for ambient concentration aerosol sampling.

Aerosol centrifuges are mechanical devices which involve a rapidly spinning body through which an aerosol sample is flowing at a constant controlled rate. They are capable of size-separating small particles because a very high centrifugal force can be applied to the particles of interest. High rotational speed centrifuges work best in the general size range from about 0.1 to 1.0 μm (aerodynamic) diameter. Classification is also accomplished with high flowrate air centrifuges used by various industries to coarse fractionate powders in the 5–50 μm size range. These devices are not normally used for the study of aerosol concentrations. Aerosol centrifuges as a class of air samplers are discussed in detail in Part I, beginning with the article by M. Tillery. Discussion of the much simpler elutriators can be found in other sources (Mercer, 1973; Silverman, Billings and First, 1971).

Cyclones are similar to centrifuges in that they depend upon centrifugal force for particle separation. Cyclones, however, are simpler mechanical devices which rely upon a rapidly spinning air stream inside a fixed housing for the separation force, as opposed to the mechanically complicated rapidly spinning body of a centrifuge. Particle size separation is not as sharp in an air sampling cyclone as it is in a centrifuge. An optimum size range for cyclones is from 3 to 30 μm (aerodynamic) diameter. Cyclones are especially useful as pre-collectors for other aerosol size measurement devices. J. M. Beeckmans' article introduces a discussion of cyclone design and calibration.

Inertial impactors are the third category of aerodynamic diameter classification devices treated in Part I in a series of articles starting with a report by V. Marple and K. Willeke. Several impaction stages connected in series are normally referred to as cascade impactors. These impactors are intermediate between cyclones and centrifuges in both mechanical complexity and particle size classification, being optimally suited to the 0.5–5 μm (aerodynamic) diameter size range. Through the use of very small dimensions and very high gas velocities, the lower size limit can be decreased below 0.1 μm with reduced pressure; conversely, through the use of very large dimensions, low velocities and high gas flowrates, the upper size limit can be increased to perhaps 50 μm. If it were not for operational problems, cascade impactors could be ideal classification devices because theoretically they can cover a very broad particle size range.

Aerosol centrifuges, cyclones, and inertial impactors are all devices which collect (or remove) gas-borne particles in a size-dependent manner. Samples of particulate matter are then available for subsequent chemical or physical analyses. In general, these devices are not aerosol monitors and are not intended for *in situ* particle size distribution measurement. D. Woods discusses one technique of automating the readout of these devices, and the uses and limitations of inertial classification devices are treated in about a dozen other articles.

PART II: LIGHT SCATTERING PARTICLE COUNTERS

Single particle optical counters provide a very different or supplementary method of aerosol measurement. These particle size distribution monitors do not collect an aerosol sample, except after the measurement. The basic measurement parameter is the amount of light scattered by a particle. This particle is then assigned an ''optical'' size which is equal to the actual size of a spherical calibration particle which scatters an equal amount of light (as measured by the instrument photosensitive detector).

Optical counters have several definite advantages. Even though some are mechanically complicated, they can be very reliable. Electronics are such that these devices can count and optically size gas-borne particles extremely rapidly and with a fair degree of reproducibility.

Although optical counters have been primarily used in clean environments, they can be adapted to fairly dirty environments. Again, there is an optimum particle size range for these devices; most operate best for particles between 0.5 and 5.0 μm diameter. Other models or modifications work well over either a slightly lower or higher size range with no theoretical upper limit, and a lower limit of less than 0.1 μm for the most sophisticated unit. This class of counters fulfills a very important function in the aerosol measurement field for the monitoring of number concentration in a preselected size interval or intervals.

The operating principle, theoretical response, calibration, limitations, and use of single particle optical counters are discussed in a comprehensive article by K. T. Whitby and K. Willeke. Counters manufactured by three different companies are treated in separate articles by A. Lieberman, J. Lepper, Jr., and L. Petralli, Jr. Details of calibration, instrument comparison, field use, and data interpretation are presented in other articles. More novel, but commercially available, optical devices are described by R. Knollenberg and discussed by R. Jeck. In all, the pros and cons of commercially available light-scattering particle counters are discussed in detail.

PART III: ELECTRICAL AEROSOL ANALYZER

The electrical aerosol analyzer is a device especially suited to the measurement of particle size distribution below the range of the optical counter. Size separation is obtained because the electric mobility of a gas-borne particle in an electric field is a function of particle size. This device collects the aerosol during measurement, but it is not a true aerosol sampler. Rather, it is a rapid response (several minutes), aerosol size distribution monitor with good size measurement capability over the general 0.01–0.5 μm (electrical) diameter size range. The electrical aerosol analyzer determines the electric mobility of an unknown particle in a known electric field and then assigns a diameter to the unknown particle equal to the diameter of a known size calibration particle having an equal electric mobility.

Although the first instrument to measure the mean size of an aerosol electrically was built over 50 years ago, it was not until about 1973 that a truly successful commercial electrical aerosol monitor was available. This device, its development, calibration, and use are discussed in detail in Part III.

The history and principle of the electrical aerosol analyzer is first discussed by B. Y. H. Liu, D. Y. H. Pui, and A. Kapadia. The calibration and performance of the instrument are detailed by D. Y. H. Pui and B. Y. H. Liu. Instrument operation and maintenance, a summary of applications, and a complete bibliography are included in G. Sem's article. Additional articles cover theory, data reduction, instrument comparison, and applications to several types of aerosol measurement situations. Comments and discussion of this measurement technique made at the aerosol measurement workshop are also included.

PART IV: CONDENSATION NUCLEI COUNTER AND DIFFUSION BATTERY

As the name implies, a condensation nuclei counter (CNC) is a device to count the number of nuclei which serve as condensation sites for a supersaturated vapor, normally water. The counting can be done manually, as it was in early devices and still is in some units, or automatically by measuring the light extinction caused by the liquid droplets formed in the condensation process. These devices, discussed in detail in Part IV by A. Hogan, T. Rich, and others, are capable of sensing or counting particles greater than about 0.005 μm (diffusional) diameter. This lower detectable particle size is somewhat uncertain but is definitely within the 0.001–0.01 μm diameter range, depending upon the composition of the particles and the instrument operating conditions (such as degree of supersaturation). CNCs are capable of particle size distribution measurements to a limited extent. If several fractions of a stable aerosol are sensed at different supersaturations and a count recorded for each sample, then the differences in number will indicate the number of particles in a calculated size interval. Operation of a CNC will be clarified after reading P. Wagner's article on droplet growth theory, J. Podzimek and J. Kassner, Jr.'s, discussion of basic calibration and E. DeMetre's and J. B. Haberl's descriptions of commercially available instruments.

A main advantage of these devices is their ability to rapidly sense or detect smaller particles at lower concentrations than can be detected by any other device or method. The devices also serve as sensing or counting mechanisms for other methods of particle size classification, namely, the diffusion battery, which preferentially removes the very smallest particles first because of a higher diffusion coefficient.

A diffusion battery is a set of circular or rectangular cross section tubes through which an aerosol stream flows. Several of these cells are normally mounted so that an aerosol stream can be deflected through various combinations of these cells to produce the differential removal as a function of particle size. A

CNC is used at the output of a diffusion battery to provide a readout or measurement of the particle number concentration. In this combination a particle size distribution device is achieved which works well over the 0.01–0.1 μm diameter range even for low aerosol concentrations. This range can be extended almost fivefold in both directions. Above 0.3 μm other techniques work far better. Below 0.01 μm the CNC is the only device capable of detecting low concentration. Therefore, this separation technique has definite applications down to a theoretical lower limit of perhaps 0.002 μm diameter. Design and operation of diffusion batteries are discussed by Sinclair and Countess.

PART V: OTHER METHODS FOR AEROSOL MEASUREMENT

Although this book is specifically concerned with four aerosol measurement methods (based upon commercially available instrumentation) it was also considered desirable to acquaint the reader with several related papers presented in a general session at the Aerosol Measurement Workshop and submitted for publication.

Several remote aerosol sensing techniques are discussed in an article by A. Green. J. Allen, R. Husar, and E. Macias present the possibility of both size and shape determination by laser light scattering. One related size distribution measurement technique based upon the electrical resistance principle is summarized in an article by S. Kinsman. Aerosol sensing based upon contact electricity is discussed by W. John, G. Reischl, and W. Devon. Performance of a new piezoelectric aerosol sensor is the subject of G. Sem and P. Daley's article.

Ideally, it would be desirable to summarize in this one volume all information related to or necessary for a complete understanding of aerosol measurement. Because that was not possible, nor even reasonable to attempt, the following discussion of published literature was prepared.

AEROSOL MEASUREMENT LITERATURE SOURCES

Many books have been published that specifically relate to the field of aerosol measurement. Several of these discuss the related problems of aerosol sampling (a representative sample is necessary in order to obtain a meaningful measurement), aerosol generation (necessary for proper instrument calibration), particle statistics (including presentation of data and calculation of measurement error), aerosol physics (physical laws governing the motion and interaction of gas-borne particles), and aerosol chemistry (formation, reaction, and composition of particles).

To help make the reader aware of books pertaining to the fine particle and aerosol fields, and acquaint him or her with the general content, the following list was prepared; it includes a short discussion of the topics covered, particularly those that relate to aerosol measurement. In addition, a list of journals is appended in which articles related to aerosol measurement would most likely be published. These lists are not complete, but we hope they include the majority of aerosol works published in English in the open literature. Bibliographies accompanying the articles in this book are quite complete for most of the measurement techniques discussed, and the reader should check these lists carefully.

ADDITIONAL REFERENCES

Allen, T., 1968: *Particle Size Measurement.* Chapman and Hall Ltd., London, 248 pp.
Topics include: Particle sampling, particle size distributions, sieving, microscopy, sedimentation, centrifugal methods, Coulter counter use, light scattering, permeametry and gas adsorption.

———, 1975: *Particle Size Measurement.* 2d ed., John Wiley and Sons, New York, 454 pp.
All aspects of particle size measurement with emphasis on powders.

American Conference of Governmental Industrial Hygienists, 1972: *Air Sampling Instruments.* 4th ed. P.O. Box 1937, Cincinnati, Ohio 45201, 587 pp.
A collection of articles dealing with many aspects of gaseous and aerosol sampling equipment and methods. Areas of concern include industrial hygiene, ambient and source sampling, biological aerosols, instrument calibration, flow systems, particle collectors and direct reading devices.

American Society for Testing Materials, 1959: *Symposium on Particle Size Measurement.* ASTM, Philadelphia, 303 pp.
Describes particle size measurement by means of sieving, gravitational and centrifugal sedimentation, light scattering, microscope counting, and using a Coulter counter.

Butcher, S. S. and R. J. Charlson, 1972: *An Introduction to Air Chemistry.* Academic Press, New York, 241 pp.
Included in this introductory text are discussions of sampling and analysis of aerosols and chemical and physical properties of atmospheric aerosols.

Cadle, R. D., 1955: *Particle Size Determination.* Interscience Publishers, Inc., New York, 303 pp.
Principal topics covered are particle statistics, sampling, optical and electron microscopy, sieve analysis, sedimentation, elutriation and surface area measurements (powders). Light scattering is also considered.

———, 1965: *Particle Size.* Reinhold Publishing Corporation, New York, 390 pp.
Text devoted to the practical importance of particle size distributions. General introductory discussion on particle behavior leads to application of basic aerosol principles to air pollution problems and particle deposition in lungs and on the skin. Also covered is the importance of size distribution in the paint, cement, powder, and paper and pulp industries.

———, 1975: *The Measurement of Airborne Particles.* Wiley-Interscience, New York, 342 pp.
Comprehensive coverage of the collection and analysis of aerosols. Topics include: Particle statistics, filtration theory, sedimentation, thermal and electrical precipitation, optical and electron microscopy, centrifugal classifiers, multistage impactors, optical measurement of aerosols, condensation nuclei counters, acoustic counters and electric aerosol analyzers.

DallaValle, J. M., 1943: *Micromeritics.* Pitman Publishing Corporation, New York, 428 pp.
Basic text on the behavior and characteristics of small particles. Includes discussion on particle size measurement, size distributions, dynamics of small particles, characteristics of packings, and physical and chemical properties of aerosols.

Dautrebande, L., 1962: *Microaerosols.* Academic Press, New York, 366 pp.
A summary of the author's and his co-workers' experiences in the generation and measurement of aerosols used to study their physiological and therapeutic effects after deposition in the respiratory tract.

Davies, C. N., Ed., 1961: *Inhaled Particles and Vapours.* Pergamon Press, New York, 495 pp.
Proceedings of an international symposium organized by the British Occupational Hygiene Society 3/29–4/1/60. It contains four papers on the size-selective sampling of airborne dust.

———, 1964: *Recent Advances in Aerosol Research.* Pergamon Press, The Macmillan Company, New York, 80 pp.
A bibliographical review of aerosol research performed between 1957 and 1962.

———, 1966: *Aerosol Science.* Academic Press, New York, 468 pp.
A detailed text composed of chapters written by the leading authorities in the field. Chapters cover aerosol generation, coagulation, electrical behavior, filtration, optical measurement, and adhesion and deposition of particles.

______, Ed., 1967: *Inhaled Particles and Vapours II*. Pergamon Press, New York.
Proceedings of an international symposium organized by the British Occupational Hygiene Society 9/28–10/1/65. It contains six papers on dust exposure evaluations and instruments.

______, 1973: *Air Filtration*. Academic Press, New York, 171 pp.
A complete analysis of the removal of aerosols by fibrous and membrane filters including descriptions of filtration theories and experimental results.

Dennis, R., Ed., 1976: *Handbook on Aerosols*. National Technical Information Service, TID-26608, Virginia, 142 pp.
Subjects discussed are aerosol generation, aerosol dynamics, optical properties, aerosol sampling and particle size measurement. Extensive reference lists are included for each subject as well as a bibliography of aerosol books and periodicals containing aerosol articles.

Drinker, P. and T. Hatch, 1954: *Industrial Dust*. McGraw-Hill Book Company, Inc., New York, 401 pp.
Comprehensive text covering both the engineering and medical aspects of the dynamics, characteristics and control of industrial aerosols.

Friedlander, S. K., 1977: *Smoke, Dust and Haze: Fundamentals of Aerosol Behavior*. John Wiley and Sons, New York, 317 pp.
Written as text for a course in particulate pollution. Topics include size distribution, transport, deposition, optical properties, measurement techniques, aerosol behavior, and gas-to-particle conversion.

Fuchs, N. A., 1959: *Evaporation and Droplet Growth in Gaseous Media*. Pergamon Press, New York, 72 pp.
A complete survey devoted to the kinetics of evaporation and growth of droplets of pure liquids in stationary and nonstationary media.

______, 1964: *The Mechanics of Aerosols*. Pergamon Press, New York, 408 pp.
This classic work reviews almost all the important papers on the subject "Mechanics of Aerosols" up through 1960. Topics covered include rectilinear, curvilinear, and Brownian motion of particles; diffusion and coagulation of aerosols; and the classification, dispersion, and turbulent motion of aerosols.

Green, H. L. and W. R. Lane, 1964: *Particulate Clouds: Dusts, Smokes and Mists*. D. Van Nostrand Company, Inc., New York, 471 pp.
A comprehensive coverage of the formation, behavior and effects of both man-made and natural particulate clouds. Descriptions of aerosol sampling, measurement, industrial collection and the dispersion and effects of particles and clouds in the atmosphere are included.

Hatch, T. F. and P. Gross, 1964: *Pulmonary Deposition and Retention of Inhaled Aerosols*. Academic Press, New York, 192 pp.
A study of aerosol deposition from a viewpoint of both aerosol mechanics and human and animal physiology.

Herdan, G., 1960: *Small Particle Statistics*. Butterworth and Company Ltd., London, 418 pp.
A complete treatment of particle statistics. This work also considers sample preparation and sizing by microscope, gravitational and centrifugal sedimentation (in liquids), surface area measurement methods and general information on powders. Some aerosol measurement discussion.

Hidy, G. M., Ed., 1972: *Aerosols and Atmospheric Chemistry*. Academic Press, New York, 348 pp.
Contains numerous articles presenting a wide variety of studies on the physical chemistry of aerosols and their relationship to atmospheric chemistry. Included is a detailed discussion of the Pasadena Smog Study.

______ and J. R. Brock, 1970: *The Dynamics of Aerocolloidal Systems*. Pergamon Press, New York, 379 pp.
A comprehensive text providing an extensive theoretical analysis of the dynamics of single and collective aerosols.

______ and ______, 1971: *Topics in Current Aerosol Research*. Pergamon Press, New York, 157 pp.
Contains two articles—the first is an analysis by N. A. Fuchs and A. G. Sutugin on the formation and physical properties of high-dispersed aerosols with particle size below 0.1 μm. The second is an article by S. L. Soo on the transport in multicomponent gas systems.

Intersociety Committee, 1972: *Methods of Air Sampling and Analysis*. American Public Health Association, Washington, D.C., 480 pp.
Describes the EPA recommended methods for the sampling of gases and aerosols.

Irani, R. R. and C. F. Callis, 1963: *Particle Size: Measurement, Interpretation and Application*. Wiley, New York, 165 pp.

Detailed descriptions of the proper use of microscopy, sieving, light scattering, turbidity, sedimentation and Coulter counters in the analysis of particle size distributions.

Junge, C. E., 1963: *Air Chemistry and Radioactivity*. Academic Press, New York, 382 pp.
Text dealing with the constituents and chemical processes in the lower 50 km of the atmosphere with an emphasis on basic and large scale phenomena. Analysis of the physical properties, chemical composition and distribution of aerosols in the troposphere is presented along with a similar analysis of the gaseous constituents and precipitation.

Kerker, M., Ed., 1963: *Electromagnetic Scattering*. Pergamon Press, The Macmillan Company, New York, 592 pp.
A detailed treatment of electromagnetic scattering with extensive references. Basic treatment is for spheres, but chapters also on stratified spheres and infinite cylinders, particle sizes and applications to polydisperse systems. Applications are primarily to physical chemistry problems in liquids rather than to the atmosphere and air pollution.

———, 1969: *The Scattering of Light and Other Electromagnetic Radiation*. Academic Press, New York, 666 pp.
Summary of the theory with applications to light scattering.

Liu, B. Y. H., Ed., 1976: *Fine Particles. Aerosol Generation, Measurement, Sampling and Analysis*. Academic Press, New York, 837 pp.
Proceedings of the 1975 Symposium on Fine Particles. Contains numerous articles on aerosol generation, aerosol sampling, aerosol measurement and analysis.

Lundgren, D. A., M. Lippmann, F. S. Harris, Jr., W. E. Clark, W. H. Marlow, and M. D. Durham, Eds., 1979: *Aerosol Measurement*. University of Florida Press.
Detailed coverage of four specific aerosol measurement techniques: inertial classification (centrifuges, cyclones, and impactors), light scattering particle counters, electrical aerosol analyzers, and condensation nuclei counters together with diffusion batteries. Numerous articles on each topic cover basic principle, mechanical operation, calibration, data gathering procedures, data interpretation, instrument limitations and specific application.

McCartney, E. J., 1976: *Optics of the Atmosphere. Scattering by Molecules and Particles*. John Wiley and Sons, New York, 408 pp.
Good summary of scattering processes and effect on light propagation.

Mednikov, E. P., 1965: *Acoustic Coagulation and Precipitation of Aerosols*. Consultants Bureau, New York, 180 pp. (Translation from Russian by C. V. Larrick.)
This text treats the problem of artificial coagulation of an aerosol through use of high-intensity sonic and ultrasonic vibrations.

Mercer, T. T., 1973: *Aerosol Technology in Hazard Evaluation*. Academic Press, New York, 394 pp.
Comprehensive text on aerosol generation, sampling and measurement. General discussion of aerosol properties, particle statistics, aerosol sampling for size and concentration measurement, instruments for aerodynamic, optical, electrical and other size measurements, respirable samplers and production of test aerosols.

Orr, C., Ed., 1977: *Filtration, Principles and Practices, Part I*. Marcel Dekker, Inc., New York.
Articles on aerosol filtration and other topics.

Orr, C., Jr. and J. M. DallaValle, 1959: *Fine Particle Measurement*. The Macmillan Co., New York, 353 pp.
Topics included are microscopy, particle statistics, sieving, sedimentation, inertial classification, light scattering, surface area measurement by flow resistance, gas adsorption and liquid-phase sorption, and pore size measurement (powders).

Shaw, D. T., Ed., 1978: *Fundamentals of Aerosol Science*. Wiley, New York, 372 pp.
Topics covered include aerosol impactors, Brownian coagulation, droplet evaporation, aerosol filtration theory, sampling of fibrous aerosol particles and electrostatics in aerosol filtration.

Silverman, L., C. E. Billings and M. W. First, 1971: *Particle Size Analysis in Industrial Hygiene*. Academic Press, New York, 317 pp.
Complete coverage of particle sampling and size analysis for both powders and aerosols. Review of sampling instruments, size analysis by optical, electron and scanning electron microscopy. Discussion of all common particle sizing apparatus (both manual and automatic), particle statistics and application areas.

Spurny, K., 1966: *Aerosols: Physical Chemistry and Applications*. Gordon and Breach, Science Publishers, New York, 943 pp.
The proceedings of an international conference in Czechoslovakia in 1962. Comprised of 943 pages—98 papers in English, 35 in German, and 7 in French. The subjects covered on aerosols are

physical chemistry, radioactivity, meteorology and astronomy, industry, agriculture, and biological effects.

Twomey, S., 1977: *Atmospheric Aerosols.* Elsevier Scientific Publishing Co., New York, 302 pp.
This text covers aspects of aerosol physics important in the formation, evolution and removal of particulate material in the atmosphere. Chapters are devoted to the influence of atmospheric particles on weather, atmospheric optics and radioactive transfer, atmospheric electricity and atmospheric energetics and climate.

van de Hulst, H. C., 1957: *Light Scattering by Small Particles.* Wiley, New York, 470 pp.
The most important and detailed book on single scattering by homogeneous spheres. A masterly presentation including mathematics, physical insight, critical discussion of literature and applications to chemistry, meteorology and astronomy.

Walton, W. H., Ed., 1971: *Inhaled Particles III. Volumes I & II.* Unwin Brothers Ltd., Surrey, England, 1090 pp.
Proceedings of an international symposium organized by the British Occupational Hygiene Society 9/14–9/23/70. Volume II contains seven papers on dust sampling and standards.

JOURNALS

American Industrial Hygiene Association Journal
Annals of Occupational Hygiene
Applied Optics
Atmospheric Environment
British Journal of Applied Physics
Environmental Science and Technology
Health Physics
Journal of Aerosol Science
Journal of the Air Pollution Control Association
Journal of Applied Meteorology
Journal of Colloid and Interface Science
Journal of Geophysical Research
Journal of the Optical Society of America
Journal of Scientific Instruments
Powder Technology
Review of Scientific Instruments
Science
Staub-Reinhalt Luft (English Translation)

CONTRIBUTORS

ALLEN, J., AeroChem Research Laboratories, Inc., P.O. Box 12, Princeton, N.J. 08540

ANDERSON, J. A., Meteorology Research, Inc., 464 Woodbury Road, Altadena, California 91001

AVOL, E. L., Rancho Los Amigos Hospital, 7601 E. Imperial Highway, Downey, California 90242

AYER, H. E., Kettering Laboratory, University of Cincinnati Medical Center, 3223 Eden Avenue, Cincinnati, Ohio 45267

BEECKMANS, J. M., Faculty of Engineering Science, The University of Western Ontario, London, Ontario, N6A 5B9, Canada

CAHILL, T. A., Physics Department, University of California, Davis, California 95616

CANTRELL, B. K., Particle Technology Laboratory, Mechanical Engineering Department, University of Minnesota, Minneapolis, Minnesota 55455

CHRISTY, D. A., Health and Safety Laboratory, U.S. Energy Research and Development Administration, New York, New York 10014

CLARK, W. E., Environmental Engineering Department, California Polytechnic State University, San Luis Obispo, California 93407

COUNTESS, R. J., Environmental Measurement Laboratory, U.S. Department of Energy, New York, New York 10014

DALEY, P. S., Major, Air Force Civil Engineering Center, Tyndall AFB, Florida 32401

DAVIS, J. G., Brookhaven National Laboratory, Department of Applied Science, Upton, New York 11973

DEMETRE, E. A., Energy Systems Programs Department, General Electric Company, 1 River Road, Schenectady, New York 12345

DEVOR, W., Air and Industrial Hygiene Laboratory, California State Department of Health, 2151 Berkeley Way, Berkeley, California 94704

ENSOR, D. S., Meteorology Research, Inc., 464 Woodbury Road, Altadena, California 91001

GERBER, H. E., Naval Research Laboratory, Washington, D.C. 20375

GREEN, A. E. S., Department of Physics and Astronomy, University of Florida, Gainesville, Florida 32611

GROBLICKI, P. J., GM Research Laboratories, Environmental Science Department, GM Technical Center, Warren, Michigan 48090

HABERL, J. B., General Electric Company, 100 Plastics Avenue, Pittsfield, Massachusetts 01201

HARRIS, JR., F. S., Rockville, Utah 84763

HO, A. T., Air Pollution Health Effects Lab., Department of Community and Environmental Medicine, University of California, Irvine, California 92717

HOCHSTRASSER, J. M., Industrial Hygiene, Tenneco Chemicals, Inc., Park 80, Plaza West, 1, Saddle Brook, New Jersey 07663

HOGAN, A. W., Atmospheric Sciences Research Center, State University of New York at Albany, 1400 Washington Avenue, Albany, New York 12222

HUSAR, R. B., Air Pollution Research Lab., Washington University, St. Louis, Missouri 65830

JECK, R. K., Naval Research Laboratory, Washington, D.C. 20375

JOHN, W., Air and Industrial Hygiene Laboratory, California State Department of Health, 2151 Berkeley Way, Berkeley, California 94704

KAPADIA, A., Particle Technology Laboratory, Mechanical Engineering Department, University of Minnesota, Minneapolis, Minnesota 55455

KASSNER, JR., J. L., Graduate Center for Cloud Physics Research, University of Missouri-Rolla, Rolla, Missouri 65401

KENOYER, J. L., Air Pollution Health Effects Laboratory, Department of Community and Environmental Medicine, University of California, Irvine, California 92717

KINSMAN, S., Coulter Electronics, Inc., 590 W. 20th Street, Hialeah, Florida 33010

KITTELSON, D. B., Mechanical Engineering Department, University of Minnesota, Minneapolis, Minnesota 55455

KNOLLENBERG, R. G., Particle Measuring Systems, Inc., 1855 S. 57th Court, Boulder, Colorado 80301

KNUTH, R. H., Environmental Measurement Laboratory, Department of Energy, 376 Hudson Street, New York, New York 10014

KOCMOND, W. C., Desert Research Institute, University of Nevada System, P.O. Box 60220, Reno, Nevada 89507

LANGER, G., National Center for Atmospheric Research, Boulder, Colorado 80302

LEPPER, JR., J. M., Climet Instruments Company, 1320 W. Colton Avenue, P.O. Box 151, Redlands, California 92373

Lieberman, A., Royco Instruments, Inc., 141 Jefferson Drive, Menlo Park, California 94025

Lippmann, M., Institute of Environmental Medicine, New York University Medical Center, 550 First Avenue, New York, New York 10016

Liu, B. Y. H., Mechanical Engineering Department, University of Minnesota, Minneapolis, Minnesota 55455

McCain, J. D., Southern Research Institute, 2000 Ninth Avenue S., Birmingham, Alabama 35205

Macias, E. S., Department of Chemistry, Washington University, St. Louis, Missouri 65830

Marlow, W., Brookhaven National Laboratory, Department of Energy and Environment, Associated Universities, Upton, New York 11973

Marple, V. A., Mechanical Engineering Department, University of Minnesota, Minneapolis, Minnesota 55455

Martens, A. E., Bausch & Lomb, 820 Linden Avenue, Rochester, New York 14625

Mewhinney, J. A., Inhalation Toxicology Research Institute, Lovelace Foundation, P.O. Box 5890, Albuquerque, New Mexico 87115

Moss, O. R., Battelle PNW-Biology Department, Battelle Boulevard, P.O. Box 999, Richland, Washington 99352

Munkelwitz, H. R., Brookhaven National Laboratory, Department of Applied Science, Upton, New York 11973

Nader, J. S., Environmental Sciences Research Laboratory, U.S. Environmental Protection Agency, Research Triangle Park, North Carolina 27711

Petralli, Jr., L. J., Met One, 154 San Lazaro Avenue, Sunnyvale, California 94086

Phalen, R. F., Air Pollution Health Effects Laboratory, Department of Community and Environmental Medicine, University of California, Irvine, California 92717

Podzimek, J., Graduate Center for Cloud Physics Research, University of Missouri-Rolla, Rolla, Missouri 65401

Pui, D. Y. H., Mechanical Engineering Department, University of Minnesota, Minneapolis, Minnesota 55455

Raabe, O. G., Radiobiology Laboratory, University of California, Davis, California 95616

Ragland, J. W., Southern Research Institute, 2000 Ninth Avenue S., Birmingham, Alabama 35205

Rao, A. K., Monsanto Company, 800 N. Lindbergh Boulevard, St. Louis, Missouri 63166

Reischl, G. P., Air and Industrial Hygiene Laboratory, California State Department of Health, 2151 Berkeley Way, Berkeley, California 94704

Rich, T. A., Deceased

Richards, L. W., Air Monitoring Center, Rockwell International, 2421 W. Hillcrest Drive, Newbury Park, California 91320

RIMBERG, D., North American Pemco, Inc., P.O. Box 655, Bardonia, New York 10954

SEM, G. J., TSI, Inc., P.O. Box 3394, St. Paul, Minnesota 55165

SENTELL, R. J., Tennecomp Systems Inc., P.O. Box J, Oak Ridge, Tennessee 37830

SINCLAIR, D., Environmental Measurement Laboratory, U.S. Department of Energy, 376 Hudson Street, New York, New York 10014

SNYDER, K., Drexel University, Philadelphia, Pennsylvania 19014

SOO, S. L., Department of Mechanical and Industrial Engineering, University of Illinois at Urbana-Champaign, Urbana, Illinois 61801

STOREY, R. W., NASA Langley Research Center, Hampton, Virginia 23665

TALLEY, D., Los Alamos Scientific Laboratory, P.O. Box 1663, Los Alamos, New Mexico 87545

TANG, I. N., Brookhaven National Laboratory, Department of Applied Science, Upton, New York 11973

TILLERY, M. I., Los Alamos Scientific Laboratory, University of California, Los Alamos, New Mexico 87545

WAGNER, P. E., Institut Fo Experimentalphysik, Der Universitat Wien, A-1090 Wien, Strudlhofgasse 4, Vienna, Austria

WHITBY, K. T., Particle Technology Laboratory, Mechanical Engineering Department, University of Minnesota, Minneapolis, Minnesota 55455

WILLEKE, K., Department of Environmental Health, University of Cincinnati, 3223 Eden Avenue, Cincinnati, Ohio 45267

WOODS, D. C., Langley Research Center, Hampton, Virginia 23665

YANG, J. Y., Union Carbide, Linde Division, 61 E. Park Drive, Tonowanda, New York 14150

PART I

Inertial Classification

AEROSOL CENTRIFUGES

MARVIN I. TILLERY

Los Alamos Scientific Laboratory, University of California, Los Alamos, New Mexico

The characteristic of primary interest in determination of airborne residence time and potential inhalation hazard of airborne particles is gravitational or terminal settling velocity. This velocity is the result of gravitational force and drag exerted on the moving particle by air. Interaction between a moving particle and air also defines the particle's dynamic or inertial properties. Thus, particle characteristics that determine sedimentation velocity can, in most cases, also be used to define the particle's inertial properties. Sedimentation velocity of spherical particles of known density can be calculated if the particle diameter is known. However, in many cases of interest, particles are irregular in shape and of unknown density. In this case, the sedimentation velocity or a characteristic related to sedimentation velocity must be measured. A hypothetical diameter can then be used to define and compare inertial properties. The aerodynamic diameter or the diameter of a unit density spherical particle with the same sedimentation velocity as the particle of interest is the most frequently used hypothetical diameter (Mercer, 1973).

Aerosols are usually encountered as a distribution of different size particles. The distribution of the parameters of interest with respect to some particle characteristic must be used to describe the behavior and potential effects of the particles. Distribution of number, mass, surface area, radioactivity, or other property of interest with respect to aerodynamic diameter is often used to describe the range and magnitude of inertial properties. Sampling instruments that separate particles according to their inertial properties permit direct determination of these distributions.

Aerosols are relatively stable suspensions of airborne particles so gravitational

Work performed at the Los Alamos Scientific Laboratory under the auspices of the U.S. Energy Research and Development Administration. Contract No. 7405-ENG-36.

sedimentation velocities are low. Accurate determination of these velocities requires long times with stable atmospheres, accurate measurement of small distances, or use of an increased deposition force. Centrifugal force is often used to increase the deposition rate and provide accurate measurements in reasonable times. Centrifugal sampling instruments normally determine deposition velocity by determining penetration distance of particles passing through a duct that is perpendicular or nearly perpendicular to the deposition force. If the inlet of the entire duct is filled with dusty air, a mixed deposit occurs with a continual

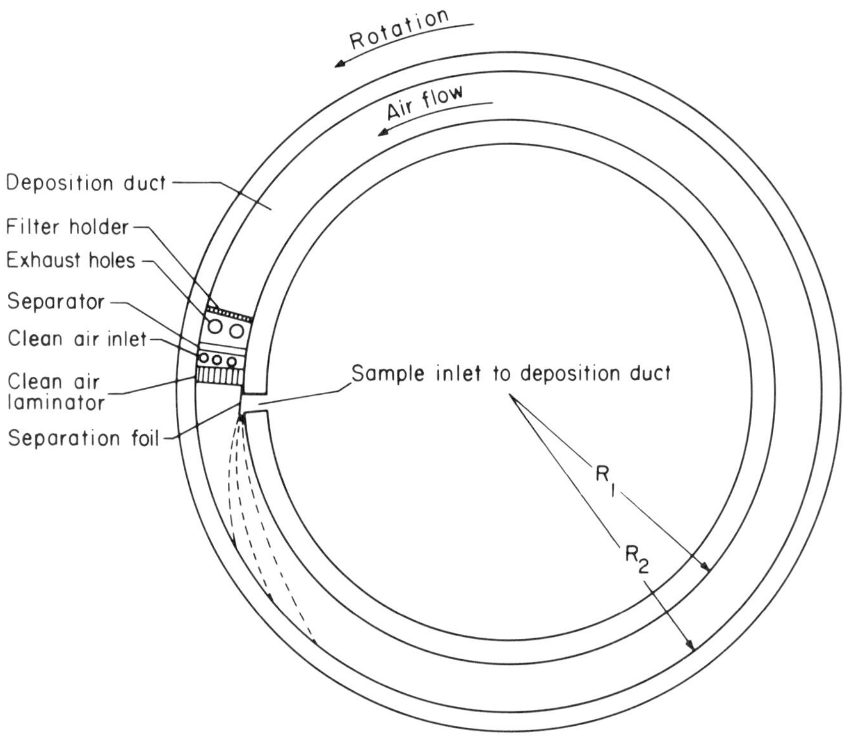

Figure 1. Top view of cylindrical duct aerosol spectrometer.

decrease in the deposit of large particles along the duct. Instruments of this type are called "semidispersive aerosol centrifuges." Centrifugal size classification instruments in which the dusty air enters the duct in a thin layer covered by a much thicker layer of clean air are called "aerosol spectrometers." Spectrometers provide a continual separation and grading of particles according to deposition velocity and are currently the most widely used aerosol centrifuges. A variety of centrifugal aerosol spectrometers have been built emphasizing specific aspects of the operating characteristics. Operating principles of all these instruments are similar. Thus, performance characteristics and operational limitation can be illustrated by considering a simple instrument.

Figure 1 shows the top view of a cylindrical duct aerosol spectrometer. Several instruments of this type have been built (Hochrainer, 1971; Matteson, Boscoe and Preining, 1974; Anderson, 1976; Tillery, 1974). In most of these instruments the depth of the duct (h) is greater than the width in the radial direction ($R_2 - R_1$). The elutriating layer of clean air is introduced into the deposition duct through a laminating device. Air is drawn down the duct by applying a vacuum behind a filter at the end of the duct. Sampling rate is regulated by drawing more air from the duct than is introduced through the clean air inlet. The deficit enters the duct as a thin layer of dusty air next to the inner wall. Flow in the duct is laminar so particles tend to remain next to the inner wall. Flow through the curving duct, and rotation of the duct results in a centrifugal force field and a radial acceleration of the particles. Instantaneous radial velocity can be represented by ignoring acceleration and equating centrifugal force to the resistance given by Stokes law:

$$\frac{dR}{dt} = \frac{\rho C D^2 \, 2\pi^2 \upsilon^2 R}{9\eta}, \qquad (1)$$

where R = radial position, t = time, ρ = particle density, C = Cunningham Slip Correction, D = particle diameter, υ = rotational rate in revolutions per unit time, and η = viscosity of duct gas.

Penetration distance (ℓ), or distance particles will travel down the duct from inlet to deposition, can be determined for particles of a given size by assuming a velocity profile across the duct given by

$$\frac{d\ell}{dt} = f(R, b), \qquad (2)$$

where b = vertical position in the duct, and ℓ = distance from the sample inlet. For tall narrow ducts of height h, penetration distance at the center line (b = h/2) can be determined by assuming that the velocity profile is parabolic across the radial dimension with a maximum at the center $(R_2 + R_1)/2$, or

$$\left(\frac{d\ell}{dt}\right)_{\frac{R_2 + R_1}{2}} = 6\bar{U}\left\{\frac{R - R_1}{R_2 - R_1} - \frac{(R - R_1)^2}{(R_2 - R_1)^2}\right\}, \qquad (2a)$$

where $\bar{U} = Q/h(R_2 - R_1)$ = average velocity in the duct, and Q = volumetric flow rate.

Maximum penetration distance (ℓ_D) for particles of diameter D results from particles entering the duct at R_1. Thus, ℓ_D is determined by solving Eqs. (1) and (2a) for dt and integrating across the velocity profile from R_1 to R_2, i.e.,

$$\ell_D = \frac{9\eta \, K \, \bar{U}}{\pi^2 \rho C D^2 \upsilon^2}\left\{\frac{R_2 + R_1}{(R_2 - R_1)} + \frac{2R_1 R_2}{(R_2 - R_1)^2} \ln\left(\frac{R_1}{R_2}\right)\right\}, \qquad (3)$$

K = ratio of center line velocity to average velocity (1.5 for tall narrow ducts). For a given instrument this can be reduced to

$$\ell_D = \frac{9\eta}{\pi^2}\left(\frac{QI}{\rho C D^2 v^2}\right), \tag{3a}$$

where I is the instrument geometrical constant given by

$$I = \frac{K}{h}\left\{\frac{R_2 + R_1}{(R_2 - R_1)^2} + \frac{2R_1R_2}{(R_2 - R_1)^3}\ln\left(\frac{R_1}{R_2}\right)\right\}. \tag{3b}$$

Figure 2 is a plot of penetration distance against $QI/\rho CD^2v^2$ and experimental data from three different cylindrical instruments. Good agreement is obtained for the instruments having rectangular deposition ducts (Hochrainer, 1971; Anderson, 1976; Tillery, 1974). In the square duct, instrument penetration is somewhat greater than predicted (Matteson, Boscoe and Preining, 1974). Eq. (3a) illustrates some of the operational limitations. The total flow rate Q must be limited to prevent particles of interest from penetrating the deposition duct. Small particle collection requires long ducts because penetration distance is inversely proportional to particle diameter squared.

Eq. (3a) indicates that high flow rates and small particle sizes can be compensated for if sufficient rotational speed is used. However, in addition to mechanical difficulties involved with high rotational rates, secondary flows develop as a result of Coriolis forces and inertia in the rotating curved duct (Stöber and Flachsbart, 1969b). These secondary flows cause localized displacement of the dusty air layer resulting in distorted deposition patterns and disagreement between deposition points and calibration curves. Secondary flows develop as vortices, with opposite directions of rotation in the top and the bottom of the duct. Depending on the relationship between direction of flow in the duct and rotation of the duct, the direction of secondary flow in the center of the channel will be toward the deposition wall or toward the center of rotation. When the period of the vortex flow is approximately equal to or greater than the time required for the sample to pass through the deposition duct, the deposition patterns on the outer wall will not be seriously disturbed. However, the deposition rate in the center of the channel will be altered depending on the direction of the vortex. This results in deviations from predicted penetration distance. This effect is most important for small particles collected at the end of long ducts. When the period of the vortex flow becomes significantly less than transit time through the duct, marked distortions can occur in the deposits, making it difficult or impossible to relate deposited and sampled concentrations as a function of particle size. The development of secondary flows that distort deposition patterns limits the range of permissible flow rates and rotational speeds.

Deposition patterns on the filter located at the end of the duct (Figure 3) illustrate the location of secondary vortices at the top and bottom of the duct (Tillery, 1974). In this case, the secondary flows did not seriously disrupt deposi-

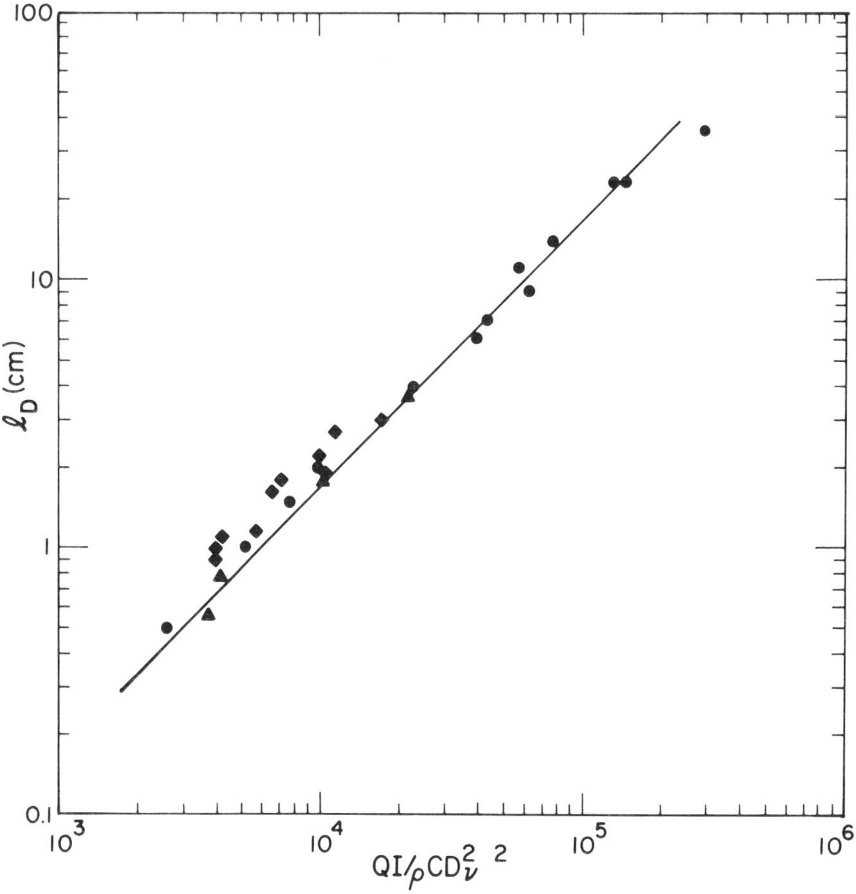

Figure 2. Penetration distance as a function of $QI/\rho CD^2\nu^2$.

Figure 3. Filter deposit from concentric instrument.

tion patterns on the outer wall. A discussion of secondary flow effects in aerosol centrifuges and illustrations of distorted deposition patterns are given in several papers (Stöber and Flachsbart, 1969b; Stöber, 1976; Stöber, Flachsbart and Boose, 1972; Stöber and Boose, 1973).

Assumptions used in derivation of Eq. (3), and failure to consider the effect of secondary flows, limit the accuracy of these analytical relationships. However, Eq. (3) defines parameters that regulate performance of these instruments and is adequate for design purposes. An empirical calibration should be used for determination of aerosol size distributions. Many instruments in common use have deposition duct shapes that prohibit mathematical analysis (Stöber and Flachsbart, 1969b; Moss, Ettinger and Coulter, 1972; Kotrappa and Light, 1972a); these instruments must be calibrated experimentally. Calibration is often accomplished by sampling an aerosol produced by nebulizing an aqueous suspension of monodisperse polystyrene latex particles. Evaporation of the droplets leaves aerosol particles consisting of multiples of the fundamental particle size.

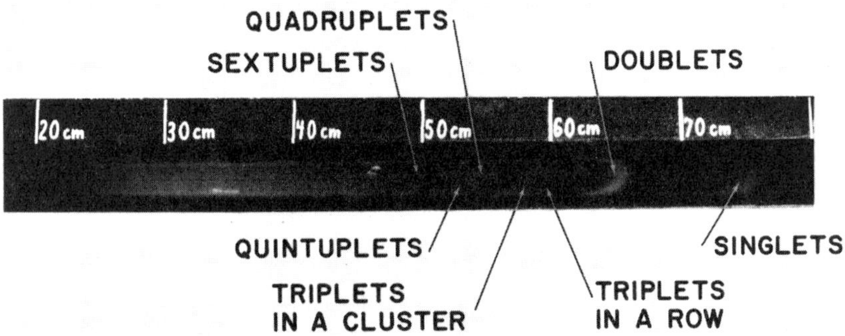

Figure 4. Deposition of polystyrene latex particles in a spinning, spiral duct spectrometer.

These particles will be collected in discrete bands in a spectrometer. Samples are normally collected on metal or paper foils lining the outer wall of the deposition duct. Figure 4 shows a foil with a collection of polystyrene latex particles. The aerodynamic diameters of aggregates of uniform sized spheres as a function of the number and configuration of particles in the aggregate have been determined by Stöber, Berner and Blaschke (1969). These values can be used with measurement of penetration distance on the foil to determine aerodynamic diameter (D_A) as a function of penetration distance. A calibration curve determined in this manner is shown in Figure 5 (Moss, Ettinger and Coulter, 1972).

The resolution of aerosol spectrometers is normally defined in terms of the ratio of the range of particle diameters found at a location to the diameter expected at the location ($\delta D/D$), and is a function of the deposit width for monodisperse particles. The deposit width for monodisperse particles results from particles having different radial positions at the duct inlet. Particles entering the duct at the inner wall ($R = R_1$) will penetrate the farthest as they have the greatest distance to travel to the deposition surface. Particles of the same size entering the deposition duct at $R = R_1 + \Delta$, where Δ is the thickness of the dusty air layer, will penetrate a shorter distance. An expression for the resolution can be obtained by differentiating Eq. (3) with respect to R_1 and D

$$\frac{\delta D_A}{D_A} = \frac{\delta \ell_D}{2\ell_D} + \frac{2R_2(R_2 - R_1) + R_2(R_2 + R_1)\ln(R_1/R_2)}{R_2^2 - R_1^2 + 2R_1R_2\ln(R_1/R_2)} \cdot \frac{\delta R_1}{(R_2 - R_1)}. \quad (4)$$

An estimate of the ratio of width of deposit to penetration distance for particles of the same size can be obtained by determining penetration distance for particles

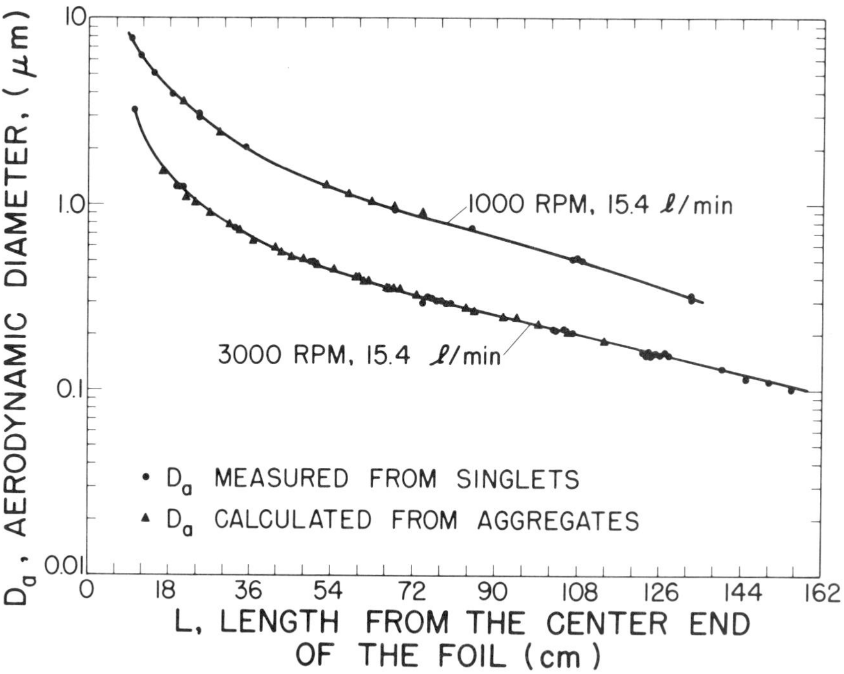

Figure 5. Calibration curve for spinning, spiral duct spectrometer.

entering the duct at R_1 and at $R_1 + \Delta$ in the same manner that Eq. (3) was determined, giving

$$\frac{\Delta \ell}{\ell_D} = \frac{2R_2\Delta - \Delta^2 + 2R_1R_2\ln\left(\frac{R_1}{R_1 + \Delta}\right)}{R_2^2 - R_1^2 + 2R_1R_2\ln(R_1/R_2)}. \quad (5)$$

The size of Δ depends on the ratio of the sampling rate S to the total flow rate Q. A reasonable approximation for the velocity profile given in Eq. (2a) over the range $0.02 \leqslant S/Q \leqslant 0.2$ is given by

$$\Delta = 0.612\left(\frac{S}{Q}\right)^{1/2}(R_2 - R_1). \quad (6)$$

Using this approximation of Δ in Eq. 5 for the instruments described by Hochrainer (1971) and Tillery (1974) indicates that

$$\frac{\Delta \ell}{\ell} D \simeq \frac{S}{Q}, \text{ when } 0.02 \leq \frac{S}{Q} \leq 0.2. \quad (7)$$

Thus, assuming δR is equal to the dusty air layer thickness, spectrometer resolution is a function of the ratio of sampling rate to total flow rate, i.e.,

$$\frac{\delta D_A}{D_A} \simeq \frac{-S}{2Q} + 0.612 \left\{ \frac{2R_2(R_2 - R_1) + R_2 (R_2 + R_1) \ln (R_1/R_2)}{R_2^2 - R_1^2 + 2R_1R_2 \ln (R_1/R_2)} \right\} \left(\frac{S}{Q}\right)^{1/2}. \quad (8)$$

A plot of resolution for two different sizes of spectrometers is given in Figure 6. Resolution improves as $R_2 - R_1$ decreases. This effect is most important for smaller instruments. The circles on the curve for $R_2 = 7.0$ cm are two relative standard deviations for particles found on a one-eighth inch diameter electron microscope grid located in the center of the collection foil (Tillery, 1974). The experimental resolution is better than predicted. This would be expected, since in the laminar flow velocity profile, fewer particles enter next to the inner wall, so the thickness of the dusty air layer (Δ) is effectively smaller than predicted.

The radial position of the particles at the end of the duct depends on deposition rate, resulting in size separation on the filter located at the end of the duct (Figure 1). This separation is not amplified by air flow normal to the deposition direction. Separation will thus be less with more overlap. An estimate of the width of the deposit can be made by calculating the radial position for particles entering the duct at the extremes of the dusty air layer (R_1, $R_1 + \Delta$). A plot of the fraction of the duct width in which particles are found as a function of particle size is given in Figure 7. Both curves are for an instrument having R_2 equal to 7 cm, operated at 3000 RPM, with a total flow rate of 5 ℓpm and a sampling rate of 0.5 ℓpm. With these operating conditions, particles having $\sqrt{\rho C}D = 0.5$ μm will be deposited on both the filter and the outer wall of the duct. The curve indicates poor separation of larger particles that are more than halfway across the duct. Particles larger than 0.25 μm in diameter will be deposited across more than 30% of the filter with the operating conditions described above. Half of the deposit of 0.3 μm particles will overlap the deposit of 0.4 μm particles. The separation of small particles found close to the inner wall will be better. The thickness of the dusty air layer for an S/Q ratio of 0.10 is about 19% of the duct width. The filter deposit thickness for equal size small particles tends to be less than this value because particles entering the duct at R_1 have a lower velocity down the duct, thus remaining in the centrifugal force field for a longer period of time than particles entering at $R_1 + \Delta$.

Penetration distance in aerosol spectrometers having ducts of more complicated geometry cannot be determined analytically. However, the parameters that affect performance of these instruments can be determined from the equations for the cylindrical instrument. The form of I in Eq. (3a) will vary with deposition duct geometry. The penetration distance for all spectrometers of this type will be directly proportional to flow rate through the duct and inversely proportional to the square of the rotational rate and particle diameter. The resolution of spectrometers is dependent on the ratio of the sampling rate to the total flow rate. Resolution also depends on the width of the deposition duct. However, this effect is not very important for instruments having R_2 greater than 5.0 cm.

The first centrifugal aerosol spectrometer was built in 1950 by Sawyer and Walton (1950). A schematic view of this instrument called the Conifuge is shown in Figure 8. The instrument consists of a narrow annulus between two right

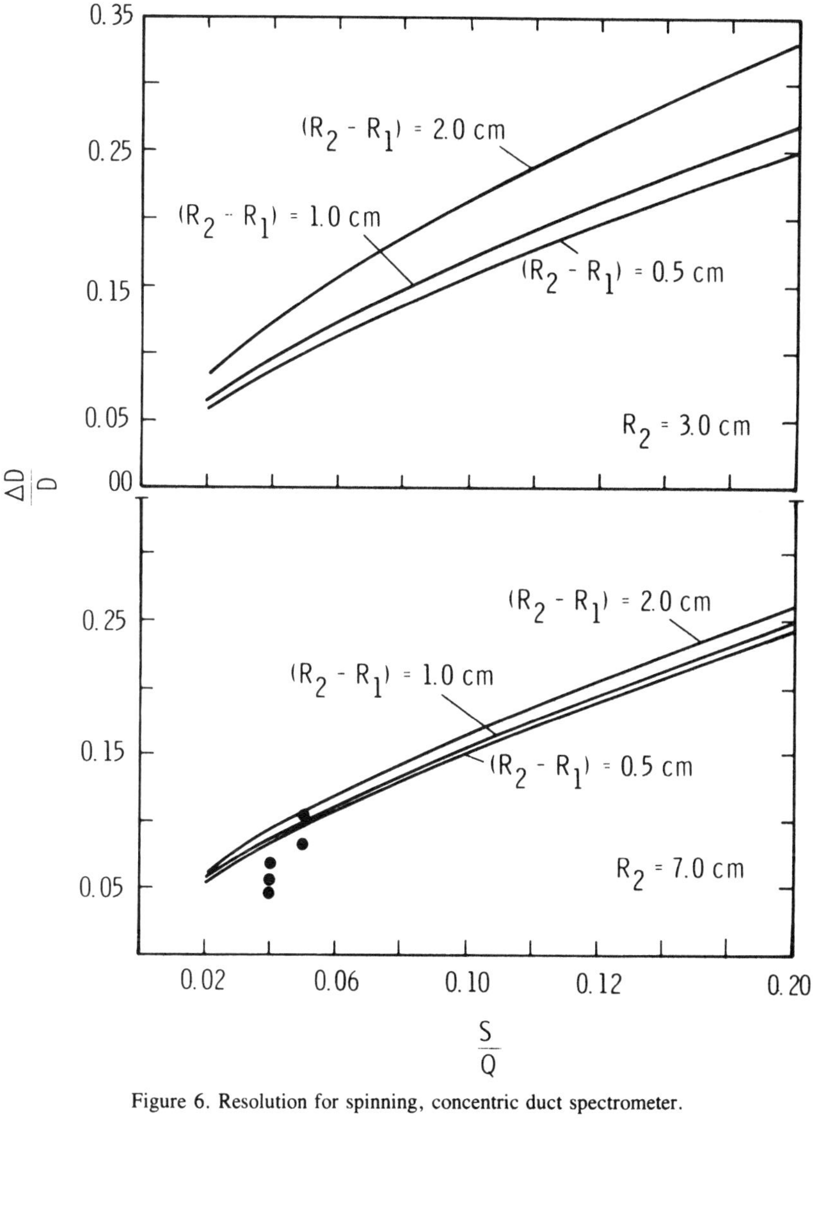

Figure 6. Resolution for spinning, concentric duct spectrometer.

cones. The cones, in a sealed chamber, are rotated by an electric motor. The rotating cones function as a centrifugal pump pulling air in the top and out through jets located around the cone base. Clean air recirculates through the sealed chamber and the top of the cones. When air is drawn from the sealed chamber, an equal amount of dusty air comes down the aerosol inlet, spreads out in a thin layer over the inner cone, and starts down the annulus covered by a much thicker layer of clean air. In the annulus, the particles are subjected to a centrifugal force accelerating the particles toward the outer wall as they pass down the channel. All particles have essentially the same distance to travel to the outer wall so they are deposited in a continuously graded spectrum according to radial velocity. Velocity down the channel decreases with distance due to the increase in cross-sectional area as centrifugal force increased due to greater

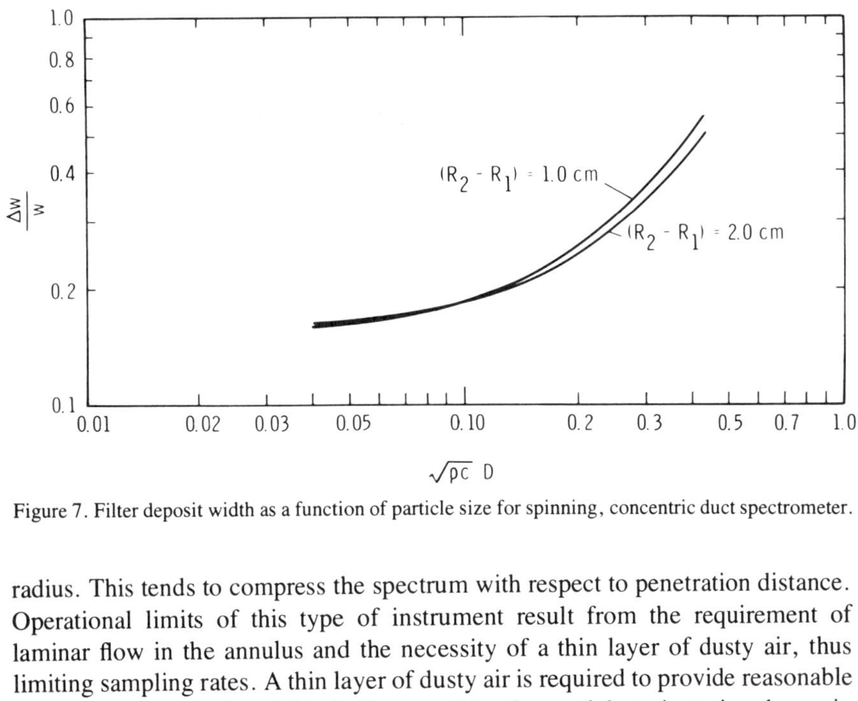

Figure 7. Filter deposit width as a function of particle size for spinning, concentric duct spectrometer.

radius. This tends to compress the spectrum with respect to penetration distance. Operational limits of this type of instrument result from the requirement of laminar flow in the annulus and the necessity of a thin layer of dusty air, thus limiting sampling rates. A thin layer of dusty air is required to provide reasonable resolution or separation. This is illustrated by the particle trajectories shown in Figure 8. Equal size particles will deposit over a length $\Delta\ell$ depending on where they enter the annulus. The thinner the dusty air layer, the better the resolution and the lower the sample rate.

An equation similar to Eq. (3) relating penetration distance to particle size, Conifuge design, and operating conditions has been developed by Stöber and Zessack (1966). This equation is

$$4(\ell_D + \ell_{-a}^3)\sin\theta + 3w\,(\ell_D^2 - \ell_a^2)\cos^2\theta = \frac{27\eta Q}{\pi^3\nu^2\rho C D^2 \cos\theta\sin\theta},$$

where ℓ_a = distance from the top of cone to sample inlet, ℓ_D = maximum penetration distance for particles of diameter D, and θ = half angle of the cone (Figure 8). Several simplifying assumptions were necessary in deriving the equation. However, Mercer (1973) has shown that the equation does a reasonable job of predicting the performance of several Conifuges.

The width of the deposit for equal size particles is dependent on the width of the dusty air layer as shown in Figure 8. The width of the dusty air layer is indicated as equal to the separation of the sample inlet funnel and the inner cone. Dusty air layer thickness is actually determined by the fraction of the annulus

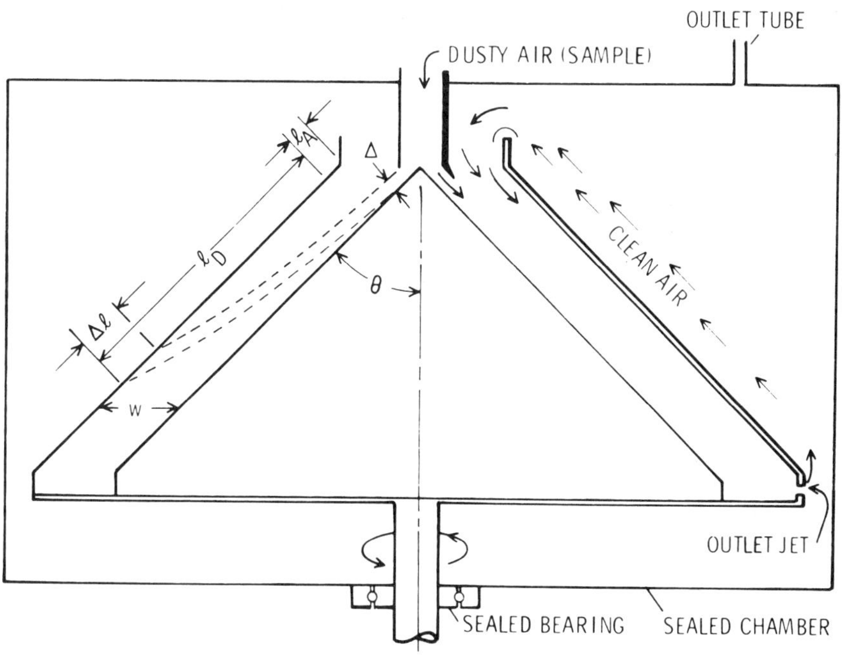

Figure 8. Schematic view of the Conifuge.

required to carry the sample flow rate. This is given in Eq. (6) as a function of the ratio of sample flow rate to total flow rate for a parabolic profile across the short dimension of a rectangular duct. The ratio of the deposit width, for uniform sized particles, to maximum penetration distance $(\Delta\ell_D/\ell_D)$ can be determined by calculating penetration distance for particles entering the annulus at the extremes of the dusty air layer. Calculations carried out assuming $\ell_a \simeq 0$ indicate a slight dependence on particle size with $(\Delta\ell_D/\ell_D)$ decreasing slightly with increasing penetration distance and decreasing particle size as shown in Table I. The most important parameters in determining $(\Delta\ell_D/\ell_D)$ are the half angle of the cones (θ), and the ratio of the sample flow rate to total flow through the annulus as shown in Figure 9. The lines in Figure 9 are the average values for the particle size range

TABLE I. Ratio of Deposit Width to Penetration Distance for 45° Conifuge Operated at 4000 RPM with a Total Flow Rate of ~13 cc/sec.

$\sqrt{\rho C}$ D (μm)	ℓ_D (cm)	$\frac{\Delta\ell_D}{\ell_D}$ S/Q=0.01	0.05	0.10	0.15	0.20
0.1	14.75	0.030	0.070	0.102	0.129	0.152
0.2	9.23	0.030	0.070	0.103	0.130	0.153
0.3	7.00	0.030	0.071	0.103	0.130	0.154
0.4	5.75	0.031	0.071	0.104	0.131	0.155
0.5	4.93	0.031	0.071	0.104	0.131	0.155
0.6	4.35	0.031	0.072	0.105	0.132	0.156
0.7	3.91	0.031	0.072	0.105	0.132	0.157
0.8	3.56	0.031	0.072	0.106	0.133	0.157
0.9	3.28	0.031	0.073	0.106	0.133	0.158
1.0	3.05	0.031	0.073	0.106	0.132	0.158
2.0	1.86	0.032	0.075	0.109	0.137	0.162
3.0	1.39	0.033	0.076	0.111	0.140	0.165
4.0	1.12	0.033	0.077	0.113	0.142	0.168
5.0	0.95	0.034	0.078	0.115	0.144	0.170

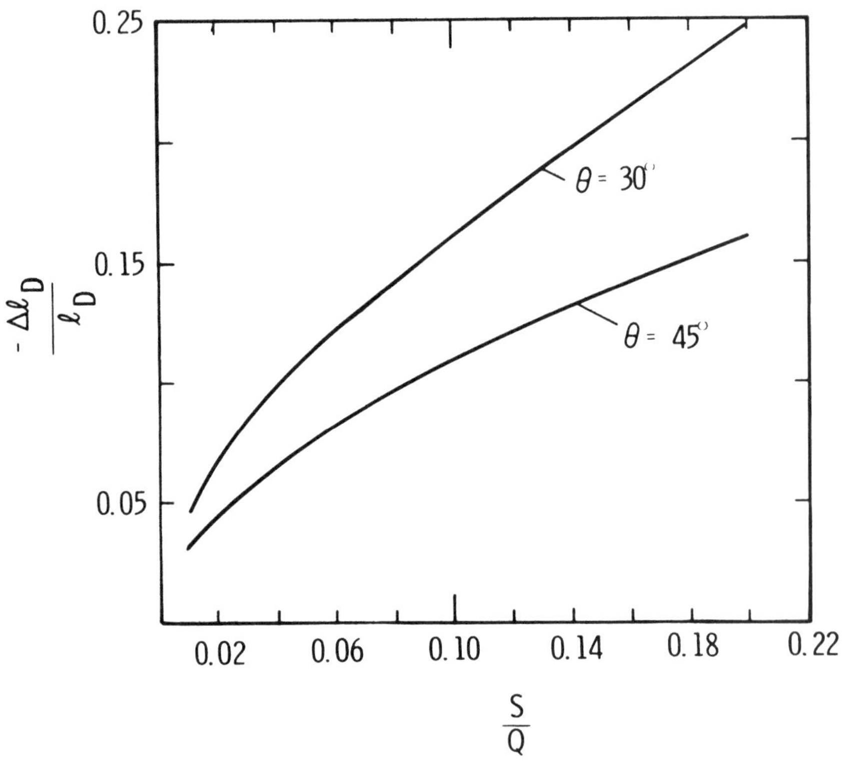

Figure 9. Length of deposit for equal size particles in two Conifuges.

given in Table I for two different "Conifuges" (45°, 30°). The relative standard deviation of $(\Delta\ell_D/\ell_D)$ over the range of diameters for each value of S/Q was less than 5%. Resolution defined as ΔD/D at any penetration distance is proportional to $(\Delta\ell_D/\ell_D)$. Some decrease in resolution results from broadening of the nonrotating, dusty air layer as it joins the rotating clean air layer at the lower edge of the inlet tube. Instruments have been designed that rotate both air flows prior to entering the deposition annulus (Hochrainer and Brown, 1969; Berner and Reichalt, 1969). These instruments have improved resolution, but they also have significant inlet losses.

Stöber and Flachsbart (1969a) designed a Conifuge having a truncated conical annulus. This design markedly increased the area of the sample inlet without increasing the thickness of the dusty air layer in the annulus. This instrument was designed to operate over a wide range of flow rates and rotational speeds and revealed the limitations of Conifuge design. At high rotational speeds and high flow rates, Coriolis forces caused secondary flows that disrupted the deposition patterns, and caused mixing that led to diffuse deposition patterns. Limiting conditions for regular deposits over the entire deposition surface were achieved at total duct flow of about nine ℓpm at 6000 RPM.

TABLE II. Conifuge Characteristics

Conifuge	θ	L(cm)	W	RPM	S(ℓpm)	Q(ℓpm)	Particle Size Range (μm)	$(S/Q)^{1/2}$
Sawyer and Walton, 1950	30°	6.0	0.58	3,000	0.025	~ 0.800	0.48–21.0	0.18
Keith and Derrick, 1960	45°	12.5	1.00	8,000	0.300	3.240	0.09–5.0	0.03
Tillery, 1967	45°	12.5	1.00	4,000	0.085	0.780	0.07–3.0	0.33
Hauck and Schedling, 1968	45°	12.5	1.00	3,000	0.674	6.740	0.50–7.0	0.31
Hauck and Schedling, 1968	45°	12.5	1.00	5,000	0.224	2.240	0.10–2.00	0.31
Hochrainer and Brown, 1969	30°	6.0	0.57	10,000	0.026	1.180	0.12–2.00	0.15
Stöber and Flachsbart, 1969b	20°	19.1	1.00	1,500	0.500	8.300	0.5–10.0	0.24
Stöber and Flachsbart, 1969b	20°	19.1	1.00	6,000	1.000	9.700	0.085–3.00	0.32
Stöber and Flachsbart, 1969b	20°	19.1	1.00	3,000	1.200	10.000	0.22–6.00	0.35
Stöber and Flachsbart, 1969b	20°	19.1	1.00	6,000	0.100	3.200	0.04–1.50	0.18
Hochrainer and Brown, 1969	90°	4.0	0.50	10,000	0.012	0.576	0.55–2.00	0.14

The characteristics of a number of Conifuges are given in Table II. The sampling rate for more than half of the instruments is less than 0.1 ℓpm. The instrument of Stöber and Flachsbart, having a sampling capability of around a liter per minute, is large and practically limited to laboratory use. The 90° instruments (Hochrainer and Brown, 1969; Berner and Reichalt, 1969) consist of an annulus between two right cylinders with flow from top to bottom similar to flow through Conifuges. A rigorous determination of penetration distance in 90° instruments has been made by Stöber and Zessack (1966) and is given by

$$\ell_D = \frac{9\eta Q\,(\ln R_2 - \ln R_1)}{2\pi^3\rho_p C D^2 (R_2^2 - R_1^2)\nu^2}, \tag{10}$$

where ℓ_D is the maximum penetration distance measured down the side of the outer cylinder, and R_1 and R_2 are the respective radii of the inner and outer wall.

Conifuge samples are normally collected on foils lining the outer cone or on slides located in slots cut into the outer cone. The foil or slide can be cut into strips and analyzed to determine the mass or particle number with respect to the diameter difference across the strip given by the calibration curve of the instrument. The use of slides or portions of the foil requires appropriate area corrections. Determination of size distribution requires corrections for inlet losses and for small particles that penetrate the annulus.

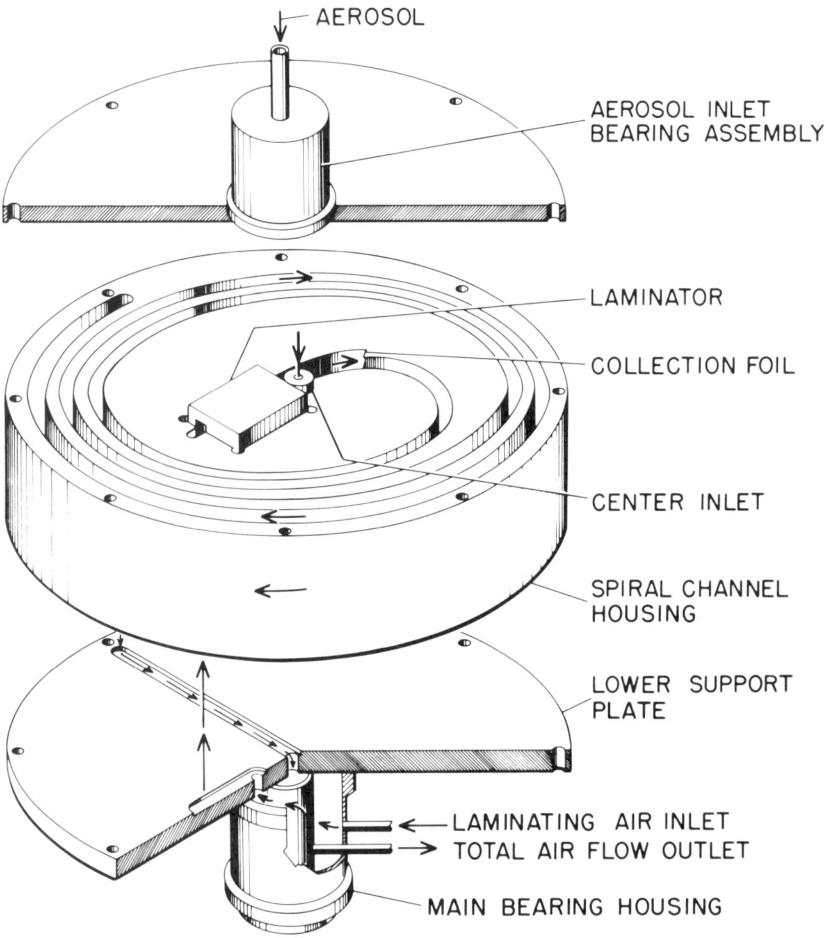

Figure 10. Blowup of spinning, spiral duct spectrometer.

The low sampling rate of the Conifuge and the instabilities encountered at high rotational speeds and flow rates indicated the need for a different design. In 1969 Stöber and Flachsbart (1969b) designed a centrifugal spectrometer with a spiral duct similar to a semidispersive centrifuge designed by Kast (1961). A blow-up view of a similar instrument is shown in Figure 10 (Moss, Ettinger and Coulter, 1972). The instrument consisted of a long (~180cm), rectangular duct (3.3 cm deep by 1.0 cm wide except at the start of the duct where the width is 1.7 cm) cut into a flat cylinder. Air flow through the instrument is controlled externally through ducts separated by rotating seals in the drive shaft of the instrument and in the base of the instrument. Filtered air is introduced into the deposition duct through a laminator consisting of several parallel foils. Suction is applied at the end of the duct. The suction flow rate is greater than the laminating air flow with the difference being drawn down the aerosol inlet, on the axis of rotation. The center inlet directs the dusty air down the deposition duct as a thin layer next to the inner wall. Particles passing down the spinning duct are subjected to an increasing centrifugal force accelerating the particles toward the outer wall where they are deposited on a collection foil. The particles are separated and deposited on the outer wall in a continuously graded spectrum according to deposition rate or aerodynamic diameter.

The complicated geometry of the deposition duct limits analytical determination of particle trajectories or penetration distances. However, the functional relationship of the parameters regulating performance is similar to the cylindrical duct instrument. Penetration distance is inversely proportional to the particle diameter squared and rotational rate squared, and is directly proportional to total flow rate. Resolution is dependent on the ratio of the thickness of the dusty air layer to the thickness of the clean air layer, or the ratio of the two flow rates. Total flow through the instrument and rotational rate are limited by the development of secondary flows that disrupt deposition patterns. Instruments of this design operate satisfactorily at 3000 RPM with total flows as high as 19 ℓpm. The particle size range deposited on the collection foil covers almost two orders of magnitude (0.09μm-5μm) under these conditions (Stöber and Flachsbart, 1969b). At 6000 RPM with a total flow of 10 ℓpm the secondary flows markedly disrupted deposition patterns (Stöber and Flachsbart, 1969b). Theoretical considerations of secondary flow development in rotating curved ducts indicate that tall, narrow ducts will tend to suppress secondary flows (Stöber, 1976). Instruments have been built with ducts as deep as 6 cm in an effort to extend the operational range. However, only marginal improvement has been noted with the deep duct instruments.

The physical size of long duct instruments and the large deposition area make these instruments somewhat impractical for portable use. Several small, single turn, spiral duct spectrometers have been built for particle separation and to provide reasonably portable sampling instruments. The first short duct instrument was designed by Kotrappa and Light (1972a). This instrument, called the Lovelace Aerosol Particle Separator, has essentially a single turn, spiral duct with approximately a one to three expansion in the radial direction over the length of the duct. The duct was expanded to utilize the Conifuge effect of decreasing velocity to enhance collection of small particles (Kotrappa and Light, 1972b). However, expansion in the radial direction increased the distance to deposition site, and decreased the average centrifugal force in the duct compared to a constant width duct with the same outer radius. These effects compromise the enhanced deposition due to increased time in the centrifugal force field. Stöber (1976) has also built several short duct instruments utilizing constant width, narrow ducts or ducts of decreasing width to suppress secondary flows. The constant duct width instrument is shown in Figure 11 with top and center inlet removed. The characteristics and performance of spiral duct spectrometers are listed in Table III. The particle size collection range for the short duct instruments is for particles collected on the deposition foil. Smaller particles are collected on the filter at the end of the duct.

Figure 11. Short, spinning, spiral duct spectrometer.

TABLE III. Characteristics of Spiral Duct Aerosol Spectrometers

Reference	Deposition Duct Dimensions (cm)			Maximum Radius (cm)	RPM	Total Duct Flow (ℓpm)	Sample Flow (ℓpm)	Particle Size Collection Range (μm)
	Length	Width	Depth					
Stöber and Flachsbart, 1969a	180	1	3.3	12	3000	10.0	0.5	0.09–5.0
Stöber and Flachsbart, 1969a	180	1	3.3	12	6000	5.0	0.3	0.18–4.0
Stöber and Flachsbart, 1969a	180	1	3.3	12	3000	19.0	1.6	0.09–5.0
Moss et al., 1972	178	1	5.08	12	3000	15.4	0.44	0.10–4.0
Kotrappa and Light, 1972	45	1.5–3.9[a]	3.9	8	3000	5.0	0.40	0.6–3.5
Kotrappa and Light, 1972	45	1.5–3.9[a]	3.9	8	3600	5.0	0.40	0.48–3.5
Kotrappa and Light, 1972	45	1.5–3.9[a]	3.9	8	4500	5.0	0.40	0.42–3.5
Stöber, 1976	55	1.67	3.3	8	3000	5.0	—	0.4–3.5
Stöber, 1976	55	1.67	6.0	8	3000	5.0	—	0.32–2.0
Stöber, 1976	55	1.6–0.8[b]	3.4	8	5000	5.0	—	0.22–2.0
Stöber, 1976	55	1.6–0.8[b]	3.4	8	3000	10.0	—	0.48–3.5

a. Expands over length from 1.5 to 3.9 cm.
b. Contracts over length from 1.6 to 0.8 cm.

In all rotating, spiral duct samplers a portion of the inlet rotates so that inlet loss corrections are necessary to determine size distributions. Losses can be calculated for well-defined rotating geometries (Mercer, 1973). In most cases the inlet geometry is not simple and is complicated by interfaces between fixed and rotating components so collection efficiencies must be determined experimentally. Correction factors relating center line deposition concentrations to airborne concentrations have been determined by Kops et al. (1974) for the long, spiral duct of Stöber. Center line deposition concentration for particles smaller than 0.2 μm was found to be enhanced by the secondary flows. Center line deposition concentration for particles larger than 0.4 μm was found to be low because of inlet losses. These losses are quite high for particles larger than 2 μm in diameter (Stöber, Flachsbart and Boose, 1972; Kops, Hermans and Van De Vate, 1974). Similar corrections are required if the entire height of the foil deposit is used to determine size distributions, as inlet losses result in lower concentrations on the lower half of the foil for large particles ($D_A > 0.4\mu m$) (Ferron and Bierhuizen, 1974). These losses are most significant for large particles, and so are important when determining mass distributions.

Centrifugal aerosol spectrometers have also been built in which the particle flow is in the radial direction (Burson, Keng and Orr, 1967/8; Redkin, 1970). In these instruments a thin stream of particles is injected into a rotating chamber of still air or a chamber in which the air is also flowing radially at a low velocity. The particles are deviated from the radial direction due to the interaction of Coriolis forces and drag. Thus, small particles deviate only slightly and deposit on the outer wall close to the intersection of the radial line from the inlet to the outer wall. Large particles deviate significantly and deposit on the outer wall at some distance from the intersection of the radial line with displacement being opposite the direction of rotation.

Several types of semidispersive aerosol centrifuges have been built. The most frequently encountered device of this type is the Goetz Aerosol "Spectrometer" (Goetz, Stevenson and Preining, 1960). The instrument shown in Figure 12 consists of two helical ducts of 2.5 turns on the outside of a 30° right cone. The radius of the cone is $\sim$ 3.8 cm at the base. The cone forming the outer wall of the ducts is removable and lined with a foil or paper for collection and subsequent analysis of the deposit. The cones are rotated by a variable speed motor at rates up to 24,000 RPM. The impeller action of the rotating ducts draws the sample down the inlet and through the ducts. Flow rate through each duct is independently controlled by jets located at the end of the duct. The farther the particles travel down the duct the greater is the centrifugal force and the deposition velocity. The deposit at any distance down the channel contains a mixture of all sizes of particles having a maximum penetration distance equal to or greater than the distance. Maximum penetration distance for a given sized particle is the distance traveled by a particle, of the given size, entering the deposition duct next to the inner wall. The increase in deposition velocity as particles travel down the duct results in a variation in deposited concentration of a given size particle out to the maximum penetration distance. A theoretical study indicates an increase of

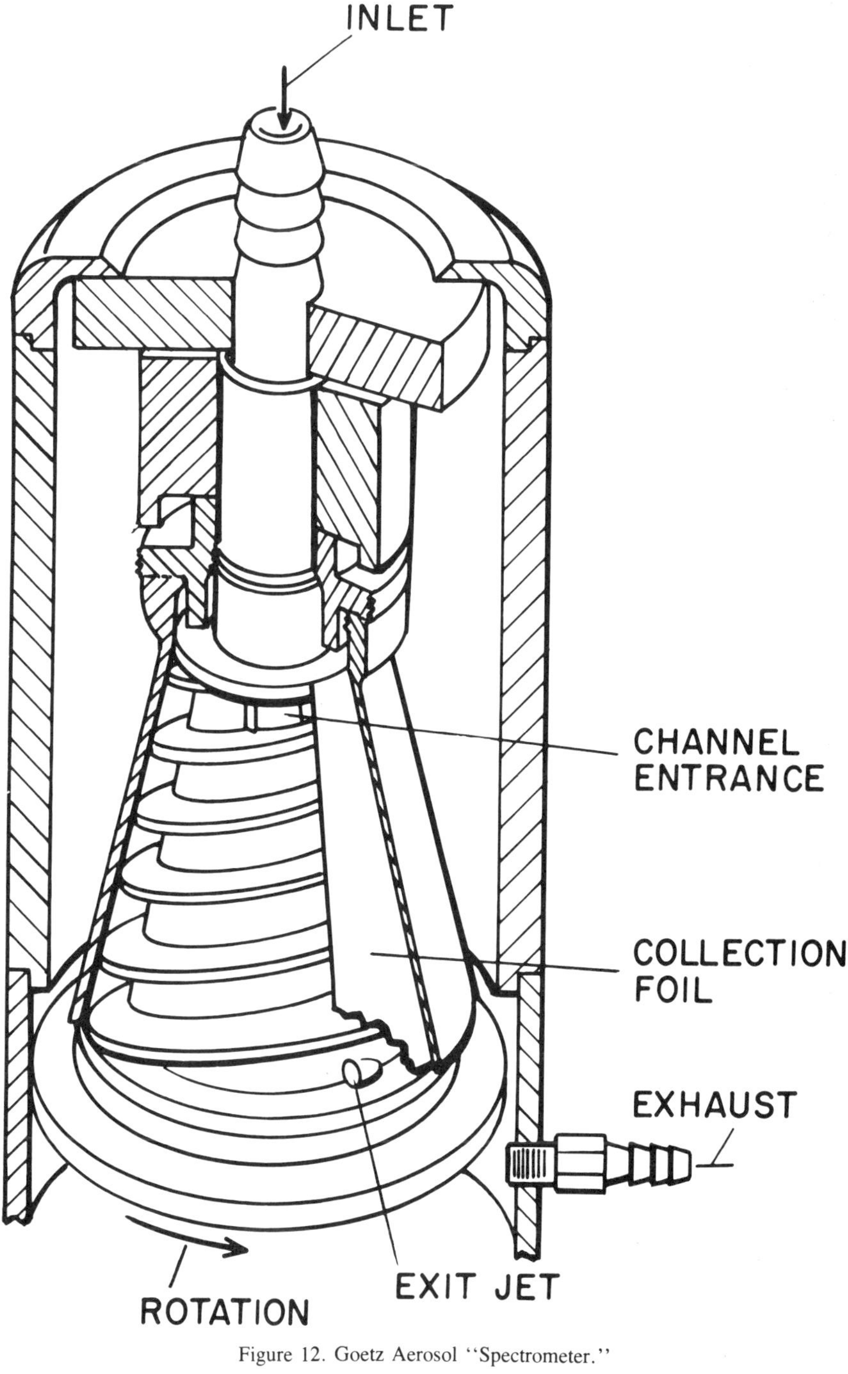

Figure 12. Goetz Aerosol ''Spectrometer.''

deposited concentration with penetration distance up to near maximum, then decreasing to zero at the maximum (Stöber and Zessack, 1964). An experimental study indicates the theory is approximately correct with cutoffs at maximum penetration distance not as sharp as indicated by theory (Baust, 1967). Turbulent flow conditions exist for a short distance below the inlet to the ducts. Theoretical treatment of deposition in the turbulent region would be difficult or impossible and has not been done. Experimental determinations of deposition in this region have been made (Baust, 1967; Raabe, 1967). Gerber has found that under limited operating conditions with a modified instrument, center line deposition concentrations are constant (Gerber, 1971; Stöber and Boose, 1973; Hochrainer, 1972). However, marked variations in deposition concentration were found across the channel requiring correction factors to determine particle size distributions from center line deposits (Gerber, 1971). Complicated deposition patterns have also been noted with a spiral duct, semidispersive aerosol centrifuge (Ruger, Maiwald and Feddersen, 1968).

The principal advantage of centrifugal aerosol spectrometers is the collection of the sample as a continuously graded spectrum in terms of aerodynamic or inertial properties. This separation simplifies the determination of particle size distributions with respect to aerodynamic diameter for most parameters of interest. The resolution, or separation of particles according to aerodynamic characteristics achieved by these instruments is not attainable by any other means. The principal disadvantages are relatively low sampling rates, collection of the sample over a large surface, and delay between the time of sampling and determination of results.

The only advantage of the semidispersive centrifuges is the high sampling rate (up to 125 ℓpm for Kast instrument [1961]) relative to the centrifugal spectrometers. The variations in deposit concentration, poorly defined maximum penetration distance, and particle losses in the turbulent flow region make determination of particle size distributions very difficult and susceptible to errors.

At the present time few of these instruments are commercially available. However, plans for almost all of the instruments are available and fabrication can be carried out by most well-equipped machine shops.

REFERENCES

Anderson, P. L., 1976: Development of a centrifuge for source sampling to determine particle size distribution. *Environ. Sci. Technol., 10,* 145.

Baust, E., 1967: Use of the Goetz Aerosol Spectrometer to measure the size spectra of polydispersed aerosol. *Staub-Reinhalt. Luft, 27,* 16.

Berner, A. and H. Reichalt, 1969: On the inlet slit systems in conifuges—part I: The ROSL system. *Staub-Reinhalt. Luft, 29,* 5.

Burson, III, J. H., E. Y. H. Keng and C. Orr, Jr., 1967/8: Particle dynamics in centrifugal fields. *Powder Technol., 1,* 305.

Ferron, G. A. and H. W. J. Bierhuizen, 1974: The measurement of polydisperse aerosols with the spiral centrifuge. Aerosol in Naturwissenschaft, Medizen und Technik, Jahreskongress der Gesellschaft für Aerosolforschung e.v.

Gerber, H. E., 1971: On the performance of the Goetz Aerosol Spectrometer. *Atmos. Environ., 5,* 1009.

Goetz, A., H. J. R. Stevenson and O. Preining, 1960: The design and performance of the aerosol spectrometer. *J. Air Pollution Control Assoc., 10,* 378.

Hauck, H. and J. A. Schedling, 1968: A modified conifuge. *Staub-Reinhalt. Luft, 28,* 18.

Hochrainer, D., 1971: A new centrifuge to measure the aerodynamic diameter of aerosol particles in the submicron range. *J. Colloid and Interface Sci., 36,* 191.

———, 1972: On the reliability of measurements with the Goetz Aerosol Centrifuge, Discussion. *Atmos. Environ., 6,* 699.

——— and P. M. Brown, 1969: Sizing of aerosol particles by centrifugation. *Environ. Sci. Technol., 3,* 830.

Kast, W., 1961: Neues staubmessgerat zur schnellbestimmung der staubkonzentration und der kornverteilung. *Staub-Reinhalt. Luft, 21,* 215.

Keith, C. H. and J. C. Derrick, 1960: Measurement of the particle size distribution and concentration of cigarette smoke by the conifuge. *J. Colloid and Interface Sci., 15,* 340.

Kops, J., H. Hermans and J. E. Van De Vate, 1974: Calibration of a Stöber Centrifugal Aerosol Spectrometer. *Aerosol Sci., 5,* 379.

Kotrappa, P. and M. E. Light, 1972a: Design and performance of the Lovelace Aerosol Particle Separator. *Rev. Sci. Instrum., 43,* 1106.

——— and ———, 1972b: Centrifuge Separator. U.S. Patent No. 3, 698, 626.

Matteson, M. J., G. F. Boscoe, and O. Preining, 1974: Design, theory and calibration of a field type aerosol spectrometer. *Aerosol Sci., 5,* 71.

Mercer, T. T., 1973: *Aerosol Technology in Hazard Evaluation.* New York, Academic Press.

Moss, O. R., H. J. Ettinger and J. R. Coulter, 1972: Aerosol density measurements using a modified spiral centrifuge aerosol spectrometer. *Environ. Sci. Technol., 6,* 614.

Raabe, O. G., 1967: Calibration and use of the Goetz Aerosol Spectrometer. *Assessment of Airborne Radioactivity,* Vienna, International Atomic Energy Agency, 417.

Redkin, J. N., 1970: Properties of atmospheric aerosol measured with a centrifugal spectrometer. *J. Geophysical Research, 75,* 3563.

Ruger, G., E. Maiwald and C. Fedderson, 1968: A centrifugal separator and its properties. *Staub-Reinhalt. Luft, 28,* 9.

Sawyer, K. F. and W. H. Walton, 1950: The "Conifuge": A size sampling device for airborne particles. *J. Sci. Instrum., 27,* 272.

Stöber, W., 1976: Design, performance and applications of spiral duct aerosol centrifuges. *Fine Particles, Aerosol Generation, Measurement, Sampling, and Analysis.* New York, Academic Press.

———, A. Berner and R. Blaschke, 1969: The aerodynamic diameter of aggregates of uniform spheres. *J. Colloid and Interface Sci., 29,* 710.

——— and C. Boose, 1973: Developing flow and particle deposition in horizontal elutriators and semidispersive aerosol centrifuges. *Atmos. Environ., 7,* 119.

——— and H. Flachsbart, 1969a: Aerosol spectrometry with a ring slit conifuge. *Environ. Sci. Technol., 3,* 641.

——— and ———, 1969b: Size separating precipitation of aerosols in a spinning spiral duct. *Environ. Sci. Technol., 3,* 1280.

———, ——— and C. Boose, 1972: Distribution analyses of the aerodynamic size and the mass of aerosol particles by means of the spiral centrifuge in comparison to other aerosol precipitators. *J. Colloid and Interface Sci., 39,* 109.

——— and U. Zessack, 1964: On the theory of a conical aerosol centrifuge. *Staub-Reinhalt. Luft, 24,* 295.

——— and ———, 1966: The measurement of aerosol particle size spectra by means of centrifugal separators. *Z. Biol. Aerosol forschung, 13,* 263.

Tillery, M. I., 1967: Design and calibration of a modified conifuge. *Assessment of Airborne Radioactivity,* Vienna, International Atomic Energy Agency, 405.

———, 1974: A concentric aerosol spectrometer. *J. Amer. Industrial Hygiene Assoc., 35,* 62.

COMMENTS ON CENTRIFUGES

Owen R. Moss

Department of Radiation Biology and Biophysics, University of Rochester, Rochester, N.Y.

These comments deal with the calibration and use of the spinning spiral aerosol spectrometer. I will first present some work of Dr. Ted Martonen (1975/76) and then I will discuss some practical points I have encountered while operating these centrifuge devices.

The overview paper on centrifuges by Tillery shows a drawing of the centrifuge built at Los Alamos. A particle enters at the center of rotation, crosses the channel, and deposits on the collection foil lining the outer wall. Martonen has recently solved the equation for the motion of a particle under the influence of the centrifugal and Coriolis forces in the channel. He assumed that there is no secondary flow. A preliminary comparison of the theory with experimental calibration points shows that the two curves diverge greatly for particles less than 0.8μm aerodynamic diameter (D_{ae}). Particles with D_{ae} of 0.25 μm are predicted to pass through the entire length of the channel. Instead, they deposit about halfway down the channel for the operating conditions used in the comparison. The instrument works much better than one would expect.

The centrifuge devices are used to sample a variety of aerosols. Air may be only a small component of the sampled gas. Martonen investigated whether the calibration of the instrument changes if the aerosol carrier gas is different from the laminating air in the channel. He used carbon dioxide, air, and helium as aerosol carrier gases. These three gases have specific gravities of approximately 1.4, 1.0 and 0.12, respectively. Figure 1 shows the results of a part of this work. Under identical operating conditions, the more dense gas tends to stream across the channel. This is very much like the results seen with cigarette smoke (Hinds, 1975; Martonen, 1976; Stöber, 1976).

This paper is based on work performed under contract with the U. S. Energy Research and Development Administration at the University of Rochester Biomedical and Environmental Research Project and has been assigned Report No. UR-3490-933.

The aerosol carrier gas can be cleaned and used for the laminating air flow in the channel. Martonen found that a laminating flow of carbon dioxide or nitrogen required that the instrument be run at lower rotational speeds in order to maintain distinct deposition patterns like those obtained during the initial calibration with dry air. It is not clear whether the conditions he used in his experiments were a

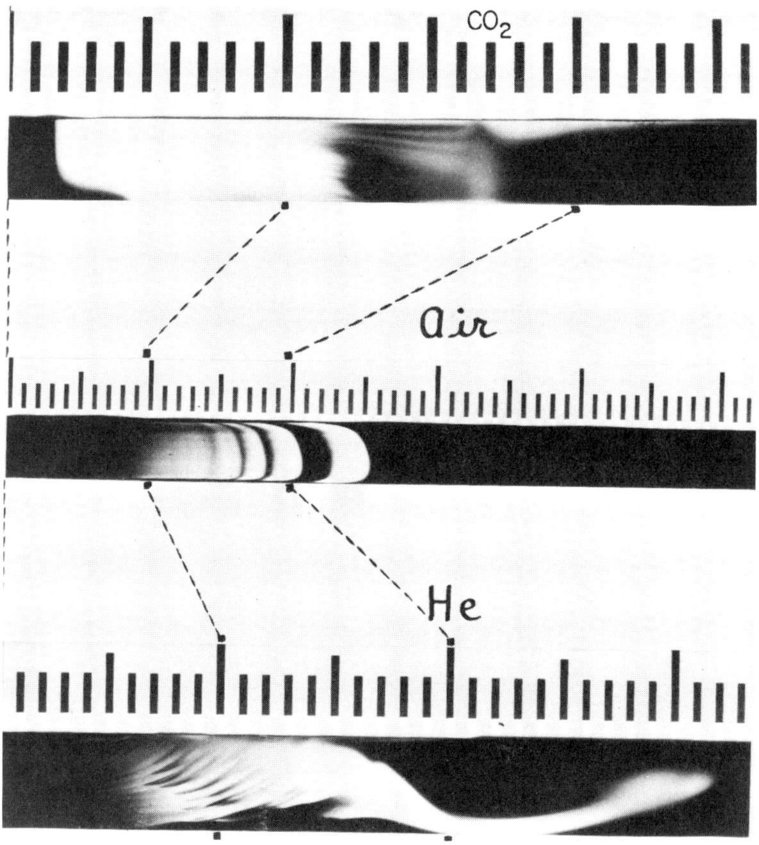

Figure 1. Centrifuge deposits with 0.822 μm diameter polystyrene latex spheres. Except for the aerosol carrier gas CO_2, air or He the sampling conditions were identical. Clean dry air was used for the laminating flow. The distance down the foil is indicated in centimeters. The centrifuge was one of the units at the Institut für Aerobiologie. The sampling conditions were 3000 rpm and 8.7 ℓpm.

good model of the aerosols the centrifuge might be used to sample. Automobile emissions might be close to what he was thinking of.

The instrument appears to keep its calibration under the various sampling conditions. The Lovelace Foundation Aerosol Particle Separator (LAPS) (Kotrappa and Light, 1972) is run at 37° C (Raabe, 1976). The specific gravity of the gas in the channel is 1.04. No change is seen in the deposition patterns of the calibration aerosol.

I have used the Stöber aerosol centrifuge at Rochester to relate the aerodynamic diameter to the projected area diameter of the particles. This type of measurement is made by gluing electron microscope (EM) grids to the collection foil. The location of the grid provides a measure of the aerodynamic diameter of the particles that land on it. A scanning electron microscope was used to observe the sample. The grid-bars contained more particles on their surface than the carbon-film-substrate, grid squares. This was very noticeable for the samples containing particles with D_{ae} greater than 5μm. The particles were very irregular in shape. For D_{ae} around 2 μm and less, there was little difference in the count density. The size distributions measured on the grid bars and grid squares were essentially the same. The grids are mechanically deformed in pulling them off the foil. The carbon film substrate would vibrate more than the adjacent metal grid bars. Particles would then be lost preferentially. This tendency to lose larger particles should be kept in mind when counting. Care should be taken in handling the foil or EM grid sample.

Obtaining a representative sample of a dust composed of particles larger than 3 μm D_{ae} is a problem with the centrifuge units. Two particle sinks located at the radius of the aerosol inlet tube from the center of rotation must be considered. One sink is the point where the walls of the stationary inlet tube match up with the rotating sides of the hole in the centrifuge head. The particle loss that occurs at this point has been reduced by machining and polishing (Kotrappa, 1971/1972). Another method that I have tried is to insert the stationary tube down into the rotor until its end opens into the spiral channel. The aerosol particles now pass from a stationary walled tube to one with a moving slit, the inlet into the channel. The loss should be less with this arrangement than that predicted for rotating tubes by Tillery (1976).

The other particle sink is the metal sheath used to form the aerosol inlet slit. If the slit is moved away from the center of rotation, the deposition patterns from the calibration aerosol become very sharp. The particle loss in such an inlet also increases (Stöber and Flachsbart, 1969). An inlet should allow the larger particles (at least up to 10 μm D_{ae}) to enter the channel. It should also allow calibration-aerosol deposition patterns of minimum overlap. The deposition patterns should be centered down the middle width of the sampling foil.

Figure 2 shows sketches of calibration aerosol deposition patterns for different aerosol inlet configurations. The first inlet is the extended sheath inlet used for high resolution studies. I have not drawn in the faint lines that tend to extend back from the deposition patterns near the top and bottom of the foil. Particles with D_{ae} greater than 4 μm are deposited inside this type of inlet and not on the collection foil. The second inlet configuration is an open one. The singlet deposition patterns become smeared.

If a slit is to be used, and one must be if the deposit is to be centered down the collection foil, then it should be located no more than one inlet-tube-radius from the center of rotation. Inlet 3 in Figure 2 shows one such configuration. Slits were constructed at the closest point to the outer wall. None of the slits designed for this position gave good deposition patterns. The aerosol appeared to have an early involvement in the development of the secondary flow. This particular inlet showed that the top aerosol stream had a larger velocity component towards the outer wall than the lower aerosol stream.

Another alternative is to place a slit at the base of a rounded step function. The example is given as inlet 4 in Figure 2. Surprisingly, the slit works fairly well. It can be shaped so that deposition occurs down the middle width of the collection foil. Particles as large as 10 μm D_{ae} appear to get in with very little loss. The deposition pattern at the center end of the foil begins to tail off visibly where the 12 to 15 μm D_{ae} particles would be expected to land (5 to 7 cm). The 10 μm D_{ae} particles then must be getting into the channel. Actually, the open inlet, number

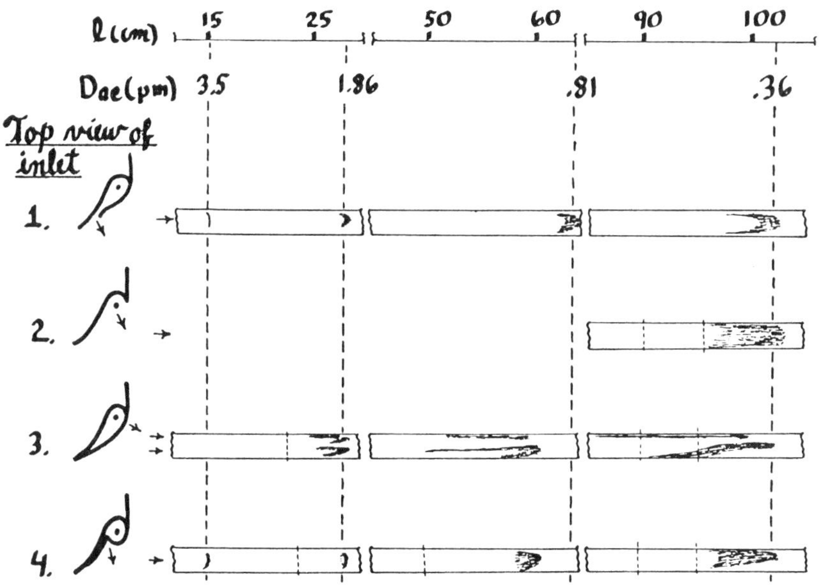

Figure 2. Sketches of deposition patterns obtained with different aerosol inlets. "1" is the distance down the foil. The small dotted lines indicate where the deposition pattern repeated due to doublets or triplets. The centrifuge was the original Stöber unit at the University of Rochester. The basic sampling conditions were 1500 rpm and 10 ℓpm.

2 in Figure 2, works fairly well with a resolution on the foil of 5 to 10 percent.

I would like to make one last comment on sampling polydisperse aerosols. In general, the sample is collected on a foil which is then cut and stored for chemical and physical analysis. I found that in cutting the foil, some of the material was lost on the scissors. When dealing with heavy deposits of dust particles having D_{ae} greater than 5 μm the loss can be more than 10%. One solution is to use foil of overlapping metal pieces. The pieces fit into a backing foil containing slots. With a little practice, selected pieces can be removed or replaced without disturbing the others. Short aluminum strips have also been used with success (Ferron and Bierhuizen, 1976).

REFERENCES

Ferron, G. A. and H. W. J. Bierhuizen, 1976: The measurement of polydisperse aerosols with the spiral centrifuge. *J. Aerosol Sci., 7*, 5-11.

Hinds, W. C., 1975: Measurement of the aerodynamic size distribution of tobacco smoke. Paper #63. Presented at Amer. Industrial Hygiene Assoc. Conf., Minneapolis.

Kotrappa, P., 1971/2: Personal communication. Lovelace Foundation for Medical Education and Research, Albuquerque.

______ and M. E. Light, 1972: Design and performance of the Lovelace Aerosol Particle Separator. *Rev. Sci. Instrum., 43*, 1106-1112.

Martonen, T., 1975/6: Personal communication. University of Rochester School of Medicine and Dentistry, Rochester.

Raabe, O., 1976: Design and use of the Mercer-style impactor for characterization of aerosol aerodynamic size distributions. Presented at Aerosol Measurement Workshop, Gainesville, Fl., March 24-26.

Stöber, W., 1976: Personal communication. Institut für Aerobiologie, Grafschaft, Schmallenberg/ Sauerland, Germany.

______ and H. Flachsbart, 1969: Size-separating precipitation of aerosols in a spinning spiral duct. *Environ. Sci. Technol., 3,* 1280-1296.

Tillery, M., 1976: Aerosol centrifuges. Presented at Aerosol Measurement Workshop, Gainesville, Florida, March 24-26.

LOVELACE AEROSOL PARTICLE SEPARATOR DESIGN MODIFICATIONS

D. TALLEY, O. G. RAABE, AND J. A. MEWHINNEY

Inhalation Toxicology Research Institute, Lovelace Foundation, Albuquerque, N. Mex.

Four years of experience employing the Lovelace Aerosol Particle Separator (LAPS) for preparation of monodisperse particles and characterization of aerosols have suggested several aspects of the engineering design which could be changed to improve performance and reliability. Specifically, the original LAPS units used three bearings on the drive shaft with the center bearing serving as a seal between the inlet for LAPS air and the exhaust. In some units the center bearing malfunctioned as a seal, leaking as much as 1.5 cc per sec of air between inlet and outlet at a pressure difference of only 0.3 psig. Also, the bearings were occasionally observed to generate a foreign aerosol which contaminated the collected aerosol. This occurred when the bearing lubricant became aerosolized and was deposited on the collection foil. Because of this loss of lubricant from the bearings and possible slight misalignment of the three bearings with respect to the LAPS shaft, premature bearing failure occurred in several LAPS units.

A complicating factor in use of the LAPS units at the Inhalation Toxicology Research Institute (ITRI) is that those used to prepare monodisperse particles are normally in glove boxes and may be highly contaminated with various long-lived radionuclides such as ^{238}Pu. Since the configuration of the bottom bearing housing prevented repair inside a glove box, extensive and costly decontamination was required before the units could be removed to a clean work area for repair.

Since a fairly major design change was required to solve the bearing and seal problem it was decided to take this opportunity to incorporate other changes in design to improve the LAPS and its reliability. A brief description of each of the design modifications follows. The original design is represented in Figure 1. The

Work performed for the Division of Biomedical and Environmental Research, U.S. Energy Research and Development Administration under Contract No. E(29-2)-1013.

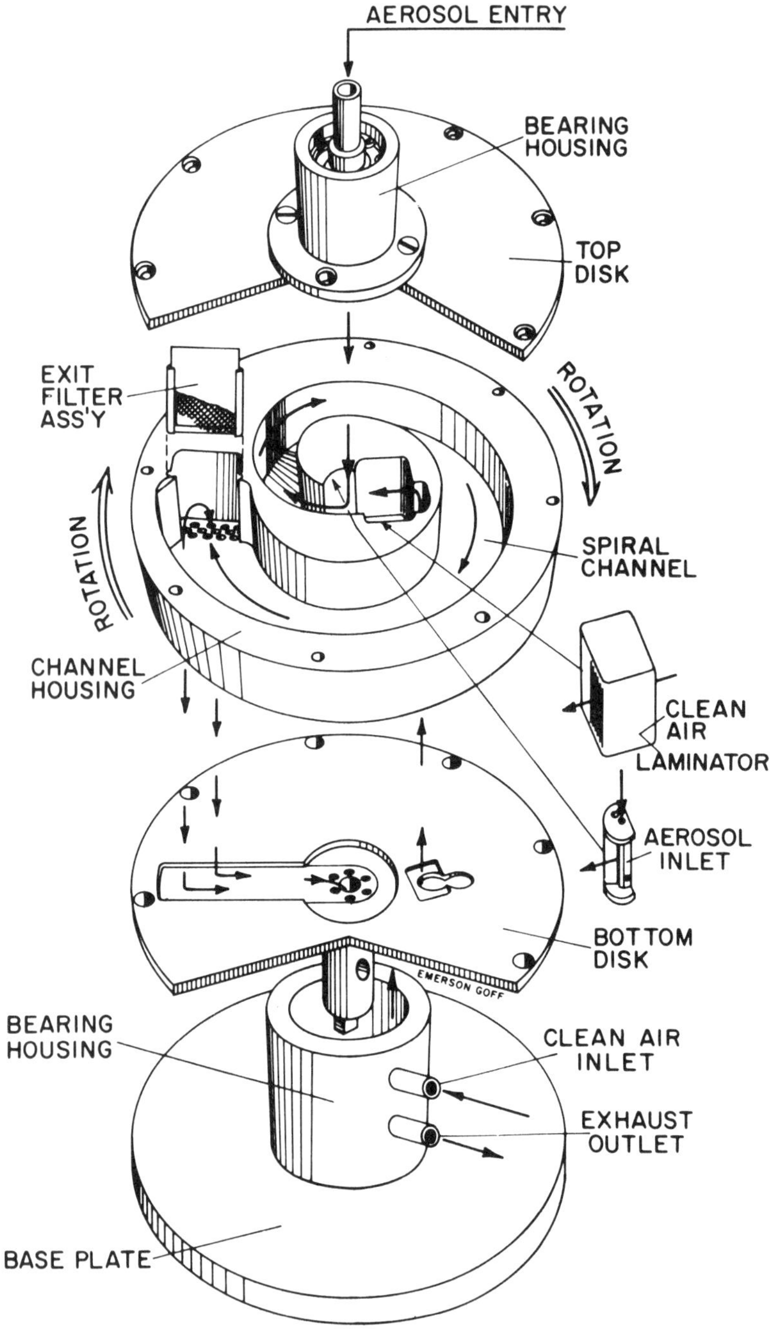

Figure 1. Exploded view of the original Lovelace Aerosol Particle Separator. Approximate scale: diameter of disk equals 17.8 cm.

modified design is represented in Figure 2. A detailed view of the drive shaft, magnetic seals and bearing housing is shown in Figure 3.

PERFORMANCE MODIFICATIONS

Magnetic Seals

Compact axial shaft seals were obtained from the Magnetic Seal Corporation, West Barrington, Rhode Island. Three of these seals were used in the drive shaft bearing housing so that the laminar air inlet and the LAPS exhaust outlet were isolated from each other and from the ambient atmosphere and bearing lubricant. Vented seal carriers were designed for effecting this separation. Another seal was used on the aerosol inlet shaft to prevent bearing lubricant contamination and aerosol dilution with ambient air.

Channel Profile

The inside wall of the spiral channel was redesigned for the final 180° so that the undesirable concave pouch at this end of the channel in the original design was eliminated and the channel made more narrow to provide an improved aspect ratio (Stöber, 1976). This allowed a square channel end measuring 3.175 cm on a side. The channel was shortened about 0.5 cm to allow for a larger filter holder, but the effective length of foil did not change due to the use of the 0.5 cm of the filter holder in front of the filter.

Channel Filters

A filter was added upstream of the laminating air to remove possible contaminants from the drive shaft seals. The spiral channel outlet filter was made to occupy the entire area of the channel outlet eliminating the constriction at the end of the original spiral.

MAINTENANCE MODIFICATIONS

Filter Holders

An injection-molded polystyrene filter holder was designed to support both the inlet and the outlet filter. The holders are inexpensive, can be loaded and stored for use as needed, and, in the case of the outlet filters, allow for convenient and safe storage of collected aerosol particles. The filter holders were designed to seal against the channel around their perimeter eliminating leakage that was sometimes observed in the original LAPS design.

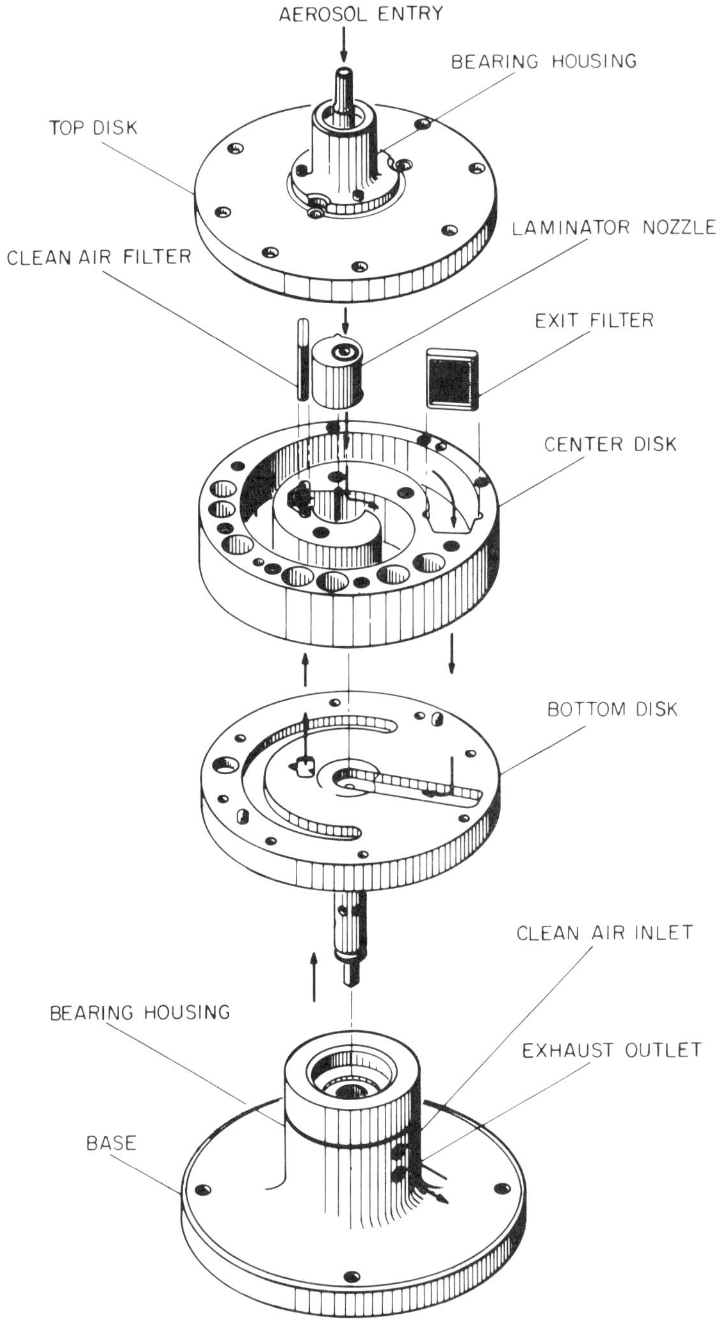

Figure 2. Exploded view of the modified Lovelace Aerosol Particle Separator. Approximate scale: diameter of disk equals 17.8 cm.

Laminator-Aerosol Inlet

The two piece clean air laminator and aerosol inlet were combined into a one piece cylindrical unit. The one piece design can be fabricated more easily with a higher degree of precision. Laminators and aerosol inlets no longer need to be supplied in matched pairs for a particular LAPS.

Drive Shaft Bearing Housing

This support base was redesigned so that assembly and disassembly could be accomplished with relative ease inside a glove box with a minimum of inexpensive special tools required. Should bearings or seals require replacement after a LAPS has been contaminated during use, expensive decontamination is not necessary.

Drive Shaft and Aerosol Inlet Shaft

These shafts were made of tool steel and brass, respectively, in the original LAPS. The modification uses 304 stainless steel for both. Rusting and corrosion as a result of decontamination procedures are thus eliminated.

Alignment of Spiral Rotor and Lid

Alignment of these mating parts in the original LAPS design was accomplished using the protruding ends of the aerosol inlet and laminator as a positioning key. This caused wear and resulting inexactness of fit between parts whose alignment was relatively critical. The modified design has dowel pins for alignment.

RELIABILITY MODIFICATIONS

Material

The original LAPS was made of 6061T6 aluminum which, while being relatively easy to machine, is soft and easily damaged. The modified LAPS is made of 7075T6 aluminum which is easier to machine and much less likely to suffer damage as a result of routine handling.

Top and Bottom Disks

The bottom disk was made thicker for greater stiffness and better sealing against the spiral rotor and for more contact with the drive shaft taper. The top disk was thickened for stiffness and for more metal support of the countersunk cap screws used in assembly.

Bolt Pattern

The modified LAPS has a new bolt pattern including three centrally located screws for more uniform clamping and a more secure seal of the top disk to the spiral rotor. The threads in the rotor are reinforced with stainless steel "Keen-sert" thread inserts.

Dynamic Balance

Dynamic balance of the original LAPS was accomplished by attaching appropriate steel counter weights to the outside of the top and bottom disks each being secured with four or five number 10-32 flathead screws which penetrated approx-

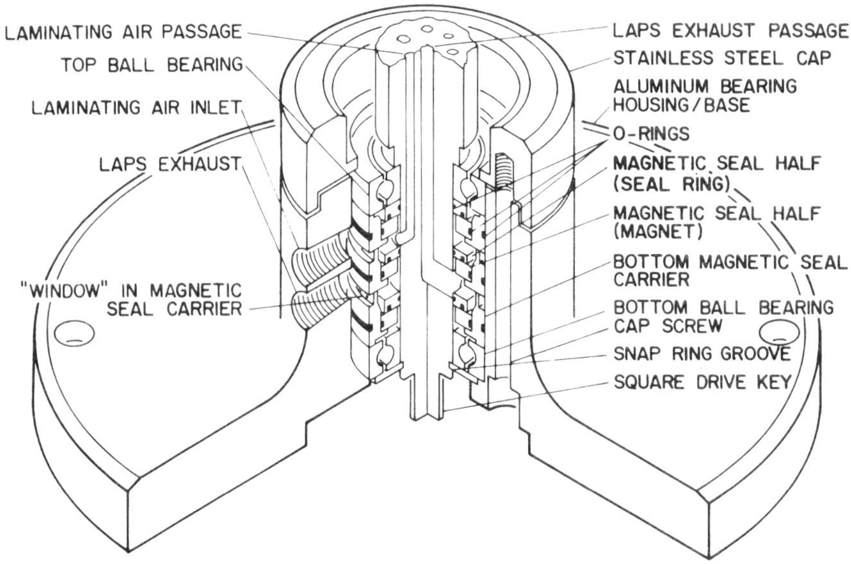

Figure 3. Cutaway section view of the driveshaft bearing housing of the modified Lovelace Aerosol Particle Separator.

imately 0.3 cm into the disk. These weights represented a potential hazard since some had been observed to loosen during LAPS operation. In the modified design appropriate quantities of metal were removed from heavy sections of the rotor to accomplish the counterweighting by metal removal rather than metal addition. In all, more than 200 g of metal were removed.

Center Bearing

Eliminating the center bearing from the drive shaft and thus eliminating any preloading of the bearings because of slight misalignment of the shaft and bearing bores should eliminate premature bearing failure. The use of three magnetic seals in place of the center bearing should protect the two remaining bearings from damage by cleaning solvents which might accidentally enter the bearing housing during decontamination. Such damage has occasionally occurred in LAPS units of the original design.

Acknowledgements—The authors acknowledge the help of Sandia Corporation, in particular John Lewis and his associates for drawings, structural calculations and configurations. Also, Mr. W. (Butch) Yaple of Continental Machine Company for his help on the design of the laminator-inlet slit.

REFERENCES

Kotrappa, P. and M. E. Light, 1972: Design and performance of the Lovelace Aerosol Particle Separator. *Rev. Sci. Instrum.*, *43*, 1106-1112.

Raabe, O. G., H. A. Boyd, G. M. Kanapilly, C. J. Wilkinson and G. J. Newton, 1975: Development and use of a system for routine production of monodisperse particles of $^{238}PuO_2$ and evaluation of gamma emitting labels. *Health Phys.*, *28*, 655-667.

Stöber, W., 1976: Design, performance and applications of spiral duct aerosol centrifuges. *Fine Particles*, B. Y. H. Liu, Ed., New York, Academic Press, pp. 351-397.

THE GOETZ AEROSOL SPECTROMETER

HERMANN E. GERBER

Naval Research Laboratory, Washington, D.C.

A major advertised advantage of the Goetz aerosol spectrometer (Preining, Stevenson and Goetz, 1959; Goetz et al., 1960; Goetz and Kallai, 1962) over its immediate predecessor, the original conifuge of Sawyer and Walton (1950), was that its sampling rate was two orders of magnitude larger. This advantage made the Goetz centrifuge popular when it first appeared commercially in the early 1960's. In the years that followed, however, confidence in the instrument declined when many papers in the literature gave inconsistent performance evaluations of the instrument. Most of the studies (Stöber and Zessack, 1964; Baust, 1967; Raabe, 1967; Stöber and Boose, 1973; Hochrainer, 1972) reported difficulties in using the instrument, although the originators and others (Gerber, 1971; Horvath, 1973; Gerber, 1974) reported acceptable performance. The present purpose is to review those evaluations and to describe those factors which must be considered when applying the instrument.

INSTRUMENT DESCRIPTION

The Goetz centrifuge is shown in the cutaway sketch of Figure 1. Two independent channels with a constant cross section of a parallelogram spiral down the outside of the conical rotor. During operation the rotor is spun rapidly, so that the airscrew action of the helical channels draws the aerosol sample into the instrument. The aerosol enters the centrifuge through the inlet tube and is deflected into the channels by a baffle at the bottom of the stationary tube. The particles upon entering the channels experience the centrifugal field and move radially across the channel flow until they hit the outside wall. This wall consists of a metal foil or other thin flexible material held tightly against the channel walls by a conical cup. After a sample is deposited, the foil is removed for analysis. A flattened collecting foil with a deposit of polydisperse latex particles is shown in Figure 2.

The flow rate F (ℓpm) through the channels depends on the rpm R of the centrifuge rotor and on the size O (cm) of the interchangeable orifice at the end of each channel. Goetz et al. (1960) gave

$$F = 2.75 \times 10^{-3}\ RO\ cm^2, \tag{1}$$

which is an approximation, since the flow rate depends to some extent on conditions external to the instrument. Flow-rate measurement must be made with a low-resistance meter, e.g., soap film meter (Barr, 1934), since the instrument is quite sensitive to pressure changes at the inlet and the outlet.

The geometry of the deposit in Figure 2 shows similarities to that of a horizontal elutriator, which is to be expected, since the operation of the instruments is similar in principle, with the force of gravity in the elutriator being replaced by the centrifugal force in the centrifuge and with the aerosol being introduced over the entire cross section of the channel in both devices. In the elutriator monodisperse particles form a deposit of uniform concentration with a length ℓ_D along the channel proportional to the inverse square of the aerodynamic diameter

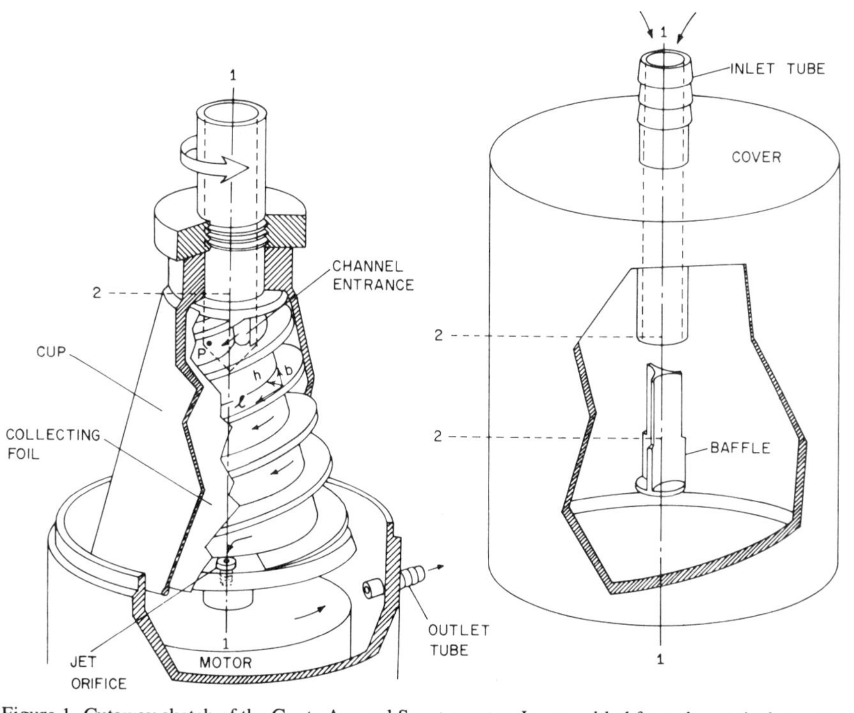

Figure 1. Cutaway sketch of the Goetz Aerosol Spectrometer. In assembled form the vertical axes (1) and the horizontal dashed lines (2) each coincide.

D of the particles. Under ideal operating conditions the concentration of those particles would also be constant on the centrifuge deposit, although ℓ_D is related to D in a more complicated manner due to the involved geometry of the centrifuge spirals. Stöber and Zessack (1964) in a theoretical study of the performance of the instrument showed that along the midline of the channel deposit

$$D \approx \left[2.29\ \eta\ F/A\ D^2\ R^2 + (\ell_a + \lambda)^{3/2}\right]^{2/3} - \lambda, \tag{2}$$

where A = particle slip factor, η = viscosity of air, λ = 16.5 cm, and $\ell_a \approx 6$ cm is defined as the distance along the deposit side of the channel from the beginning of the channel to the point opposite the channel entrance.

The simplest technique of measuring ℓ_D along the flattened channel deposits which describe Archimedean spirals is with accessory equipment (Goetz and Kallai, 1962). It can also be obtained by measuring the length r of the line perpendicular to the top edge of the foil and extending to the deposit end, which is formed by the arc of radius r + $\bar{r}_o$ (Figure 2). Goetz and Kallai (1962) gave the relationship between ℓ_D and r as

$$\ell_D = \frac{1}{2}\left\{\frac{r(\bar{a}^2 + r^2)^{1/2}}{\bar{a}} + \bar{a}\ \ln\left[r + (\bar{a}^2 + r^2)^{1/2}\right]\right\}_{r_0}^{r + \bar{r}_0}, \tag{3}$$

where $\bar{r}_0$ = 7.0 cm, r_0 = 7.4 cm, and $\bar{a}$ = 1.77 cm/rad is the pitch of the deposit spirals. An approximate expression,

$$\ell_D \approx 3.0\ r^{1.355}, \tag{4}$$

accurate for $\ell_D \geqslant 4.0$ cm, can also be used.

Under the assumption that the deposit concentration of monodisperse particles on the centrifuge deposit is uniform and ends abruptly at each ℓ_D, a deposit of polydisperse particles is evaluated in the same way as an elutriator deposit. The discrete size distribution of the particles is given by

$$\frac{\Delta N}{\Delta D} = \frac{\Delta C}{\Delta \ell_D}\ \frac{\Delta \ell_D}{\Delta D}\ \frac{A_D}{F\ t}. \tag{5}$$

The number C of particles per unit area is counted at different ℓ_D values near the midline of the deposit, and the resulting curve of C vs. ℓ_D is differentiated to yield $\Delta C/\Delta \ell_D$. The ratio $\Delta \ell_D/\Delta D$ is obtained from calibration data which relates D of monodisperse particles to their ℓ_D value (Figure 3). The last ratio is a weighting factor needed to give the size distribution an absolute value; it includes the area A_D on the deposit covered by particles of size D, the flow rate through the channels, and the time t during which the aerosol sample is collected.

The cumulative size distribution of polydisperse particles is simply

$$\frac{N(<D)}{M} = \frac{1}{M\,F\,t} \sum_{C=0}^{C} A_D\,\Delta C, \tag{6}$$

where N(<D) is the number of particles per volume of aerosol with sizes smaller than D and M is their total number.

INSTRUMENT EVALUATIONS

The purpose of almost all experimental evaluations of the Goetz aerosol centrifuge was to establish the geometry of the centrifuge deposit with monodisperse latex particles and thus verify the accuracy of Eq. (5) for sizing polydisperse aerosol. All the evaluations showed a more or less sharp dropoff in the latex particle concentration curves, which is the first requirement for an elutriatorlike deposit. The values of ℓ_D and D corresponding to the dropoffs are shown in the calibration curves provided by the builders of the centrifuge (Figure 3). These curves, apparently for an early centrifuge model, differ somewhat from those

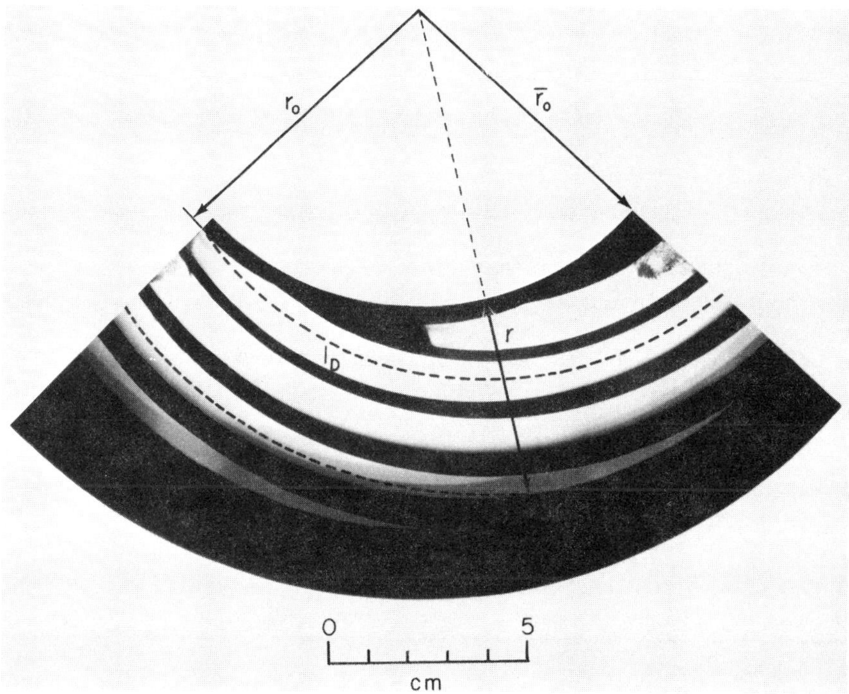

Figure 2. Flattened collecting foil with a deposit of polydisperse latex particles. The deposit length of the monodisperse fraction is ℓ_D.

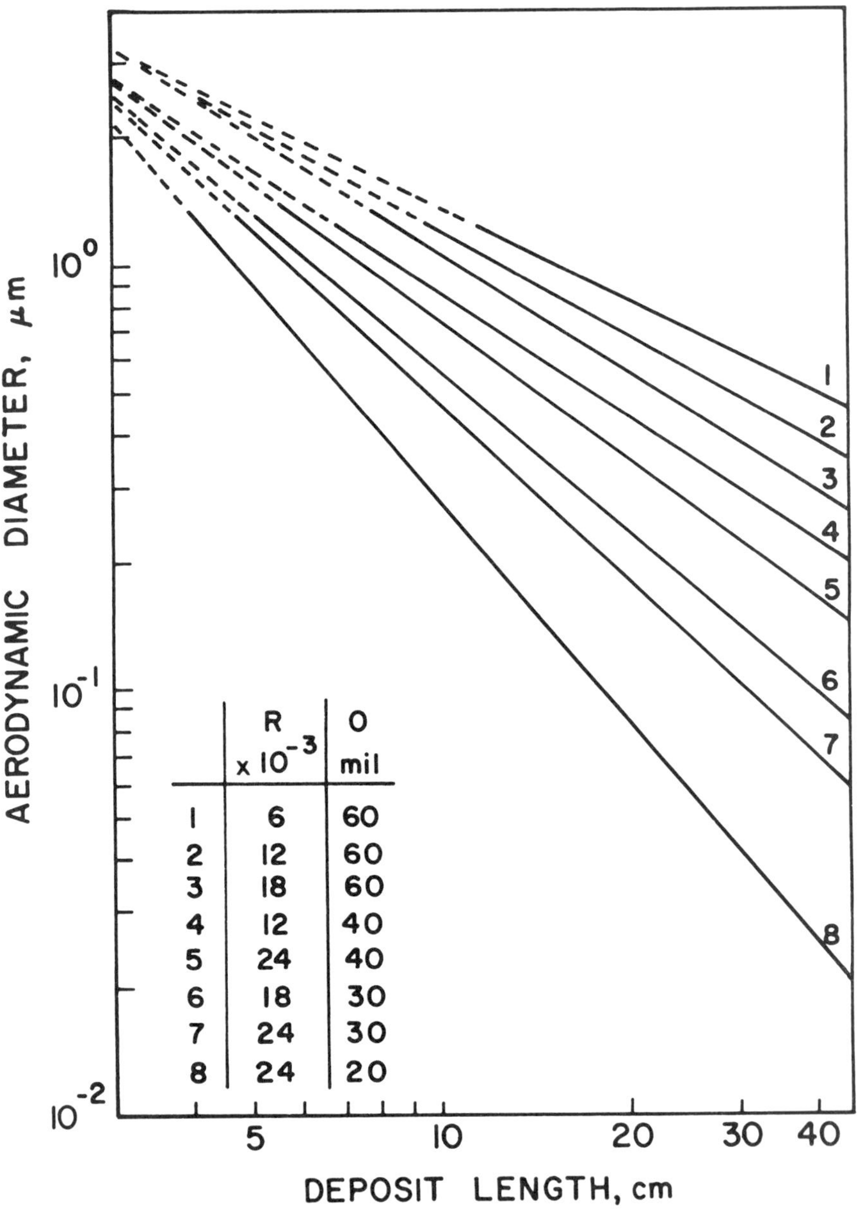

Figure 3. Deposit length for particles of a given aerodynamic diameter as a function of centrifuge rpm R and the size O of the flow-limiting orifice, according to Goetz, Stevenson and Preining (1960).

corresponding to the commercial instrument. The curves for the commercial instrument have proved to be reproducible.

Evidence for the second requirement, that of a uniform deposit concentration for monodisperse particles, was given by Preining, Stevenson, and Goetz (1959). Their two concentration curves measured along the midline of the deposit for two values of O, and one value each of R and D (Figure 4) did indeed show the required uniform deposit concentration as well as the sharp dropoff. However, the originators of the centrifuge neglected to give concentration curves for other possible combinations of D, R, and O (D/R/O), thus leaving the impression that the instrument operated ideally for at least all D/R/O described by the curves in Figure 3. Subsequent evaluations (Stöber and Zessack, 1964; Baust, 1967; Raabe, 1967; Stöber and Boose, 1973; Gerber, 1971) of the

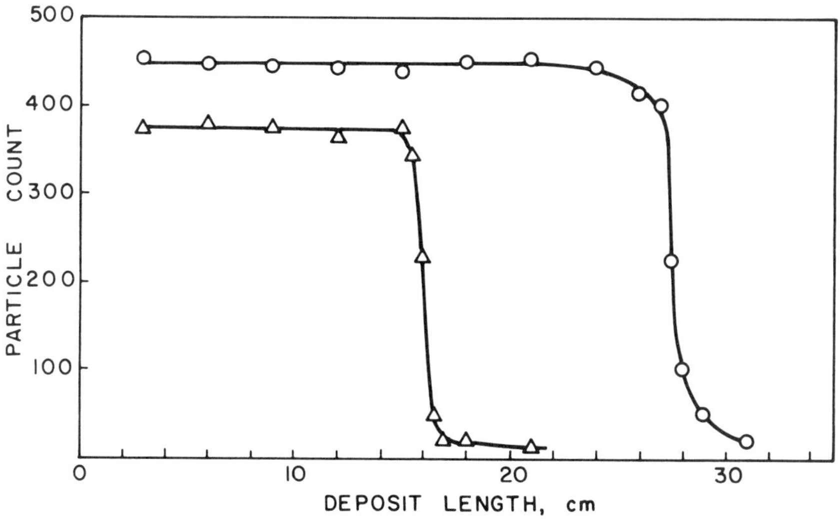

Figure 4. Concentration of latex particles 0.56 μm in diameter along the midline of the foil deposit for R = 12,000, O = 40 (Δ), and O = 60 (o), from Preining, Stevenson and Goetz (1959).

centrifuge tested other D/R/O and found concentration curves which usually were non-uniform and sometimes showed broad cutoffs. The reputation of the centrifuge for erratic behavior came about because investigators usually tested their own choice of D/R/O and found shapes of concentration curves that had not been seen previously.

The last (Gerber, 1971) of the series of evaluations found a consistent pattern in all of the previous calibration work. All latex-particle concentration curves, measured near the midline of the deposit, had three distinct regions as shown in the example in Figure 5. The curves increased over the distance ℓ_a, had a nearly constant slope S_1 (with both positive and negative values) from ℓ_a to ℓ_b, and dropped off with an average slope of S_2 past ℓ_b. It was discovered that for any given operating condition of the centrifuge, i.e., any given values of R and O, the shape of the concentration curves for various values of D was nearly invariant when the length ℓ_b - ℓ_a was normalized (Figure 6). The previous calibration data as a whole reflected that behavior (Figure 7). Some concentration curves showed the required $S_1 = 0$, although most showed $S_1 < 0$, which behavior was also found by Baust (1967).

Figure 8 shows a wide range in the slopes S_2 of the dropoff region; however an overall pattern in the data is not as evident as for S_1, perhaps due to the greater sensitivity of the cutoff to the technique of latex particle generation. As a general rule large values of $1/S_1$ are found when $R \gtrsim 20000$ rpm and when flow rates greater than 5 ℓpm are used. Although the operating conditions exceeding those limits still give a Reynold's number reflecting laminar flow in the channels, it appears that those limits mark a transition to turbulence in the channel entrances.

A series of calibration runs was repeated with the baffle on the inlet tube removed (Gerber, 1971). Those concentration curves consistently showed values of $S_1 \geqslant 0$ (Figure 7). Velocity-profile measurements near a channel entrance in the centrifuge (point P, Figure 1) run with and without the baffle at the same values of R and O showed major differences, suggesting that the inlet geometry

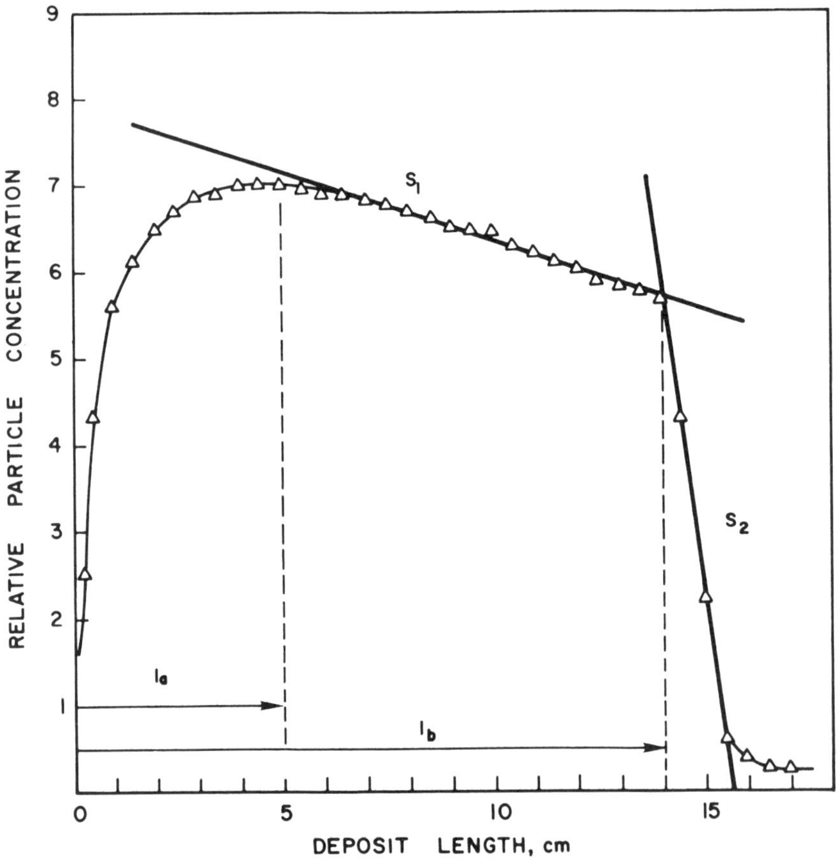

Figure 5. Typical latex particle concentration curve (D = 0.357 μm, R = 18,000, O = 30 mil).

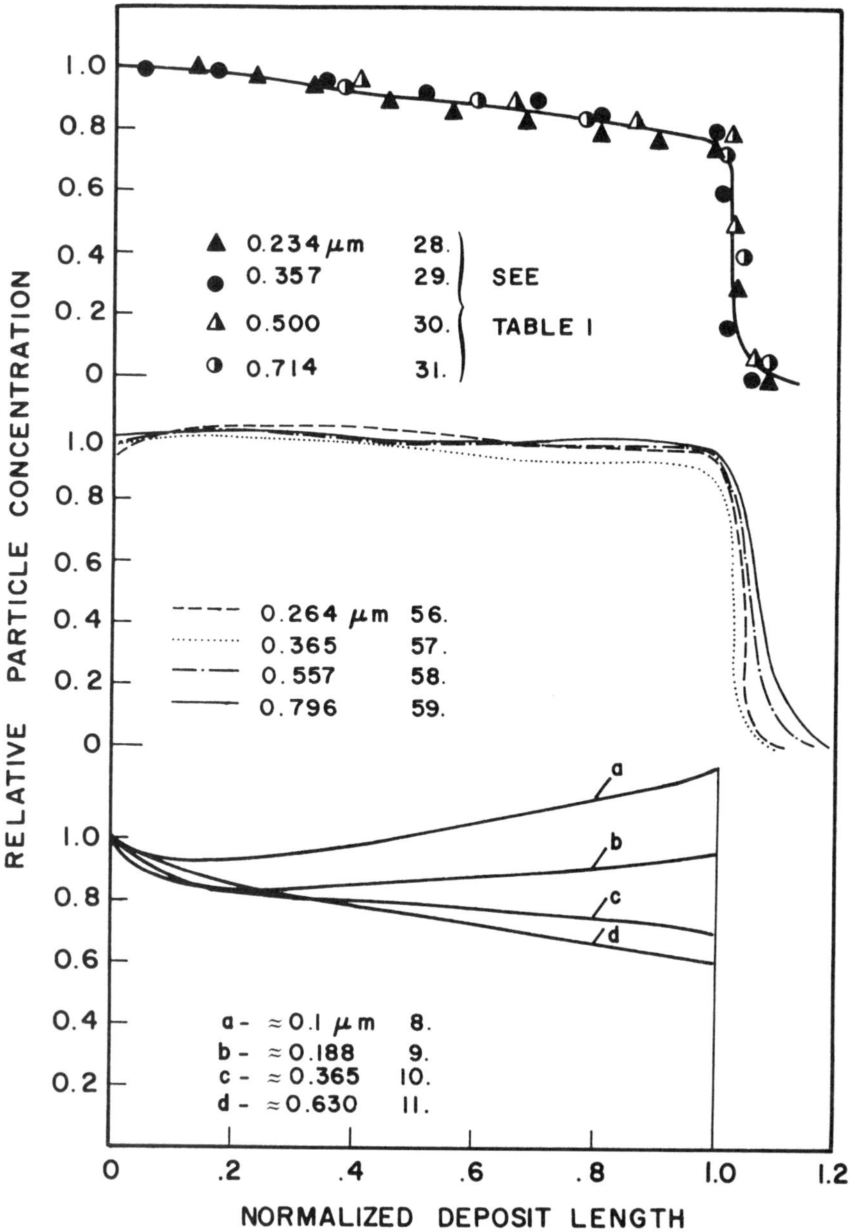

Figure 6. Experimental latex particle concentration curves (Gerber, 1971, upper curve; Horvath, 1973, middle curve) and theoretical predictions by Stöber and Zessack (1964), lower curve. All curves were normalized over the distance $\ell_b - \ell_a$ and they reflect the same operating conditions of the centrifuge. The instrument used by Horvath (1973) differed slightly from the Goetz centrifuge in baffle geometry and channel dimensions.

influences the slope of the concentration curves by causing strongly developing velocity profiles in the first few centimeters of the channels.

Stöber and Zessack (1964) calculated the concentration of monodisperse particles along the channel deposit for a given set of operating conditions under the assumptions that the particles were uniformly distributed across the channel at point ℓ_a, that laminar, fully developed flow existed beyond ℓ_a, and that no velocity gradients existed between the inner and outer channel walls. Their

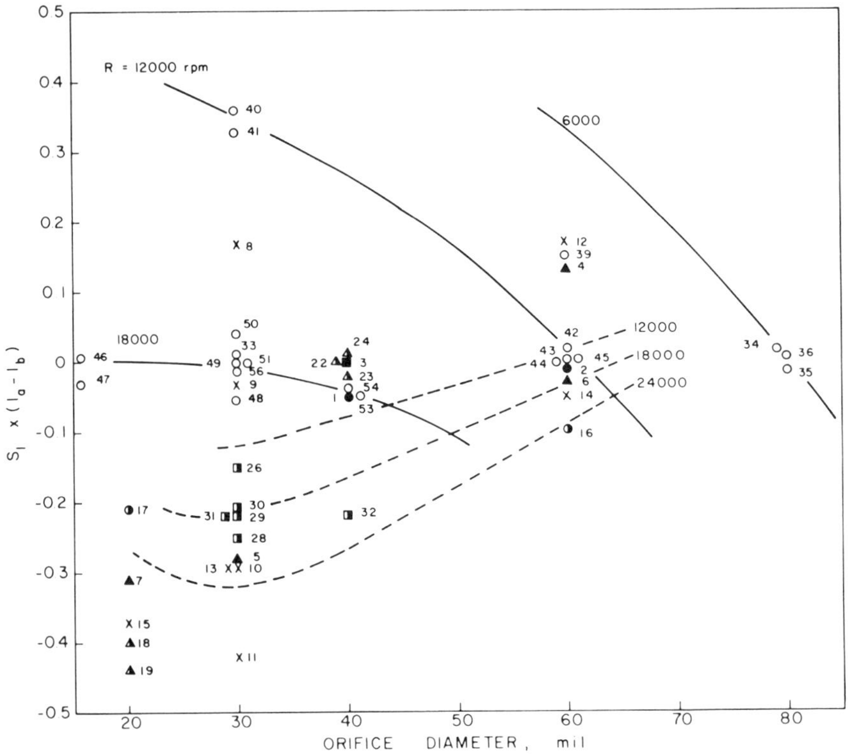

Figure 7. Normalized slope S_1 of latex-particle concentration curves as a function of the operating conditions of the centrifuge. The stock instrument is indicated by dashed lines, and the instrument with baffle removed is indicated by solid lines. See Table I for the key.

theoretical results do not compare favorably with the pattern demonstrated by the experimental concentration curves of others (Figures 6 and 7), or with their own experimental curves included in Figure 7. Stöber and Boose (1973) qualitatively attributed the disagreement to the effect of the developing velocity profile as well as secondary circulation in the channel induced by the Coriolis force.

Figures 7 and 8 show that operating conditions exist which give the required elutriatorlike deposit at least along the center line of the deposit. Horvath (1973) pointed out, however, that the instrument operating with nonuniform concentration curves could still give the correct size distribution of aerosols if recurrence correction formulas are applied to Eq. (5).

TABLE I. Key for Figures 7 and 8. Each D(μm)/R(rpm × 10^{-3})/O(mil) Represents a Calibration of the Goetz Centrifuge with Latex Particles. For Some Tests D Is Expressed as a Function of ℓ_D

● Preining et al. (1959)	▲ Baust (1967)	37. 1.275/6/80
1. 0.560/12/40	18. 0.163/22/20	38. 1.417/6/80
2. 0.560/12/40	19. 0.365/22/20	39. 1.090/8/60
	20. 0.796/22/20	40. 0.357/12/30
■ Preining (1962)	21. 1.305/22/20	41. 0.357/12/60
3. 0.557/10/40	22. 0.163/22/40	42. 0.500/12/60
	23. 0.211/22/40	43. 0.714/12/60
Stöber and Zessack (1964)	24. 0.365/22/40	44. 1.090/12/60
▲ Experimental	25. 0.796/22/40	45. 0.109/18/16
4. 0.365/6/60		46. 0.234/18/16
5. 0.365/18/30	Gerber (1971)	47. 0.109/18/30
6. 0.365/18/60	◨ centrifuge A	48. 0.234/18/30
7. 0.365/24/20	26. 0.357/12/30	49. 0.357/18/30
	27. 1.090/12/60	50. 0.500/18/30
× Theoretical	28. 0.234/18/30	51. 0.714/18/30
8. l_D(60cm)/18/30	29. 0.357/18/30	52. 0.375/18/40
9. l_D(32cm)/18/30	30. 0.500/18/30	53. 0.500/18/40
10. l_D(16cm)/18/30	31. 0.714/18/30	54. 0.714/18/40
11. l_D(10cm)/18/30	32. 0.357/22/40	55. 0.357/22/30
12. 0.365/6/60		
13. 0.365/18/30	○ centrifuge A,	◒ Horvath (1973)
14. 0.365/18/60	baffle removed	56. 0.264/18/30
15. 0.365/24/20	33. 0.357/18/30	57. 0.365/18/30
		58. 0.557/18/30
◐ Raabe (1967)	○ centrifuge B,	59. 0.796/18/30
16. 1_D(60cm)/24/60	baffle removed	
17. 1_D(32cm)/24/20	34. 0.357/6/80	
	35. 0.714/6/80	
	36. 1.090/6/80	

ACCURACY OF PARTICLE SIZING

Deposit Area

The accuracy of Eq. (5) also depends on the values of A_D, which are only approximately equivalent to the area $\ell_D \times b_0$ available for particle deposition on the collecting surface (b_0 is the distance between the upper and lower channel walls). Goetz and coworkers (Preining, Stevenson and Goetz, 1959; Goetz and Kallai, 1962; Preining, 1962) realized early that deposits of monodisperse particles were not of width b_0 but instead gradually narrowed to a fraction of that value for long deposits. Further they found that the deposit concentrations away from the midline of the deposit were nonuniform to some degree (e.g., Figure 9), even though the midline concentration was constant. They measured the areal extent of the latex particle deposits with a planimeter and divided that area by the maximum possible coverage, $\ell_D \times b_0$. The resulting ratio K_1 is applied to Eq. (5) to correct the assumption that $A_D = \ell_D \times b_0$. Similar measurements (Gerber, 1971) which in addition took into account the deposit inhomogeneities away from the deposit midline gave values of K_1 which agreed well with the earlier values (Figure 10).

Entrance Losses

Baust (1967) determined particle losses in the centrifuge inlet by measuring the activity of radioactively tagged particles on the centrifuge deposit as well as on a parallel reference filter. His results are given in Figure 11 along with the results of recent work based on the counts of latex particles. No obvious explanation was

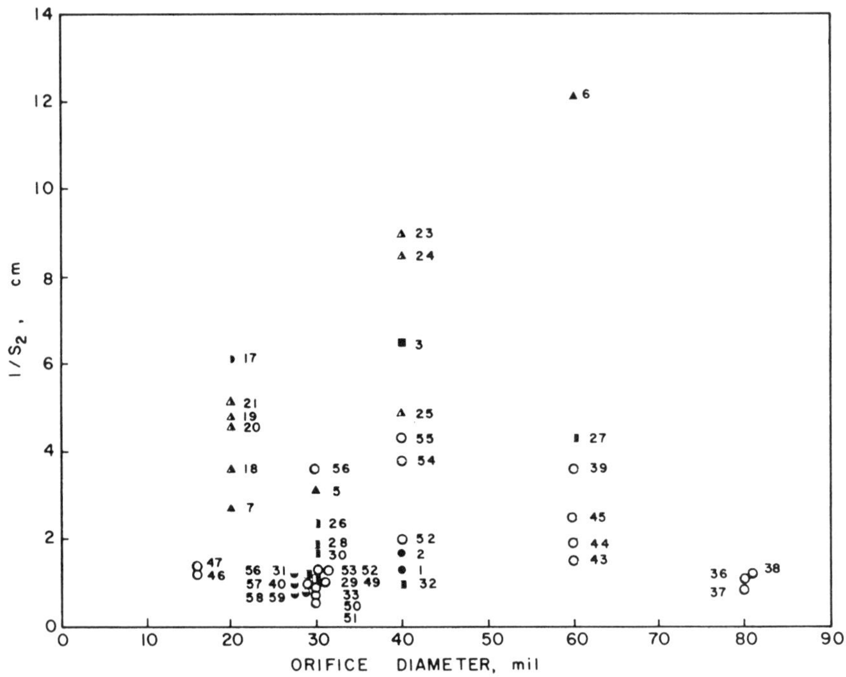

Figure 8. Slope S_2 of the cutoff region for latex particle concentration curves. $S_2 = \Delta C/\Delta \ell$, where $\Delta C = 1.0$.

found for the significant differences seen between the data of Baust and the other data. The recent work shows that the losses depend on the operating conditions of the centrifuge. Also, the losses are small for particles with sizes that correspond to $\ell_D > 6$ cm, which is the location on the deposit about opposite the channel entrance. The large loss of particles with smaller ℓ_D values suggests that those particles are lost in the short connection between the inlet tube and the channels.

Pressure Sensitivity

The reduction of atmospheric pressure from the sea-level value shortens the values of ℓ_D and requires that a correction be applied to the calibration curves

(Figure 3) and to the value of F in Eq. (5) (Gerber, 1964). Two factors contribute to shorten the deposits: The particle slip correction becomes more pronounced at lower pressures, and the flow rate through the channels decreases. The correction factor K_2 applied to values of D in Figure 3 is given with an accuracy of 5% by the empirical relationship

$$K_2 = \left(\frac{D}{C_1}\right)^{2 \exp - \left[\left(\frac{76}{76 - P}\right)^{C_2}\right]}, \tag{7}$$

where P is atmospheric pressure in cm Hg, D is given in micrometers, $C_1 = 3.0$ and $C_2 = 0.95$ for operating conditions with $O \leqslant 0.75$ mm, and $C_1 = 2.0$ and $C_2 = 0.98$ for $O > 0.75$ mm. After the calibration curves are corrected, new values of $\Delta\ell_D/\Delta D$ must be calculated. The value of F which in addition affects the absolute value of the size distribution through Eq. (5) was found to require a factor of 0.8 at P = 30 cm Hg.

Resolution Limit

The ability of the centrifuge to resolve fine structure in a polydisperse size distribution depends on the sharpness of the deposit dropoff for particles of each size in that distribution. A measure of the sharpness is the distance on the deposit over which the dropoffs of monodisperse latex particles are smeared. Such measurements were made by fitting cumulative normal curves to the dropoff regions of latex particle concentration curves (Gerber, 1971). Subtracting the known value of the variance in the size of the latex particles from the variance of the normal curves gave the dispersion in the dropoff inherent to the instrument. The resolution limit of the centrifuge is defined as the relative standard deviation (σ/D) of that dispersion.

The resolution limits shown in Figure 12 for several operating conditions of the centrifuge show that exceptionally precise particle sizing is possible. Similar resolutions should exist for other operating conditions as long as $R \leqslant 20000$ rpm and $F \leqslant 5$ ℓpm, since under those limits the average slope S_1 in the dropoff region was found to be steep (Figure 8). The precise resolution suggests that the flow throughout the channels is laminar for those operating conditions, since otherwise the sharp border between aerosol and particle free air in the channels which creates the fine resolution would not exist.

Statistical Sampling Error

The particle counts per unit area along the deposit which form the basis of the derived particle-size distribution introduce a statistical sampling error governed by the Poisson distribution. Some have suggested that the magnitude of that error was sufficient to make use of the centrifuge impractical (Hochrainer, 1972; Horvath, 1973). The relative statistical sampling error was calculated for a range of particle-size distributions to place that suggestion in proper perspective. The size distributions were described by the Junge distribution

$$dN/d\log D = C/D^{\gamma}, \tag{8}$$

for which $-2 \lesssim \gamma \lesssim 3$, thus encompassing most possibilities. In the evaluation of those hypothetical distributions on the centrifuge deposit, eleven equal areas

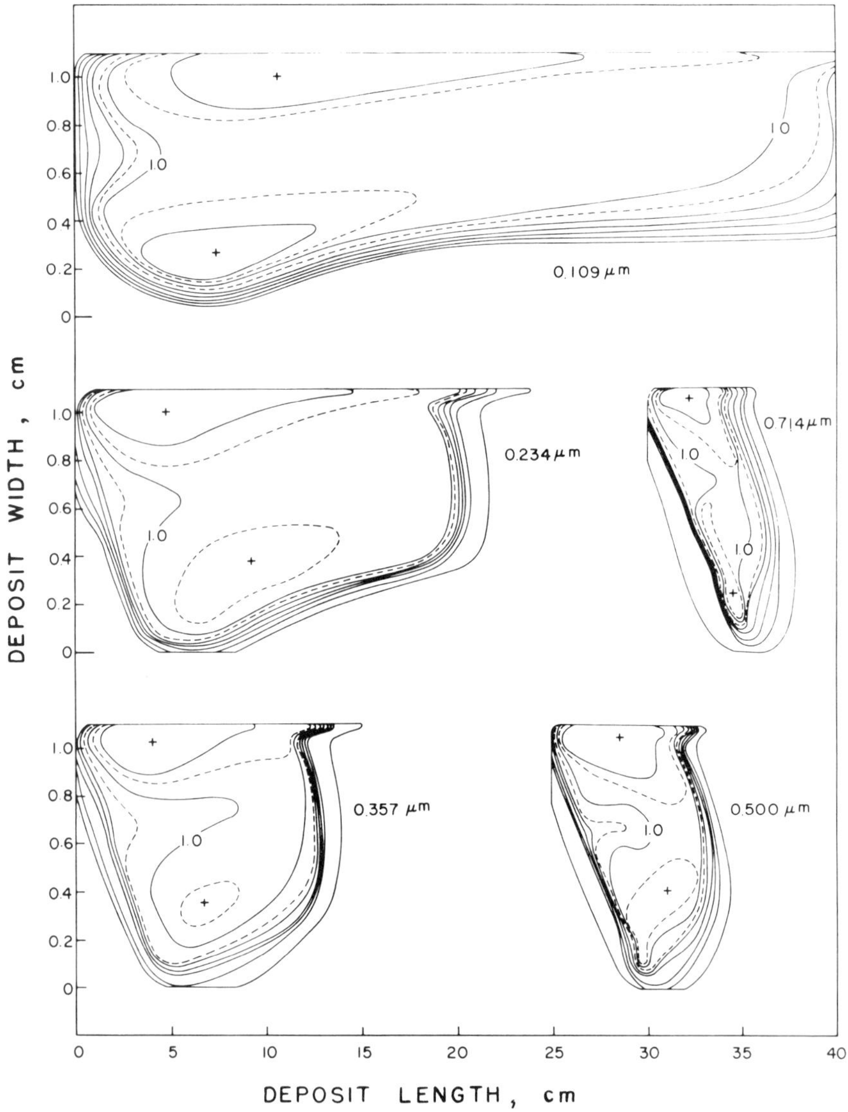

Figure 9. Deposit concentration of monodisperse latex particles for the centrifuge with the baffle removed (R = 18,000, O = 30 mil). Separation between solid concentration isolines is 0.2, and that between solid and dashed lines is 0.1. The outermost line is 0.0, and areas labeled (+) have values larger than 1.0.

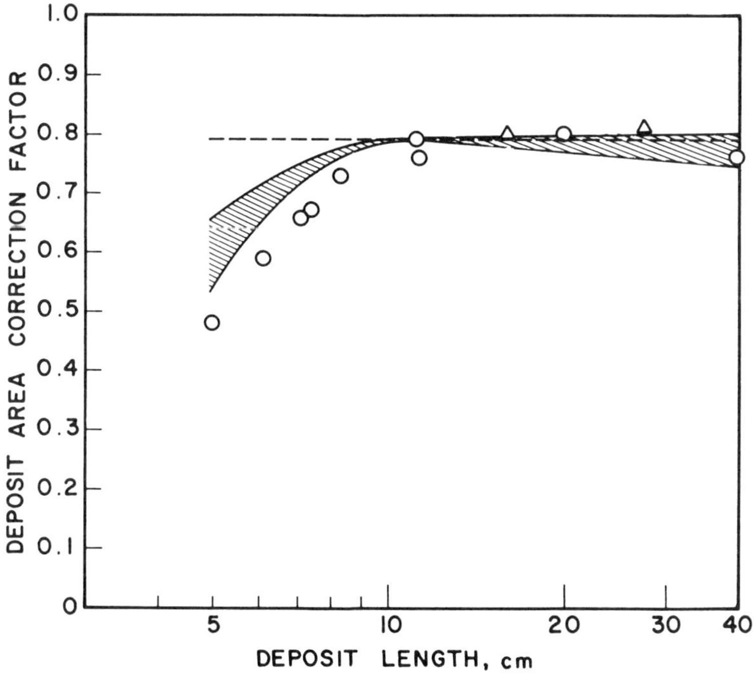

Figure 10. Deposit area correction factor K_1 vs. ℓ_D to correct for the assumption that the deposit width is a constant 1.15 cm.

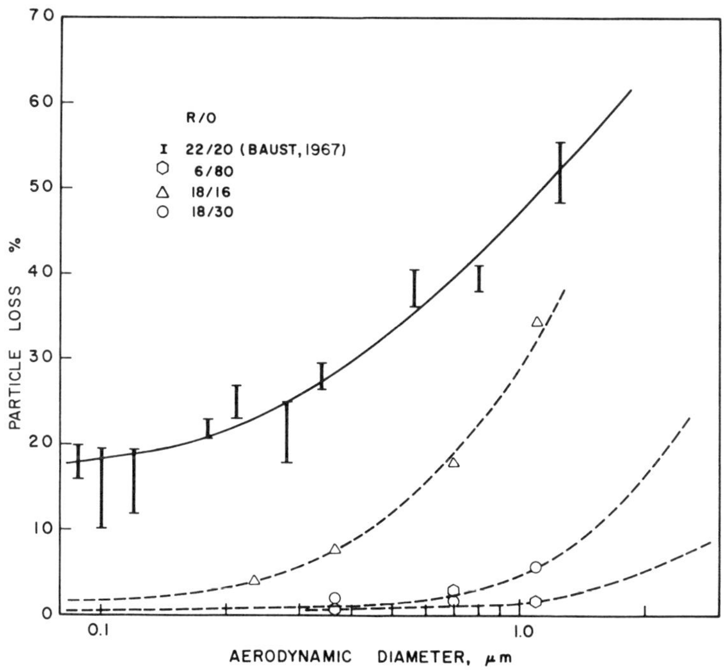

Figure 11. Loss of particles in the inlet of the centrifuge.

were counted to give a ten-point size distribution with the points separated by equal geometric intervals over one decade of particle size. The total count was 25000 particles. The average relative statistical sampling error was calculated for discrete and cumulative size distributions as well as for the total number of particles (Figure 13). As expected, the error is small for all cumulative distributions, although it is also reasonable for discrete distributions with $\beta \lessapprox 2$. However, for size distributions with a predominance of small particles (e.g., in the atmosphere $\beta \approx 3$), a substantial error is found. By reducing the number of

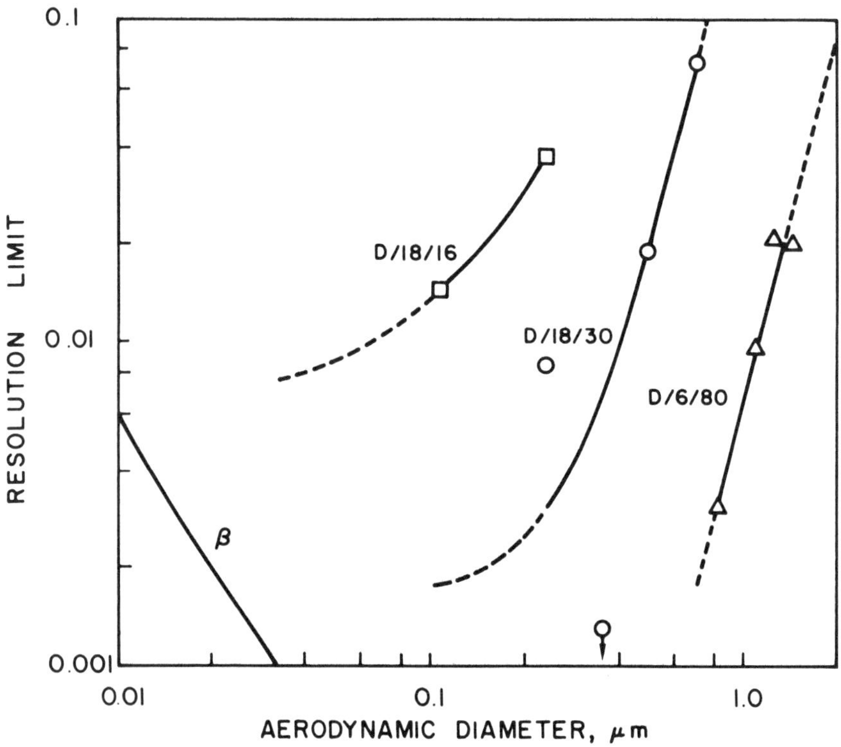

Figure 12. Resolution limit σ/D of the Goetz centrifuge. The contribution from the Brownian diffusion of the particles is β.

data points in the distribution or increasing the particle counts per point, the errors shown in Figure 13 can be significantly reduced.

CONCLUSIONS

In retrospect the lack of sufficient calibrations of the Goetz centrifuge by its builders was the principal reason the instrument developed a poor reputation. The few data they showed indicated that the centrifuge operated ideally in that it gave elutriatorlike deposits for monodisperse particles. Most of the subsequent calibrations showed deposits of other shapes. When all calibrations were compared, however, they showed a consistent pattern, with the same deviations from the ideal elutriator behavior. By taking those deviations into account, as has been described here, the instrument can be successfully applied to measure the size distribution of aerosol particles.

The weakness of the instrument lies in the large statistical sampling error which is unavoidable when trying to measure the discrete size distribution of aerosols with a much greater number of small than large particles.

The strengths of the instrument are the relatively high sampling rate, accurate

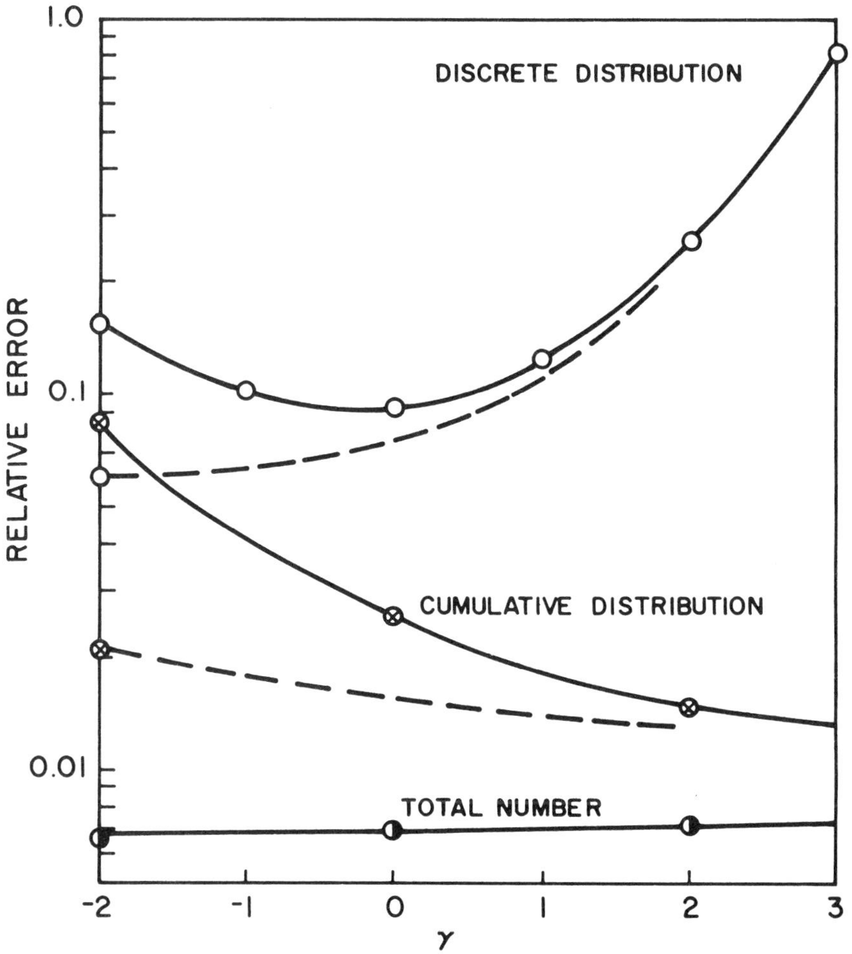

Figure 13. Average statistical sampling error for ten-point size distribution described by the Junge distribution, $dN/d \log D = C/D^{\gamma}$, over the particle-diameter interval 0.1 to 1.0 μm and for a total particle count of 25,000 particles. The dashed lines give the error when the counting areas on the centrifuge deposit are optimized.

measurement of the cumulative size distribution, and ability to measure particles with diameters as small as a few hundred angstroms (Gerber et al., 1970; Gerber, 1976). Operation of the Goetz centrifuge at much higher rpm values than those of other aerosol centrifuges makes the measurement of such particles possible and a unique capability.

REFERENCES

Barr, G., 1934: Two designs of flow meter, and a method of calibration. *J. Sci. Instrum., 11*, 321–324.

Baust, E., 1967: Use of the Goetz Aerosol Spectrometer to measure the size spectra of polydisperse aerosols. *Staub-Reinhalt. Luft, 27*, 16–23.

Gerber, H. E., 1964: Pressure calibration of the Goetz Aerosol Spectrometer. Ft. Monmouth, N.J., U.S. Army Electronics Command Technical Report No. 2456.

——, 1971: On the performance of the Goetz Aerosol Spectrometer. *Atmos. Environ., 5*, 1009–1031.

——, 1974: Discussions: Developing flow and particle deposition in horizontal elutriators and semi-dispersive aerosol centrifuges. *Atmos. Environ., 8*, 1344–1346.

——, 1976: Relationship of size and activity for AgI smoke particles. *J. Atmos. Sci., 33*, 667–677.

——, U. Katz, C. I. Davis, and L. O. Grant, 1970: Some size distribution measurements of AgI nuclei with an aerosol spectrometer. *J. Atmos. Sci., 27*, 1060–1067.

Goetz, A., H. J. R. Stevenson and O. Preining, 1960: The design and performance of the aerosol spectrometer. *J. Air Pollution Control Assoc., 10*, 378–383.

—— and T. Kallai, 1962: Instrumentation for determining size- and mass-distribution of submicron aerosols. *J. Air Pollution Control Assoc., 12*, 470–486.

Hochrainer, D., 1972: Discussions: On the reliability of measurements with the Goetz Aerosol Centrifuge. *Atmos. Environ., 6*, 699.

Horvath, H., 1973: Discussions: On the performance of the Goetz Aerosol Spectrometer. *Atmos. Environ., 7*, 1003–1011.

Preining, O., 1962: Das Goetzsche Aerosolspektrometer, probleme bei betrieb und auswertung. *Staub-Reinhalt. Luft, 22*, 129–133.

——, H. J. R. Stevenson and A. Goetz, 1959: The analysis of aerosol spectra. 52nd Annual Air Pollution Association Meeting, Los Angeles. Preprint 59-42, 29 pp.

Raabe, O. G., 1967: Calibration and use of the Goetz Aerosol Spectrometer. *Assessment of Airborne Radioactivity*. Vienna, International Atomic Energy Agency.

Sawyer, K. F. and W. H. Walton, 1950: The 'Conifuge'—a size-separating sampling device for airborne particles. *J. Sci. Instrum., 27*, 16–23.

Stöber, W. and U. Zessack, 1964: On the theory of a conical aerosol centrifuge. *Staub-Reinhalt. Luft, 24*, 295–305.

—— and C. Boose, 1973: Developing flow and particle deposition in horizontal elutriators and semi-dispersive aerosol centrifuges. *Atmos. Environ., 7*, 119–130.

QUESTIONS

Q: I'd like to know the physical size of the last instrument described by Dr. Stöber?

Stöber: While the prototype design followed the conventional rotor design of several years ago, the new rotor model which we are working on now has a diameter of about 33 centimeters and the geometry of the duct will be somewhat different. There are only 1½ turns on the spiral duct, and the width of the duct has been reduced to something like 7 mm, instead of the usual depth of 35 mm.

Q: What about the humidity response and the difficulties with using this type of instrument?

Stöber: One has to maintain constant relative humidity at least to a reasonable degree. The shift of the frequency in this case, where the basic frequency is at 10 megacycles, is about 1 cycle per percent relative humidity and we can adjust to that. The sampling rate is up to 10% of the total flow. If we maintain a clean air flow at 0% humidity and we introduce aerosol at 100% humidity then we have only 10% humidity in the air mixture. Within 0 and 10% we can adjust for the differences caused by changes in humidity. We can also provide the possibility of just using humid air at a certain level and maintaining that humidity within the duct. Some particles, particularly hydrophilic ones, grow and shrink with humidity; this is something one has to take into consideration.

Q: What can you do to prevent water condensation when you are sampling a hot humid air stream?

Stöber: We have used this instrument for sampling diesel exhaust which has a lot of humidity. We use a silica gel column with a hollow center core so the diesel exhaust aerosol goes through that column and the water diffuses to the surrounding silica gel wall. That works very well and we have no dewpoint problems whatsoever. Of course, again, particles might shrink and swell with the different humidity levels but I think that is something one has to keep in mind and can live with.

Q: I have a question for Otto Raabe. He spoke about the importance of discharging particles. If we discharge an aerosol, we change its deposition properties. Shouldn't we really be sampling the aerosol in its natural charged state?

Raabe: There are two times when you have to worry about charge. First, if you are calibrating with monodispersed polystyrene latex particles you have to discharge them, or else you are going to get some anomalous deposit positions because of the charge and you won't be able to accurately define the physical size and density of those particles. The second case is in field sampling, for example, plutonium aerosols in glove boxes. We have found a high electrostatic charge which did lead to anomalous results with respect to the deposition both in cascade impactors and in the spiral disk centrifuge if we didn't first pass the aerosol through a Krypton 85 discharger. Now I say we must discharge because there is no way to characterize this complicated phenomenon that relates both to electrostatic charge and aerodynamic diameter. They should be characterized separately.

Study the charge in one set of experiments. Study the actual aerodynamic properties of a particle without charge in another set.

Q: What do you mean by discharging?

Raabe: Discharging here in the broadest meaning means that we have reduced the charge to the Boltzmann equilibrium. This is not actually completely discharging, but reducing the charge to such a relatively small amount that the behaviour of the particles is not markedly affected by their charge. We are talking about reducing charges from hundreds or thousands of charges per particle to Boltzmann equilibrium which involves some uncharged, some with positive, some with negative; but small charges do exist.

Stöber: Maybe I should add some remarks to that. I think from our experience with the spiral centrifuge the only difference we obtained in the deposit of charged or uncharged particles is that the arrangement of the particles deposited close to the foil is different when the particles are charged than when they are uncharged. The reason is that a particle which is settling in a location and cannot get rid of its charge throws up a field and influences an approaching particle which is settling in the same location. The electron micrographs are different depending on whether particles were charged or not, or whether they settled on an insulating foil which we use very often, such as Formvar film, or on a metal foil. The location, as far as my experience goes, is not changed, only the picture of the deposit.

Moss: I'd like to make a comment about the question of charge. Again, this comes back to the basic thing we'll be talking about in this Workshop and that is what are we trying to measure? With the centrifuge devices we're measuring a very specific thing, terminal settling velocity. And why are we measuring that? Because it is one of the mechanisms by which the lung takes out particles from the air. So the question on charge, in terms of these instruments, reduces to how long does a charge remain on a particle when it's in 100% humid atmosphere such as you have when you are inhaling it into your lungs? In that sense you want to discharge your particle when you are looking at this particular property.

Tillery: I think the important thing here is that latex aerosols that are generated from generators have very high charges, extremely high charges, and such charge levels will not normally be encountered in sampling conditions. The enhanced deposition is a result of image forces which require fairly high electric charges. So the type of aerosol you're normally going to encounter isn't going to be a problem, and it's not going to be a problem, I think, in relating the deposition in these instruments to lung deposition.

Q: If you are sampling a natural unknown aerosol, what is the accuracy with which you can determine size distribution and how does this relate to optical size measurement?

Raabe: That's an exceptionally difficult question. Size distribution measured with a centrifuge depends on the density and shape of the particle. Optical counter sizes depend on the refractive index of the particles, while centrifuges do not. That is the crux of the problem, and I don't believe there is an answer yet.

Stöber: I think the problem really is whether you can relate mass distribution measurement in terms of aerodynamic diameter to number distribution measurement. You can do both analyses with the instrument. You can put Formvar film along the sampling foil, take it off, and weigh your deposits. Immediately you get a mass distribution in terms of aerodynamic diameter. This is not necessarily related to a number distribution because if you sample an unknown aerosol having particles with all kinds of shape factors, then you are at a loss. There is no direct relationship to the number distribution and vice versa. If you look at electron micrographs and count your particles in certain locations then you come up with a number distribution but you cannot transfer that into a mass distribution without knowing shape factors. Of course you can determine shape factors from electron micrographs, but it's a messy procedure. This is a basic problem which we cannot get away from. It's always worthwhile to state what you are referring to; a primary mass distribution measurement, or a primary number distribution measurement. The other kind of distribution cannot directly be derived from the first measurement.

ANALYSIS OF THE CYCLONE AS A SIZE SELECTIVE AEROSOL SAMPLER

J. M. Beeckmans

Faculty of Engineering Science, The University of Western Ontario, London, Ontario, Canada

The reverse flow cyclone has long been used as a cheap and reliable device for removing relatively coarse particles from air streams. Although all cyclones will collect particles in all sizes with at least some measure of efficiency, cyclones are not generally considered to be efficient devices for removing particles smaller than 2.0 to 3.0 μm. Recent experimental studies on cyclone efficiency have, however, shown that quite high collection efficiencies can be obtained with submicron particles in some circumstances (Chan and Lippmann, 1975; Blachman and Lippmann, 1974; Knuth, 1969; Lippmann and Chan, 1974; Lippmann and Kydonieus, 1970). A wide range of cut sizes (also called effective cut-off aerodynamic diameters (ECAD), defined as the aerodynamic diameter of particles having a collection efficiency of 0.5 at the stated operating conditions) may be obtained by suitable selection of cyclone geometry, cyclone diameter and inlet velocity, which raises the possibility of using calibrated cyclones for determining size distributions in heterodisperse aerosols. A multistage sampler with six identical cyclones in parallel, all operating at a different flow rate, was developed by Lippmann and Kydonieus (1970) for this purpose. A multistage sampler based on cyclones is inherently able to collect much larger samples than a cascade impactor, and may be capable of yielding much more accurate information than the impactor, particularly in determining particle size distributions by mass weight. The design of efficient cyclone samplers, and the extrapolation of data to conditions outside of the calibration range requires some understanding of cyclone theory and cyclone operation, and it is our purpose to review existing data and theories to assist in this objective.

DIMENSIONAL SIMILARITY IN CYCLONES

Dimensional similarity between two geometrically similar cyclones, or between two operating conditions in a given cyclone, requires, as a minimum, similarity in fluid flow and in particle trajectories. These conditions will be satisfied by the equality of the cyclone Reynolds numbers (*Re*) and of the Stokes numbers (*Stk*). There may, however, be other mechanisms or forces at play beyond those scaled by equating the Reynolds numbers and Stokes numbers. These include electrostatic forces, turbulent velocity, fluctuation, induced inertial impingement of particles on the cyclone walls, reentrainment of deposited particles, and elastic recoil of particles on deposition. Moreover, in some cases the particle Reynolds number based on centrifugal drift velocity may be outside the Stokes law range, in which case the inertial forces will not be correctly scaled by the Stokes number. All of these complications should be borne in mind in evaluating cyclone data, and in extrapolating such data outside the calibration interval.

REVIEW OF CYCLONE THEORIES

Early cyclone theories, such as those of Rosin, Rammler and Intelman (1932), and Barth (1956), considered only inertial forces and resulted in models which predict capture efficiencies of unity for particles slightly larger than the cut size. This conclusion, which is contary to observation, resulted from neglect of the role of turbulent diffusion. Nevertheless, although the older theories failed to predict a reasonable shape for the efficiency curve, they did predict the cut size of most industrial cyclones reasonably well. The cut size may be expressed in terms of *Stk* and of the cyclone geometrical parameters in these theories.

Subsequent theoretical developments have included effects attributable to turbulent diffusion. Leith and Licht (1972) assumed complete mixing of particles in a horizontal plane, and a deposition velocity at the wall which was a function of the axial distance coordinate. This velocity was taken as equal to the centrifugal drift velocity which a hypothetical particle initially released near the center of the cyclone would have attained by the time it had descended to the level of the horizontal plane in question. The model is physically and conceptually unsound; nevertheless its predictions are in good agreement with limited data on industrial cyclones obtained using heterogeneous test dusts. Such data are inherently suspect, however, because it is not clear to what extent the dust had been deagglomerated prior to reaching the cyclone. The agreement between the theory and the data of Lippmann and his collaborators, who used dilute monodisperse aerosols, is generally poor.

Turbulent diffusion was quantitatively taken into account in the theories of Beeckmans (1972, 1973). The results were obtained as solutions of differential equations, whose parameters could be expressed in terms of the Stokes number and of the cyclone geometrical parameters. It was assumed that the turbulent diffusivity is uniform throughout the cyclone, and proportional to the Reynolds number. The expression for turbulent diffusivity was obtained by a momentum balance on the spinning gas, and it is based on the assumption that the tangential gas velocity is inversely proportional to $r^{0.5}$. Data on gas velocity distribution in a cyclone are scarce, and it is generally considered that the exponent varies between 0.5 and 0.7. No data are available on the effect of *Re* on the exponent, but if the exponent were to depend on *Re* the parameters of the differential equation would depend on *Re* as well as on *Stk*. Beeckmans' theories are not particularly successful in correlating experimental cyclone efficiency data.

ANALYSIS OF CYCLONE EFFICIENCY DATA

Data obtained with redispersed powder cannot be considered reliable, and will not be considered further. The data of Lippmann and collaborators were obtained with dilute monodisperse aerosols, and will be considered here, together with limited, unpublished data from our laboratories obtained with monodisperse, uranine-methylene blue aerosol.

The least restrictive assumption for correlating the data for a given cyclone is that the efficiency is a function of the particle size over the domain of *Re* and *Stk*, as well as of the electrical parameters, the elastic constants of the particles, and of the cyclone walls, etc. Such a broad generalization clearly has little utilitarian value in our present state of knowledge of cyclone operation.

The somewhat more restrictive assumption that efficiency may be correlated over the domain of *Re* and *Stk* cannot be verified for Lippmann's data, because the only source of variation for *Re* is the inlet velocity V_i. Thus, although the data cover a certain area of the domain of *Re* and *Stk*, there are no data in which both *Re* and *Stk* were respectively equal in two independent experiments. A reliable test of the assumption in question would require experiments in which cyclone diameter D, gas viscosity μ, gas density ρ_f or particle density ρ_p were varied, as well as particle size and inlet velocity.

While Lippmann's data do not furnish a proof of the validity of the assumption $\eta = \eta(Stk, Re)$, it would seem logical to seek an empirical correlation in terms of these variables. A linear relationship is clearly unsuitable, but a variety of other functions have been found which correlate the data more or less well. Functions which were tested include the following:

$$\ln N_T = a_0 + a_1 \ln (Re) + a_2 \ln (Stk) \quad (1)$$

$$\ln N_T = a_o + a_1 \ln(Re\ Stk) + a_2 \ln^2 (Re\ Stk) \quad (2)$$

$$\eta = F (X) \quad (3a)$$

where

$$X = \ln[(Stk^{0.5}\ Re^b)/(Stk^{0.5}\ Re^b)_{50}]/\ln \sigma_g \quad (3b)$$

TABLE I. Ranges of the Experimental Variables

Cyclone	D (mm)	V_i (m/sec)	d_{ae} (μm)	Re $\times 10^{-3}$	Stk $\times 10^3$	n
Unico	25.4	1.82–2.39	2.3–5.1	3.6–4.7	1.4–7.9	30
Aerotec 3/4	19	1.94–2.47	2.2–6.6	2.9–3.7	1.8–21	30
Aerotec 2	55.2	6.0–8.6	2.7–6.6	26–37	3.0–25	20
Dorr-Oliver $Q \leq 5$ ℓpm	10	3.7–20.8	1.07–9.27	2.9–16	6.3–230	47
Dorr-Oliver $Q > 5$ ℓpm	10	24.3–133	0.24–2.0	19–100	1.9–79	56
Beeckmans	76–152	6.1–21.3	0.71–5.6	54–150	0.054–16	21

NOTE: For all cyclones, it was assumed that the kinematic viscosity had a value corresponding to air at 20°C, 1 atm ($\nu = 1.51 \times 10^{-5}$ m^2 sec^{-1}).

N_T is the cyclone efficiency expressed in transfer units, F is the normalized cumulative log probability function, $(Stk^{0.5}\ Re^{b})_{50}$ is the value of the quantity in brackets at $\eta = 0.5$, σ_g is the geometric standard deviation, and b is an arbitrary constant. The cyclones used by Lippmann and collaborators are described in the literature. Our own experiments were performed with two geometrically similar cyclones with diameters equal to 3 inches and 6 inches. Details of these experiments will be published elsewhere. Experimental variables and their ranges for all cyclones considered are given in Table I. Results of the regression analyses are shown in Tables II and III. Note that the data for the Dorr-Oliver 10 mm cyclone were split into two groups, one for which the flow Q was less than or equal to 5 ℓpm, and one for which Q exceeded this value. As can be seen in Table II, this procedure resulted in a considerable reduction in the standard error of estimate ϵ. The last column in these tables may be

TABLE II. Linear Regression Analyses on the Function $\ell n\ N_T = a_0 + a_1\ \ell n\ Re + a_2\ \ell n\ Stk$

Cyclone	a_0	a_1	a_2	ϵ [a]	e^{ϵ}
Unico	−2.90	1.093	1.179	0.098	1.10
Aerotec 3/4	−6.52	1.58	1.29	0.153	1.16
Aerotec 2	0.205	0.526	1.202	0.133	1.14
Dorr-Oliver, all data	−6.81	1.09	1.10	0.705	2.02
Dorr-Oliver $Q \leq 5$ ℓpm	−3.03	0.96	1.90	0.387	1.47
Dorr-Oliver $Q > 5$ ℓpm	−11.65	1.36	0.67	0.246	1.28
Beeckmans	−3.86	0.584	0.516	0.195	1.22

a. ϵ equals standard estimate of error. ($\epsilon = \Sigma_i(\ln N_{Ti} - Y_i)^2/(n-1)$, Y being the value of ln N_T calculated from the regression equation using appropriate values of Re and Stk corresponding to the i'th observation, and N_{Ti} being the transfer unit efficiency corresponding to the i'th experiment).

interpreted as follows. Let z_i equal the ratio of the observed N_T to that calculated using the regression equation, and suppose that whenever $z_i < 1$ it is replaced by its reciprocal, so that for all z_i, $z_i > 1$. It may be shown that geometric mean value of z_i equals

$$\bar{z}_g = e^{\epsilon\sqrt{(n-1)/n}}$$

Thus for $n \gg 1$, the last column in the tables approximates the geometric mean value of z_i.

The data could not be fitted by linear regression to a function of the form

TABLE III. Linear Regression Analyses on the Function
$\ell n\ N_T = a_0 + a_1\ \ell n\ Re\ Stk + \ell n^2\ Re\ Stk$

Cyclone	a_o	a_1	a_2	ϵ	e^ϵ
Unico 240	− 4.0	1.51	−0.063	0.096	1.10
Aerotec 3/4	− 5.92	2.53	−0.206	0.10	1.11
Aerotec 2	−12.6	3.36	−0.195	0.122	1.13
Dorr-Oliver $Q \leq 5\ \ell pm$	−12.9	2.82	−0.11	0.611	1.84
Dorr-Oliver $Q > 5\ \ell pm$	− 6.8	1.41	−0.052	0.44	1.55
Beeckmans	− 3.2	0.59	0.0073	0.196	1.21

TABLE IV. Non-linear Regression Analyses Based on the Function S_2

Cyclone	b	$(Stk^{0.5}\ Re^b)_{50}$	σ_g	$\epsilon(\%)$
Unico 240	0.50	4.0	1.64	2.58
Aerotec 3/4	0.725	26.5	1.53	1.95
Aerotec 2	0.225	0.83	1.56	2.11
Dorr-Oliver $Q \leq 5\ \ell pm$	0.225	1.51	1.30	5.84
Dorr-Oliver $Q > 5\ \ell pm$	1.05	6058	2.13	3.79
Beeckmans	0.5	13.7	3.15	5.83

TABLE V. Non-linear Regression Analyses Based on the Function S_1

Cyclone	b	$(Stk^{0.5}\ Re^b)_{50}$	σ_g	ϵ	e^ϵ
Unico 240	0.475	3.4	1.68	0.103	1.11
Aerotec 3/4	0.50	4.24	1.55	0.107	1.11
Aerotec 2	0.475	11.0	1.57	0.117	1.12
Dorr-Oliver $Q \leq 5\ \ell pm$	0.25	1.88	1.46	0.442	1.56
Dorr-Oliver $Q > 5\ \ell pm$	1.1	10,400	2.36	0.191	1.21
Beeckmans	0.60	43.0	3.39	0.200	1.22

represented in Eqs. (3a) and (3b), but instead a non-linear minimization routine was used to find the minimum value of sums of squares based on two functions of the dependent variable η, on the domain of the three disposable parameters contained in the equations. The functions minimized were

$$S_1 = \Sigma\, ln^2(N_{Ti}/Y_i) \tag{4}$$

and

$$S_2 = \Sigma(\eta_i - \eta_{ci})^2 \tag{5}$$

Y_i equals the value of N_T calculated using the fitting function with the values of *Stk* and *Re* corresponding to the *i'th* data point, and η_{ci} is the corresponding calculated value of the efficiency. Results of these analyses are given in Tables IV and V. Figures 1 and 2 show plots of some of the data using probability and logarithmic coordinates.

DISCUSSION AND CONCLUSIONS

A comparison of Tables IV and V shows that changing the minimization criterion has an appreciable effect on the parameters, even when the form of the function remains unchanged. This is particularly significant with respect to the parameter b, which determines the relative effects of the Reynolds number and of the Stokes number. Note that b changed appreciably for both AEROTEC cyclones as the minimization function changed. Changes in $(Stk^{0.5}\,Re^{b})_{50}$ are less significant, since they reflect primarily changes in b. The ratio $b/0.5$ is compared in Table VI with a_1/a_2, the ratio of regression coefficients relating to the functions *ln(Re)* and *ln(Stk)* in Table II. Agreement between columns 2 and 3 is within 20% for all cyclones, and with the exception of the AEROTEC 2 cyclone there is equally good agreement between columns 1 and 3. We conclude that data based on monodisperse aerosol studies indicate that cyclone efficiency cannot be correlated in terms of the Stokes number alone, and that the Reynolds number must also be included in the correlation. Furthermore, for a given cyclone geometry this type of correlation may only be valid for fixed values of cyclone diameter, gas viscosity and gas density. Changes in the proportions of the cyclone will, of course, have a profound effect on the parameters of the correlation.

COMPARISON OF CYCLONES AND IMPACTORS AS SIZE-SELECTIVE SAMPLERS

The ability of the cyclone to sample large volumes without saturation effects has been alluded to earlier, and this is an important advantage relative to impactors. On the other hand, as can be seen in Tables IV and V, the cyclone efficiency curve does not have a particularly sharp cut-off, as evidenced by relatively large values of σ_g. Mercer (1974) has compared values of σ_g found by plotting impactor efficiency curves on log probability coordinates against $Stk^{0.5}$. In order to compare these data with cyclone data, we must reduce both sets of data to a common basis. This presents no problem in comparing impactor data with cyclone data obtained by varying particle size at constant flowrate, because in this case *Re* is constant and consequently, in Eq. (3b) the normalized variable becomes $ln(Stk/Stk_{50})^{0.5}$. The slope of an efficiency plot on log probability paper,

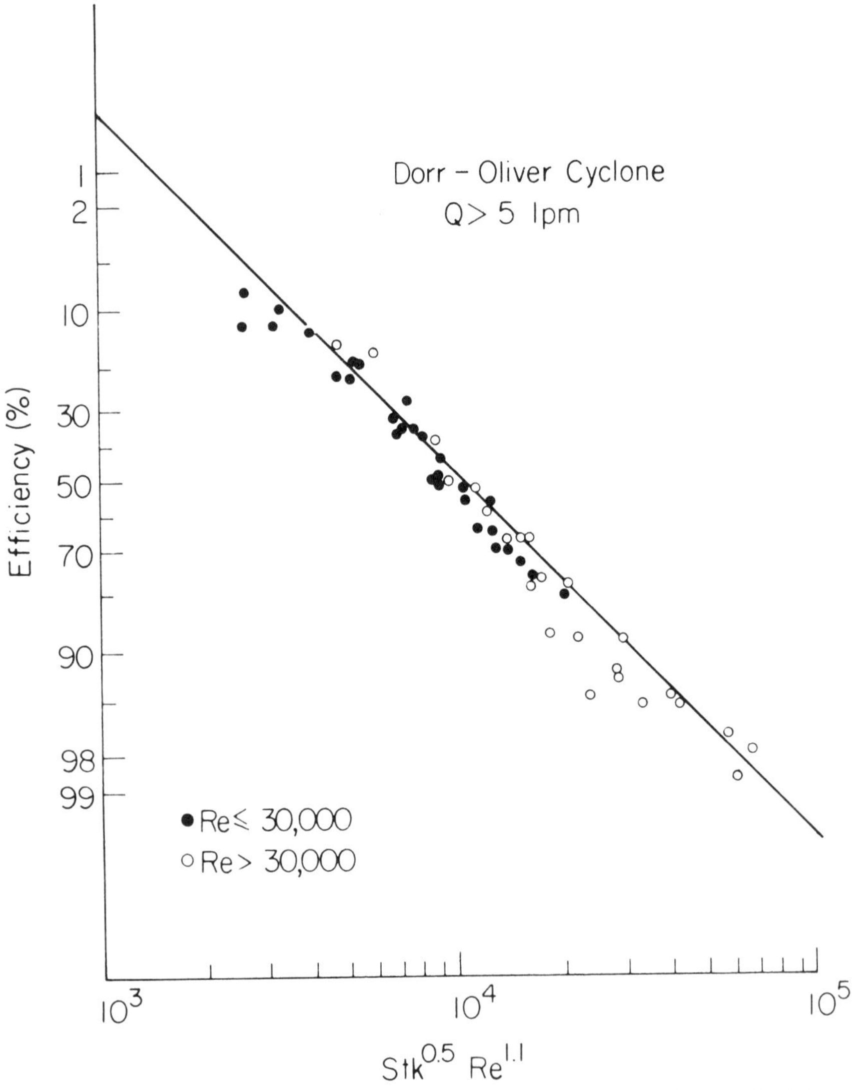

Figure 1. Efficiency of the 10 mm Dorr-Oliver cyclone for flows in excess of 5 ℓpm, vs. $Stk^{0.5}Re^{1.1}$. Line shows correlation using cumulative log-normal distribution function (Table V).

which determines σ_g, is the same whether the abscissa is $Stk^{0.5}$ or $(Stk/Stk_{50})^{0.5}$, hence Mercer's calculated σ_g's for impactors may be compared directly with σ_g's for cyclones operated at constant flow.

If the velocity of the cyclone is changed at constant particle size, then both *Re* and *Stk* will be affected. It is possible, however, to plot efficiency vs. *Stk* at a single particle size, and to determine σ_g for such a plot. If the efficiency is in fact a function of both *Re* and *Stk*, a different curve will be obtained for each particle size. It is easy to show that the following relationship exists between $\sigma_g^{(1)}$, the dispersion parameter for the argument $Stk^{0.5}$, and $\sigma_g^{(2)}$, the dispersion parameter for the argument $Stk^{0.5}Re^b$:

$$\sigma_g^{(1)} = (\sigma_g^{(2)})^{1/(1 + b)}. \tag{6}$$

Eq. (6) shows that $\sigma_g^{(1)} < \sigma_g^{(2)}$ whenever $b > 0$. Thus for the cyclones studied, since b is always positive, the variance of a plot of efficiency vs. $Stk^{0.5}$ for

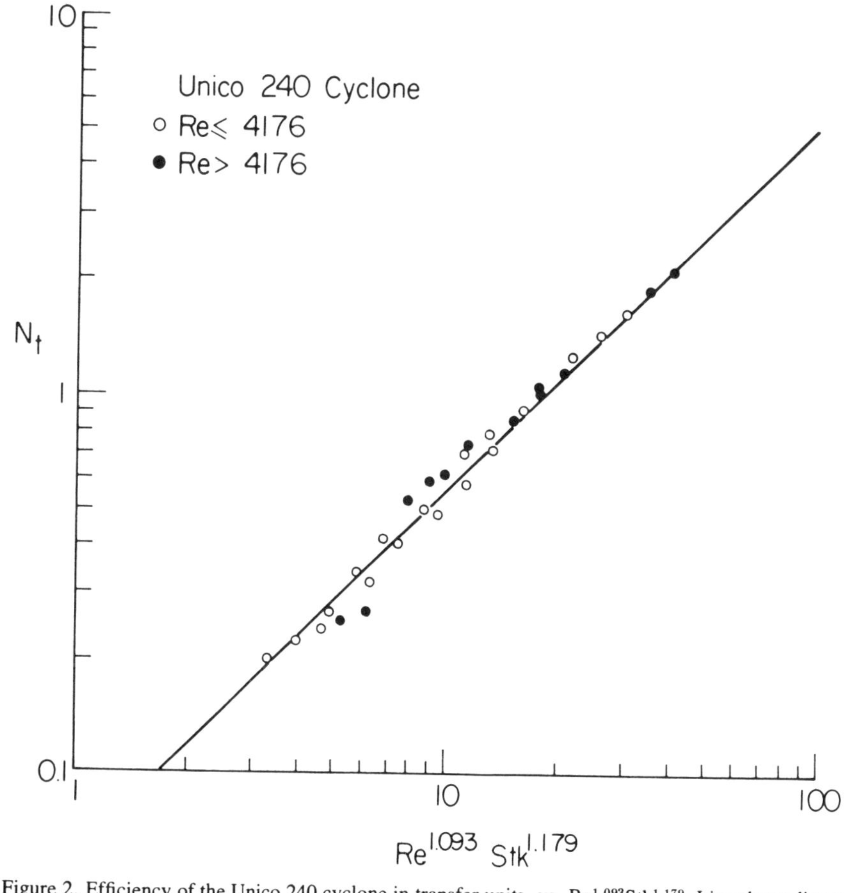

Figure 2. Efficiency of the Unico 240 cyclone in transfer units, vs. $Re^{1.093}Stk^{1.179}$. Line shows linear regression equation (Table II) on the arguments ℓn Re and ℓn Stk.

experiments at constant flowrate and variable particle size is invariably smaller than a similar plot obtained by varying flowrate at constant particle size. The fact that Mercer found the same phenomenon with impactor data suggests that these might be better correlated as a function of the form $Stk^{0.5}Re^{b}$ rather than as a function of Stk only.

The data correlated in Tables IV and V were obtained over a range of flows and particle sizes. The slope of an efficiency curve plotted on logarithmic and probability coordinates using as argument the function $Stk^{0.5}Re^{b}$ must equal the slope of a similar plot of efficiency versus $Stk^{0.5}$ at constant inlet velocity. This follows from the fact that if Re is constant the abscissas of two curves are merely displaced by a constant value on a logarithmic scale. Thus the σ_g's in Tables IV and V may be directly compared with σ_g's for impactor data obtained at constant flow. Mercer calculated values for these for round jets in the range 1.10 to 1.54, with a mean value of 1.27. For cyclones, the lowest value found for σ_g was 1.3

TABLE VI. Comparison of (b/0.5) with Ratio of Coefficients a_1/a_2 in Regression Analysis on Variables $\ell n(Re)$ and $\ell n(Stk)$

Cyclone	b/0.5 (Table V)	b/0.5 (Table IV)	a_1/a_2 (Table II)
Unico 240	0.95	1.0	0.93
Aerotec 3/4	1.0	1.45	1.22
Aerotec 2	0.95	0.45	0.438
Dorr-Oliver $Q \leq 5$ ℓpm	0.5	0.45	0.505
Dorr-Oliver $Q > 5$ ℓpm	2.2	2.1	2.03
Beeckmans	1.2	1.0	1.13

for the 10 mm Dorr-Oliver cyclone in the low flow range (Table IV). Several cyclones had σ_g's in the range 1.5–1.7, and some cyclones had σ_g's considerably less sharp than this. In general, therefore, the cyclones had cut-off characteristics appreciably higher than most impactors. The same conclusions may be reached by considering the slope of the plot of η versus $Stk^{0.5}/(Stk^{0.5})_{50}$. For both round and rectangular impactors at constant inlet velocity the value of this slope is approximately 1.6 (Mercer, 1965). It may be shown that for a device whose efficiency may be correlated by a function of the form $Stk^{0.5}Re^{b}$, the slope s of efficiency vs. $Stk^{0.5}/(Stk^{0.5})_{50}$ at $\eta = 0.5$ is given by the following expressions:

a. Re constant, Stk varied by changing particle size

$$s = 1/\sqrt{2\pi}\, ln\, \sigma_g. \tag{7}$$

b. Particle size constant, inlet velocity varied

$$s = (2b + 1)/\sqrt{2\pi}\, ln\, \sigma_g. \tag{8}$$

Slopes computed for the cyclones, using Eq. (7), ranged between 0.33 and 1.05. These values are less than the corresponding impactor slopes, in agreement with the previous conclusion based on σ_g values for cyclones and impactors.

The more diffuse cut-off characteristics of cyclones relative to impactors means that there will inevitably be a greater loss of definition in the reconstruction of the particle-size distribution of the aerosol under test from cyclone selective sampler data relative to cascaded impactor data. It is difficult to generalize quantitatively about the relative errors in the computed particle size distributions obtained by the two methods. It seems likely that the error obtained using a given set of cyclones or impactors is dependent on the ratio of the dispersion parameters of the particle population and of the size-selective samplers, as well as on how well the ECAD's of the set of samplers is distributed over the effective range of particle sizes in the aerosol. In any case, the dispersion parameter (σ_g) of the aerosol population should be considerably larger than that of any single sampler if a reasonable estimate is to be obtained for the population parameters of the aerosol. Thus if the particle sizes of the aerosol cover a broad spectrum it is likely that a set of size-selective cyclones would give almost as good an estimate of particle population as an impactor cascade. On the other hand, should the aerosol contain a relatively narrow range of particle sizes, the error in determining the aerosol particle size distribution obtained using cyclones would probably be appreciably larger than the corresponding error in using impactors.

REFERENCES

Barth, K., 1956: *Brennst.-Wärme-Kraft, 18*, 1.

Beeckmans, J. M., 1972: A steady-state model of reverse-flow cyclone. *J. Aerosol Sci., 3*, 491–500.

———, 1973: A two-dimensional turbulent diffusion model of the reverse flow. *J. Aerosol Sci., 4*, 329–336.

Blachman, M. W. and M. Lippmann, 1974: Performance characteristics of the multicyclone aerosol sampler. *J. Amer. Industrial Hygiene Assoc., 35*, 311–326.

Chan, T. L. and M. Lippmann, 1975: Presented at the 25th Canadian Chemical Engineering Conference, Montreal.

Knuth, R. H., 1969: Recalibration of size-selective samplers. *J. Amer. Industrial Hygiene Assoc., 30*, 379–385.

Leith, D. and W. Licht, 1972: A.I.Ch.E. Symp. Ser. *68*, No. 126, 196–206.

Lippmann, M. and T. L. Chan, 1974: Calibration of dual-inlet cyclones for "respirable" mass sampling. *J. Amer. Industrial Hygiene Assoc., 35*, 189–200.

——— and A. Kydonieus, 1970: A multi-stage aerosol sampler for extended sampling intervals. *J. Amer. Industrial Hygiene Assoc., 31*, 730–737.

Mercer, T. T., 1974: Presented at a meeting of Gesellschaft für Aerosol Forschung, Bad Soden.

Rosin, P., E. Rammler and W. Intelman, 1932: *Z. Ver. Dt. Ing.*, 76, 433.

USE OF CYCLONES FOR SIZE-SELECTIVE AEROSOL SAMPLING

MORTON LIPPMANN

Institute of Environmental Medicine, New York University Medical Center, N.Y., N.Y.

The penetrating analysis presented by Dr. Beeckmans in his review paper in these Proceedings has, at long last, provided a scientifically sound basis for characterizing the performance of small cyclones in air sampling applications. Chan and Lippmann (1976) have shown that previously proposed predictive relations for cyclone efficiency (Rosin et al., 1932; Lapple, 1951; Barth, 1956; Leith and Licht, 1972; Beeckmans, 1973) are inconsistent with experimental observations in tests using low concentrations of monodisperse test aerosols. Using a hyperbolic tangent relation for cyclone performance first developed by Blachman and Lippmann (1974), they also described the empirical constants of this equation for four commercially available, small cyclones which have been used in air sampling, and for several Stairmand-type cyclones. It should be noted that the previous predictive theories were developed from test data on, and have primarily been applied to, cyclones used for air cleaning applications.

The discrepancies between the predictive theories and experimental performance could be due to any or all of the following differences between the conditions within cyclones used in air cleaning as opposed to cyclones used in air sampling: 1) In air cleaning, the concentrations are much higher than in the ambient aerosols which are drawn into sampling cyclones, and may lead to particle agglomeration within the cyclone; 2) In air sampling, the particles usually remain on the cyclone wall where they were initially deposited, while in air cleaning, deposited particles are continually scoured from the initial deposition sites and migrate to the collecting hopper; 3) The flow patterns in air cleaning cyclones are always turbulent, while in the smaller cyclones used in air sampling, the flows may be completely or partially laminar. The presentation of Ayer and Hochstrasser in these Proceedings clearly demonstrates transitions in pressure-flow relationships in cyclones which imply that laminar flow conditions can exist within small cyclones.

APPLICATIONS IN TWO-STAGE SAMPLING

By far the greatest usage of small cyclones in aerosol sampling has been as the first stage of a two-stage sampler. In most cases, they have been operated at flowrates which were believed to permit cyclone penetrations equivalent to the penetration of inhaled dust to the non-ciliated deep lung spaces in humans. Such applications have been called "respirable" dust sampling, and their rationale has previously been discussed (Lippmann, 1970, 1976; AIHA Aerosol Technology Committee, 1970).

In "respirable" dust sampling, it is essential that the cyclone cut-off characteristics be known and constant, and that each cyclone is operated at a flowrate which produces the desired cut-off. In the absence of a usable theory, it has been necessary to design and calibrate air sampling cyclones empirically. Unfortunately, dependence on an empirical calibration can lead to controversy and/or errors. The original calibration of the Dorr-Oliver 10 mm nylon cyclone (Lippmann and Harris, 1962) failed to include an appropriate aerodynamic shape factor. With Kotrappa's (1971) shape factor correction, the recommended sampling rate was reduced to 1.8 ℓpm, which is in good agreement with most of the more recent empirical calibrations (Ettinger et al., 1970; Lippmann and Kydonieus, 1970; Caplan et al., 1977; Seltzer et al., 1971) which recommend either 1.7 or 1.8 ℓpm. However, MESA, which specifies acceptable samplers and their operating characteristics in mine dust evaluations, still insists that the 10 mm cyclone be operated at 2.0 ℓpm on the basis of its own calibration (Tomb and Raymond, 1969).

There are conflicting calibration data for several of the 12.5 to 50 mm diameter stainless steel cyclones which have also been used for size-selective sampling. Some of the differences can be attributed to the same causes as for the 10 mm cyclone, i.e., differences in calibration techniques, data interpretation, and failure to use appropriate correction factors. Another factor is the dimensional variations between nominally identical samplers which were documented by Lippmann and Chan (1974). The steel cyclones are individually assembled, and the quality control exercised by the manufacturers has not always been adequate. The 10 mm nylon cyclones, which are injection molded, have not been observed or reported to vary significantly from one to another.

Finally, empirical calibrations can only be valid if the field applications are made without significant changes in the sampling train configurations. The paper by Ayer and Hochstrasser in these symposium proceedings demonstrates that the cut-off characteristics of a given cyclone can be significantly changed by altering the coupling between the cyclone and back-up filter and thereby altering the flow pattern within the outlet pipe of the cyclone. These observations appear to explain many of the current discrepancies between different laboratories' calibrations.

APPLICATIONS IN MULTI-STAGE SAMPLING

Size-mass distribution analyses of aerosol constituents have been made using cascade impactor samplers for more than thirty years, beginning with the original May (1945) impactor and the Laskin (1950) Univ. of Rochester modification. Analyses have also been performed with aerosol centrifuges which deposit particles on a collection foil according to their aerodynamic size. However, both types of sample collectors have significant limitations when used for such applications. Solid particles tend to bounce off or be reentrained from impaction surfaces which lack an adhesive layer or whose adhesive becomes saturated. Unfortunately, saturation usually is reached either before the investigator is aware of it, or before the accumulated sample is large enough for the subsequent analysis. Applications of centrifuges are limited by their low sampling rates, and large physical size and cost.

The limitations of the other inertial classifiers for compositional analysis by particle size fraction have led to the development of samplers using cyclones in either parallel or series arrays. The first development of a portable sampler using a parallel array of cyclone-filter series samplers was described by Lippmann and Kydonieus (1970). Each cyclone had a different cut-size and the overall size-mass distribution of the total aerosol or any of its chemical constituents could be determined from analyses of the collection on the filter following each cyclone and a parallel filter sampler operated without a cyclone pre-collector. The performance characteristics of an improved version of this sampler were described by Blachman and Lippmann (1974). A much larger version of this sampler for fixed station ambient air pollution sampling was described by Bernstein et al. (1976). A parallel multicyclone sampling train for stack sampling applications was described by Chang (1974).

The major advantage in using cyclones instead of impactors for such applications is that their performance is not significantly affected by the amount of sample collected (Blachman and Lippmann, 1974). They share with impactors the advantage of relatively low fabrication cost.

Acknowledgements.—These investigations were supported by research contracts HSM-099-71-48 and CPE-70-NEG-128 from the National Institute for Occupational Safety and Health, by RP 117-1 from the Electric Power Research Institute, and is part of a center program supported by Grant ES 00260 from the National Institute of Environmental Health Sciences.

REFERENCES

AIHA Aerosol Technology Committee, 1970: Guide for respirable mass sampling. *J. Amer. Industrial Hygiene Assoc.*, *31*, 133–137.

Barth, W., 1956: Calculation and design of cyclones on the basis of recent tests. *Brennstoff-Warme-Kraft*, *8*, 1–9.

Beeckmans, J. M., 1973: A two-dimensional turbulent diffusion model of the reverse flow cyclone. *J. Aerosol Sci.*, *4*, 329–336.

Bernstein, D., M. T. Kleinman, T. J. Kneip, T. L. Chan and M. Lippmann, 1976: Development of a high-volume sampler for the determination of particle size distributions in ambient air. *J. Air Pollution Control Assoc., 26,* 1069–1072.

Blachman, M. W. and M. Lippmann, 1974: Performance characteristics of the multicyclone aerosol sampler. *J. Amer. Industrial Hygiene Assoc., 35,* 311–316.

Caplan, K. J., L. J. Doemeny and S. D. Sorenson, 1977: Performance characteristics of the 10 mm cyclone respirable mass sampler, Part I—monodisperse studies. *J. Amer. Industrial Hygiene Assoc. 38,* 83–95.

Chan, T. L. and M. Lippmann, 1976: Particle collection efficiencies of air sampling cyclones. *Environ. Sci. Technol., 11,* 372–382.

Chang, H. C., 1974: A parallel multicyclone size-selective particulate sampling train. *J. Amer. Industrial Hygiene Assoc., 35,* 538–545.

Ettinger, H. J., J. E. Partridge and G. W. Royer, 1970: Calibration of two-stage air samplers. *J. Amer. Industrial Hygiene Assoc., 31,* 537–545.

Kotrappa, P., 1971: Revision of Lippmann-Harris calibration for two-stage sampler using shape factor. *Health Phys., 20,* 350–351.

Lapple, C. E., 1963: Dust and mist collection. *Chemical Engineers Handbook,* J. H. Perry, Ed. New York, McGraw-Hill.

Laskin, S., 1950: The modified cascade impactor. UR-129, Rochester, New York, The University of Rochester Atomic Energy Product.

Leith, D. and W. Licht, 1972: The collection efficiency of cyclone type particle collectors—A new theoretical approach. *AIChE Symposium Series: Air Pollution and its Control, 68,* 196–206.

Lippmann, M., 1970: "Respirable" dust sampling. *J. Amer. Industrial Hygiene Assoc., 31,* 138–159.

Lippmann, M., 1976: Size-selective sampling for inhalation hazard evaluations. *Fine Particles,* B. Y. H. Liu, Ed., New York, Academic Press, 287–310.

Lippmann, M. and W. B. Harris, 1962: Size-selective samplers for estimating "respirable" dust concentrations. *Health Phys., 8,* 155–163.

Lippmann, M. and A. Kydonieus, 1970: A multi-stage aerosol sampler for extended sampling intervals. *J. Amer. Industrial Hygiene Assoc., 31,* 730–738.

Lippmann, M. and T. L. Chan, 1974: Calibration of dual-inlet cyclones for "respirable" mass sampling. *J. Amer. Industrial Hygiene Assoc., 35,* 189–200.

May, K. R., 1945: The cascade impactor, an instrument for sampling coarse aerosols. *J. Sci. Instrum., 22,* 187–195.

Rosin, P., E. Rammler and W. Intelman, 1932: Principles and limits of cyclone dust removal. *Z. Ver. dt. Ing., 76,* 433–437.

Seltzer, D. F., W. Bernaski and J. R. Lynch, 1971: Evaluation of size selective presamplers. II. Efficiency of the 10 mm nylon cyclone. *J. Amer. Industrial Hygiene Assoc., 32,* 441–446.

Tomb, T. F. and L. D. Raymond, 1969: Evaluation of collecting characteristics of horizontal elutriator and 10 mm nylon cyclone gravimetric dust samplers. Presented at the 1969 Annual Meeting of the American Industrial Hygiene Association, Denver, May.

CYCLONE DISCUSSION

HOWARD E. AYER AND JOHN M. HOCHSTRASSER

Kettering Laboratory, University of Cincinnati, Cincinnati, Ohio

Our interest in cyclones has been related to the two-stage sampling of dust to provide a "respirable" fraction for correlation with health effects. The first portion of our comments, therefore, will discuss the use of small cyclones for this purpose, and the interpretation of results which are obtained from the cyclone sampling. The balance of the talk will discuss results of laboratory experiments with miniature cyclones used for two-stage sampling and give some of our theories as to the performance of the cyclones.

The small cyclones are routinely used to sample for dust which may produce silicosis or coal workers' pneumoconiosis. Because this is a routine procedure, the main requirements are convenience and low cost with accuracy. It is these factors of convenience, cost and availability which have led to the widespread use of cyclones, in particular the 10-mm Dorr-Oliver cyclone, in North America. Our comments are directed toward the accuracy with which we represent the relative health hazards by these measurements. In addition to the 10-mm cyclone, we have used the Health and Safety Laboratory (HASL) ½"-cyclone (Unico 18) and the Aerotec 2 most often, but we have also used the Aerotec ¾" and the HASL 1" (Unico 240). Based primarily on results of comparative "respirable" mass and impinger sampling in the Vermont granite sheds, a threshold limit value for quartz based upon two-stage sampling was recommended (Ayer, 1969). A subsequent trip to the granite sheds, this time a pilot study for granite cutters operating their equipment without exhaust ventilation, produced the average data of Table I (Ayer et al., 1973). From this table it may be noted that for practical purposes the 10-ℓpm elutriator made in the NIOSH shops, the Isleworth sampler, the HASL ½" cyclone at 10-ℓpm, and the 10-mm nylon cyclone at 1.7-ℓpm were equivalent in average performance.

Lippmann and Chan (1974) have presented a calibration of dual inlet cyclones including the Aerotec ¾'' and 2'' models, and the HASL ½'' and 1'' models. Others attempting to use these cyclones at the flow rates recommended by Lippmann and Chan have encountered inconsistencies, however. Yablonsky et al. (1975) noted that the concentration of a clay dust cloud passing the Unico 240 (HASL 1'') was 1.3 times the concentration passing the Aerotec ¾'' when both were operated at the flow rates recommended by Lippmann and Chan. Thompson et al. (1976) in using the Unico 240 and the Aerotec ¾'' at the recommended flow rates, noted that the Unico 240 passed 1.24 times as much dust as the Aerotec ¾''. Using an Andersen Impactor, Yablonsky determined that the Aerotec ¾'' was passing essentially that fraction of the total dust which would be expected for a cyclone with characteristics which corresponded to the retention criteria specified by the Threshold Limit Values Committee of the ACGIH (1970). Thompson et al. (1976) used a Coulter counter to determine the efficiency of the cyclone for various size fractions and determined that the Aerotec ¾'' operating at its recommended flow rate performed very close to the criteria specified in the Threshold Limit Values. Thus, independent analyses by differing methods confirmed the calibration of Lippmann and Chan for the Aerotec ¾'', but produced

TABLE I. Average Respirable Dust Concentrations

	mg/m^3
Horizontal Elutriator Samplers	
Isleworth (MRE) sampler (2.5-ℓpm)	10.7
10-ℓpm elutriator	11.6
Hexhlet (50-ℓpm)	14.4
Cyclone Samplers	
½-inch steel cyclone at 10-ℓpm	10.9
10-mm nylon cyclone at 1.7-ℓpm	10.7

rather different results for the Unico 240. We believe that data obtained in a study of the performance of miniature cyclones (Hochstrasser, 1976) can possibly explain some of these discrepancies.

The study was performed using the ''high efficiency'' Stairmand design (1951). Stairmand's design was varied by using two shorter cone lengths, three different outlet inside diameters, and three different outlet outside diameters giving a total of 15 configurations of cone and outlet diameters. The basic body diameter was 19 mm. For the cyclones with the shorter cones, the pressure drop across the cyclone varied regularly with the flow as shown in Figure 1. However, with the long cone configuration, it was found that there were two definite pressure-flow regimens which could be defined by whether or not the flow in the outlet pipe was laminar or turbulent, i.e., whether the Reynolds number was less than 2,000 or greater than 4,000. When the Reynolds number was between 2,000 and 4,000 either the laminar or the turbulent flow pressure drop could exist, with the turbulent condition being the more stable. This is illustrated in Figure 2. The prediction of cyclone efficiency for a Stairmand design cyclone by Leith and Licht (1972), Lapple (1950), and Barth (1956) are shown in Figure 3 for one of the higher flow rates. The predicted efficiency according to Blachman

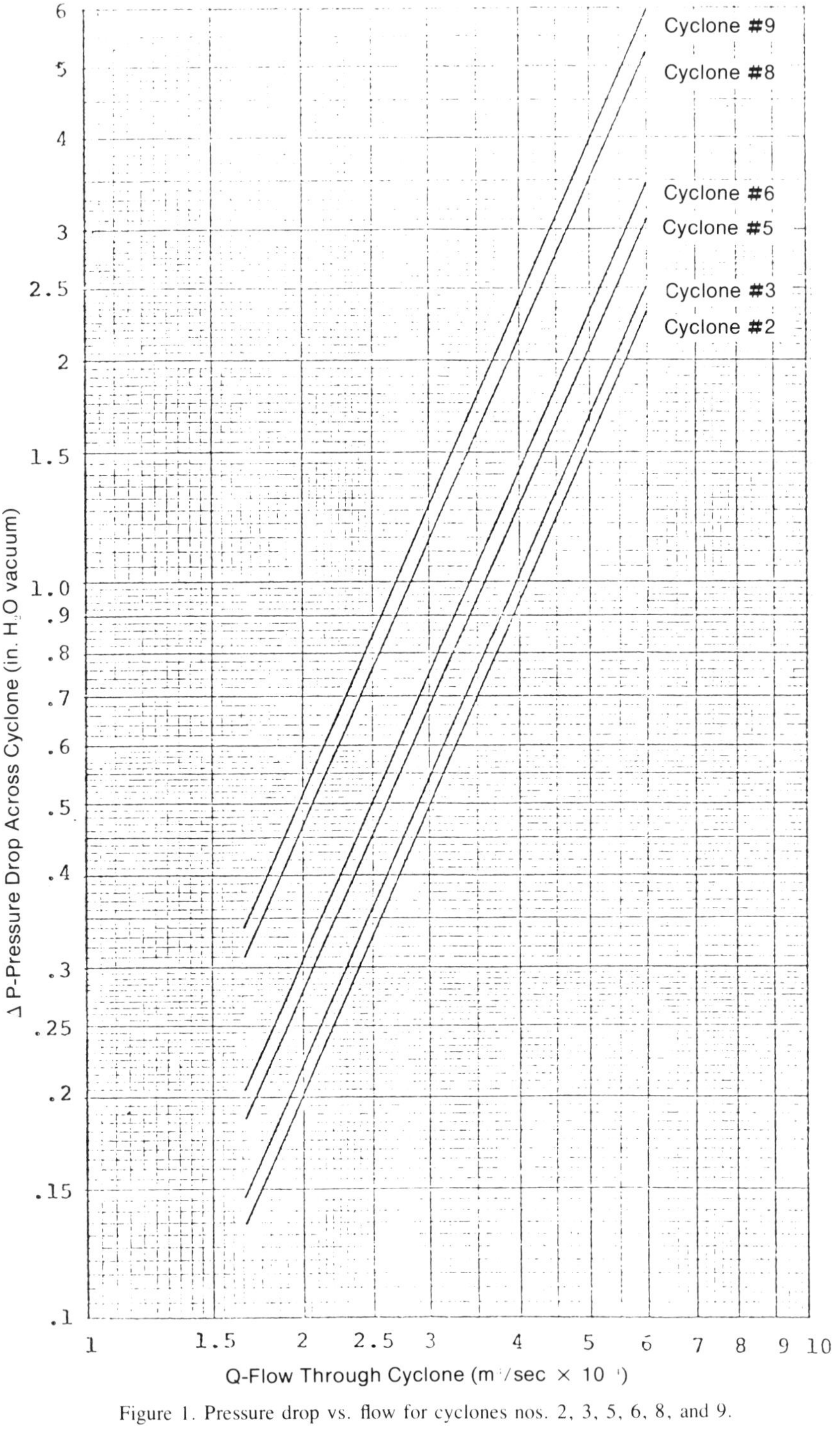

Figure 1. Pressure drop vs. flow for cyclones nos. 2, 3, 5, 6, 8, and 9.

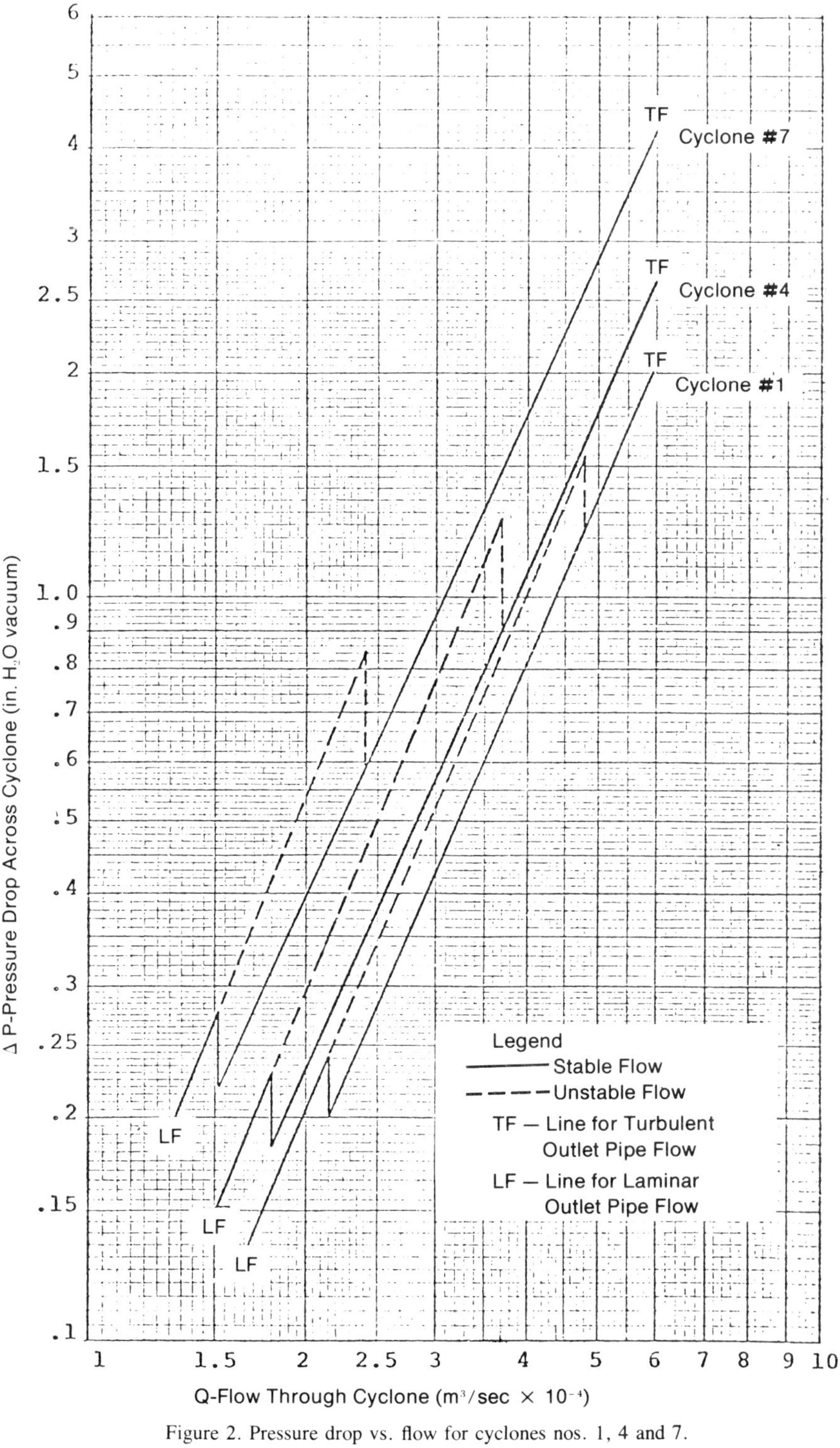

Figure 2. Pressure drop vs. flow for cyclones nos. 1, 4 and 7.

and Lippmann (1974) is also shown, along with the experimental efficiency data. It may be seen that the observed efficiency is predicted reasonably well by either the Barth equation or that of Blachman and Lippmann. In this case there appeared to be little difference in efficiency for large particles whether the outlet

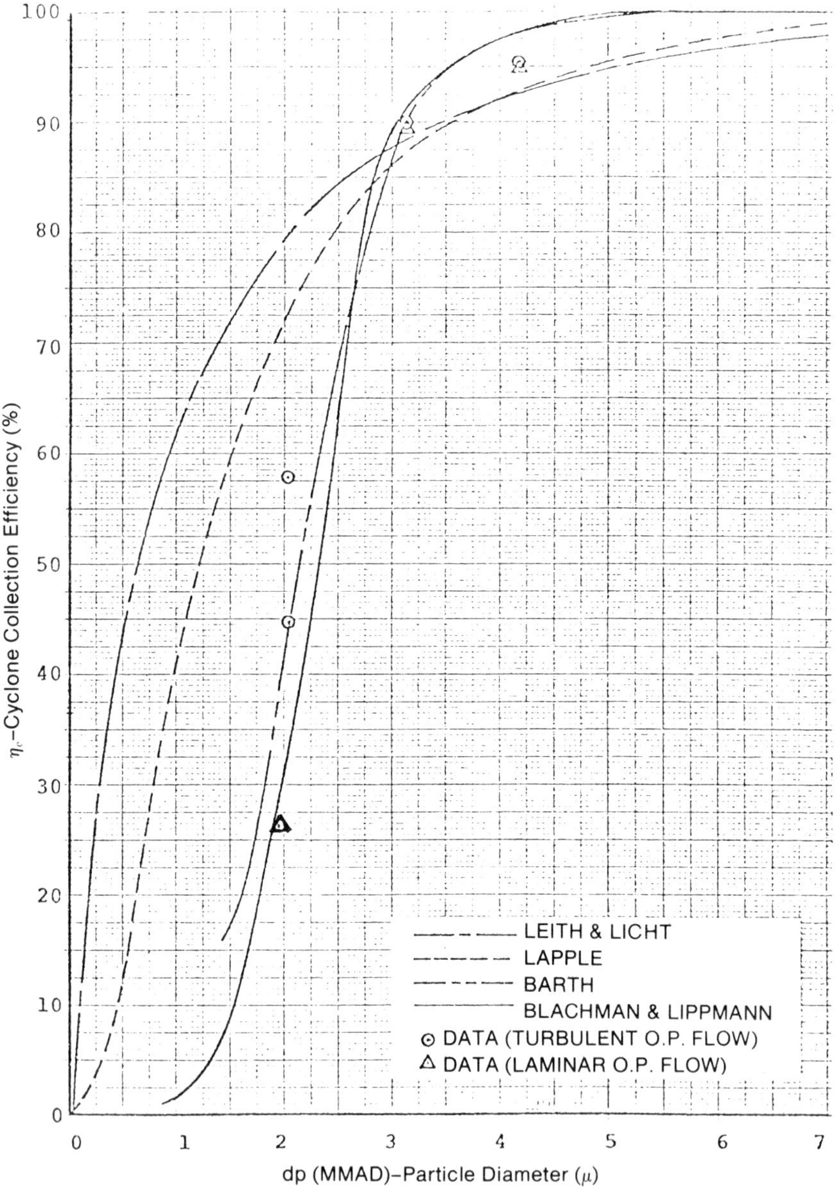

Figure 3. Cyclone efficiency vs. particle size for cyclone no. 1 at Q = 3.32 × 10^{-4} m^3/sec.

pipe flow was laminar or turbulent. At the lowest flow for cyclone configuration No. 1, (illustrated in Figure 4) only the Blachman and Lippmann prediction reasonably matched the observed efficiencies. It may be noted at this lowest flow the 50% efficiency point was roughly 3.5 μm aerodynamic diameter, the 50% efficiency specified in the TLV criteria. It is theorized that when the outlet pipe

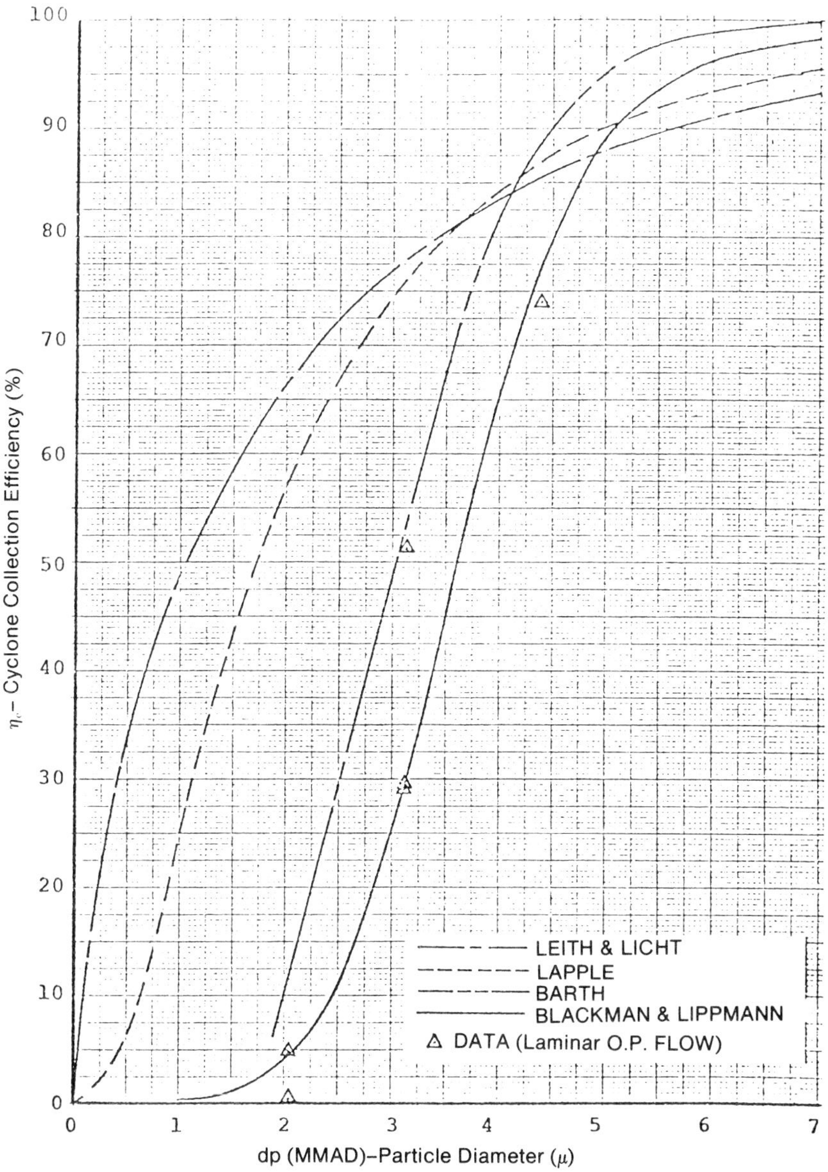

Figure 4. Cyclone efficiency vs. particle size for cyclone no. 1 at $Q = 1.66 \times 10^{-4}$ m^3/sec.

flow is turbulent, the cyclone is operating in the conventional manner as illustrated in Figure 5. However, when the outlet pipe flow is laminar, the flow pattern in the cyclone is different; the outlet vortex does not form in the normal manner, but rather the flow pattern is as shown in Figure 6. The existence of the "zone of stagnation" shown in this diagram is confirmed by both a ring of particles collecting at the plane of transition with none below this point, and by

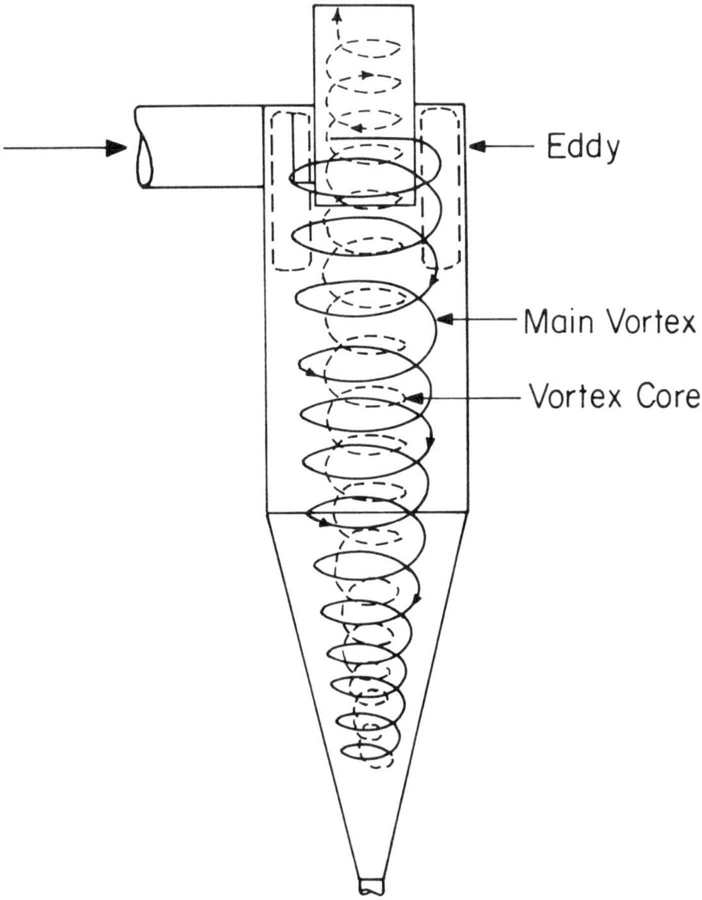

Figure 5. Eddy and vortex flows in a cyclone collector.

the fact that there is no further change of pressure inside the cyclone below the plane of transition. The cyclones calibrated by Lippmann and Chan came reasonably close to the ACGIH retention criteria only when the cyclones were operating with the outlet pipe in the laminar flow condition.

It thus appears that the operation for "respirable" dust sampling cyclones is dependent upon the laminar flow condition existing in the outlet pipe and the short circuited flow condition. If this is so, then it is imperative that the outlet conditions be maintained similar to those under which the cyclone was calibrated. Conditions immediately downstream from a cyclone outlet may be reflected back, as it were, into a flow system changing the pressure conditions inside the system. The location of the plane of transition may then shift, drasti-

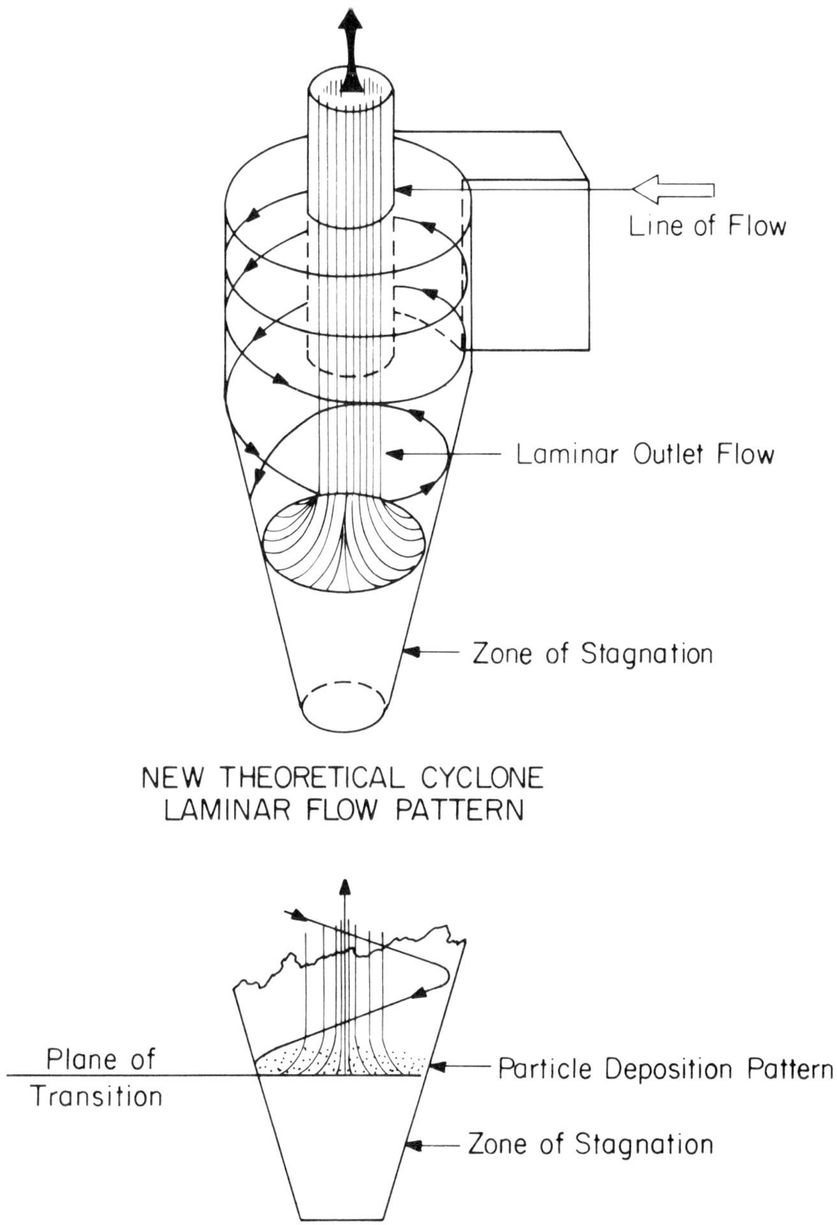

Figure 6. Particle deposition pattern (laminar flow cyclone).

cally altering cyclone efficiency. It is suggested that many of the inconsistent results which have been observed in "respirable" dust cyclone calibration are not the result of experimental error or because of minor variations in construction between the experimental cyclones. Lippmann and Chan carefully described the way in which each cyclone outlet was fitted to the filter holder in their calibration setup. When this same configuration was used by others, e.g., Thompson of the Bureau of Mines and Yablonsky of our laboratory with the Aerotec ¾" cyclone, results comparable to those of Lippmann were observed although measurement techniques and the aerosol used were quite different. Likewise, the data of Thompson et al. for the carbon steel Aerotec 2 cyclone very closely matched those of Lippmann and Chan for the carbon steel cyclone at 430 ℓpm as the outlet configurations were quite similar in the two experiments. However, whereas Lippmann and Chan used a close-coupled filter for both the Unico 240 and the Unico 18, this configuration was not used by Thompson et al. in their experiments and the results were not comparable to those of Lippmann. Likewise, Yablonsky in our laboratory did not use a close-coupled filter for the Unico 240 and again the penetration was some 25% greater than would have been predicted by Lippmann's characteristic. These findings are consistent with those of Knight (1976) who has observed different penetration through the 10-mm nylon cyclone in the two MSA configurations, i.e., the coal mine dust cassette and the silica dust apparatus which uses the Millipore field monitor cassette.

The relationship of the outlet configuration of miniature cyclones to efficiency has not yet been investigated in detail. The foregoing observations are adequate, we believe, to demonstrate that these outlet configurations are important and that standardizations of pump and cyclone for a personal sampler will not characterize the performance of the sampler unless the outlet configuration through the filter is also standardized. It must be recognized that not all cyclones can match the ACGIH TLV Committee criteria for retention. Neither the Stairmand design investigated by Hochstrasser nor the 10-mm nylon cyclone is a close match for the TLV curve under the conditions of calibration; both have a cutoff characteristic which is too sharp, though the 50% penetration can be adjusted to an aerodynamic diameter of 3.5 μm.

The use of methods giving aerodynamic diameter vs. efficiency for miniature cyclone calibrations has eliminated many of the differences between laboratories. Elimination of pulsating flows in field units has further removed inconsistencies. It remains for the important area of outlet configuration to be specifically investigated and standardized, so that remaining conflicts may be resolved.

REFERENCES

Ayer, H. E., 1969: The proposed ACGIH mass limits for quartz: review and evaluation. *J. Amer. Industrial Hygiene Assoc., 30,* 117–125.

———, J. M. Dement, K. A. Busch, H. B. Ashe, B. T. H. Levadie, W. A. Burgess and L. DiBernardinis, 1973: A monumental study—reconstruction of a 1920 granite shed. *J. Amer. Industrial Hygiene Assoc., 34,* 206–211.

Barth, W., 1956: Design and layout of the cyclone separator on the basis of new investigations. *Brenstoff-Warme-Kraft, 8,* 1–9.

Blachman, M. W. and M. Lippmann, 1974: Performance characteristics of the multicyclone aerosol sampler. *J. Amer. Industrial Hygiene Assoc., 35*, 311–326.

Hochstrasser, J. M., 1976: The investigation and development of cyclone dust collector theories for application to miniature cyclone presamplers, Ph.D. Dissertation, University of Cincinnati.

Knight, G., 1976: Tentative specifications for respirable dust sampling. In correspondence to ACGIH Aerosol Hazards Committee, January 28.

Lapple, C. E., 1950: Gravity and centrifugal separation. *J. Industrial Hygiene Quart., 11,* 41.

Leith, D. and W. Licht, 1972: The collection efficiency of cyclone type particle collectors—a new theoretical approach. Reprint from *AIChE Symposium Series, Air Pollution and Its Control, 68,* 196.

Lippmann, M. E. and T. L. Chan, 1974: Calibration of dual inlet cyclones for "respirable" mass sampling. *J. Amer. Industrial Hygiene Assoc., 35,* 189–200.

Stairmand, C. J., 1951: The design and performance of cyclone separators. *Trans. Inst. Chem. Engr.* (London), *29*, 356.

Thompson, E. M., H. N. Treaftis and T. F. Tomb, 1976: Comparison of recommended respirable mass dust sampling devices. Presented at American Industrial Hygiene Conference, Atlanta.

Threshold Limit Committee, 1970: Threshold limit values (TLV's) of airborne contaminants and intended changes by ACGIH. Amer. Conf. Governmental Industrial Hygienists, P. O. Box 1937, Cincinnati, Ohio 45201.

Yablonsky, J., H. E. Ayer, J. Svetlik and S. W. Horstman, 1975: Calibration system for dust sampling: Final report. Contract No. DAMD 17-74-C-4204, Cincinnati, University of Cincinnati.

AEROSOL MEASUREMENT BY CYCLONE COLLECTORS

S. L. Soo

Department of Mechanical and Industrial Engineering, University of Illinois at Urbana-Champaign, Urbana

A previous study (Soo, 1973) showed that via a rigorous integration procedure the collection efficiency η_c of a cyclone separator with a steep cone is given by

$$\eta_c = 1 - \exp\left[- \sigma\left(\frac{2\pi\, C^2\, L}{F\, Q\, R^2} + 8\pi\, \frac{\rho_{p1}\, q^2\, R^2\, L}{\epsilon_0\, m^2\, F\, Q} + 2\pi\, E_0\, \frac{q}{m}\, \frac{RL}{FQ}\right)\right], \quad (1)$$

where σ is the sticking probability of the particles to the wall or the layer of collected particles; C is the maximum vorticity of the system; L is the length and R is the maximum radius of the cyclone; Q is the volume flow rate; $F = 3\pi\, \mu\, d/m$ is the inverse relaxation time of particle-fluid momentum transfer for particle mass m, size d, and fluid viscosity μ; ρ_{p1} is the density of the particle cloud per unit volume of the mixture; ϵ_0 is the permittivity of free space, q is the charge per particle; and E_0 is the applied electric field. The first term in parenthesis is the parameter for collection by centrifugal force exerted by the vortex; the second term is that by space charge of the particles; and the third term is that by the applied electric field. The magnitude of σ depends on contacting surfaces and depositing forces. It was shown that for continuous removal of particles at the bottom of the collector cone, a high value of sticking probability is often not desirable because it would hamper continuous removal of collected particles. However, this is not a concern when a cyclone separator is used for collecting aerosol samples and sticking probability near 1 will be desirable. Consistency of results is achieved via controlled test conditions. The flow relations are shown in Figure 1.

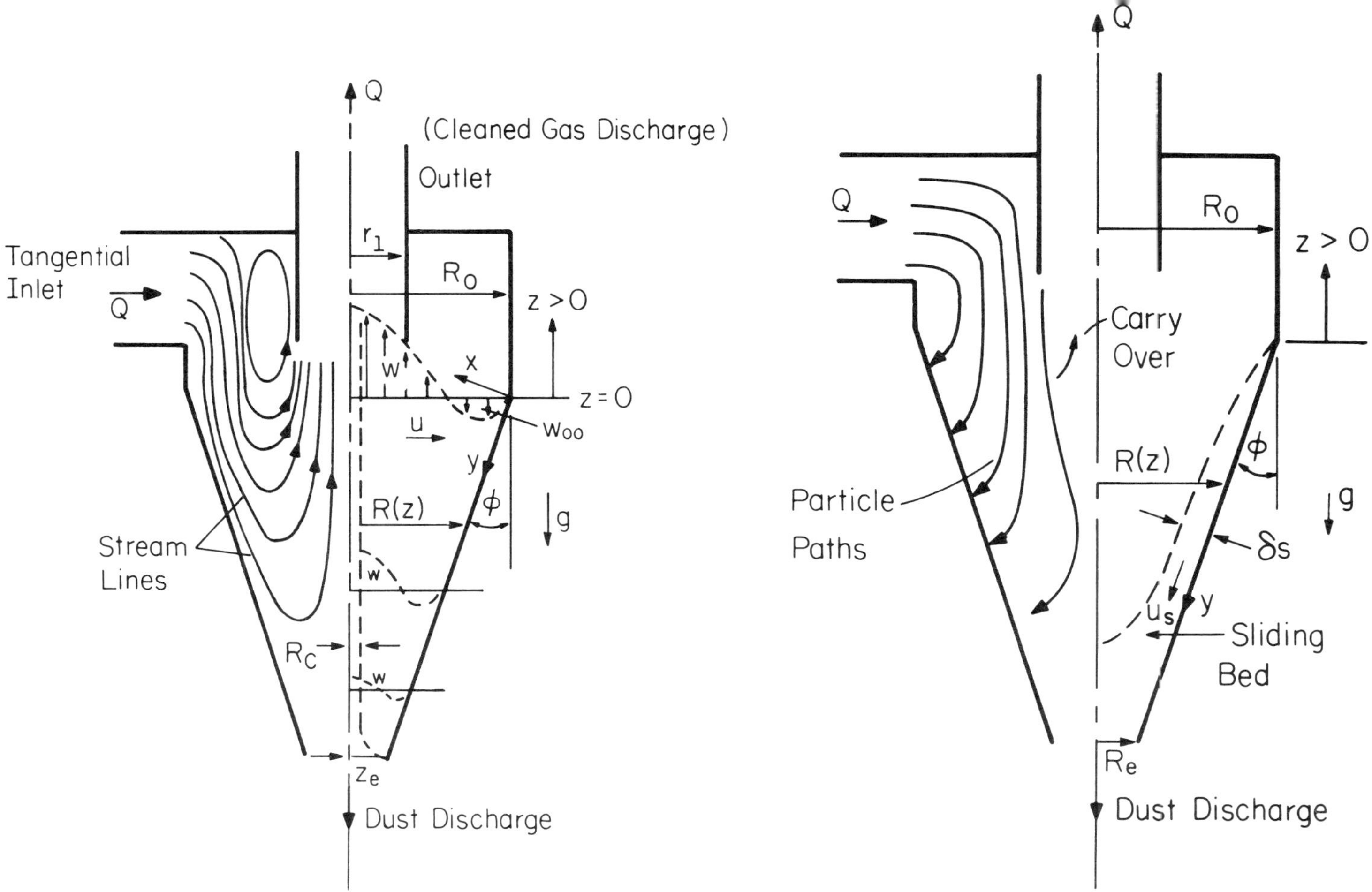

Figure 1. Conditions inside a cyclone separator and coordinates (Soo, 1973). Left: Fluid velocity distribution. Right: Particle path and sliding bed.

VORTEX ACTION

When considering vortex action alone, the first term in the parentheses of Eq. (1) shows that large collection efficiency of a given size and mass of particles is given by a large value of

$$\frac{2\pi\ C^2\ L}{Q\ R^2}\ \frac{m}{3\pi\ \mu\ d} = \frac{\pi\ C^2\ L\ d^2\ \bar{\rho}_p}{9Q\ R^2\ \mu}\ , \tag{2}$$

where $\bar{\rho}_p$ is the density of the particle material, and the relation is reduced to the

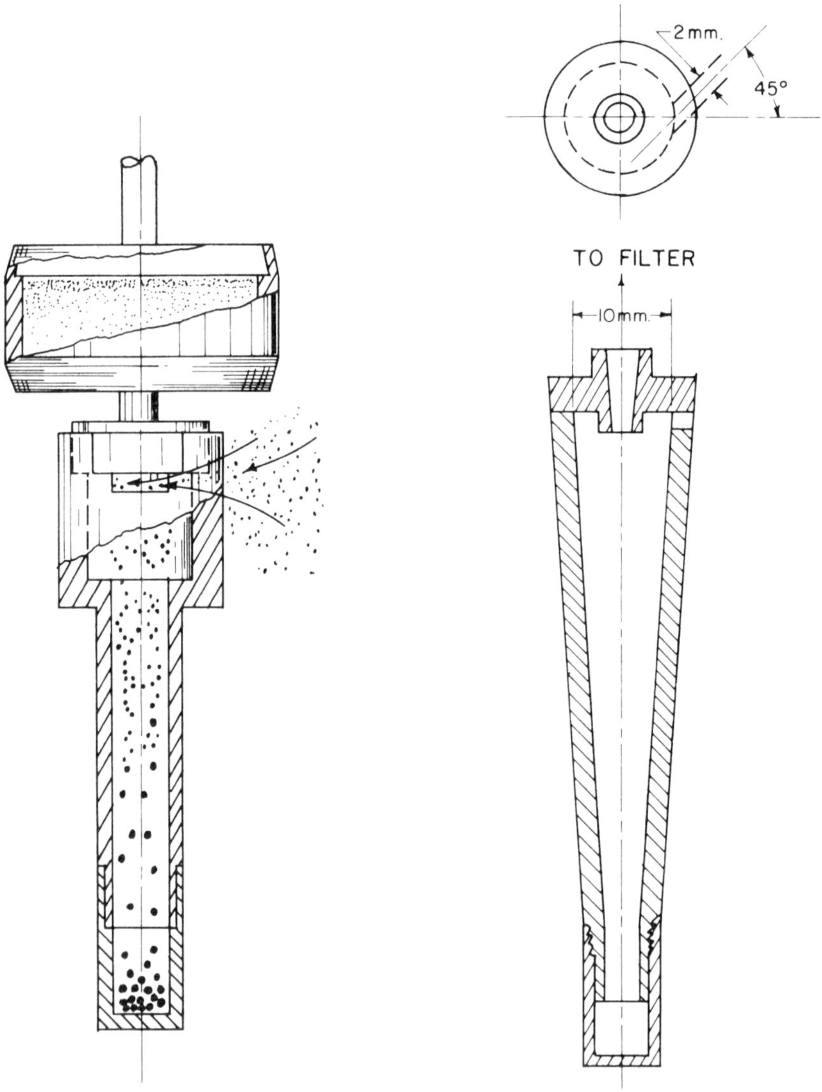

Figure 2. 10-mm Cyclone with filter (AIHA Aerosol Technology Comm., 1970). A recent design courtesy of Pittsburgh Mining Center, Bureau of Mines (Morse, 1970).

right-hand side of the expression in Eq. (2) for spherical particles. Classification is accomplished by choosing a given vorticity C such that particles of a given value of $d^2\bar{\rho}_p$ are efficiently collected. That is, corresponding to cascade impactors, cyclones in series with increasing vorticity can classify as well as collect particles of various sizes.

Wide application is in cyclone elutriators for sampling and evaluating respirable coal mine dusts such as given in Figure 2 showing a 10 mm diameter nylon cyclone with filter and a recent design. The function and performance of this nylon cyclone are expressed in terms of penetration with the expressed aim of simulating pulmonary deposition with relations as shown in Figure 3 in comparison to AEC and British Medical Research Council (BMRC) standards (AIHA Aerosol Technology Comm., 1970). Test data given by Tomb and Raymond (1970) (Figure 4) and Seltzer et al. (1971) showed that by operating at different flow rates in liters per minute (or vorticity), a sampling cyclone can be operated to collect over different ranges of sizes of aerosols (between 4.0 and 8.0 μm at 2.5 ℓpm and between 2.0 and 4.0 μm at 2.7 ℓpm). The lack of greater degree of selection via flow rate (or vorticity) shows the existence of other influences.

ELECTRIC EFFECT

The influence of the electric effect is seen in the experimental results of Stein et al. (1973). Measurements of concentrations vs. time of dispersed aerosols of NJZ zinc cadmium sulfide fluorescent powder were made in a polystyrene chamber of

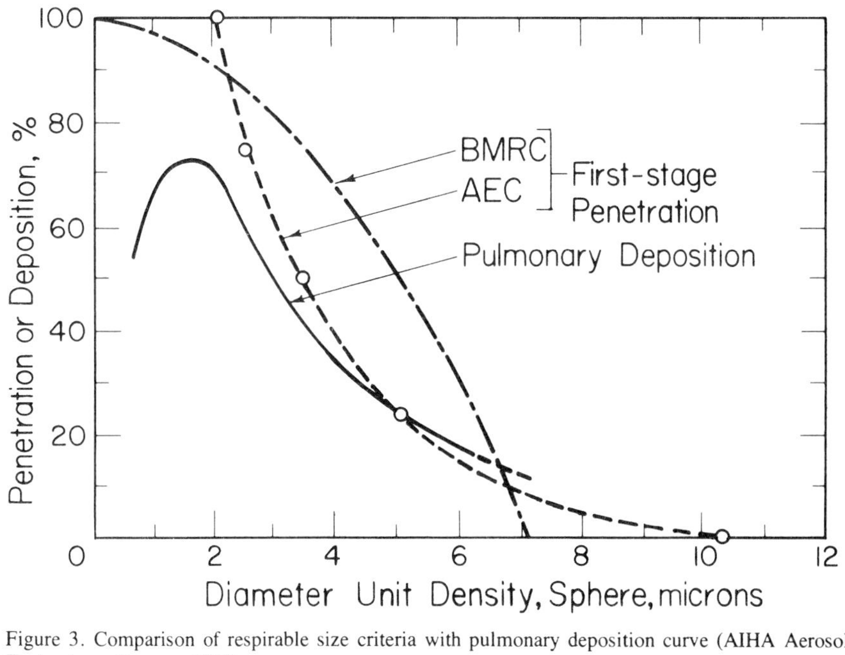

Figure 3. Comparison of respirable size criteria with pulmonary deposition curve (AIHA Aerosol Technology Comm., 1970).

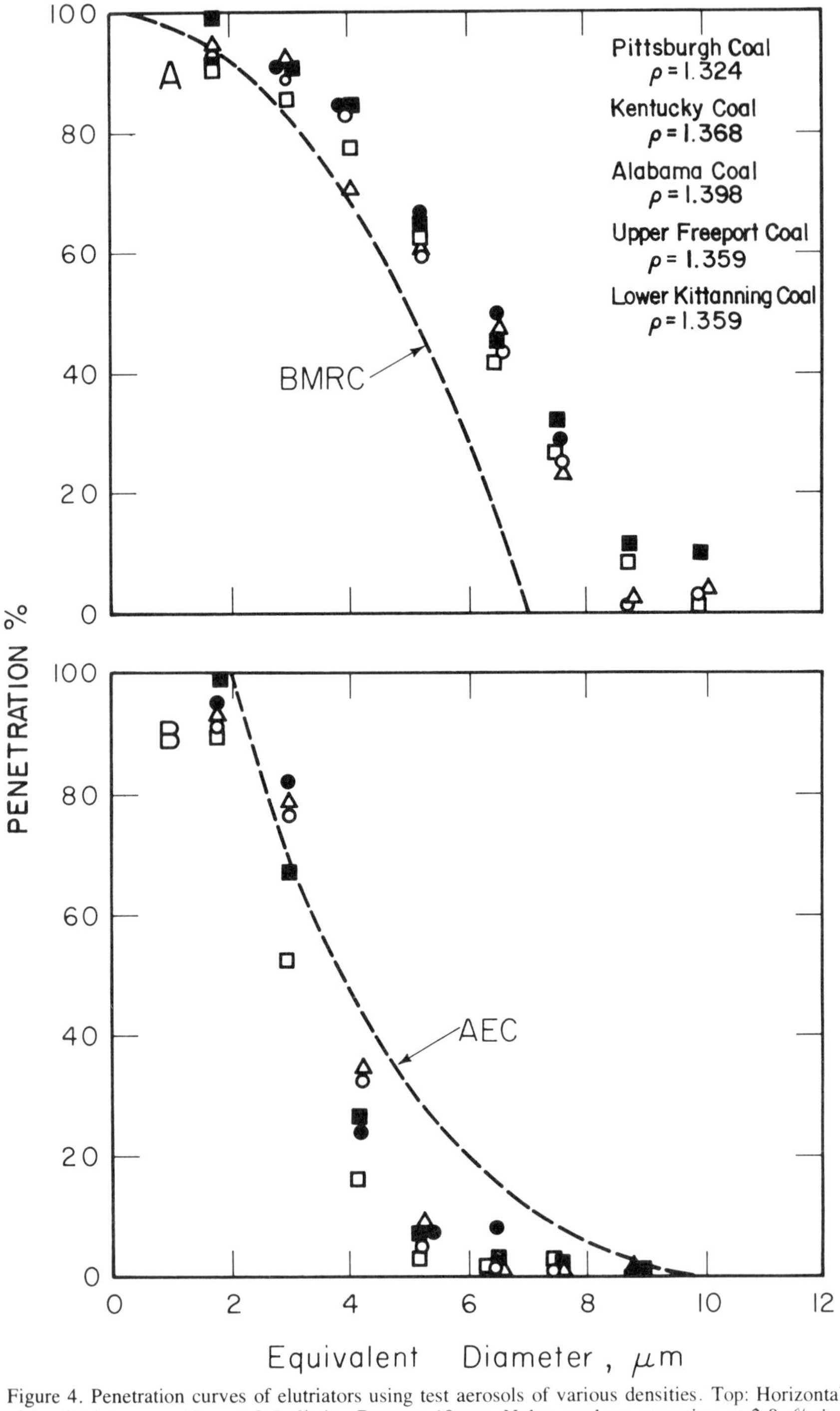

Figure 4. Penetration curves of elutriators using test aerosols of various densities. Top: Horizontal plate elutriator operating at 2.5 ℓ/min. Bottom: 10-mm Nylon cyclone operating at 2.8 ℓ/min. Collected by membrane filter followed by Mines microprojection technique for size distribution 2000X magnification or Coulter Counter (Tomb and Raymond, 1970).

86 cm diameter at the conditions of charged and uncharged inner surface. Charges were built up by triboelectric effect when the untreated interior was cleaned. The surface potentials in the charged chamber varied from zero volt to 10 kv within charge islands varying from 1 mm to 10 cm in diameter. This type of charge buildup was eliminated when the interior was coated with a film of antistatic agent (Arquad 12-50—Armour Chemical Co.). A comparison of ex perimental results of these charged and uncharged conditions is shown in Figure 5 where the charged chamber (radius R) wall collected the aerosols rapidly according to a time constant proportional to (E/R) (q/m)/F where E is the electric field. Collection by the uncharged chamber wall occurred slowly with a time

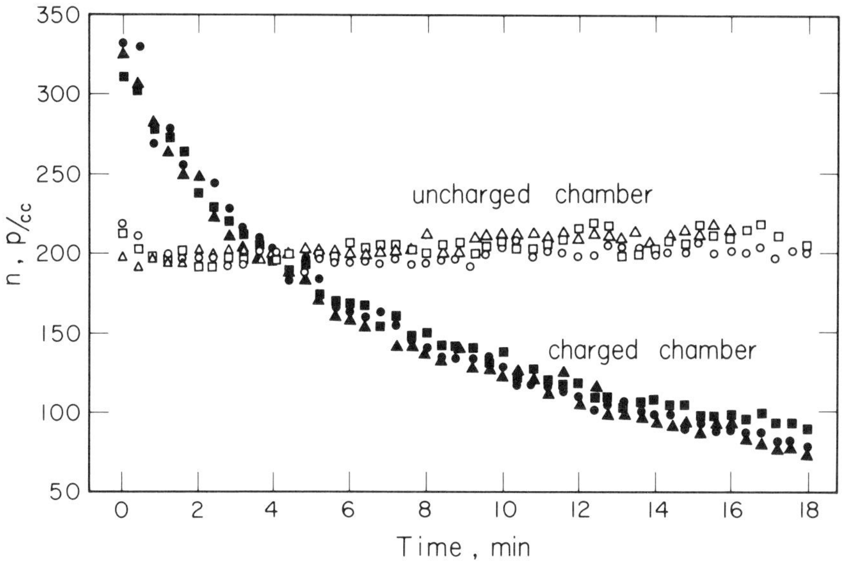

Figure 5. Comparision of concentration vs. time for an uncoated (charged) and coated (uncharged) chamber with dry dispersed aerosol. (When using a coated chamber, the results for dry and spray dispersed aerosols were similar [Stein et al., 1973].)

constant proportional to v_d/R, where v_d is a deposition velocity (Sehmel, 1968) of aerosols diffused to the wall. The magnitude of v_d depends on the properties of the contacting surfaces.

The nylon cyclones cited in the above were not tested under the condition of controlled surface charges, and the influence of the last term in the parenthesis of Eq. (1) could have affected the test results. This internal electric field can be either utilized or eliminated by careful surface treatment (Stein et al., 1973). The space charge effect in a 10 mm cyclone is expected to be small.

Utilization of surface charges or other forms of applied electric field on a particle cloud will improve collection efficiency for particle sampling purposes. Other geometries for controlled electric field are readily visualized.

OTHER EFFECTS

We may also expect significant effects of surface forces such as van der Waals force and liquid force especially at high humidity. Their effect on collection of particles has been dealt with (Soo and Tung, 1972). The fact that surface charges have an important influence is seen in that with similar cyclones of nylon and stainless steel of similar geometry, the steel cyclone collects less. The need for control of test conditions is also seen in tests with similar nylon cyclones; similar penetration characteristics were obtained by Knuth at 1.4 ℓpm and by Tomb at 2.0 ℓpm (Tomb and Raymond, 1970).

The effect of agglomeration is probably small for low dust concentration, say on the order of 1 mg per cubic meter, at most of the test conditions.

DISCUSSION

The previous study (Soo, 1973) of the three-dimensional velocity distribution as given in Figure 1 shows that the velocity distribution influences the collection and density distribution of dust particles of various sizes and the elutriation of dust particles near the exit whose terminal velocity is below the local air velocity. The electrostatic and surface forces are also readily seen in the density distribution and the deposition. The general relation such as Eq. (1) permits scaling of designs to convenient sizes for detailed measurements.

REFERENCES

AIHA Aerosol Technology Comm., 1970: Guide for respirable mass sampling. *J. Amer. Industrial Hygiene Assoc., 31,* 133–137; also U.S. Bu. Mines Info. Circular 8503.

Morse, K. M., 1970: Problems in the gravimetric measurements of respirable coal mine dust. *J. Occupational Medicine, 12,* 400–409.

Sehmel, G. A., 1968: Aerosol deposition from turbulent airstreams in vertical conduits. Report BNWL-578, Battelle Northwest, Richland, Virginia, WA99352, March.

Seltzer, D. F., W. K. Bernaski and J. R. Lynch, 1971: Evaluation of size-selective presamplers, II: efficiency of the 10 mm nylon cyclone. *J. Amer. Industrial Hygiene Assoc., 32,* 441–446.

Soo, S. L., 1973: Some basic aspects of cyclone separators. *Proc. of 1st Internat'l. Conf. on Particle Technol.,* IIT Research Inst., 9–16.

______ and S. K. Tung, 1972: Deposition and entrainment in pipe flow of a suspension. *J. Powder Technol., 6,* 283–294.

Stein, R. L., W. H. Ryback and A. W. Sparks, 1973: Deposition of aerosol in a plastic chamber. *J. Colloid and Interface Sci., 42,* 441–447.

Tomb, T. F. and L. D. Raymond, 1970: Evaluation of the penetration characteristics of a horizontal plate elutriator and of a 10 mm nylon cyclone elutriator. Bureau of Mines Report of Investigations 7367, March.

QUESTIONS

Q: Can you separate 30 μm from 0.1 μm particles with a cyclone?

Lippmann: Obviously you can, but the difference in size is so great that I wonder if it's feasible. With 30 μm particles I'd be worried about the inlet conditions and getting enough of them into the inlet of the cyclone. With a difference in size that great you would have your choice of other convenient instruments, such as elutriators and impactors.

Beeckmans: I think there is one big danger in trying to make such a separation with a cyclone; 30 μm droplets are going to have quite a high velocity in the body of the cyclone due to centrifugal forces, and are going to impact rather efficiently with the small aerosol particles. I think you'd be better off using a horizontal elutriator with that particular problem.

Lippmann: The small particles could be trapped in the internal circulation and perhaps deposit by turbulent diffusion as well. I think Dr. Beeckmans' suggestion is probably the most practical. A horizontal elutriator for a 30 μm separation wouldn't be very big, and it would provide a very easy and gentle separation which would be nearly ideal.

Q: What is the Reynolds number range for your work?

Beeckmans: It varies quite a bit. The low is around 2900. Two or three of the cyclones have had roughly this lower point (of 2900). The high on that group of data was 150,000.

Q: Have there been studies in recent years on the inlet conditions? Will the Reynolds number used cause a boundary layer build up? The transition from laminar to turbulent is very much affected by what goes on upstream of the inlet. It may be that some differences in results are caused by differing inlet conditions.

Beeckmans: That could well be. I think there is a difference of opinion between myself and some of the other panelists here. I don't believe there is such a thing as a laminar cyclone, at least not in ones that are being studied. I base that on the fact that if you look at the pressure drop characteristics of the cyclones the exponent of pressure drop against flow is always 2 or a little greater. That to me doesn't indicate a laminar flow condition.

Whitby: I'd like to make one comment. About 15 years ago we did some work on much bigger cyclones; these were approximately 10 to 20 cm cyclones. We were trying to see what the interaction of different size particles was. This work was never really completed and never was published. What it did show was that small aerosols, i.e., 3 μm aerosols, got scrubbed off the wall very, very readily. There was no question but that particles over 10 μm scoured the wall and removed particles which would otherwise stick. In other words it would cause considerable difference in removal or efficiency if larger (10 μm) particles were or were not present. This is probably an area that needs further study.

Lippmann: We have lab data on that question. We just have to keep in mind that there is a very basic difference, as has been stated, between air cleaning cyclones and sampling cyclones. It's hard to construct a hypothetical situation where you are going to build up enough mass of dust to get into scouring when you're using cyclones for sampling. If you do you will get reentrainment. When we built up very thick wall deposits purposely, we started to get into reentrainment problems. However, in terms of realistic industrial atmospheres and conceivable sampling times, I don't think reentrainment is going to happen.

Soo: We have treated this problem of reentrainment in a paper in the *International Journal of Multiphase Flow* (1973: *1,* 89–101) in relation to the larger particle impinging on the smaller particle actually causing reentrainment.

Q: What about theory vs. experimental performances?

Lippmann: I leave the theory to Dr. Beeckmans who can describe it much better than I can. I would hope that Dr. Whitby would send all those data he has buried in his laboratory to Dr. Beeckmans who has asked me many times if I know of anybody who has good laboratory calibration data with monodisperse aerosols. He hasn't been able to dig enough of it up himself and has been restricted, as you can see, to some of his own and some of ours.

Q: Some people report that some cyclones have a much sharper cut-off than do others. Is there any suspicion that these cyclones are operating in a different manner than the normal cyclones?

Lippmann: Well, yes. The small sampling cyclones do have different sharpness of cut-off from one to the other. For the intended purpose of 2-stage sampling to match the ACGIH curves the steel cyclones fortuitously, and I'm sure it's entirely fortuitous, match the entire shape of the curve quite well, whereas the nylon cyclone in the low flow range where it is used for this purpose is much too sharp. You can't match the entire curve. You can only match it at one point, and the 50% cut-off point is usually selected as the point to match. For industrial hygiene applications, in my opinion, the sharpness of the cut-off is not a critical factor.

Olin: I have a partial answer to Dr. Beeckmans' search for data. There's been a lot of work done by Southern Research Institute on cyclones as precutters for stack sampling. We have developed a cyclone precutter for that particular purpose and have calibrated it for monodisperse aerosols. The precutter that we have has a σ_g as good as any commercial impactor (σ_g = 1.2 to 1.25). Also, we have just begun to make a cyclone preseparator for the high volume sampler, cut-off of about 10 μm to remove large particles before they enter our high volume cascade impactor.

Q: I want to address this question to Dr. Soo. How can one change the slope of the particle collection efficiency curve on the one hand, and on the other hand, will it be really a constant or does the particle sticking coefficient probably change with loading?

Soo: It is an important characteristic quantity. It is one of the quantities that has not been sufficiently determined and identified. I think any determination of cyclone efficiency without taking into effect particle surface interaction is just not going to represent the true situation. The sticking probability if you have a higher centrifugal force, namely a greater vorticity, would tend to increase because it is more driving force toward the wall. Also the electrostatic effect actually has a greater input into the sticking probability. We find some chemicals when applied to cyclones actually form a rather smooth deposit layer. This then shows the significance of the difference in the electrostatic deposition effect and the centrifugal effect. The centrifugal effect actually has less of an effect on the variation sticking probability because inside the boundary layer the particle has a soft landing. But with electrostatic effect the particles are driven all the way into the boundary layer.

Lippmann: If the sticking coefficient is other than 1.0, you should not be using cyclones as size-selective samplers. They must retain what they deposit.

INERTIAL IMPACTORS

VIRGIL A. MARPLE AND KLAUS WILLEKE

Mechanical Engineering Department, University of Minnesota, Minneapolis
Kettering Laboratory, University of Cincinnati Medical Center, Cincinnati, Ohio

Cascade impactors are instruments which have been extensively used for the aerodynamic size classification of aerosol particles. The basic configuration, shown in Figure 1, is simple. A jet of aerosol from a nozzle is directed at an impaction surface: the high inertia, large particles cross the air streamlines and strike the impaction plate, while the smaller particles with less inertia are not collected but remain airborne and pass on to a second nozzle-impactor plate arrangement. The second nozzle is of a smaller diameter than the first nozzle, which increases the velocity and inertia of all the remaining aerosol particles. The larger of the previously uncollected particles thus have sufficient inertia to be collected by the second stage. This process continues as the aerosol stream enters successive stages of the cascade impactor. The final element of a cascade impactor is usually a filter which collects all particles too small to be collected by the last impaction stage.

The nozzles of each stage can be of any shape, but are normally rectangular or round. Many cascade impactors have several nozzles of an identical size at a given stage with the aerosol jets from the nozzles impinging upon a common impaction surface.

The most important characteristic of an impactor stage is the collection efficiency curve, which gives the fraction of particles of a given size collected from the incident aerosol stream as a function of particle size. Ideally, an impactor would collect all particles larger than a certain size and none of the smaller ones, which would correspond to the sharp ideal cutoff illustrated in Figure 2. The efficiency curve of a typical real impactor, also shown in Figure 2, spans over a range of particle sizes but still has a good "sharpness of cut".

It has been shown (Marple and Liu, 1975) that ideal impactor behavior can be obtained if two requirements are met:

1. In the region between the jet exit plane and the impaction plate, the Y-component of fluid velocity is a function of Y only (Y is parallel to the center line of the nozzle, see Figure 3).
2. The Y-component of the velocity of the particles at the jet exit plane is uniform across the jet.

These two requirements are reduced to only the first requirement, if the velocity of the particles is equal to the fluid velocity at the exit plane of the jet. Such requirements are nearly met in a real impactor, as can be seen by the velocity profiles in Figure 3 for a rectangular impactor. It can be seen that the velocity component perpendicular to the jet exit plane, V_y, is independent of X over a major portion of the jet, where X is the distance from the jet center line. However, this criterion is not met in the boundary layer near the wall. Particles passing through this region are the cause for the nonideal cutoff characteristics at the larger efficiencies of the real impactor performance curve shown in Figure 2.

THEORETICAL ANALYSIS

The collection efficiency curves of real impactors have been the object of many studies reported in the literature. Although most of the studies have been experimental in nature (e.g., Ranz and Wong, 1952; Mercer and Chow, 1968; Mercer and Stafford, 1969; Andersen, 1966; Lundgren, 1967), several theoretical studies have been made as well (e.g., Ranz and Wong, 1952; Mercer and Chow, 1968; Mercer and Stafford, 1969; Marple, 1970; Davies et al., 1951).

The theoretical studies show that the impaction characteristics of the impactor are governed by the fluid's velocity flow field, which in turn is specified by the Navier-Stokes equations. A solution of the Navier-Stokes equations (Marple et al., 1974a, b) reveals that the velocity flow field is a function of the physical configuration of the impactor and the Reynolds number, Re, of the flow passing through the nozzle. The Reynolds number, based on the hydraulic diameter of the nozzle, is defined as

$$\mathrm{Re} = \frac{\rho V_0 W}{\mu} \tag{1}$$

for round impactors, and

$$\mathrm{Re} = \frac{\rho V_0 2W}{\mu} \tag{2}$$

for rectangular impactors, where ρ is the fluid density, V_0 is the mean velocity at the nozzle throat, W is the nozzle diameter or width and μ is the fluid viscosity. Furthermore, the trajectory of the particles in the flow field, which determines whether or not the particles are collected, is found by solving the particles' equation of motion (Marple and Liu, 1974), the solution of which is a function of

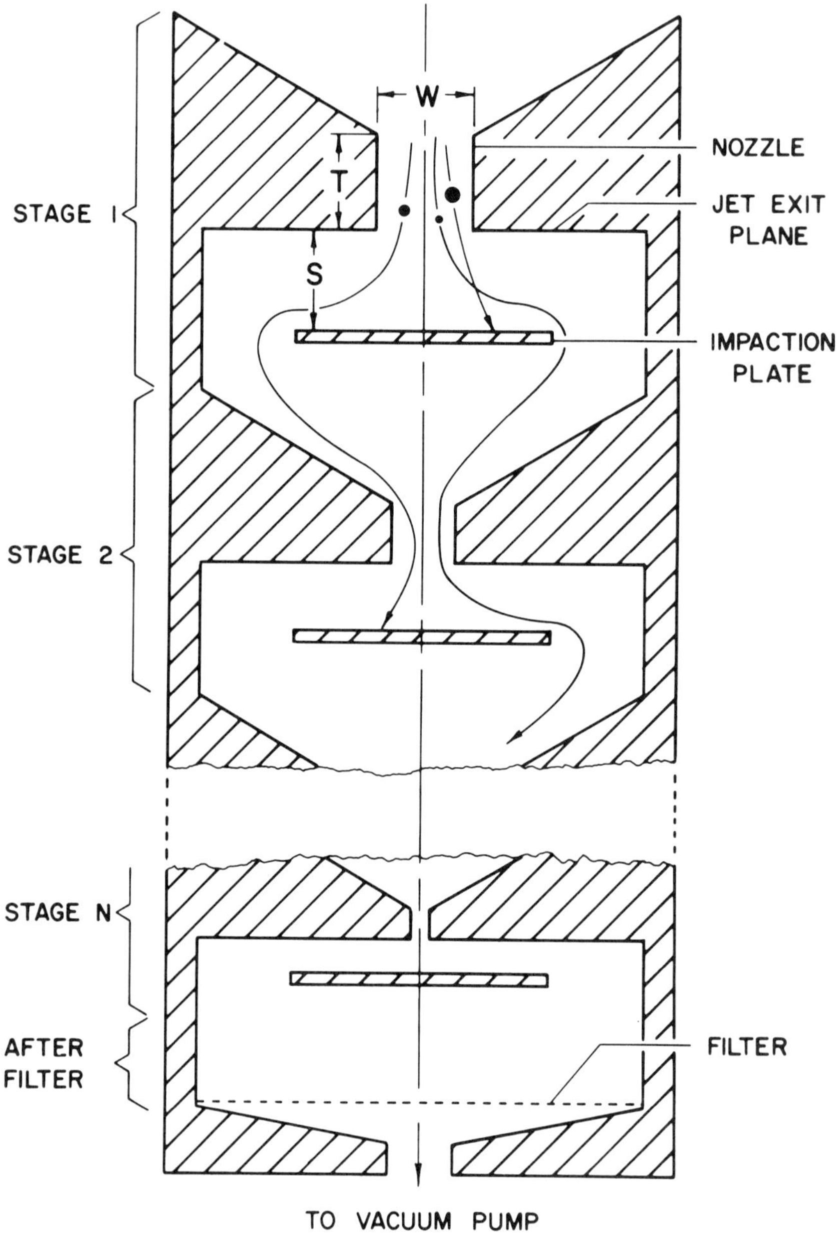

Figure 1. Schematic of cascade impactor with trajectories shown for particles of three different sizes.

the particles' Stokes number and of the parameters which characterize the flow field. The Stokes number is defined as the ratio of the aerodynamic particle stopping distance to the radius, or half-width, of the impactor throat

$$\mathrm{Stk} = \frac{\rho_p \, C V_0 D_p^2 / 18\mu}{W/2}, \tag{3}$$

where ρ_p is the particle density, C is the Cunningham slip correction factor, and D_p is the particle diameter. The square root of the Stokes number, $\sqrt{\mathrm{Stk}}$, is a dimensionless particle size.

The collection efficiency of an impactor is therefore a function of the Stokes number, the Reynolds number, and the physical configuration of the impactor. It

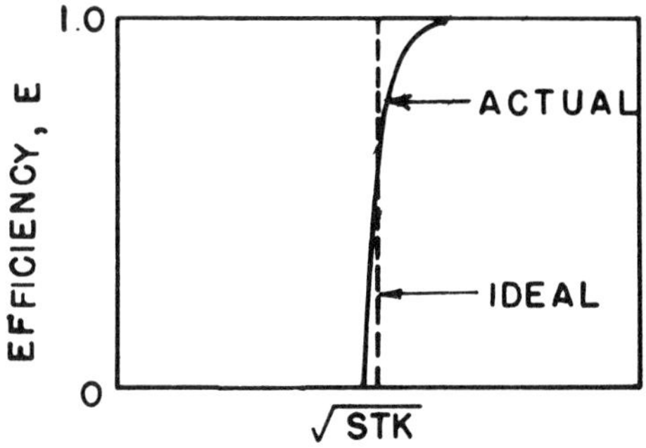

Figure 2. Efficiency curve for a typical impactor.

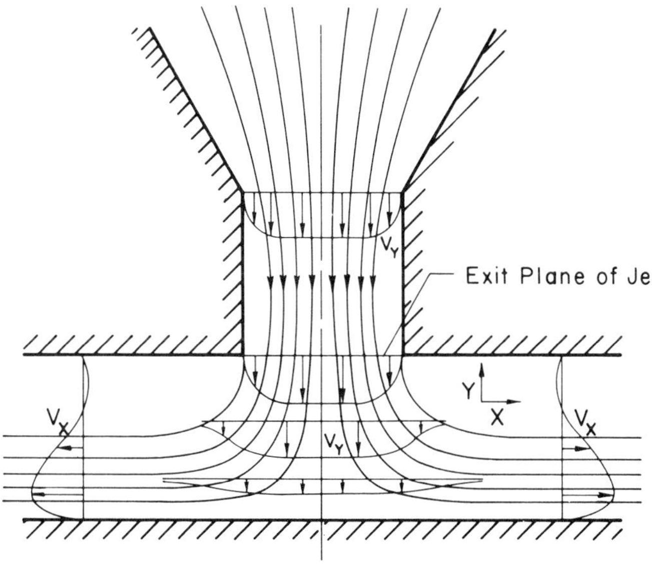

Figure 3. Flow fields in a rectangular impactor (S/W = 1, T/W = 1, Re = 3000).

has been found that the physical configuration for either a round or rectangular nozzle can be sufficiently specified by the value of the nozzle's diameter or width, W, the nozzle-to-plate distance, S, and the nozzle throat length, T (Marple, 1970). Since the Stokes number and Reynolds number are dimensionless, it is convenient to express the physical dimensions in dimensionless form by expressing all dimensions in units of the jet width, such as S/W and T/W.

Figure 4 shows the influence of these various parameters on the collection efficiency. In this figure, the collection efficiency, E, is the collected fraction of all particles of a particular value of $\sqrt{Stk}$ passing through the nozzle. The influence of S/W, Re, and T/W on the collection efficiencies is shown in parametric form in Figures 4a, b, and c, respectively.

Several interesting conclusions can be made by examining the curves in Figure 4. First, all the efficiency curves are similar in shape except for the cases of low Reynolds number (Re $<$ 500) or extremely high Reynolds numbers (Re $=$ 25,000). For low Reynolds numbers, the poor cutoff characteristics are caused by the thick viscous boundary layer near the impactor nozzle wall. For high Reynolds numbers the knee in the efficiency curve at the low values of efficiency appears to be caused by a very thin boundary layer over portions of the impaction plate adjacent to the stagnation point. This thin boundary layer, having a thickness about equal to the particle diameter, allows smaller particles to impact than in areas where the boundary layer is thicker.

Second, the efficiency curve's position on the efficiency vs $\sqrt{Stk}$ plot is relatively independent of S/W, T/W, and Re, except for small values of S/W and Re. This can be seen more clearly in Figures 5 and 6, where the value of $\sqrt{Stk_{50}}$ is cross-plotted as a function of S/W and Re, respectively. In Figure 5, one should note that $\sqrt{Stk_{50}}$ drops sharply with decreasing nozzle-to-plate distance for S/W $<$ 1.0 for rectangular impactors and S/W $<$ 0.5 for round impactors. $\sqrt{Stk_{50}}$ is defined as the value of $\sqrt{Stk}$ at which 50% of the particles are collected.

COMMERCIAL IMPACTORS

Due to the simplicity of construction and use, and the relatively sharp cut-off characteristics, cascade impactors have been widely used for the size classification (and size-classified chemical analysis) of aerosols. Figure 7 lists some of these impactors and the approximate cutoff size (50% collection efficiency points) for unit density spheres (equivalent aerodynamic diameter) of the various stages at the given flow rate. Also note in Figure 7 that a distinction is made between single or multiple nozzle impactors, where multiple nozzles refers to the use of several nozzles in each individual stage.

The cutoff sizes of each stage provided by the manufacturer will generally be for a given set of operating conditions of flow rate Q (which determines the air flow velocity through the nozzle, V_0), particle density ρ_p, gas viscosity μ, and gas density ρ. If any of these operating parameters are different from those

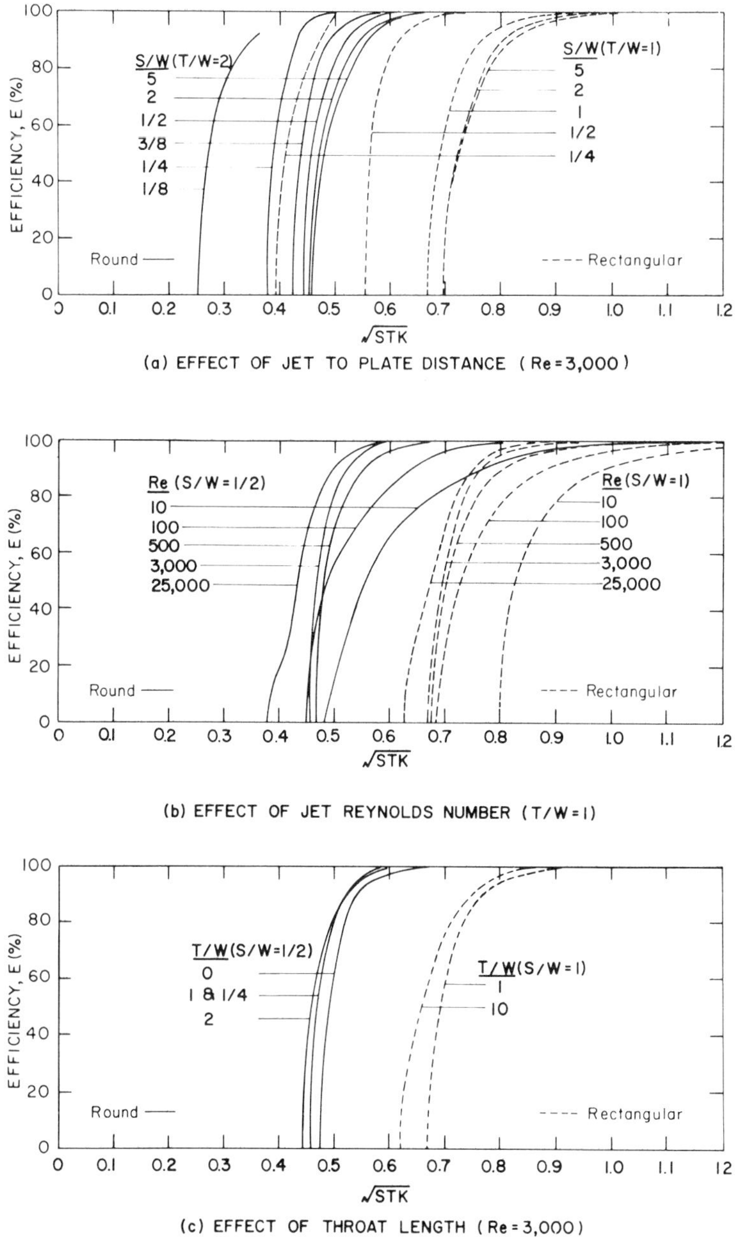

Figure 4. Impactor efficiency curves for rectangular and round impactors showing the effect of jet-to-plate distance S, Reynolds number Re, and throat length T.

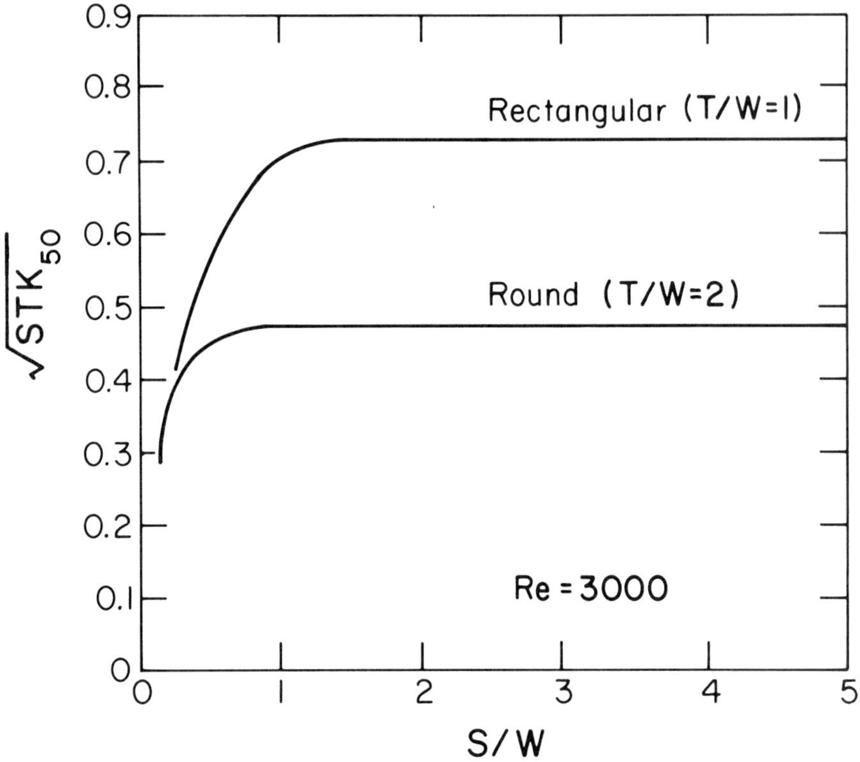

Figure 5. Impactor 50% cutoff size as a function of the jet-to-plate distance.

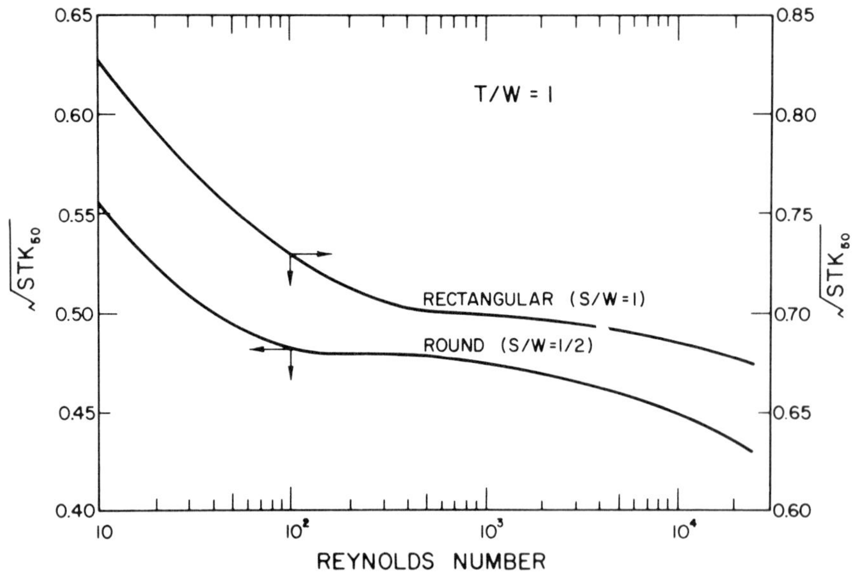

Figure 6. Impactor 50% cutoff size as a function of the Reynolds number.

specified by the manufacturer, the cutoff size may be different due to changes in the values of Re and Stk.

For example, if the Reynolds number is not greatly different from the one assumed by the manufacturer, the influence of varying these parameters can be found by analyzing Eq. (3) and noting that $\sqrt{Stk_{50}}$ is constant. Thus,

$$\sqrt{C}\,D_{50} = \sqrt{C'}\,D_{50}{}'\sqrt{\left(\frac{\rho_p{}'}{\rho_p}\right)\cdot\left(\frac{Q'}{Q}\right)\cdot\left(\frac{\mu}{\mu'}\right)}, \tag{4}$$

where the primed values are operating conditions specified by the manufacturer and the unprimed values are the actual operating conditions. D_{50} denotes the particle diameter at 50% collection efficiency. If the actual Reynolds number is greatly different from the Reynolds number calculated for the manufacturer's specified conditions, the change in $\sqrt{Stk_{50}}$ should be considered. Figure 4 can be used to approximate this effect.

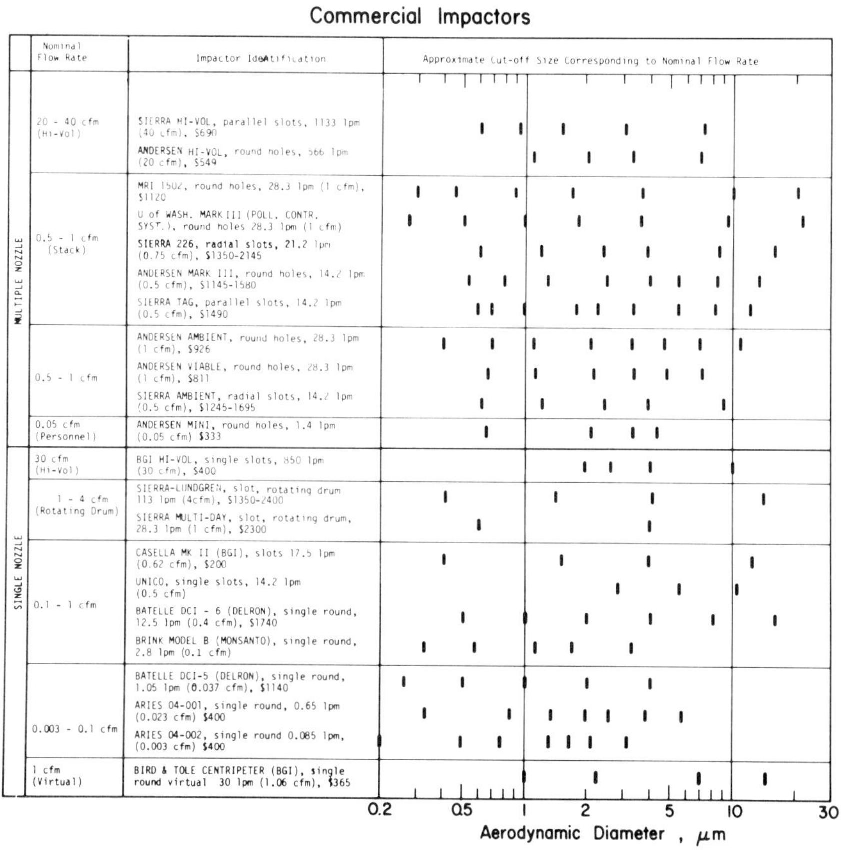

Figure 7. Some commercially available cascade impactors.

IMPACTOR DESIGN

Although there are many commercial impactors available, investigators often design and build their own impactors. Many of these special impactors have been reported in the literature (e.g., Schott, 1973; Cooper and Spielman, 1974; May, 1975; Marple and Rubow, 1976).

A procedure for designing impactors (Marple and Willeke, 1976) is presented below in Steps 1 through 7. This procedure makes use of data from the efficiency curves in Figure 4 and from two design charts shown in Figures 8 and 9. The design chart of Figure 8 shows graphically the relationship between the Reynolds number, flow rate, cutoff size, nozzle diameter and number of nozzles in each stage for multiple nozzle round impactors. The design chart of Figure 9 shows

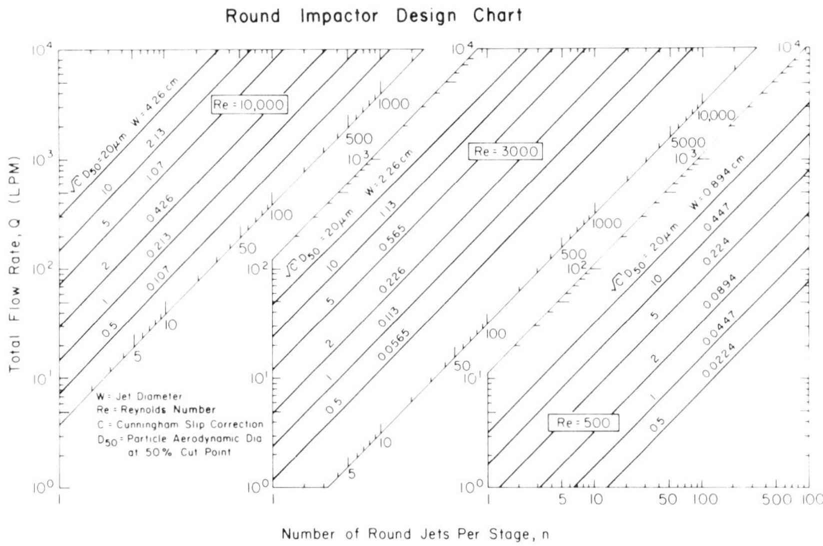

Figure 8. Design chart for round impactors (D_{50} = aerodynamic diameter at 50% cut point).

the relationship between the Reynolds number, flow rate per unit nozzle length, cutoff size and nozzle width for rectangular impactors. These design charts give the number of round nozzles (Figure 8), or the flow rate per unit nozzle length (Figure 9) necessary to achieve a desired Reynolds number in the nozzle of the impactor. The recommended steps are:

1. Choose a desired cutoff diameter, D_p. If the density, ρ_p, is different from unity, calculate the value $\sqrt{C}\,D_{50}$ (where D_{50} is the equivalent aerodynamic diameter of a unit density sphere) from:

$$\sqrt{\rho_{ae}}\,\sqrt{C}\,D_{50} = \sqrt{\rho_p}\,\sqrt{C}\,D_p, \tag{5}$$

where ρ_{ae} = 1 gm/cm³. The variation in the Cunningham slip correction is usually negligible for small diameter differences. Thus, the values of C on both sides of Eq. (5) are approximately equal to each other in most cases.

2a. (*Round*) - Use Figure 8 and determine the number and size of jets required for the desired operating Reynolds number, Re, and aerodynamic cutoff size, $\sqrt{C}\ D_{50}$. Since the cutoff characteristics are fairly constant and sharp over a Reynolds number range from 500 to several thousand, a Reynolds number of 3,000 may be assumed for the calculations.

2b. *(Rectangular)* - Use Figure 9 and choose a desirable Reynolds number (again Re = 3,000 is satisfactory) and determine the Q/L value and jet width, W, for the desired aerodynamic cutoff size $\sqrt{C}\ D_{50}$. Special note must be made of the resulting value of the jet aspect ratio, L/W, since end effects can become detrimental to the sharpness of cut for nozzles with small L/W values.

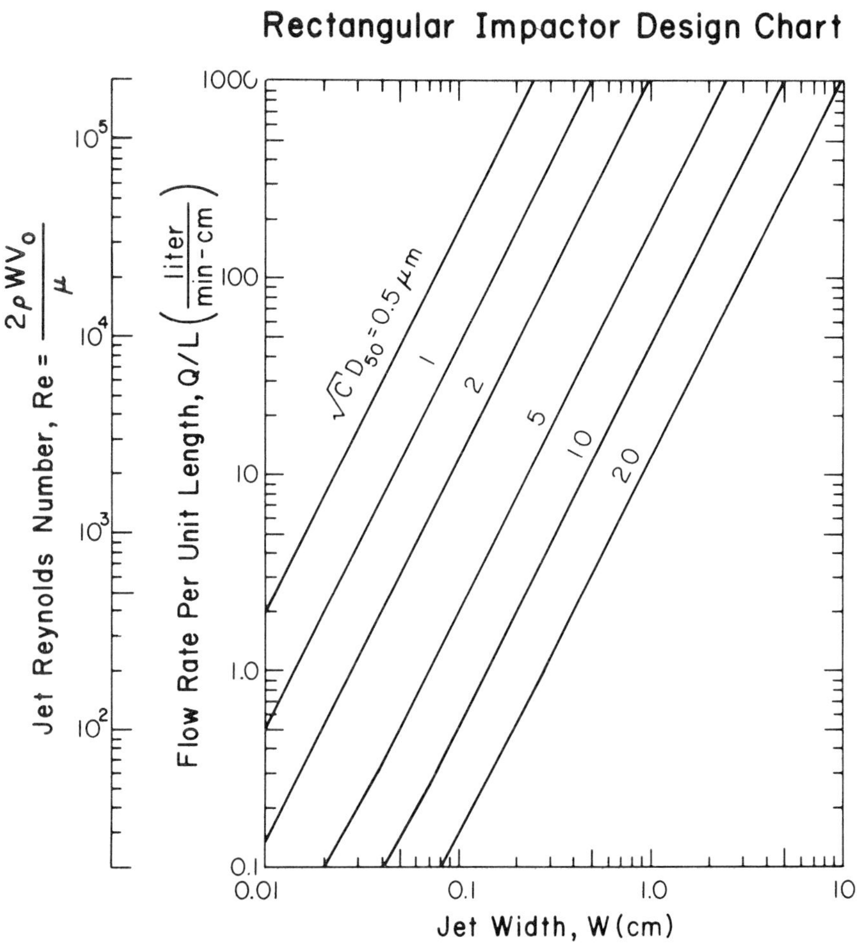

Figure 9. Design chart for rectangular impactors (D_{50} = aerodynamic diameter at 50% cut point).

3. Select a convenient size of jet diameter or jet width which is close to the value found in Step 1. For a round impactor this would be a standard reamer size.

4. For this value of jet diameter or jet width, check the Reynolds number using Eqs. (1) and (2).

5. Determine the value of $\sqrt{Stk_{50}}$ from Figure 4 and calculate the cutoff size, $\sqrt{C}\, D_p$, by using Eq. (3).

6. Determine the pressure in the impaction region, P_2, by assuming that the pressure drop in the stage is equal to the dynamic pressure of the jet. Thus

$$P_2 = P_1 - \tfrac{1}{2}\,\rho V_o^{\,2}, \tag{6}$$

where P_1 = static pressure at the impactor stage inlet
P_2 = static pressure at the impaction plate
ρ = fluid density
V_o = average velocity in the jet

7. Determine the cutoff particle size, by calculating the Cunningham slip correction (Wahi and Liu, 1971) from the equation:

$$C = 1 + \frac{0.163}{D_p P_2} + \frac{0.0549}{D_p P_2} \exp\left(-6.66\, D_p P_2\right), \tag{7}$$

where P_2 = static pressure (atmospheres)
D_p = particle diameter (μm).

OPERATIONAL PRECAUTIONS

As has been shown, an impactor has the ability to sharply classify particles into distinct ranges of aerodynamic size. However, to obtain reliable data from an impactor, one must be aware of its limitations. For example, precautions must be taken in the use of the design charts and of the calculated efficiency curves for conditions other than those specified. Also, an impactor cannot size particles which are not sampled. Thus, the inlet collection efficiency must be known. Once the particles are sampled, the particles must be deposited only upon the impaction plates and, once deposited, must remain on the plates. Particle loss and particle reentrainment may result in indicated size distributions which are significantly different from the actual size distributions.

High Velocity Effects

In using impactors, it is often desirable to be able to collect as small a particle as possible. As Eq. (3) implies, small particles can be collected by increasing the nozzle velocity, V_o. However, the theory previously presented for impactors assumes that the fluid is incompressible. The assumption of incompressibility (Olson, 1961) is normally assumed valid for air flows with Mach numbers, M, less than 0.2, for which the density is in error by only 2%. However, M = ⅓, for which the density is in error by about 5%, may be considered a practical upper limit (Cohen and Montan, 1967).

High velocity through the nozzle may also bring about more particle losses between stages and from the impaction plate due to reentrainment (discussed later). These losses, however, can be minimized by the use of a proper impaction plate adhesive.

Inlet Losses

As is the case with any size analyzing instrument, the efficiency with which the impactor inlet is sampling the aerosol particles must be known before accurate size distribution and particle concentration data can be obtained (Fuchs, 1975).

Recent work (Agarwal, 1975) has shown that large particles (up to about 100 μm) can be drawn into an inlet from still air with efficiencies of nearly 100%. However, there may still be considerable losses on the internal surfaces of the inlet due to impaction and settling. If sampling from a moving air stream is required, isokinetic conditions at the inlet will aid in efficient sampling of large particles. Small crosswind velocities may significantly reduce the sampling efficiency in ambient air sampling situations.

One method which has been used to aid in the sampling of large particles is to place the impaction surface on a rotating arm (Jaenicke and Junge, 1967; Noll and Pilat, 1971; Gruber and Harms, 1974). Since these devices have no inlet, this source of losses is eliminated.

Interstage Losses

Particle losses in an impactor, generally referred to as wall losses or interstage losses, are the deposition of particles on surfaces other than the impaction plate. Currently, no theory exists to predict these losses, and thus, they must be determined experimentally.

Recently, three commercial cascade impactors have been investigated experimentally to determine the interstage losses. In a detailed evaluation (Rao, 1975) of the Lundgren impactor (rotating drum impaction plate), the Andersen Viable aerosol sampler (28.3 ℓpm, multiple round jets) and also in a similar evaluation, of the Sierra high-volume cascade impactor (Willeke, 1975; Willeke and McFeters, 1975) (1133 ℓpm, multiple rectangular jets), the interstage losses were determined as a function of particle aerodynamic size, and were expressed as the fraction of the total number of particles entering the impactor (Figure 10).

A study of Figure 10 shows that for the Lundgren and the Andersen Viable impactors, interstage losses are not a serious problem for particles smaller than about 5 μm, but increase rapidly for larger particles. The Sierra impactor, however, has appreciable losses for particles smaller than 5 μm. It was found (Willeke, 1975; Willeke and McFeters, 1975) that the interstage losses occurred primarily on both inner walls of the rectangular nozzle, apparently due to lateral impaction of the entering flow. Tapering of the nozzle inlets reduced these losses.

Particle Reentrainment (Bounce-off and Blow-off)

The limitations of impactors due to particles bouncing off the impaction plate or being blown off the impaction plate after collection are essentially the same: in both cases, particles which should have been collected are reentrained into the airstream. Thus, we shall use the term reentrainment for both blow-off and bounce-off. For single stage impactors, this means that the concentration of

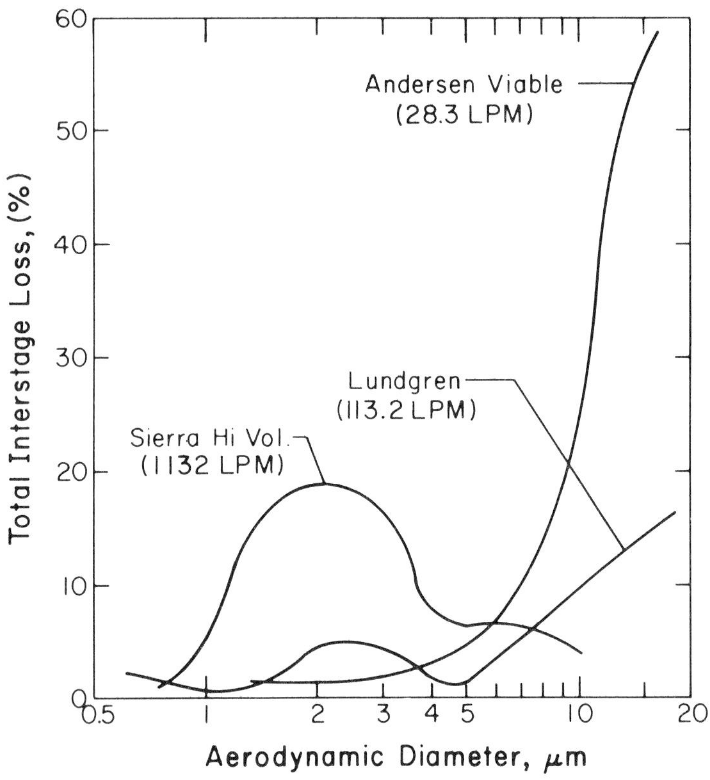

Figure 10. Total interstage losses for the Andersen (multiple round holes), Lundgren (single slots), and Sierra (multiple slots) impactors.

particles collected on the impaction plate will be too small. For cascade impactors, the resulting size distribution will be shifted to smaller sizes, since reentrained particles will be collected on stages intended to collect smaller particle sizes.

The degree of particle reentrainment is a function of the type of particle and the nature of the impaction surface. Liquid particles will impact on any type of surface with very little reentrainment. The reentrainment of solid particles, however, is a strong function of the type of impaction surface used. The maximum amount of reentrainment is experienced with a dry, smooth surface and the least with a very sticky surface. Table I lists some of the sticky surfaces that have been used.

A detailed experimental study (Rao, 1975) to determine the degree of reentrainment from several types of impaction surfaces was made using single-stage impactors (single round jets) designed according to the theoretical criteria de-

TABLE I. Impaction Surface Adhesives Used in Impactor Studies

Reference	Adhesive
May (1945)	Flypaper mixture—one part rosin to three parts castor oil (benzene solution) Polyisobutane (chloroform solution) One part methylated starch to three parts tricresyl phosphate (alcohol or amyl acetate solution) Oil or vaseline
Davies et al. (1951)	Glycerin jelly
Stern et al. (1962)	Dow-Corning Hi-Vac silicone grease
McFarland and Zeller (1963)	Dow Corning silicone oil (DC-200 fluid, 2% solution in hexane)
Zeller (1965)	One part dibutyl phthalate, 10 parts cellulose acetate butyrate and 90 parts cylohexanone Scotch tape and other films
Lundgren (1967)	Viscous oils and sticky tapes Dow Corning Hi-Vac silicone grease Minnesota Mining Kel-F Polymer wax
McFarland and Husar (1967)	100,000 cs. Dow DC-200 silicone oil (1% solution in hexane)
Mercer and Chow (1968)	Dow Corning antifoam A
Hogan (1971)	G.E. silicone resin SR-516
Berner (1972)	Vaseline dissolved in carbon tetrachloride or toluene
Wesolowski et al. (1973)	Sticky film Dow Corning Hi-Vac grease
McCain et al. (1974)	10 to 15% solution in benzene of high vacuum silicone grease
Walkenhorst (1974)	Several oils, $CaCl_2$ solution

scribed above. The results for one of the impactors are presented in Figure 11. It should be noted that the test aerosols were polystyrene latex spheres which are quite resilient. Thus, the curves of Figure 11 can be considered to be approximately the worst obtainable.

Figure 11 shows that with an oil-coated glass plate there was essentially no reentrainment, since the efficiency curve agreed well with the theoretical curve and attained nearly 100% efficiency. At the other extreme, the uncoated glass plate had severe reentrainment for particles larger than about $\sqrt{\mathrm{Stk}} = 0.47$ and had a maximum collection efficiency of about 30%. When a filter paper was used as a collection surface, the efficiency curves were found to be between the two extreme cases of the oil-coated and the dry glass plates, and tended to have poor sharpness-of-cut characteristics.

In order to avoid the problems of particle reentrainment, a dichotomous (or virtual) impactor may be used (Dzubay and Stevens, 1975). In such a device the particles are impacted into a slowly pumped void and are then collected onto a membrane filter. The virtual impactor, however, is considerably more complicated than the normal type of impactor, and consists of only one classification stage.

DATA ANALYSIS

The particles collected on the impaction plates are analyzed after the sampling is terminated. In the most common type of analysis, the concentration and size distribution (by number or mass) of the aerosol particles is determined from the deposits by one of the several methods of analysis of which only the more widely used ones will be discussed here.

The size distribution of the aerosol can be determined either from a single stage impactor or from a cascade impactor. For a single stage impactor the deposit must be analyzed with a microscope to get a number distribution. With a cascade impactor, the particle deposits of the various stages can also be analyzed with a microscope, but are more commonly analyzed gravimetrically to obtain a mass distribution.

The number or mass concentration of the aerosols can be determined from the size distribution data. If the size distribution is found from a portion of the deposit by microscope analysis, the concentration is calculated from the quantity of particle-laden air sampled and from the fraction of the total deposit analyzed. If the gravimetric analysis is used, only the quantity of air flowing through the impactor needs to be known.

Methods other than the gravimetric analysis are also used to determine the mass of particles collected on an impaction plate. In one instrument (Chuan, 1970) the particles are impacted onto a piezoelectric crystal which determines the mass by monitoring the change in the resonant frequency of the crystal. Another instrument (Lilienfeld and Dulchinos, 1972) uses beta attenuation to detect the mass of particles collected.

Impactors also have been used in conjunction with an optical particle counter to obtain the number size distribution and particle concentration (Cooper and Spielman, 1974). The aerosols were passed through an impactor with a variable jet width. The particles not collected on the impaction plate were counted with an optical particle counter. By varying the jet width in a step-wise manner, a size distribution was obtained.

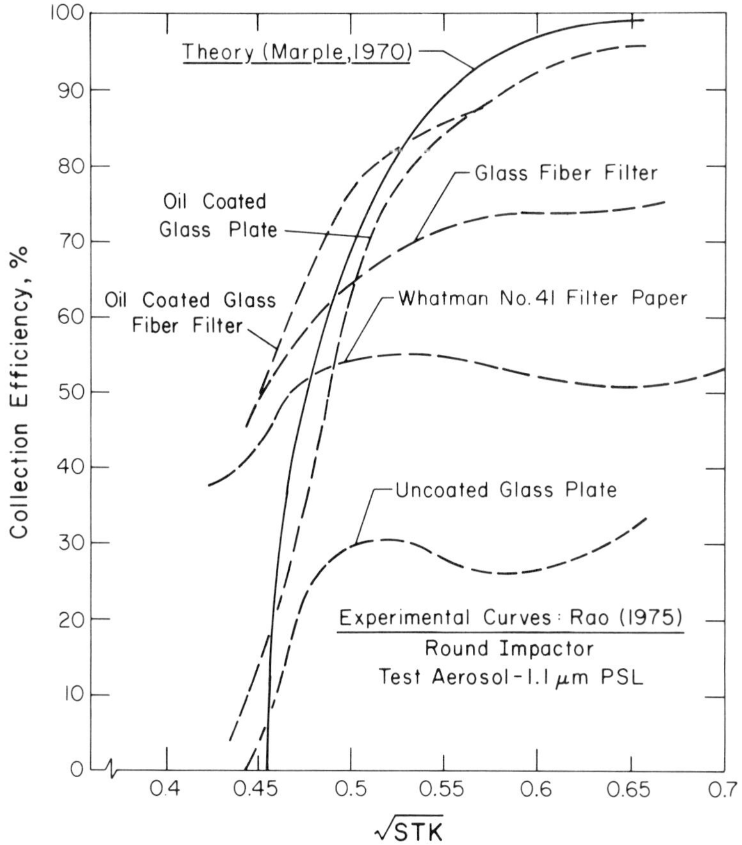

Figure 11. Collection efficiency of a typical single stage impactor for various collection surface media.

The elemental composition of the deposits can be determined by techniques such as atomic absorption, neutron activation and x-ray fluorescence. The collection of the particles should be on a medium having a very low level of impurities. For instance, it has been found (Dzubay and Stevens, 1975) that glass fiber filter paper is unacceptable for x-ray fluorescence analysis, but membrane filters, made of mixed esters of cellulose and having 0.8 μm pore size, are satisfactory.

REFERENCES

Agarwal, J. K., 1975: Aerosol sampling and transport, Ph.D. Dissertation, University of Minnesota, Particle Technology Laboratory, Publ. No. 265.

Andersen, A. A., 1966: A sampler for respiratory health hazard assessment. *J. Amer. Industrial Hygiene Assoc.*, *27,* 160.

Berner, A., 1972: Practical experience with 20-stage impactor. *Staub-Reinhalt. Luft*, *32,* 1.

Chuan, R. L., 1970: An instrument for the direct measurement of particulate mass. *J. Aerosol Sci.*, *1,* 111–114.

Cohen, J. J. and D. N. Montan, 1967: Theoretical considerations, design, and evaluation of a cascade impactor. *J. Amer. Industrial Hygiene Assoc., 28,* 95–104

Cooper, D. W. and L. A. Spielman, 1974: A new particle size classifier: Variable slit impactor with photo-counting. *Atmos. Environ., 8,* 122.

Davies, C. N., M. Aylward and D. Leacey, 1951: Impingement of dust from air jets. *AMA Arch. Ind. Hyg. Occup. Med., 4,* 354

Dzubay, T. G. and R. K. Stevens, 1975: Ambient air analysis with dichotomous sampler and x-ray fluorescence spectrometer. *Environ. Sci. Technol., 9,* 663.

Fuchs, N. A., 1975: Sampling of aerosols. *Atmos. Environ., 9,* 697–707.

Gruber, C. W. and G. Harms, 1974: Fugitive dust measurement by the rotary ambient adhesive impactor. Presented at 67th Annual APCA Meeting, Denver, Colorado.

Hogan, A. W., 1971: Evaluation of a silicone adhesive as an aerosol collecting medium. *J. Appl. Meteor., 10,* 592.

Jaenicke, R. and C. Junge, 1967: Studien zur oberen grenzgrösse des natürlichen aerosoles (Studies of the upper size limit of natural aerosols). *Beitrage zur Physik der Atmosphäre, 40,* 129–143.

Lilienfeld, P. and J. Dulchinos, 1972: Portable instantaneous mass monitor for coal mines dusts. *J. Amer. Industrial Hygiene Assoc., 33,* 136–145.

Lundgren, D. A., 1967: An aerosol sampler for determination of particle concentration of size and time. *J. Air Pollution Control Assoc., 17,* 225.

Marple, V. A., 1970: A fundamental study of inertial impactors, Ph.D. Dissertation, University of Minnesota, Particle Technology Laboratory Publ. No. 144.

——— and B. Y. H. Liu, 1974: Characteristics of laminar jet impactors. *Environ. Sci. Technol., 8,* 648–654.

——— and ———, 1975: On fluid flow and aerosol impaction in inertial impactors. *J. Colloid and Interface Sci., 53,* 31–34.

———, ——— and K. T. Whitby: 1974a: On the flow fields of inertial impactors. *ASME J. Fluid Eng., 96,* 394–403.

———, ——— and ———, 1974b: Fluid mechanics of the laminar flow aerosol impactor. *J. Aerosol Sci., 5,* 1–16.

——— and K. Rubow, 1976: Aerodynamic particle size calibration of optical particle counters. *J. Aerosol Sci., 7,* 425–433.

——— and K. Willeke, 1976: Impactor design. *Atmos. Environ., 10,* 891–896.

May, K. R., 1975: An "ultimate" cascade impactor for aerosol assessment. *J. Aerosol Sci., 6,* 413–419.

———, 1945: The cascade impactor: An instrument for sampling coarse aerosols. *J. Sci. Instrum., 22,* 187.

McCain, J. D., K. M. Cushing and W. B. Smith, 1974: Methods for determining particulate mass and size properties: Laboratory and field measurements. *J. Air Pollution Control Assoc., 24,* 1173.

McFarland, A. R. and H. W. Zeller, 1963: Study of a large-volume impactor for high-altitude aerosol collection. General Mills Inc. Electronics Division, Minneapolis, Mn. Report No. 2391, Contract AT (11-1), p. 401.

——— and R. B. Husar, 1967: Development of a multistage inertial impactor, University of Minnesota, Particle Technology Laboratory Publ. No. 120.

Mercer, T. T. and H. Y. Chow, 1968: Impaction from rectangular jets. *J. Colloid and Interface Sci., 27,* 75.

——— and R. G. Stafford, 1969: Impaction from round jets. *Ann. Occup. Hyg., 12,* 41.

Noll, K. E. and M. J. Pilat, 1971: Size distribution of atmospheric giant particles. *Atmos. Environ., 5,* 527–540.

Olson, R. M., 1961: *Essentials of Engineering Fluid Mechanics.* Scranton, Pennsylvania, International Textbooks Co., p. 165.

Ranz, W. E. and J. B. Wong, 1952: Impaction of dust and smoke particles. *Ind. Eng. Chem., 44,* 1371.

Rao, A. K., 1975: An experimental study of inertial impactors, Ph.D. Dissertation, University of Minnesota, Particle Technology Laboratory, Publ. No. 269.

Schott, J. H., 1973: Jet-cone impactors as aerosol particle separators, M. S. Thesis, University of Minnesota.

Stern, S. C., H. W. Zeller and A. I. Schekman, 1962: Collection efficiency of jet impactors at reduced pressures. *I & EC Fundamentals, 1,* 273.

Wahi, B. and B. Y. H. Liu, 1971: The mobility of polystyrene latex particles in the transition and the free molecular regimes. *J. Colloid and Interface Sci.*, *37*, 374–381.

Walkenhorst, W., 1974: Untersuchungen über den haftgrad von staubteilchen. *Staub-Reinhalt. Luft*, *34*, 182.

Wesolowski, J. J., 1973: Ambient air aerosol sampling. *Proc. Second Joint Conf. in Sensing of Environmental Pollutants*, Instrument Society of America.

Willeke, K., 1975: Performance of the slotted impactor. *J. Amer. Industrial Hygiene Assoc.*, *36*, 683–691.

——— and J. J. McFeters, 1975: The influence of flow entry and collecting surface on the impaction efficiency of inertial impactors. *J. Colloid and Interface Sci.*, *53*, 121–127.

Zeller, H. W., 1965: Large volume impactor collector. Litton Systems Inc., Applied Science Division, 2295 Walnut St., St. Paul, Minnesota 55113, Report No. 2893, Project No. 89125.

COMMENTS ON INERTIAL IMPACTOR CALIBRATION AND USE

RONALD H. KNUTH

Health and Safety Laboratory, U.S. Energy Research and Development Administration, N.Y., N.Y.*

Work performed several years ago at the Health and Safety Laboratory (HASL) revealed the serious effects particle rebound and varying stage collection efficiency have on determining cut-off diameters in the Andersen 1 cfm "Ambient Fractionating Sampler."[1] This experimental work was undertaken initially to verify the manufacturer's stage calibration but it was found almost immediately that (1) impacting particles tended to rebound from the glass impaction surfaces and then penetrate through succeeding stages to the final filter, and (2) stage collection efficiencies often increased or decreased with time depending on the relative change in the impaction surface with particle accumulation directly below the jet. Coating the impaction surface with a thin layer of rubber cement altered both the impaction pattern and the removal efficiency of the stage.

In these tests, Polystyrene Latex (PSL)[2] particles, 1.099 μm in diameter, were sampled at the standard flow rate of 1 cfm using glass impaction plates provided with the instrument. According to the manufacturer's calibration, particles of that size should be removed by the 5th and 6th stages of the impactor.

The possibility of inefficient particle collection due to rebound arose when microscopic examination of the impaction areas revealed too few particles and unusual deposition patterns. A series of tests was run to investigate the rebound characteristics of coated and uncoated impaction plates. The concentration of latex particles as generated in these experiments with a Laskin nozzle (Echols and Young, 1963) is quite constant with time and for the 15 minute tests the number of particles sampled was assumed to be constant. Therefore, the total

*Now the Environmental Measurements Laboratory, U.S. Department of Energy, N.Y., N.Y.

1. Andersen 2000, Inc., P.O. Box 20769, Atlanta, Ga. 30320.
2. Dow Chemical Co., Midland, Mich. 48640.

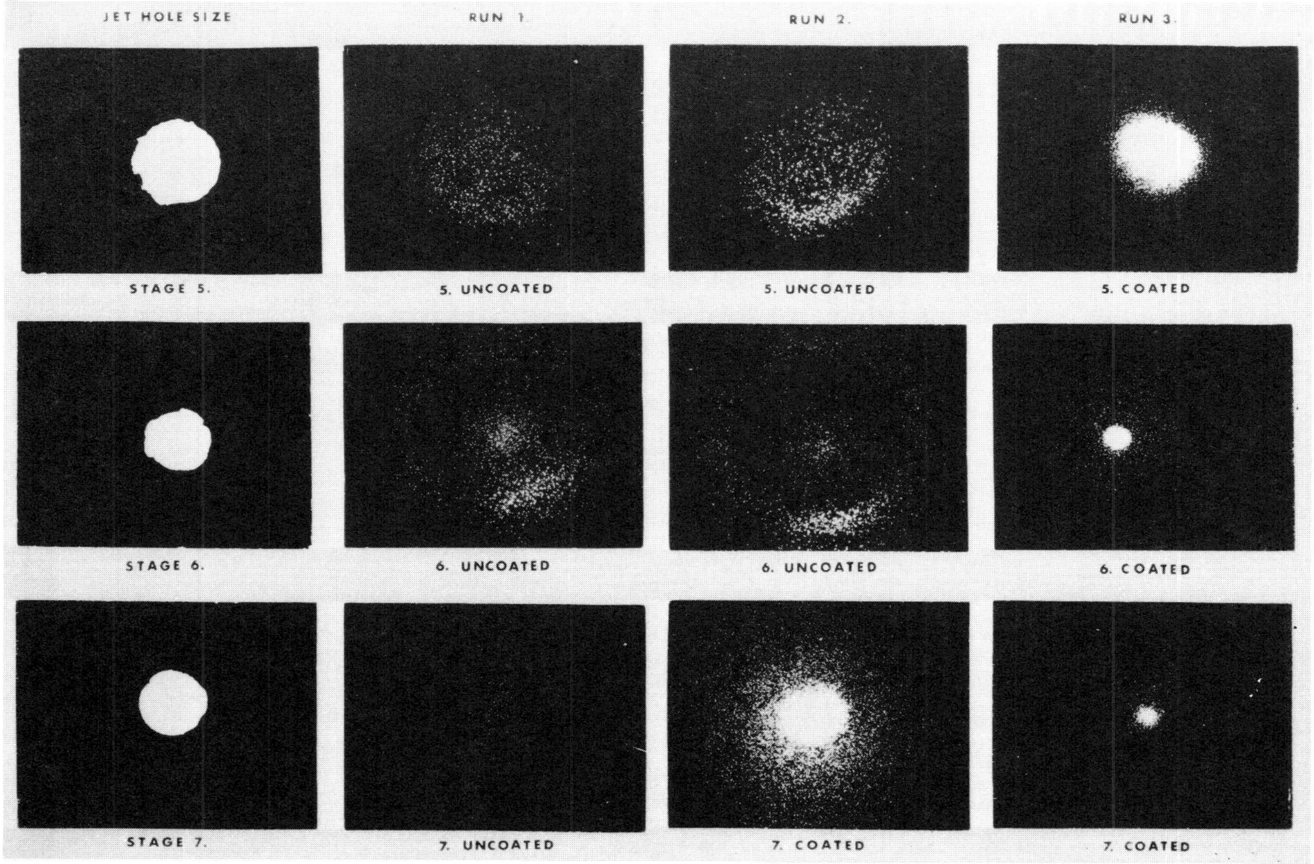

Figure 1. Latex particle impaction patterns in the Andersen Sampler.

number of particles retained by the stages should have been approximately equal in successive tests if there was no difference in the removal efficiency of the coated and uncoated plates.

Figure 1 shows representative photographs of the impaction areas beneath the jets of stages 5, 6, and 7 from a series of three runs. In run 1, all plates were uncoated; in run 2, stages 5 and 6 were uncoated and stage 7 was coated with rubber cement; in run 3, all three stages were coated with rubber cement. These photographs reveal that uncoated plates allow some of the particles to rebound and continue through succeeding stages by a cascade effect eventually reaching the final filter. In run 1, particles rebounded from the uncoated plates and continued through to the final filter. In run 2, the rubber cement coating on stage 7 retained many of the particles that had passed stages 5 and 6. In run 3, with all three stages coated with rubber cement, most of the particles were retained by stages 5 and 6 in accordance with the manufacturer's calibration. A typical rebound pattern for an uncoated impaction plate is illustrated by stage 6, run 2 in Figure 1. This ring of particles encircling the impaction area directly below the

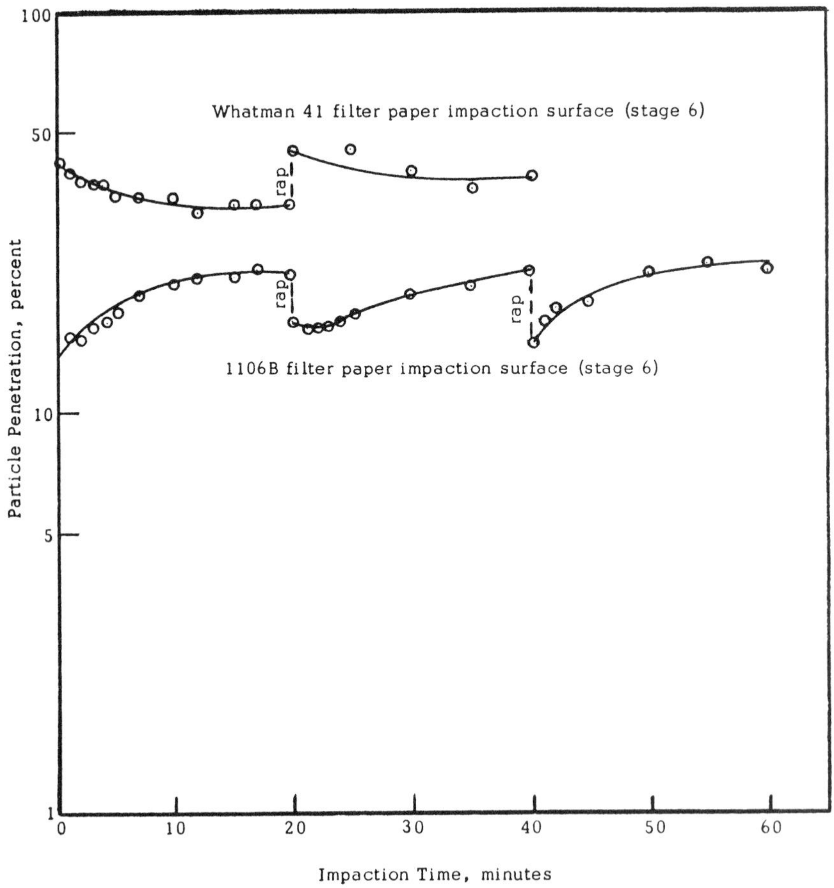

Figure 2. Impaction removal efficiency with time in the Andersen Sampler.

jet was found at the higher velocity stages when impaction plates were not coated. It is suspected that it consists of particles that either are swept by the air flow from the central deposit or rebound slightly from the impaction plate and are precipitated again away from the jet where the velocity is lower. Particles that rebound higher above the impaction plate are carried to the next stage. The dense deposit of particles in the ring is always located on the side closest to the edge of the impaction plate. In impaction areas near the edge of the plate, the pattern is in the form of a wide crescent with very few particles deposited in the opposite arc of the ring.

Additional tests were run using 5.0 μm diameter iron oxide particles produced with a spinning disc aerosol generator. According to the manufacturer's calibration, an aerosol of that size should be removed by stages 2 and 3. In general, impaction patterns for the 5.0 μm particles were similar to those previously shown for the latex particles. Very few of the 5.0 μm aerodynamic diameter iron oxide

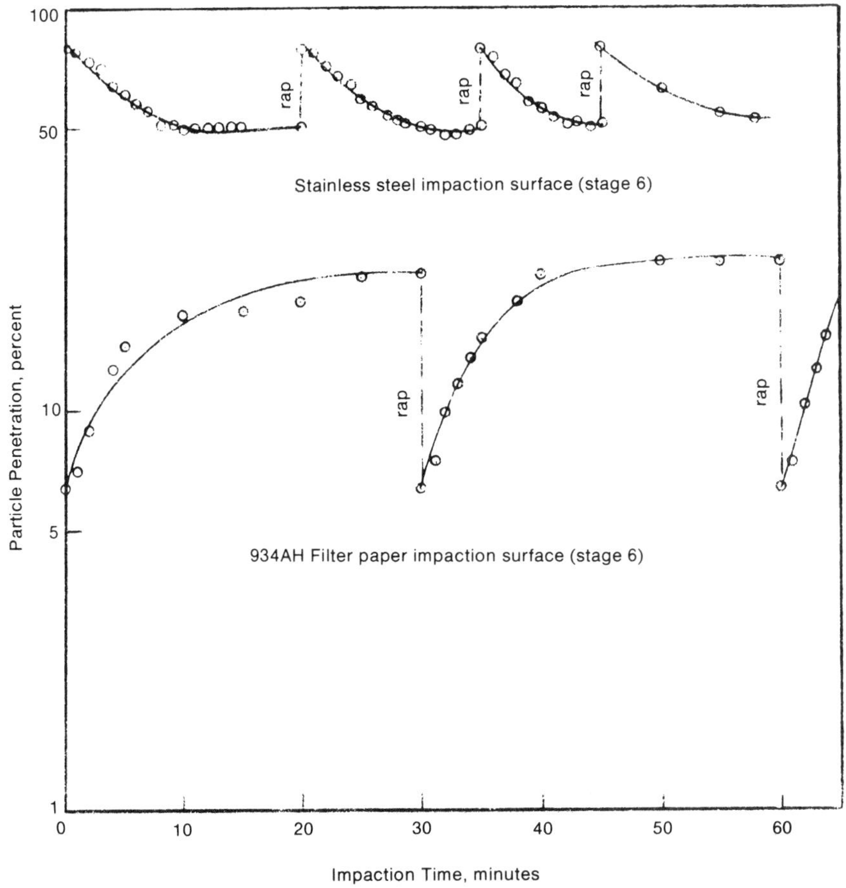

Figure 3. Impaction removal efficiency with time in the Andersen Sampler.

particles were found impacted beyond stage 5 when the impaction plates were coated. When plates were not coated, characteristic rebound patterns similar to those previously shown for the latex aerosol particles were found on stages 4 through 7.

Attempts to determine the impaction efficiency of a single stage using a photometer to measure the concentration of monodisperse latex particles upstream and downstream from the stage were inconclusive. Penetration values varied from test to test for seemingly identical conditions. The variations could

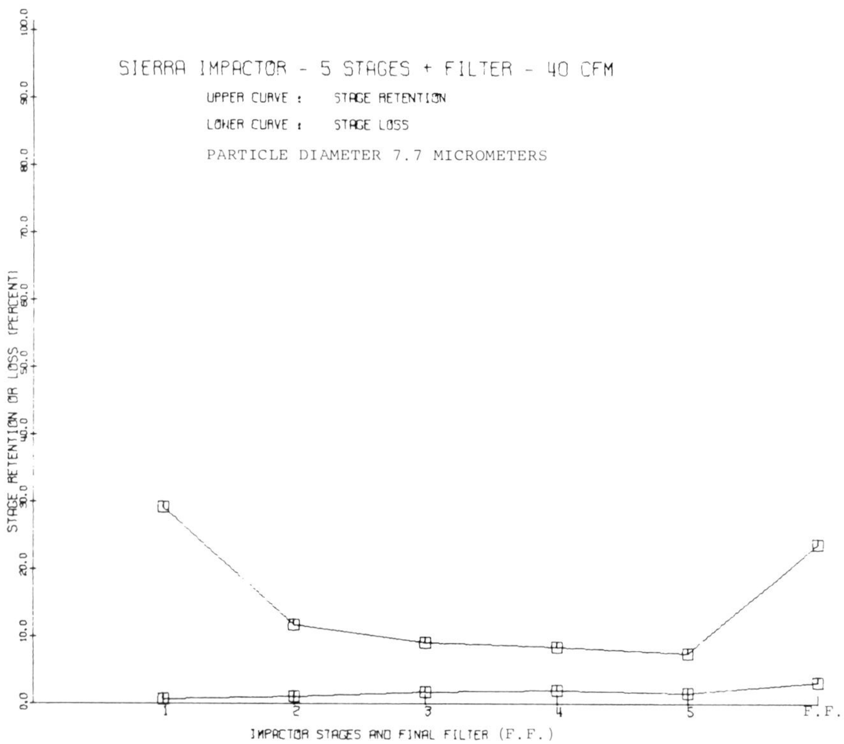

Figure 4. Effect of particle rebound on stage collection.

not be correlated with changes in particle concentration, humidity, or temperature. However, a time dependence of impaction efficiency did seem possible. This was verified by plotting particle penetration versus time for the 1.099 μm PSL particles. Figures 2 and 3 show the time variation of the collection efficiency for different impaction surfaces and the effect of succesive sharp raps on the impactor. It is presumed that each rap moved the impaction plate slightly, exposing clean impaction areas to the impinging air jets. In every test, a sharp rap on the sampler caused the penetration to return to initial conditions, resulting in the cyclic curves shown. The reproducibility of the individual curves varied depending on the impaction surface and the amount of displacement of the impaction surface caused by the sharp rap.

The change in penetration is not as great for Whatman 41 and 1106B filter papers, shown in Figure 2, as for the bare stainless steel and for 934AH filter surfaces shown in Figure 3. However, in the case of Whatman 41 and bare stainless steel, penetration decreases with time in contrast to the 934AH and 1106B filter surfaces which show increasing penetration with sampling time.

Recent calibration work on the Sierra 40 cfm Cascade Impactor[3] at this labora-

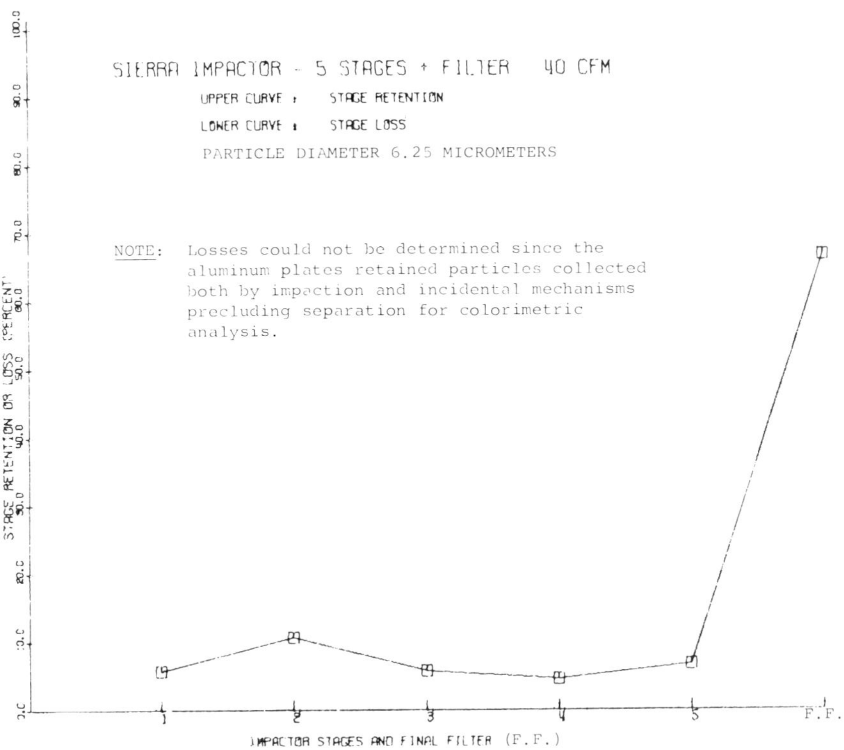

Figure 5. Effect of 6.25 micrometer diameter particle rebound from aluminum plate impaction surface on stage collection.

tory has again shown the serious effects particle rebound can have on measurements of particle size. The impactor was tested at the manufacturer's suggested flow rate of 40 scfm using the recommended Type A (binderless) glass fiber filter paper as an impaction surface. The aerosol used for the calibration consisted of solid, dry, monodisperse spheres of methylene blue, generated by a spinning disc. The effect of rebound was again seen in scanning electron micrographs that were made of all five impaction stages and the final filter. A test with an aerosol 7.7 μm in diameter revealed similar particle sizes on each stage and the

3. Sierra Instruments, Inc., P.O. Box 909, Carmel Valley, Calif. 92924.

final filter and no evidence of fractured particles. Figure 4 is a graph of the experimental data from this test which indicates the significant effect of particle rebound on stage collection efficiency. According to the manufacturer's (D_p) 0.50 values, 50% or more of these seven micrometer particles should be collected by stage 1, and the remainder by stage 2. However, the figure shows that the particles rebounding from stage 1 continue on through the impactor with approximately 10% of the total aerosol being removed by each succeeding stage.

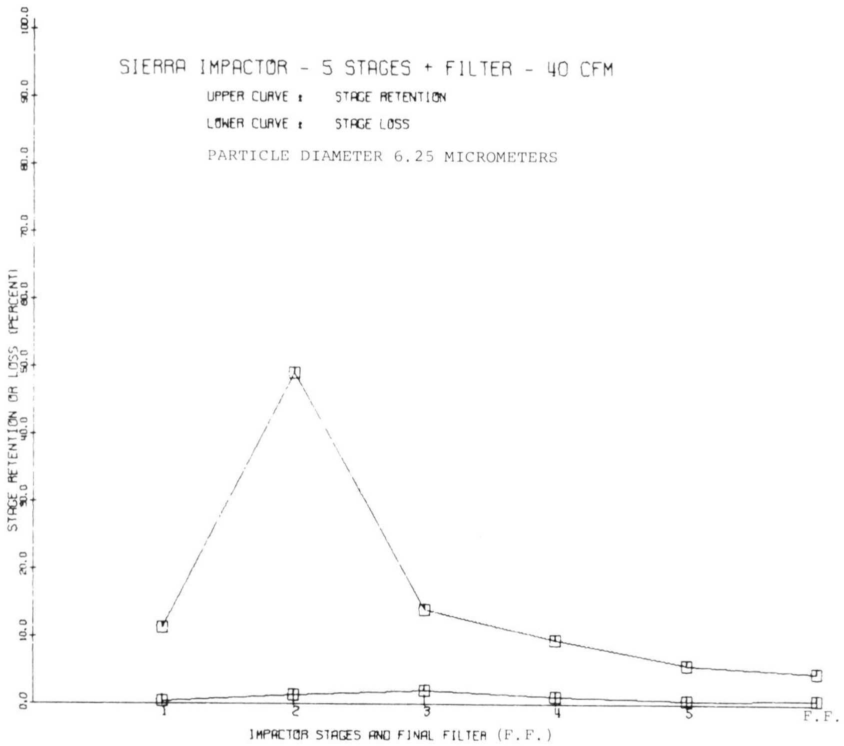

Figure 6. Stage collection of 6.25 micrometer diameter particles with filter paper impaction surface.

The final filter captures all remaining particles which in this test amounted to 24% of the total collected.

The effect of impaction surface was investigated by operating the impactor without the stage filter papers. Figure 5 shows the result of a test with a 6.25μm aerosol; 67% of the aerosol passed through the bare impaction stages to the final filter. By contrast, the result of operating the impactor with filter paper impaction surfaces gave the test results of Figure 6. The failure of stage 2 to retain all of the particles not retained by stage 1 indicates that some rebounding was still taking place in this test. The difference in stage collection efficiency with and without the filter paper impaction surfaces is quite evident. Photographic analysis of the stages and final filter again revealed no evidence of particles fracturing and the size distributions of the particles on all stages correspond to the aerosol collected on the input reference filter.

The rebound of large particles through the impaction stages to the final filter would tend to produce an underestimate of the mass median diameter (mmd) values. To illustrate the amount of deviation from the true values of mmd and geometric standard deviation (σ_g) which could occur, a hypothctical ideal log-normal aerosol having a mmd of 7.5 and a σ_g of 1.5 was replotted using the

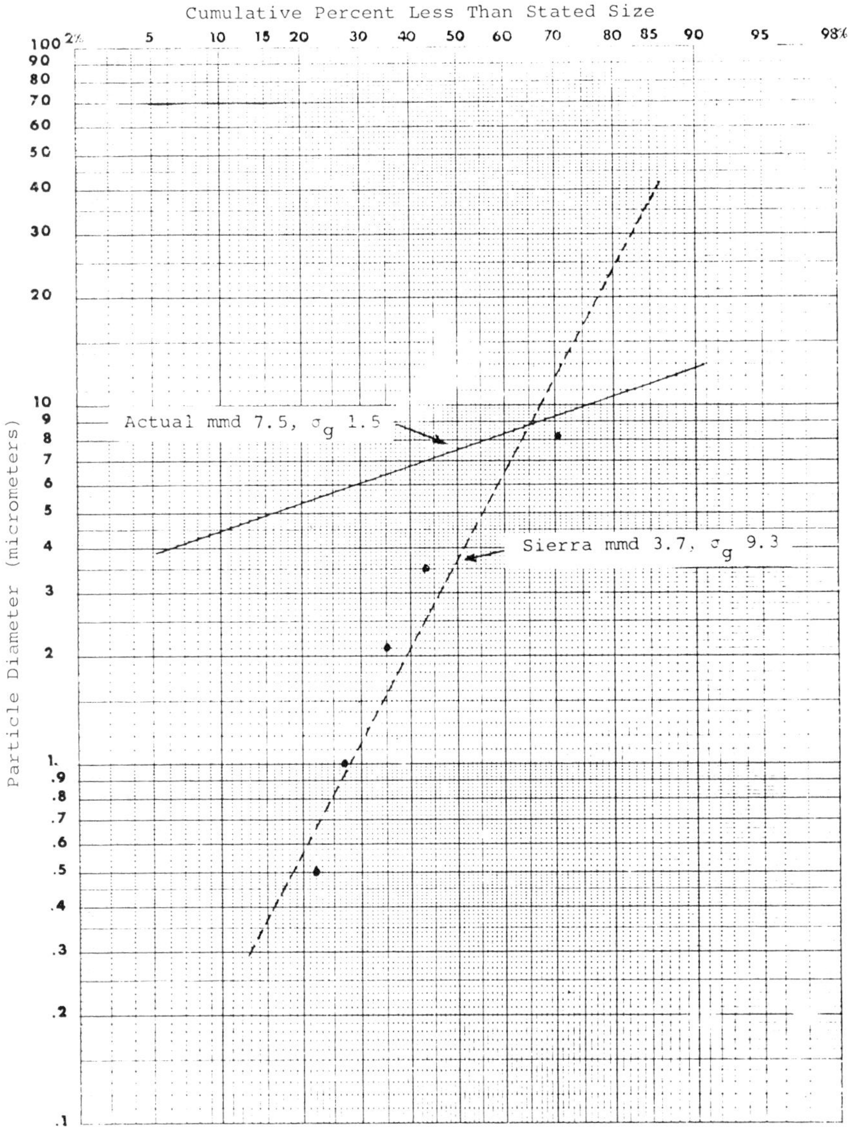

Figure 7. Particle size distribution analysis.

experimental collection curves developed from the methylene blue calibration of the Sierra. Figure 7 shows log-probability plots of the actual aerosol and the stage values that would be obtained from measurement with the Sierra impactor. The Sierra mmd and σ_g values were read from the line of best fit which was calculated by a least squares method.

Miller (1974) has reported disagreement in mass median diameters determined experimentally by three high volume cascade impactors, the Andersen (20 cfm), Sierra (40 cfm), and BGI (30 cfm). His data show that mmd's obtained with the Sierra and BGI are consistently much lower than the mmd's obtained with the Andersen when these instruments sampled the same ambient aerosol. The rebound of larger solid, dry particles through the impaction stages to the final filter of the Sierra and BGI would tend to produce an underestimate of the mmd values. It is possible that the disagreement in mmd's reported by Miller was caused by rebound characteristics of the Sierra and BGI impactors.

Particle rebound and reentrainment from impaction surfaces have not been quantified on theoretical grounds. These effects can cause serious errors in the size distribution calculated from theoretical efficiency curves. Moreover, experimental calibrations are often performed with aerosols of insufficient size range or with atypical properties (e.g., oil droplets) that fail to characterize performance accurately under actual conditions of use.

Therefore, it is prudent for prospective users of particle size classifiers to verify calibrations and other information reported by vendors before selecting a commercial instrument for application to environmental studies.

REFERENCES

Echols, W. H. and J. A. Young, 1963: Studies of portable air-operated aerosol generators. NRL Report 5929.

Miller, K. and H. W. DeKoning, 1974: Particle sizing instrumentation. 67th APCA Annual Meeting, Denver, Colorado.

SOME COMMENTS ON THE SELECTION AND USE OF AN IMPACTOR

A. KISHAN RAO*

Midwest Research Institute, Kansas City, Mo.

The suitability of a given impactor for sampling an aerosol depends upon the impactor flow rate, the aerosol concentration and the amount of sample required. The impactor selected should provide sufficient sample in a reasonable length of time.

For sampling solid particles (which could bounce), a multi-jet impactor is preferable over the single-jet impactor. In a multi-jet impactor, the jet velocity to impact a given sized particle is low. An adhesive coating of the collection surface is preferable. Glass-fiber filter papers, if used as collection surfaces, are effective in preventing bounce, but modify the impactor cutoff characteristics.

When the aerosol size range of interest is too wide, say from 0.1 to 30 μm, as in the case of inlet of control devices, a multiple cascade impactor approach is suggested instead of a single cascade impactor with several stages in series. This will increase the effort and to some extent the cost, but minimize the impactor problems. Measurement of multi-modal aerosols will be greatly facilitated by this approach. The overloading of the first few stages can be avoided this way.

CALIBRATE THE IMPACTOR

The impactor chosen must be calibrated. The impactor theory can satisfactorily predict the performance of an impactor if its design and operation satisfy the assumption of the theory. A multi-jet impactor such as the Andersen Sampler does not satisfy the theory, and theoretical prediction of its performance can lead to considerable error. Furthermore, the nature of the collection surface affects the performance significantly. The impactor stage loading also affects the performance. Therefore, experimental calibrations under conditions of use (i.e., with the substrate and flow rates to be used) are essential if reliable data are required.

*Present Address: Monsanto Company, St. Louis, Mo.

OBSERVE THE FIRST STAGE AND FINAL FILTER CAREFULLY

The loading of the first stage and the final filter and the wall losses can speak of the quality of sampling. If excessive loading of the first stage occurs, a precutter of some sort should be used. The cutoff characteristics of this precutter should be known so that the first stage data could be used. If a precutter is used the impactor should be calibrated with the precutter. Excessive loading of last stage and wall losses are an indication of particle bounce. Ignoring the filter catch in the size distribution calculations does not improve the sampling accuracy. It is best to discard such a run and employ bounce preventative measures in subsequent runs.

The density of aerosol is an important parameter of the aerosol and should be experimentally measured.

MEASUREMENT OF PARTICULATE AEROSOL MASS CONCENTRATION USING A PIEZOELECTRIC CRYSTAL MICROBALANCE

DAVID C. WOODS

Langley Research Center, Hampton, Va.

The mass concentrations of particulate aerosols have been measured either directly or indirectly using various techniques including light scattering, filtration, and others. These methods suffer some disadvantages in that assumptions have to be made regarding size distributions, refractive index, and density in the case of light scattering, and the filtration method lacks the time resolution needed for rapidly varying conditions. A technique for measuring mass concentration directly by impacting the particulates on an oscillating quartz crystal and observing the change in the crystal's frequency has been employed. The frequency change is proportional to the mass collected on the crystal and the rate of change in frequency is proportional to the mass concentration. This technique allows measurements to be made in real time. In addition to mass concentration measurements the particles collected on the sensing crystal can be examined by scanning electron microscopy. This identifies the elemental constituents of the particles and in some cases gives an indication of particle size and shape. A commercially available quartz crystal microbalance (QCM) employing this technique is described by Chuan (1970). Chuan has also reported results from field measurements using the QCM (1971). An early model of the QCM was evaluated extensively by Daley and Lundgren (1974). In this report some samples of data taken with the QCM that illustrate the possible applications and limitations of this measurement technique are presented. Included are data from rocket exhaust measurements from both a ground monitoring station and from aboard an aircraft and data from an environmental quality study made in Norfolk, Virginia.

DESCRIPTION OF QCM

An electronic schematic (Chuan, 1970) of the QCM is shown in Figure 1 to illustrate the principle of operation. The frequencies from the two oscillators are mixed to produce a beat frequency. The beat frequency increases in proportion to the mass of particulates collected on the sensing crystal. The beat frequency is converted to an analog signal (saturation) which can be measured on a strip chart recorder. The saturation signal is then differentiated to give a signal (conc.) proportional to the rate of increase in mass collected on the sensing crystal. Since the air flow through the instrument is constant the conc. signal is proportional to the mass concentration of particulates in the air. Usually both the sat. and conc. signals are recorded during measurements since there are times when it is helpful to have both signals to interpret the data.

A photograph of the QCM is shown in Figure 2. The sensing head cover was removed to show the location of the crystals. There is a digital display as well as the analog outputs. There is also a section containing a disc to collect particles for

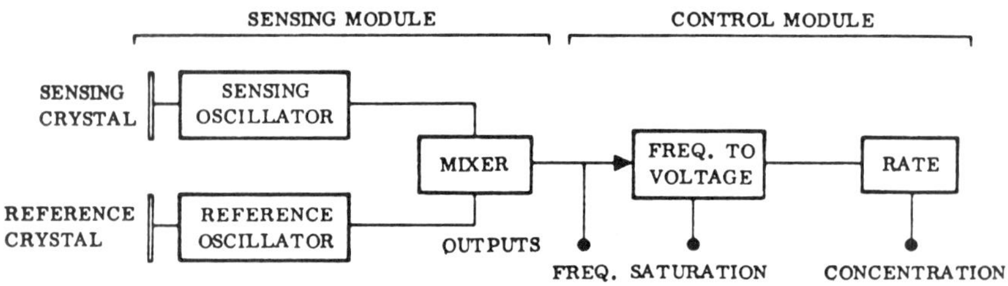

Figure 1. Electronic schematic of QCM.

time-resolved elemental analysis. Particles collected on the sensing crystal may also be analyzed by scanning electron microscopy. Figure 3 shows a photograph of a modified version of the QCM which was designed for remote field operation. It records data on a strip chart and on cassett tape for computer processing.

The particles impact on the surface of the sensing crystal after flowing through one of four small jet holes (0.04 cm diameter), thus forming four impaction sites on the crystal (see Figure 4). Although an adhesive coating is applied to the crystal surface the collection efficiency decreases when the particles begin to pile up at the impaction sites. This causes a nonlinear response in the QCM. When high mass concentrations (several hundred micrograms per cubic meter [$\mu g/m^3$]) are measured this pile up problem occurs within a short time. As a rough guide, the crystal can collect for about 1 hour if the average mass concentration is 100 $\mu g/m^3$. When the efficiency starts to drop off because of the pile up it is necessary to move to a new set of impaction sites on the crystal. This can easily be done by simply turning the crystal over to the opposite side or interchanging the reference crystal with the sensing crystal. Used crystals can be cleaned, recoated and used again (Daley and Lundgren, 1974).

The QCM will indicate a negative mass concentration if there is a loss of material from the sensing crystal, since the concentration signal is the differential of the saturation or accumulated mass signal. This loss of mass may be evaporation of moisture or some other volatile material. The effect of evaporation is illustrated in Figure 5. The sharp increases in the concentration signals are caused by moisture depositing on the sensing crystal indicated by the accompanying increase in the saturation signal. Evaporation is indicated by the loss in mass or drop in the saturation signal which causes an apparent negative mass concentration. This figure illustrates the importance of having both the concentration and saturation signals to properly interpret the data.

SAMPLE DATA

In deploying instruments for the rocket exhaust measurements the ground sampling stations are set up downwind of the launch pad along the predicted path of the exhaust cloud. Figure 6 shows a plot of mass concentration (averaged from a

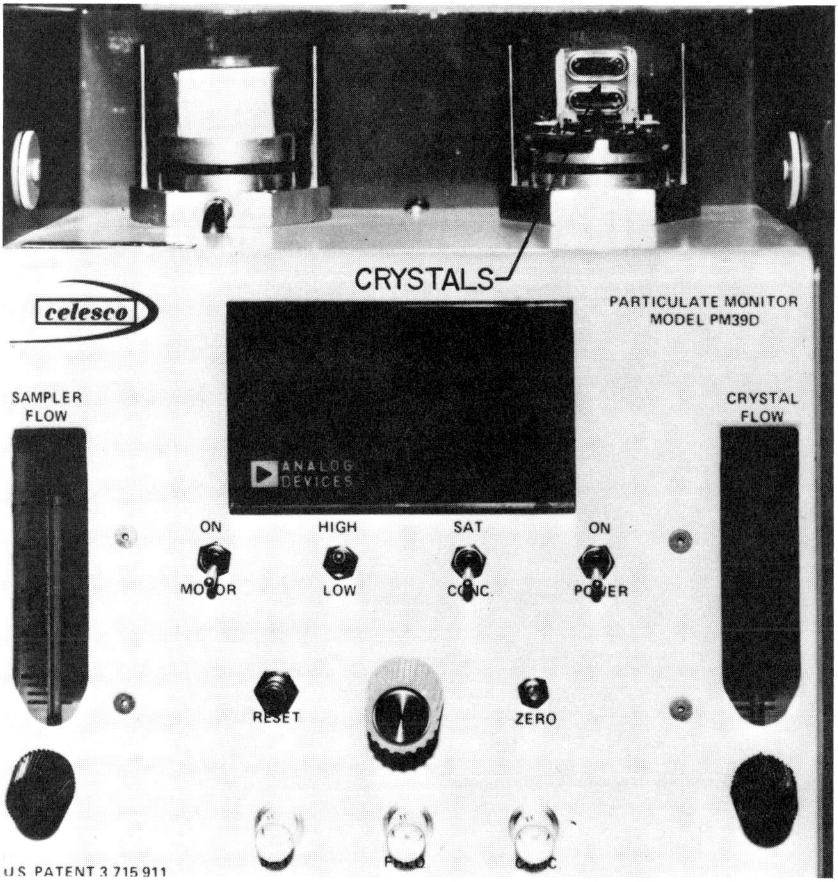

Figure 2. Photograph of QCM.

Figure 3. Photograph of QCM for remote operation.

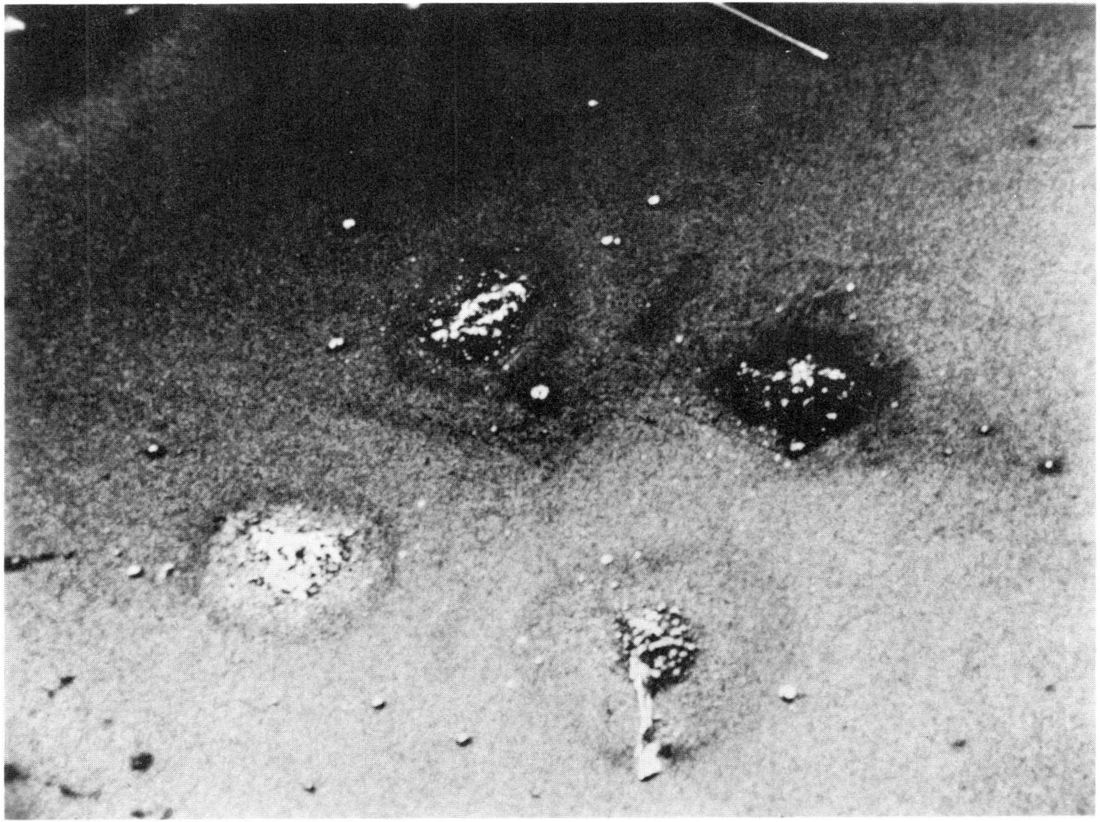

Figure 4. Scanning electron photomicrograph showing particles collected at the four impaction sites.

QCM trace) as a function of time in minutes after launch of a solid propellent rocket at Kennedy Space Center (Hulten et al., 1974). The QCM was located about 3 km downwind from the launch point and about 5° off the center line of the exhaust cloud path. The mass loading prior to the predicted cloud arrival time was typical of the background levels. As the cloud passed over the monitoring station the mass loading increased by a factor of four above the background level and dropped back to the background level as the cloud moved on.

Measurements have been made in the rocket exhaust cloud from aboard a Cessna 402 twin engine plane. To reduce the particle buildup problem on the

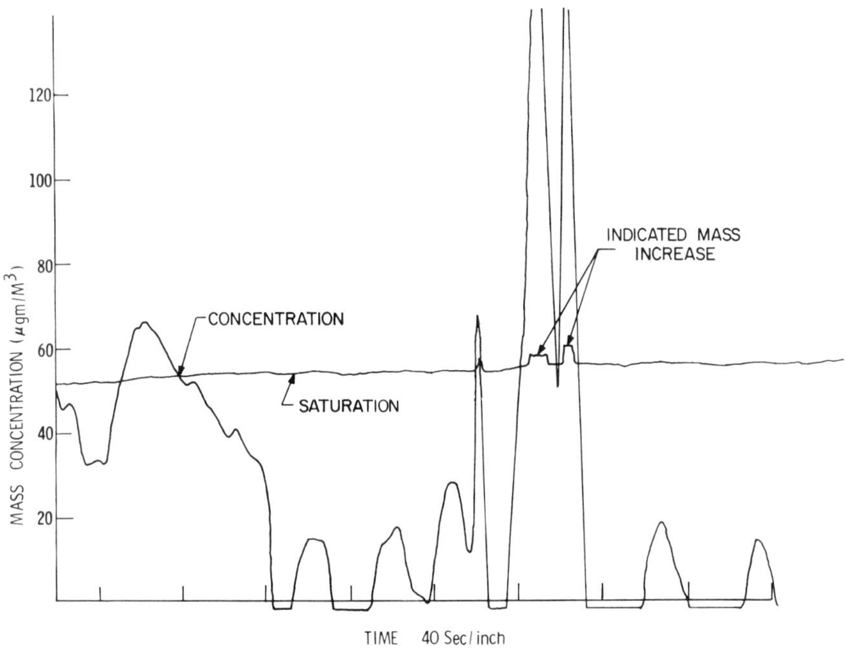

Figure 5. QCM trace showing the effect of evaporation from the crystal surface.

crystal mentioned above, three separate sensing heads were mounted in the nose of the aircraft (Figure 7) with a means of switching from one to the other. The electronics for the QCM were mounted in the aircraft cabin. Figures 8 and 9 show the mass concentration as a function of time in seconds in the exhaust cloud from a Titan III rocket. The QCM measurements (solid line) are compared with measurements made with an integrating nephelometer (dashed line) which was also aboard the aircraft but had a separate sampling probe. In Figure 10, from a different launch, the nephelometer has a normal response but the QCM signal increases very rapidly and goes negative several times. This is the same kind of response as in Figure 5 where evaporation on the crystal surface caused an apparent negative concentration. This suggests that there are volatile substances in the exhaust cloud which either deposit directly on the sensing crystal or attach to other particles which impact on the crystal. This could perhaps be eliminated by heating the inlet air and evaporating the volatile material before it reaches the sensor as is done in the nephelometer.

Figure 11 shows a representative set of measurements taken over a 24-hour period near downtown Norfolk, Virginia (Storey et al., 1975). The mass concentrations as measured with several different instruments are plotted as a function of time of day. The high volume filters had a 12-hour collection period, the nuclepore filters collected over 2-hour periods, Royco sampled once every 10 minutes and the QCM sampled for 10 minutes once every hour. Although the trends are similar the QCM values are substantially lower than the filter measurements. The differences can probably be accounted for by different size

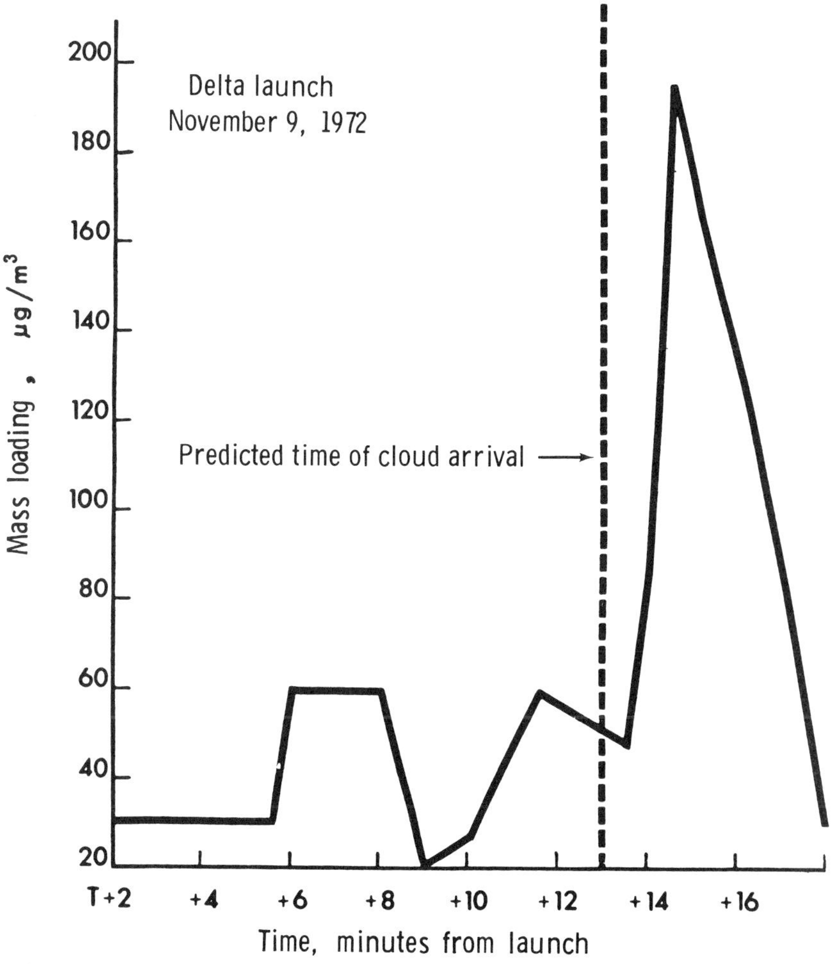

Figure 6. Mass loading as a function of time in minutes after launch, measured with a QCM.

Figure 7. Sampling aircraft.

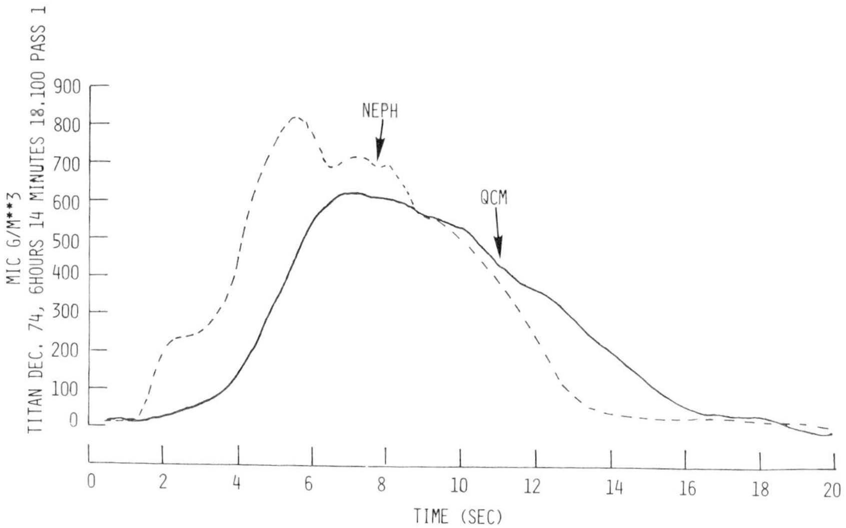

Figure 8. Mass concentration as a function of time in seconds in the exhaust cloud.

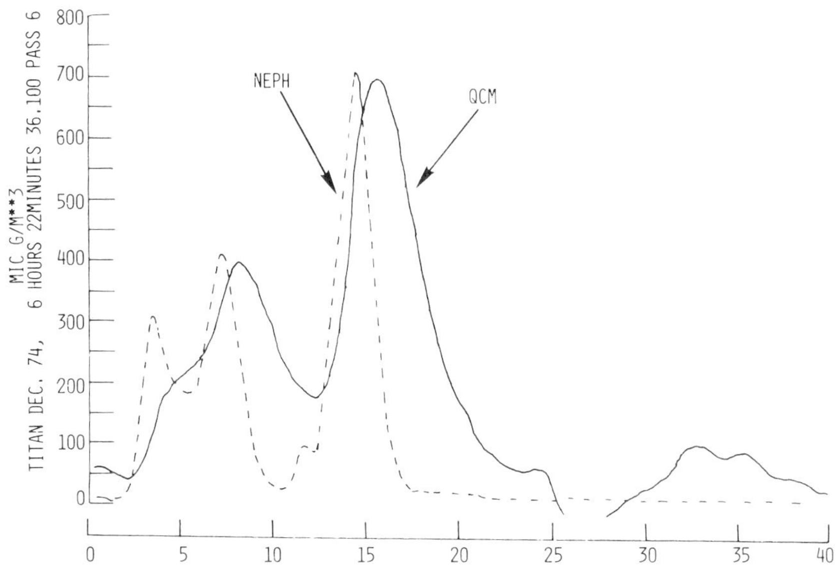

Figure 9. Mass concentration as a function of time in seconds in the exhaust cloud.

ranges of the samplers, different collection efficiencies, and settling of large particles in the inlets to the QCM.

For the most part we have relied on the manufacturer's calibration of the QCM in which it is compared with a high volume sampler. Some preliminary calibrations on the QCM have been made at Langley Research Center using monodisperse latex spheres with diameters of 0.76 μm, 1.01 μm, 2.02 μm, and 5.7 μm. The particles were generated with a Royco model 256 nebulizer and counted with a Royco 220 light scattering photometer to determine a number density from which a mass concentration was calculated. A sensing efficiency was defined as the ratio of the indicated mass concentration from the QCM to the concentration as determined from the Royco. The results of the measurements using both coated and uncoated sensing crystals are shown in Figure 12. The efficiency for the coated crystal decreases with larger particle size to about 50% at 5.7 μm. The uncoated crystal is less efficient for all sizes but very inefficient for particles greater than 1.0 μm in diameter. It is essential that a coating be used to collect the larger particles.

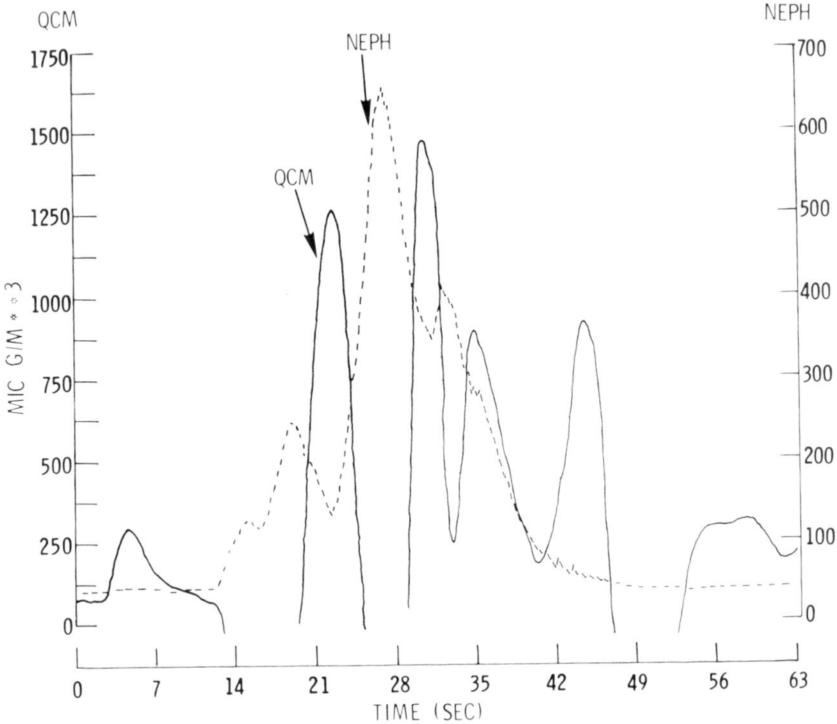

Figure 10. Mass concentration as a function of time in seconds in the exhaust cloud.

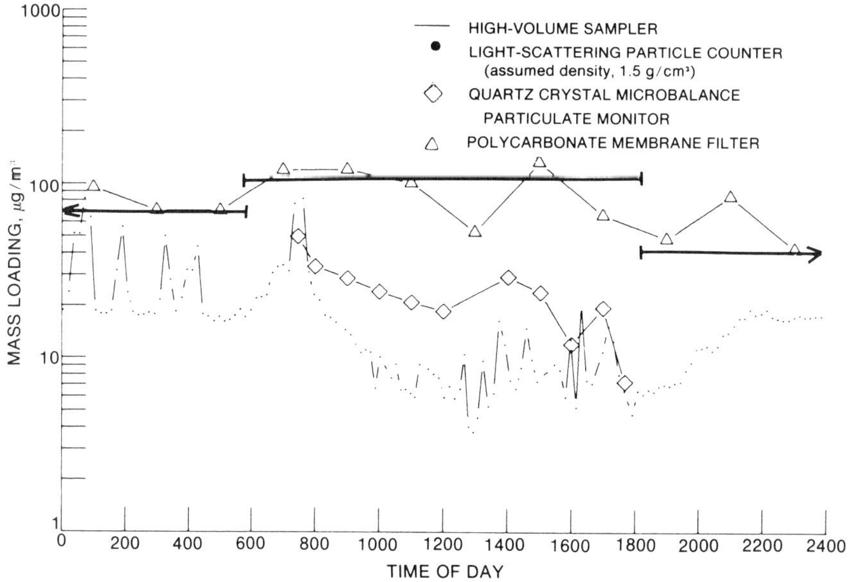

Figure 11. Atmospheric mass loading as a function of time of day.

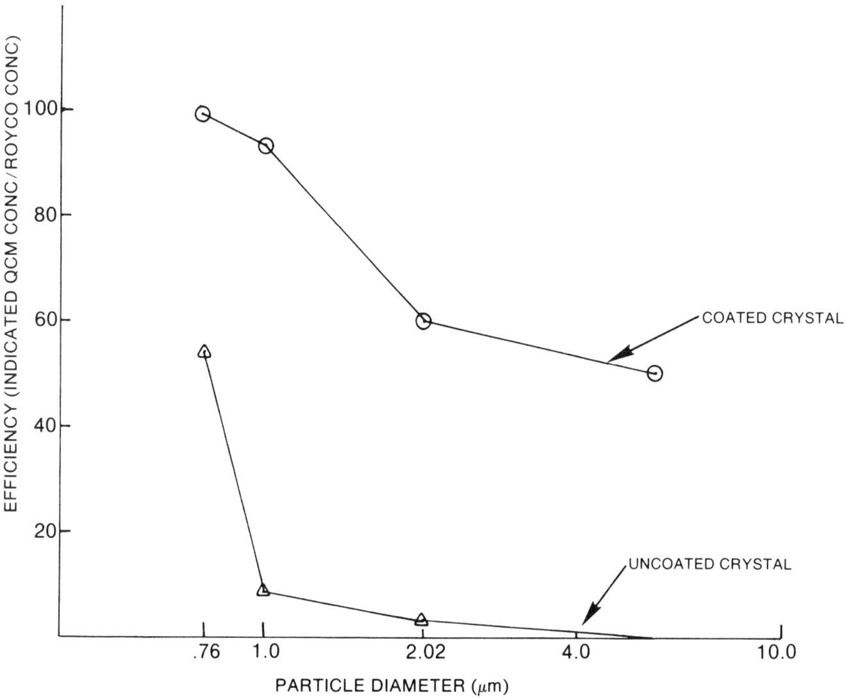

Figure 12. QCM sampling efficiency as a function of particle (latex spheres) diameter.

SUMMARY

The quartz crystal mass monitor has been used to measure real-time mass concentration under various field conditions. The measurement technique allows for direct mass measurement as opposed to other techniques where mass is inferred. The accuracy of measurements made with the QCM probably depends on the size range of the particles measured and the amount of material collected on the sensing crystal. However, for general trends in mass concentration under rapidly varying conditions, the QCM can be a very useful tool.

REFERENCES

Chuan, R. L., 1970: An instrument for the direct measurement of particulate mass. *Aerosol Sci.,* 1, 111–114.

Chuan, R. L., 1971: Measurement of particulate pollutants in the atmosphere. Joint Conference on Sensing of Environmental Pollutants, Palo Alto, California, Nov. 8–10.

Daley, P. S. and D. A. Lundgren, 1975: The performance of piezoelectric crystal sensors used to determine aerosol mass concentrations. *J. Amer. Industrial Hygiene Assoc., 36,* 518–532.

Hulten, W. C., R. W. Storey, G. L. Gergory, D. C. Woods, and F. S. Harris, 1974: Effluent sampling of Scout "D" and Delta launch vehicle exhausts. NASA TM X-2987.

Storey, R. W., R. J. Sentell, D. C. Woods, J. R. Smith, and F. S. Harris, 1975: Atmospheric particulate measurements in Norfolk, Virginia. NASA TM X-3285.

COMMENTS ON SURFACE COATINGS FOR LUNDGREN-TYPE IMPACTORS

THOMAS A. CAHILL

*Department of Physics, Florida State University, Tallahassee**

Quantitative chemical analyses of particulate samples collected by inertial impactors are of little use if errors in particle collection efficiency are large. Thus, there has been a continuing effort to develop adhesive coatings designed to suppress bounce-off and other phenomena that will result in particles leaving the impaction surface. Most of this effort has centered around aerosols generated in the laboratory, and evaluation in field conditions has been limited in extent. As part of the aerosol monitoring efforts of the California Air Resources Board, the Air Quality Group of the Crocker Nuclear Laboratory, University of California, Davis, has utilized the Lundgren-type Sierra Multiday Impactor for extensive studies of the size and elemental composition of atmospheric particulates at numerous sites and on a continuous basis since 1973 (Flocchini et al., 1976). Evaluations of various coatings have been made during this study. These evaluations led to the selection of certain greases as the standard coating used for all impaction stages on the basis of particle collection efficiency and lack of chemical interferences with analytical processes based on ion-excited x-rays.

Tests were performed in the dry, hot weather characteristic of much of California during the summer, conditions known to maximize particle loss phenomena such as bounce-off for soil-derived aerosols. The tests were designed to determine the relative collection efficiency for various coatings for ambient aerosols in realistic conditions, and the total collection efficiency of the system as measured by comparisons between impactors and total filters without previous particle sizing. Typically, 4 multiday impactors with total filters were run side by side on the roofs of buildings in Sacramento, California, and Davis, a small town 25 km away. Care was taken to use similar intake manifolds, as the effect of the manifold-induced cut point on soil derived aerosols is major. Identical manifolds

Work supported in part by the California Air Resources Board.

*This work was performed while the author was on leave from the Physics Department, University of California, Davis.

were used in the tests in Sacramento, but this was not possible at Davis. A correction had to be applied to the soil-derived particles due to this cause, based on unit comparisons with identical substrates and filters. Impaction surfaces were all of mylar type S, from DuPont, Inc. Three coatings were used in these tests. The paraffin and Apiezon grease coatings were applied by a 2 second dip into saturated (~5%) solutions of each material in toluene at room temperature, which gave reproducible coatings of around 65 $\mu g/cm^2$. Apiezon type L is the preferred grease due to chemical purity considerations. The silicone coating was derived from Dow-Corning high vacuum grease via a smearing technique which gave rather thick coatings compared to the other methods. Samples were collected in 24 hour increments, and analyses were made for elements sodium through lead. Many elements were present in large amounts on all days. and these were used to test particle collection efficiency on an element by element

TABLE I. Relative Collection Efficiency vs. Coatings

Location	Date	No. Days	Particles, Size	None (%)	Coating Type (%) Paraffin	Coating Type (%) Silicone	Apiezon (%)
Sacramento	6/75	3	Soil,[a] 3.6–20 μm	5 ± 1	44 ± 5	93 ± 4	100
Davis	7/75	2	Soil, 3.6–20 μm	13	37	97	100 ± 3
Sacramento	6/75	3	Na, Cl, 0.65–20 μm	38	89	89	100
Sacramento	6/75	3	S, 0.65–20 μm	<9	35	70	100
Sacramento	6/75	3	Br, Pb, 0.65–20 μm	9 ± 2	51 ± 23	92 ± 27	100 ± 12

a. Soil = Al, Si, K, Ca, Ti, Mn, Fe.

and size by size basis. Soil-derived elements including Al, Si, K, Ca, Ti, Mn, and Fe behaved in a statistically identical manner in all cases and, from other data at these sites (Flocchini et al., 1976) they could be treated as a class. In cases when enough data were present, variances in the values could be calculated, and these are included in the results. Analytical errors were typically 5% absolute.

In summary, Apiezon type L grease and silicone grease both exhibited high collection efficiencies for many types of particles and a wide variety of size ranges. Apiezon type L has been chosen for our programs due to its chemical purity, as the silicon in silicones interferes with the use of atmospheric silicon as a tracer of soil derived materials. Impaction units treated with Apiezon type L coatings proved to have statistically identical collection efficiencies as simple total filters of known high efficiency such as Millipore AA. Other tests verified that no more than 1% of the large particles present in the largest size ranges crossed over to the smallest size range when these coatings were used. Thus, the combination of Lundgren-type impactors, proper greases and substrates, and care not to exceed a single particle monolayer in the deposition resulted in quantitative collection efficiencies, with typical variances of ±10%.

TABLE II. Total Collection Efficiency vs. Coatings

				Total Filters (%)		Coating Type (%)			
Location	Date	No. Days	Particles	MP,AA	N,0.4[a]	None (%)	Paraffin	Silicone	Apiezon (%)
Sacramento	6/75	3	Soil, <20 μm	100	—	48 ± 12	79 ± 4	121 ± 12	117 ± 8
Davis	9/75	7	Soil, <20 μm	100	119	—	—	—	116[b]
Sacramento	6/75	3	S, Br, Pb, <20 μm	100	—	96	101	—	114
Davis	9/75	7	S, <20 μm	100	103	—	—	—	98
Davis	9/75	7	Br, Pb, <20 μm	100	101	—	—	—	97

a. Nuclepore 0.4 μm pore, 3 times normal pore density.
b. Includes correction for different large particle cut point due to manifold design. Uncorrected value: 0.63 ± 0.01.

REFERENCE

Flocchini, R. G., T. A. Cahill, D. J. Shadoan, S. J. Lange, R. A. Eldred, P. J. Feeney, G. W. Wolfe, D. C. Simmeroth and J. Suder, 1976: Monitoring California's aerosols by size and elemental composition. *Environ. Sci. Technol., 10,* 76.

DESIGN AND USE OF THE MERCER-STYLE IMPACTOR FOR CHARACTERIZATION OF AEROSOL AERODYNAMIC SIZE DISTRIBUTIONS

Otto G. Raabe

Radiobiology Laboratory, University of California, Davis

The use of an impactor with a single circular orifice for each jet provides a convenient method for determining the aerodynamic size distribution of an easily assayed aerosol such as of radioactive or fluorescent material. This type of impactor is preferred when small sample rates and short sampling periods are desired (and practical) since the stage characteristics can be accurately prescribed and particle bounce or reentrainment can be minimized with the use of an appropriate liquid film on the collectors.

Following the lead of Ranz and Wong (1952) and Mitchell and Pilcher (1958), Mercer, Tillery, and Ballew (1962) designed a miniature five-stage cascade impactor using four individual circular jets and a filter stage for collecting the small particles that negotiate the impaction stages. Each successive jet hole was smaller than the previous jet and the ratio of collector separation to hole size was maintained similar for each jet. This impactor was specifically designed to sample radioactive aerosols in an inhalation research laboratory operated by Lovelace Foundation, Albuquerque, New Mexico. Continuing in this line Mercer, Tillery, and Newton (1970) described the improved version having seven impaction stages.[1] This design and its various modifications are commonly called the Mercer-style impactor. However, this impactor is called the Lovelace impactor by Mercer (1973) in his book, and other names by other sources. It should be noted that the current recommended design of this impactor as described herein is somewhat different from the design described by Mercer et al. (1970). The Mercer-style impactor was developed under various Energy Research and De-

Prepared under Energy Research and Development Administration Contract (ERDA) E(04-3)-472 and based in part on work performed at Lovelace Foundation under ERDA Contract E(29-2)-1013.

1. Raabe (1970) described the similar Lovelace impactor.

velopment Administration, ERDA (formerly Atomic Energy Commission) contracts.

BASIC CONSIDERATIONS

Impactor Operation

The collection efficiency of each stage of any cascade impactor for various particle sizes depends upon a number of geometric and operational factors including (a) the linear velocities of the air streams (usually described by a hypothetical average, U_0, equal to the volumetric flow rate divided by the area of the opening); (b) the jet opening shape and size (conveniently described by the jet width for slits or the jet diameter for holes, either of which may be called W); (c) the aerosol particle shape; (d) the particle size (geometric diameter, D, if the particle is spherical); (e) the particle density, ρ; (f) the slip correction, C(D) (important for smaller particles); (g) the air viscosity, η; (h) the jet-to-collector separation distance, S; (i) the jet throat length, T; and the character of the jet airflow as can be conveniently described by the Reynolds' number, N_{Re}.

Aerodynamic separation for a given particle based upon inertial momentum in an airstream which changes direction can be characterized by the particle stopping distance. This is defined as the distance traveled in the forward direction before coming to rest with respect to the surrounding air for a particle traveling initially at a speed U_o (in the forward direction) with respect to the air (Fuchs, 1964). The stopping distance for a spherical particle of geometric diameter, D, can be expressed as:

$$L = \frac{U_o \rho C(D) D^2}{18\eta} = \frac{\rho^* U_o D_{ar}{}^2}{18\eta} \tag{1}$$

with L the stopping distance, ρ the physical density of the particle (assuming the density of air is essentially nil), ρ^* the standardizing density equal exactly to one g/cm^3, C(D) the slip correction for correcting Stokes' law of viscous resistance for smaller particles, η the viscosity of air and D_{ar} the aerodynamic (resistance) diameter equal to $D\sqrt{[\rho C(D)/\rho^*]}$ defined by Raabe (1976).

Ranz and Wong (1952) used a dimensionless parameter ψ for describing the operation of an impactor stage given by dividing the stopping distance for a given particle by the jet size W:

$$\psi = L/W. \tag{2}$$

Others including Davies and Aylward (1951), Mercer (1963), Marple (1970), and Marple and Willeke (1976) have used the dimensionless Stokes number, P, for a particle given by dividing the stopping distance by half the jet size:

$$P = 2L/W. \tag{3}$$

Based on the principles of physical and dynamical similarity and dimensional analysis (Duncan, 1953), an efficiency curve for an impactor stage given as a function of either of these dimensionless parameters will be applicable to another impactor stage of the same type but of a different size which has the same characteristic ratios S/W and T/W and conforms with respect to flow similarity of pressure drop and Reynolds' number. The flow Reynolds' number, N_{Re}, for a circular jet is given by:

$$N_{Re} = \frac{\rho_a U_o W}{\eta} \tag{4}$$

with ρ_a the physical density of the air.

IMPACTOR DESIGN AND OPERATIONS

The design of the impactor is illustrated in Figure 1. Each stage of the impactor has a collector cover-slip holder which can accommodate a standard 22 mm diameter glass cover slip as is commonly used in light microscopy. The cover slip on each stage must be coated with a suitable liquid film to prevent particle bounce during impaction. Anti-foam A spray (Dow Chemical Co.) available in spray cans from most scientific supply companies has been found to be a satisfactory coating for impactor stages. The cover slip holders are spring-fit with wire legs so that they automatically are positioned against the metal pins which act as jet-collector spacers (determining exactly the S/W ratios).

Each stage is O-ring fit into a metallic cylindrical body so that interstage leaks are prevented. The nose inlet cap of the device contains the first impactor jet orifice. The nose inlet is chamfered and O-ringed to fit a one-half inch diameter opening from which samples can be taken. The back end-cap contains a built-in filter holder to accommodate a 25 mm diameter filter which serves as the eighth collection stage. The back end-cap also has a pipe thread outlet to be connected to the flow control and vacuum system used to draw samples through the impactor. Nominal pressure drop through the impactor varies up to about 1.5 psi depending upon the chosen flow rate.

The current design features of the impactor are summarized in Tables I and II, respectively, for the two units with different flow-rate ranges. The so-called 100 cm^3/min impactor is normally operated at flow rates from 50 to 150 cm^3/min. The one-liter-per-minute unit is normally operated from 300 to 1000 cm^3/min. When $\sqrt{\psi}$ is plotted versus the efficiency, Mercer and Stafford (1969) found that the 50% collection point for round jets with S/W = 1 at a Reynolds' number between 600 and 3500 had a value for $\sqrt{\psi} \simeq 0.32$. Therefore, the effective cutoff aerodynamic resistance diameter, ECD_{ar}, can be calculated as

$$ECD_{ar}\ (\mu m) = 1.257 \times 10^3 \left(\frac{W^3}{F}\right)^{0.5} \tag{5}$$

where W is the jet diameter in cm and F is the volumetric flow rate in cm^3/min. It is generally assumed that the equation (5) satisfactorily describes the ECD of stages for Mercer-Style impactors. Based on Marple (1970) the error in ECD values associated with this assumption probably does not exceed 10% in most cases, although the constant can be corrected as required in individual cases.

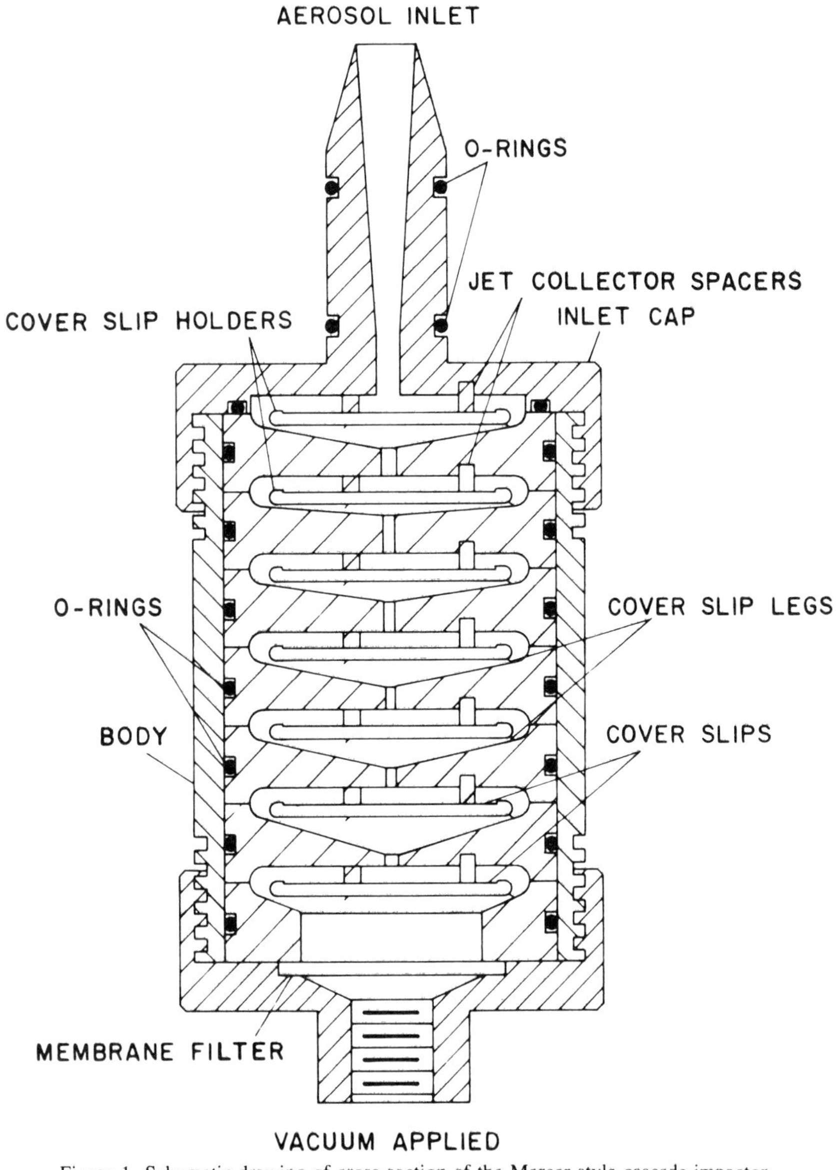

Figure 1. Schematic drawing of cross section of the Mercer-style cascade impactor.

Operation of the impactor depends critically on maintenance of a constant sampling rate. Normally, the impactor should be operated for no more than 10 minutes to avoid alteration of the effectiveness of the liquid film on the stage collectors in preventing bounce and to avoid reentrainment caused by sample build-up depending on the aerosol concentration.

TABLE I. Cascade Impactor Characteristics Including Jet Diameter (W), Ratio of Collector Distance to Jet Diameter (S/W), Ratio of Jet Throat Depth to Diameter (T/W) and Operational Reynolds Number (N_{Re}) and Effective Cutoff (Aerodynamic) Diameter (ECD_{ar}) at 100 cm^3/min. (21°C and Barometric Pressure 76 cm Hg)

Stage Number	W (cm)	S/W	T/W	N_{Re}	ECD_{ar} (μm)
1	0.081	1	3[a]	170	2.90
2	0.066	1	3	210	2.13
3	0.053	1	3	260	1.53
4	0.043	2	3	320	1.12
5	0.033	2	3	420	0.75
6	0.025	2	3	550	0.50
7	0.015	2	3	920	0.23
Filter	—	—	—	—	0

a. Plus tapered entrance.

TABLE II. Seven-Stage Round-Jet Cascade Impactor Characteristics Including Jet Diameter (W), Ratio of Collector Distance to Jet Diameter (S/W), Ratio of Jet Throat Depth to Diameter (T/W) and Operational Reynolds Number (N_{Re}) with Effective Cutoff (Aerodynamic) Diameter (ECD_{ar}) at Sampling Rate of 800 cm^3/min. (21°C and Barometric Pressure 76 cm Hg)

Stage Number	W (cm)	S/W	T/W	N_{Re}	ECD_{ar} (μm)
1	0.236	1	3.0[a]	500	5.10
2	0.178	1	1.8	600	3.34
3	0.142	1	3.0	800	2.38
4	0.117	1	3.0	950	1.78
5	0.089	2	3.0	1200	1.18
6	0.069	2	3.0	1600	0.81
7	0.041	3	3.0	1700	0.37
Filter	—	—	—	—	0

a. Plus tapered entrance.

The samples are analyzed by determining the relative mass of material collected on each stage and the back-up filter. For radioactive aerosols this is conveniently accomplished by counting the emissions. Since the dead-space in the impactor is about 15 cm^3 from the inlet to the filter, a dead space correction should be made for each sample on each stage if the dead space volume is a significant fraction of the total sample volume. The data consists of the mass (or corrected mass) of material on each of eight stages including the filter.

Using a method of weighted least-squares, described by Raabe and Tillery (1968), a log-normal function describing the aerosol mass distribution with respect to aerodynamic (resistance) diameter has been routinely fit to the data. It is also possible to approximate a log-normal function for the mass distribution with use of log-probability paper. The distribution is then describable with a mass (or activity) median aerodynamic diameter (MMAD and AMAD) and associated geometric standard deviation (σ_g).

REFERENCES

Davies, C. N. and M. Aylward, 1951: The trajectories of heavy solid particles in a two-dimensional jet of ideal fluid impinging normally upon a plate. *Proc. Phys. Soc. B., 64,* 889-991.

Duncan, W. J., 1953: *Physical Similarity and Dimensional Analysis.* Edward Arnold and Co., London, 57-80.

Fuchs, N. A., 1964: *The Mechanics of Aerosols.* Pergamon Press, New York, 23-73.

Marple, V. A., 1970: *A Fundamental Study of Inertial Impactors.* Ph.D. Dissertation, University of Minnesota, 74-125.

______ and K. Willeke, 1976: Inertial impactors: Theory, design, and use. *Fine Particles,* B. Y. H. Liu, Ed., Academic Press, 411-466.

Mercer, T. T., 1963: On the calibration of cascade impactors. *Ann. Occup. Hyg., 6,* 1-14.

______, 1973: *Aerosol Technology in Hazard Evaluation.* Academic Press, 237-239.

______, M. I. Tillery and C. W. Ballew, 1962: *A Cascade Impactor Operating at Low Volumetric Flow Rates.* LF-5, Lovelace Foundation, Albuquerque, N. M., 1-19.

______ and R. G. Stafford, 1969: Impaction from round jets. *Ann. Occup. Hyg., 12,* 41-45.

______, M. I. Tillery and G. J. Newton, 1970: A multi-stage, low flow rate cascade impactor. *J. Aerosol Sci., 1,* 9-15.

Mitchell, R. I. and J. M. Pilcher, 1958: Design and calibration of an improved cascade impactor for size analysis of aerosols. *Proc. Fifth AEC Air Cleaning Conference,* TIP-7551, OTS, Dept. of Commerce, Washington, D.C., 67-84.

Raabe, O. G., 1970: Generation and characteristics of aerosols. *Inhalation Carcinogenesis,* Hanna, Nettesheim and Gilbert, Eds., U.S. Atomic Energy Commission, Oak Ridge, Tennessee, 123-172.

______, 1976: Aerosol aerodynamic size conventions for inertial sampler calibration. *J. Air Poll. Contr. Assoc., 26,* 856-860.

______ and M. I. Tillery, 1968: Analysis of cascade impactor data by a weighted least-squares method. *Fission Product Inhalation Program Annual Report 1968-1968,* Lovelace Foundation, Albuquerque, N.M., 165-171, and *J. Amer. Indust. Hyg. Assoc., 29,* 102 (Abstract).

Ranz, W. E. and J. B. Wong, 1952: Impaction of dust and smoke particles on surface and body collectors. *Industr. Eng. Chem., 44,* 1371-1381.

QUESTIONS

Jeck: There is one thing I think the user should be aware of, especially those who use several single stage impactors and hope to get a size spectrum analysis for particles from a polydispersed source. I've seen cases where the person sits by the microscope and examines the slide through the microscope and judges particle diameter. Even though the impactor may have an efficiency cut-off of around 2μm he often doesn't realize that there are particles there below maybe 0.5 μm because they are harder to see. The bigger particles are very easy to recognize and unless you know lots of small particles are there you tend to overlook them because they look like the background. Another thing is that although most of the interest here seems to be in regard to sampling solid particles, there may be some interest in sampling for size analysis of volatile particles such as water droplets or perhaps some other liquids. I haven't heard anybody address the problem of how much evaporation can go on in these aerosol samplers; presumably it is worse in the multi-stage impactors than in the single stage ones. Also, there hasn't been much interest expressed in determining the actual number density of these particles; I think this is a difficult item to measure with impactors. You get relative size spectra but the actual number density for each size integral seems to be difficult to do.

Cahill: This is particularly true when you are trying to sample sulfuric acids that may exist in the atmosphere. In some experiments we used dichotomous samplers and collected on filters which were then stored for a month or so. In dry conditions we experienced 30% loss on the filter itself. When we kept the filter at dry ice temperatures we experienced no loss. So evaporation can occur not only during the process of collection but also after the process of collection. If the particles are small and are below the optical range, below 0.1 μm, this is particularly true.

Q: In regard to Wood's statement on piezoelectrics, two things are worth mentioning. One is that the expected weight response for piezoelectric crystals and their use is often not the effect to the crystal; it is the effect of the aerosol on the crystal which is somewhat in line with the statement just made.

Woods: There is one thing that relates to evaporation. I had three different sampling heads and I noticed that the situation got worse after having sampled on one head for several passes through the cloud. For the first pass I never had this problem, sometimes on the second pass and then on the third and fourth pass through the cloud on that same crystal the evaporation problem got worse. It could be due to absorption of hydroscopic particles on the crystal.

Q: What about calibration curves from different people being different?''

Rao: The impactor design should obey theory. For example, the Lundgren impactor satisfies theory, so its calibration is good. The calibration of the Andersen sampler does not satisfy the theory. The highest jet Reynolds number is about 390 which is less than 500. The lowest jet Reynolds number is less than 100. The theory is good for single jet impactors. Andersen's is a multi-jet impactor and we assume that the flow through each hole is the same but since the pressure drop across the stage is very low the flow is not equal through each hole and you can see this from the deposits. So the calibration will be off. Moreover since the collection surface, the loading, and everything else affects the calibration, it's better to calibrate experimentally.

Whitby: I'd just like to point out that EPA has sponsored a great deal of work on the various types of impactors. Southern Research Institute has found that some of the units are definitely not performing as claimed, and even at the cut-off point they have very low efficiency. If you use a lower stage and somehow the big particles get through, the curves are not as predicted in the theory. Many of them break back over and you'll find that if you, say, double the size of the stage 50% collection efficiency size, it will actually have very poor collection. So one of the things you need to do is make sure you set up the device as the manufacturer specifies. If you try to change it around by dropping out stages or running it at higher or lower flow you will run into problems.

Marple: One thing I want to say about the theory is that if your impactor boundary condition corresponds to that of the theory they do agree very well. I believe Dr. Willeke has looked at the Sierra, and I think he also looked at the Andersen, and he found that it agreed very well to the theory for the first stages. Further into the impactor it may have changed.

Lippmann: I think the customer or the user is frequently at fault (if the manufacturer's calibration is correct and the laboratory calibrations have confirmed it) by trying to collect more sample than the slide can hold. Effective retention is limited to a monolayer of collection. If you have any doubt, you ought to calculate beforehand, on the basis of the size of the spot you can expect to collect in a monolayer thickness, how much mass you will get. If it's not enough for your analysis, please don't go out and try to sample.

Whitby: I'd like to comment on things that were not brought up. Dr. Berner in Vienna has worked on this. If you have a grease or oil that wets the particles and the rate of buildup is not too high, he was able to demonstrate to our satisfaction that you could get some pretty big piles under the right conditions without causing the particles to fall off. This is only if the oil works itself up through the particles and provides a wet surface at the top. The effective Stokes number of collection for the wall is decreased. What that means is that the losses of these particles (that bounce off the stage) onto the wall builds up, and so if one has particle bounce, the wall losses can increase quite drastically. This can give some very peculiar size distributions and results.

Cahill: We have been finding for very dry, bouncy aerosols that you lose about half; about half goes to the next stage. The next stage is summed over them. If half is captured normally you will still collect 75% and 25% will then be collected by the wall in a rather efficient manner. What you say is true. Once you are out of the slip stream, all hell can break loose pretty rapidly.

PART II

Light Scattering Particle Counters

SINGLE PARTICLE OPTICAL COUNTERS: PRINCIPLES AND FIELD USE

KENNETH T. WHITBY AND KLAUS WILLEKE

Mechanical Engineering Department, University of Minnesota, Minneapolis
University of Cincinnati Medical Center, Cincinnati, Ohio

Single particle optical counters (Whitby and Vomela, 1967; Liu et al., 1974a, b; Gucker and Rose, 1954; Lieberman and Allen, 1969; Martens and Fuss, 1968) (OPC) have found widespread use in clean room monitoring, pollution research and laboratory aerosol studies because of their ability to make *in situ* measurements (Whitby et al., 1972; Willeke et al., 1974; Willeke and Whitby, 1975) of the concentrations and size distributions of particles suspended in the air. A beam of light inside the instrument is focused onto a ''view volume'' through which the airborne particles pass one at a time. The amount of light scattered or absorbed from each individual particle is measured by a photosensitive detector. The signal amplitudes registered by this detector are then stored in the channels of a Multi-Channel Analyzer (MCA) from which the particle size distribution is determined. The range for particle sizing extends from about 0.5 μm to about 10 μm in diameter for most commercial OPC's with some going as low as 0.3 μm and some as high as 40 μm.

Aerosol photometers and transmissometers, not discussed here, measure light scattering or extinction from a cloud of particulates, and do not resolve the individual particle sizes. These measurements will provide accurate concentration measurements provided the size distribution is invariant with time and is the same as the one of the calibration aerosol.

There have been many discussions of the theory and principles of operation of single particle counters. Detailed descriptions of all commercial counters and most experimental ones are available. However, not as much information on their calibration, adjustment, and non-ideal behavior is available. Also not as much has been published on utilization of these counters for the measurement of the size distribution of atmospheric aerosols in the field. Therefore, this paper will emphasize the application of OPC's rather than the theory of these instruments. It has been the experience of our laboratory that proper maintenance, calibration and understanding of the non-ideal characteristics of an optical counter may be more important to its successful application in the field than the theoretical virtues of its optical system. A point sometimes neglected by instrument manufacturers is that unless an instrument is reliable in operation and in the maintenance of its calibration, its virtues of principle are of little value.

For completeness, a brief review of the principles of operation and the main characteristics of most commercially available counters will be given first. Then instrument calibration and useful modifications will be discussed followed by a discussion of nonideal characteristics. Applications of OPC's for the measurement of atmospheric aerosol size distributions will be given last.

OPERATING PRINCIPLE

The light extinction system, shown in Figure 1a, will be used to exemplify the operating principle of the single particle optical counter. Light from a lamp, usually an incandescent lamp, sometimes a laser, is condensed onto an aperture and is then recondensed into the view volume where a sharp image of the defining aperture is formed. The aerosol flow is ducted to the light beam in this plane of focus. A view volume may then be defined as the region bounded by the cross-sectional area of the aerosol stream and the height of the aperture image or the height of the detector image, whichever is smaller. The light emanating from the view volume is re-imaged onto a photosensitive detector, such as a photomultiplier or a photodiode.

In a single particle optical counter with an incandescent light source, a light extinction system, such as shown in Figure 1a, is only practical for particles larger than several micrometers in diameter. Since most particles in an ambient or industrial environment are smaller in size, this light extinction technique is rarely used for aerosols, but is frequently used for the characterization of particles in a liquid.

Most commercially available single particle optical counters measure the light scattered out of the incident beam rather than the light intensity reduction of that beam. The basic geometrical arrangement of common scattering instruments is, in general, as shown in Figures 1b and 1c. The details of light collection and detection, however, may differ in how the scattered light is focused onto the photodiode or photomultiplier. Each commercial system has its own arrangement of illumination and acceptance angles. Specification of the following four angles (graphically represented in Figures 1b and 1c), fully describes the light interaction in the view volume: γ is the half angle of the illuminating cone, α, the light trap half angle, β, the collecting aperture half angle, and ψ, the inclination between illuminating and collecting cone axis.

The optical system in an OPC is sometimes characterized by the scattering angle limits. Referring to the axi-symmetric system of Figure 1c, scattering of light from illuminating angle γ to collection angle β gives an upper limit of $\gamma + \beta$; scattering from γ to light trap angle α gives a lower limit $\alpha - \gamma$. For instance,

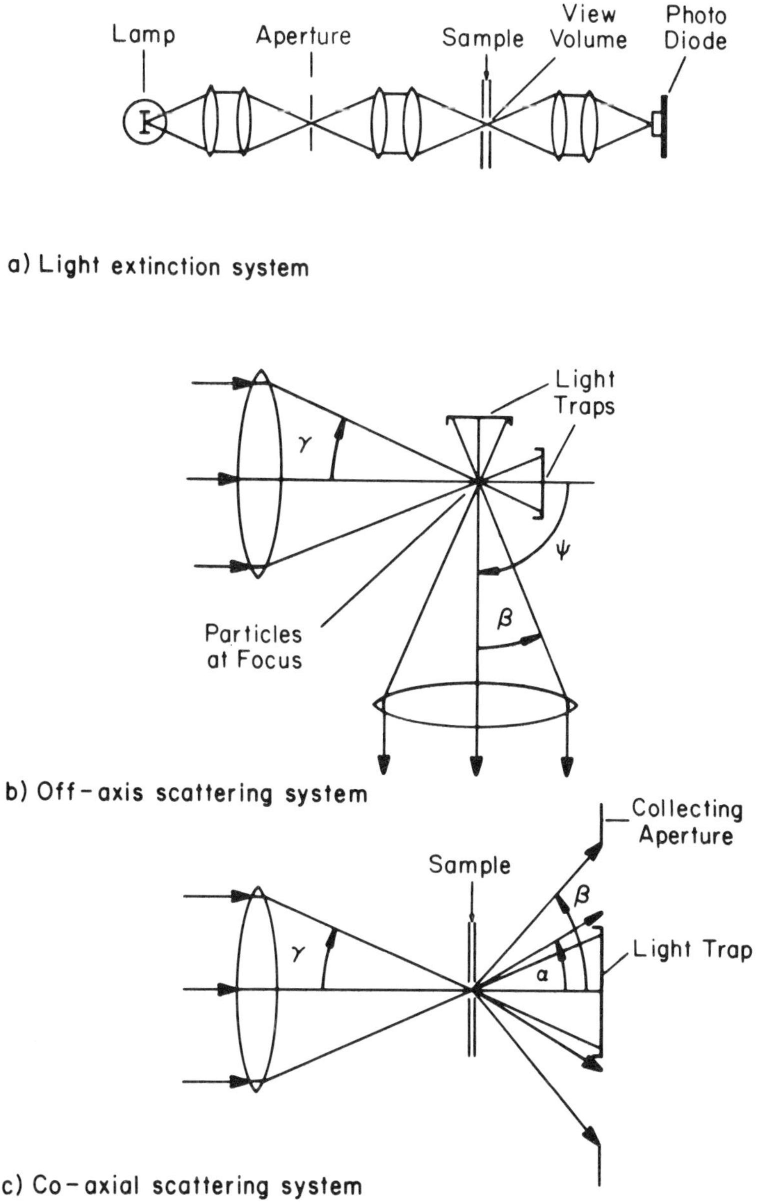

Figure 1. Geometries of illumination and collection in conventional single particle optical counters.

an instrument with $\gamma = 15°$ and viewing angles $\alpha = 35°$ to $\beta = 90°$ has a scattering range of 20° to 150°.

When particles are illuminated by a monochromatic light source, such as a laser, the Mie scattering curve of scattered light flux shows periodic amplitude oscillations (Quenzel, 1969; Gebhardt et al., 1970) with respect to particle size parameter $\alpha = \pi D_p/\lambda$, where D_p is the diameter of the particle (assumed spherical) and λ is the wavelength of illumination. The counter response may, therefore, not be a single-valued function of particle size. Incandescent light usually washes out these oscillations and is, therefore, used in most commercial OPC's. When a powerful light source is used, e.g., a high intensity laser light, volatile particles may partially evaporate. However, the microsecond exposure of the view volume is generally too short for evaporation to occur.

COMMERCIAL INSTRUMENTS

The significant characteristics of several commercial OPC's are listed in Table I. All of these instruments have incandescent light illumination.

The Royco 220[1] is a right-angle scattering instrument with the axis of the collecting aperture inclined at angle $\psi = 90°$ to the axis of the illuminating cone (Figure 1b). All other instruments, listed in Table I and schematically shown in Figure 1c, have a common illumination and collecting aperture axis, i.e., $\psi = 0°$ and collect the light scattered into a hollow cone.

The Royco 218 and 245, TechEcology 200 and 208[2] and Climet CI-250[3] collect light from the forward scattering region and may be termed forward or near-forward scattering instruments. Their half-angle of illumination is about 5° to 12°, their light-trap half-angle ranges from about 8° to 18° and their collecting-aperture half-angle ranges from about 20° to 30°. The scattered light in these instruments is condensed and focused onto a photosensitive detector by lenses with a central dark stop. The light collection system is therefore much like a dark field microscope in which essentially all of the illuminating beam is rejected after passage through the view volume and the scattered light is focused onto a detector. The Climet CI-201 and the Bausch & Lomb 40-1A[4] have viewing angles which extend from about 30° to as high as 90°. In the Bausch & Lomb instrument, the view volume is located at the focal point of a parabolic mirror with a light trap at its center. The mirror reflects and collimates the scattered light onto a lens which condenses and focuses the light onto a photomultiplier. In the Climet CI-201 instrument, the use of lenses for light collection is avoided by locating the view volume at one of the two focal points of an elliptical mirror which reflects the scattered light to a photomultiplier at its other focal point.

The view volume varies by about one decade from 0.25 to 4 mm^3 and the sampling rate Q_s varies by over two decades from 170 $cm^3\ min^{-1}$ (0.006 cfm) to

1. Royco Instruments, 41 Jefferson Drive, Menlo Park, Calif. 94025.
2. TechEcology, Inc., 645 No. Mary Avenue, Sunnyvale, Calif. 94086.
3. Climet Instruments Co., 1620 W. Colton Ave., Redlands, Calif. 92373.
4. 820 Linden Avenue, Rochester, N.Y. 14625.

28,300 $cm^3\ min^{-1}$ (1 cfm) for the tabulated counters. Battery-operated portable instruments are generally limited to pumps handling low flow rates.

A counter which utilizes the light extinction techniques is manufactured by HIAC.[5] The Company claims that the instrument sizes airborne particles from 150 μm to 5 μm and from 60 μm to 2 μm depending on the sensor used.

In a counter (Schuster and Knollenberg, 1972; Knollenberg, 1974) developed by Particle Measuring Systems[6] the particles are injected into the radiation field of an open cavity He – Ne gas laser. The light extinction by the particles perturbs the resonance in the cavity with a resulting power loss of the laser, i.e., the laser behaves as a nonlinear amplifier of perturbations in the resonant cavity. It is claimed that this counter, designated as an "Active Scattering Aerosol Spectrometer," sizes particles from 5 μm to 0.15 μm.

TABLE I. Characteristics of Commercial Single Particle Optical Counters[a]

Optical Counter	γ^b (degree)	α^c (degree)	β^d (degree)	V (mm^3)	Q_s ($cm^3 min^{-1}$)	Q_s (cfm)
Bausch & Lomb 40-1	13	33	53	0.5	170	0.006
Climet CI-201	15	35	90	0.5	7,080	0.25
Climet CI-250	12	18	28	0.4	472	0.017
Royco 218	5	11	30	0.25	283	0.01
Royco 220 (ψ = 90°)[e]	24	—	24	2.63	2,830	0.1
Royco 245	5	16	25	4.0	28,300	1.0
Tech Ecology 200	5	8	20	0.46	283	0.01
Tech Ecology 280	5	10	20	2.5	2,830	0.1

a. Values given are approximate, for use as guidelines only.
b. γ = illuminating cone half angle.
c. α = light trap half angle.
d. β = collecting aperture half angle.
e. ψ = inclination between illuminating and collecting cone axis.

THEORETICAL RESPONSE CALCULATIONS

The response of single particle optical counters, which typically measure particle sizes from 0.5 μm to 10 μm in diameter, can be calculated through the use of the Mie-scattering theory (Quenzel, 1969; Gebhardt et al., 1970; Cooke and Kerker, 1975; Martens and Doonan, 1970; Hodkinson, 1966; Hodkinson and Greenfield, 1965; Hodkinson and Greenleaves, 1963; Gucker and Ross, 1955; Gucker et al., 1962; Gucker and Tuma, 1968; van der Hulst, 1957; Mie, 1908; Kerker, 1969; Cassatt and Maddock, 1974; Pinnick and Hoffman, 1973; Oeseburg, 1972). This assumes that the particles are spherical, ideally smooth, and are illuminated by light of unit flux per unit beam cross-sectional area, and calculates the scattering intensity as a function of scattering angle θ (measured with respect to the incident ray), particle refractive index m and particle size parameter $\alpha = \pi D_p/\lambda$. The results of Mie theory calculations are usually presented as angular intensity

5. 4719 West Brook Street, Montclair, Calif. 91763.
6. 1855 South 57th Court, Boulder, Colo. 80301.

functions i_1 (α,m,θ) and i_2 (α,m,θ), the components of scattered light polarized in and normal to the plane containing the directions of illumination and observation. The quantity $(\lambda^2/8\pi^2) \times (i_1 + i_2)$ is then the flux per unit solid angle scattered in the direction θ by a particle of given size and refractive index. The intensity functions are represented by infinite series whose terms are functions of spherical Bessel and Legendre polynomials. The scattering angle θ is calculated with respect to the incident ray which is inclined at angle ϕ to the axis of illumination. The response calculation for each particular optical design, therefore, includes a geometrical factor $F(\theta,\phi)$. An additional factor $f(\lambda)$ accounts for the wavelength distribution of the emissive power of the light source and the spectral sensitivity of the photosensitive detector. The relative counter response R for spherical particles, in the absence of coincidence and cross sensitivity, is then

$$R(m,D_p) = \int \int \int (\lambda^2/8\pi^2)\,(i_1 + i_2)\,F(\theta,\phi)\,f(\lambda)\,d\lambda\,d\theta\,d\phi. \qquad (1)$$

As an example, the relative, theoretical amount of scatter (Cooke and Kerker, 1975) of one particular commercial counter is shown as a function of particle size in Figure 2. Calculations for other commercial counters are given elsewhere (Quenzel, 1969; Gebhardt et al., 1970; Cooke and Kerker, 1975; Martens and Doonan, 1970; Hodkinson, 1966; Hodkinson and Greenfield, 1965; Hodkinson and Greenleaves, 1963; Gucker and Ross, 1955). A preamplifier amplifies the photoelectric signal. If the amplification is linear, the counter response will closely follow the functional relationship given by these calculations.

The counter response may also be represented as light flux received per unit particle projected area for unit luminance in the source (e.g., Hodkinson, 1966). The resulting curve shows the flux increasing by about 4 decades per decade increase of α when $\alpha \leqslant 1$ (Rayleigh or dipole scattering). The curve reaches a peak at about α = 2 to 5, oscillates (Mie scattering) and settles to an approximately constant flux value with small or no oscillations when $\alpha > 50$, i.e., the flux per unit particle-projected area is approximately constant for large particles (geometric scattering). The rapid decrease of flux for $\alpha < 2$ imposes a lower size limit on conventional OPC's, ca. 0.3 μm for white light illumination.

Stray light from optical elements such as lenses and light traps, Rayleigh scattering of the air molecules and electrical noise result in background scattering intensities which must be lower than the scattered light obtained by passage of the smallest particle through the view volume.

INSTRUMENT CALIBRATION

Calibration with Ideal Aerosols

Relative, theoretical calculations may be used to determine how monotonic the response of a counter will be with respect to particle diameter, and how much the response will depend on particle refractive index within the size range of interest.

However, theoretical calculations are unable to predict the effects of particle shape. Also, individual instruments, either due to normal variations in manufacturing tolerances or due to the wear and tear of field use, are likely to have calibrations significantly different from the theory. For all of these reasons an experimental calibration (Whitby and Vomela, 1967; Liu et al., 1974a, b; Gucker and Rose, 1954; Lieberman and Allen, 1969; Martens and Fuss, 1968; Lundgren and Cooper, 1969; Rimberg and Thomas, 1970; Boden and Coenen, 1970; Cooke

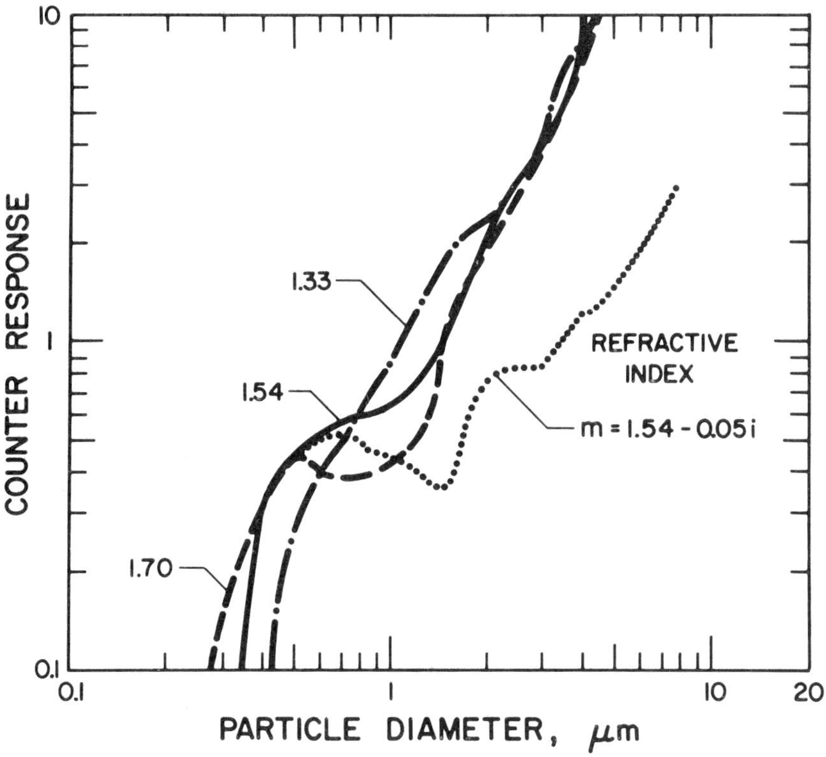

Figure 2. Theoretical response of the Bausch and Lomb 40-1A particle counter (adapted from Cooke and Kerker, 1975).

and Kerker, 1973; Bakhanova and Ivanchenko, 1973) is always necessary to establish the actual voltage output of the counter as a function of particle size for a given aerosol.

The least expensive technique is one in which monodisperse polystyrene latex (PSL) or polyvinyl toluene latex suspensions[7] are nebulized (Whitby and Liu, 1968), dried, and sampled into the OPC. They are available in sizes from about 0.1 μm to about 3 μm, and have a refractive index of 1.6. They may be considered

7. Dow Chemical Company, Midland, Mich.

ideal[8] because of their spherical, smooth shape, i.e., they fit the assumptions for the theoretical response calculations outlined above.

An alternate and more elaborate technique utilizes either the spinning disc[9] or the vibrating orifice monodisperse aerosol generator.[10] The latter (Berglund and Liu, 1972) allows the calculation of particle diameter to an accuracy of about 2 percent using the experimental values of signal frequency, liquid flow rate and concentration of aerosol material in solution with a volatile substance. Aerosols generated with these instruments approximately cover the entire size range of conventional OPC's. The effect of particle refractive index may be studied by generating ideal particles of di-octyl phthalate (DOP, m = 1.49) and Cargille index-of-refraction liquids with m = 1.4, 1.5, and 1.6.[11] The Cargille 1.7 refractive index liquid was found (Liu et al., 1974a) to be unstable and sufficiently volatile at room temperature to cause significant uncertainties in the calculated particle diameter.

An example (Liu et al., 1974b) of an experimental calibration with DOP aerosols is shown in Figure 3. The instrument's theoretical response (Knollenberg, 1974) is shown in Figure 2. Experimental calibration curves for other commercial optical particle counters are shown in Figures 4 through 6, and will be discussed in the next chapter. The effect of particle shape and light absorption may be studied by generating particles such as methylene blue, sodium chloride and india ink (Whitby and Vomela, 1967).

Calibration on Non-Ideal Aerosols

The refractive index and shape of airborne particles found in environments such as mines and metallurgical processing may be quite different from these artificial aerosols. The main difficulty in calibrating OPC's for such dusts is the preparation of monodisperse calibrating aerosols. Conventional aerodynamic dispersion of powders, such as coal, does not give sufficiently monodisperse and deagglomerated aerosols. It has been found that a fluidized bed (Liu et al., 1974b; Guichard and Magne, 1967; Willeke and Whitby, 1974) serves well in the deagglomeration of powders. The airborne particle may then be classified to a known size by means of a differential mobility analyzer (Liu et al., 1974b).

8. There is some evidence that polystyrene latex aerosols are not as optically ideal as DOP aerosols. In our calibration work with both aerosols, the standard deviation of the response from DOP aerosols of equivalent monodispersity usually seems to be somewhat smaller than for PSL, especially for the Royco 220 with its 90° optical system. Furthermore, without an independent check, the sizes furnished with PSL latex suspensions cannot be assumed to be correct to better than about 6%. It must be remembered that the Dow calibrations are performed under vacuum in an electron microscope, whereas the PSL aerosols are used to calibrate OPC's at atmospheric pressure and temperature, and are the residue of a droplet produced from an atomized suspension of latex and surface active materials. Variations in dilution, droplet size, atomizer characteristics and aerosol drying can affect the physical and optical characteristic of the final aerosol. For all of these reasons it is not reasonable to assume that PSL calibration sizes are inherently more accurate than about ±5% unless extra precautions are taken to verify the PSL aerosol sizes.

9. Sierra Instruments, P.O. Box 909, Village Sq., Carmel Valley, Calif., 93924.

10. Thermo-Systems, Inc., 2500 Cleveland Avenue, St. Paul, Minn. 55113.

11. R. P. Cargille Laboratories, Cedar Grove, N.J.

However, classification by electrical mobility is limited to submicron particles.

Unclassified larger particles may be sized by optical microscopy after passage through the OPC. However, the counter's response is spread over a range of channels for a particle of a given size so that the original size distribution of the dust must be deduced by numerical inversion (Liu et al., 1974b). This broadening effect limits the resolving power of the instrument as discussed in a later chapter.

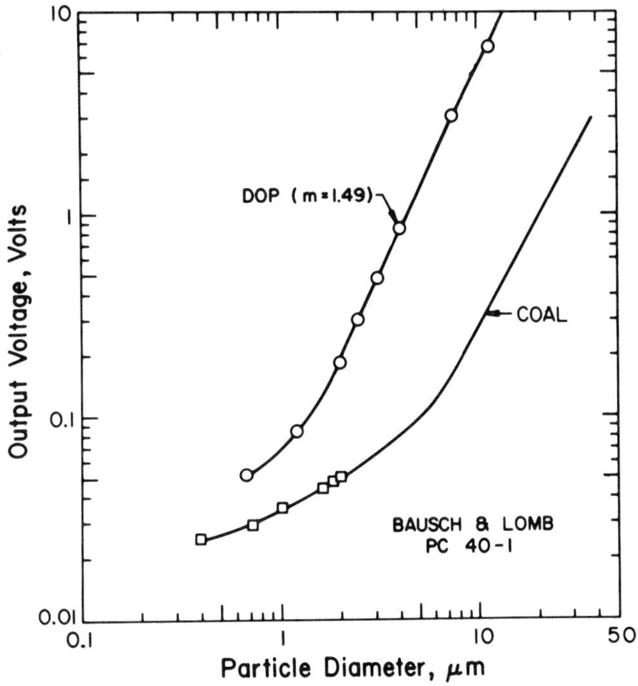

Figure 3. Experimental calibration curve for the Bausch & Lomb 40-1A particle counter.

Aerodynamic Calibration of Optical Particle Counters

Marple and Rubow (1976) have recently developed a set of special impactors and a procedure for the aerodynamic calibration of OPC's on very nonideal aerosols such as coal dust. For many applications such as those in industrial hygiene or in measuring mine dust, the aerodynamic size distribution is the measurement of choice.

Since this technique was described in detail by Marple in another paper during this symposium, the technique will be described here only briefly.

In this technique a steady, well-dispersed aerosol that has been made moderately monodisperse by preclassification is generated by a fluidized bed. This aerosol is sampled both with and without a special sharp-cutoff impactor attached to the inlet of the OPC. By using a series of these impactors a curve of response vs. aerodynamic sizes corresponding to the cutoff sizes of the impactors may be constructed. Figure 7 is a calibration made by this technique.

INSTRUMENT ADJUSTMENT AND MODIFICATION

Work with several dozen different OPC's of most of the commercially available makes has shown that proper alignment of the optical and inlet system is of great importance, so that the aerosol stream passes through a uniform portion of the

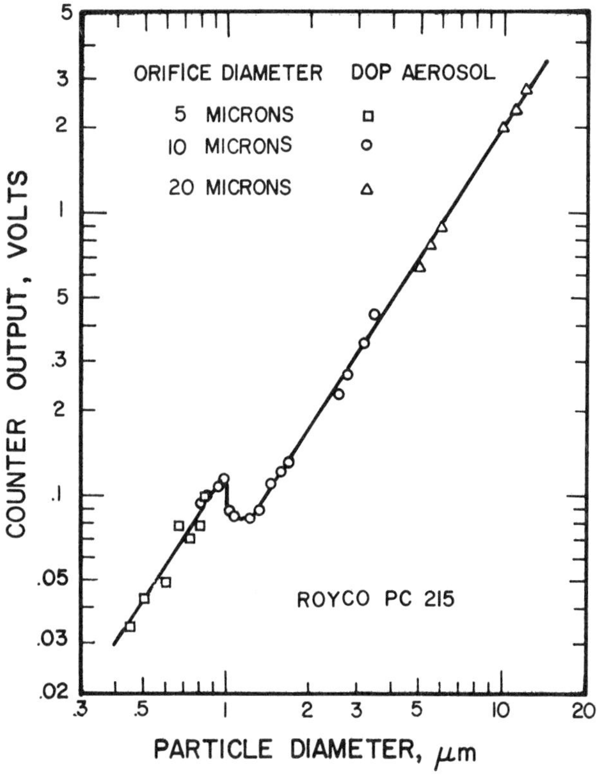

Figure 4. Vibrating orifice generated calibration curve for the Royco 215 particle counter. The optical system of the Royco PC 218 is smaller than the older Royco PC 215.

illuminated beam. Although the design of some counters makes it possible to make these adjustments without an aerosol generator, the best procedure is to use a good monodisperse aerosol and then look at the response on a multichannel analyzer having enough channels so that both the height and the width of the pulse distribution produced by the counter can be observed. While observing the pulse distribution, the lamp, and/or the inlet position should be adjusted for the maximum pulse distribution voltage and the minimum pulse width.

Special attention should be paid to the uniform illumination of the view volume. The light source of which a defocused image is formed in the view volume should produce the most homogeneous radiation density possible. For instance, band-like tungsten filaments were found to be superior to spiral tungsten filaments (Jaenicke, 1972).

Inlet System Modifications

Some commercial OPC's have sheath air inlet systems and some do not. Some companies have been prevented from using sheath air systems for patent reasons. Also, the largest market for OPC's has been for clean room monitoring where a large view volume is necessary to obtain sufficient counts. For clean room monitoring where all of the counts are usually grouped into just a few channels, good resolution is less important than a large sampling rate. However, when one of these "clean room counters" is to be used for aerosol size distribution mea-

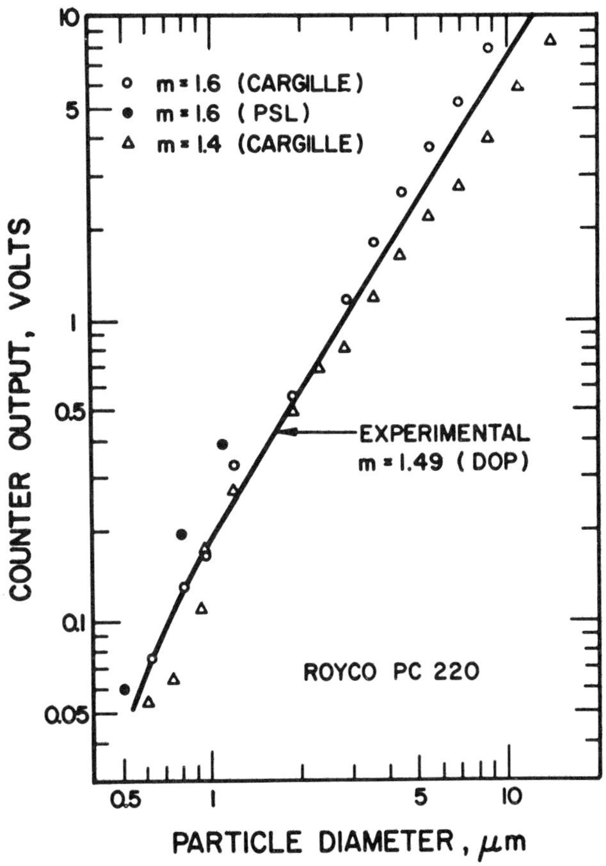

Figure 5. Experimental calibration curve for the Royco 220 particle counter.

surement in the laboratory or field, some modification to the inlet system is often necessary to provide better resolution and to decrease the effective sampling rate so that the higher aerosol concentrations encountered in urban aerosol sampling will not create coincidence and spurious pulse generation problems.

A sheath-air inlet further improves the quality of the view volume. In a typical sheath-air inlet (Liu et al., 1974a) shown in Figure 8, about 90 percent of the incoming aerosol flow is diverted through an external filter which separates the particles from the air stream. The filtered air is then reintroduced around the

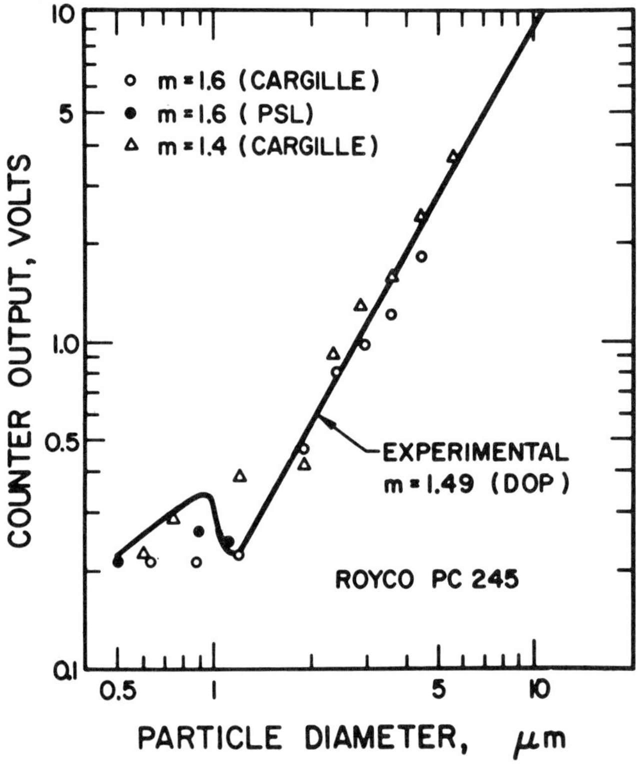

Figure 6. Experimental calibration curve for the Royco 245 particle counter.

remaining 10 percent of the original aerosol flow which passes unimpeded through a straight, inner tube. The combined stream is then introduced into the optical view volume where the particles are counted. The flow is laminar and the aerosol stream flows as a central core within the clean-air sheath into the view volume.

A sheath-air inlet increases the resolution of the counter by focusing the aerosol sample stream into a more uniformly illuminated portion of the view volume, and permits a higher aerosol concentration to be counted because of the decrease in size of that view volume. A sheath-air inlet also prevents or reduces the recirculation of aerosols into the view volume. The signal-to-noise ratio may be improved if the aerosol sample stream is focused into a portion of the view volume which has a higher mean intensity than the original view volume.

The sheath-air inlet shown in Figure 8 is a passive system requiring no pumps or valves. The bypass ratio is independent of inlet static pressure and if the bypass filter is large enough, filter loading will have little effect on the bypass ratio.

The measured aerosol concentration will be lower than the true aerosol concentration, if particles are lost in the OPC inlet. The ratio of these two concentrations defines the sampling efficiency. If particles of sizes larger than about 5 μm in diameter are to be sampled, the inlet tube should preferably be short without any bends. An example is given in Figure 9 which shows a conical inlet design (Liu et al., 1974a) for the Royco 245 high-flow-rate counter. With this inlet, the original efficiency for sampling from a calm air environment was improved to about 80% for 10 μm particles and about 43% for 30 μm particles as shown in Figure 10. One may conclude from Figure 10 that measured concentrations

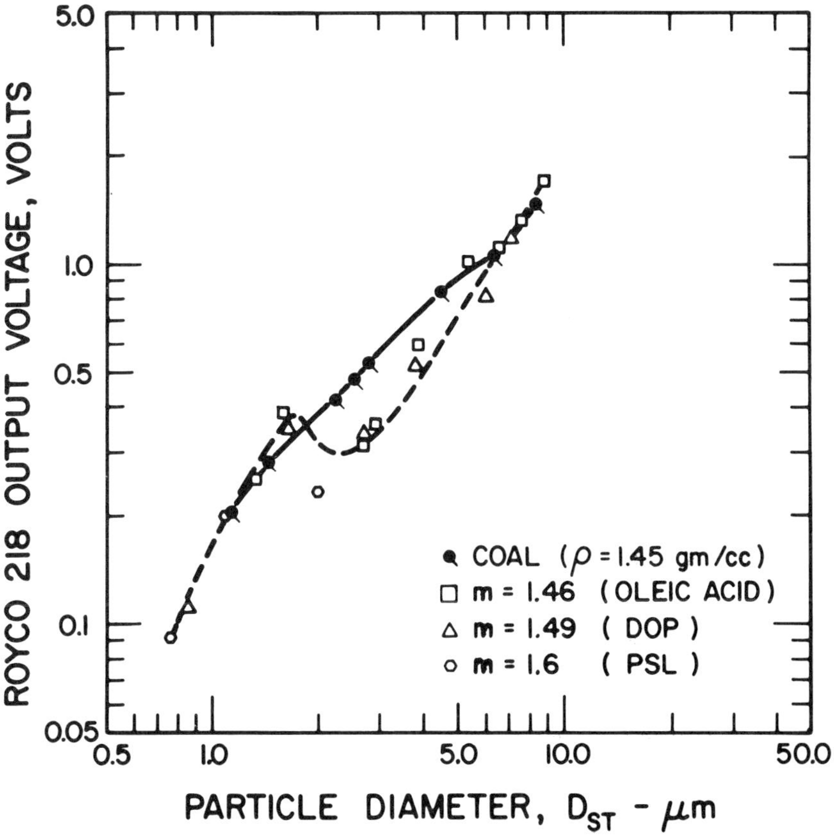

Figure 7. Calibration curves for Royco 218 optical particle counter. D_{ST} is the aerodynamic diameter.

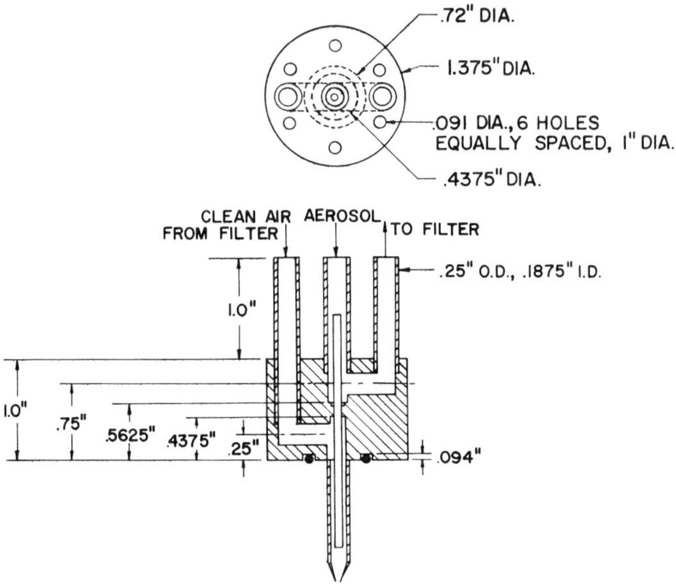

Figure 8. Sheath-air inlet for the Royco 220 particle counter.

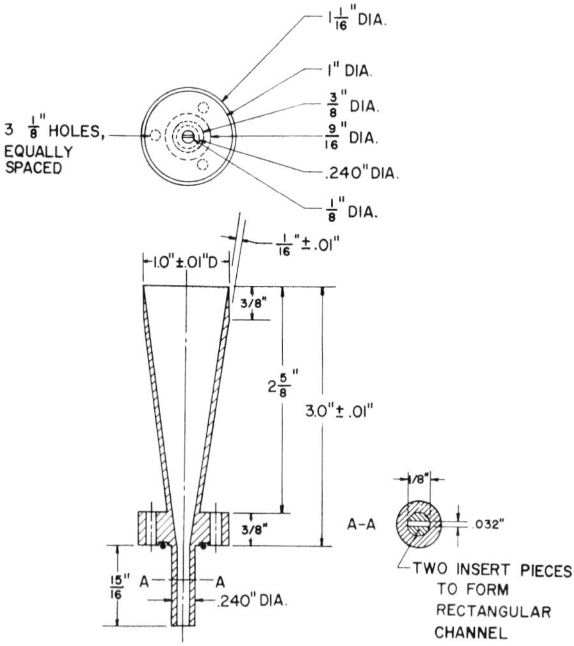

Figure 9. Conical inlet for the Royco 245 particle counter.

should always be modified by the sampling efficiency when particles larger than 5 to 10 μm in diameter are sampled.

NONIDEAL CHARACTERISTICS OF OPTICAL PARTICLE COUNTERS

With good manufacture, maintenance, and calibration, the performance of OPC's on ideal monodisperse aerosols is found to agree quite well with theory. However, when one of these counters is used to measure real atmospheric aerosols, the following characteristics of atmospheric aerosols create problems which must be taken into account if useful data is to be obtained:

1. Refractive index variations from 1.33 (water) to 1.7 (minerals)
2. Shape variations which spread out the counter response from a $\sigma_g = 1.1$ up to 1.8
3. Absorption which may cause response shifts to smaller sizes of up to a factor of 5.
4. The number of particles in a size range decreases about as D_p^{-4}.
5. Concentration variations of 1000/1 in the atmosphere.
6. Atmospheric aerosols usually contain appreciable water so that they may evaporate in the inlet system of the counter.
7. Urban and source concentrations are usually high enough so that some degree of dilution is necessary before the aerosol reaches the sensing zone of the counter.

In this section the practical consequences of the above characteristics will be discussed, as well as some of the solutions that have been found to be useful.

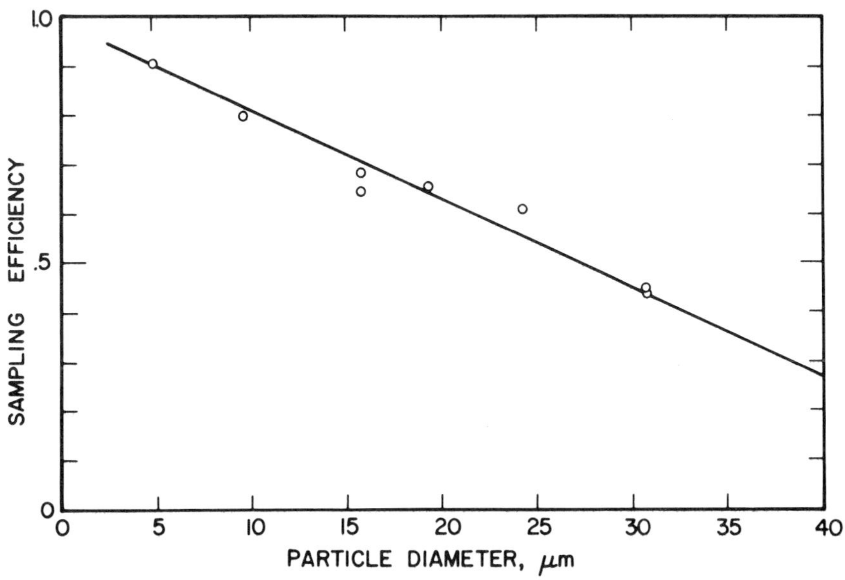

Figure 10. Sampling efficiency as a function of particle diameter for the modified conical inlet (Figure 8) of the Royco 245 high-flow-rate particle counter.

Aerosol Refractive Index

The chemical composition of natural aerosols may vary from particle to particle. The actual particle index of refraction, m, is therefore unknown. Many ambient aerosols contain a substantial quantity of water because of their hygroscopicity.

The real refractive index atmospheric aerosols may therefore range from n = 1.33 (water) to about n = 1.7. Because of the importance of refractive index to atmospheric optics calculations, the relationship between atmospheric aerosol

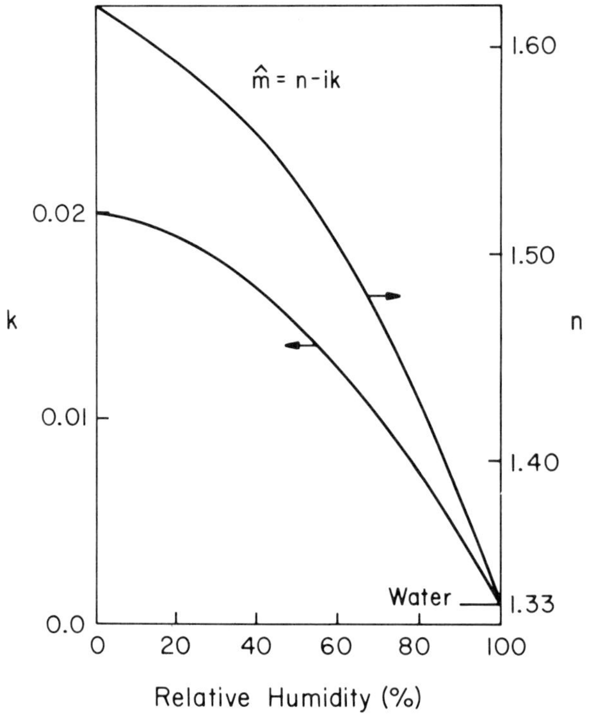

Figure 11. Atmospheric complex refractive index model derived from the work of Winkler and Junge (1972) by Sverdrup (1976). The complex refractive index, m, is a combination of the real part, n, and the absorption coefficient, k.

refractive index, composition and relative humidity has been extensively studied. Sverdrup (1976) has recently summarized this work (Winkler and Junge, 1972; Hanel, 1971). Figure 11 shows the relationship between relative humidity and refractive index for an average atmospheric aerosol. From this figure it is clear that the refractive index of atmospheric aerosols at average humidities is less than the 1.6 of PSL that is usually used for calibration of OPC's. It is best to calibrate OPC's using aerosols having a refractive index close to 1.5. For practical purposes, OPC's are usually calibrated with ideal particles of refractive index 1.5 to 1.6. The size measured by an OPC is then (Whitby and Vomela, 1967) an "Equivalent latex sphere diameter" or an "Equivalent DOP sphere diameter" depending on the calibrating aerosol used. The theoretical response curve for the example given in Figure 2 (Bausch & Lomb 40-1A Counter) shows that the counter response is a strong function of refractive index, particularly for $D_p < 2$ μm. The degree of dependence on refractive index varies considerably from one type of counter to another because of the difference in optical configuration.

An experimental calibration (Liu et al., 1974b) of the Bausch & Lomb 40-1A counter for DOP particles of refractive index m = 1.49 is shown in Figure 3. The counter output voltage is seen to monotonically increase with particle size. When calibrated with DOP particles, a similar monotonic increase was found for the Royco 220 particle counter, Figure 5, whereas the calibration curves for the Royco 215, Figure 4, and the Royco 245, Figure 6, show a dip for particle sizes of about 1 μm in diameter. No useful size distribution information can be obtained in the size range where counters display such a dip in the calibration curve. A counter operating with monochromatic laser light may have a multivalued curve over a large range of particle sizes. It is hoped that this Laboratory will soon be able to furnish calibration curves for the Climet and the TechEcology counters as well.

Agreement between theoretical prediction and experimental calibration is generally very good (Liu et al., 1974a). After an experimental calibration curve has been established for a given refractive index, the theoretical response for this refractive index and for other refractive indices may be drawn onto the experimental plot. However, the reference size through which the theoretical response curve is dimensionalized to counter output voltage should preferably be from a monotonic portion of the response curve where the response is least affected by variations in refractive index, i.e., $D_p > 2$ μm for the example in Figure 2.

Special aerosols may have refractive indices which greatly differ from those generally found in the ambient air. Figure 2 shows the relative theoretical counter response (Cooke and Kerker, 1975) to particles of refractive index 1.33 to 1.70 and to absorbing particles with m = 1.54 - 0.05i. A typical light absorber is coal dust for which an experimental calibration curve (Liu et al., 1974b) is shown in Figure 3. As seen, the output signal amplitude from the OPC ranges from a factor of 2 to 25 lower for coal than for ideal transparent DOP particles of the same size. One may define a size shift factor (Whitby and Vomela, 1967) as the ratio of the coal particle size to that of a DOP particle which generates the same output voltage from the OPC. This size shift increases monotonically from about 3 for small particles to about 5 for large particles in Figure 3. Submicron particles may have unity size shift; e.g., the curves for m = 1.54 and for m = 1.54 - 0.05i merge at a particle diameter of about 0.7 μm in the example of Figure 2. The degree of the size shift depends again on the optical configuration of the counter.

Effect of Particle Shape

It has been shown previously (Whitby and Vomela, 1976) that only a very small departure from sphericity is sufficient to cause a significant loss in resolution for counters, such as the Royco 220, which accept light through only a small angle on only one side of the axis. However, all counters are affected to some extent, with those accepting light through the largest solid angle in the forward direction being affected the least.

The coal dust for the calibration curve, Figure 3, was produced in a ball mill so that the particles were nodular in shape. However, coal dust particles produced by a cutting tool were found to have very irregular shapes (Marple, 1974), such as spheres or platelets. This difference in shape does not necessarily change the calibration curve but will decrease the resolution of the counter because the scattered light received at a fixed angle from the optical axis will depend on the orientation of the particle in the view volume and will therefore have a larger range of intensities than one would expect from a spherical particle.

The effect of particle shape is expected to be less for a counter with a light collection aperture axis coaxial with rather than at some angle to the illumination axis, because the scattered light is integrated through 360° around the optical axis. A broad range of viewing angles (β - α) for a coaxial system as found, e.g., on the Climet Model 201, will further smooth out geometrical variations.

How important is the effect of refractive index and particle shape on the usefulness of size distribution data obtained from OPC's when measuring airborne aerosols and dusts? Let us consider two applications, atmospheric aerosols and mine dusts.

Our current understanding of the physical state of atmospheric aerosols is that most of the time the particles smaller than a few μm are liquid solution droplets, spherical in shape with a refractive index less than 1.5. The mechanically produced particles in the atmosphere larger than a few μm are usually nearly transparent, irregular in shape, and have a great variety of refractive indices, mostly 1.5 or larger. The combination of the fine particles ($D_p < 2$ μm) being of low refractive index and nearly ideal in shape with the coarse particles ($D_p > 2$ μm) being of somewhat higher refractive index and irregular in shape means that theoretically predicted dips in the responsive curve shown in Figure 2 for $n = 1.7$ are either not actually observed or are much weaker than predicted. The irregularity of the coarse particles also tends to smooth out the response curves. This is illustrated in Figure 7.

Comparisons between instruments suggests that if OPC's are calibrated using ideal monodisperse aerosols of $n = 1.5$, that measured size distribution will be within the experimental error of the methods.

Tables II and III show the ratio of volume concentrations measured by the Electrical Aerosol Analyzer (EAA) to the Royco 220 during the GM sulfate study (Table II) and for the Royco 220 to 245 (Table III). Both of these counters were calibrated with $n = 1.5$ DOP. Note that although the EAA/Royco 220 ratio is 1.21, the σ is over 1. This is due primarily to current fluctuations in the last range of the EAA. Recent modifications to the EAA by TSI have greatly reduced these fluctuations. A recent repeat of the EAA to Royco 220 comparison using one of the modified EAA's gave a ratio of 0.89 and a σ of only 0.12 for 10 runs.

Our conclusion is that properly calibrated, properly applied OPC's are just as accurate for *in situ* sizing of atmospheric aerosols as any other measurement technique and that the resulting size distribution data is adequate to resolve all of the important features of the distributions.

Mineral dusts such as those produced by dust storms or in mines consist of irregular particles mostly in the coarse particle range, and sometimes absorbing as in the case of coal dust. For such dusts, the imaginary part of the refractive index and the shape effects have such a strong effect that a calibration for each dust on each counter is necessary to obtain anywhere near the accuracy achieved on atmospheric aerosols.

TABLE II. Ratio of $\Delta V/\Delta \log D_p$ as Measured by the EAA to the Royco 220 OPC in the Overlap between 0.562 and 1 μm for GM Proving Grounds Sulfate Study

Aerosol	N	Mean $\frac{\text{EAA}}{\text{Royco 220}}$	σ
Background before	69	1.36	1.52
Background after	49	1.25	1.49
During test	153	1.03	.73
		Avg. = 1.21 ±	

TABLE III. Ratio of $\Delta V/\Delta \log D_p$ as Measured by the Royco 220 OPC to the Royco 245 OPC in the Overlap between 5.62 and 7.48 μm for the GM Proving Grounds Sulfate Study

Aersol	N	Mean $\frac{\text{Royco 220}}{\text{Royco 245}}$	σ
Background before	58	0.80	.54
Background after	43	1.13	.64
During test	121	1.03	.54
		Avg. = 1.21 ±	

Coincidence

Coincidence (Jaenicke, 1972; Bader et al., 1972; Nikiforova and Sedunov, 1972) refers to the simultaneous presence of more than one particle in the optical view volume. This determines the maximum particle concentration, which an OPC may measure accurately. In the case of coincidence, the indicated particle number concentration, N_i, is lower than the true concentration, N_t, and the shape of the size distribution is weighted towards larger particles because the pulse height analyzer registers two or more particles in the view volume as one particle of a larger size.

The statistical probability P, that no particle (p = 0), one particle (p = 1) or more than one particle (p = 2, 3, 4, ...) is present at the same time in the view volume v is given by the Poisson distribution.

$$P_p = \frac{\mu^p e^{-\mu}}{p!} \tag{2}$$

where

$$\mu = N_t v \tag{3}$$

is the average number of detectable particles in v. The probability of finding no particles in the view volume is then

$$P_o = e^{-\mu} = e^{-N_t v}, \tag{4}$$

and the probability of finding one, two or more particles in v is

$$(1 - P_o) = 1 - e^{-N_t v} = N_i v \tag{5}$$

so that the ratio of indicated to true particle concentration becomes

$$\frac{N_i}{N_t} = \frac{1 - e^{-N_t v}}{N_t v}. \tag{6}$$

This derivation, however, considers only the static situation of finding particles traversing the view volume. A particle will trigger a pulse only if the view volume is not already occupied by particles in which case the indicated count is (Jaenicke, 1972; Princen and Kwolek, 1965; Pisani and Thomson, 1971; Mercer, 1973)

$$\frac{N_i}{N_t} = e^{-N_t v}. \tag{7}$$

Equations (6) and (7) are plotted in Figure 12. For small values of $N_t v$, Equation (6) reduces to $1 - N_t v/2$, and Equation (7) to $1 - N_t v$, i.e., twice the coincidence of Equation (6). Since the view volume can usually not be defined accurately and the scattered light intensity of the particle entering the view volume may not immediately exceed the threshold value, either Equation (6) or (7) can be used to predict the limiting number concentration for which coincidence may become significant. Equation (7) should be used for a conservative estimate.

Generation of Countable Pulses by Subcountable Sized Particles

When we purchased our first OPC in 1964 (a Royco 200), we observed that when it was exposed to high concentrations of aerosols near the lower end of its counting range, the counter indicated a rather monodisperse aerosol of much larger size. In fact, if one loaded up the optical system with concentrations of 0.3 μm particles on the order of $10^6/\text{cm}^3$ and then put a filter on the inlet so that the optical system (it had no sheath-air system) could clean itself out, one could observe an apparently monodisperse aerosol with an initial apparent mean size of about 10 μm, which would gradually decrease in size until it disappeared in the 0.3 μm to 0.4 μm size range.

Subsequent investigation of this phenomenon showed that it was due to statistical generation of pulses by subcountable particles in high concentrations (Whitby and Liu, 1967).

The characteristics of these spurious pulse distributions are illustrated in Table IV and Figure 13. To form these distributions polystyrene latex aerosols generated by atomization were mixed with monodisperse DOP aerosols generated by a condensation generator. Data on the mixtures is given in Table IV. When the DOP aerosols concentration was less than 100/cm^3, there were only a few counts

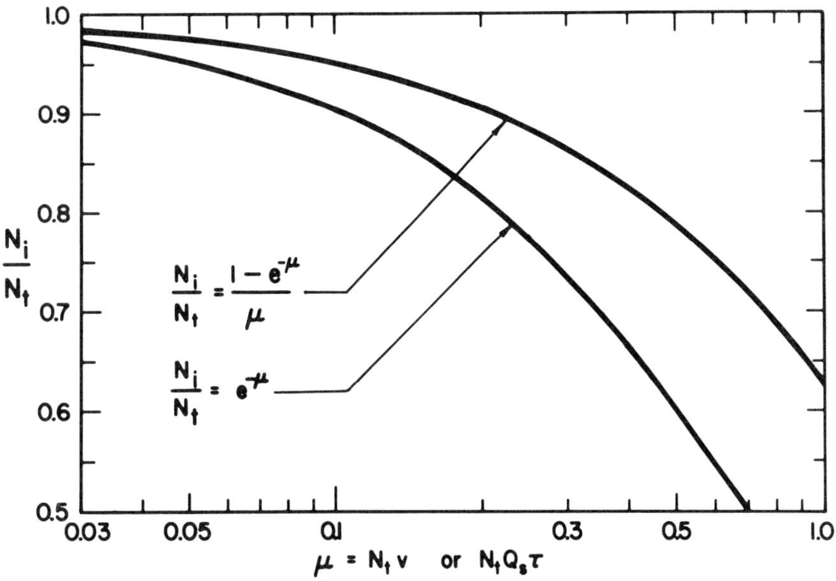

Figure 12. Ratio of indicated to true particle concentration as a function of average particle number concentration $N_t v$ in view volume v, or as a function of average particle count rate $N_t Q_s \tau$ during stretched pulse duration τ.

in the 0.3 to 0.4 μm channel. Thus the distribution observed in run 163 for only DOP aerosol is due only to the DOP aerosol.

For runs 157, 158, and 159, the PSL concentration was held constant and the DOP concentration increased. As the DOP concentration was increased, three main effects on the indicated distribution may be noted in Figure 13.

1. The apparent size of the PSL mode increased from the true value of 0.786 μm to 1.6 μm.
2. The number of particles counted in the 0.3 to 0.4 μm channel decreased by almost two orders of magnitude.
3. The counts in the PSL mode increased from 188 to 297/cm^3.

Figure 14 is a graph calculated from the theory which gives the median apparent size as a function of the average number of particles in the view volumn, vN, where v is the view volume and N is the aerosol number concentration. The number concentration scale, N, is calculated for the Royco PC 200 which has a view volume of 0.002 cm^3. From this graph it is seen that spurious pulses will begin to be observed when the number of particles in the view volume in the size range, just below the lower size limit of the lowest channel, exceeds a few particles. As a practical criteria, if the average number of particles in the view volume in the quarter decade of size just below the lower size boundary of the smallest size interval is less than 1, then spurious pulse generation will not have a significant effect on the indicated size distribution. For example, Table V

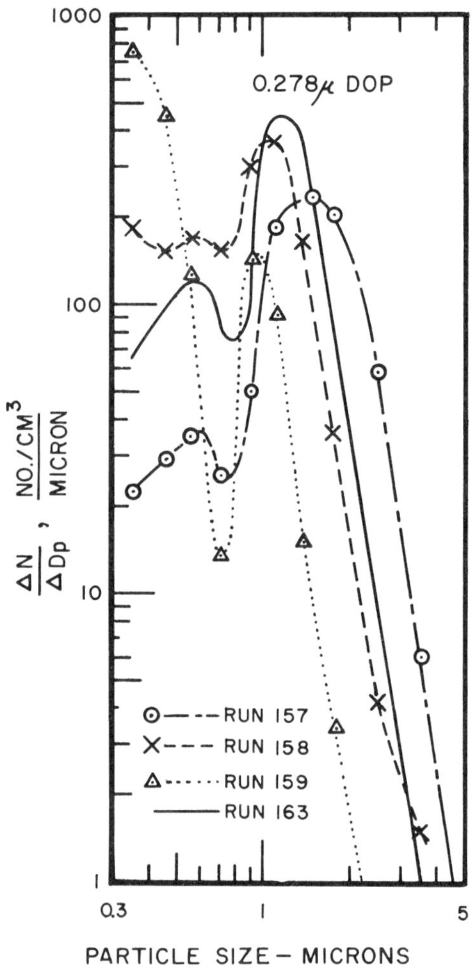

Figure 13. Indicated size distributions generated by a Royco 220 when sampling mixtures of 0.278 μm DOP and 0.796 μm PSL aerosols. Run 163 is a distribution generated when sampling the 0.278 μm DOP only. Note that it has a mode at about 1.2 μm.

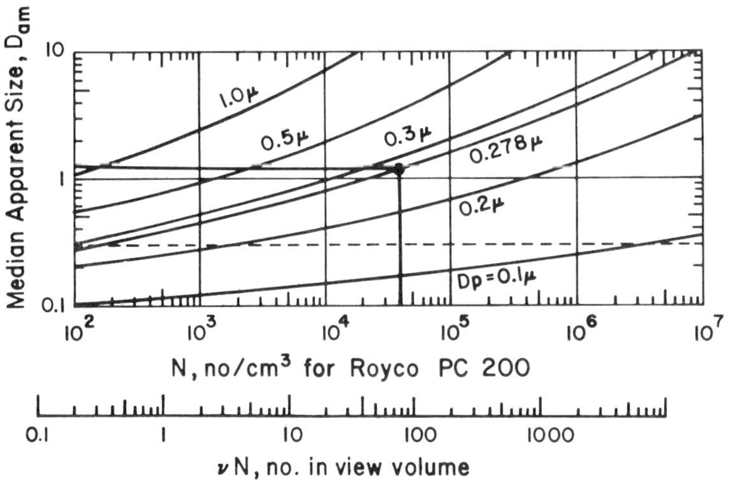

Figure 14. Median apparent size of the spurious size distribution generated by subcountable particles in the view volume as a function of νN and N, the number concentration for the Royco PC 200 having a view volume of 0.002 cm^3.

gives the number of particles in a typical trimodal urban aerosol. Channel 8, which covers the quarter size decade from 0.18 to 0.32 μm, has 1300 particles/cm^3 in it. If the Royco PC 200 with its 0.002 cm^3 view volume were measuring this aerosol undiluted, there would be an average of 2.4 particles from this size range in the view volume at all times. From Figure 14 it is seen that this would produce a small distortion of the size distribution in the 0.3 to 0.4 μm channel. However, if the accumulation mode volume concentration were to increase to 300 μm^3/cm^3, as has been observed in both St. Louis and Los Angeles during heavy pollution episodes, the number of particles in the 0.18 to 0.32 μm channel would be about 25, high enough to cause a serious distortion of the size distribution indicated by the OPC.

Spurious pulse generation and the consequent distortion of the indicated size

TABLE IV. Effect of 0.278 μm Monodisperse DOP Aerosol on the Royco PC 200 Response to 0.796 PSL Aerosol Prepared by Atomization

	Run Number			
	157	158	159	163
DOP conc., no./cm^3	4.3×10^4	9.8×10^4	6.1×10^2	4.1×10^4
PSL conc., no./cm^3	52	77	56	None
Residue conc., no./cm^3 [a]	2.6×10^4	6.4×10^4	2.7×10^4	None
Indicated conc. from Royco, no./cm^3	297	288	188	163

a. Residue particles from the atomization of the PSL. These particles are smaller than 0.1 μm.

distributions may be avoided by reducing the view volume or by diluting the aerosol ahead of the counter. From Table I it is seen that there is a large difference in the view volumes of the different commercial counters.

The sheath-air inlet system shown in Figure 8 has an effective dilution factor, or reduction in the view volume, of about a factor of 10. When using such an inlet, spurious pulse generation will not be a problem. Single particle OPC's used to size urban aerosols should not have an effective view volume of greater than about 0.0002 cm^3 if they are to be used without dilution.

Count Variance, σ_r^2

When any counting device is used to make successive counts, it will be found that the relative standard deviation σ_r will vary as:

$$\sigma_r = (n)^{1/2}. \tag{8}$$

The particle count is given by

$$n = N_t Q_s t \tag{9}$$

where Q_s is the volumetric sample flow rate and t is the time of sampling.

Equation (8) is plotted in Figure 15. For instance, a count of 100 particles has a relative standard deviation of 10% and a count of 1000 particles has a relative standard deviation of about 3%.

Real aerosol size distributions are characterized by a rapid decrease in the number concentration of the larger particles relative to the smaller ones usually about as D_p^{-4}. If the concentration to the view volume is kept small enough to avoid coincidence and spurious pulse generation, then often the total count in the larger size channels is so small and the count variance so large that it effectively limits the maximum size which the counter can count with acceptable accuracy to a size about 10 times the lower size limit of the counter. Although the count may be increased by extending the counting time, there are practical limits to extending the time especially if the counter is used as a part of a sizing system such as the Minnesota Aerosol Analyzing System (MAAS) where the counting time is limited by the desired cycle time of other instruments to a maximum of about 10 min.

To illustrate the magnitudes of this effect, σ_r was calculated for a typical trimodel urban aerosol, for the Royco 220 and 245 as they have been used in our field work. The number concentration for the model distribution is given in Table V and the volume distribution with $\pm 1\ \sigma_r$ plotted for the two counters is shown in Figure 16. Also shown are the size limits at which 1 count in 10 min. would be expected. Since the aerosol concentration in an urban area might be only 1/10 of the average value on a clean day, the actual usable upper limit for these counters is only about 10 μm for the Royco 220 and 40 μm for the Royco 245. We have found that a maximum ratio of the upper size limit of a given counter to the lower

TABLE V. Trimodal Model Size Distribution of a Typical Urban Aerosol

Channel	Interval Boundary D_p (μm)	Number in Interval ΔN (no./cm^3)	Channel	Interval Boundary D_p (μm)	Number in Interval ΔN (no./cm^3)
	.003				
1		2.58×10^4	10		26.2
	.056			1	
2		1.25×10^5	11		2.71
	.01			1.8	
3		1.36×10^6	12		0.82
	.018			3.2	
4		3.3×10^4	13		0.22
	.032			5.6	
5		2.96×10^3	14		0.035
	.056			10	
6		2.75×10^3	15		0.0032
	.1			18	
7		2.79×10^3	16		0.0018
	.18			32	
8		1.3×10^3	17		5.8×10^{-6}
	.32			56	
9		264	18		1.09×10^{-7}
	.56			100	

Mode	Volume (V) μm^3/cm^3	Mode Size DGV (μm)	Mode σ_g
Nuclei	0.5	0.02	1.6
Accumulation	30.0	0.35	1.9
Coarse particle	30.0	5.0	2.2

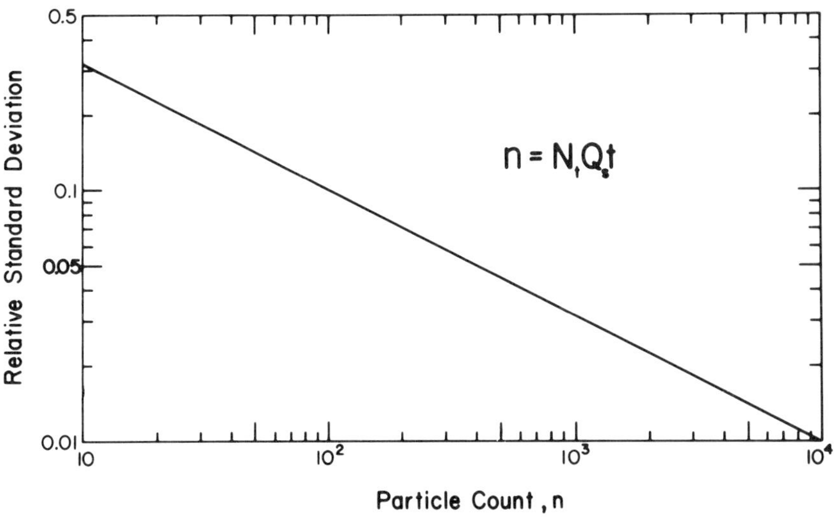

Figure 15. Particle count accuracy of OPCs.

limit of 10/1 is all that is practical for atmospheric aerosol sizing. To size more than one decade of size, one must use several counters in parallel, each optimized for a different size range. For example in the MAAS we have used a Royco 220 operating at an effective flow rate of 300 cc/min for the 0.4 to 7.5 μm range and a Royco 245 operating at a 28 ℓpm flow rate for the 5.6 to 40 μm range.

Aerosol Evaporation

It is now well known that atmospheric aerosols, especially those in or near urban areas, are quite hygroscopic due to their content of such substances as sulfuric acid and ammonium sulfate. In California, for instance, aerosols measured during the ACHEX program had on the average 50% water by weight at 75% relative humidity. At these high water contents, changes in temperature between

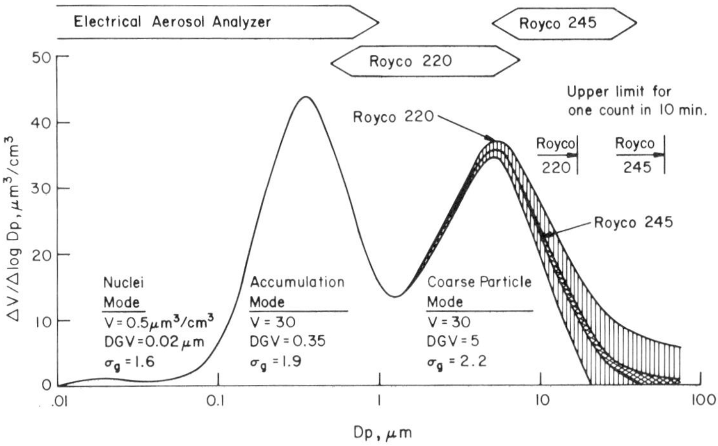

Figure 16. A $\pm\,\sigma$ count superimposed on the volume size distribution of an average urban trimodal size distribution. The upper limits are for one count in 10 minutes of sampling for the respective counters. The bars at the top show the actual count range over which the counters have been used in the MAAS system.

the sampling point in the atmosphere and the sensing zone of the OPC may cause significant changes in aerosol size.

This effect is illustrated in Figure 17 where three runs made with the MAAS at three different temperature differences between the inside and outside of the mobile laboratory are shown. The Royco 245 was located outside the trailer so that its measurements were at the outside temperature, TO, whereas the EAA and the Royco 220 were located in the trailer and made their measurements at a few degrees above the inside temperature, TI. For Run 56 where the inside and outside temperatures were nearly the same, the match between the 220 and 245 is good. However, for Run 52 where the inside temperature was 7.9°C warmer than the outside temperature, the volume measured by the 245 is several times larger than the 220.

This mismatch data was analyzed for a number of sites and was found to be a useful index of the amount of water in the aerosol. The mass fraction of water in the aerosol, M_W/M_T, as a function of relative humidity is shown in Figure 18 compared to measurements made by other investigators. Agreement is within the scatter from the other methods.

If OPC's cannot be operated so that the sensing zone is at the sampling temperature, then some correction for relative humidity and temperature is desirable.

We have measured the temperature of the inlet tubes entering the sensing zone of the Royco 220 and found that it is usually about 2°C above the ambient temperature. Thus, even when the counter is sampling ambient air just outside its case, there will be a small amount of aerosol evaporation.

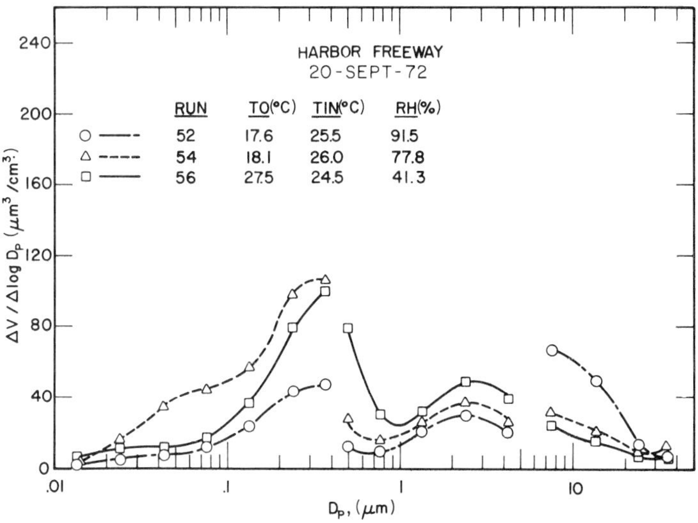

Figure 17. The effect of changing relative humidity on aerosols from the Harbor Freeway in Los Angeles (Sverdrup and Whitby, in press). The mismatch between the Royco 245 measuring the 5.6 to 40 μm range and the Royco 220 measuring the 0.42 to 5.6 μm range is greatest for the higher RH where the aerosol evaporated more in being transported to the inside of the mobile lab.

Resolution

An optical counter in combination with a good multichannel analyzer (MCA) can in principle resolve exceedingly narrow size distributions of ideal particles. In practice, however, instrument defects such as noise, nonuniformity of illumination of the view volume, imperfect discriminators, and the non-ideality of the aerosols with respect to refractive index, particle shape, coincidence, and spurious pulse distributions all tend to place limits on the achievable resolution.

Resolution may be expressed in several ways. One way is shown in Figure 19 where singlets, doublets and triplets produced by a vibrating orifice generator were measured by a Bausch and Lomb 40-1 counter connected to a Hewlett Packard MCA. Here the doublets, which have 1.26 times the volume of the singlets, are clearly resolved. Another quantitative and therefore more generally useful method is to determine the relative standard deviation or geometric standard deviation of the indicated size distribution when a very monodisperse aerosol is sampled by the counter.

Some typical values of the σ_g for several different counters and aerosols are shown in Table VI. The best σ_g that we have observed for the best counters in the best adjustment on ideal aerosols is in the range from 1.04 to 1.07. More typically, counters in good to average adjustment give σ_g's of about 1.1 to 1.15.

As was discussed earlier, irregular or absorbing aerosols degrade the resolution of OPC's seriously, the amount depending on the aerosol and the particular counter. The shape of the counter output signal is approximately Gaussian and therefore may be characterized by the variance of the pulse distribution σ_o^2, where σ_0 = the output voltage pulse height standard deviation. If each distributed effect that contributes to the output standard deviation, σ_0 is of great importance, so that the aerosol stream passes through a uniform portion of the illuminated beam:

$$\sigma_{ro}^2 = \sigma_{rm}^2 + \sigma_{rn}^2 + \sigma_{ri}^2 + \ldots \sigma_{rj}^2 \tag{10}$$

where

$\sigma_{rm} = \dfrac{\sigma_m}{D_{pm}}$ = relative standard deviation of the monodisperse size distribution

$\sigma_{rn} = \dfrac{\sigma_n}{V_o}$ = standard deviation of the noise relative to the output voltage V_o

$\sigma_{ri} = \dfrac{\sigma_I}{I}$ = standard deviation of the illumination in the view volume relative to the average intensity I

$\sigma_{rj} = \dfrac{\sigma_j}{E_j}$ = relative standard deviation of the j'th effect E_j

In Equation 10, only a few of the factors that can contribute to the spreading of the output pulse distribution have been identified because it is usually not possible to separate them anyway. To measure σ_m of a very monodisperse aerosol is not easy, nor is it easy to determine the variability of the illumination in the view volume. Although it is not too difficult to determine σ_n, noise has a significant effect only for particle sizes near the lower end of the counter range.

In practice, the best policy is to determine σ_{rm} using a very monodisperse aerosol such as PSL or DOP generated with the vibrating orifice aerosol generator for which σ_{rm} is typically less than 2%. Then σ_{ro}^2 is assumed to be equal to the sum of σ_{rm}^2 and all of the variances of the other effects. Typically, these residual σ_{rr} may be as much as 0.15.

How important are counter residual dispersion or cross-channel effects during actual use? If the counter output is being used to measure the size distribution of nearly monodisperse aerosols in the laboratory, obviously these effects are very important and must be evaluated for the particular counter being used. However, for atmospheric aerosol size distribution measurements where the σ_g's of the distribution modes in the optical range are rarely less than 1.9 (Figure 16), the percentage error is only about $(0.1)^2/(0.9)^2 \times 100 = 1.23\%$. Thus for applications such as our use in the MAAS, the resolution of most commercial counters in average adjustment is adequate to resolve the accumulation and coarse particle modes in atmospheric aerosols, unless the aerosol is highly irregular in shape.

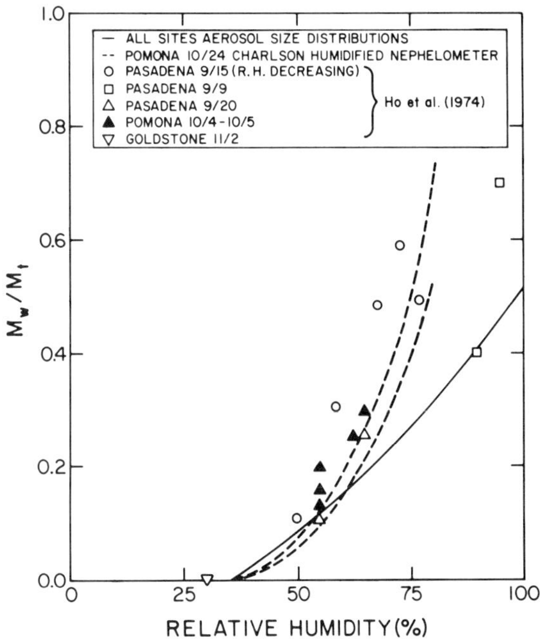

Figure 18. Estimates of the liquid water content of aerosols measured in the ACHEX program by three different techniques. The mismatch technique, which measures the water content at a size of about 6 μm, is a little bit lower than that of the other techniques but in general agreement.

For irregular aerosols, counters such as the Royco 245 and 218, Bausch & Lomb 40-1, and Climet counters have advantages over the Royco 220.

Pulse Processing

The photodetector in an optical counter receives a pulse of photons when a particle passes through the optical view volume. The photon pulse is then converted to a current or voltage pulse by the counter electronics. This voltage pulse must be classified and stored as a count by a multichannel analyzer until it is read as a pulse height distribution by an output device. Many conventional MCA's which were designed for the processing of the fast rise time pulses generated by nuclear counters often will not accept the longer pulses generated by OPC's without pulse stretching.

The function of the pulse stretcher will be illustrated by describing one specific device, the University of Minnesota Model 170-1 pulse converter which was developed from the Royco 170-1 pulse converter (PC). This PC first removes low frequency signals and D-C offset. The incoming pulses generally are Gaussian in shape and have a pulse width from a few to several hundred microseconds depending on the OPC. The pulse amplitudes are then linearly or linear-log amplified depending on a panel switch setting. In the linear-log mode, pulses with amplitudes less than one volt are amplified linearly, and those above, logarithmically. This is advantageous over straight log amplification because log amplification also decreases the signal to noise ratio by the log of the s/n ratio at the low end.

When the amplified signal reaches its peak, the system stretches the pulse at

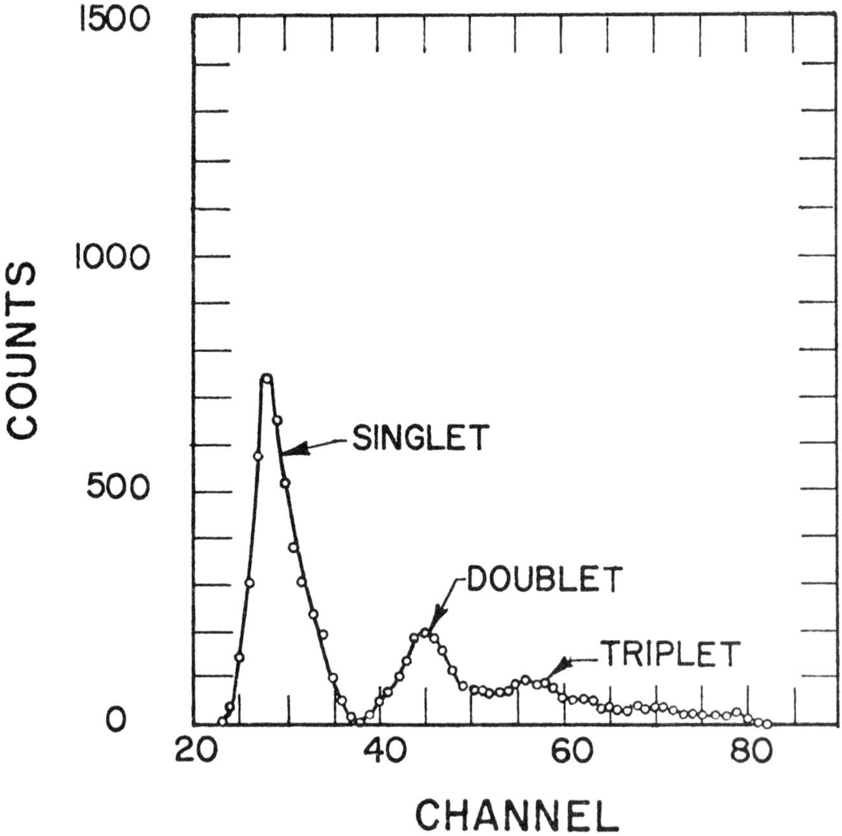

Figure 19. Pulse height distribution from a B & L 40-1 optical counter when sampling a very monodisperse DOP aerosol under conditions of poor dispersion where doublet and triplet particles are generated. The doublets which have 1.26 times the volume of the singlets are clearly resolved.

the voltage level for a preset period of time at the end of which a "gating" pulse commands the MCA to read the signal amplitude. The signal voltage is held as a charge on a capacitor which is quickly discharged and reset to its baseline value after the MCA has read its amplitude.

When the pulse stretching technique is used, the pulse has to be stretched beyond the duration of the signal which introduces a dead time during which no additional signals are processed. The calculation of particle count loss is similar to the one for particle coincidence loss in the view volume. Referring to the Poisson distribution, Equation (2), we have now:

$$\mu = \dot{n}_t\tau = N_tQ_s\tau \tag{11}$$

where $\dot{n}_t$ is the true particle count rate (Equation 9) and τ is the time duration of

TABLE VI. Indicated Monodisperse Particle Uniformity for Various Counters

Counter	Aerosol	Indicated σ_g
B & L 40-1 with sheath air inlet, best	PSL	1.04
B & L 40-1 average adjustment	PSL	1.1
Royco 220 with sheath air average adjustment	PSL or DOP	1.1
Royco 220 with sheath air average adjustment	Irregular transport $D_p = 2.5\ \mu m$	1.65
B & L 40-1 average adjustment	Irregular transport	1.3–1.4
B & L 40-1 average adjustment	Coal dust	1.8

the stretched pulse, i.e. μ is equal to the average number of counts made during pulse duration τ. The expression for the indicated particle count n_i is then similar to Equation (5)

$$\frac{n_i}{n_t} = e^{-N_tQ_s\tau} = \frac{N_i}{N_t} \tag{12}$$

which is plotted in Figure 10. As seen from Figure 10, a less than 5% loss in particle count approximately requires that

$$N_tQ_s\tau < 0.05. \tag{13}$$

Similarly, a less than 5% particle coincidence loss in the view volume requires

$$N_tv < 0.05. \tag{14}$$

The loss of signal count may be reduced by dilution of the aerosol concentration. However, dilution techniques may incur a loss of large particles.

An alternate technique of peak detection makes use of pulse differentiation. With this technique, a comparator looks at the slope of the signal and gives a "gating" signal for amplitude reading by the MCA only when the differentiator output changes from positive to negative at the peak of the signal. However, fluctuations on the signal due to noise may produce local maxima so that the MCA may record more than one output pulse from a single input pulse.

Multichannel Analyzers (MCA's)

The voltage pulses leaving the OPC sensor or the pulse shaper must be discriminated, sorted into voltage ranges, and counted. Most manufacturers of OPC's also market an MCA compatible with their sensors. Many users have also used nuclear-type MCA's having a very large number of channels and some form of digital readout. One serious deficiency of the nuclear-type MCA's for field work is that they have too many channels and thus require too much labor to reduce the counts to a usable size distribution. This is not much of a problem if the counter is connected directly to a computer, but is serious if the numbers must be read off a teletype printout.

To avoid these problems, our laboratory has developed a special eight-channel MCA which has a two-line analog output. One line is a three-digit number and the other is a one-digit exponent. This MCA has all the resolution that is required for most field work and its output signals can be read off of a digital display or recorded with a conventional analog data logger. This MCA has been used successfully in aircraft, in a car, and in mines when powered by a portable battery pack.

For laboratory research or where the maximum flexibility is required, a nuclear-type MCA connected to a teletype or to a computer offers the maximum flexibility and resolution.

CONCLUSIONS AND RECOMMENDATIONS FOR THE APPLICATION OF SINGLE PARTICLE OPTICAL COUNTERS TO THE MEASUREMENT OF ATMOSPHERIC AEROSOLS

Measurements by single particle optical counters are superior to manual sampling and optical microscope counting techniques when the *in situ* particle size distribution and concentration of air environments is to be determined, because OPC's measure particles automatically and in situ. Detailed information may thus be provided on the time variations of these quantities, which cannot be achieved by microscope techniques with any degree of statistical counting accuracy without great expenditure in time. In addition, human error and bias are minimized, and possible errors resulting from sample collection onto a substrate and from transfer of the sample to a proper measurement location are excluded.

Most OPC's can accurately count particles over a range of at least four decades of particle concentration. The limits are set by coincidence resulting from electronic dead time of simultaneous occurrence of more than one particle in the view volume, when sampling high concentrations, and by particle count accuracy, when sampling low concentrations. However, any range of particle concentration can usually be measured by proper choice of particle counter(s), air dilution and sampling time.

The lower limit of particle size detected by available commercial OPC's with white light illumination is approximately 0.3 μm in diameter. The upper limit is set by the sampling efficiency of the counter inlet tube, and the probability of sampling enough particles during the counting period.

A number of experimental counters which use lasers for illumination, very small view volumes, and extra high sensitivity detectors, have sized particles as small as 0.08 μm. However, none of these counters is commercially available.

Our laboratory has found that the Electrical Aerosol Analyzer is superior to the OPC for the measurement of atmospheric aerosols smaller than 0.5 μm.

Atmospheric Aerosols

This section summarizes the conclusions and recommendations for applying OPC's to the measurement of atmospheric aerosols in the ambient atmosphere as has been done in the many air pollution field projects.

1. OPC's should be calibrated on monodisperse, transparent aerosols of n = 1.5 rather than PSL of n = 1.6. The refractive index of atmospheric aerosols is rarely above 1.5, especially in the fine particle size range at normal humidities.

Most OPC measurement of atmospheric aerosols is done under such conditions that the effective refractive index is between 1.33 and 1.5 - 0.02i, and the fine particles are usually spherical. When evaluating the theoretical response curves of a counter, curves for $n > 1.5$ and for $k > 0.02$ have little relevance. Highly absorbing particles or particles of $n > 1.6$ are usually so irregular in shape that the curves calculated for spherical particles are not applicable.

2. Aerosol dilution must be sufficient so that no more than one particle is in the view volume at any one time to prevent spurious pulse distribution generation and coincidence loss. For most counters to be used at urban atmospheric aerosols concentrations, this requires the use of some form of inlet dilution and/or the use of a sheath-air inlet system.

3. Because of the range of aerosol concentrations encountered in the normal atmosphere, it is rarely possible to use a single OPC operating at a given flow rate with a given dilution ratio over more than one decade of particle size. To cover the size range from 0.3 to 50 μm would require three counters and to cover the range from 0.5 to 40 μm, two counters.

4. For sampling and sizing of aerosols larger than about 7 μm, good inlet design coupled with inlet sampling efficiency evaluation is required to obtain valid data. Most commercial counters designed for the clean room market have poor sampling efficiencies for particles larger than 10 μm and therefore require modification and evaluation.

Dusts

Typical dusts that might be sized by OPC's are solid, quite irregular in particle shape and may have significant optical absorption. Although the refractive index is usually unknown, it is quite likely to be greater than 1.5. The irregular shape causes the response on even relatively monodisperse dusts to be spread over enough channels so that resolution of a distribution with a $\sigma_g < 1.6$ is usually impossible. This spreading effect tends to wash out dips in the response curves (see Figure 7).

Following are some general recommendations for the application of OPC's for the sizing of dust.

1. Although difficult, calibration on the dust to be measured, preferably by the aerodynamic technique developed by Marple, is necessary if better than a factor of two agreement between the monodisperse ideal aerosol calibration and the equivalent sphere size of the dust is to be obtained. If the dust is also absorbing, then calibration is necessary to avoid very large errors.

2. OPC's should not be used to attempt to resolve dust distributions having $\sigma_g < 1.6$ unless it can be shown experimentally on the particular dust and with the particular counter that the resolution is better.

3. Forward-scattering type counters such as the Royco 218 have better resolution on dusts than counters which accept light at larger angles and are therefore to be preferred.

REFERENCES

Bader, H., H. R. Gordon and O. B. Brown, 1972: Theory of coincidence counts and simple practical methods of coincidence count correction for optical and resistive pulse particle counters. *Rev. Sci. Instru., 43,* 1407–1412.

Bakhanova, R. A. and L. V. Ivanchenko, 1973: The calibration curve of photoelectric counters and computation of particle size distribution when the relationship between particle size and electrical pulse amplitude is ambiguous. *J. Aerosol Sci., 4,* 485–490.

Berglund, R. N. and B. Y. H. Liu, 1972: Generation of monodisperse aerosol standards. *Environ. Sci. Technol., 7,* 147–153.

Boden, A. and W. Coenen, 1970: Darstellung von teilchengroessenvert-lungen mittels eines einfachen specktrometers als zusatz zu streulichtmessgeraeten. *Staub-Reinhalt. Luft, 30,* 483–487.

Cassatt, W. A. and R. S. Maddock, Eds., 1974: *Aerosol Measurements* (includes several articles on optical measurements), NBS Special Publ. No. 412, National Bureau of Standards (U.S.).

Cooke, D. D. and M. Kerker, 1973: Particle size distribution of colloidal suspensions by light scattering based upon single particle counts—polystyrene latex. *J. Colloid Interface Sci., 42,* 150–155.

______ and ______, 1975: Response calculations for light scattering aerosol particle counters. *Appl. Optics, 14,* 734–739.

Gebhardt, J., J. Bol, W. Heinze and W. Lebschert, 1970: A particle-size spectrometer for aerosols utilizing low-angle scattering of the particles in a laser beam. *Staub-Reinhalt. Luft, 30,* 5–14.

Graedel, T. W., 1974: Channel width determination and electronic pulse processing losses in optical particle counters. *J. Aerosol Sci., 5,* 125–131.

Gucker, F. T., R. L. Rowell and G. Chui, 1965: A study of intensity functions for dielectric spheres calculated from the Mie theory. *Aerosols Physical Chemistry and Applications,* K. Spurny, Ed., Czechoslovak Acad. of Sci., 59–113.

——— and D. G. Rose, 1954: A photoelectronic instrument for counting and sizing aerosol particles. *Br. J. Appl. Phys. Supplement No. 3*, 138–143.

——— and ——— 1955: The response curves of aerosol particle counters. Presented at the 3rd Nat'l. Air Pollution Symp., Pasadena, Calif., April 18–20.

——— and J. Tuma, 1968: Influence of the collecting lens aperture on the light-scattering diagrams from single aerosol particles. *J. Colloid Interface Sci., 27,* 402–411.

Guichard, J. C. and J. L. Magne, 1967: Application du lit fluidise en milieu gaseau a la generation des aerosols. Inst. Nat. Rech. Chim. Appl., Vert-le-Petit, France, Note interieure No. 51.

Haenel, G., 1971: New results concerning the dependence of visibility on relative humidity and their significance in a model for visibility forecast. *Beitr. Phys. Atmos., 44,* 137–167.

Hemenway, D. R., 1974: Pulse processing from particle detectors, Ph.D. Dissertation, Univ. of North Carolina at Chapel Hill.

Ho, W., G. M. Hidy, and R. M. Govan, 1974: Microwave measurements of the liquid water content of atmospheric aerosols. *J. Appl. Meteorol., 13,* 871.

Hodkinson, J. R., 1966: The optical measurement of aerosols. *Aerosol Science,* C. N. Davies, Ed., Academic Press, 468 pp.

——— and J. R. Greenfield, 1965: Response calculations for light-scattering aerosol counters. *Appl. Optics, 4*, 1463–1474.

——— and I. Greenleaves, 1963: Computations of light-scattering and extinction by spheres according to defraction and geometrical optics, and some comparisons with the Mie theory. *J. Optical Soc. of Amer., 53,* 577–588.

Jaenicke, R., 1972: The optical particle counter: Cross-sensitivity and coincidence. *J. Aerosol Sci., 3,* 95–111.

Junge, C. E., 1963: *Air Chemistry and Radioactivity*. Academic Press, 382 pp.

Kerker, M., 1969: *The Scattering of Light and Other Electromagnetic Radiation*. Academic Press.

Knollenberg, R. G., 1974: Active scattering aerosol spectrometry. *Aerosol Measurements,* Cassatt and Maddock, Eds., NBS Special Publ. No. 412, National Bureau of Standards (U.S.), 57–64.

Lieberman, A. and R. J. Allen, 1969: Theoretical and experimental light scattering data for a near forward system. Presented at the Amer. Assoc. for Contamination Control, May.

Liu, B. Y. H., R. N. Berglund and J. K. Agarwal, 1974a: Experimental studies of optical particle counters. *Atmos. Environ., 8,* 717–732.

———, ——— and ———, 1974b: Author's reply. *Atmos. Environ., 8,* 1344.

———, V. A. Marple, K. T. Whitby and N. J. Barsic, 1974c: Size distribution measurement of airborne coal dust by optical particle counters. *J. Amer. Industrial Hygiene Assoc., 8,* 443–451.

Lundgren, D. A. and D. W. Cooper, 1969: Effect of humidity on light-scattering methods of measuring particle concentrations. *J. Air Pollution Control Assoc., 19,* 243–247.

Marple, V. A., 1974: Development, calibration and application of size distribution instruments at the University of Minnesota. *Aerosol Measurements,* Cassatt and Maddock, Eds., NBS, Special Publ. No. 412, National Bureau of Standards (U.S.), 176.

——— and K. L. Rubow, 1976: Aerodynamic particle size calibration of optical particle counters. *J. Aerosol Sci., 7.*

Martens, A. E. and K. H. Fuss, 1968: An optical counter for dust particles. *Staub-Reinhalt. Luft, 28,* 14–18.

——— and D. D. Doonan, 1970: Comments on: Influence of refractive index on the accuracy of size determination of aerosol particles with light-scattering aerosol counters. *Appl. Optics, 9,* 1930–1931.

Mercer, T. T., 1973: *Aerosol Technology in Hazard Evaluation,* Academic Press.

Mie, G., 1908: Beitraege zur optik trueber medien. *Ann. Physik, 25,* 377–445.

Nikiforova, N. K. and Y. S. Sedunov, 1972: An investigation of errors arising in photoelectric counters—aerosol particle analyzer. *J. Aerosol Sci., 3,* 441–453.

Oeseburg, F., 1972: The influence of the aperture on the optical system of aerosol particle counters on the responsive curve. *J. Aerosol Sci., 3,* 307–311.

Pinnick, R. G. and D. J. Hoffmann, 1973: Efficiency of light-scattering aerosol particle counters. *Appl. Optics, 12,* 2593–2597.

Pisani, J. F. and G. H. Thomson, 1971: Coincidence errors in automatic particle counters. *J. Phys. E.*

Princen, L. H. and W. F. Kwolek, 1965: Coincidence corrections for particle size determinations with the Coulter Counters. *RSI, 36,* 646–653.

Quenzel, H., 1969: Influence of refractive index on the accuracy of size determination of aerosol particles with light-scattering aerosol counters. *Appl. Optics, 8,* 165–169.

Rimberg, D. and J. W. Thomas, 1970: Response of an optical counter to monodisperse aerosols. *Atmos. Environ., 4,* 681–688.

Schuster, B. G. and R. Knollenberg, 1972: Detection and sizing of small particles in an open cavity gas laser. *Appl. Optics, 11,* 1515–1520.

Sverdrup, G. M., 1976: Parametric measurement of submicron atmospheric aerosol size distributions, Ph.D. Dissertation, Univ. of Minnesota.

——— and K. T. Whitby, The effect of changing relative humidity on aerosol size distribution measurement. *Hutchinson Memorial Volume on the ACHEX Program,* in press.

van der Hulst, H. C., 1957: *Light Scattering by Small Particles,* John Wiley.

Wales, M. and J. N. Wilson, 1961: Theory of coincidence in Coulter particle counters. *RSI, 32,* 1132–1136.

Whitby, K. T., 1975: Private communication, unpublished work of the authors.

——— and B. Y. H. Liu, 1967: Generation of countable pulses by high concentrations of subcountable sized particles in the sensing volume of optical counters. *J. Colloid Interface Sci., 25,* 537–546.

——— and ———, 1968: Polystyrene aerosols—electrical charge and residue size distribution *Atmos. Environ., 2,* 103–116

———, ———, R. B. Husar and N. J. Barsic, 1972: The Minnesota Aerosol-Analyzing System used in the Los Angeles Smog Project. *J. Colloid Interface Sci., 39,* 136–164.

——— and R. A. Vomela, 1967: Response of single particle optical counters to nonideal particles. *Environ. Sci. Technol., 1,* 801–814.

———, W. E. Clark, V. A. Marple, G. M. Sverdrup, G. J. Sem, K. Willeke, B. Y. H. Liu and D. Y. H. Pui, 1975: Characterization of California aerosols. I. Size distributions of freeway aerosol. *Atmos. Environ., 9,* 463–482.

Willeke, K., S. K. Lo and K. T. Whitby, 1974: Dispersion characteristics of a fluidized bed. *J. Aerosol Sci., 5,* 449–455.

———, K. T. Whitby, W. E. Clark and V. A. Marple, 1974: Size distributions of Denver aerosols—a comparison of two sites. *Atmos. Environ., 8,* 609–633.

——— and ———, 1975: Atmospheric aerosols: Size distribution interpretation. *J. Air Pollution Control Assoc., 25,* 529–534.

Winkler, P. and C. Junge, 1972: The growth of atmospheric aerosol particles as a function of the relative humidity. Part I—method and measurement at different locations. *J. Research Atmos., 6,* 617–638.

QUESTIONS

Q: There was discussion about inlet efficiencies for particle sizes of 5 to 40 μm. What about smaller particles' inlet efficiencies?

Whitby: Generally speaking, the counters are ok from 6 or 7 μm on down. The only problems are those that sometimes arise at very small flows. There have been quite a few efforts both in the United States and in Europe to reduce the lower size limit in counters. These usually require very small flow rates, very small holes in the plumbing, and very small inlet volumes. These counters can suffer from inlet problems also, but generally problems begin at about 5 to 7 μm. We've had a couple of counters where the original inlets had only 50 to 60% inlet efficiency at 10 μm. Most on the market now are ok up to 10. Troubles begin at 10 or sometimes 5 or 6 μm. My feeling is that one must look into the inlet efficiency problem before 10 μm or else the results are not good.

Q: I didn't understand one thing in the presentation of the data. Shortly after the spinning disk particle data, you showed some data for irregular absorbing particles. It wasn't clear whether in the calibration of the optical particle counter you were attempting to take into account the absorbing properties in the particle calibration.

Whitby: Irregularities in the transparent particles effectively reduce the resolution of the right angle system to a σ_g of 1.5. If you have a 45° system that integrates all around the axis (like the Royco, the Bausch & Lomb, or the Climet) then the spreading effect is much less. There is a shift in response due to absorption but the spreading effect on the size distribution (say you are calibrating with a monodispersed aerosol) is about the same as for a transparent particle. This is not theory, this is just plain experimental measurements.

Q: Are you saying that it's not practical to use an optical counter to measure stack gas aerosols or dry aerosols?

Whitby: If you wish to do that you have to calibrate on the kinds of aerosols you wish to measure. I do not believe that you could take the ideal aerosol calibrations and predict what a particular counter would do. The ideal thing to do would be to do a calibration with that kind of aerosol. We measure aerosols like that and we could see smearing out, deterioration in the quality of the data which we think was due to these effects. Please understand that transparent, irregular particles only destroy the resolution but don't shift the size. Practically speaking, the coarse particle mode in the size distribution has a geometric standard deviation of about 2 to 2½. The shift in the resolution is about from 1 to 1.6. Therefore, the resolution is still good enough to resolve the coarse particle mode of the dust reasonably well. Problems arise when you have a fairly monodispersed, irregular aerosol. Then the counter spreads it out quite severely, or if it's absorbing as well as irregular, it moves the size distribution over sometimes by a factor of 4 in size.

Harris: What Dr. Whitby is saying is that some of us for years have given median size distributions expressed in terms of equivalent latex sphere sizes.

Q: How did you measure the size of your DOP and oil aerosols?

Whitby: Where we used DOP aerosols, generated with a spinning disk, we sized them microscopically. I didn't discuss the calibration aerosols at all because Drs. Pui and Liu are coming in tomorrow and they can tell exactly how this is done. Since the development of the vibrating orifice generator, the accuracy with which you can calculate the size of the aerosol is so much better than any method of measurement that ordinarily we don't recheck it.

ROYCO INSTRUMENTS PARTICLE COUNTERS: CAPABILITIES AND LIMITATIONS

A. LIEBERMAN

Royco Instruments, Inc., Menlo Park, Calif.

Royco Instruments manufactures a variety of light scattering single particle counters for airborne particle measurement, along with a parallel line of liquidborne particle counters which are not included in this discussion. The airborne particle counter line covers several instruments designed for a wide variety of applications. All of the instruments are single particle counting devices which count and size individual particles as they are passed through an illuminated sensing volume. Light scattering is measured at either right angles or at near forward scattering angles. Particle size measurement capability ranges from 0.2 μm to greater than 100 μm. Sample flow rates vary from 0.01 cfm to 1.0 cfm. A variety of data readout capabilities is available, some interchangeable as part of the instrument. In addition, most instruments are supplied with electrical connections for data take off points. There is also a range of peripheral components that can be used with the airborne particle counters. The components include aerosol generators for calibration purposes, sample acquisition devices and additional data processing circuitry.

The range of operating capabilities summarized briefly above does not apply to each and every one of the instruments. Table I shows operational data for the several airborne particle counters that Royco manufactures. Models 203 and 220 are right angle scattering instruments, while models 218, 225 and 245 are near forward scattering instruments, as indicated by the scattering angle range data. Generally speaking, right angle scattering instruments normally have better full-range resolution than near forward scattering angle instruments, but are more affected by differences in refractive index. Figure 1 shows the optics for right angle scattering instruments and Figure 2 shows optics for near forward scattering instruments. Figure 3 shows the response curve for a right angle scattering optical system and Figure 4 shows the response for a near forward scattering angle optical system.

The Model 218, Figure 5, is a portable instrument designed primarily for observation of relatively high concentrations of aerosols in specific size ranges. It is relatively light, can be operated by battery, but does not have available data take off points for operation of multi-channel analyzers or data processing equipment. The built-in size detection system allows measurement of only one particle size range at a time. Internal construction is shown in Figure 6.

Model 225 is a high concentration measurement instrument, shown in Figure 7, designed primarily for observation of polluted environments where concentrations may range to several milligrams per cubic meter. The Model 245 was

TABLE I. Optical Design Parameters for Various Counters

	Model				
Optical Design	203	218	220	225	245
Light half-angle	24°	5°	24°	5°	5°
Trap half-angle	—	9°	—	16°	16°
Collector half-angle	24°	25°	24°	25°	25°
Scattering angle range	42–138	4–30	42–138	11–30	11–30
Sensitivity, μm	0.2	0.5	0.3	0.3	0.5
Dynamic range, standard	40:1	40:1	40:1	40:1	40:1
Sensing volume, cu mm	1.3	0.15	2.6	0.4	4
Coincidence error, %	4	6	9	15	4.5
Concentration limit, pcf	10^6	10^7	10^6	10^8	3×10^5
Resolution, % of mean at 1 sigma	3	15	5	5	15
Flow rate, cfm	0.01	0.01	0.1	0.1/0.01	1
Data	Remote, Selectable	Digital, Fixed	Remote, Selectable	Integral, Selectable	Integral, Selectable
Dimensions, inches					
Main frame	16×10×18	15×10×4	16×10×18	16×8×21	16×8×21
Sensor	—	—	—	5×4×21	5×4×21
Weight, pounds	32	18(23)[a]	32	42	42
Main frame	32	18(23)[a]	32	42	42
Sensor	—	—	—	12	12
Power, watts	150	55(25)[a]	150	250	250
Environment	7°C–52°C operating, −40°C–70°C, non-operating 95% Relative Humidity operating, 98% non-operating 1G operating, 5G non-operating				

a. Battery operation.

designed primarily for measurement in industrial clean room applications where low concentrations can be encountered. Thus the high flow rate is used to provide sufficient sample to allow reasonable statistical validity in areas where particle concentration may be in the less than 100 particles per cubic foot range. Figure 8 shows the optical assembly for the near forward scattering sensors.

One of the problems with any single particle counting instrument is the possibility of recirculation of particles through the sensing volume. Models 203 and 225 have a sheath air flow system which avoids this problem. The remaining models have a sufficiently high flow rate to sensing volume ratio so that the effect of recirculated particles is insignificant.

Calibration procedure for all instruments involves generation of monodisperse latex particles within the dynamic size range capabilities of the instrument, setting the response of the detector to a fixed level so that the signal to noise ratio for the smallest particle will be on the order of 2 to 1 or more, then establishing comparator levels for each of the several size ranges that are to be measured over the dynamic range of the instrument. This procedure will normally establish the size and the accuracy to a level of approximately 5%. A final touchup is then carried out by comparing the response of each instrument with that of an in-house standard that is more carefully calibrated with a range of particle sizes. This

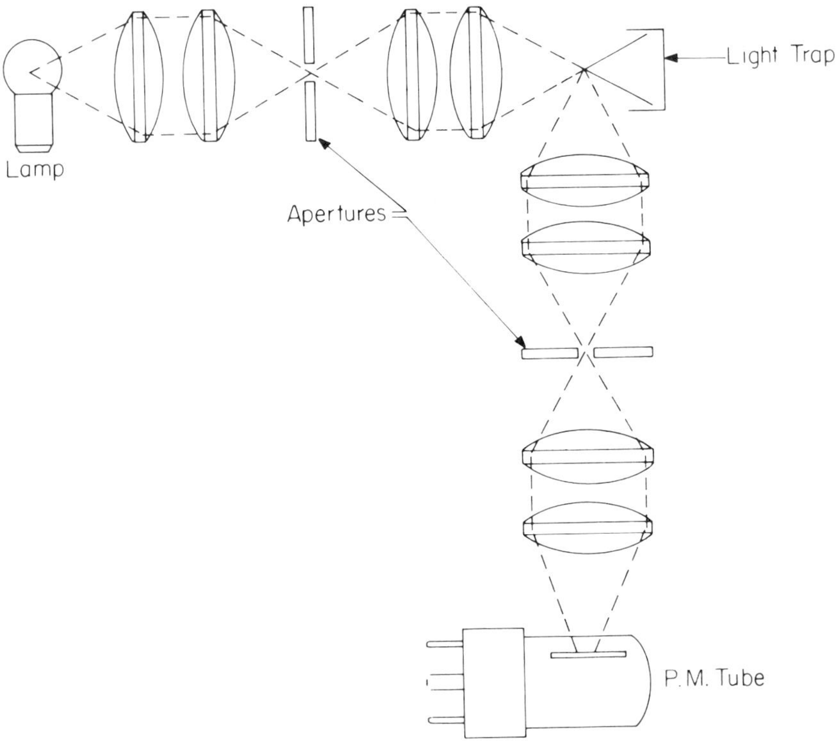

Figure 1. Layout of optical components for right angle scattering.

procedure is also used to establish number correlation from one instrument to another. Following the primary calibration a field standardization procedure is employed which permits establishing instrument operation at the same base line level used for primary calibration. The field standardization procedure involves provision for generating constant light level pulses and setting the instrument gain at a reference level based on the light pulse amplitude.

A series of available plug-in data readout modules can be used with any of the instruments except for the Model 218. The plug-in modules permit display of the rate at which particles in either one or two selected size ranges are being accumulated in real time. Alternately, a plug-in module may be used that indicates the number of particles counted in one of five selected size ranges. The size range may be advanced manually or automatically. Alternately, particles counted in five size ranges can be shown simultaneously. In addition, pulse information can be obtained from a rear panel coaxial cable connector for pulse height analyzer interface systems. Note that the pulse durations are typically 10 to 20 microseconds (exception: Model 203 at 400 microseconds and Model 220 at 200 microseconds) with rise times of 2 to 3 microseconds. A standard nuclear pulse height analyzer requires pulse shaping circuits in order to handle these pulses.

Available peripheral systems include a Model 256 nebulizer type aerosol generator, Figure 9, for generation of latex aerosols from suspension, and a Model 258 submerged orifice atomizer, Figure 10, that will generate milligram

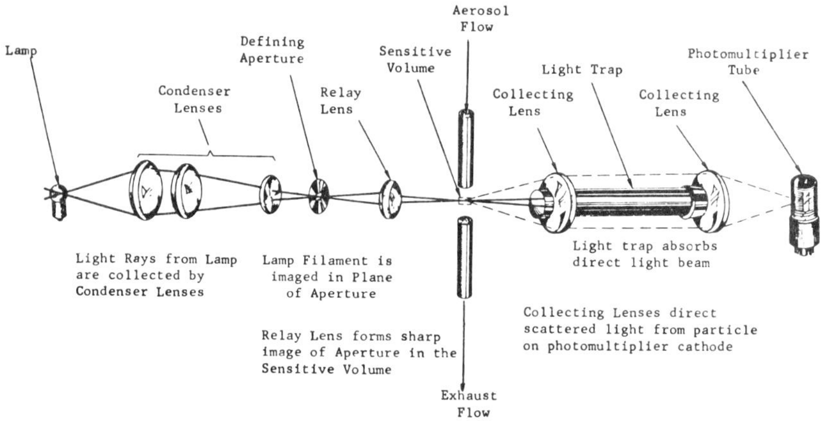

Figure 2. Layout of optics components.

per liter quantities of submicron oil or water droplet aerosol for filter testing purposes. Sample handling peripherals include a Model 173 Sample Probe for point sampling at filter face systems. The sample probe inlet is designed for isokinetic sampling a 1 ft^3/min flow in a 100 ft/min field. The probe is fitted with a repeater meter that indicates the rate of particle accumulation so that the operator can observe the exact point of leak. Another sample handling system is a Model 252 Aerosol Dilutor. This is a gas ejector system which uses a high speed stream of filtered clean air to aspirate a 1:100 sample for dilution. This system should not be used for measurement of particles any larger than 2 to 3 μm in diameter. A Model 160/161 Valve Manifold System, Figure 11, allows programmed transfer of valves connected to a manifold which is then connected to the particle counter inlet. Up to 10 valves can be selected with lines normally from 10 to 15 meters in length for sequential measurement at a number of sample points. Care is required to ensure equivalent sample flow rates with different line lengths or configurations. In addition, it is necessary to realize that particle loss

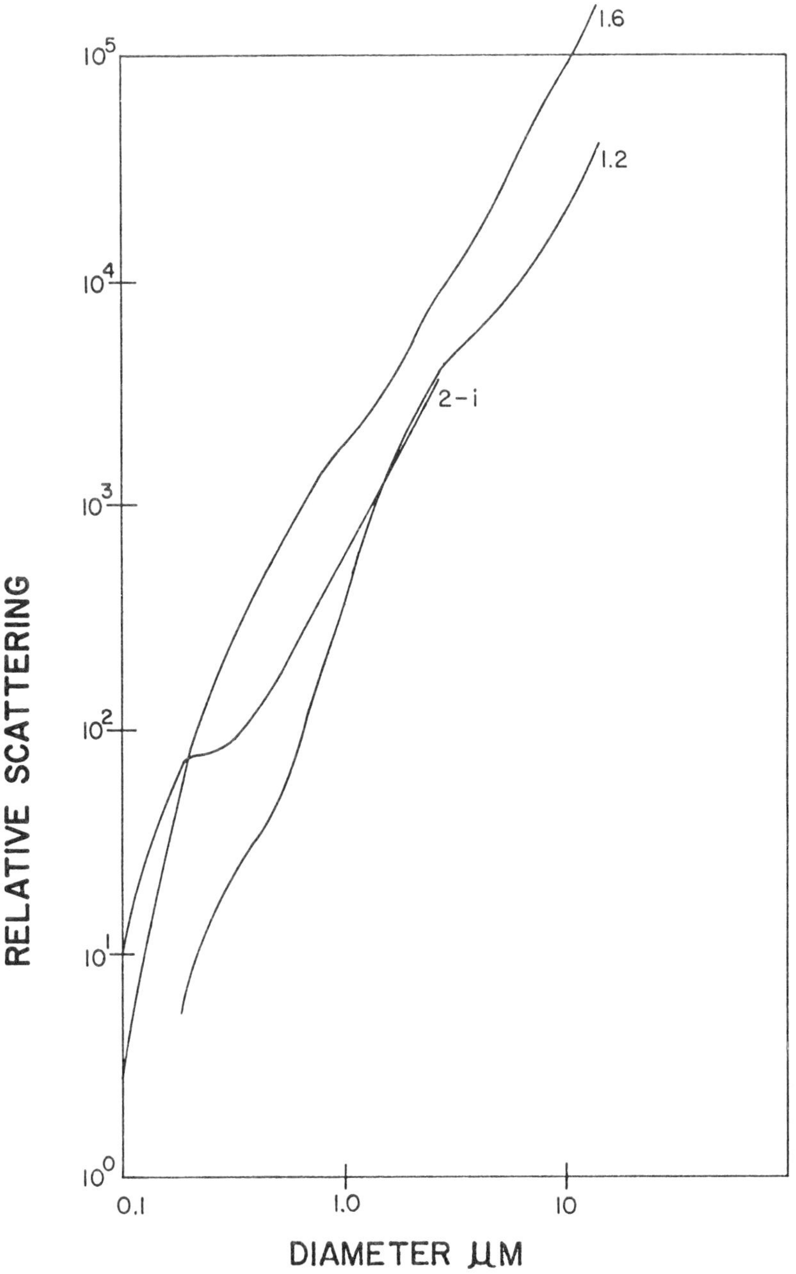

Figure 3. Light scattering vs. particle size for Royco model 200/220 optics.

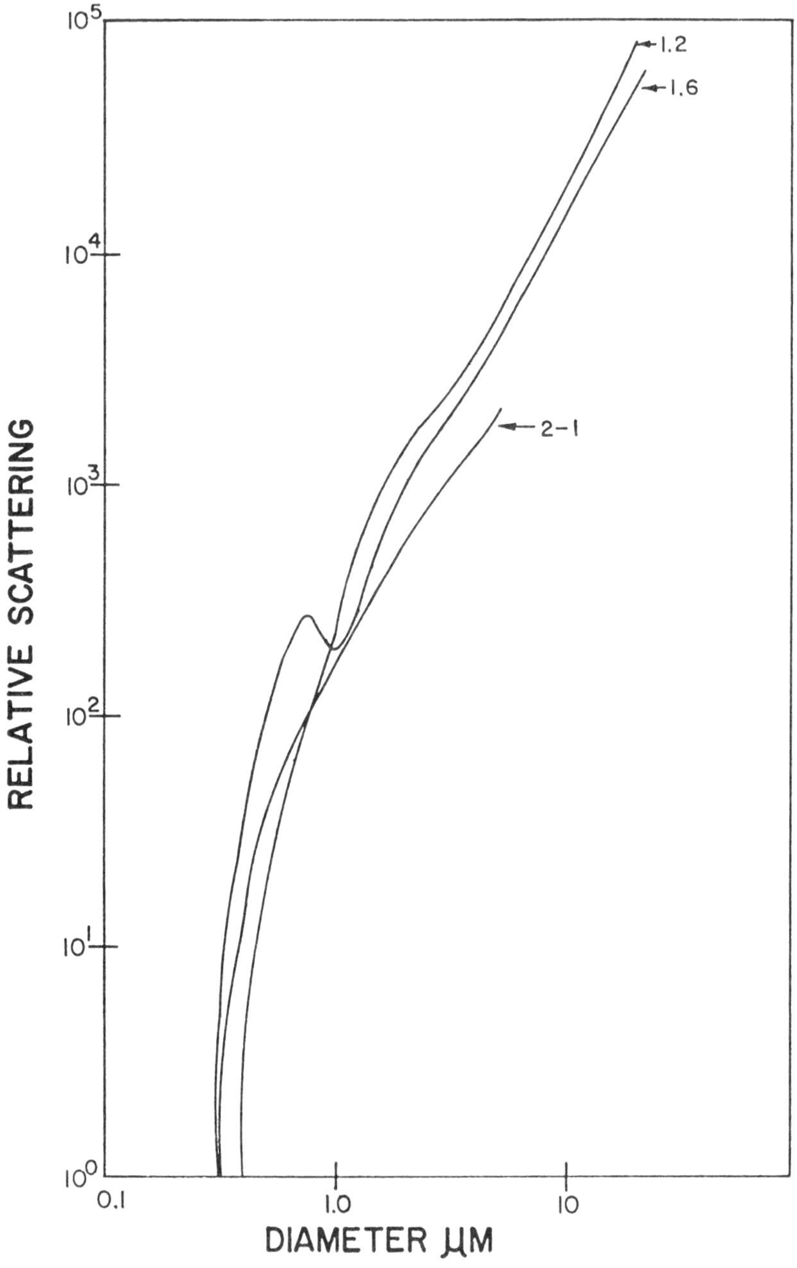

Figure 4. Light scattering vs. particle size for Royco model 245 optics.

in lines may occur due to wall loss or gravitational deposition. Also, pressure drop through the additional line and valve must not exceed the capability of the sample pump.

Note that the dynamic range indicated for each of the instruments is 40 to 1. This dynamic range limit was selected to provide reasonable sizing resolution

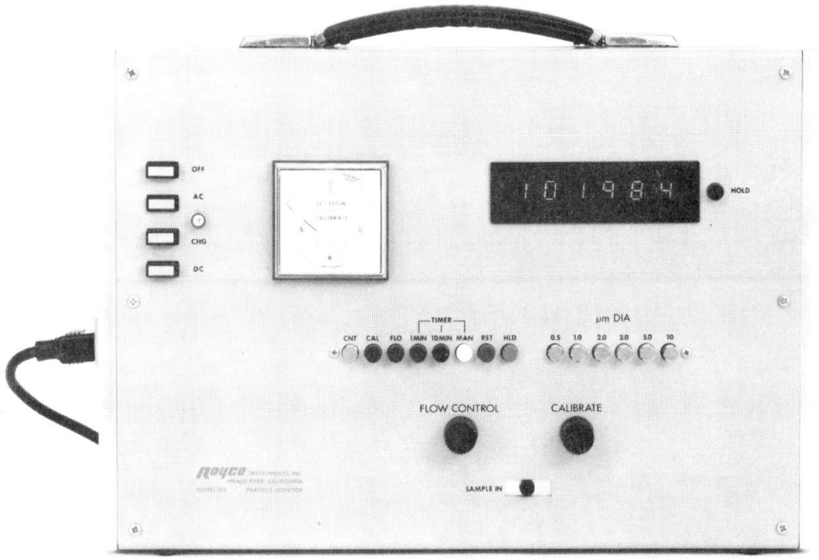

Figure 5. Model 218, front view.

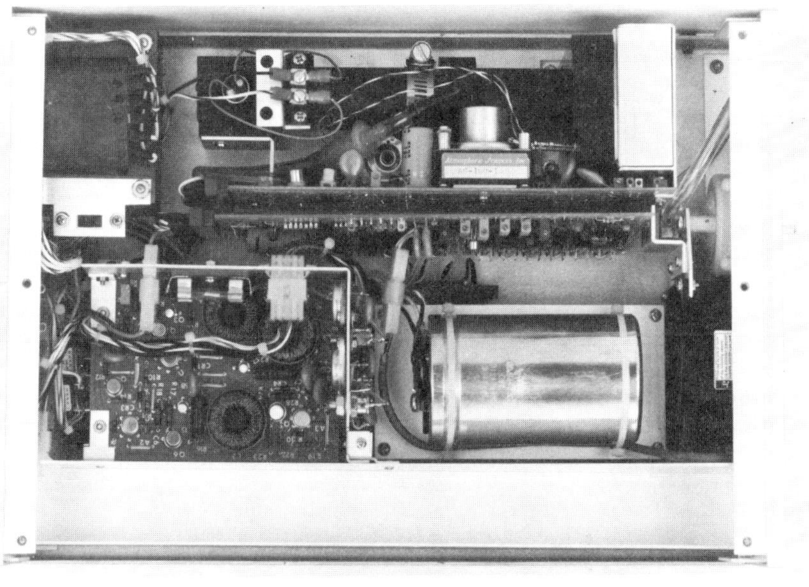

Figure 6. Model 218, rear view.

over the entire dynamic range of the electronic and pulse analyzing system. It is possible to calibrate any individual sensor over a much wider dynamic range. The photodetector/preamplifier combination is linear over a range of at least 100 to 1. It is, therefore, possible to shift the dynamic range to a different portion of the entire spectrum by changing system gain or changing illumination intensity. In either case it is necessary to recalibrate, especially if the illumination wave length is affected by the method selected to change illumination intensity.

If the sample flow rate is changed then the rise time of the pulses which are measured will also change. Accommodation of a wide range of pulse rise times requires an electronic system with wide band width. Since electronic noise levels increase with band width, then sizing sensitivity will be decreased if a wide band width system is used. The net result then is that large changes in sample flow will result in attenuation of fast pulses, thus producing erroneous data, both in terms of size and count if an electronic system with too wide band width is used.

Figure 7. Model 225/241, front view.

Figure 8. Model 241, optics assembly.

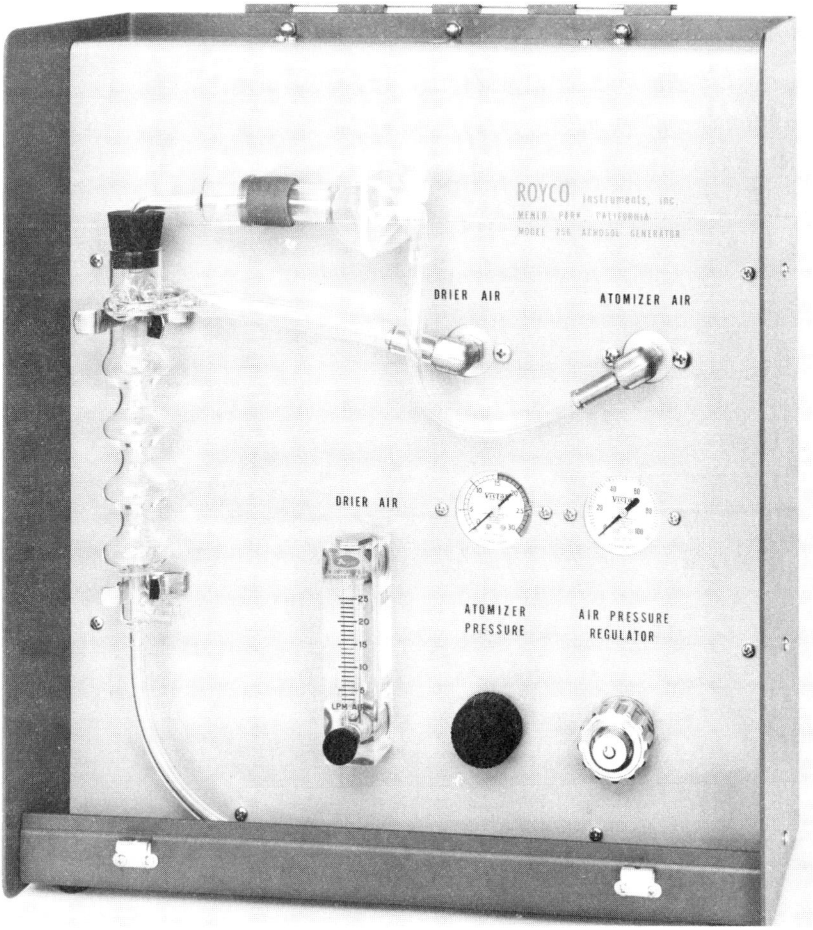

Figure 9. Model 256, aerosol generator.

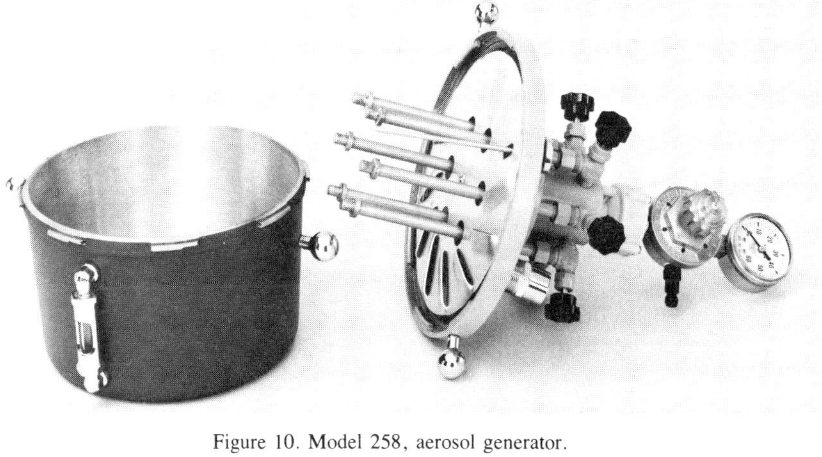

Figure 10. Model 258, aerosol generator.

Environmental temperature limits of 7° - 52° C are specified. Within these ranges the electronic system drift is essentially zero. At much lower or higher temperatures the semiconductors in the electronic system may malfunction. The relative humidity limit arises from the requirement to pass gas through a system with a restriction. If the relative humidity approaches saturation then adiabatic expansion immediately after the inlet tube may produce droplets of condensable vapor in the sensing volume. Excessive vibration levels will result in displacements of optical assemblies sufficient to produce erroneous information.

Figure 11. Model 160/161, valve manifold.

Field problems in application include a number of areas. These include incorrect sizing as the chemical and physical nature of the aerosol to be sampled varies appreciably from that of the calibration material. Another problem area is that sampling or inlet efficiency decreases progressively with particle size, resulting in significant losses or even non-appearance of larger particles. This can result in difficulty in correlating data from two or more particle counters sampling from the same environment. Differences in resolution and signal to noise ratio from instrument to instrument may affect the indicated particle size distribution as well as the concentration of particles measured in the smallest size range.

QUESTIONS

Q: When you want to sample ambient aerosols do you have to worry about non-isokinetic sampling conditions?

Lieberman: If you're sampling in the wind and you know what the wind velocity is, and you want to sample particles larger than about 5 μm you do have to worry about isokinetic sampling. There are a couple of ways to avoid the isokinetic sampling problem. One way is to use a large duct, six inches plus in diameter, and aspirate an air sample into this duct at a fixed velocity. The sampling area in this large duct will give major measurement errors at the periphery and much less at the center. You then sample isokinetically from the center at the velocity. Another way to avoid the problem is to put a sample gooseneck on a weather vane, sample through the wind superisokinetically, and then calculate the sampling error.

SYNOPTIC ANALYSIS OF AEROSOL PARTICLES

JAMES M. LEPPER, JR.

Climet Instruments Company, Redlands, Calif.

Analysis of natural aerosols involves the sampling of a very wide variety of aerosol particles. The particles may consist of liquids such as water or hydrocarbons. . . they might consist of solids such as minerals and pulverized materials, and biological debris consisting of insect parts like scales and hairs, or single-celled biota like algae and bacteria.

The two opposite views of these particles involve the detailed analysis of a few particles, or a synoptic analysis consisting of analysis of many particles divided into a few categories. The extreme of the detailed analysis category is the Science Spectrum Differential II, and at the generalized extreme there is the Climet CI-208. The 208 collects and measures thousands or millions of particles then gives you just one orange, NIXIE number. You don't learn how many arms or legs each particle has, but on the other hand you have a small number of data that can be used to control some small number of feedback loops.

To use feedback control of any system, etc., there must be one feedback loop for each variable controlled. If one has thermostats for both an air conditioner and a heater, they will work against each other, both heating and cooling at the same time. Thus, if a full spectrum of data is gathered about a particle's properties, all of the data cannot be used to control one knob—it is best suited for a research application where the data are searched for features.

Thus, the problem solved by the CI-208 sensor is how to make a wide variety of particulates have an output signal that is a function of a single degree of freedom (size: diameter, area, or volume). This is described by the following four topics.

LIGHT COLLECTION GEOMETRY

The variety of particles to be measured by the CI-208 sensor makes it obvious that light scattered by the particulates is going to go in different directions depending on a number of properties of each particle. This simple requirement directly leads us to the wide-angle optics system.

With an optical sensor the natural method of measurement is some function of the particle silhouette area. The area is in turn converted to the equal shadow area of a sphere, and the sphere's diameter is the reported size, "equivalent spherical diameter." Further, an amount of light intercepted by a particle, equal to its shadow in dimensions, is either scattered or absorbed. For particles in the range of 0.3 to

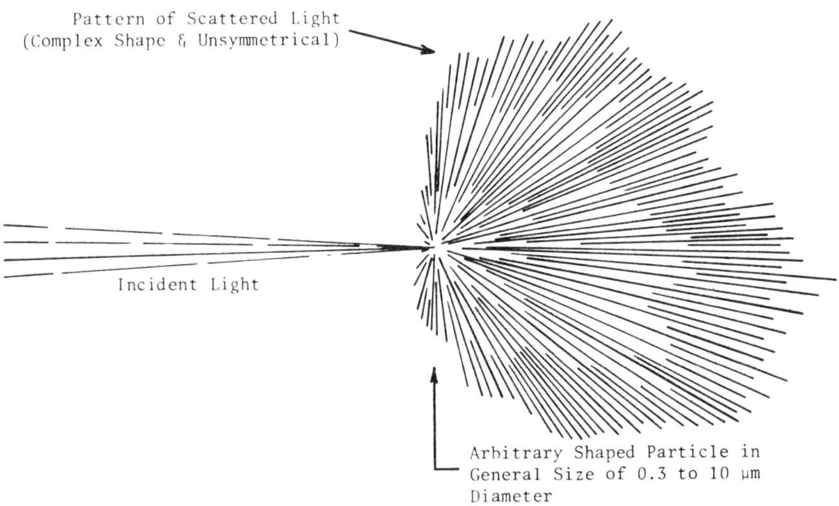

Figure 1. The incoming light beam strikes an arbitrarily shaped particle, scattering light in various directions, and virtually all in the forward hemisphere.

10 μm in diameter, incident light is scattered with only a small amount absorbed in the short path available.[1]

Since the light is scattered in random directions (Figure 1), it is necessary to surround the particle with the collection optics.

For the special cases of spherical particles, calculations can be made of the scattering angles due to various sized particles. The curves that are generated for specific angles are not monotonic functions of scatter vs. size. The curves show a resonance when the size of the light waves are simple multiples of the particle size. Differing combinations of particle size and wavelength size change the shape of the scattered light distribution, but the total light scattered remains essentially the same.

Light scattering patterns are usually predicted with the use of the old Mie

1. The major exception is for semiconductor materials. Free carbon is the only common material encountered with such properties, thus coal appears $\approx$ 20% smaller in diameter.

scattering formulas. But they are only formulas without closed solutions. They require computerized numerical solutions even for the simple cases of spheres. For complex shapes and heterogeneous particles, there is little hope of solution. In the meanwhile, hardware to do the job of collecting the scattered light can still be built. The plan is elementary: just collect all of the scattered light, irrespective of the direction. The implementation is harder: how do you make optics that surround an object? Graphically, it is clear. The solution must take the form of an ellipse, where all of the rays leaving one focus reflect from the elliptical surface and converge on the second focus. Accurate production of such an elliptical surface is the problem which Climet solved.

The work of Derry Cooke and Milton Kerker (Cooke and Kerker, 1975) is a beautiful job of thorough analysis. Building on the various attempts to incorporate all of the significant variables of particle light scattering, they succeeded in creating the definitive analysis of the various existing instrument configurations. Yet there still remains the question of particle shapes. At that point, I fall back on the heuristic argument of the wide angle and wide spectrum optics: wherever the light is scattered, a mirror with a large solid angle is sure to get most of it. And, since there is only so much light per unit area, the collected light is proportional to the silhouette area.

The Climet elliptical mirror[2] surrounds the particles in the sensing region of the instrument (Figure 2). Of the light scattered in the forward hemisphere, about 5.1 steradians out of 2π in the forward hemisphere are collected. Back-scattered light in the rear hemisphere is usually less than a percent of the total scattered light, and it is not collected. Because of the incoming light coming in a cone of about f/2, and a light trap in the center, rays from about 15° to 105° are captured (Figure 3).

Thus, instead of finding the best compromise angle to work at, the Climet sensor avoids compromising the signal, and collects virtually all of the scattered light.

THE ELLIPTICAL MIRROR

The light-collecting mirror used in the Climet 201/208 series of particle analyzers is in a very unusual shape for precision optics. It has a diameter of 7.06 cm and a focal distance of 5 cm. By definition, the f number would be 0.708 (focal length divided by diameter), which is a very good light collector. Very good glass optics are limited to light gathering power of about f/1 to f/0.8. But the deep elliptical shape of the mirror defies the definition of f number. The sensor uses an annular region forward of the focal point, not the round region behind the focal point. It thus collects light 2.55 times better than the f number indicates. It is better described by the amount of solid angle[3] through which it collects light. It collects

2. Actually, a prolate hemispheroid.
3. Solid angle is measured in steradians, the angle subtended by a region of one square unit at a spherical radius of one unit.

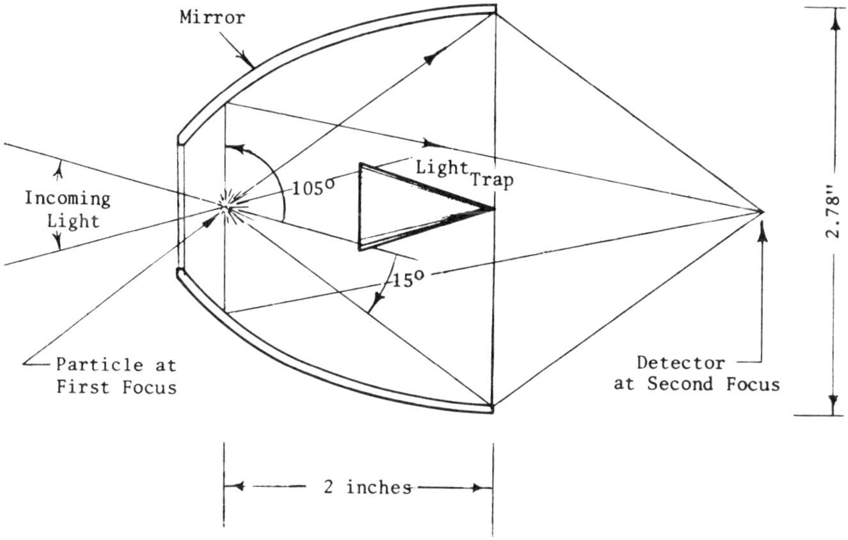

Figure 2. A cross section of the elliptical mirror shows the manner in which the scattered light is collected from the first focus and directed to the second focus. Scattered light is collected at angles ranging from a minimum of 15 degrees to a maximum of about 105 degrees.

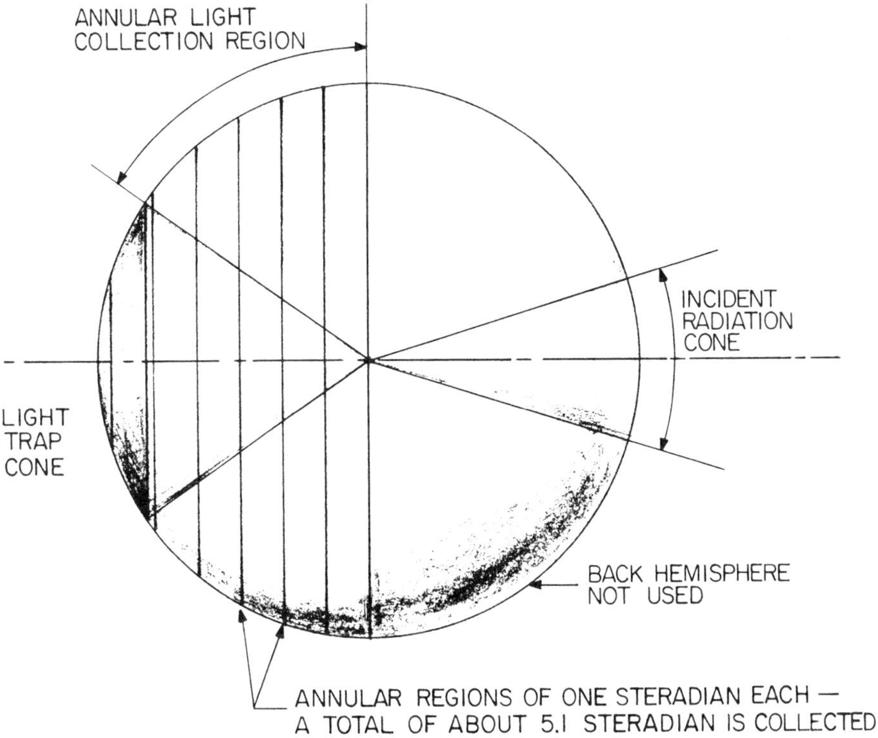

Figure 3. This mathematical sphere represents the solid angles through which light approaches and leaves a particle in the field of view. It is a unit sphere near the particle and is centered at the focus of the elliptical mirror.

light over a solid angle of about 5.1 steradians, while the same dollars spent on glass optics would collect only about one steradian.

It cannot be made by conventional optical techniques. It is formed by electroplating nickel onto a stainless steel mandrel. The mandrel was formed and the mirror tested by special optical instruments made by Climet.

SPECTRAL DISTRIBUTION

The second most important means of creating a generalized sensor, thus obtaining a synoptic view of aerosols, lies in the lamp and detector combination.

The light source is an important contributor to the uniformity of the 208 sensor. Just as different-sized particles give different scattering distributions, different wavelengths have the same effect. It is, therefore, desirable to have a very broad spectrum of light used to further smooth out the effects of particle size tuning to the incident wavelengths.

The spectrum is controlled by both the detector and the source. The Climet source is a quartz-halogen cycle lamp with a 3000°K color temperature. The small amount of blue light added by a temperature above the common 2875°K makes an important contribution for the small-sized particles in the 0.5 to 0.3 μm range.

The other contributor to the total spectral response is the detector. In the 208 sensor a photomultiplier tube is used with a spectral cutoff at about 600 nm. The effective spectrum covers a range of about 2:1; just enough to smooth out the curves of wavelength dependence (Figure 4).

In the Climet sensor CI-250, the lack of a mirror is somewhat compensated for by the use of an extended spectrum. A silicon photodiode detector is used instead of a photomultiplier tube. The extended spectral range is enough to smooth the calibration curve to a monotonic shape. It doesn't curve back down as do the curves for other forward-scatter sensors.

A laser-illuminated sensor is something that immediately comes to mind. For particles much larger than the wavelength, the forward-scattered light is very close to the line of the incident beam. The laser's narrow beam, however, makes it possible to mask off easily the incident beam and capture energy scattered at narrow angles, the narrow angles only exaggerating the particle resonance artifacts. The wide-angle mirror looks ideal to eliminate those effects. We tried an Argon laser and found that the calibration curve was nearly monotonic, but with a slope of less than one (log-log) for the range of about 0.2 to 1 μm size. It does allow one to detect particles down to about 0.1 μm with its 10:1 signal-to-noise ratio increase; but there is little discrimination between 0.1 and 0.5 μm. Only a few runs were made, so an exact curve was not generated.

The small beam made possible by the laser's small divergence could be useful in extending the concentration range of particles; but this is accomplished much less expensively by our dilution system. The increased signal-to-noise ratio can be used to count faster; but we are already sampling near Mach 0.5 and can't go faster without sonic choking.

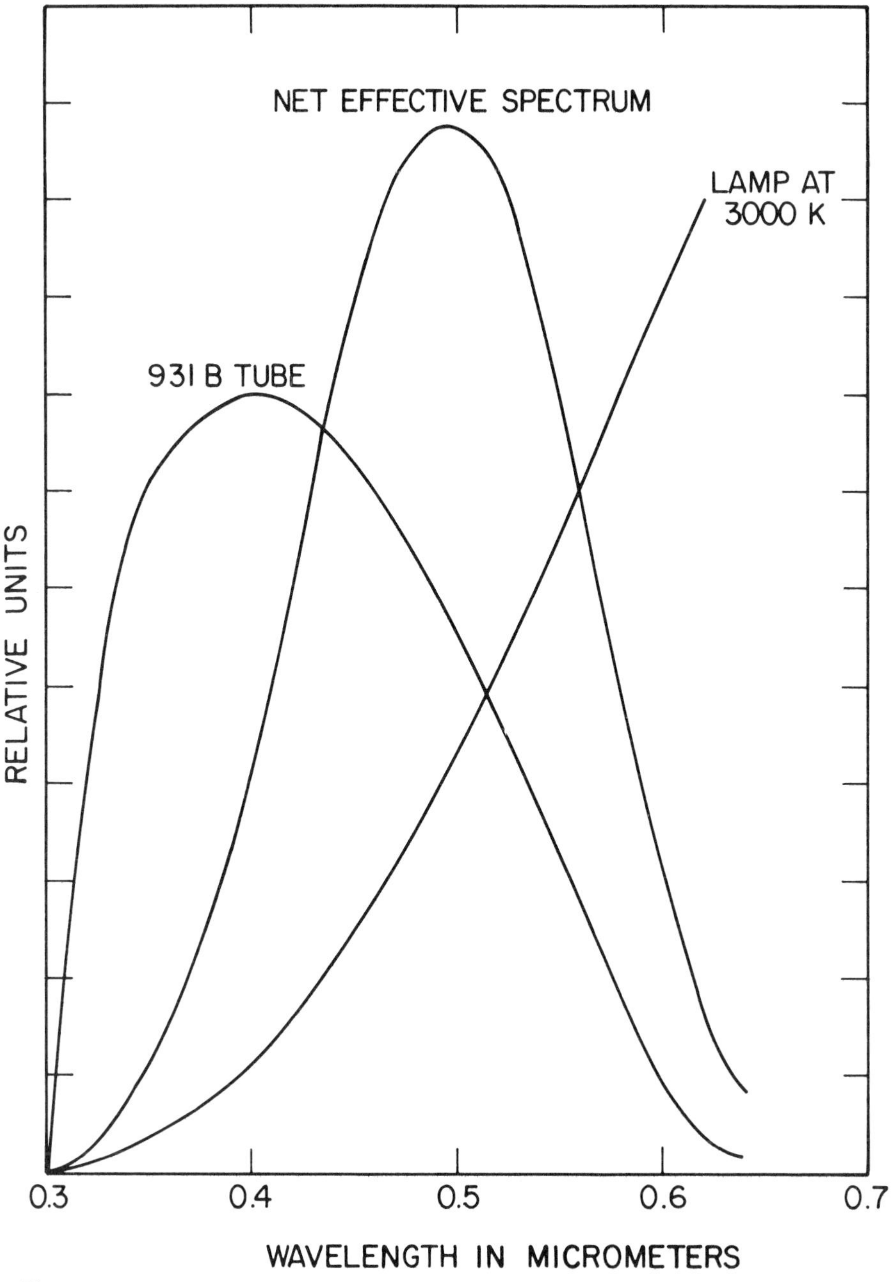

Figure 4. Optical spectral sensitivity for the CI-208 is the product of the lamp emission and detector sensitivity spectra. Use of a 3000°K quartz-bromine lamp provides a range of over 2:1 to smooth wavelength tuning effects.

SAMPLE FLOW SYSTEM

The third most important aspect of a particle sensor which allows a generalized view of the particle jumble is a straight line inlet with gentle acceleration of the particles.

What is a particle and what is not? A single size measurement cannot answer the question of the bond strength between two particles. Do you count a pair as two small particles or one large particle? The best that an instrument can do is measure what is presented to it, without altering the appearance of it. The experimentalist, then, must prepare the aerosol to be in the desired state before it enters the instrument. Usually the hardest part of aerosol measurement is getting an honest sample to the instrument.

The flow in particle analyzers looks like one of the simpler aspects of the design, and has been neglected in the past. Accuracy of the flow rate measurement has been another weak area. The accuracy of a sensor giving counts per unit volume is just as dependent on the flow rate as the count rate for the accuracy of the data.

Getting a representative sample to the instrument is the first problem. Flow rates are frequently set to suit some particular sampling problem, usually clean rooms. The many measurements with aerosols to find the optimum in terms of accurate sampling go unheeded.

Measurements at Climet were made, and they generally verified those in the literature. Flow rates of 0.25 and 1.5 CFM were used in a test of atmospheric aerosol for sizes 1.0 μm and greater, with various sized tubes 279 cm (about 9′) long. The high flow rate consistently lost significant numbers of particles. Losses were about 25% more for larger sizes while about half of the particles were lost for ¼ inch OD tubing.

We must admit this is a desirable finding since our nozzle flow is near sonic, and our field of view is limited to a region about twice that of the sample jet; thus our ¼ CFM and ¼-inch tubing.

The inlet nozzle does a special job for us. It is shaped like the classical DeLaval converging nozzle (Figure 5), and accelerates the boundary of the flow to a speed approximately matching the flow speed of the center lamina. This avoids a speed distribution of the particles as they pass through the field of view. The pulses generated by the particles are very uniform, with widths varying about ± 10–20%. The particles are smoothly accelerated to about Mach 0.39 at the nozzle tip. The smooth acceleration, and the rapid trip to the instrument field of view (about 40–50 microseconds) prevent changes in the particles before they are measured.

The sample travels as a free jet for about a centimeter before it is measured. The gap is necessary to minimize light scattering by the nozzle tip.

When a free jet leaves a nozzle, the boundary becomes turbulent, increasing with distance. There is a cone of calm inside the jet until the turbulence fills the entire jet (Figure 6). The turbulence entrains air from inside the sensor, and the cross section of the jet increases with distance. At the exit nozzle a portion of the mixture cannot enter the exit, and will recirculate inside of the sensor. If a co-axial sheath of clean air is provided, the center cone is extended a slight amount. To be effective, it should be isokinetic (speed matched) with the jet.

If the air protecting the jet is allowed to enter the sensor some distance from the jet, it will converge on the exiting jet. Converging flow naturally dampens turbulence, just as it does in the converging section of the inlet nozzle. This effect is seen in wind tunnel nozzles, laminar flow wings, etc. Oscilloscope photos of the pulses show high uniformity in pulse width and shape, and force the conclusion that the flow through the view volume is essentially laminar, and that the jet is smaller than the field of view. This uniformity of pulses is in marked contrast with most other particle sensors.

The converging flow appears to do an unusual thing: accelerate the particles in an open flow. They leave the nozzle at approximately Mach 0.39, implying a

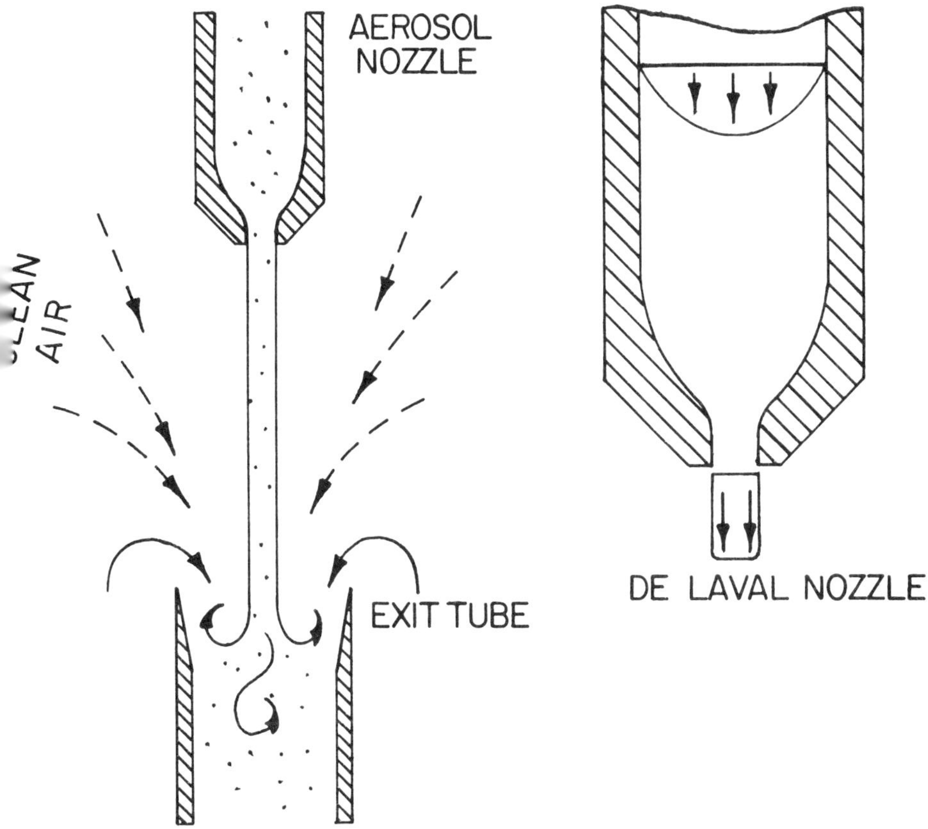

Figure 5. The converging flow of clean air stabilizes the jet so that little mixing takes place. The exit flow is larger than the volume of the mixed clean and loaded air. The smooth column of the jet viewed in a smoke test shows a nearly perfect boundary as illustrated. The De Laval nozzle accelerates the slow boundary layer air in the laminar inlet flow. The flow leaving the nozzle then has a very uniform velocity distribution.

pulse width of 5.2 microseconds. The actual pulse width is only 3 microseconds, giving a speed of about Mach 0.66.

The effectiveness of this approach is shown by the air flow required. With the converging system we find that anything over a flow equal to the sample is sufficient; while with the co-axial flow sheath proportions of 5:1 or 10:1 are needed.

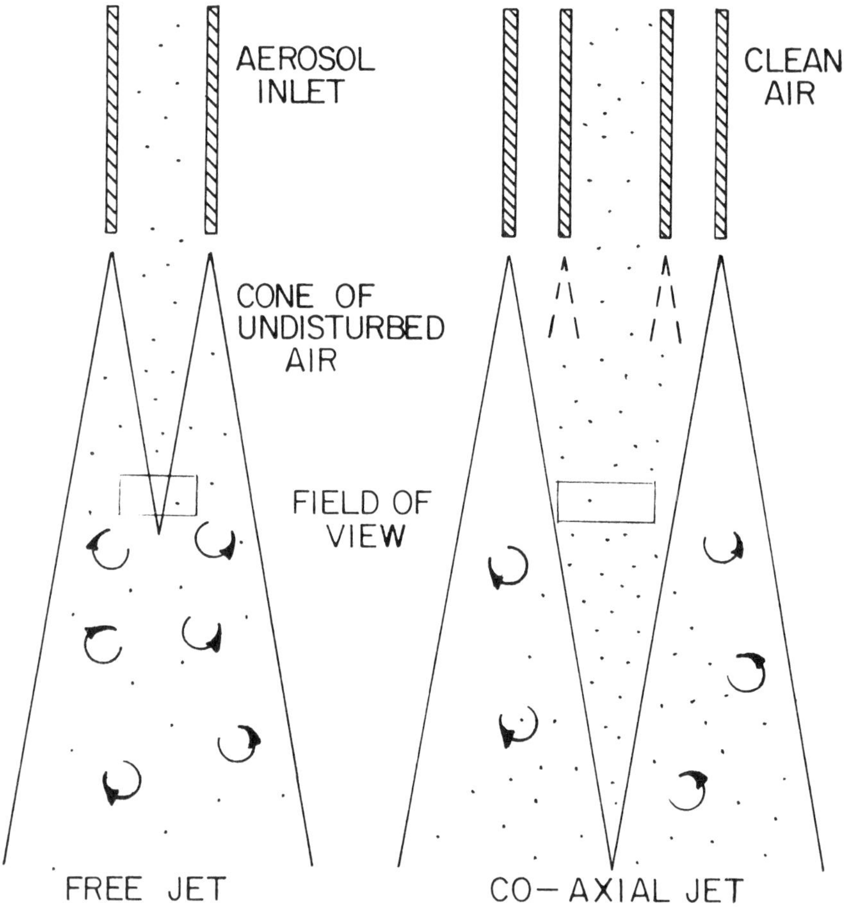

Figure 6. Either the free jet or the co-axially sheathed jet entrains large amounts of air in the sensor chamber. Since the volume of the mixed air exceeds the exit flow, part of the mixture is recirculated in the chamber. Note that while the sheath air extends the cone of undisturbed air to the field of view, it does not prevent mixing and the recirculation of particles.

A dilution system[4] is available which extends the basic flow scheme of the 208 sensor. Usually dilution involves mixing and the attendant loss of particle accuracy. The Climet CI-294 system dilutes by using a small sample stream, surrounding it with 100 times as much filtered air, then sending the coaxial system

4. Patent No. 3,791,196.

through the instrument without mixing until the flow exits the sensor (Figure 7). By using a ratio of 100:1, with convergent flow below the sample injection point, the flow does not have to be isokinetic. The sample flow can be further decreased (with decrease from full scale accuracy) so that the ratio goes to 1000:1 or greater. That extends coverage from about 3×10^{10} to 3×10^{11} particles per m^3 (10^9 to 10^{10} per ft^3).

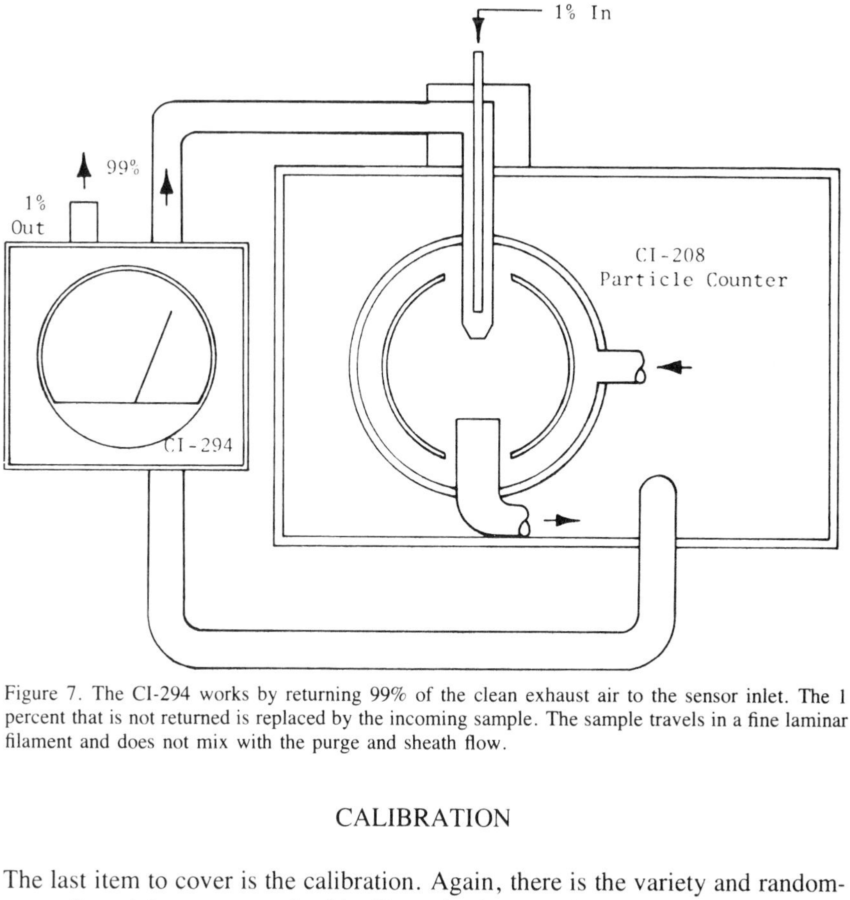

Figure 7. The CI-294 works by returning 99% of the clean exhaust air to the sensor inlet. The 1 percent that is not returned is replaced by the incoming sample. The sample travels in a fine laminar filament and does not mix with the purge and sheath flow.

CALIBRATION

The last item to cover is the calibration. Again, there is the variety and randomness of particles to contend with. Even the best particles available have some measurable variation. Since particles arrive in random order, and at random times, it is necessary to use statistical methods to describe or characterize a particular aerosol. The very objective of aerosol counters is to obtain a representative sample of some larger group of particles. That, too, implies statistics.

The statistics that are involved are very straightforward and commonly used in other fields. A group or aliquot of particles is represented by some simple model, and the aerosol is described in the parameters of the model. Otherwise, all of the data would have to be used as the descriptor of the aerosol.

The best known model is the Gaussian distribution (Figure 8). While it poorly fits data for natural aerosols, it very well fits the preferred portion of the preferred particles. That is, it fits for the large particles of a batch, but does not describe the small debris particles that you get for free. Note that there is no fundamental reason for the particles to fit the Gaussian model; it merely depends on the technique to prepare and separate the particles. Each batch must be examined to see if it suits the model; otherwise some other descriptor must be used.

To describe a particle distribution in terms of a Gaussian model, only two data

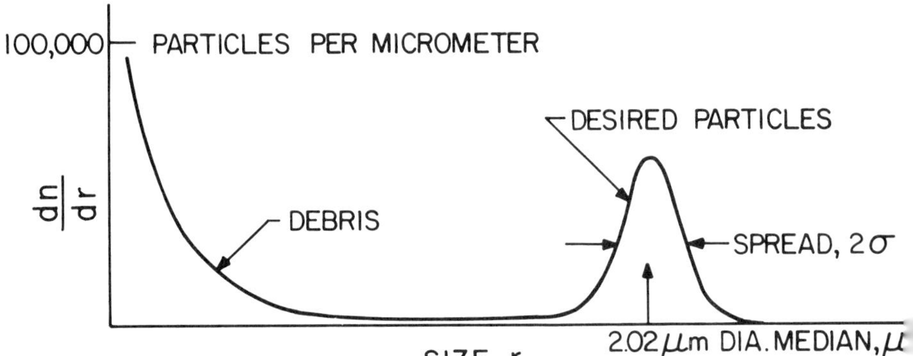

Figure 8. The distribution of calibration particles.

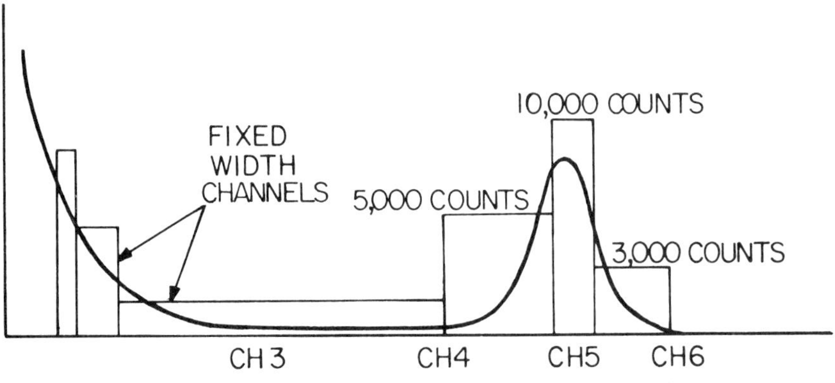

Figure 9. If the calibration (amplifier gain) is low, it is shown as an imbalance of counts in Ch. 4 and Ch. 6.

are necessary: the median, μ, and the spread or standard distribution, σ. More measurements can be made such as skewness, etc., until there are as many degrees of freedom as there are data, but they are really measures of how bad the model's fit is.

In production testing it is sufficient to use three size-channels to define very precisely the calibration point, and the instrument's spread, σ (Figure 9). The central channel covers voltages from just below the mean size of the calibration particles to just above it. The amount of particles that get counted in the channels to either side of the central channel is a measure of the distribution spread. Sigma is read from a graph relating the proportion of particles in the side channels to the amount in the central channel (Figure 10). To normalize the data, the counter stops when the central channel has counted 10,000 particles. For example, if the central channel stops at 10,000 counts, and the side channels have 3000 counts, that is a narrow distribution. If the count were 8000, the distribution is near the maximum we allow.

To get the distribution of particles centered on the channel set to the calibrated size, one could just maximize the count in it while adjusting the sensor gain. It is more accurate, however, to match the count in the two side channels. Then, the wings of the distribution curve spread equally on either side of the central

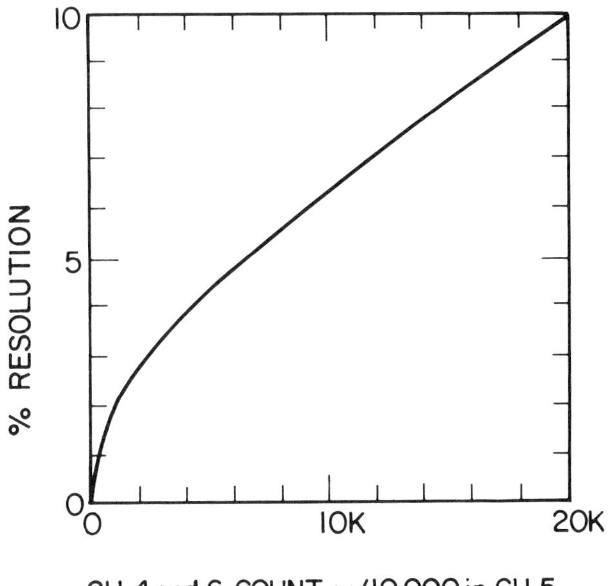

Figure 10. Conversion chart allows translation of percentage of side channels (4 and 6) to standard deviation.

channel. The accuracy of the centering is just the difference of side channels counts times a scale factor of μ/σ. For example, the side channels might read 3600 counts, each. They would be matched with an uncertainty of the square root of 3600, 60 counts or 1.67%. The accuracy scale factor for a worst case instrument at 5% would be 1.67%/20. The resulting small number, 0.083%, is much smaller than the uncertainty of the particle size, and an essentially perfect calibration is done.

To double check the calibration, three other channels are monitored, and the no-particle rms noise is measured. The three channels are below the calibration channels. The lower three, 1, 2, and 3, monitor the small debris that comes with the calibration particles. When the particles are turned off, and only clean air is in the sensor, channel 1 monitors the noise below the 0.3 μm threshold. It should vary in direct proportion to the rms noise. If it doesn't, there is non-random noise (interference), and that can be eliminated. Channel 2 monitors from the 0.3 μm threshold and above, and should have no counts when the particles are off. Counts there mean noise spikes which can be fixed, or dust in the sensor chamber which will be soon carried out. Dust can enter during assembly, or during turn-on and turn-off transients when the purge air flow is not smooth.

Calibration quality must be maintained in the field. A variation on the factory calibration is used for the field calibration technique. It is necessary for the instrument to have very narrow channel discrimination to achieve accurate calibration. The normal channels are ranged typically 0.3, 0.5, 1.0, 3.0, 5.0 and 10.0 μm. These are too broad for calibration. A simple circuit extender card is used to connect the normal logic threshold card. The extender card has switches and a resistor network to create narrow channels around the size of the calibration particles. For example, with 2.02 μm particles, the channels could be set at 1.72, 2.02, and 2.32 μm. The number of particles coming from the particle generator is measured at the "greater than 1.72" level. That will include virtually all of the 2.02's. Then the count at "greater than 2.02" is set to one half of the previous count. When half of the particles fall below 2.02, and half above, the sensor is calibrated. The upper channel is used to be sure that there are not a lot of doubles or other operational problems. Note that the particle generator must remain constant during the test. Particle generators do vary, and it is necessary to repeat the test several times to get the highest accuracy (better than 1%). Again, however, the μ/σ factor improves the accuracy. If the aerosol generator varies 20% and μ/σ is 1/20, the location of μ is accurate to about 1%. Thus, for the price of an inexpensive circuit card and a particle generator, a prime calibration can be done in the field.

REFERENCE

Cooke, D. D., and M. Kerker, 1975: Response calculations for light-scattering aerosol particle counters. *Applied Optics, 14,* 734–739.

THE AERODYNAMIC SIZE CALIBRATION OF OPTICAL PARTICLE COUNTERS BY INERTIAL IMPACTORS

VIRGIL A. MARPLE

Mechanical Engineering Department, University of Minnesota, Minneapolis

At the University of Minnesota, we have been using optical particle counters to size nonideal aerosols (i.e., light absorbing, irregular particles). However, to obtain reliable data, the optical particle counters must be calibrated on the specific material which they will be sizing. Also, since the main reason for studying these aerosols is to determine the quantity of respirable particles present, a calibration in terms of the aerodynamic or Stokes' diameter is preferred. Thus, a method using an especially designed calibration impactor was developed, which provides a calibration of an optical particle counter based upon either the Stokes' or aerodynamic diameter (Marple and Rubow, 1976).

To obtain a single calibration point, two runs are made with the optical particle counter. For one of the runs, the calibration impactor is attached to the inlet of the optical particle counter, and for the other run, the impactor is removed. An analysis of the data from the two runs yields a calibration point corresponding to the 50% cut point size of the impactor. By changing the size of the impactor nozzle, different calibration points can be obtained.

CALIBRATION IMPACTOR

The calibration impactor is shown in Figures 1 and 2. Basically, the impactor consists of a body, impaction plate, cover, and inlet nozzle. Two bodies were designed so as to make the calibration impactor easily adaptable to any instrument inlet (body A) or to any interfacing plenum chamber (body B). The inlet and cover of the impactor have been made into two parts so that inlets with nozzles of different diameters can be easily interchanged to change the particle cutoff point of the impactor. The cover can be removed from the body for easy access to the impaction plate. The inlet shown on the impactor in Figure 1 is

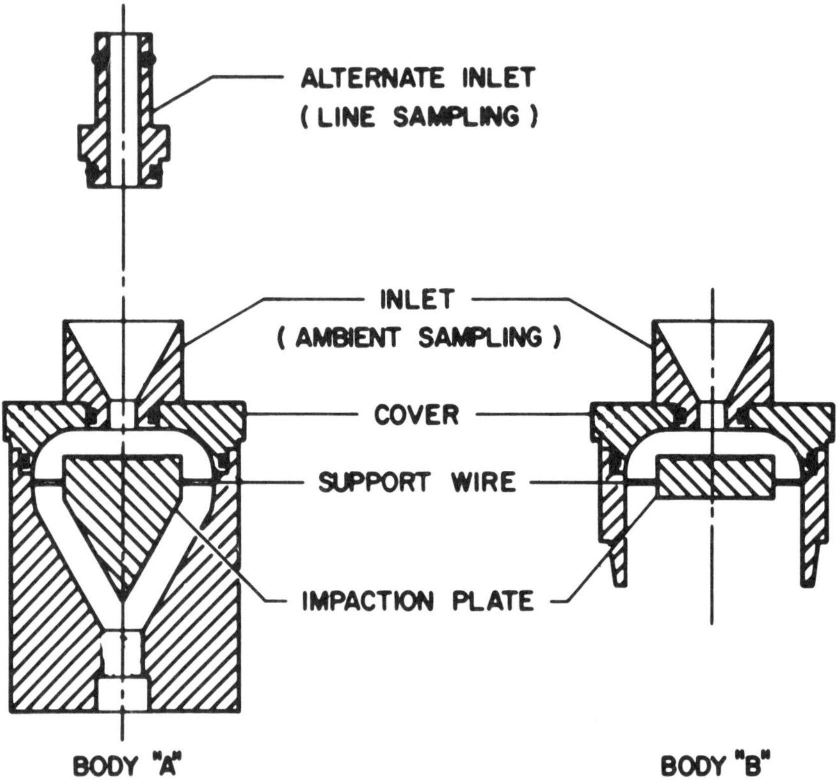

Figure 1. Schematic of calibration impactor.

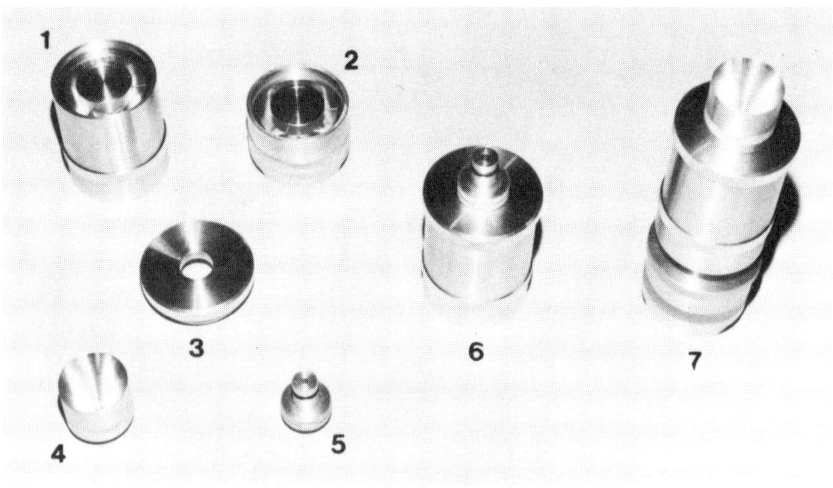

Figure 2. Calibration impactor. 1. Body "A" with impaction plate. 2. Body "B" with impaction plate. 3. Cover. 4. Inlet nozzle (ambient sampling). 5. Inlet nozzle (line sampling). 6. Assembled single impactor. 7. Assembled double impactor.

generally used when sampling from the ambient surroundings. The alternate inlet is used whenever particles must be transported to the impactor by tubing, or when two impactors are used in series. For the latter case, the alternate inlet serves both as a nozzle for the second impactor and as a connector between two impactors, since the I.D. of the recess at the exit of impactor body A is the same as for the O.D. of the alternate inlet. For either inlet, the impaction characteristics of the impactor will be identical as long as the nozzle diameters are the same. Also for all nozzles used, the criterion of S/W ≥ 1.0, necessary for the cutpoint

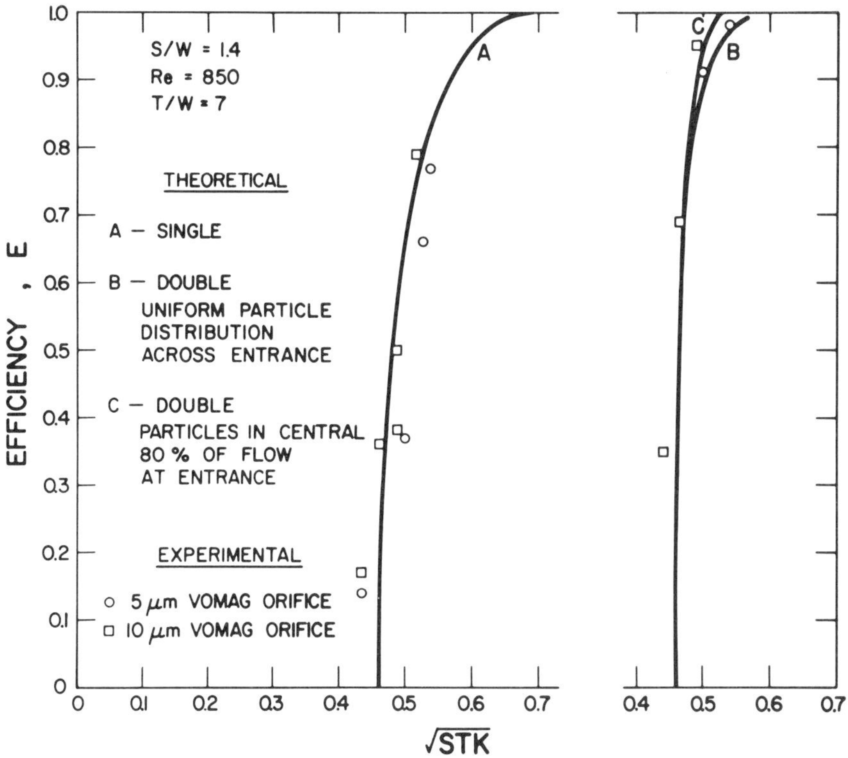

Figure 3. Comparison of theoretical and experimental efficiency curves for the single and double calibration impactor.

to be insensitive to small variations in S, has been followed (Marple and Liu, 1974), where S is the nozzle-to-plate distance and W is the nozzle diameter.

Studies of inertial impactors have shown that if impactors are properly designed and operated, the cutoff characteristics of an impactor can be theoretically predicted (Marple and Liu, 1974; Rao, 1975; Jaenicke and Blifford, 1974). Also, the sharpness of cut between the aerodynamic size of particles which are collected and those which are not is quite sharp. For example, the theoretically predicted calibration curve for the calibration impactor is shown as curve A in Figure 3. The experimental data points shown in Figure 3 substantiate the validity of the theoretically predicted calibration curve for that impactor. The details of this calibration are given in Marple and Rubow (1976).

The curves B and C in Figure 3 represent a range of combined calibration curves that would be expected if the aerosol is passed through two impactors in series with identical nozzle diameter, and thus, cutoff points. For curve B, it was assumed that there was a uniform particle distribution across the entrance entering the second impactor, and for curve C, it was assumed that the particles were in the central 80% of the flow at the entrance to the second impactor. The concentration of the particles in the central portion of the flow of the second impactor would be due to particle inertial effects at the entrance. It can be seen in

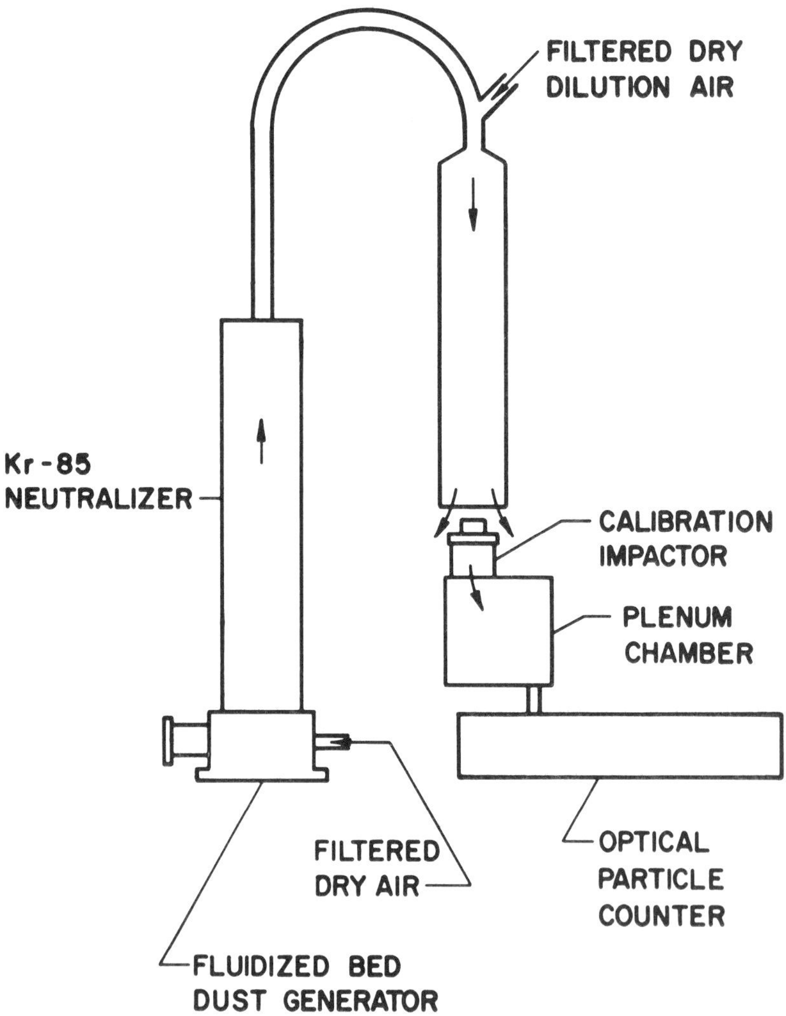

Figure 4. System for laboratory calibration of optical particle counters with calibration impactors.

Figure 3 that using two impactors in series does give a combined efficiency curve which is sharper than using a single impactor. However, for the purposes of calibrating an optical particle counter, the sharpness of cut of a single impactor is sufficiently sharp for a good calibration.

CALIBRATION OF OPTICAL PARTICLE COUNTERS

The testing facility for the laboratory calibration of the Optical Particle Counter (OPC) is shown in Figure 4. The dust is generated in a fluidized bed dust generator which disperses the dust particles as individual particles with essentially no agglomerates. Incorporated in the dust generator is a Kr-85 source which ionizes the air and neutralizes the particles' static charge (Whitby and Liu, 1968). The particles then pass through the calibration impactor and into the OPC being calibrated. A plenum chamber is placed between the impactor and OPC to allow the particles to become thoroughly mixed before reaching the OPC.

To obtain a calibration point for the OPC, two runs must be made. One is made with the OPC sampling through the calibration impactor and the other is made with the OPC sampling through just the inlet (i.e., no impaction plate in the impactor body). The data from the OPC during these runs is recorded by a multichannel analyzer (MCA), and must be plotted as particle concentration vs. the voltage levels of the signals from the OPC corresponding to specific MCA channels. Next, the particle concentration for the impactor case is divided by the particle concentration for the inlet-only case for the voltage level of each MCA channel and the resulting particle concentration ratios, R, are plotted vs. the OPC voltage output. Two such curves shown in Figure 5 vary from 1.0 for particles too small to be collected in a calibration impactor, to zero for particles large enough to be collected with 100% efficiency. The voltage level where the particle concentration ratio is 0.5 will correspond to the 50% cutoff size of the impactor.

If the particle concentration for the calibration impactor run should be different from that of the inlet-only run, the ratio of the concentration for particles too small to be collected will differ from 1.0. However, by adjusting this final curve by a constant factor such that the concentration ratio will be 1.0 for small particles, a correct calibration can still be obtained.

The two curves shown in Figure 5 illustrate the ratio R vs. the OPC output voltage for two different types of aerosol particles. The calibration of the Bausch and Lomb 40-1A OPC is with polystyrene latex (PSL) particles, which are considered to be ideal transparent spheres. The calibration point of 1.8 volts corresponds to a 5.80 μm diameter PSL particle. The calibration of the Royco 218 OPC is with coal dust particles, which are light absorbing and irregular in shape. However, even for this case, the calibration point of 0.42 volts corresponding to a 2.27 μm diameter cut size of the impactor is easily found. Since the density of the coal particles is 1.45 gm/cc, the aerodynamic diameter of this calibration point is 2.73 μm.

To determine the validity of this calibration technique, polydisperse 6 μm to 14

μm diameter PSL particles were aerosolized and used with a single impactor to obtain three calibration points for a Bausch and Lomb 40-1A OPC. These calibration points were then compared to a calibration curve which had been determined previously by using a vibrating orifice monodisperse aerosol (VOMA) generator to generate monodisperse DOP aerosols (Berglund and Liu, 1973). The comparison, Figure 6, shows that the calibration points determined with the impactor are in excellent agreement with the previously determined calibration curve.

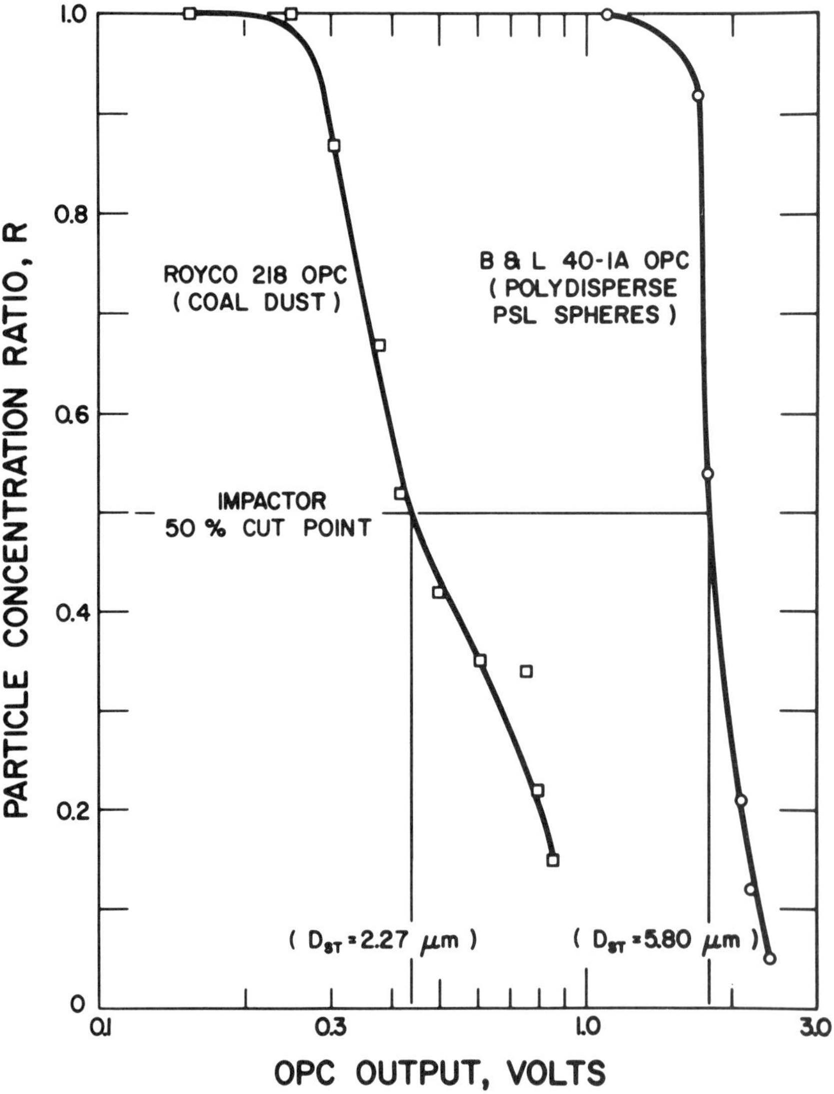

Figure 5. Typical particle concentration ratio curves.

Having demonstrated that the single impactor can be successfully used with ideal aerosols, it was then used to calibrate a Royco Model 218 OPC with coal dust. The resulting calibration curve is shown in Figure 7. Eight calibration points were obtained by using eight different size nozzles on the impactor. Details of the inlet sizes are given in Marple and Rubow (1976). The flow rate through the Royco 218 OPC was 2 ℓpm, which is larger than normal due to the incorporation of a sheath air inlet (Liu, Berglund and Agarwal, 1974) into the flow system.

Besides the coal calibration curve in Figure 7, a calibration as determined with

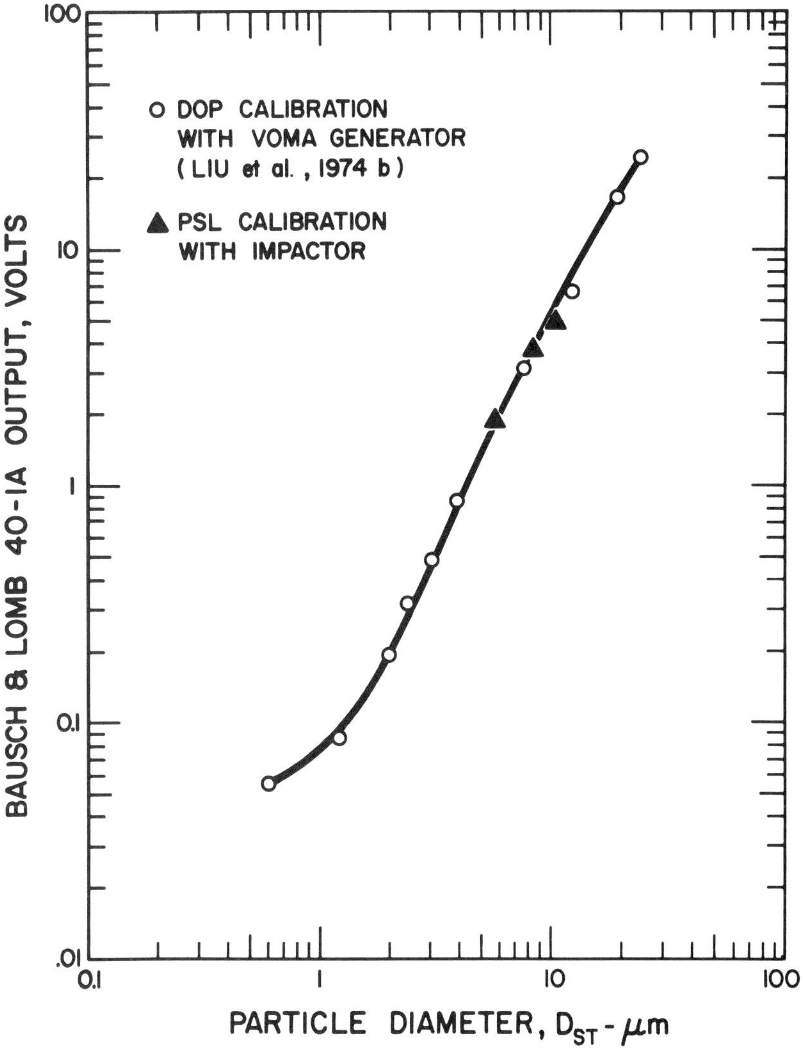

Figure 6. Calibration curve for a Bausch & Lomb 40-1A optical particle counter.

monodisperse ideal aerosols is also shown. Figure 7 shows that both calibration curves are nearly identical at sizes larger than about 6 μm. At the smaller sizes, the two efficiency curves are different primarily due to a dip being found in the calibration curve for the ideal aerosols. A dip of this nature in optical particle counters with forward light-scattering optics is expected for ideal aerosols with indices of refraction, m, of 1.46 and 1.6 (Liu, Berglund and Agarwal, 1974; Lieberman and Allen, 1969). The severity of the dip is a function of the particle's index of refraction, with the dip being more severe for particles of m = 1.6 than for particles of m = 1.46. For light absorbing coal particles the dip is not seen in the calibration curve. This also was predicted by Lieberman and Allen (1969), who determined the general shape of calibration curves from theoretical results of Hodkinson and Greenfield (1965).

CONCLUSIONS

An impactor has been developed and used to calibrate optical particle counters on irregular particles. By using an impactor for the calibration, the optical particle counter will have a calibration curve which corresponds to the aerodynamic or

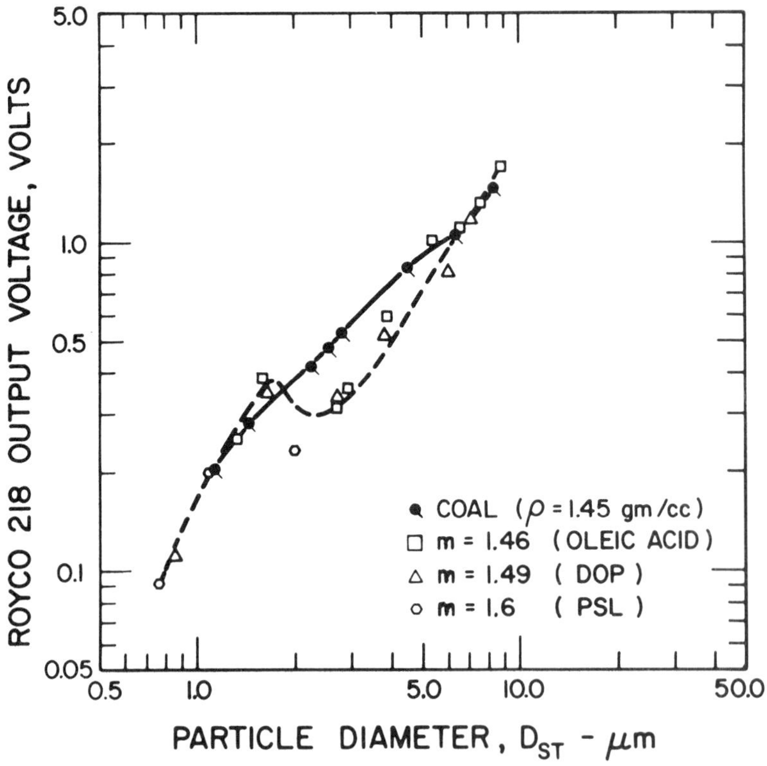

Figure 7. Calibration curves for Royco 218 optical particle counter.

Stokes' diameter of the particle. Also, since this calibration technique merely requires that two runs be made with an optical particle counter, one run sampling through the impactor and the other run sampling without the impactor on the inlet, the entire calibration curve can be generated in the field on the same aerosol which the optical particle counter will be sizing. Thus, there needs to be no artificial generation of the aerosol in the laboratory to obtain the calibration.

REFERENCES

Berglund, R. N. and B. Y. H. Liu, 1973: Generation of monodisperse aerosol standards. *Environ. Sci. Technol., 7,* 147–153.

Hodkinson, J. R. and J. R. Greenfield, 1965: Response calculations for light-scattering aerosol counters and photometers. *Appl. Optics, 4,* 1463–1474.

Jaenicke, R. and I. H. Blifford, 1974: The influence of aerosol characteristics on the calibration of impactors. *J. Aerosol Sci., 5,* 457–464.

Lieberman, A. and R. J. Allen, 1969: Presented at Amer. Assoc. Contamination Control, May 19-22.

Liu, B. Y. H., R. N. Berglund and J. K. Agarwal, 1974a: Experimental studies of optical particle counters. *Atmos. Environ., 8,* 717–732.

———, V. A. Marple, K. T. Whitby, and N. J. Barsic, 1974b: *Amer. Indust. Hygiene Assoc. J., 8,* 443.

Marple, V. A. and B. Y. H. Liu, 1974: Characteristics of laminar jet impactors. *Environ. Sci. Technol., 8,* 648–654.

——— and K. Rubow, 1976: Aerodynamic particle size calibration of optical counters. *J. Aerosol Sci., 7,* Part 5.

Rao, A. K., 1975: An experimental study of inertial impactors, Ph.D. Dissertation, University of Minnesota, Particle Technology Laboratory Publ. No. 269.

Whitby, K. T. and B. Y. H. Liu, 1968: Polystyrene aerosols—electrical charge and residue size distribution. *Atmos. Environ., 2,* 103–116.

DESIGN CONSIDERATIONS OF THE TECHECOLOGY PARTICLE COUNTERS

LOUIS J. PETRALLI, JR.

TechEcology, Inc., Sunnyvale, Calif.

Particle counters will operate satisfactorily under ideal laboratory conditions but have a high failure rate when operated under normal field use. This high failure rate is caused by poor instrument design.

We concentrated on solving instrument operational maintenance and reliability problems when we designed the particle counters presently manufactured by TechEcology.

Typical reliability failures:

1. Mechanical damage caused by shock or vibration during shipping or handling.
2. Change of the internal secondary calibrate system without a change in the calibrate "status."

Typical operational problems:

1. Erroneous counts caused by local electrical line noise.
2. Counts caused by recirculating particles.

Typical maintenance problems:

1. Replacement of light source.
2. Field repair of instrument by non-technical help.

The TechEcology Particle Counters have been designed to eliminate these problems.

Secondary Calibrate System

A primary instrument calibration using particles of a known size is done at the factory. At the same time an internal system is set up so that a secondary calibration can be performed at any time. The secondary system should be able to measure any change in the instrument that would affect the primary calibration. The ideal secondary system should not by-pass or add any components to the overall system. In addition, calibration should be performed at the same light level as when counting small particles.

Mechanical approaches inherently add components to a ''calibrate system'' that are not part of the normal operating system.

Mechanical Calibrate Systems

One method utilizes a mechanical probe that is inserted in the light path and scatters light. The scattered light is measured and used to establish a secondary calibrate standard. This method has several technical and operational disadvantages.

1. Any contamination or other physical change of the mechanical probe will change the amount of scattered light and the calibration ''status''.
2. The mechanical probe is not part of the sensor during prime calibration.
3. The mechanical probe will also scatter many times more light than particles and does not represent a normal light level operating condition.

Another mechanical method uses fiber optics and a chopper to produce light pulses that are focused on the detector. This method completely by-passes the collection optics; in other words, the collection optical system could be removed and the instrument would still ''calibrate.'' This system also uses components that are not part of the normal operating system and operates at high calibrate light levels.

At least 50% of the field failures that normally occur are caused by a change in the secondary calibrate system because of shock or vibration during shipment. To prevent this failure we decided to eliminate the conventional mechanical calibrate system, and we developed an electronic system.

TechEcology Electronic Calibrate System

A small amount of diffused light from the sensor's lamp is optically focused on the solid state detector. The signal is amplified and the calibrate status indicated on a front panel meter. This status test completely checks the optical and electronic systems without using any new components. Any change in electronic gain, lamp intensity, or lens contamination can be detected and a proper adjustment made. By eliminating choppers, fiber optics or mirrors used in mechanical calibrate systems the reliability of the Particle Counter to survive physical misuse is greatly increased. The disadvantage of this system is that the signal to noise ratio is reduced.

Erroneous Counts

The particle counter's electrical output amplitude varies as a function of particle size, and the instrument's electronics detect the pulse height. Any electrical pulse of the right amplitude will be recorded as a particle.

Particle counters will record electrical noise that is either self induced or on the A. C. line. All particle counters should "zero count". In other words, with particle free air being sampled zero particles should be counted.

This problem is minimized by careful electronic design. Every instrument shipped by TechEcology must operate with a test of 500 volts of SCR line noise and perform a zero count.

Field Maintenance

Field maintenance is time consuming, expensive, and frustrating to users not disciplined in electronics, fluid flow, and optics. As a matter of fact anyone should be able to perform field maintenance if the instrument is properly designed.

To facilitate field repair of TechEcology's Particle Counters a unitized modular construction design was used. Various subsystems are used and can be replaced easily in the field. The only tool needed is a screw driver. The sensor, pump assembly, power supply, counter or front panel control can be exchanged in five minutes.

A good example of how careful design can reduce field repair time is the approach used in replacement of the TechEcology Sensor Lamp Assembly.

A mechanical assembly was designed that would allow movement of the lamp filament with respect to the lamp mount. Every lamp is adjusted during assembly so that the filament is optically in the exact same place for every lamp assembly.

The lamp used as a light source for the particle counter has a rated life of 1500 hours. When a lamp failure occurs, a new, factory pre-aligned lamp assembly can be installed in a few minutes. The user does not have to worry about re-focusing the filament at the "view volume," wherever that may be.

Past particle counter designs have concentrated on designing a new and better optical system. We used a standard optical system and developed a reliable instrument that will operate under field use and is easy to maintain.

AN EVALUATION OF THE CLIMET 208 AND ROYCO 220 LIGHT-SCATTERING OPTICAL PARTICLE COUNTERS

WILLIAM E. CLARK* AND EDWARD L. AVOL**

Air Monitoring Center, Rockwell International, Newbury Park, Calif.

Two light-scattering optical particle counters, the Royco Model 220 and the Climet Model 208, were experimentally evaluated as part of an EPA sponsored program to install aerosol concentration and size distribution monitoring instruments into EPA Community Health Air Monitoring Program (CHAMP) field stations. The purpose of the evaluation was to provide comparative data between the two instruments for such performance characteristics as signal and noise level and particle size resolution ability. The experimental evaluation consisted of measuring the response of each instrument to a number of monodisperse particle sizes in the range from 0.3 to 6.0 μm in diameter using a 256 channel pulse height analysis system.

EQUIPMENT AND PROCEDURE

Royco Model 220

The Royco Model 220 optical particle counter manufactured by Royco Instruments, Inc., Menlo Park, California, utilizes a right angle light-scattering optical system to detect particles. Aerosol is continuously drawn through a viewing volume illuminated by a tungsten lamp. A photo-multiplier tube located 90° from the axis of the focused light beam detects the light scattered from the particles and provides a voltage pulse for each particle which passes through the viewing volume. The amplitude of the voltage pulse is a measure of the particle size. The particular instrument used in this evaluation was equipped to measure particles as small as 0.3 μm in diameter.

*Present Address: Environmental Engineering Department, California Polytech State University, San Luis Obispo, Calif. 93407.

**Present Address: Rancho Los Amigos Hospital, 7601 E. Imperial Highway, Downey, Calif. 90242.

The Royco 220 was evaluated as it was received from the factory and also after it had been modified by the addition of a sheath-air inlet system similar to the one developed at the University of Minnesota and described by Liu (1974). Figure 1 is a schematic diagram of this device. The sheath-air inlet replaces the inlet normally supplied with the instrument and operates by directing a portion of the aerosol through a filter to remove the particles and back around the remaining incoming aerosol stream as an annular sheath of clean air. Liu reports that this modification significantly enhances the performance of the Royco 220 by decreasing the size of the viewing volume so that higher concentrations can be measured, preventing the recirculation of aerosols into the viewing volume, increasing the resolution of the counter, and also increasing the signal-to-noise ratio. For this experimental evaluation the sheath-air inlet was operated with a total flow of $47.2 cm^3 sec^{-1}$ and an aerosol flow of $0.668 cm^3 sec^{-1}$.

Climet Model 208

The Climet Model 208 optical particle counter manufactured by Climet Instruments Co., Redlands, California, utilizes an elliptical mirror optical system to detect particles. The sample viewing volume is located at the primary focal point of the mirror and a photomultiplier tube is located at the secondary focal point. Light scattered by particles passing through the viewing volume is collected over a solid angle in the forward direction which extends from 15 to 105 degrees with respect to the incident light path. The photomultiplier tube provides a voltage pulse for each particle which passes through the viewing volume and the amplitude of the pulse is a measure of the particle size. The flow system of the Climet 208 is designed so that the particle passing through the viewing volume is surrounded by a sheath of clean air. As with the Royco counter, the particular Climet counter used in this evaluation was equipped with the 0.3 μm sensitivity option.

The Climet 208 was also evaluated using the Climet Model 294-1 dilution system. This dilution system, shown schematically in Figure 2, works in conjunction with the Climet 208 flow system to provide an additional clean air sheath around the incoming stream of aerosol. The dilution system is designed to provide an adjustable dilution rate of from 100 to 1 to 1000 to 1. For this evaluation the aerosol flow rate was adjusted to $1.67 cm^3 sec^{-1}$.

Pulse Height Analysis System

A Nuclear Data, Inc. Model 1100, 256 channel pulse height analyzer (PHA) was used to measure voltage pulse height distributions from the two optical counters. The PHA was equipped with a variable conversion gain which allowed pulse amplitudes to be measured with a voltage resolution of from 1.95 to 31.3 mv. The Climet 208 counter provided voltage pulses which had a rise time of about 2μsec. These short pulses could be fed directly to the PHA analog to digital converter. The Royco 220, however, provided pulses which had a rise time of

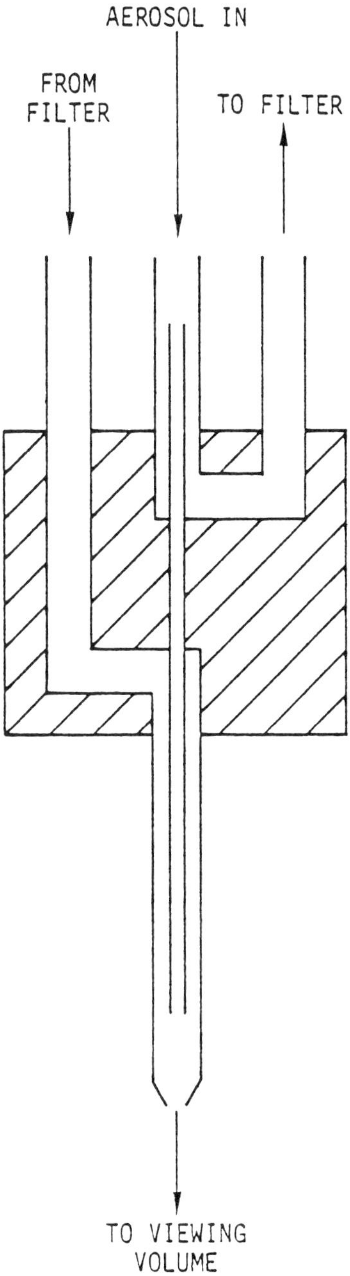

Figure 1. Schematic diagram of the sheath air dilution system for the Royco 220 counter.

about 30 μsec. For these longer pulses it was necessary to use the Nuclear Data Particle Sizing Amplifier (PSA) ahead of the analog to digital converter. The calibration of the pulse height analysis system depended to some extent on the input pulse width especially for low voltage pulses. For this reason the system was calibrated for each conversion gain used in the evaluation with pulses produced by a waveform generator which were shaped similar to the pulses provided by the Royco and Climet counters.

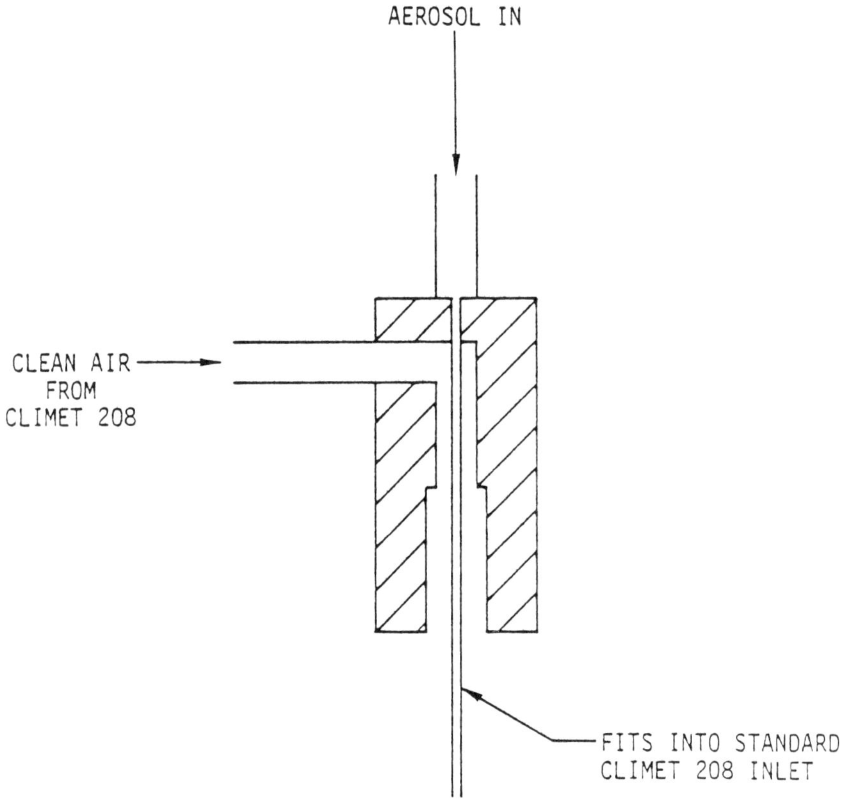

Figure 2. Schematic diagram of the Climet model 294-1 dilution system inlet.

Aerosol Generation

A Royco Model 256 aerosol generator was used to produce monodisperse polystyrene latex aerosol particles over a range of sizes from 0.3 to 2.0 μm. The Royco 256 generator incorporates the use of an atomizer to produce an aerosol from dilute aqueous solutions of uniformly sized polystyrene latex particles. In addition, a Berglund-Liu vibrating orifice monodisperse aerosol generator was used to produce two monodisperse particle sizes in the range from 2.0 to 6.0 μm in diameter. The Berglund-Liu generator described by Berglund (1973) produces a monodisperse aerosol by forcing a liquid solution of the desired aerosol material dissolved in alcohol through a vibrating orifice by a syringe pump at a predetermined rate. For this evaluation, monodisperse dioctyl phthalate (DOP) particles were produced from solutions of DOP in methyl alcohol.

Reduction of Experimental Data

Monodisperse aerosol particles were introduced to the optical counters, and the voltage pulse amplitude distribution was measured with the Nuclear Data pulse height analysis system. Figure 3 shows the pulse height distribution for the Climet counter sampling 1.01 μm particles. The horizontal axis of the display

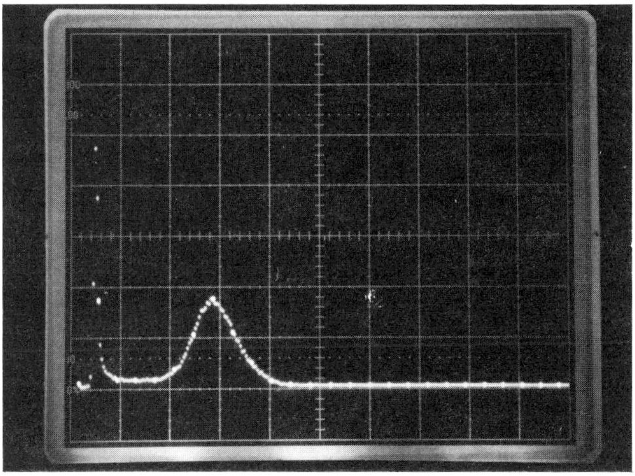

Figure 3. Pulse height distribution. Counter: Climet. Particles: 1.01 micron. Horizontal scale: ≈ 2mv/channel. Vertical scale: 800 counts/division.

shows the 256 channels of the PHA. For the particular conversion gain used for this run, each channel represents approximately 2mv and the full scale voltage is 0.5 volt. The vertical axis displays the number of pulses counted in a prescribed time interval for each voltage channel. The large number count in the first few channels represents instrument noise pulses.

The median voltage and the standard deviation of measured pulse height distributions was determined by plotting the cumulative frequency distribution for each run on normal probability graph paper and fitting the data with a straight line. Figure 4 shows the data displayed in Figure 3 plotted in this manner. The cumulative percent of the total count less than the upper voltage boundary of the indicated channel number was plotted against channel number. A straight line on this type of plot represents a normal distribution function. The pulse height distributions measured with both counters were approximately normal in shape over the central 95 percent of the distribution ($\bar{v} \pm 2\sigma$). The median and standard deviation of the pulse height distribution were determined by reading the median channel and the channel corresponding to 84% from the cumulative frequency plots and then applying the appropriate voltage vs. channel number calibration curve depending on the particular conversion gain utilized.

The pulse height distribution for the smaller monodispersed particles extended into the noise level for both counters. Data of this type was reduced by first subtracting a background noise count made with just water in the aerosol generator and then determining the cumulative frequency distribution over the central portion of the distribution extending essentially from the channel corresponding to the saddle point on the left side to a channel on the right side which had a similar number count.

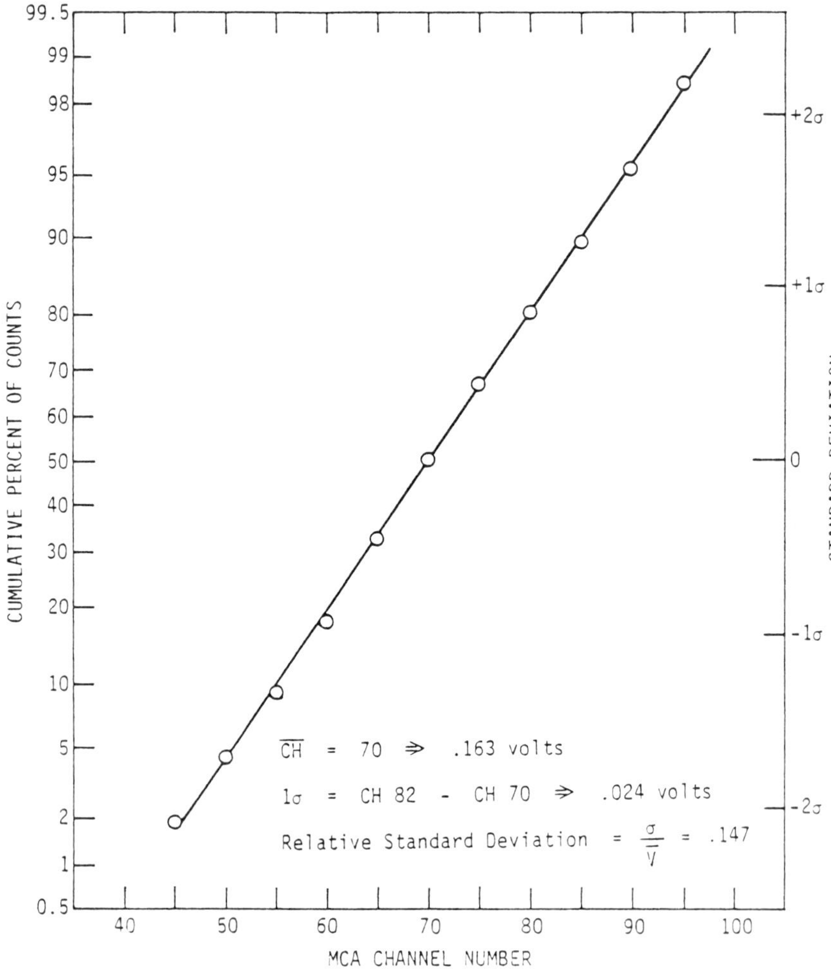

Figure 4. Cumulative frequency plot for Climet 208 counter and 1.01 μm particles.

RESULTS AND DISCUSSION

Pulse Shape

The Climet 208 counter produced pulses which were approximately Gaussian in shape and which had a pulse width of approximately 4 μsec at one half the pulse amplitude. The basic shape and width of the majority of the Climet pulses appeared to be fairly uniform over the complete range of the instrument. Some spurious pulses were also observed from the counter which had about the same amplitude as the 4 μsec pulses but which were from about 10 to 50 μsec in width. It is thought that these pulses resulted from particles recirculating through the edge of the viewing volume at a slower rate than the primary aerosol stream. The addition of the Climet dilution system to the basic Climet 208 counter did not reduce the frequency of the spurious pulses.

The Royco 220 counter produced pulses which were approximately Gaussian in shape but which had a width which depended on the pulse amplitude. The majority of the pulses measured had rise times from about 20 to 40 μsec with the larger pulses displaying the longer rise times. A considerable number of spurious pulses were also observed which had considerably longer rise times of about 100 to 200 μsec. As with the Climet counter, it is thought that the longer rise time pulses result from particles recirculated into the edge of the viewing volume. The frequency of the spurious pulses was significantly reduced when the Royco was operated with the sheath-air inlet system.

Pulse Amplitude vs. Particle Size

The median voltage pulse amplitudes for each monodisperse particle size for the Climet 208 counter are tabulated in Table I and shown graphically in Figure 5. The factory calibration curve shown is from the Climet 208 operation and maintenance manual. Experimentally determined points agreed with this curve to within ± 22%. Approximately the same median voltage amplitudes were measured for each particle size when the Climet counter was operated with the Climet Model 294 dilution system.

The median voltage pulse amplitudes for each monodisperse particle size introduced to the Royco 220 counter and the Royco modified with sheath-air inlet are tabulated in Table I and shown in Figure 6. The factory calibration points shown are for the specific instrument evaluated. The experimental data for the unmodified Royco agreed with the factory calibration at 2.02 μm. Below this size, however, the measured pulse amplitude decreased more rapidly than the factory data and deviated by a factor of 1.77 at 0.312 μm. Modification of the Royco counter with the sheath-air inlet resulted in increasing the median pulse amplitude measured for each particle size by 21 to 43 percent.

Resolution

In order to determine the resolution of the two optical counters, the relative standard deviation of the pulse amplitude distribution measured for each monodisperse particle size was calculated by dividing the standard deviation of

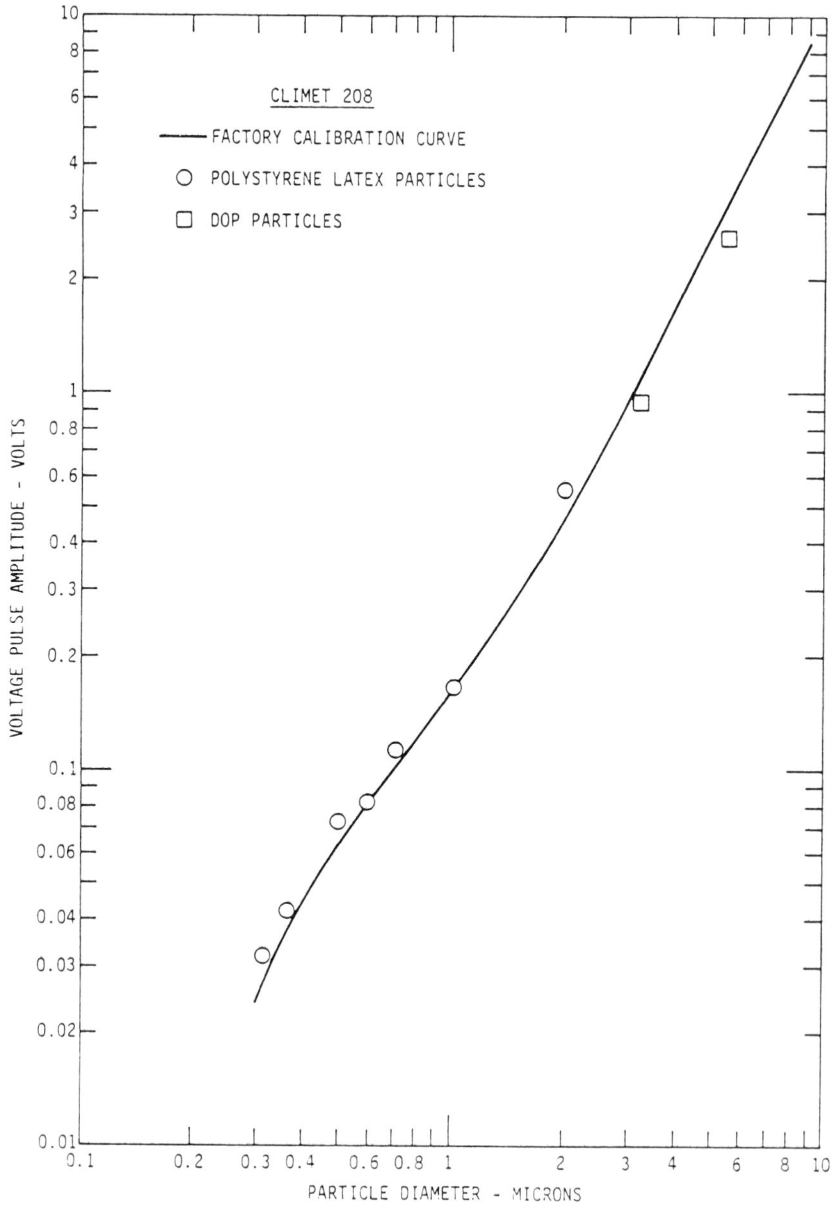

Figure 5. Voltage pulse amplitude vs. particle size for the Climet 208 counter.

the pulse height distribution by the median pulse height. Table II summarizes this data for both instruments. For both the Climet and the Royco counters the relative standard deviation was greater in value for the smaller particle sizes and progressively less in value for the larger particle sizes. The relative standard

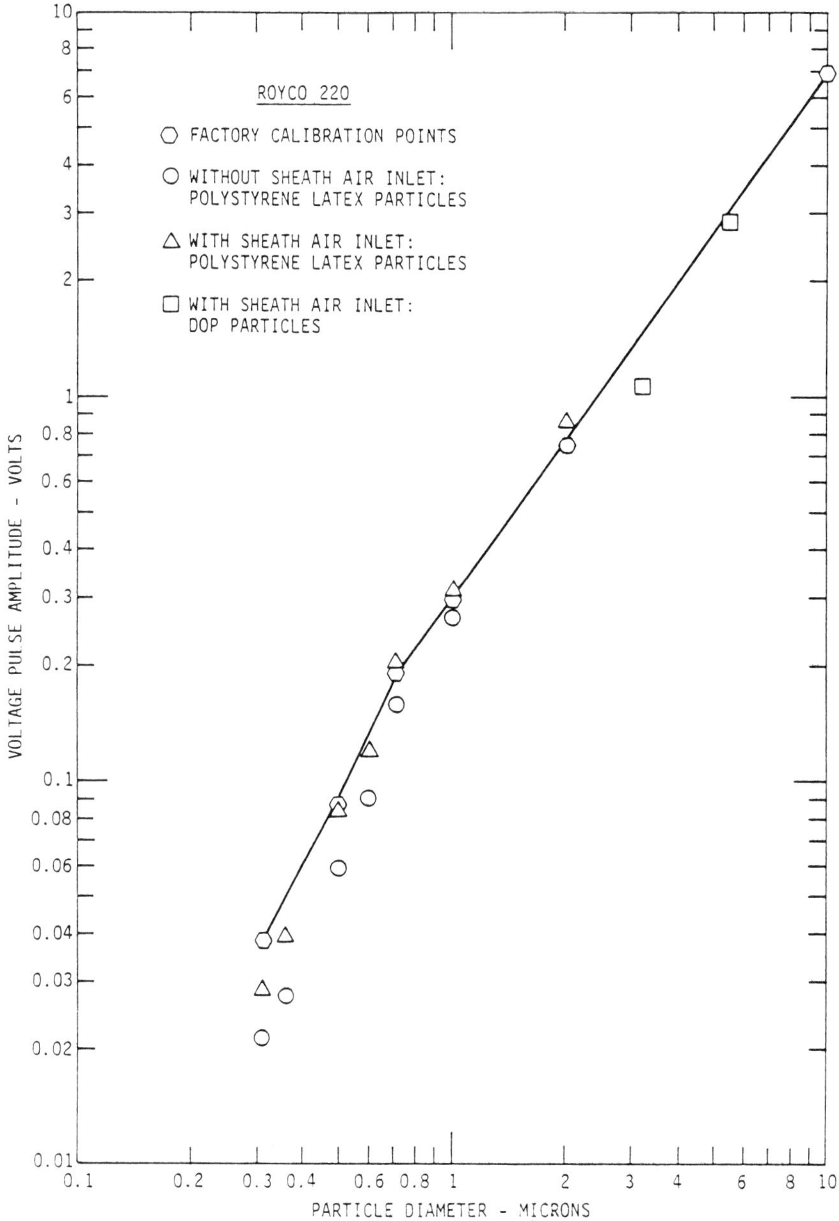

Figure 6. Voltage pulse amplitude vs. particle size for the Royco 220 counter.

deviation of pulse height distribution was approximately the same for the Climet counter when the instrument was operated with or without the Climet 194 dilution system. For the Royco counter, however, the relative standard deviation was reduced significantly when the instrument was operated with the sheath air dilution system.

TABLE I. Median Pulse Amplitudes for Monodisperse Particles Introduced to the Royco 220 and Climet 208 Optical Counters

Monodisperse Particle Size (μm)	Median Voltage (Climet 208)	Median Voltage (Royco 220)	Median Voltage (Royco 220 with Sheath-Air Inlet)
0.312	0.032	0.0215	0.0288
0.364	0.042	0.0275	0.0392
0.5	0.072	0.0590	0.0820
0.6	0.081	0.0895	0.119
0.714	0.112	0.158	0.202
1.01	0.163	0.261	0.315
2.02	0.542	0.640	0.862
3.23	0.942	—	1.06
5.49	2.56	—	2.82

TABLE II. Relative Standard Deviation of Pulse Amplitudes for Monodisperse Particles for the Climet 208 and Royco 220 Counters

Monodisperse Particle Size (μm)	Relative Standard Deviation of Pulse Amplitude			
	Climet 208 (%)	Climet with Climet 294 dilution system (%)	Royco 220 (%)	Royco with sheath-air inlet (%)
0.5	23.6	23.4	27.0	15.9
0.6	22.2	—	25.3	13.4
0.714	18.8	17.3	24.7	9.90
1.01	14.7	13.8	22.8	9.21
2.02	7.38	7.69	20.7	6.73
3.23	5.40	—	—	5.66
5.49	5.50	—	—	4.26

Noise Level

Particle-free air was introduced to the two counters and the voltage pulse height distribution of the noise signals was measured with the Nuclear Data system. Figure 7 shows the noise count distribution with voltage for both instruments for an accumulation time of one minute. Also shown on Figure 7 are the voltages corresponding to 0.3 μm particles as determined in this evaluation and shown in Figures 5 and 6 for the Climet, and the Royco with and without the sheath air inlet. The noise level for both instruments extends past the voltage pulse height corresponding to 0.3 μm particles. Table III shows the noise count rate for voltage pulses greater in magnitude than voltage pulses corresponding to 0.3, 0.325 and 0.35 μm particles. The noise count rate for the Climet 208 was considerably higher

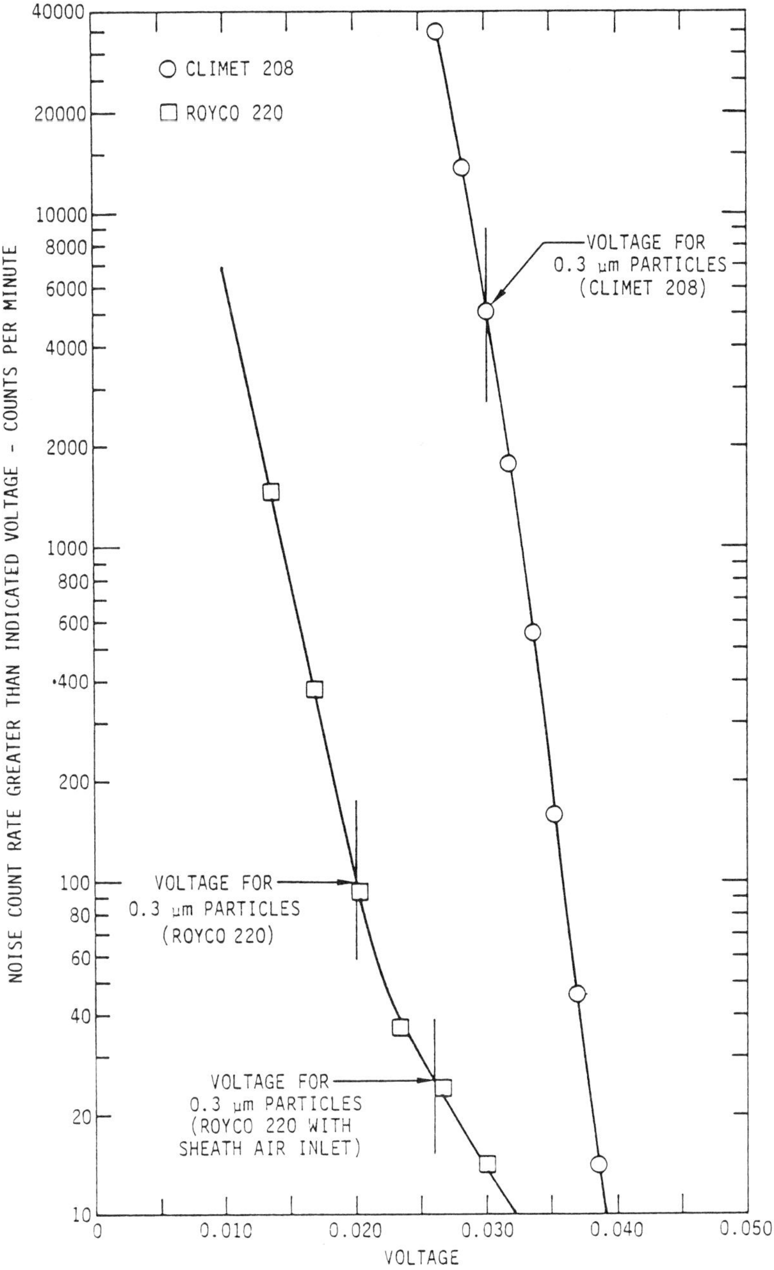

Figure 7. Noise count distribution vs. voltage for Climet 208 and Royco 220 counters.

than for the Royco 220 for a lower particle size limit of 0.3 μm. However, because the noise count rate curve for the Climet 208 as shown in Figure 7 dropped off sharply with the voltage pulse height, the noise count rate was approximately the same for the two counters for a lower particle size limit of 0.35 μm.

Although subtracting the noise count reduces the error associated with instrument noise, the noise is also superimposed on the voltage pulses resulting from particles passing through the viewing volume. In this regard, the higher noise level of the Climet counter evaluated in this study is reflected in the experimental

TABLE III. The Noise Count Rate of Voltage Pulses Greater in Magnitude than the Voltages Corresponding to 0.3, 0.325, and 0.35 μm Particles

Optical Counter	0.3 μm	0.325 μm	0.35 μm
Climet 208	5000 min^{-1}	400 min^{-1}	10 min^{-1}
Royco 220	100	18	25
Royco 220 with sheath-air inlet	25	12	<10

result that the resolution of the Climet counter for submicron particles was somewhat less than the resolution of the Royco counter operated with the sheath-air inlet.

Following this experimental study, it was discovered by the manufacturer that the photomultiplier installed in the Climet 208 used had been heated to an abnormally high temperature during the instrument assembly process. This may have been the cause of the high noise count rate associated with the particular Climet counter evaluated.

REFERENCES

Berglund, R. N. and B. Y. H. Liu, 1973: Generation of monodisperse aerosol standards. *Environ. Sci. Technol., 7,* 147–153.

Liu, B. Y. H., R. N. Berglund and J. K. Agarwal, 1974: Experimental studies of optical particle counters. *Atmos. Environ., 8,* 717–732.

AEROSOL GROWTH MEASUREMENTS USING THE DIFFERENTIAL II AND CLIMET 208 LIGHT SCATTERING SPECTROMETERS

I. N. TANG, H. R. MUNKELWITZ, AND J. G. DAVIS

Department of Applied Science, Brookhaven National Laboratory, Upton, N.Y.

The measurement of angular light scattering is a common method of particle-size determination from about 0.2 to over 10 μm. In general the light scattering method has the advantage of allowing particles to be sized and counted *in situ* without the necessity of sample collection and preparation. This is particularly desirable when sampling liquid droplets, where such collection and preparation may drastically alter the size and other physical properties of the original aerosols. In aerosol growth studies, light scattering techniques offer the further advantage of continuously monitoring the size changes of airborne particles undergoing phase transition, vapor condensation, or chemical reaction.

This paper describes our experience with two commercially available optical instruments, namely, the Climet 208 Particle Analyzer and the Differential II Particle spectrometer. It is shown that, with appropriate modifications and calibration, these instruments are ideally suited for laboratory studies of particle growth.

EXPERIMENTAL

Two different systems were designed and constructed. In the first, a continuous flow growth chamber enabled the size changes of a salt aerosol having a well defined initial size distribution to be investigated as a function of relative humidity. In the second system, a static growth chamber allowed the phase transformation and subsequent growth of a single particle to be followed at a predetermined relative humidity. A description of the two systems follows.

Work performed under the auspices of the U.S. Energy Research and Development Administration.

The Continuous Flow Aerosol Growth Apparatus

Figure 1 is a schematic of the continuous flow apparatus which incorporates a modified Climet 208 Particle Analyzer for monitoring particle size distributions. This instrument is in many ways different from the standard version. The basic elliptical mirror, light source and photo-multiplier sensor have been retained but the sheath gas recirculating system and electronics have been replaced as follows.

In order to insure a low aerosol background, both sample carrier gas and sheath gas are now obtained from the controlled boil off of liquid nitrogen. Gas

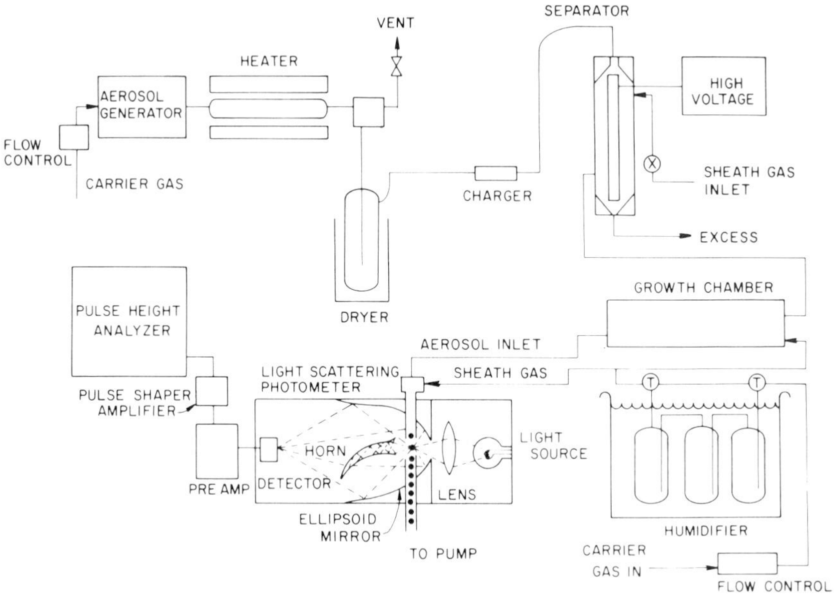

Figure 1. Schematic diagram of the continuous flow aerosol growth apparatus.

flow through the sensing volume is continuously monitored with precision flow meters and controlled with micrometer valves. A slightly negative pressure with respect to ambient is maintained in the measuring chamber by an external, continuously pumped ballast volume.

The output of the photomultiplier tube is fed directly into an external charge-sensitive preamplifier for conversion into voltage pulses. The pulses are subsequently amplified, shaped and sorted through the use of a linear amplifier and a 100 MHz pulse height analyzer. A 1024-channel data storage and retrieval system is employed to allow detailed, on-line analysis of the size distribution data. This technique, together with the associated electronic devices, is essentially standard practice in pulse-counting applications (Tang and Castleman, 1972, 1975) and has been employed by other investigators (Liu et al., 1974; Husar, 1974).

The response of the optical detector and associated electronics was characterized in the usual manner. First, the linearity of pulse amplification was established by substituting a variable pulse-height generator for the Climet photodetector. As shown in Figure 2, a plot of pulse height in mV vs. channel number for the Multichannel Analyzer gives a straight line. Next, the detector and electronics were calibrated to correspond to particles of a known size and refractive index. For this purpose, dilute aqueous suspensions of polystyrene latex spheres were nebulized, dryed and counted with the complete instrument. The observed relative response which is a function of refractive index and the cross-sectional area of the sphere was plotted vs. size for a number of different sizes. The result

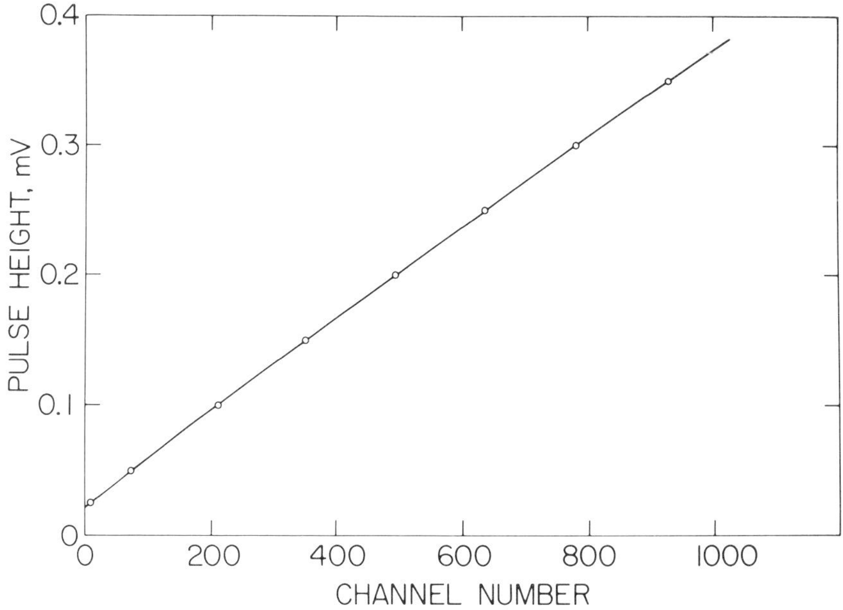

Figure 2. Calibration of pulse height analyzer.

of this calibration is shown in Figure 3 compared with data taken from Husar (1974) for a similar instrument. In order to make this comparison the data were normalized to give a relative response of 1.0 at a diameter of 1.0 μm. The solid line represents the calculated relative response expected from such a system and is in good agreement with both sets of data points.

In order to further characterize the system, the growth of sodium chloride aerosols by water vapor condensation was studied. Referring to Figure 1, the experimental apparatus consists of an aerosol generator, an electrostatic size-separator, a growth chamber and the modified Climet 208. Aerosols were generated from aqueous salt solutions by either nebulizing or bubbling. The droplets were passed through a heater and dryer where water was evaporated and removed. The dry gas stream containing the aerosol passed through a charger and subsequently into the mobility separator where the charged aerosols were sorted and a distinct size fraction selected either for observation in the dry gas or for reaction in the growth chamber.

The growth of sodium chloride aerosols with an initial size distribution peaked at 0.3 μm in diameter was studied. The onset of particle growth at 76% relative

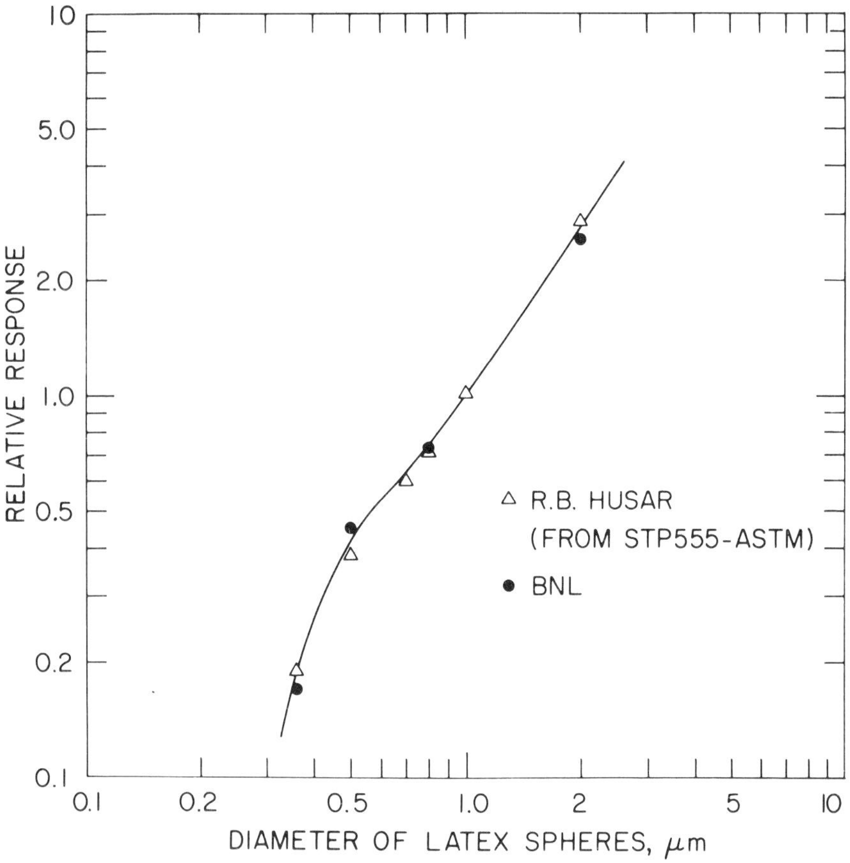

Figure 3. Relative response of the Climet 208 for latex spheres of different sizes.

humidity (R.H.) was well defined and the subsequent droplet growth with increasing humidity was found to be in good agreement with theoretical predictions. Figure 4 is a composite picture of the size distribution of a NaCl aerosol at various stages of growth. In each picture, the number of particles per channel is recorded on the axis at each channel of the Multichannel Analyzer, the channel number being proportional to particle size. The size distribution at 75.7% R.H. is identical to that of the initial aerosol produced in a dry atmosphere, indicating that no particle growth has yet taken place. The two peaks in the picture correspond to two discrete size ranges. The first peak represents particles of 0.3 μm in diameter, each carrying a single unit of electrostatic charge, while the second peak represents particles exactly twice the size of the first but each carrying two units of charge. Since the charge-to-mass ratio, and hence the electric mobility, was the same for the particles of these two size ranges, they were taken out as a group from the electrostatic size separator but resolved into different sizes by the optical particle counter.

The onset of growth at the phase transition point was observed at ~76% R.H. when the particle size distribution suddenly shifted to the right as shown in the

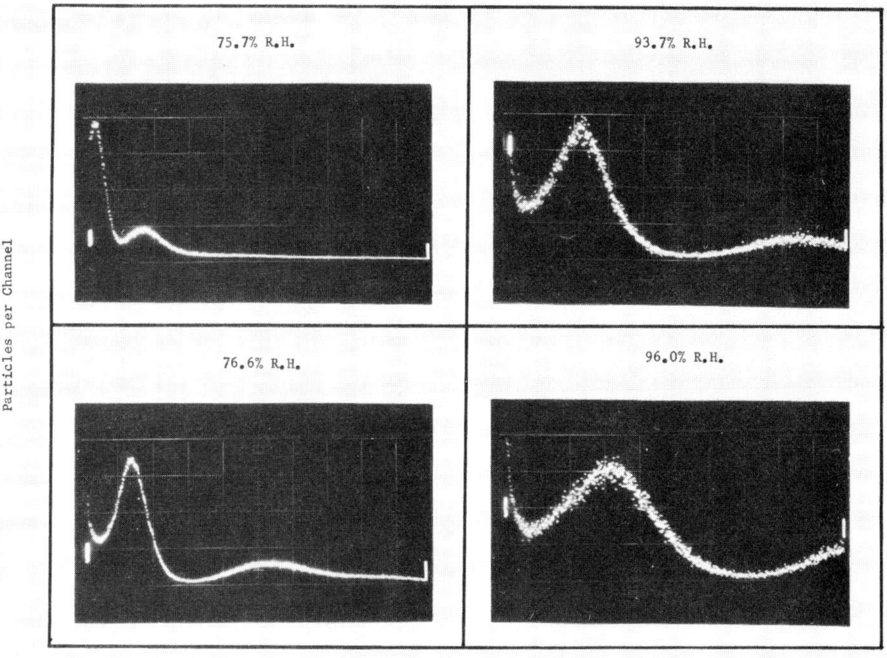

Figure 4. Growth of NaCl aerosol in a moist environment.

picture. The continuing growth of particles at higher humidities is also evident from the pictures taken at 93.7% and 96.0% R.H. as shown in Figure 4. The experiments allow a quantitative study to be made of the growth of aerosol particles as a function of relative humidity.

The Single Particle Growth Apparatus

The apparatus consists of an aerosol production and size-selection system similar to the one described above. However, a Differential II light-scattering spectrometer was used to observe the growth of a single particle exposed to a specific relative humidity. A schematic diagram of the light-scattering cell of the instrument is shown in Figure 5. Wyatt and Phillips (1972) have described the operation of the Differential II. In the present study, the light-scattering cell was modified to become an integral part of a growth chamber. This was accomplished by machining out an indentation on the bottom of the cell to accommodate a stainless steel dish containing a saturated aqueous salt solution.

Briefly, the experimental procedure is as follows: An aerosol stream from the electrostatic size separator is further diluted with carrier gas and injected into the light-scattering cell. A single particle which happens to drift into the laser beam is selected and maneuvered into the center of the electrical field by careful manipulation of pneumatic and electrostatic controls. After a few moments of such manipulation the chosen particle is trapped by the radial component of the electric field and other particles in the chamber are swept away because of a different mass to charge ratio and unfavorable position with respect to this field.

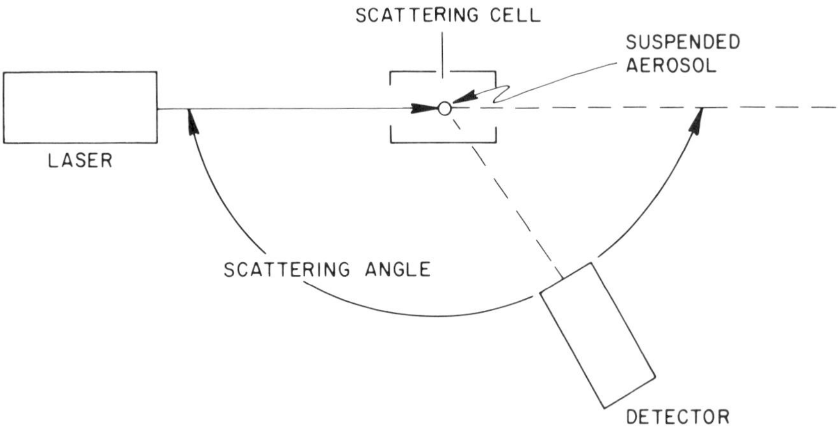

Figure 5. Schematic of single particle light scattering spectrometer.

Once an aerosol particle is properly positioned, a photoelectrically controlled levitation servosystem takes over and automatically holds the particle stationary during measurement.

The light scattered by the particle is measured as a function of angle with respect to the laser beam by the detector which traverses approximately 160° around the cell. In this manner, the size change of the aerosol particle is monitored continuously *in situ* during growth. Particle growth is deduced from the recorded angular dependence of the scattered light intensity. The relative humidity in the cell is determined by the water vapor pressure of a saturated salt solution above which the aerosol particle is suspended.

The effect of changes in relative humidity on the scattered light intensity from a single sodium chloride aerosol particle is shown in Figure 6. A deliquescent salt such as sodium chloride is ideally suited to such an experiment since the evaporation and condensation processes are extremely fast and the onset of growth is characterized by a sharp transition point at 76% R.H. In Figure 6, the first scan

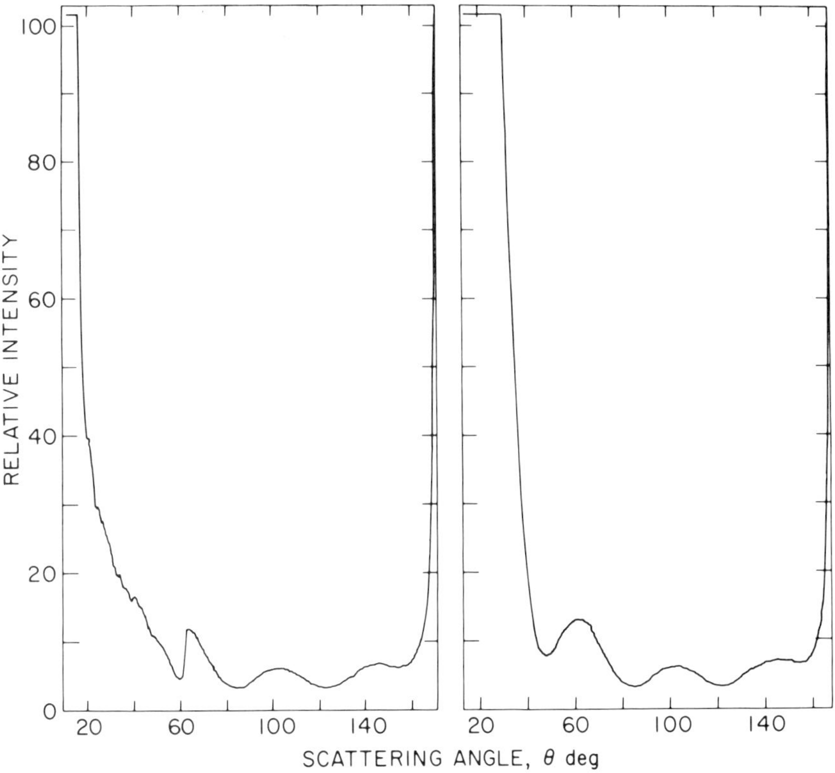

Figure 6. Successive scans of a single NaCl particle during and after growth at 79% relative humidity.

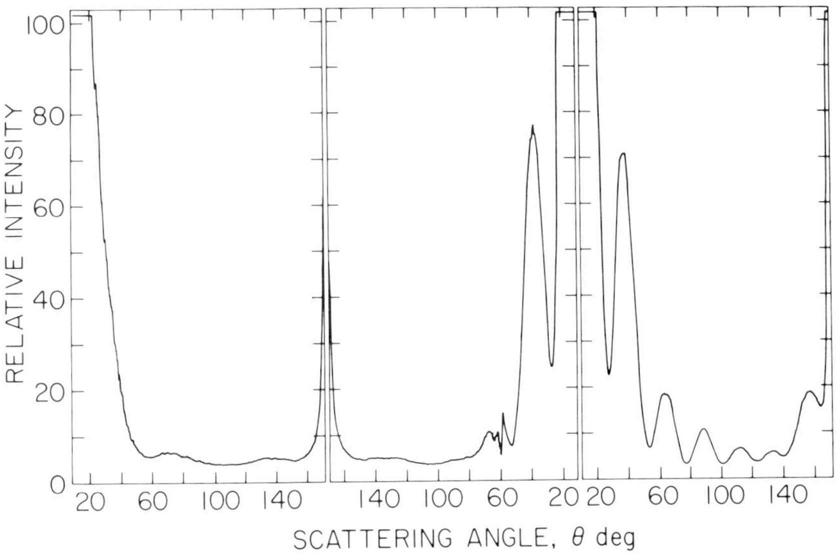

Figure 7. Successive scans of a single NaCl particle before, during and after growth.

shows that this transition was taking place when the detector moved to an angle of about 60° with respect to the incident laser beam. The sharp increase in relative intensity which was completed in about 0.8 seconds in this particular event serves to illustrate the rapidity of the change. The next trace shows the scattering pattern exhibited by this particle fully grown at 79% R.H.

The light scattering data from another single sodium chloride aerosol particle is shown in Figure 7. The first frame shows the initial aerosol in an environment where the relative humidity is less than 76%. In the following trace a sharp increase in the scattered light intensity is seen to occur at about 60°, indicating that the aerosol particle is undergoing a rapid size change. The final trace again shows the scattered light spectrum of the same particle completely grown at 79% R.H. If spherical shape is assumed for the initial aerosol an estimate of the size

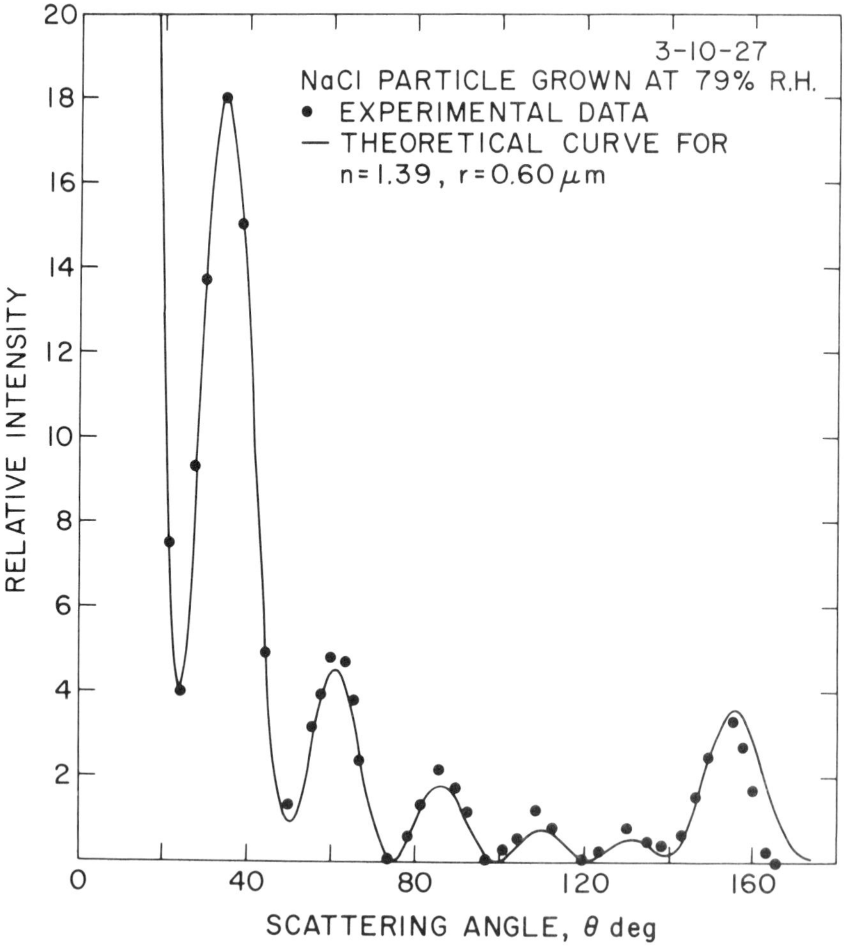

Figure 8. A preliminary comparison of experimental light scattering data with theoretical curve for droplet size estimation.

change can be made from Mie theory by comparing the initial and final traces.

Figure 8 shows light scattering data taken for a suspended NaCl aerosol grown from an initial size of 0.3 μm in radius to its equilibrium size at 79% R.H. The closed circles represent the background corrected light scattering data while the solid line represents the curve calculated from Mie theory for a spherical particle of refractive index 1.39 and radius 0.6 μm. The theoretical curve is taken from the Atlas of Light Scattering Curves (Wyatt and Phillips, 1972) and gives only a preliminary fit of the data for size estimation. A precise computer fit using the exact refractive index is expected to produce a better estimate of particle size required in growth studies.

CONCLUSION

Light scattering techniques are well suited to the study of liquid aerosol systems. Commercially available instruments such as the Climet 208 and the Differential II with appropriate modifications can serve as useful research tools for aerosol studies in the laboratory. While a study of the growth of sodium chloride particles is academic, the methods described in the present study may be readily applied to aerosol systems composed of mixed salts (Tang, 1976). Such multicomponent systems are characteristic of natural aerosols and a systematic investigation will prove to be fruitful in elucidating the behavior of these environmentally important aerosol systems.

REFERENCES

Husar, R. B., 1974: Recent developments in IN SITU size spectrum measurement of submicron aerosols. Special Technical Publ. 555, Amer. Soc. for Testing and Materials.

Liu, B. Y. H., R. N. Berglund and J. K. Agarwal, 1974: Experimental studies of optical particle counters. *Atmos. Environ., 8,* 717–732.

Tang, I. N., 1976: Phase transformation and growth of aerosol particles composed of mixed salts. *J. Aerosol Sci., 1,* 361.

——— and A. W. Castleman, Jr., 1972: Mass spectrometric studies of the gas-phase hydration of the monovalent lead ion. *J. Chem. Phys., 57,* 3638–3644.

——— and ———, 1975: Gas phase solvation of the ammonium ion in ammonia. *J. Chem. Phys., 62,* 4576–4578.

Wyatt, P. J. and D. T. Phillips, 1972: A new instrument for the study of individual aerosol particles. *J. Colloid Interface Sci., 39,* 125–135.

QUESTIONS

Q: How would you define the humidities in your test chamber to such a high degree of precision?

Munkelwitz: We're using saturated salt solutions to control the vapor pressure.

Q: What is the interpretation of the sudden build-up in light intensity?

Munkelwitz: There are two things happening, the refracted index is changing. It's going from the higher value of 1.55 to about 1.38. Also, the particles are becoming larger and this is what actually gives you the intensity change. In this particular case, there is about a factor of 2 change in size.

AEROSOL FIELD MEASUREMENTS USING LIGHT-SCATTERING PHOTOMETERS

RICHARD W. STOREY

NASA Langley Research Center, Hampton, Va.

Personnel of the Aerosol Measurement Research Branch in the Instrument Research Division at NASA Langley Research Center have been obtaining field measurements of aerosols during the past several years (Hulten et al., 1974; Gregory and Storey, 1975; Storey et al., 1974, 1975). The measurements have been obtained in the process of conducting several projects as follows: (1) Launch vehicle exhaust effluent monitoring, (2) local characterization of atmospheric particulates, and (3) tethered balloon-borne light-scattering particle counter. The objectives of these projects are, briefly stated, as follows. For (1) above, accomplish ground-level measurements of particulate concentrations and dosages along the direction of rocket exhaust effluent cloud travel and either side of the centerline. For (2) above, establish a mean level of air quality and deviation about this mean, ascertain diurnal changes or special events in air quality, and evaluate instrumentation and sampling procedures. For (3) above, obtain measurements of particulates in the inversion and boundary layer using a tethered balloon-borne light-scattering particle counter.

In accomplishing the objectives of the first two projects many different particle measuring instruments were used. Among these were the quartz crystal mass monitor, high-volume samplers, multi-stage impactors, integrating nephelometer, particulate concentrator, Nuclepore filters and light scattering particle counters. The light scattering particle counters are the instruments that will be discussed in this paper.

INSTRUMENTS

Two commercially available instruments were used for the first two projects. They were the Climet CI-201 sensor with data acquisition system and the Royco model 220 particle analyzer with data acquisition system.

The Climet particle measuring system consists of three units (Figure 1): CI-201 particle analyzer, CI-209 counter printer, and a model 1600 incremental magnetic tape recorder. In operation, the aerosol or particle is drawn into the view volume (Figure 2) by a vane-type pump which provides a constant air flow rate through the view volume. In addition, the Climet has a sheath air system to maintain the particles in the center of the view volume and to prevent them from drifting back into the view volume. The light scattered in the forward direction by the particle is focused by an elliptical mirror onto a photo-multiplier, the unscattered light passing through the view volume is absorbed by the cone. The

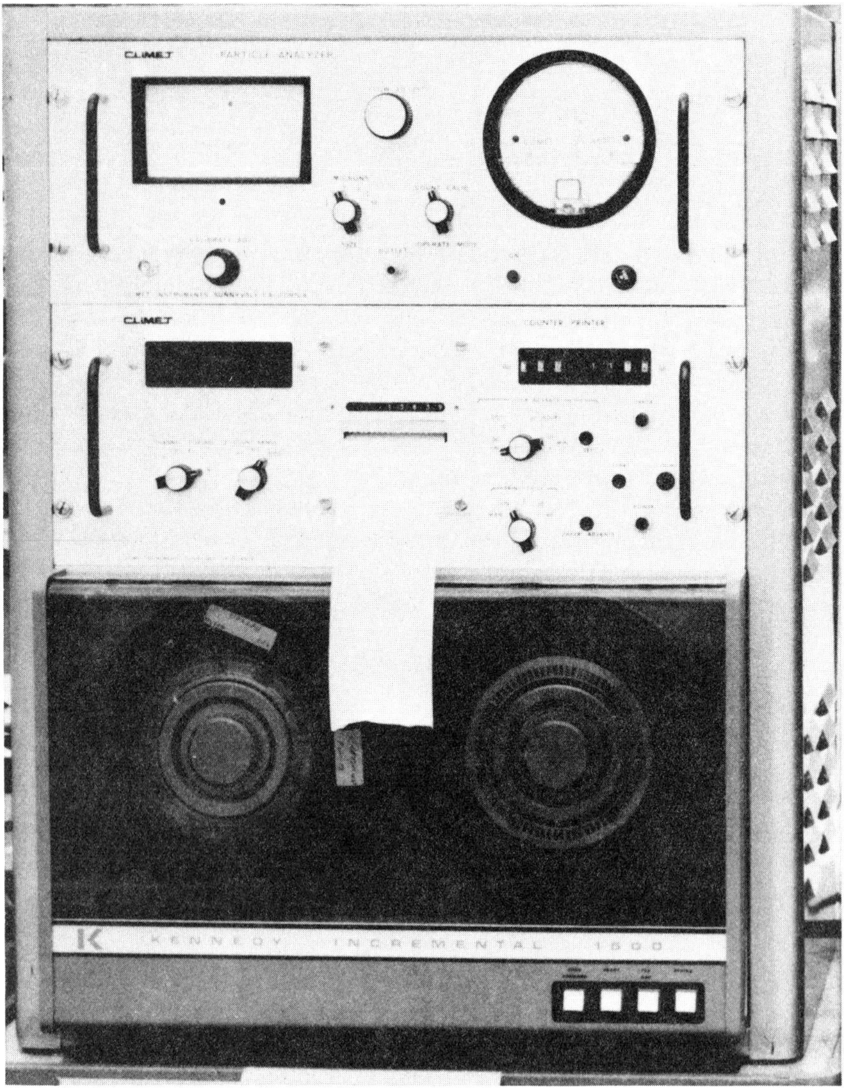

Figure 1. Climet particle measuring system.

photo-multiplier output current pulse is determined by the size of the particle. The output pulses are fed into a counter-printer which divides the particles into six size ranges, five of which are from 0.3 to 10 μm, with the sixth range being 10 μm and greater. The size range can be increased by a factor of 10 switch in the sensor preamplifier. The number of particles in the six size ranges (channels) are

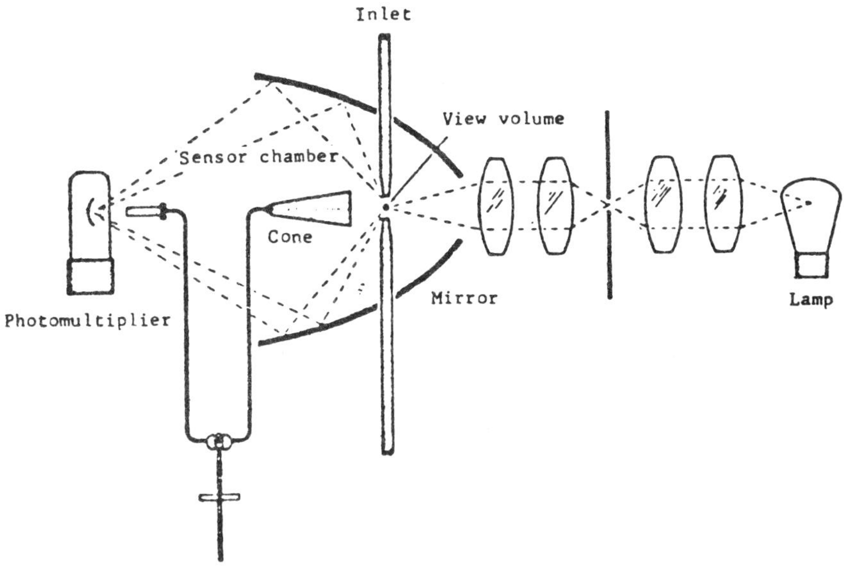

Figure 2. Climet optical system.

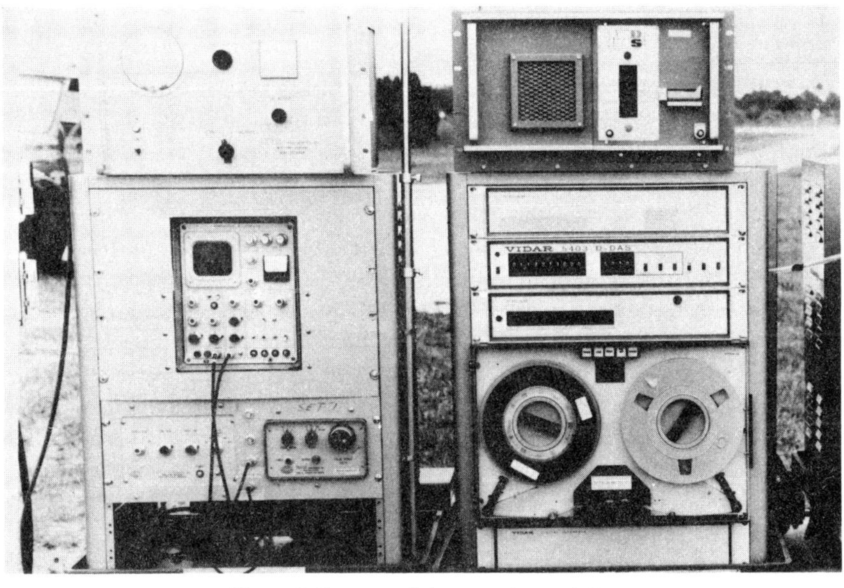

Figure 3. Royco particle measuring system.

printed out on a paper tape and a magnetic tape recorder to yield number of particles per size range.

The Royco particle measuring system (Figure 3) consists of a Royco model 220 sensor, pulse converter, pulse height analyzer, and pulse generator. The data acquisition system consists of a digital interface, controller, paper tape printer, seven-track magnetic tape recorder, digital clock, and power supplies. In operation, the aerosol or particle to be measured is drawn in at the intake to the model 220 sensor and passes through an illuminated view volume. Light scattered at 90° by a particle is focused onto a photomultiplier by a lens-slit configuration (Figure 4). The photomultiplier then yields a current pulse determined by the size particle occupying the view volume at that instant. The output pulses are fed into the pulse converter for conversion to the proper pulse shape for acceptance by the pulse height analyzer, which sorts, counts and displays them. The particle size range of the instrument is 0.5 μm diameter to 6.5 μm diameter; this size range is divided into 100 channels by the pulse height analyzer (PHA). The pulse generator generates various level (amplitude) pulses at a rate of 120 hertz to calibrate the pulse height analyzer.

The output from the pulse height analyzer is fed to a digital interface and then to a controller where it is recorded simultaneously on the paper tape digital printer and the seven-track magnetic tape recorder. Thumb wheels on the controller permit manual input of various test parameters to the two recorders. Real time from the digital clock is also recorded on both printer and tape recorder.

FIELD MEASUREMENTS

Launch Vehicles

The measurement of particulate fallout from rocket exhaust clouds requires the ability to deploy or re-deploy the measuring stations at times as close as 2 hours prior to launch. For a land deployment, the instruments are moved in a van and placed on pallets. However, the need to re-deploy on short notice resulted in configuring the systems so that they could be operated directly from the vans. If, as is the case quite frequently, the cloud went to sea, then the instruments are deployed on sea craft with all of the problems associated with instrument protection, such as salt spray, accelerations due to rough water, etc.

A typical deployment of measuring stations is shown in Figure 5, where station R had a Climet light-scattering photometer, station S a quartz crystal mass monitor and station 1808 the Royco light-scattering photometer. All of the sites were downwind of the predicted cloud path. Figure 6 shows the Climet light-scattering photometer measurements before, during, and after launch plotted in terms of mass loading in each of five channels, where each channel represents a specific particle size. The number of particles is converted to mass by assuming a density of 2.3 g/cc (for Al_2O_3), multiplying by the mean volume of each particle in a size interval, and the number of particles in that size. The right-hand side of the curves are plotted in more detail for the times during and after which the cloud, moving with a predicted path direction of 215°, was expected to be over site R (2.76 km at 225°). Since no significant change in the data occurred, it was concluded that the cloud did not pass over site R.

Figure 7 shows the mass monitor results from site S, 3.18 km from the launch point in a downwind direction of 209°. The time change of the mass loading (μg/m^3) indicates the cloud arrival at site S approximately T + 14 minutes from launch, as indicated by a greater than factor 3 increase in mass. The model predicted the cloud would arrive 3-km downwind on the 209° path at T + 13 minutes after launch. The natural atmospheric background before launch was in the range of 20 to 60 μg/m^3. Elemental analysis of the particles collected on the

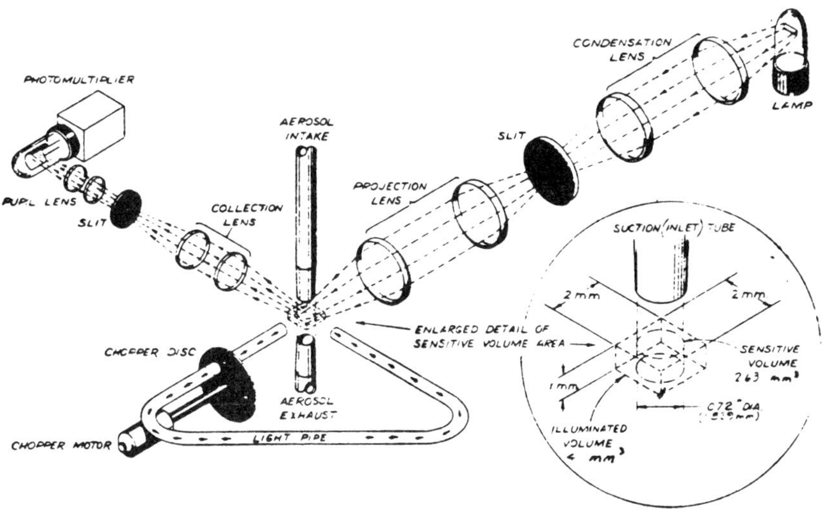

Figure 4. Royco 220 sensor optical system.

mass monitor crystal show significant amounts of aluminum, which is known to be a major constituent of the exhaust effluent.

Figure 8 shows the Royco light-scattering photometer measurements conducted at site 1808, 5.25 km from the launch point at 217°. The 100 channels of data are grouped into four major particle size ranges and the mass loading calculated and plotted as a function of time. The arrival of the cloud at site 1808, approximately 23 minutes after launch, is clearly evident by the change in particles which, it is to be noted at this site, was much more marked in the smaller sizes. Measurements were made for an hour after launch, although by that time the air at the site (as seen from the plots) had not yet returned to normal prelaunch background. Prelaunch model prediction data indicated that the cloud would be 5-km downwind on the 215° course at T + 23 minutes. Figure 9 shows the mass loading data from the Royco as a function of time, as determined from the total number of particles measured in the size range of the instrument (0.6 to 6 μm).

The Royco light-scattering photometer was used to monitor a Scout rocket launch at Wallops Island, Va., and provided the only real-time particle-size distribution data during launch, as shown in Figure 10. The time shown for each curve is the start of a particle sampling collection interval of 1 minute. The solid curve at time 11:07:54 is representative of prelaunch background conditions where no appreciable change in the natural background data was noted for several hours before the launch. The size distribution curve for the time 11:10:32 shows a sudden radical change at 32 seconds after the launch. The curves for times up to 9 minutes after launch show successive decreases toward a return to natural atmospheric background.

Because of the large number of particles produced in the atmosphere during

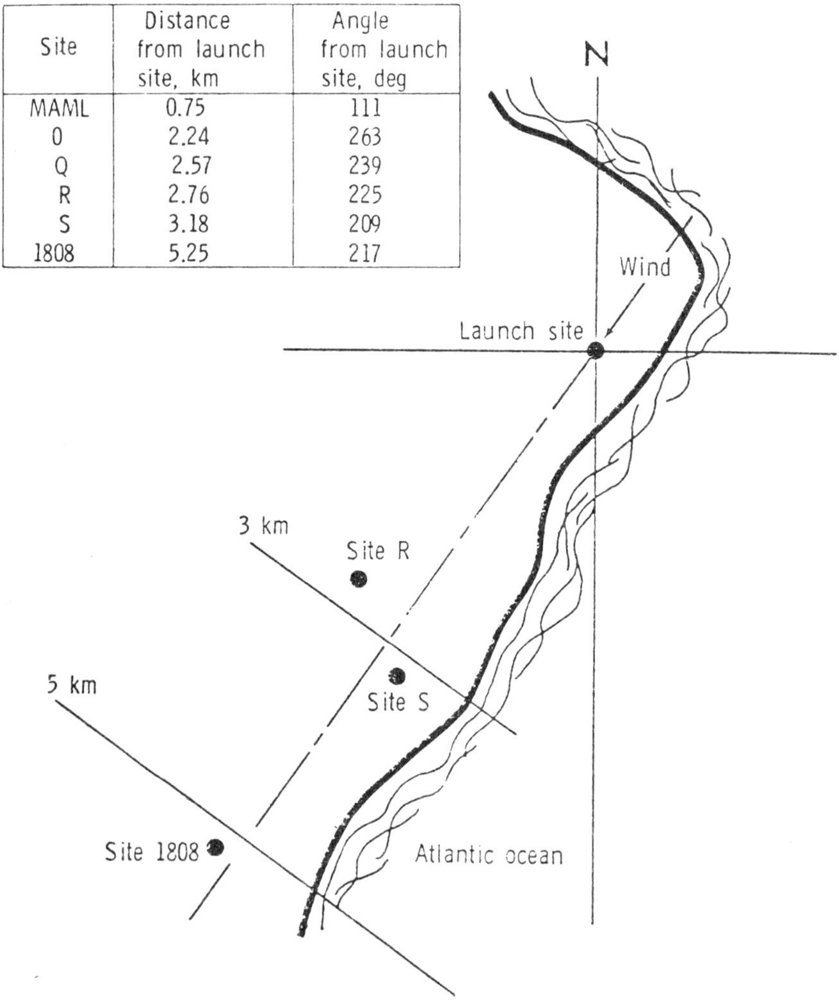

Site	Distance from launch site, km	Angle from launch site, deg
MAML	0.75	111
0	2.24	263
Q	2.57	239
R	2.76	225
S	3.18	209
1808	5.25	217

Figure 5. Delta effluent monitoring sites at Kennedy Space Center.

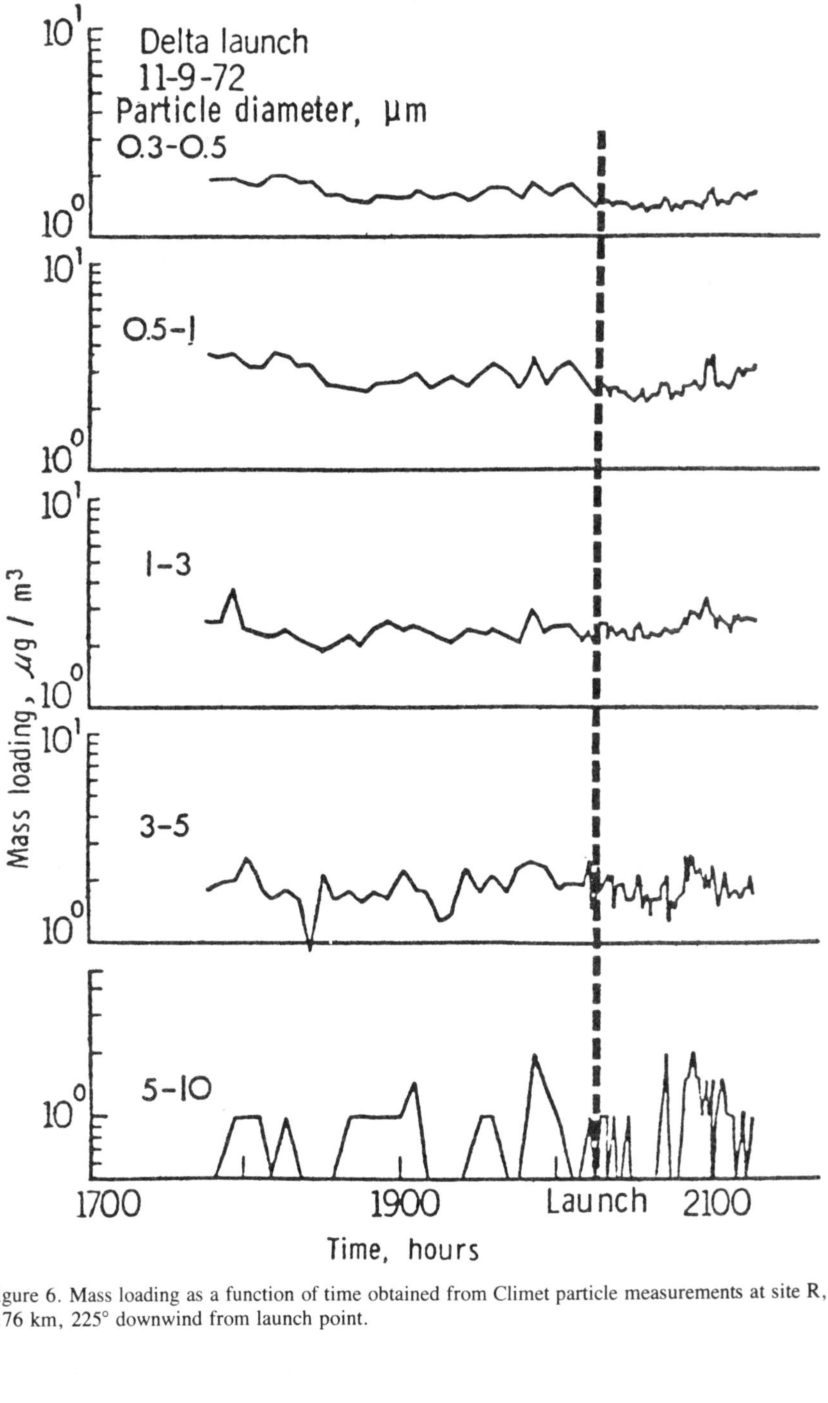

Figure 6. Mass loading as a function of time obtained from Climet particle measurements at site R, 2.76 km, 225° downwind from launch point.

the launch, the Royco instrument was providing a distorted size distribution, brought on by coincidence problems due to the presence of more than one particle in the counting volume at the same time. This condition results in values too low for the number of small particles. The dip in the curve for time 11:10:32 around 1.3 μm indicates this coincidence condition, and that many of the smaller particles were coincident with other particles and hence not individually counted. The horizontal part of the same curve does not show definite indications of convergence or dropping off; therefore, there must have been large particles above 6 μm diameter which were not measured because of the instrumentation upper-sizing limitations of 6 μm. Without a coincidence correction, one may

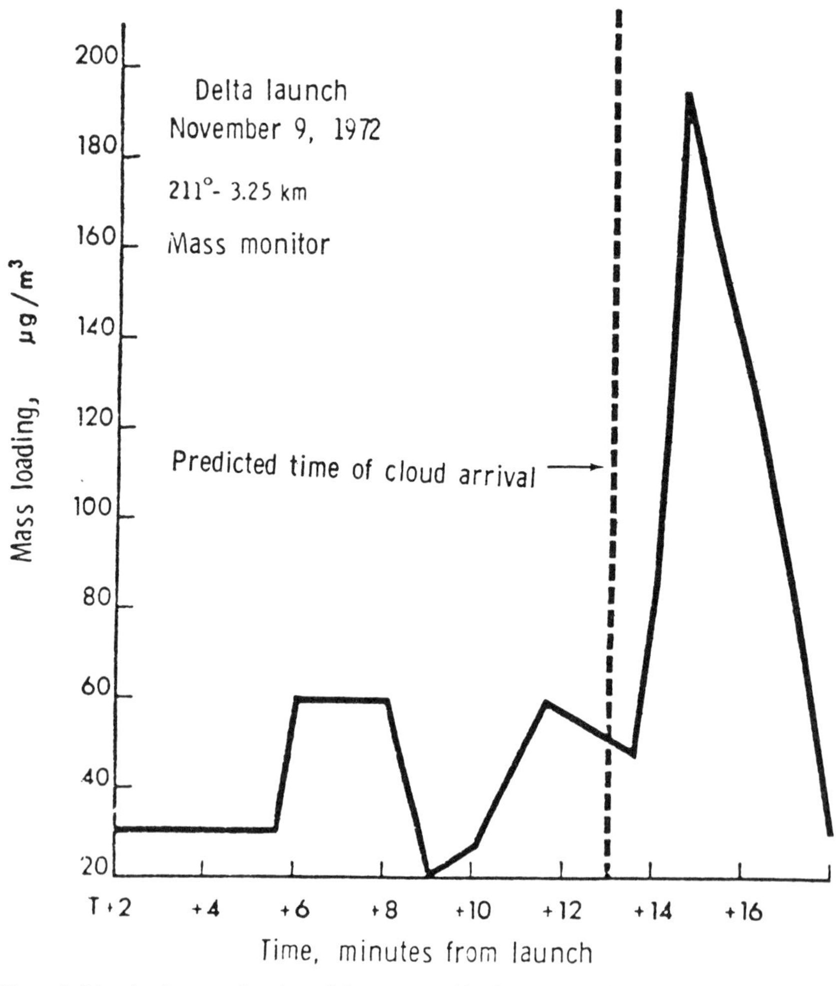

Figure 7. Mass loading as a function of time measured by the mass monitor at site S, 3.18 km, 209° downwind from launch point.

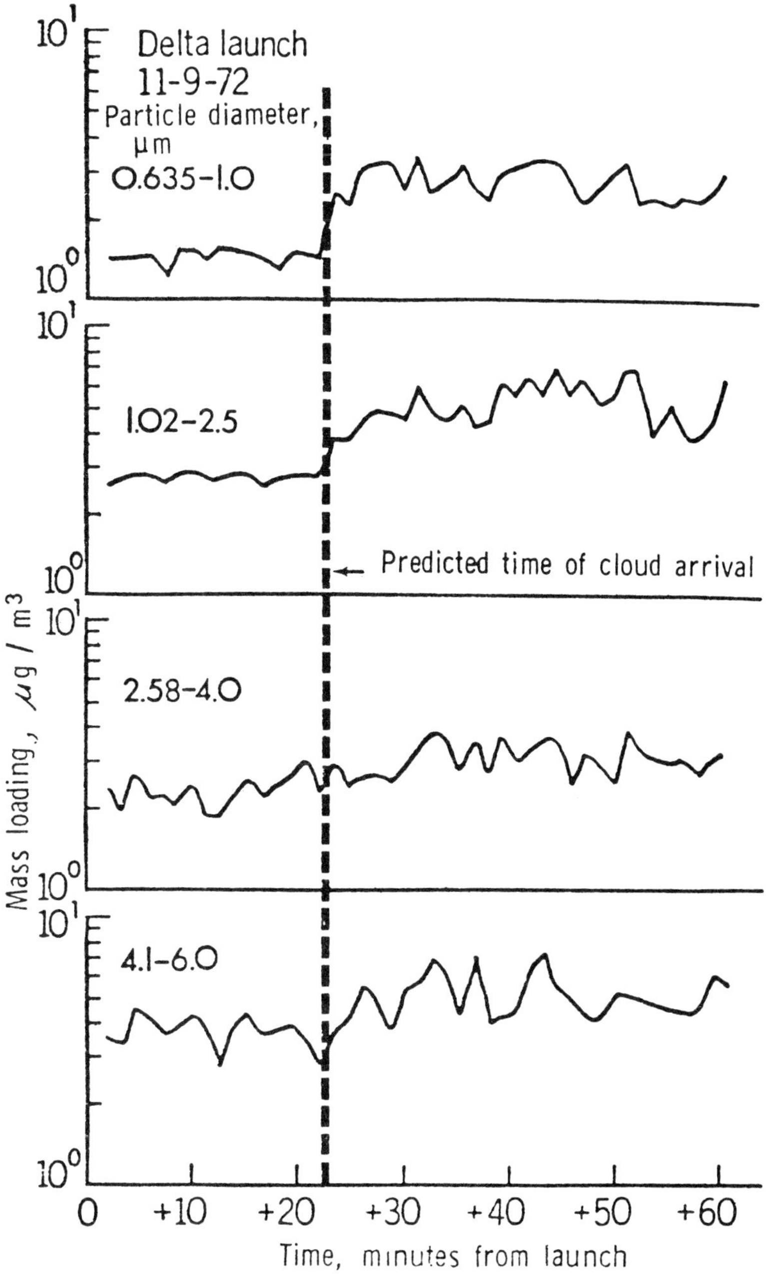

Figure 8. Mass loading as a function of time obtained from Royco particle measurements at site 1808, 5.25 km, 217° downwind from launch point.

conclude an average of three orders of magnitude increase in the number of particles 3 to 6 μm in diameter resulting from the launch.

Aerosol Characterization

A characterization of atmospheric particulates was conducted in cooperation with the Virginia State Air Pollution Control Board, Region VI, and Old Dominion University, from August 6 to August 19, 1973, at a site near the center of Norfolk, Virginia. Simultaneous measurements were made with the following instruments: a quartz crystal microbalance particulate monitor, a light-scattering multirange particle counter, a high-volume air sampler, and Nuclepore filters. To assess the impact of meteorological conditions on air quality variations continuous data on temperature, relative humidity, wind speed, and wind direction were

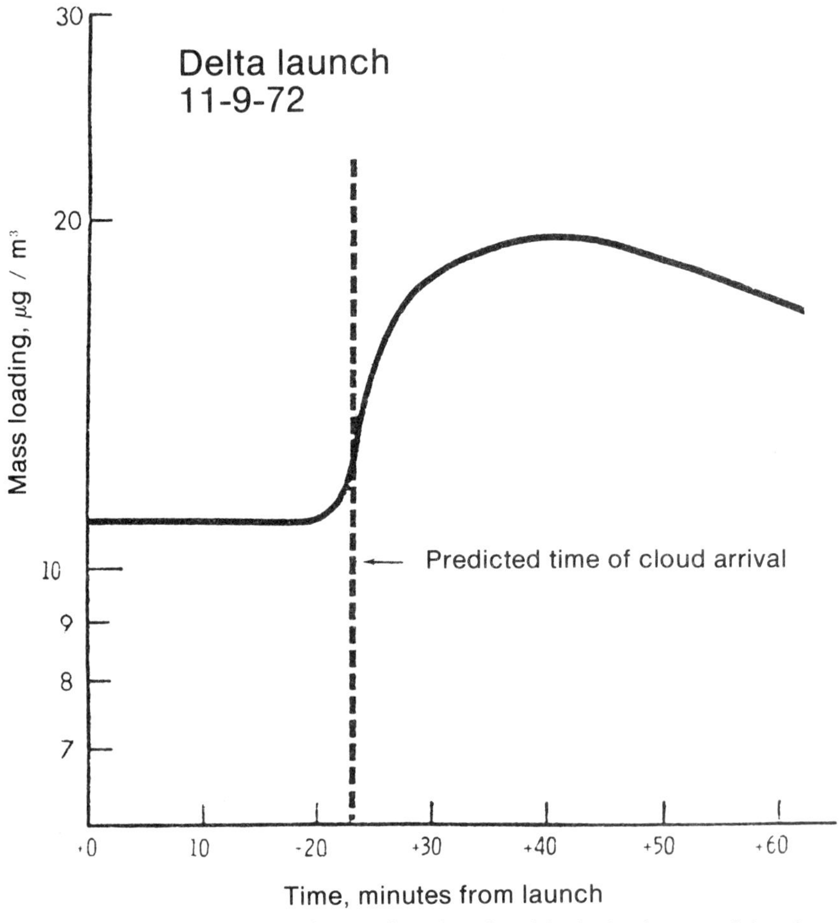

Figure 9. Mass loading determined from total number of particles in the size range 0.6 to 6 μm, measured at site 1808, 5.25 km, 217° downwind from launch point.

recorded. Particulate elemental composition was obtained by neutron activation analysis and scanning electron microscopy of filter samples.

The instruments were installed in an air-conditioned, mobile laboratory (Figure 11). In addition to the instrumentation, a Class 100 clean modular unit (bench), which permits filter loading and unloading on site, also was installed in the mobile laboratory.

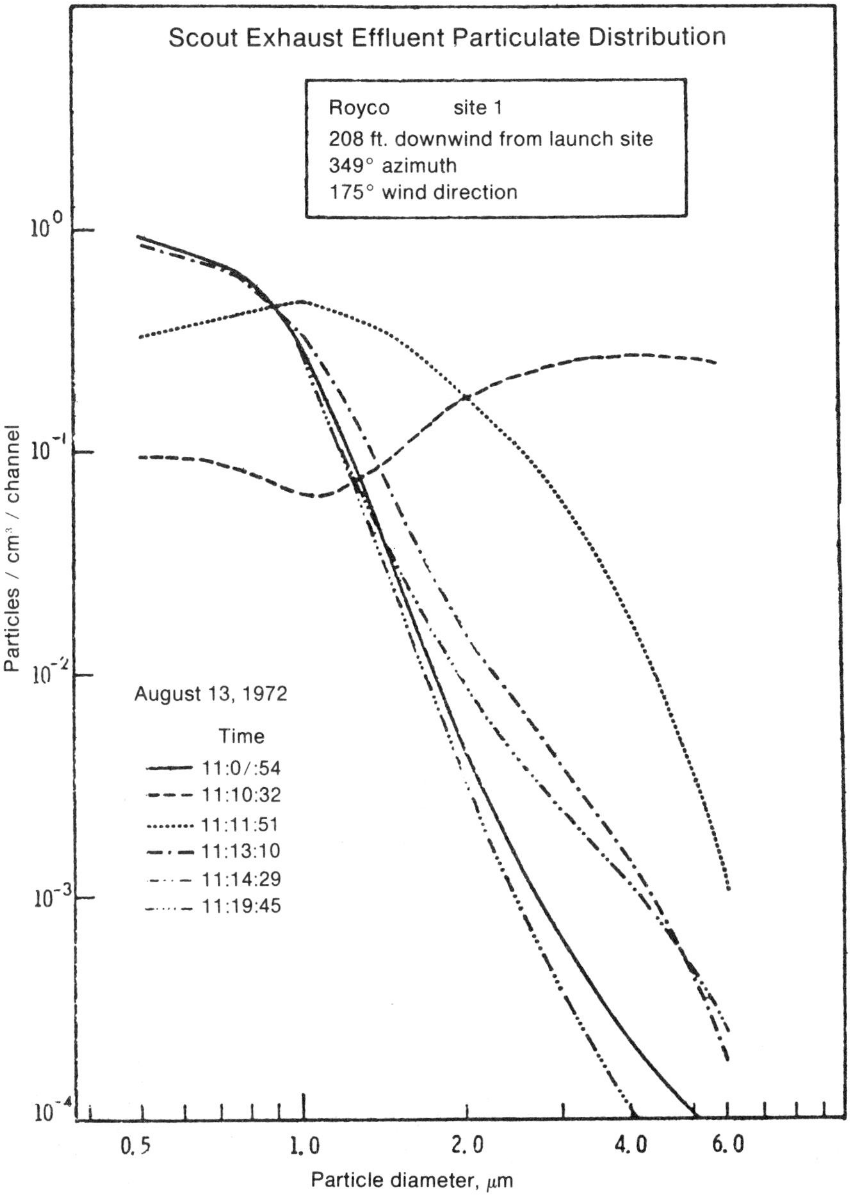

Figure 10. Particulate size distribution measured at site 1 before, during and after launch of the Scout rocket.

The main sampling tube consisted of two sections of smooth-wall aluminum tubing, 2-meters long by 15-cm diameter. This 4-meter-long tube was connected by an O-ring seal to a feed-through port on the roof of the mobile lab, which placed the inlet of the main sampling tube at a height of 6 meters above the ground. Various diameter inlet tubes, 25-cm long, were contained in a circular flange and connected by an O-ring seal to the underside of the port. A blower in the port maintained a flow rate of approximately 1400 ℓpm through the sampling tube. The inlet tubes were connected by short (30 to 60 cm) lengths of Tygon

Figure 11. Instrumentation van.

tubing (0.64-cm diameter) to the particle counter and the quartz crystal microbalance (mass monitor).

The light-scattering particle counter sampled particulates continuously, with a 10-minute sampling cycle. The data were checked periodically to determine if the particle concentration was large enough to cause coincidence effects (more than one particle in the scattering volume at any time) in the photometer. The counts seem to indicate that coincidence errors were not a significant problem. This procedure yielded continuous data on particle size distribution and concentration for the 2-week period, except for interruptions caused by instrument problems.

Due to limitations of the quartz crystal microbalance existing at the time of the experiment (which have subsequently been eliminated), a sample was collected for a 10-minute period once every hour from 0600 to 1800 hours. The sample periods were chosen to coincide with a 10-minute sample period of the light-scattering photometer.

The high-volume sampler was placed on a platform adjacent to and at the same height as the sampling tube on the mobile lab. The sampling was done on a 12-hour basis from 0600 to 1800 and 1800 to 0600 hours. The filters (collector and backing) were sheets of 20- by 25-cm fired glass fibers. The backup filter was used to prevent loss of collector filter weight due to loss of fibers and to prevent contamination by the sampler support screen.

The 47-mm diameter polycarbonate membrane filters, in their holders, were mounted on a rack on top of the mobile lab. A vane pump was used to draw ambient air through the filters. The pump was connected to a sequential sampler, and separate filters sampled for a 2-hour period during the 24 hours.

Figure 12 shows the particulate mass loading as a function of time of day over a 24-hour period for August 16, 1973. The measurements were made with four different instruments, as indicated. The particle counter data were continuous with a time resolution of 10 minutes. The mass monitor sampled for 10 minutes once each hour during the day, typically from 6:00 a.m. to 6:00 p.m. Data from each instrument represent an average over the 10-minute period. Each Nuclepore filter collected for a 2-hour period. The data points are plotted at the midpoints of the sampling period. The high-volume sampler sampled for 12-hour periods, from 0600 to 1800 and from 1800 to 0600. In general, trends in data agree for the different instruments. However, these instruments measure different size ranges of particles, have different collection efficiencies, and should not be expected to agree completely. In going from the 10-minute time resolution of the particle counter data to the 12-hour resolution of the high-volume sampler data, the increasing loss of information becomes quite apparent. Although the longer sample time is advantageous in smoothing random variations which may be of no interest, events which are of interest and have been shown as diurnally occurring are lost. From these data, it would appear that a sample time of ½ hour should be sufficient to resolve most events of interest.

For subsequent projects involving characterization of aerosols an additional mobile laboratory was obtained (Figure 13). In this laboratory the sampling tubes are brought in separately to each instrument, thus avoiding horizontal runs and 90° bends in the inlet lines going to the various instruments.

Instrument Problems

Some of the *typical* problems experienced with the light-scattering photometers while making these measurements are discussed below:

(1) "O" ring failure between the drive motor and the inlet pump on the Royco 220 sensor results in loss of data, since no particles are drawn into the view volume. This "O" ring should be inspected and/or replaced before each series of measurements.

(2) In both the Climet and Royco, the outlets became clogged with sea salt deposits with the attendant decrease in flow through the instrument. The inlet and outlet were cleared by passing a wire through the outlet.

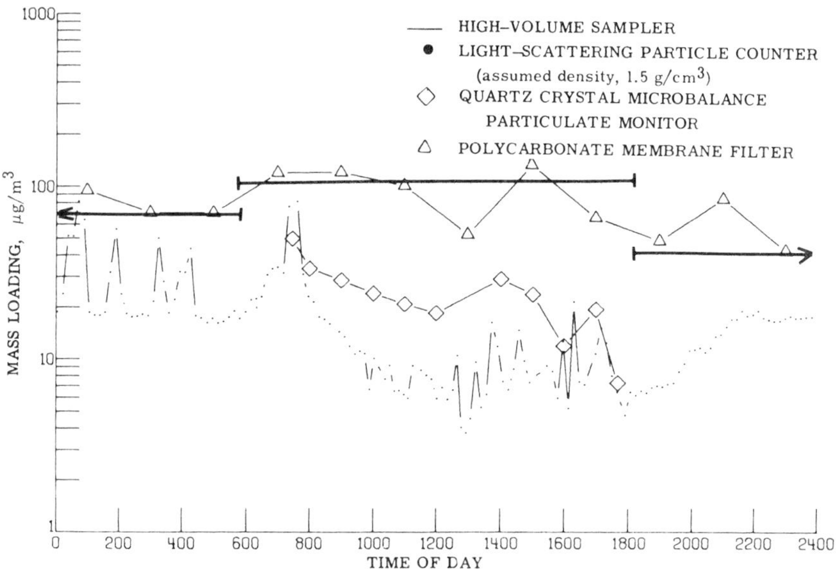

Figure 12. Atmospheric mass loading for August 16, 1973.

(3) The Climet paper tape would not feed through, which meant 6 channels printed in one tape position. The problem was due to malfunction of the paper advance solenoid.

(4) The Climet printer would print 1, 2, or 3 channels, and the print wheels would spin until the instrument power was removed. This problem occurred intermittently and could be corrected in some cases by reseating the P.C. boards.

(5) The lamp intensity would vary on the Royco 220 sensor which prevented proper setting of the calibrate adjust control. The problem was traced to loose mechanical connections on the lamp-voltage leads. The mechanical connections were replaced with solder connections.

(6) The pulse height analyzer, used with the Royco 220 sensor, would shift

the channel location of the particles as the ambient temperature changed. This was caused by temperature sensitive transistors on the P.C. boards. The boards were reworked with silicon transistors, and this problem was corrected.

(7) Coincidence problems occurred with the Royco 220 sensor and to a lesser extent with the Climet sensor. Coincidence causes the instruments to give a

Figure 13. Mobile laboratory.

Figure 14. Tethered balloon-borne flight package.

distorted number and size distribution plot, for example, two smaller particles are counted as one large particle. Dilution of the incoming particles is used to reduce the coincidence. What is needed is a commercially available, off-the-shelf, relatively inexpensive dilution system with dilution ratios of $\times 10$ and $\times 100$ that does not size discriminate during the dilution process.

(8) While condensation in the lines going to the instruments is not strictly an instrument problem, it is a problem connected with the measurements. It occurred when the ambient air temperature and relative humidity were high compared to these parameters inside the mobile labs. The solution was to heat slightly the lines inside the mobile labs by using heat lamps controlled by variacs. This appeared to correct the problem.

Tethered Balloon-Borne Light-Scattering Particle Counter

The light scattering particle counter flight package with its lifting, tethered balloon is shown in Figure 14. The particle counter measures three particle sizes (of equivalent polystyrene spheres). The sizes are from $\geqslant$0.314-μm diameter to 8.0-μm diameter; $\geqslant$ 0.484-μm diameter to 8.0-μm diameter; and $\geqslant$3.16-μm to 8.0-μm diameter. In addition, the following parameters are measured: pressure, temperature (fast), temperature (slow), relative humidity, wind speed, and ozone. These ten (one spare) channels of data are telemetered in parallel (simultaneously) to the ground station shown in Figure 15. The ground station demodulates the ten channels of information and feeds them to a magnetic tape recorder. Also, any one of the channels can be plotted as a function of altitude, on the x-y plotter, in real time.

A Nuclear Data, 4410 mini-computer based data acquisition system is presently being interfaced to the ground station. This system filters, averages and converts to engineering units the ten channels of information from the demodulator. Also, the data is recorded on seven-track magnetic tape and plotted on a point plotter in real time. First initial flights to check out the system were made at Wallops Island in 1975. This system will be used in making atmospheric aerosol measurements jointly with lidar and acoustic probing of the inversion and boundary layers.

Calibration

The three instruments available for calibrating these specific three light-scattering instruments are the Royco model 255 aerosol generator, a vibrating orifice, and a spinning disk aerosol generator. Because of the ease of setting the instrument up, the routine calibrations are done with the Royco model 255 aerosol generator.

Figure 16 shows the effect on the size distribution plot of the approaching failure of the light source in the Royco 220 sensor during a calibration. The two upper curves were obtained with polystyrene latex spheres of 1.01 and 2.02 μm diameter, generated from a Royco model 255 generator. The shape of the upper curves stayed the same even though the suspension (containing the spheres) was diluted further (2 to 1) and various adjustments of atomizer and drier air were tried. At this point the 220 sensor light source failed. A new lamp was installed and its alignment checked. No adjustment of the lamp mount was required. The

Figure 15. Ground station.

particles were then generated in the original suspension (no 2 to 1 dilution) and the two curves for the respective size particles were obtained (lower curves).

Another problem with this particular Royco 220 sensor was that it indicated spurious counts in channels 1 and 2 with a few counts out to channel number 8 on the pulse height analyzer even after 2 hours of sampling with a membrane filter on the 220 inlet. These spurious counts were eliminated by having the exterior of the inlet tube flashed with copper and then "blued" (made black).

CONCLUDING REMARKS

The field-monitoring experiments of the rocket launches served to establish techniques, procedures, and necessary instrumentation requirements for more comprehensive experiments. The exhaust cloud particulate size number distribution

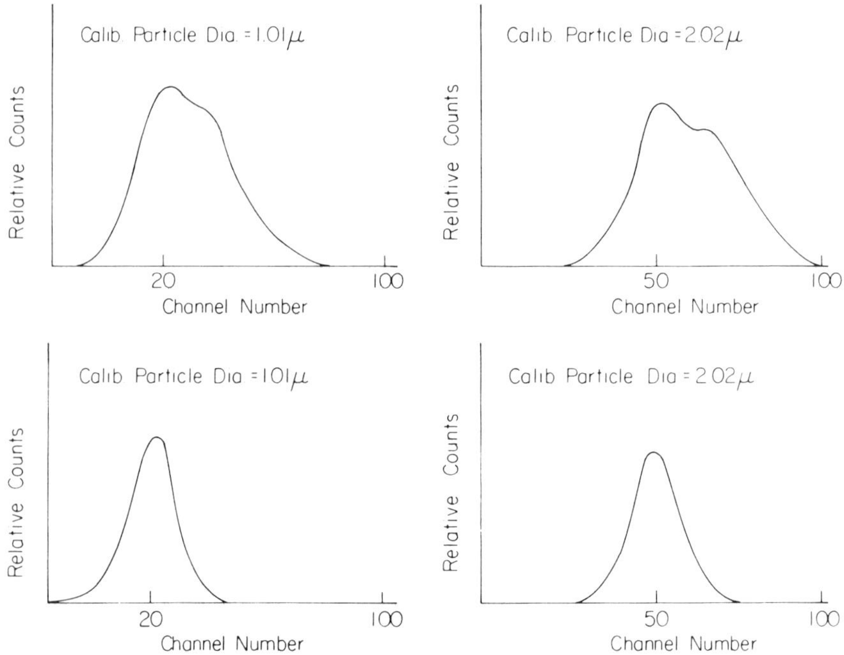

Figure 16. Royco size distribution.

(total number of particles against particle diameter), mass loading, morphology, and elemental composition have been determined within limitations.

Suitable dilution systems will be employed to avoid coincidence problems with the light-scattering photometers. Real-time measurements of particle size number distributions using the light-scattering photometers will be expanded to include the larger particles, greater than 6 μm.

During the 2-week period in which air quality measurements were made at one site in Norfolk, Virginia, a set of instruments measuring the mass loading, particle size distribution, and elemental composition of the aerosols produced data requested by the Virginia State Air Pollution Control Board, Region VI. The light-scattering multirange particle counter and quartz crystal microbalance particulate monitor gave consistent results. These results indicate that short-term (event-type) variations in the atmospheric mass loading can sometimes be missed by the 2-hour Nuclepore filter averaging periods. A time resolution of ½ hour seems necessary to detect such significant events.

The instrument problems discussed do not include all of the instrument measurement problems experienced but are typical of the problems encountered in obtaining day-to-day field measurements for the projects described.

Within limitations inherent in the light-scattering particle counters, the instruments yield time resolved, rapidly acquired, aerosol size distribution data. These data can be correlated with other measurements to detect variations of particulate characteristics including short-term events.

The tethered balloon-borne light scattering instrument package shows promise for obtaining aerosol characteristics as well as various meteorological parameters of the boundary and inversion layers.

REFERENCES

Gregory, G. L. and R. W. Storey, Jr., 1975: Effluent sampling of Titan IIIC vehicle exhaust. NASA Report No. TMX-3228.

Hulten, W. C., R. W. Storey, G. L. Gregory, D. C. Woods, and F. S. Harris, Jr., 1974: Effluent sampling of Scout "D" and Delta launch vehicle exhausts. NASA Report No. TMX-2987.

Jaenicke, R., 1972: The optical particle counter: Cross-sensitivity and coincidence. *J. Aerosol Sci.*, *3*, 95–111.

Storey, R. W., R. J. Sentell, D. C. Woods, W. C. Hulten, J. R. Smith, and F. S. Harris, Jr., 1974: Particulate air quality measurements in Norfolk, Virginia. APCA Paper 74-69, presented at the 67th Annual Meeting of the Air Pollution Control Assoc.

Storey, R. W., Jr., R. J. Sentell, D. C. Woods, J. R. Smith, and F. S. Harris, Jr., 1975: Atmospheric particulate measurements in Norfolk, Virginia. NASA Report No. TMX-3285.

COMPUTER GRAPHICS AND ANALYSIS OF ATMOSPHERIC AEROSOL SIZE DISTRIBUTIONS

RONALD J. SENTELL

NASA Langley Research Center, Hampton, Va.

The Aerosol Measurements Research Branch of Langley Research Center has participated in air quality studies with Region VI of the Virginia State Air Pollution Control Board and in atmospheric diffusion studies at Kennedy Space Center since 1970. Thousands of aerosol size distributions have been measured and processed. Without computer processing and computer graphics, analysis of such a volume of data would be impractical. To facilitate analysis of the data, the approach was adopted to examine first an overview of size distribution as a function of time of day in terms of a three-dimensional projection (Storey et al., 1975). From this projection, one easily can determine size distribution stability or change. Size distributions for given times then can be selected and examined in more detail.

The aerosol size distributions presented in this paper were measured with a right-angle scattering photometer. A fundamental problem with any commercial, single-angle, scattering-type photometer in measuring atmospheric aerosol size distributions is the dependency of the scattered light on aerosol index of refraction. In calibration of such instruments or for measuring a known aerosol in the laboratory, the index of refraction is known, and an accurate size distribution may be obtained. However, in the ambient atmosphere, the index of refraction of the sample is never known with any degree of certainty. Although an accurate size distribution is not measured, valuable data still are obtained in terms of a fingerprint for a particular species of air parcel. Fingerprints, such as typical urban pollution, North-Atlantic background, continental background, and serious smog episodes, are reported in the literature (Slinn, 1975; Willeke and Whitby, 1975; Harris, 1976).

Aerosol Measurement System

The aerosol measurement system, which collected the data presented in this paper, is composed of a right-angle scattering photometer (Royco model 220), a pulse height analyzer (TMC model 102), and a data acquisition system (Vidar model 5403). The aerosol size range measurable with this system is approximately 0.5 μm to 6.5 μm in diameter, with this size range being divided into 100 size intervals by the pulse height analyzer. Data from the pulse height analyzer are recorded on paper tape by a high-speed printer and/or on a 7-track magnetic tape. Additional data, such as time of day from a digital clock and various experimental parameters, also are recorded. This system is shown in Figure 1. A sampling schedule of one sample per 10 minutes normally is used for ambient measurements.

Figure 1. Aerosol measurement system.

Meteorological Station

The Meteorological Research, Inc. weather station measures and records the following meteorological parameters: wind speed, wind direction, temperature, and relative humidity. The station is a self-contained, battery-operated unit with accuracy of all measurements on the order of $\pm 2\%$. The chart drive is a clockwork mechanism with an accuracy of ± 60 seconds per 24 hours.

Data Reduction

Aerosol size distributions are examined in several perspectives. First, a three-dimensional projection of aerosol number concentration divided by the size window (dN/dLog R) as a function of size and time of day from midnight to midnight is examined for general trends and anomalies. Second, to further examine trends, each size distribution for a given time is (1) integrated over size to give a total number concentration, and (2) fitted in a least-squares sense to the Junge (1963) power-law distribution, which is often a good, first approximation, to give an indicator of the shape of the size distribution. Each parameter is then plotted as a function of time of day from midnight to midnight. Finally, on the basis of

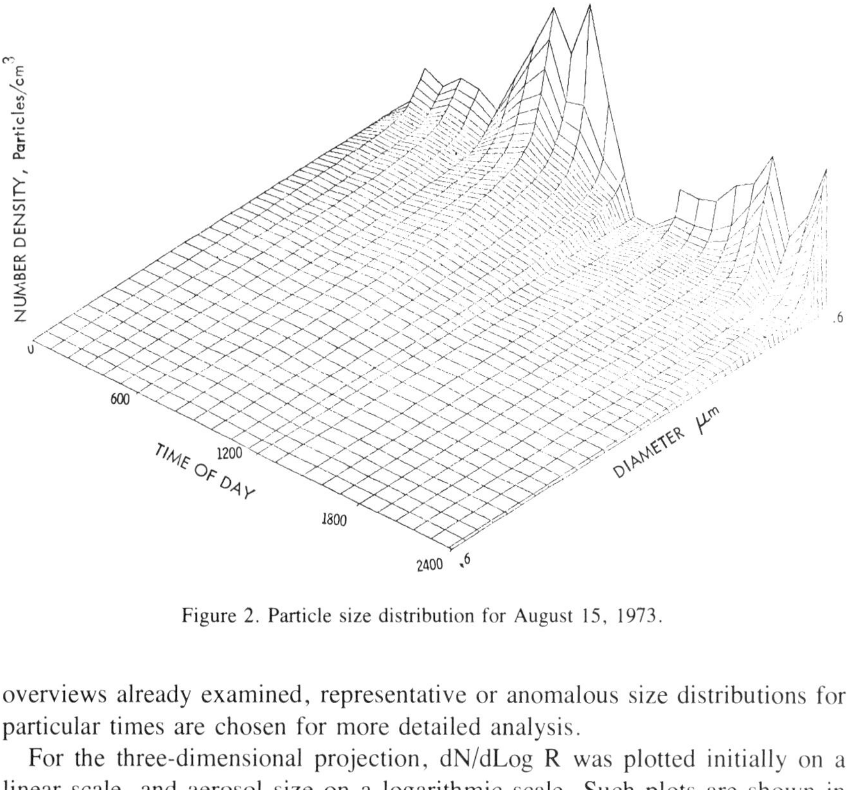

Figure 2. Particle size distribution for August 15, 1973.

overviews already examined, representative or anomalous size distributions for particular times are chosen for more detailed analysis.

For the three-dimensional projection, dN/dLog R was plotted initially on a linear scale, and aerosol size on a logarithmic scale. Such plots are shown in Figures 2 and 3. However, to gain a better view of how well-behaved a distribution is with respect to a power-law model, dN/dLog R should be plotted on a logarithmic scale. Then, a straight line indicates an agreement with the model. Such a plot is shown in Figure 4.

Aerosol size distributions divided into 100 channels and taken every 10 minutes for several weeks quickly accumulate into a large volume of data. Without computer processing and computer graphics, analysis is an unmanageable task. Magnetic tapes from the aerosol measurement system are read directly into a

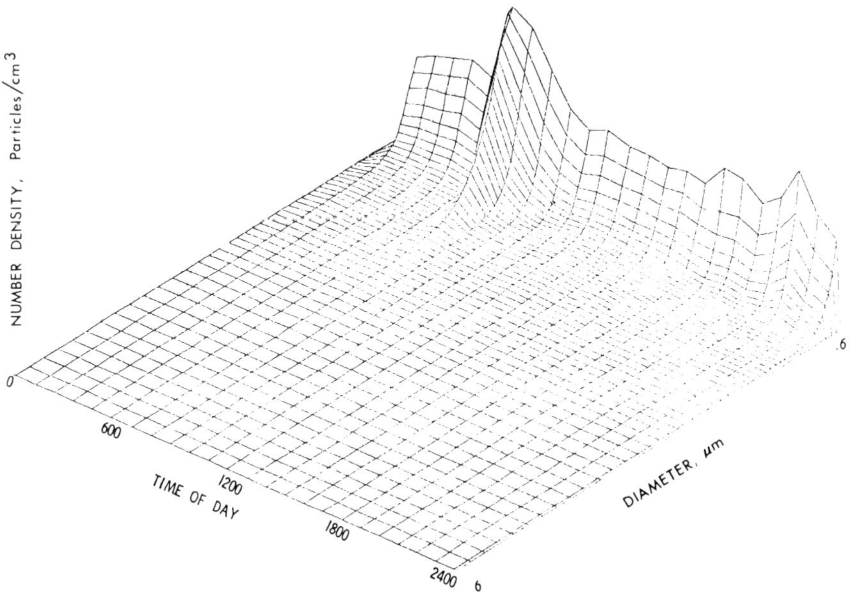

Figure 3. Particle size distribution for August 18, 1973.

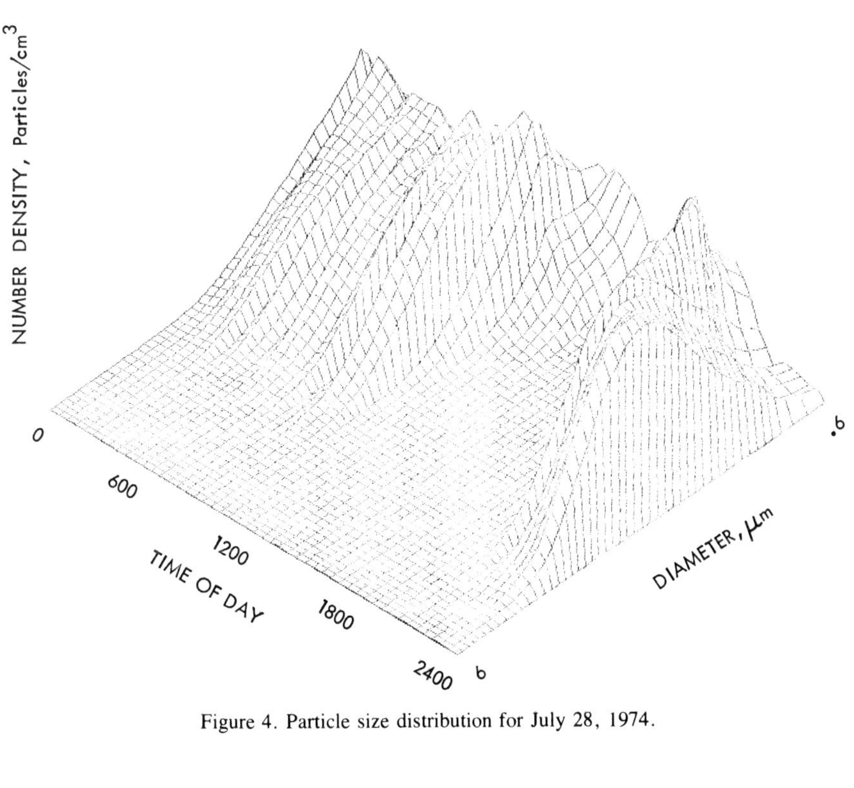

Figure 4. Particle size distribution for July 28, 1974.

CDC 6600 computer. The data are analyzed, and then plotted on a Calcomp plotter. Even with computer processing, some data must be averaged or discarded to save computer time and cost. For the three-dimensional projection-type plotting with hidden lines removed, the cost is proportional to the number of points cubed. For a 50 by 50 array, the cost is about \$10. For a 150 by 100 point array, the cost would be \$2160. For the projection plots in Figures 2, 3, and 4, every third distribution and every other size are plotted. In the single-parameter plots shown in Figures 5 and 6, all data are plotted.

Supporting meteorological data are recorded on a strip chart recorder, aver-

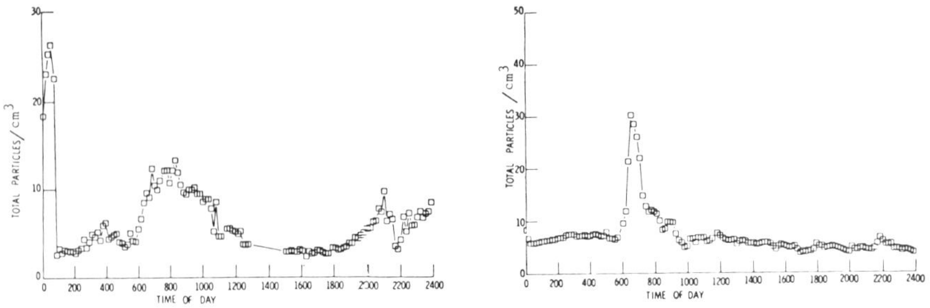

(a) August 15, 1973.

(b) August 18, 1973.

Figure 5. Particle number density.

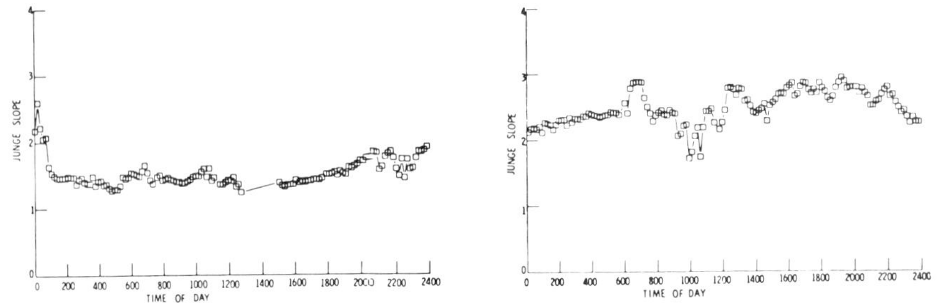

(a) August 15, 1973.

(b) August 18, 1973.

Figure 6. Particle size distribution parameter.

aged, and punched on cards for computer input. For data shown in Figures 7 and 8, data are averaged over an hour with each point plotted being the midpoint of the average.

DISCUSSION

Specific examples illustrate the usefulness of the different plotting techniques for identifying general trends and events in aerosol size distribution. During an air quality study from August 6 to August 19, 1973, with Region VI of the Virginia State Air Pollution Control Board in Norfolk, Virginia, a consistent, positive excursion in aerosol number concentration was observed in the early morning. Projection plots for August 15 and August 18, 1973, are shown in Figures 2 and 3. Figure 5 further shows this event for integrated number concentration. Figure 6 shows variation in power-law exponent or Junge slope.

Two observations are made. First, the early morning excursion on August 15 is not associated with a definite excursion in Junge slope; while on August 18, the excursion in number concentration definitely is associated with an excursion in Junge slope. Second, the average slope for each day is different, suggesting a different source of aerosols. Examining the meteorological data for these 2 days (Figures 7 and 8), one finds the wind direction from the southeast on August 15 and from the northwest on August 18. The possibility of different sources for the

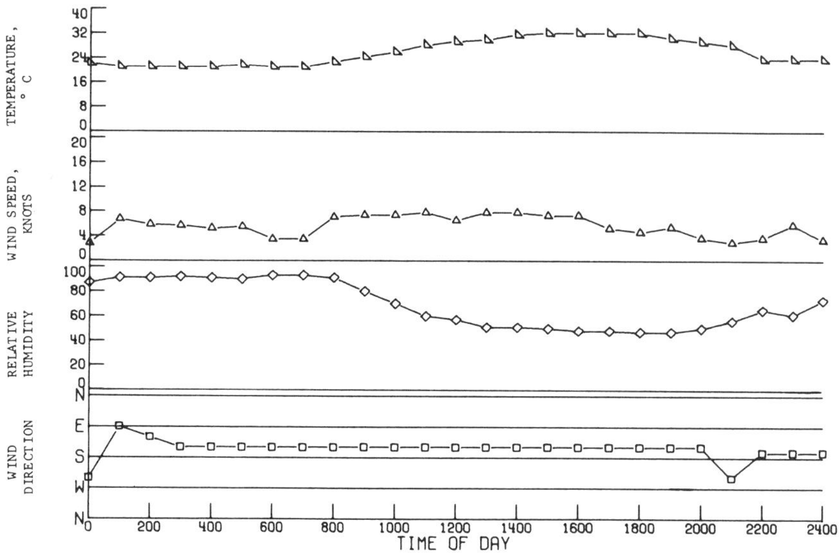

August 15, 1973.

Figure 7. Meteorological data.

2 days is strengthened. Also, the possibility of the morning excursion being connected to an atmospheric mechanism, such as fumigation, rather than a particular source at a given location is strengthened. In exploring the possibility of a change in aerosol concentration in the morning being associated with a change in relative humidity, one finds a change in relative humidity at about the right time on August 15. However, on August 18, relative humidity is constant all day. From these 2 days, relative humidity does not appear to influence the morning event. Thus, on these 2 days, an excursion in aerosol concentration is observed with respect to a day's time frame. Differences in size distribution are easily observed between the 2 days.

A projection plot for another air quality study in Norfolk on July 28, 1974, is shown in Figure 4. One quickly discovers a power-law-type aerosol for the morning which changes throughout the day. This example further illustrates the importance of observing the complete size distribution as a function of time. Without this type of plot, one easily might assume two or three distributions as being representative of the whole day.

CONCLUSIONS

Computer processing and plotting of atmospheric, aerosol size distributions greatly facilitate the otherwise impractical task of analyzing time variations in size distributions for a long period of time. A three-dimensional projection of

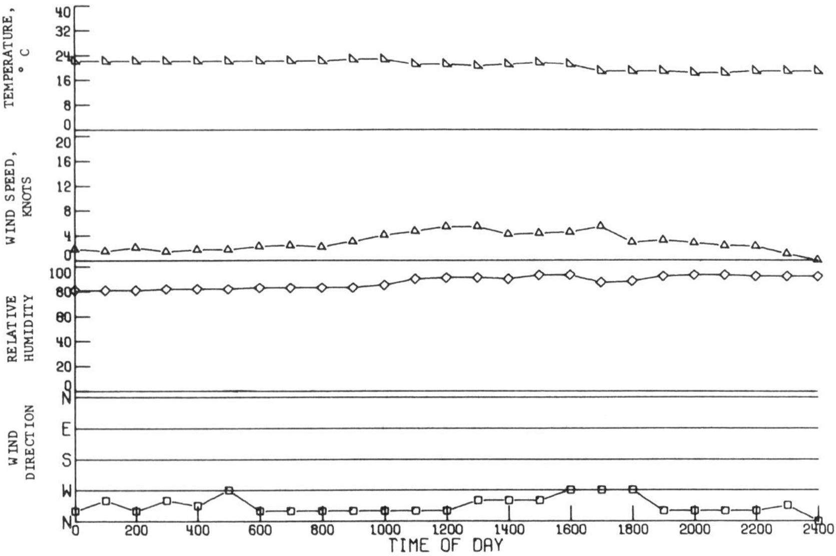

August 18, 1973.

Figure 8. Meteorological data.

aerosol concentration as a function of size and time of day from midnight to midnight quickly shows the behavior of the aerosol, whether stable or changing. Integrated number concentration over the measured size range, and a distribution parameter from a power-law distribution model show trends in the data. Based on these plots, a decision can be made about whether a sample for a particular time is representative or anomalous. One then may choose a particular time frame to be studied in more detail.

REFERENCES

Harris, F. S., Jr., 1976: Atmospheric aerosols: A literature summary of their physical characteristics and chemical composition. NASA Report No. CR-2626.

Junge, C. E., 1963: *Air Chemistry and Radioactivity,* Academic Press, 115–117.

Slinn, W. G. N., 1975: Atmospheric aerosol particles in surface-level air. *Atmos. Environ., 9,* 763–764.

Storey, R. W., Jr., R. J. Sentell, D. C. Woods, W. C. Hulten, J. R. Smith and F. S. Harris, Jr., 1975: Particulate air quality measurements in Norfolk, Virginia. APCA paper 74-59, presented at the 67th Annual Meeting of the Air Pollution Control Association, June 1974, NASA TMX-3285.

Willeke, K. and K. T. Whitby, 1975: Atmospheric aerosols: Size distribution interpretation. *J. Air Pollution Control Assoc., 25,* 529–534.

QUESTIONS

Q: What kind of count rates did you have with the Royco? Did you have a diluter on it?

Sentell: No, we did not.

Whitby: The only way that you can get that kind of a size distribution towards the end of that plot is to have these spurious pulses that I referred to. In other words, the count rate is overloading. That depresses the counts of the small particles and shifts the modes so the size distribution you're fitting may be the spurious portion of the distribution and not really the actual distribution. It's a very strange distribution.

Jeck: When I went through cloud physics the teacher told me that clouds were similar to fogs. The fact is that clouds and fogs aren't a bit the same. As we find out, the average cloud size spectrum contains many, many more small particles. In fact, after you think about it, it has to make sense because visibility in fogs is always better than it is at a cloud base. In this particular instrument we used, there is a natural bias against small particles. If you take the average, cloud-based spectrum you will find it has a very small sigma and is a log-Gaussian distribution. It will, in effect, have no distortion because the distribution is quite narrow. If you go to an exponential distribution there is a natural bias against small particles. My main point is that much of this information exists now through customer and factory experience but, as far as I know, is not generally available except upon request. If a customer doesn't know what to request he runs into these problems without knowing it. Some kind of bibliography of brief documentation on application would certainly be very helpful to customers.

Knollenberg: There's not a wide application for that kind of instrument. You have more experience with that than I do, and all of the customers certainly have more experience. We learn as much from you as you do from us.

HOLOGRAPHY AND IMAGE ANALYSIS

ALEXANDER E. MARTENS

Bausch & Lomb, Rochester, N.Y.

Holographic techniques augmented by automatic image analysis have been used to derive a variety of data on aerosols, including natural and artificial fog, cosmetic and industrial sprays, liquid sprays in the combustion chambers of rockets, and others (Thompson and Ward, 1966; Thompson, Ward and Zinky, 1967; Thompson, 1974). The information which can be derived in this manner includes particle size distributions, particle velocity distribution, spray patterns and their change in time. Currently, the holographic methods are the only means available for comprehensive *in situ* analysis of aerosols. The applications of holography to particle analysis have been described by Thompson, Ward, and Zinky (1967). The use of image analysis in the evaluation of restored images or holograms themselves reduces time and improves accuracy of the measurements.

Investigation of aerosols via holography involves a three-step process. In the first step, a hologram of the volume in space containing the aerosol is generated. In step two, the holographic record is used to restore the original image. In the third step the image is analyzed. Under certain conditions, the second step can be omitted.

Although two methods for making holographic records of the aerosols are available, on-axis Frauenhoffer holography, and off-axis holography, the former has several important advantages which led to the commercial utilization of that technique. Basically, the aerosol is illuminated by an intense, pulsed, coherent light source, usually a ruby laser or one of several types of gas lasers. The light is diffracted at the particles. The diffracted light forms an interference pattern with the original collimated illuminating beam, which in this case acts as the reference beam, forming a hologram. The hologram can be and usually is recorded photographically, or by producing thermal deformation in special polymers. The size resolution is a function of the illuminating light's wavelength, hence UV has an advantage. For example, nitrogen laser operating at 3371Å has been successfully used (Labrum, 1975).

Particle velocity has an important effect on the system's ability to record particle sizes. For example, Thompson (1974) has calculated that to measure the size of a 10 μm particle moving at 1 m/sec correctly, the laser pulse duration should be under one microsecond. By using two light pulses in rapid succession, with an accurately known time interval between them, the velocity vectors for individual particles can be determined by noting the displacement of the two particle images.

Image analysis can be carried out directly on in-line holograms to determine particle size distribution. This, however, is limited to spherical particles. Much more data can be obtained by analyzing restored images, layer-by-layer. In this manner, using advanced image analysis equipment, it is possible to determine size, shapes, orientation, coordinates, and velocities of individual particles.

Finally, in some applications it is possible to by-pass the hologram altogether by forming, through a microscope objective, images of particles on the target of a Vidicon, and recording on video tape the dynamic behaviour of the aerosol with time. The tape is played back into an image analysis system, and applying certain in-focus criterion to the particle images, a variety of data can be extracted. Pulsed laser is used as an illumination source (Labrum, 1975). The particle images or diffraction patterns from, respectively, reconstructed images or directly from holograms are optically relayed onto the photo-sensitive target of a scanner similar to those used in television. Video signals are generated by scanning the images which correspond to the spatial distribution of brightness. The signals are electronically processed to extract automatically a great variety of information on the aerosol, including particle sizes and shapes, instantaneous position, velocity vectors, etc.

The combination of holography with automatic image analysis provides a powerful analytical tool for investigation of aerosols.

REFERENCES

Labrum, J. H., 1975: Size and velocity measurement techniques of high speed microscopic particles. *Electro-Optics 1975 Conference Proc.*, Anaheim, California, *6*, 1–9.

Thompson, B. J., 1974: Holographic particle sizing techniques. *Sci. Instrum.*, *7*, 781–788.

——— and J. H. Ward, 1966: Particle sizing—the first direct use of holography. *Scientific Research, 1.*

———, ———, and W. R. Zinky, 1967: Application of hologram techniques for particle size analysis. *Appl. Optics*, 519–526.

SINGLE PARTICLE LIGHT SCATTERING SPECTROMETERS

ROBERT G. KNOLLENBERG

Particle Measuring Systems, Inc., Boulder, Colo.

The work at Particle Measuring Systems (PMS) relating to aerosol particle size measurements began with quite unusual design objectives. The first carrier medium we were exposed to was a hard vacuum. The second application was for aircraft measurements. Both of these environments necessitated *in situ* measurements with high spectral resolution. The former requirement is obvious in a vacuum since particle trajectories cannot be controlled. High spectral resolution implies that the size distribution function itself is of fundamental importance. High spectral resolution is often a requirement for research studies in particulate problems but may be an unnecessary elaboration in many cases. In fact, the bulk of all single particle optical aerosol counters sold do not have sufficient resolution for true particle size spectrometry. Many simply operate with presettable lower and upper size thresholds and provide simple yes or no answers; e.g., is the air more contaminated than yesterday, or are there sufficient, large aerosol particles to cause problems in this chemical process or to contaminate that pharmaceutical product? For such uses, a simple particle counter is often found sufficient. However, there remains a host of problems wherein the particle size distribution is itself of greatest importance, providing clues as to physical processes, formation mechanisms, hints as to origin, as well as greater precision in determining such integrated properties as particle mass density, extinction coefficient, atmospheric visibility and those properties which are sensitive to higher moments of the distribution function.

A corollary to sufficient particle size resolution is the problem of satisfying the sampling statistics at the large infrequent particle sizes. Because of the wide range of applications and environments, an ideal instrument generally exists only for a particular size and concentration. Attempts to cover much more than an order of magnitude particle size most often leave measurements at either end of the size range compromised. Because of the many variants in designing for the wide size range and concentrations found in aerosol work, we will not attempt to cover all aspects of aerosol measurements using light scattering instruments. Rather we will restrict ourselves to describing a few instruments we have designed which are gradually becoming available with major emphasis on research needs wherein particle size resolution, *in situ* sampling, and high sensitivity are required. These requirements characterize the particle size spectrometer—differentiating it from the particle counter.

ESTABLISHING THE DESIGN CRITERIA FOR PARTICLE SIZE SPECTROMETERS

A Discussion of Imaging Spectrometers and Their Limitations

Any general discussion of particle sizing techniques should not omit imaging systems. The bulk of our current knowledge of particles has its roots in many hours of painful viewing through optical microscopes, and in recent years, the more exact measurements from electron microscopy. It is not the purpose of this paper to readdress those arguments detailing the shortcomings in the practical use of such imaging instruments. They are well known and have consumed volumes in discussion. Rather it is our purpose to show how certain attributes of high resolution imaging can complement particle size measurements even where the particle is too small to resolve optically and light scattering methods are employed. First, however, we must identify the inherent limitations of pure imaging techniques.

In microscopy (whether at optical wavelengths or with electron beam) it is possible to size particles without the troublesome aspects of shape, and particle refractive index, and recently Knollenberg (1970, 1972, 1976) has designed automated imaging instrument systems which can readily size particles *in situ*. There are in fact, only three aspects of imaging particle size spectrometers which ordinarily restrict their application to particles larger than a few microns:

1. Fundamental resolution limit (approximately 0.4 μm at optical wavelengths)
2. Sample volume becomes vanishingly small as size approaches the wavelength
3. Cost.

It is our experience that if the size is large enough (greater than a few microns) an imaging particle size spectrometer will invariably be superior to any technique employing only the measurement of scattered light. Probably the best example that we can offer of the capability of imaging instruments is the 2-D Spectrometer developed by PMS and used for airborne *in situ* particle spectrometry of cloud elements. The 2-D Spectrometer has all of the characteristics of high resolution shadowgraphy but with complete data compression through its inherently digital image data acquisition scheme.

The 2-D Optical Array Probe utilizes a photodiode array and photodetection electronics similar to standard optical array probes (Knollenberg, 1970, 1972); however, the 2-D Probe contains a high speed front end memory enabling each photodetector element to encode many bits of shadow information from each particle. The particle's transit scans the array and image slices are recorded across the shadow to develop a two-dimensional image. It is thus a true parallel processor of image data as opposed to serial scanning devices common to vidicons and the Reticon array. The probe can collect particle image information at the rate of 128 million bits per second. However, data is recorded only when particles are present resulting in automatic data compression.

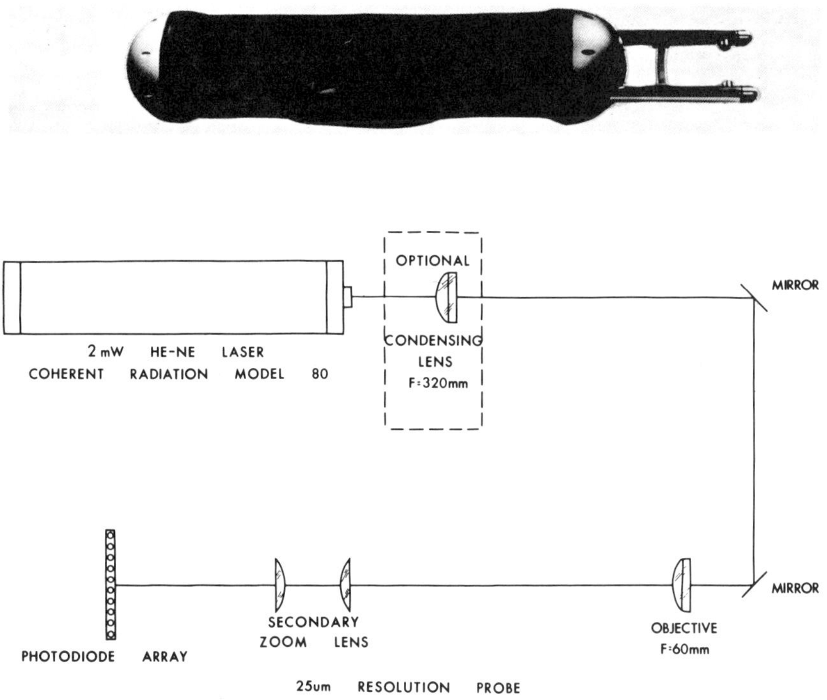

Figure 1. 2-D Optical array probe and optical system.

Two types of 2-D probes are manufactured, covering size ranges of 25–800 μm and 200–6400 μm. A photograph and optical system for the 25–800 μm instrument are shown in Figure 1. In operation, particles passing through the laser beam, in the vicinity of the object plane, are imaged by a two-stage imaging system onto a photodiode array. Total magnification can be varied from 5×–10× by setting the zoom. A magnification of 8× provides 25 μm resolution. The 25 μm resolution instrument employs 32 active photodiode elements in the array and takes image slices at a rate of up to 4 million per second when a particle passes through. At air speed of 100 m sec^{-1}, 250 nsec corresponds to 25 μm of distance, which is the effective diameter of each element in the array at 8× magnification. At this air speed the size resolution across the array (particle width) is identical to the image slice resolution through the array (particle length) resulting in square raw image data.

Data from each image slice is stored in a static MOS shift register (see Figure 2) which serves as a buffer for writing onto computer-compatible tape via the Data Acquisition System (DAS). Two MOS buffers are employed in a ping-pong fashion so one can be available for loading (at a rate determined by incoming particles) while the other is used to unload a previous sampling of particles (at a rate necessary for writing onto tape). As long as unloading is accomplished earlier than loading in each case, no indigestion is encountered and the buffer roles can be alternately reversed with no loss of data.

Examples of plotted image data are shown in Figures 3A and 3B generated by a printer/plotter. Shadowed elements are plotted dark (zeros) and unshadowed elements are plotted light (ones) with each particle image separated from adjacent images by a coded time slice and an all zeros slice. In this particular system, the top 8 bits of the time slice and all zeros slice, are fixed alternate bits used for identification purposes. Time progresses from left to right on the plots, with the time slice following each image.

The 2-D Probe as described has 25 μm resolution; however, this resolution limit is dictated by circuit response (rather than optical limits) imposed by the 100 m sec^{-1} relative particle velocity. A reduction to 10 m sec^{-1} would thus allow 2.5 μm resolution with appropriate optical magnification scaling. In the more recent gray scale 2-D Spectrometer (4 levels of shadow density), we employ an adjustable zoom lens giving continuous size ranging capabilities from 1–64 μm up to 2–128 μm. Particle size detection in the neighborhood of 1 μm is not emphasized as much as having 1 μm resolution on a 10 μm particle. The overriding factor here is the vanishing, small depth-of-field as one approaches the wavelength in size. For example, with 64 array elements the sample volume cross section is 64 μ wide.[1] The depth-of-field for a 1 μm particle is only 10 μm whereas for a 10 μm particle it is 1000 μm ($D_f \propto r^2$). The corresponding sample volume cross sections of 6.4×10^{-4} mm^2 and 6.4×10^{-2} mm^2 are found to bound that which is normally employed in single particle light scattering spectrometers (100 μm $\times$ 100 μm = 10^{-2} mm^2). Because of the rapid change of sample cross section with size, one normally finds that somewhere between 1 and 10 μm there is a reasonable lower limit for an imaging system because of sampling considerations. About the only instance in which the sample volume is adequate at the limit of resolution in array instruments is in particle source characterization (e.g., atomization nozzles) where concentrations are much higher.

It should be pointed out that it is possible to increase the sample area of most light scattering instruments; however, at similar resolutions, the imaging system will rapidly surpass a light scattering instrument at sizes much larger than 10 μm. It should also be borne in mind that at any equivalent sample area, the imaging

1. The sample volume cross section or sample cross section is the cross sectional area through which particles flow and are sampled. The sample volume is the illuminated volume where size measurements are made and accepted. The product of the sample volume cross section and flow velocity is the volume sampling rate.

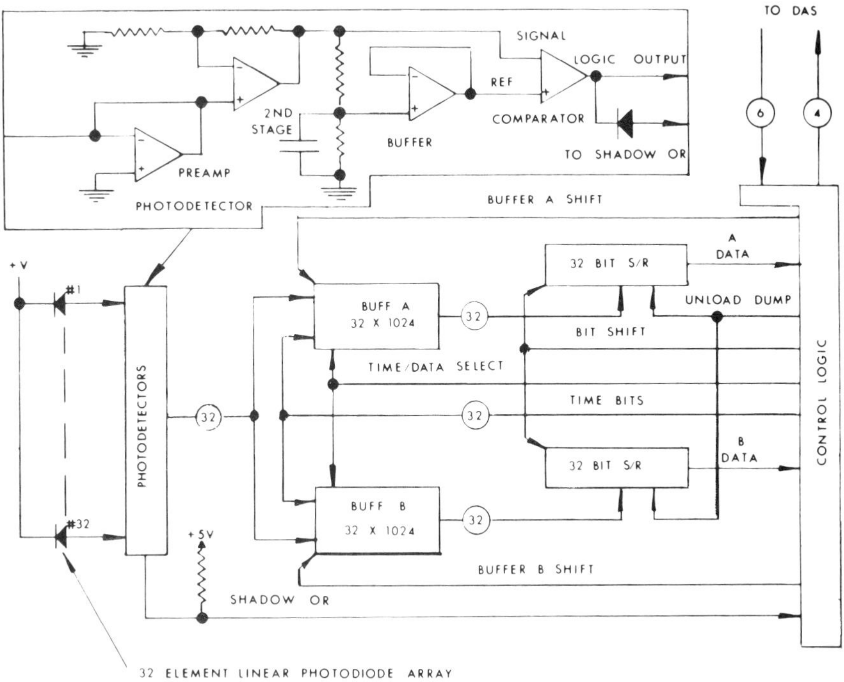

Figure 2. Simplified 2-D optical array probe circuit diagram.

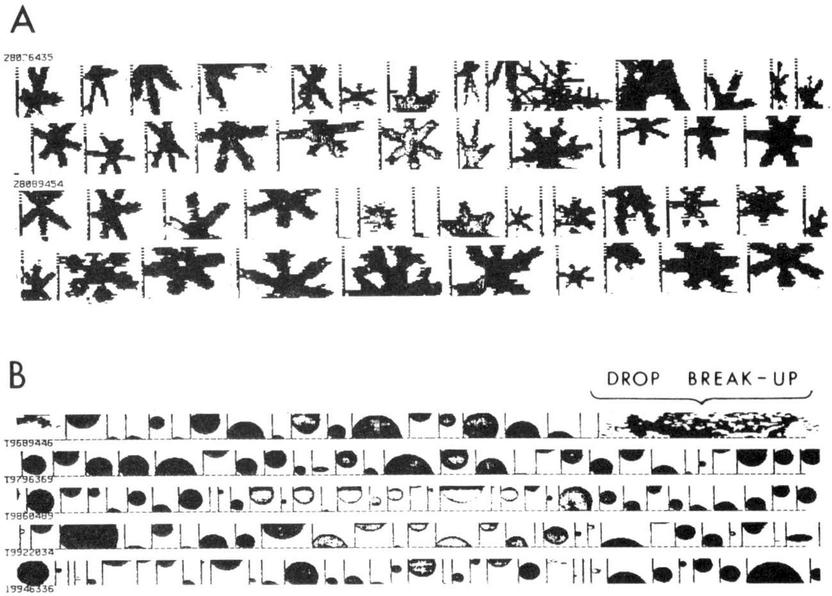

Figure 3. Samples of 2-D imagery in snow and rain taken at 80 m sec^{-1} airspeed.

spectrometer will have significantly less coincidence error because of the intrinsic thinness of its associated view volume.[2]

Discussion of Design Criteria for Single Particle Light Scattering Spectrometers

From the above discussion it is apparent that even if cost were not a factor, light scattering spectrometers would be necessary. They have no fundamental resolution limit provided one can achieve the necessary energy density. On the other hand, it has been difficult to design single particle light scattering spectrometers that are truly *in situ* instruments. We have found it only possible to achieve *in situ* sampling with single particle light scattering spectrometers when the measurement is combined with an imaging system. In this case, the imaging system allows for an independent means of sample volume determination through particle trajectory analysis. In the remainder of this paper, we will discuss only instrument designs in which high resolution imaging systems are used to collect the scattered light and provide the needed resolution to perform trajectory analysis. Two somewhat different techniques will be discussed wherein sample volume limits can be established using double pulse height analysis of image signal characteristics produced by masked beam-splitters.

Resolution

In the imaging spectrometer discussed, up to 64 array elements were employed giving roughly one part in 64 resolution. In most cases this high resolution is useful in particle sizing applications. It is also necessary to provide sample area. With regard to light scattering instruments, resolution is not as important since, because of various added uncertainties, increased accuracy is not obtained with added resolution. What is sufficient resolution? One can find frequent reference in the literature to the use of multichannel analyzers with up to several thousand size channels. However, multichannel analyzers were designed for nuclear event transient analysis and not particle size measurements. In general, there is little necessity to provide greater resolution than that dictated by the accuracy and precision of the instrument. In our single particle scattering spectrometers, we typically provide 15 size channels and 4 size ranges which when combined allow for up to 60 size classes. Greater resolution (more size classes) can not generally be supported by the inherent instrument accuracy. We have also found size classes of equal width to be preferable to nonlinear weighting of size intervals, e.g., logarithmic. This is particularly true when the size ranges overlap and one is trying to compare data from adjacent ranges. It should also be remarked that higher resolution can be more easily supported at submicron sizes than at larger sizes. The slope of the scattering signal with particle size is so steep in the submicron range that effects of refractive index are not so severe. We typically provide 100 Å resolution in the 0.1–0.3 μm range on our most sensitive spectrometers. The greatest uncertainty in particle sizing accuracy occurs in the range

2. View volume is the total illuminated volume over which particle interaction is observed.

of 0.8–1.5 μm which happens to coincide with the first resonant dip in the calibration curve calculated from MIE theory. More will be said in this regard; however, because of the general increase in uncertainty in this range, high resolution is not as meaningful since it does not carry with it high accuracy.

Optical Configuration

Next, we should establish the optical configuration most useful for single particle light scattering spectrometry. In recent years there has been a great deal of discussion about the various attributes of monochromatic vs. white light sources, forward scatter vs other collecting geometries, the choice of solid state photodiodes vs photomultiplier detectors. The debate is still in progress.

Light Source.—At PMS, our first systems used white light sources (both tungsten filament and high pressure arc lamps). However, today we invariably employ gas lasers in most of our spectrometers. The line of reasoning developed to support this trend lies in several areas. First, *in situ* sampling demands unusual requirements that generally dictate monochromatic collimated illumination. Monochromatic light is required for filtering out background light. Collimated illumination is desirable in order to maintain uniform energy density over the sample volume dimensions. A laser adequately meets both requirements.

Collecting Geometry.—It is also desirable that a scattering spectrometer collect a larger percentage of the forward scattered light. Forward scattering instruments are least sensitive to refractive index and particle morphology for sizes up to 5 μm if the diffraction peak is collected. Of course, it would be desirable to collect all of the scattered light, but this is not possible without giving up the *in situ* sampling requirement. What we are saying is that if one is limited to a certain solid angle, the best choice is to place it to collect the forward scattered components. One should always remember that the light diffracted by the particle is a function only of the particle cross sectional area. The remainder of the light scattered by a particle is either reflected, refracted or absorbed, all of which are also sensitive to the particle refractive index. For particles larger than 1 μm radius, 85% of the diffracted light is contained at angles smaller than 20°. It can be assured that if one could indeed separate the diffracted light components, he could design a nearly ideal instrument. Unfortunately this can only be done if the particles are totally absorbing. For particles with less than total absorption, a significant fraction of the refracted light overlaps with the diffracted components and separation is impossible.

Resonant Scattering.—With sufficient temporal and spatial coherence, the refracted and diffracted forward scattered components can constructively or destructively interfere (resonance) making the combined particle scattering signal a nonunique function of particle size. Herein lies the only disadvantage in the use of monochromatic radiation. Most lasers possess spatial coherence (uniphase wave front), perhaps its only value being the ease with which theoretical MIE calculations can be applied. The greatest undesirable effect of spatial coherence is the pronounced resonance in the scattering, giving rise to nonunique calibration. However, the real effects are generally not as severe as those computed. What then are practical implications of resonant scattering?

Formal solutions to light scattering require MIE computations. However, treating diffracted, refracted and reflected components as separable wave functions provides additional insight. If we assume no absorption and strong forward scatter, we can consider only the diffracted and refracted light as significant. The phenomenon of resonant scattering results from interference of diffracted and refracted light (so-called anamolous diffraction). It results because of the phase shift in the refracted wave proportional to the optical path through the particle. Refracted light diverges in a cone of nearly constant angular substance regardless of particle size whereas the diffracted light distribution is size dependent. Maximum possible interference (constructive or destructive) occurs when waves have similar divergence and amplitudes. Eighty five percent of the diffracted light is within the airy disk central maximum, its half angle being $\sin \theta = 0.61\ \lambda$. Refracted light for a sphere at $m = 1.5$ is contained within a half angle of 45°. Similar half angles for diffraction occur where $r = 0.6\ \mu m$. The phase shift is 180° so we have destructive interference and minimal scatter in forward directions. This simple model is less appropriate as one moves off axis where phase shifts in diffracted light require consideration but it generally describes the behavior in forward scattering. Note that absorption reduces refracted light resulting in less interference.

An examination of assumptions underlying MIE scattering in the above examples is also worthwhile. The critical assumption is that of the uniphase wavefront which in the limit is impossible to produce. MIE computations produce maximum resonance in scattering. All real situations result in reduced resonance. Lasers are often thought of as uniphase fronts but this is approximately true only for TEM_{00} mode. The effects of nonuniform phase across the beam must be compared with the particle size to see if the source region for diffracted light (requiring several Fresnel zones) is a region of nearly constant phase. It is easy to create nonuniform phase fronts even with lasers. Multimode tubes serve the purpose quite well. A high order multimode laser beam may have 10 complete phase reversals across its diameter. Such a beam focused to 30 μm diameter would illuminate particles in the resonance range with 180° phase shifts. The effects of phase variation reduce resonance and result in a more desirable calibration curve. The effect can be seen in Figure 4 where the beam diameter is varied with respect to the particle size with predictable changes in resonance affected calibration.

In summary one sees that the actual resonant scattering is not as pronounced as that calculated. Even if it were, the laser would still be necessary for true *in situ* sampling. Additional arguments can also be presented which are of a subtle nature. The average aerosol particle is generally an agglomerate of many smaller particles, each having a different refractive index. This is particularly true of the larger aerosol particles where resonance effects are most pronounced. Their net effect would be to produce suppression of resonance through randomization of the phase of the refracted components.

One cannot perform calculations for particles of amorphous shape; however,

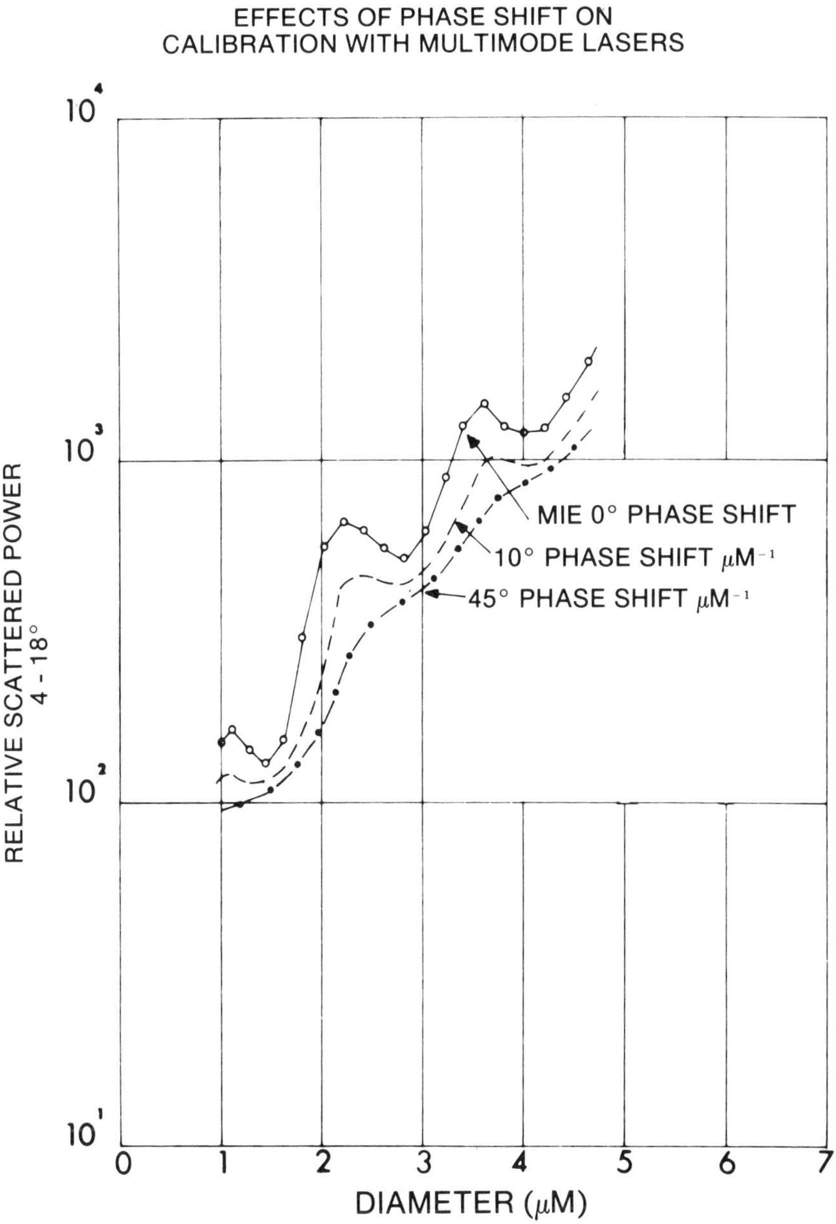

Figure 4. The suppression of resonance effects when multimode lasers are used as illumination sources. The top curve is the theoretical MIE response. The lower curves show the reduction in resonance amplitude due to variations in phase across the wave front.

certain effects would be inferred. Particles having crystalline faces begin to scintillate at sizes between 5 and 10 μm and the collection of scattered light at favorable orientations can provide widely varying response. The collection of forward scattered light avoids most of this uncertainty. Also, consider the extreme example of a 2 μm diameter totally absorbing spherical particle and two collecting geometries, one from 4°–22° and a second from 22°–180°. Since there is only diffracted light, and 85% of the diffracted light is collected within the first 22°, the 4°–22° collecting geometry will collect seven times the amount of light as the 22°–180° collecting geometry with 1/20 the effective solid angle.

We conclude that only a collimated laser gives one sufficient steerable light power to enable strong forward scattering light collection. Other arrangements where the resonance effects can be suppressed using lasers include active scattering, combined scattering extinction and multiwavelength laser scattering.

Detectors

Most of our particle size spectrometers employ solid state photodiode detectors whether operating as imaging or light scattering instruments. There is no question that ultimate sensitivity is greatest with photomultiplier tubes; however, with *in situ* sampling one generally also has greater background light. In such situations the ultimate detection of the photomultiplier tubes can never be realized and the signal-to-noise ratio becomes governed entirely by detector quantum efficiency which is several times higher for photodiodes. Knollenberg and Luehr (1976) have provided detailed calculations of signal-to-noise ratios comparing photomultiplier tubes and photodiodes which the reader can refer to for more detailed discussion.

SAMPLE VOLUME DEFINITION THROUGH TRAJECTORY ANALYSIS

In making *in situ* measurements, the greatest problem lies in establishing sample cross section limits. In standard single particle counters the illuminated field is large enough that the particle flow intersects with a small enough region to be considered of uniform intensity. However, lasers (particularly TEM_{00} mode) produce highest intensity only at the beam center and one must limit measurements to regions of uniform intensity without restricting particle flow.

Two methods are now employed at PMS in single particle scattering spectrometers which can be used to establish a region of accepted measurement within the laser beam and provide control over sample volume size and total view volume. The first method (hereafter Method I) utilizes only the comparison of particle scattering signal amplitudes. The second method (hereafter Method II) utilizes both the comparison of optical scattering signal amplitudes and pulse width analysis. Both methods utilize masked beam-splitters to develop paired scattering signals from refocused images within a dark field.

Sample Volume Definition Using Method I

In Method I, depicted in Figure 5, a high resolution optical system collects the scattered light and reimages at 10× magnification or greater within a dark field. The transmitted laser beam is dumped at a central stop on the first lens element. A 50% beam-splitter produces two image planes for two detectors. The reflected image prism face is masked with a vertical slit to block central transmission during transit. The other detector prism face is unmasked. The masked beam-splitter derives two signals which, in conjunction with double pulse height analysis, provide a means of determining if a particle's position is in the desired sample volume. The relative size of the sample volume cross section with respect to the laser beam is depicted in Figure 6. The sample cross section includes only the region near the center of the laser beam. The sample cross section is noticeably diamond shaped. The center of this diamond cross section coincides with the object plane of the collecting optics. The points of this diamond cross section define the limiting depth-of-field. Both the width of the sample cross section and the depth-of-field vary inversely with the magnification used in the collecting

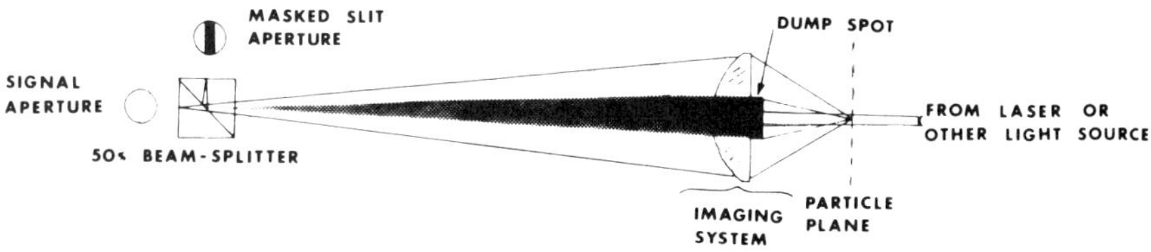

Figure 5. Simplified optical system for Method I.

optics. The diamond shaped sample cross section results from the accept/reject criteria used after comparing signals at the masked and unmasked detectors.

The light transmitted on axis through the beam splitter is the signal used to size the particles. The light reflected at 90° forms its image on a circular aperture with a central opaque vertical mask. Shown in Figure 7 are the size and positions of images formed by particles at various positions in the illuminated volume. It is important to understand that for small particles (essentially point objects) the image size is only a function of its axial displacement from the object plane. The image size is linearly related to the numerical aperture of the collecting optics and is given approximately by: image size = N.A. × displacement from object plane. It is apparent that only images that are near the object plane and the center of the sample volume form images with light concentrated on the opaque slit on the masked detector. Thus, such images transmit little signal through the masked aperture. We normally use a gain ratio of masked aperture detector to signal aperture detector of 4× for best noise immunity. Particles whose pulse amplitudes, as seen by the masked aperture detector, are greater than those seen by the signal aperture detector are rejected. The diamond shaped sample cross section results. The sample cross section is the product of the sample volume width and one-half the depth-of-field.

This method of trajectory analysis has several unique advantages. First it is possible to localize as small a region as necessary to achieve uniform illumination by increasing magnification. The sample cross section is an inverse function of the square of the magnification ($\propto M^{-2}$) while the sample volume and view volume are $\propto M^{-3}$. Thus, coincidence errors can be reduced by increasing

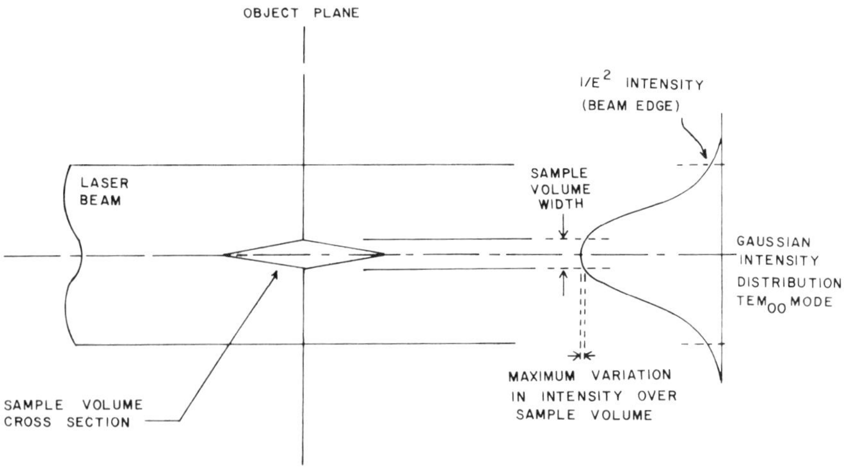

Figure 6. Relative size and position of sample volume cross section.

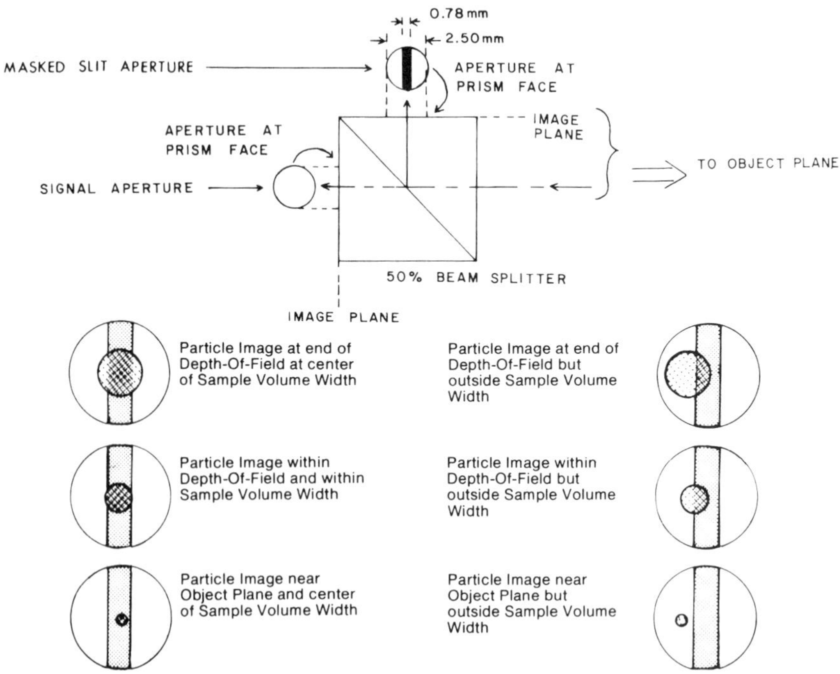

Figure 7. Image sizes and positions on masked aperture detector.

magnification to most any desired limit. A third result of the use of Method I is that stray light will also be controlled in the same proportion as coincidence error. We have used this type of system to localize central regions of a laser beam less than 20 μm wide. Thus, with a highly focused He-Ne laser and 1 mW output power, one can achieve 0.1 μm sensitivities. The only disadvantage lies in the very small sample cross section that results. Method I is limited primarily to measurements at small sizes (less than 5 μm) which behave essentially as point objects. At larger sizes the image size becomes a function of particle size if high magnification is employed and increases the probability of being rejected (cannot differentiate large particle images from images of out-of-focus unresolved particles). Correction for the loss of sample volume with increasing size can be made; however, Method II is more commonly employed at those sizes where noticeable size dependency exists. Method I is equally applicable to white light sources.

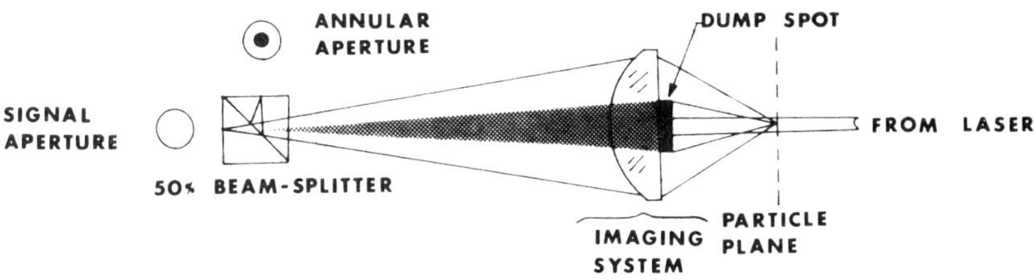

Figure 8. Simplified optical system for Method II.

Sample Volume Definition Using Method II

Whenever it is desirable to maintain a somewhat larger sample cross section and where measurements are desired for larger particles, another method of imaging trajectory analysis is used which utilizes scattering signal pulse pair analysis using a masked detector to limit depth-of-field and transit time analysis to determine sample cross section width.

In instruments employing Method II (depicted in Figure 8) the laser is a high order multimode 5 mW He-Ne tube and focused to approximately a few hundred microns diameter using condensing optics. The point of focus coincides with the center of the sampling aperture. Particles passing through the laser beam in the sampling aperture scatter energy into the optics. The amount of scattering for a given particle size is a function of the laser mode, as well as the exact chordal transect through the beam. With a high order multimode laser, the probability of intersecting the peak isophote is as high as 90% (typically 80%). In fixed flow systems, the particle transit times can be measured precisely, allowing for edge effect rejection using pulse-width comparison techniques.

The scattered energy is relayed from the object plane and focused through an interference filter to a masked beam-splitting prism. The masking on the prism is a central circular dump spot on the 90° transmitting face creating an annular aperture. This serves to allow only light transmission from particle images when the particles are displaced a certain distance in either direction from the object plane and no transmission when at or very near the object plane. As long as the magnification is low, the image size is not an appreciable function of particle size and all particle images will give the same percentage of transmission through the masked aperture at similar depth-of-field limits.

The total depth-of-field over which a particle is viewed extends over about a 5 mm path. It is usually necessary to truncate the actual depth-of-field to considerably less than this distance to avoid errors associated with the loss of collecting aperture and scattering signal. This is accomplished by comparing signal levels from each detector using peak readers (see Figure 9). Because of the annular aperture of the masked prism face, the light from a particle near the object plane is greatly reduced and is, in general, less because of the loss of collecting aperture. The gain ratio of annulus signal to scattering signal must be greater than unity and is normally 2.5:1. The depth-of-field is defined as that region where the signal pulses are of greater amplitude than the annulus signal pulses. The depth-

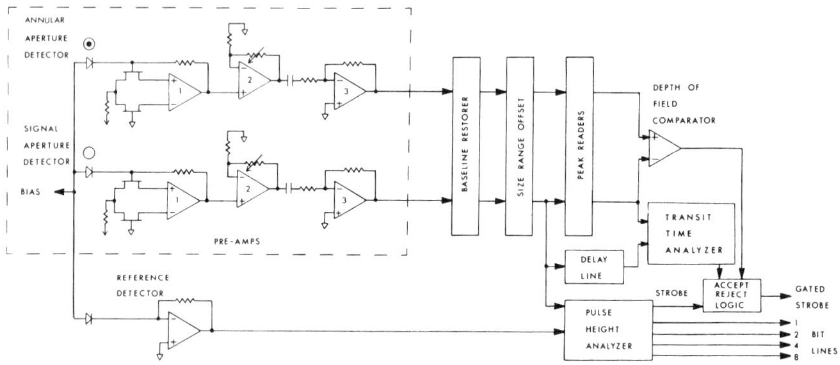

Figure 9. Simplified circuit block diagram of instruments using Method II sample volume definition.

of-field may be varied by increasing or decreasing the gain in the annulus signal path. The increase in gain decreases depth-of-field while a decrease in gain increases depth-of-field. The instrument is normally set with a depth-of-field between 2 and 3 mm for several reasons. First, the depth-of-field cannot be reduced much below 2 mm because of optical considerations. Secondly, reduction of depth-of-field requires increase of annulus gain and deteriorates annulus S/N. Thirdly, the illumination is uniform enough to allow a 3 mm depth-of-field. The small changes in angular distribution of scattered light with differing drop size have negligible effect on depth-of-field as long as the gain ratio is sufficiently high.

In order to reject pulses from particles passing through the edge of the laser beam, the transit time of the particle pulses must be analyzed. At constant velocity the transit time is determined by the particular chordal transect through the laser beam. When the particle passes through the center of the laser beam, it passes through the widest section and produces the longest transit time. Near the beam edges, the transit times are very short. The beam edges are also regions of weaker intensity, and particles passing through the edges are thus undersized.

In order to avoid pulse amplitude dependency on pulse width, measurements must be made at a fixed percentage of peak amplitude. The signal pulse train is first delayed by a delay line for several microseconds. The output of the signal peak reader is divided by a factor of two to derive a comparator threshold at one-half the peak amplitude. The half-width of the delayed pulses are then measured to derive transit time pulses. Particle transit times are measured and compared with stored average values (the average transit time for about 1000 pulses is stored in an up-down counter which maintains a running mean of the average particle transit time). Pulses greater than average in width are accepted.

GENERAL DESCRIPTION OF PARTICLE SIZE SCATTERING SPECTROMETERS USING METHOD I AND II

As was mentioned in the previous section the use of Method I is generally restricted to particles smaller than a few microns and Method II is employed at larger sizes. It should also be pointed out that Method II generally results in a much larger sample cross section as well. In this section we will describe several spectrometers which use Method I and Method II. These are summarized in Table I along with several other instruments which are not fully described due to space limitations.

Instruments Using Method I

The Active Scattering Aerosol Spectrometer (ASAS).—The ASAS is designed with four overlapping size ranges with each size range divided into fifteen linear size intervals providing up to 60 size channels. It is packaged in two separate enclosures, housing the probe with its optics and pre-amps and an electronics console (See Figure 10). The ASAS Probe utilizes aluminum extrusions for mechanical stability. Laser and detector alignment is achieved with spring-loaded x-y screw adjustments. The ASAS laser is a hybrid He-Ne (632.8 nm) tube with an aluminum envelope specially designed for the ASAS. The ASAS pulse height analyzer has its reference voltage derived from the source of illumination providing effective automatic gain control (AGC). A programmable amplifier is used to gain switch and provide the size ranging to accommodate the large dynamic range of the instrument.

The primary method of aspiration is unlike any used in conventional aerosol counters. It is in essence a miniature wind tunnel. A 10:1 accelerator produces a flow rate of approximately 6.15 cm sec^{-1} intersecting with a 3 cm length of the laser beam. The sample volume is positioned at the center of flow. The instrument relies entirely upon optical definition of the sample volume using full trajectory analysis by Method I.

The data acquisition system within the electronics console has a MOS memory with twenty addresses. Fifteen of the addresses are used for particle size distribution storage. The remaining addresses are normally used for other housekeeping

TABLE I. Optical Parameter Values for Various Aerosol Spectrometers

	Active Scattering Aerosol Spectrometer	Classical Scattering Aerosol Spectrometers			
	ASAS-300	LAS-100	CSASP-100	CSAS high volume	FSSP-100
		Method I			Method II
Size ranges	(4) 0.15–0.3, 0.2–0.6, 0.4–1.0, 0.6–3.0 μm	(1) 0.3–5.0 or 0.6–10.0 μm	(2) 0.3–3.0, 1–20 μm	(1) 1–30 μm	(4) 0.4–6.0, 1–15, 2–30, 3–45 μm
Size channels/size range	15	15	15	15	15
Minimum detectable size	0.15 μm[a]	0.3 μm	0.3 μm	1 μm	0.4 μm[b]
Size resolution	0.01 μm typical in 0.15–0.3 μm range	0.2 μm typical	0.1 μm typical in 0.3–3.0 μm range	1 μm	0.1 μm typical in most sensitive range
Sample cross section	0.018 mm^2 @ 10X magnification	0.018 mm^2 @ 10X magnification	0.018 mm^2 @ 10X magnification	1.7 mm^2	0.2 mm^2
Flow velocity	6.15 m sec^{-1}	8.0 m sec^{-1}	6.15 m sec^{-1}	30 m sec^{-1}	100 m sec^{-1} typical
Sample rate (volume)	0.1 cm^3 sec^{-1} @ 10X magnification	0.14 cm^3 sec^{-1} @ 10X magnification	0.1 cm^3 sec^{-1} @ 10X magnification	50 cm^3 sec^{-1}	20 cm^3 sec^{-1}

a. 0.05 μm possible with photomultiplier tube detectors.
b. 0.2 μm possible with photomultiplier tube detectors.

information such as selected size range and time of day. An eight channel subcommutated analog input is also available. A selectable digital display and graphical CRT display are provided for real-time data monitoring requirements of precise particle counts or distribution functions. Magnetic tape and line printer interfaces are standard with the ASAS system.

The optical diagram shown in Figure 11 is the typical ASAS laser cavity configuration. The plasma tube is sealed with a high reflectivity curved mirror and a Brewster's window. An externally adjustable, high reflectivity mirror completes a hemispherical cavity laser operating in TEM_{00} mode. At the sample volume location the laser beam is approximately 500 μm diameter and produces a power density in excess of 500 W cm^{-2}.

The high resolution, multi-ranging capabilities of the ASAS have greatest importance in the submicron size range. Figure 12 illustrates the spectral resolu-

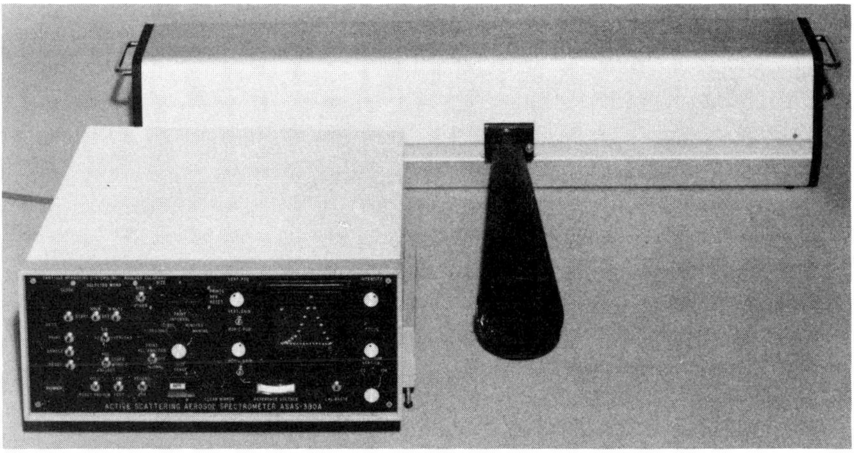

Figure 10. Photograph of active scattering aerosol spectrometer system, probe and electronics console.

tion of the ASAS with a mixture of three DOW lattices on Range #2. All 15 size channels are sampled simultaneously. Size ranges may be manually selected; however, in the AUTO mode, the ASAS is programmed to sequence through all size ranges. Size range and size channel information is encoded in the data outputs.

Air flow rates in the ASAS are invariant and determined to a greater accuracy than in plumbed systems. The constant speed axial blower varies less than 1% with a ±10 volt change in line voltage. To provide airflow adaptation to existing plumbing, a test section insert is provided.

Classical Scattering Aerosol Spectrometers.—The newest classical scattering spectrometer is the Laser Aerosol Spectrometer (LAS) shown in Figure 13, which will be described here. Additional classical scattering spectrometers are described in Table I. The standard LAS has 15 size intervals covering the 0.3–5.0 μm size range. The laser is prealigned. Detector alignment is achieved with two spring-loaded x-y screw adjustments. The LAS also uses a reference voltage derived from the source of illumination to provide AGC. A 100:1 accelerator in the LAS produces a flow rate of approximately 8.0 m sec^{-1} intersecting with a 1 cm length of the laser beam. The sample volume is positioned at the center of flow.

The LAS has a MOS memory with sixteen addresses. Fifteen of the addresses are used for particle size distribution storage. The remaining address is normally used for elapsed time. A selectable digital display is provided for real-time data monitoring requirements of precise particle counts. The LAS provides outputs to drive a laboratory oscilloscope for real-time displays of particle distribution functions and a line printer interface with automatic print functions.

The LAS has a simplified optical system. The diagram in Figure 13 shows the typical LAS optical system. The laser tube is a 5 mW unit operating high order multimode. A microscope objective focuses the laser beam down to approximately 60 μm diameter at the sample volume and produces a power density in

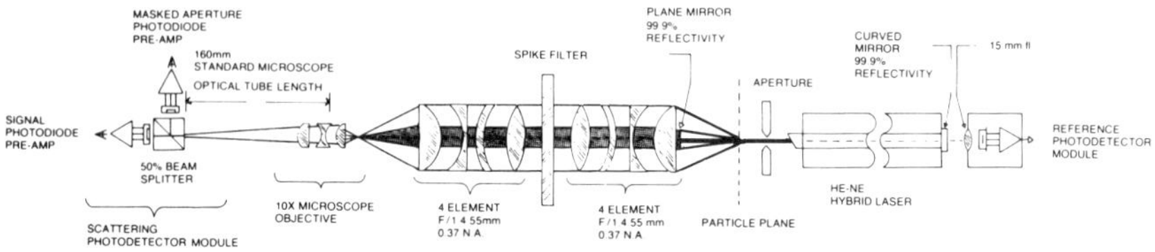

Figure 11. Active scattering aerosol spectrometer optical system.

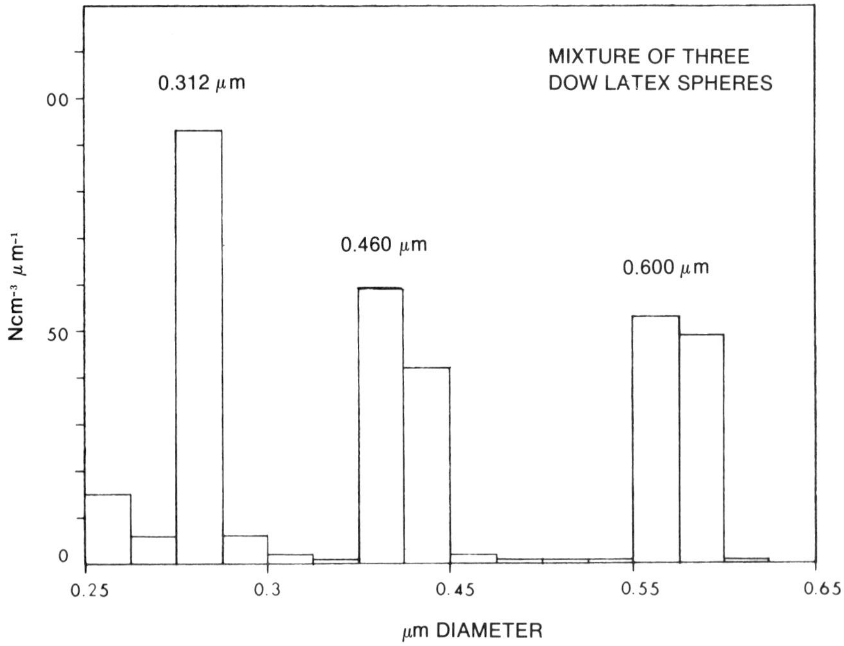

Figure 12. Particle size spectrum from a mixture of three DOW latices illustrating resolution of the active scattering aerosol spectrometer.

excess of 100 W cm^{-2}. The LAS uses standard microscope objectives in the imaging collecting optics with sample volume definition by Method I.

Instruments Using Method II

Forward Scattering Spectrometer Probe (FSSP).—The FSSP is a new airborne instrument designed for *in situ* particle size measurements using Method II which can also be used aboard ground vehicles (see Figure 14). The FSSP has four overlapping size ranges with each size range divided into fifteen linear size

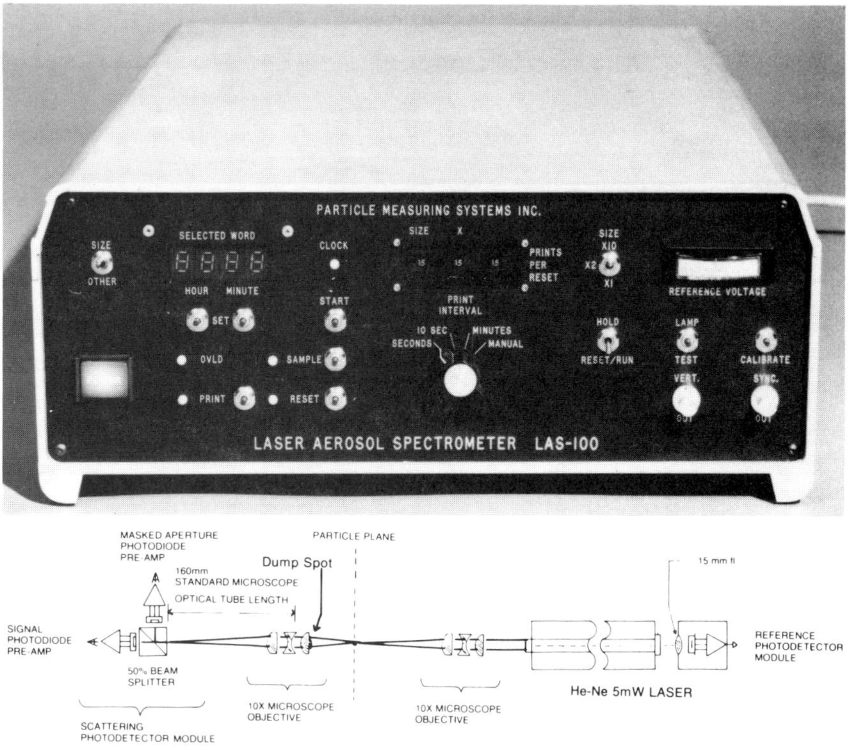

Figure 13. Top: laser aerosol spectrometer. Bottom: Optical system.

intervals providing up to 60 size channels in the 0.5–45 μm range. Detection to 0.2 μm is possible with photomultiplier tubes. The FSSP is packaged in a cylindrical pod as a plug-in assembly. The FSSP has similar AGC circuitry and programmable amplifiers for gain switching. The probe electronics are self-contained, the output being the particle size in binary code accompanied by a strobe pulse to increment an appropriately addressed memory channel in a data acquisition system. The FSSP interfaces directly to minicomputers, computers or other data acquisition systems with memory storage.

The FSSP optical system is shown in Figure 15. The laser used in the FSSP is a high order multimode 5 mW He-Ne tube. The beam is focused to approximately

200 μm diameter using condensing optics with a 60 mm effective focal length. The single most important feature of the FSSP increasing its performance over earlier ASSPs[3] is its improved optical system which reduces the sample volume to where coincidence errors are less than 10% at concentrations of 10^3 cm^{-3}. Figure 16A illustrates comparative cloud droplet spectra from flights using the PMS 206 aircraft and a high framing rate data system having one-fifth second resolution. Fig. 16B illustrates pairs of droplet spectra from two different FSSPs. The results are typical of the intercomparisons of two identical instruments.

For cloud physics work the FSSP normally has a primary size range of 2–30 μm diameter, resolving particles into 15 size channels. Four switchable size ranges

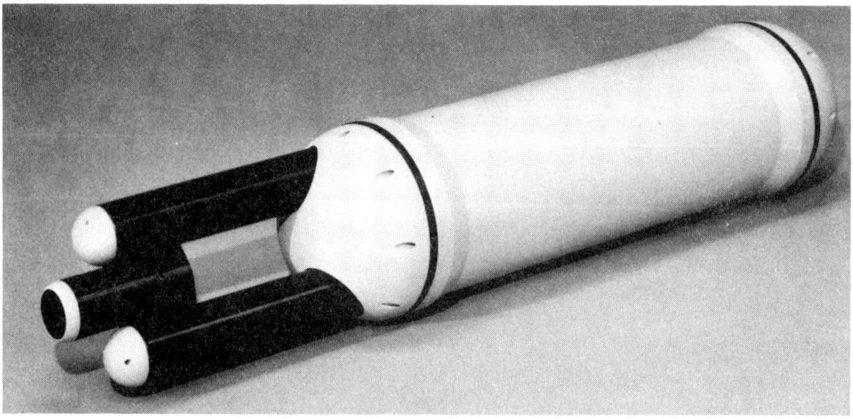

Figure 14. The forward scattering spectrometer probe.

cover 0.5–45 μm. It is normally set up to size particles having velocities from 10–125 m sec^{-1}.

For aerosol work photomultiplier tube detectors normally replace the photodiode detectors. The total size range covered is generally 0.2–10 μm using four overlapping size ranges.

DISCUSSION AND CONCLUSIONS

We have described single particle light scattering spectrometers covering particle sizes from 0.1–45 μm. These instruments all employ forward scattering and rely upon particle trajectory analysis to establish sample volume in relatively undisturbed flow. Of the two types of trajectory analysis presented, Method I is the more powerful technique. Using this method, one is limited only by the illumination densities he can achieve. Background light, whether arising from stray illumination or molecular scattering, can be reduced to nonlimiting values. With

3. The ASSP (Axially Scattering Spectrometer Probe) is the original instrument designed by PMS using Method II. Today it is built only for users who cannot accommodate the higher performing FSSP because of aircraft hardpoint mounting problems.

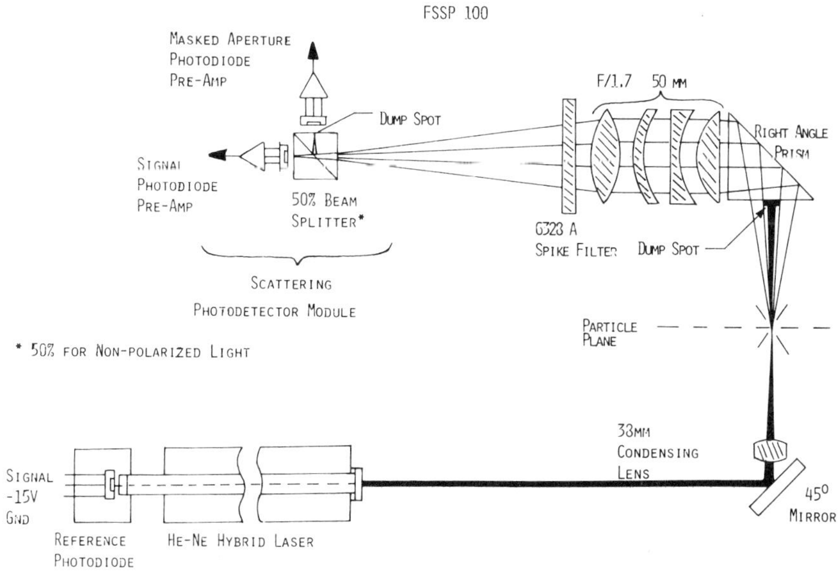

Figure 15. Forward scattering spectrometer optical system.

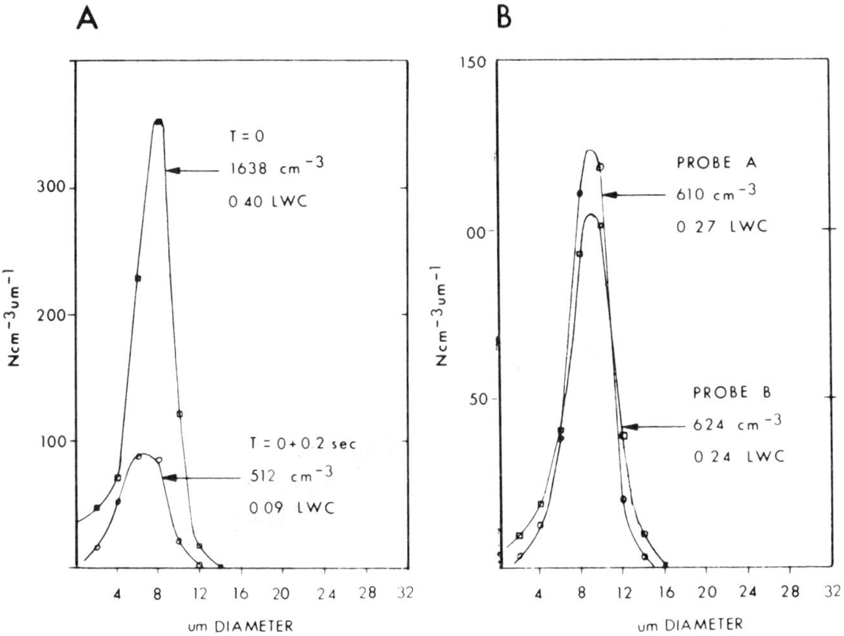

Figure 16. Flight data sampled near cloud base with forward scattering spectrometer probes.

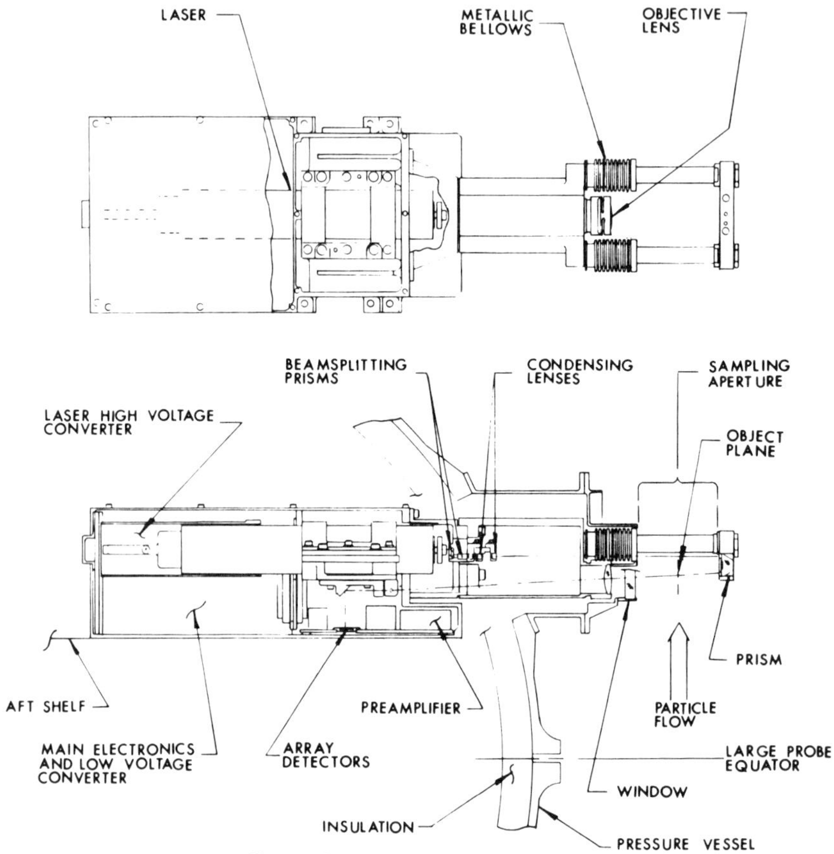

Figure 17. Pioneer Venus instrument.

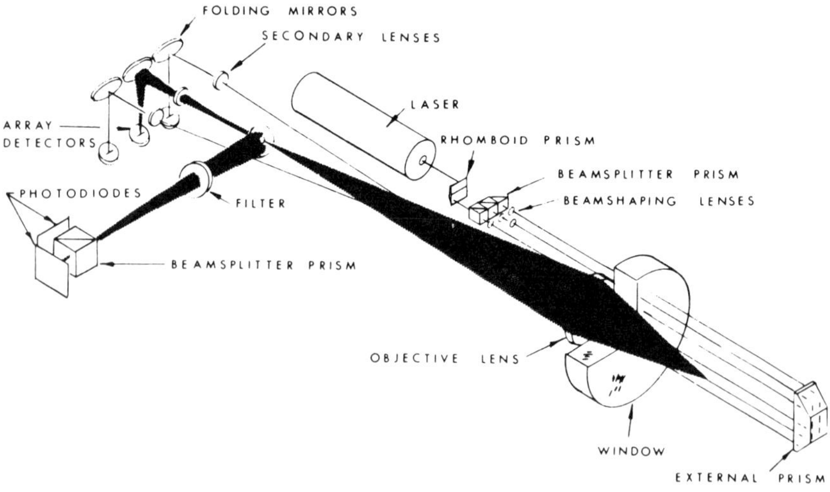

Figure 18. Pioneer Venus instrument optical diagram.

photomultiplier tube detectors and high magnification (to achieve near zero background) the ultimate limit of detectivity using electron counting can be realized. This limit is obviously determined for a particle size which generates at least one photoelectron from the collected scattering signal. We have not presented a case for electron counting techniques because they result in extremely small sample volumes (unless in a vacuum) and fit an unusual category of measurements. The limit is below 100 Å using high powered lasers in a vacuum.

We have tried to show attributes of imaging systems when combined with light scattering techniques. An instrument that we have configured which sizes both by shadowgraph imaging and light scattering is the Cloud Particle Size Spectrometer for Pioneer Venus in 1978. The instrument contains a single primary imaging system and secondary optics to produce three shadowgraph imaging optical array spectrometer size ranges and one light scattering size range (see Figures 17 and 18). The light scattering size range covers particle sizes from 0.5–5.0 μm while the optical array covers sizes from 5.0–500 μm. The instrument will make measurements from 68 km to the planetary surface where temperatures exceed 800°F and pressures 100 atmospheres. The instrument represents the result of optimizing imaging as well as light scattering methods to cover a wide dynamic range.

REFERENCES

Knollenberg, R. G., 1970: The optical array: an alternative to extinction and scattering for particle size measurements. *J. Appl. Meteor., 9.*

———, 1972: Measurements of the growth of the ice budget in a persisting contrail. *J. Atmos. Sci., 29,* 1367–1374.

———, 1976: Three new instruments for cloud physics measurements: The 2-D Spectrometer, The Forward Scattering Spectrometer Probe, and The Active Scattering Aerosol Spectrometer. Presented at *The International Cloud Physics Conference,* July 26–August 6.

——— and R. E. Luehr, 1976: Open cavity laser "active" scattering particle spectrometry from 0.05 to 5 microns. *Fine Particles: Aerosol Generation, Measurement, Sampling, and Analysis,* B. Y. H. Liu, Ed., Academic Press, 669–697.

PERFORMANCE OF THE PMS AXIALLY SCATTERING SPECTROMETER PROBE

RICHARD K. JECK

Naval Research Laboratory, Washington, D.C.

The Atmospheric Physics Branch at the Naval Research Laboratory (NRL) was one of the first customers of Particle Measuring Systems, Inc., (PMS), of Boulder, Colorado. An Axially Scattering Spectrometer Probe (ASSP) and an Optical Array Probe (OAP) were delivered to NRL in June of 1973. Since that time we have used both probes in a number of experiments in the laboratory, aboard aircraft and ship, and in other field applications involving the size measurement of ambient aerosols, and cloud and fog droplets.

The principle selling points which influenced our decision to buy the PMS probes were the following:

1) The probes are designed for *in situ* measurements with the sampling volume essentially open to the environment. This design avoids the possibility of inadvertent loss or evaporation of volatile particles in inlet tubing and makes it easier to accomplish isokinetic sampling conditions during airborne measurements.

2) The system provides real time display and digital readout of the measured size spectra. This is a great advantage over impactors, for example, where post-sampling analysis is tedious and time consuming and thereby discourages one from taking more than the minimally required number of data exposures.

3) The design incorporates electro-optical features to overcome the "edge effect" problem by detecting and eliminating from the count those particles that are only partially illuminated by the light beam and which would therefore be subject to undersizing.

4) The design promises good size resolution by dividing each of several ranges into 15 channels or size categories.

While the instruments do advance the state of the art in these areas, we have nevertheless discovered several sources of systematic error which can easily go undetected. These errors can significantly affect the indicated particle size spectrum and/or number density (number of particles per unit volume) without being apparent to the operator. Therefore we feel it is important to make available to current and potential users the results of our experience.

Some of the difficulties arise from anisokinetic sampling conditions when the probes are deployed in other than the airborne situations for which they were designed. In particular the ASSP (and it is to this probe exclusively that we'll be

TABLE I. Problems Encountered during Operation of the ASSP

		Source of Problem			Data Affected		
Problem	Effect	Field problem	Design problem	Inadequate factory documentation	Clear air	Haze	Fog
Electronics module overheated	Instrument failure	X	X	X	X	X	X
Water in optics	Loss of cal.	X	?	X	X	X	X
Unsecured lens cell	Variable d.o.f.		X	X	X	X	X
	Errors in sampling volume			X	X	X	X
Improperly set response times	Saturation of particle count	X		X	X	X	X
	Errors in number density	X		X	X	X	X
Side effects of edge rej. circuitry	Distorted drop size spectrum		X	X		X	X
	Errors in readout of velocity & vol.		X	X	X	X	?
Anisokinetic sampling							
Fast aspiration	Errors in number density	X		X	X	X	X
	Distorted drop size spectrum	?		X		?	?
Use of horn (with fast aspiration)	Errors in number density	X		X	X	X	X
	Distorted drop size spectrum	X		X		?	X

referring in the remainder of this article), due to its apparent versatility, invites one to use it in a variety of laboratory and other stationary applications which can lead to the difficulties described below. The unrecognized measurement errors result at least in confusing data which can require literally weeks of extra work to interpret and to correct for the effects of the errors. At worst the data may be completely invalidated even though during operation the counts appear to be registering normally on the display. In either case the errors may not be discovered until later analysis or until the data is compared with that from other particle sizing instruments or perhaps with transmissometers, visiometers or liquid water content sensors, for example, that are employed simultaneously with the PMS probe.

We have perhaps encountered more problems than most users, both because we were one of the first customers of PMS and did not have the benefits of previous customers' experience, and because we have tried to use the ASSP in quite a variety of applications. The problems that have been encountered so far are summarized in Table I. The first column lists these problems while the second column lists the effects of the problems on instrument performance. The third and fourth columns indicate whether the problem occurs only under certain application conditions or whether it is inherent in the design. The last three columns indicate, from our experience to date, which types of data will be affected by the malfunction or problem listed in the first column. Taken in order, a description of the problems and effects is given in the following paragraphs.

Item 1: Overheating of the ASSP Electronics Module

The ASSP is accompanied by several printed circuit cards, power supplies, etc., which are contained in a metal enclosure called the electronics module. The module can be seen in Figure 1. Since noise pickup considerations require the module to be connected to the ASSP with a short cable, a problem in properly shading the module may arise when the ASSP is deployed outdoors in the open. The module is not ventilated and solar heating can raise the interior temperature to a point where some of the semiconductor devices will fail. The LM311 (integrated circuit) comparators on the pulse height analyzer (PHA) cards are particularly susceptible. The owners manual specifies the operating environmental temperature to be between −40°C and +40°C but no cautionary information is given about shading the module from the sun.

Item 2: Water Collecting in the ASSP Optics

In Figure 1 the ASSP, electronics module, and OAP are shown mounted on bombracks beneath the wing of an S-2D type of aircraft. Unfortunately, deployment of the ASSP in this "nose down" orientation evidently permits water to accumulate in the collecting optics in the "nose" of the probe whenever the plane flies through clouds. There is apparently enough sedimentation of droplets into the exit aperture provided for the laser beam and the scattered light that the accumulated water will seriously degrade the performance of the probe. The net result is a shift in the calibration and an underestimation of particle number density, neither of which may be apparent from watching the real time data display. Errors such as these are often detected only if the post-measurement analyses of the data reveal size spectra and/or number densities that deviate sufficiently from the expected results to make the data suspect. Again, as was mentioned in the introduction, routine comparison of the ASSP data with simultaneous records from transmissometers, impactors, or liquid water content sensors is recommended in order to check for inadvertent measuring errors with the ASSP.

The possible accumulation of water in the optics is a serious problem also because water has been known to collect there and remain for days. This could jeopardize the validity of ASSP data from an entire field trip if the presence of the water goes undetected. Even worse, collected water has been known to degrade the optical thin film filter which is sandwiched between two lenses in the "nose" of the probe. This additionally attenuates the scattered light and can cause a gross

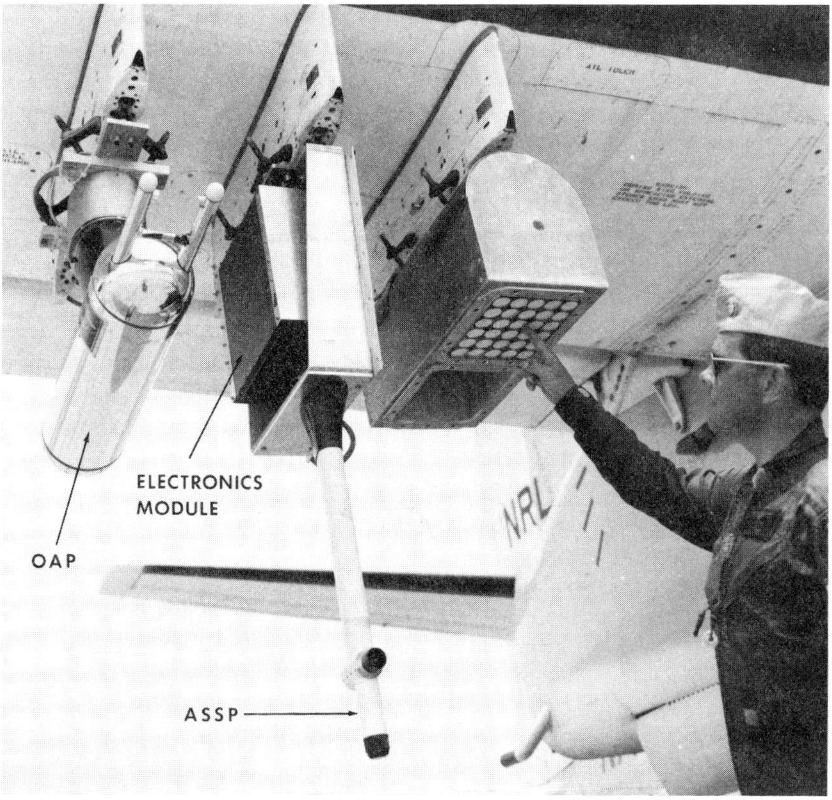

Figure 1. View of the ASSP and OAP as originally installed on the NRL S-2D research aircraft.

shift of the calibration toward larger particle sizes due to the resultant reduction in sensitivity of the optical system. Such optical degradation may be progressive and unless it is discovered by timely calibration checks, data from more than one field trip may be lost.

Aircraft which regularly fly over or near the ocean at low altitudes are usually washed periodically to reduce corrosion from salt spray. Water under pressure from washing apparatus can also force its way into both the ASSP and OAP with similar disastrous effects. The aircraft maintenance crew should be alerted to this danger and instructed not to carry out a washdown with the probes installed.

Item 3: Unsecured Lens Cell

The collecting optics in the "nose" of the ASSP consist of two lenses with an optical filter sandwiched between them. These components are packaged in a short cylindrical sleeve referred to as the lens cell. Both the lens cell and another cylindrical package containing a prism and the photodetectors slide into an interior, axial access tube extending the length of the "nose." Since the optical train has a definite focal length which is designed to view about 4mm of laser beam length in the center of the sampling aperture, the position of the lens cell within the access tube is critical in defining the depth of field and the location of the sensitive volume. In some of the earliest ASSP models the lens cell was not adequately secured in place and vibration or handling the probe could cause the cell to slide within the access tube. This would result in a variable depth of field and therefore in a variable sensitive volume which in turn leads to errors in the indicated particle number density. Since a displaced lens cell may not have any obvious effects on the observed size spectrum during routine operation, the resulting errors in number density can not only go undetected but the errors may be different from measurement to measurement if the probe is moved during the experiments.

PMS has since added a set screw to all probes to hold the lens cell in position. However, all the problems just described can still occur if the set screw loosens or is not tightened after performing maintenance or repair operations which require removal of the lens cell.

An itemized check list to be used before and after each use of the probe is probably the best way to avoid some of the problems described so far. Otherwise one may be faced with the difficult problem of having to redetermine what the actual depth of field was during some previous experiments where the lens cell was dislocated, for example.

Item 4: Improper Circuit Response Times

The analog and digital circuits in the ASSP system make use of several time constants and delay times, respectively, to provide optimum response to signals generated by particles passing through the laser beam with velocities in some anticipated range. In addition, certain clock rates which are vital to the proper performance of the edge rejection circuitry are preset according to the anticipated particle velocity range. For example, for airborne use the response times and clock rates are set to short values at the factory. For situations where the probe is dedicated to stationary applications, and thus should be aspirated at much slower airspeeds, the factory recommends lengthening these values to be compatible with lower particle velocities.

Problems can arise, however, if the available particle velocities end up being too high for the given response times or if one is attempting to sample air with a high particle number density. In such cases of high particle rates (a function of both the particle velocity and number density), the indicated number density will be artificially decreased with increases in particle velocity or in actual number density. This is a saturation effect associated with the finite dead time following the transit of a particle. The dead time is established by one of the digital delays in the gating circuitry. For sampling at aircraft speeds this dead time is set at 10μsec, thus enabling the probe to count up to 100,000 particles per second as long as the particles are neither coincident nor enter the laser beam less than 10μsec apart. For low speed aspiration, however, the dead time in one case was set at 170μsec by the factory. It was found in this case that with fog droplets being aspirated through the ASSP at about 30kt, saturation began to occur whenever the actual number density exceeded a modest value of about 100 cm^{-3}

This effect can easily go undetected by the operator since the counts accumulated per unit time on the readout/display will probably show an actual increase with increasing particle rate in many cases. It is only when a computation of indicated number density is performed, by dividing the particle count accumulated during a given time interval by the volume of air sampled during that interval, that the real trend in indicated number density can be determined.

Since each particle passing through the laser beam retriggers the dead time delay whether the previously triggered dead time has expired or not, it is possible that for sufficiently high particle rates the probe may hardly register a count. In practice it is more likely that just a finite decrease in the indicated number density will occur as the particle rate exceeds the limits dictated by the delay time.

In most laboratory set ups, testing for this effect is complicated by difficulties in maintaining a constant, actual number density in the aerosol samples drawn into the probe at varying aspiration rates. Attempts to avoid these problems by keeping the time constants fixed at short values and aspirating a stationary probe at correspondingly high airspeeds may improve this problem, but will probably lead to additional problems which also cause artificial depressions in the indicated number density as described in Item 6.

Even with normally fast delay times this saturation problem can apparently occur when trying to size fog droplets or other relatively large aerosols in the presence of smoke. If the smoke particles are large enough to contribute to at least the first size channel *and* also exist in numbers large enough to keep the dead time retriggered at a high duty rate, then the indicated number density for all particles will be erroneously low.

Other problems to keep in mind that are associated with adverse particle rates are particle coincidences at high actual number densities and failure of the edge rejection circuitry when the particle velocity either exceeds or falls below the range dictated by the preset clock rates. Both of these will affect the apparent size distribution of the particles.

To help guard against such inadvertent errors it is again suggested that whenever possible, data from other instruments such as transmissometers and impactors be used to check the ASSP response until its performance has been tested for all variations of the sampling conditions expected in any given application.

As with the other items described in this article, the owners manual contains no helpful information on these practical application problems and the inexperienced user is likely to apply the ASSP to error prone situations without being aware of the limitation of the instrument.

Item 5: Side Effects of the Edge Rejection Circuitry

The edge rejection feature is an option which, in the ASSP, uses the measured transit times of particles in the laser beam to determine which particles have transected the beam edge and are therefore subject to being undersized. For the multimode lasers normally installed now in the ASSP, PMS reports that the illumination intensity is distributed trapezoidally across the beam diameter with all the rolloff in intensity occurring in the outer 20% of any beam radius. The average transect for all particles across the beam corresponds to a chord section equal to about two thirds of the beam diameter. For a given flow rate there is an average transit time, derived from the pulse widths of the scattered light signals, corresponding to this average chord length. By means of digital counting and gating techniques the transit time of each particle is compared with the running average. Particles traversing the laser beam along a chord shorter than the average will have shorter than average transit times and will not activate the gating circuitry needed to register a count in any of the size channels. Thus, when operating properly, only "valid" scattering events corresponding to particles passing through the central two-thirds of the beam diameter are counted.

Also, a related option allows the running average transit time to be processed to determine particle velocity which, when multiplied by the fixed sampling area of the beam, can be used to compute particle number density.

We have encountered two problems with these options. The circuitry appears to be at least partly responsible for a distortion of the size spectrum measured in hazes and fogs due to an inherent bias against small particles in the presence of larger ones. Secondly, there is a gross overestimation of particle velocity when the ASSP is used for sizing clear air aerosol, i.e., in the absence of fog or cloud droplets or other particles larger than about 3 μm diameter.

The first problem is illustrated in Figure 2 where the size spectrum of a maritime fog is presented as measured on two size ranges of the ASSP. The loss of particle counts in the first few size channels regularly occurs in fogs and is obviously an artifact on the 3–45 μm range shown here.

The explanation of the effect is, according to PMS, due to the fact that on all ranges the transit time information for each particle is derived from the pulse width of the scattering signal at the threshold of the first size channel. It is argued that the relatively weak scattering signals associated with the first few size channels will also be narrower, as measured at the threshold of channel 1, than the stronger signals from larger particles. Thus, when "large" particles are present, their wider pulse widths are reported to dominate the transit time averaging circuitry and many smaller, valid particles are rejected because they are mistaken for edge grazing particles.

The factory has recommended several alternatives for coping with this problem. One alternative is to adjust the clock rates in the velocity averaging circuitry to bias the average velocity towards the apparently faster particles and thus prevent fewer of them from being rejected. This would discriminate less against particles in the first few size channels but unfortunately it would also allow more edge grazing particles to be counted too.

A second possibility is to eliminate the edge rejection circuitry altogether and do a statistical correction on the indicated size spectrum during the data analysis. The owners manual describes a method for correcting the measured spectrum by

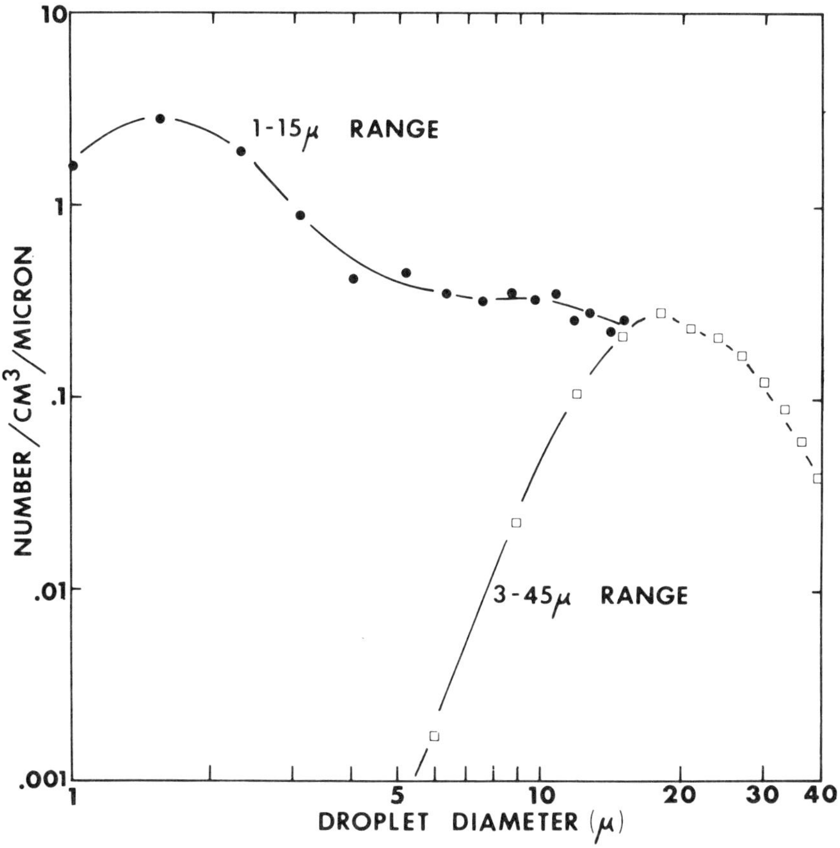

Figure 2. Example of the underestimation of number densities by the ASSP in the first several size channels during fog droplet measurements.

systematically removing a certain fraction of the counts from each of the smaller size channels and distributing them over the larger channels while conserving the total number of particles. The statistical method is deficient for the first few size channels, however, where there are not enough smaller size channels to draw from. Also the correction is valid only if the size spectrum does not extend beyond the size range in use during the measurements. Except for clear air aerosol measurements the size spectrum will usually be wider than the most appropriate range on the instrument and some assumption about the upper end of the distribution will have to be made. There may be some situations where the actual size spectrum is sufficiently narrow that one can select a size range which keeps the spectral peak away from the lower size channels and thus minimize the effect of the edge rejection errors. Also, as is pointed out by PMS, if one is interested mainly in the total extinction coefficient associated with a wide distribution of particle sizes, the contribution from the smaller particles may be negligible in certain cases anyway so that errors due to the edge rejection circuitry will not be important. In general, however, the owner would like to have confidence in the indicated size spectra in all circumstances and the lack of published information by the manufacturer on the limitations of this circuitry and on all the conditional requirements is unfortunate.

According to PMS the edge rejection circuitry is not expected to cause any spectral distortion in the measurement of clear air aerosols or where the number of particles increases exponentially with decreasing particle size. Indeed, our measurements so far have shown no clear cut distortion in clear air conditions but it does begin to show up under hazy conditions where the relative humidity is high and the visibility is down to 5km or less. In Figure 3 a clear day, continental aerosol size spectrum is depicted as measured simultaneously by our ASSP and by one of PMS' newer instruments, the Active Scattering Aerosol Spectrometer (ASAS), courtesy of Dr. Gary L. Trusty of NRL.

The data in Figure 3 are shown not only to demonstrate the proper performance of the edge rejection circuitry in the measurement of clear air aerosols, but also to show that two PMS probes of somewhat different design and which are calibrated differently will give satisfactory agreement, for clear air measurements at least, when both probes are operated under favorable conditions. In this case the ASSP was aspirated by suction at 10 cfm (particle velocity $\approx$ 16 kt). The actual sampling volume swept out by the sensitive volume of the laser beam was 200 cc per minute. (Greater aspiration rates give rise to sizing errors described in Item 6.)

Although the edge rejection function appears to operate properly in clear air aerosol measurements, the associated particle velocity processing circuitry does not. Either with natural aspiration by the wind or with forced aspiration by suction, the indicated average particle velocity is three to five times greater than the actual airspeed through the probe. For example, with a measured 15 kt wind aspirating the probe, the indicated particle velocity is about 80 kt. The accuracy is much improved when larger particles such as fog droplets are being measured, but in general the electronic velocity indication does not appear to be reliable. Since the computed number density depends on knowledge of the particle velocity, some other means should be used for measuring the speed of the sampled air.

Item 6: Errors Associated with Anisokinetic Sampling

When the ASSP is deployed as a stationary probe the particles must be fed through the sensitive volume either by being carried on the wind or other stream of air, or by being sucked through by means of forced aspiration with a suction pump. For outdoors applications the steady suction method can avoid the problem of variable and/or low winds and consequently of variable sampling volumes. But it does introduce problems associated with anisokinetic sampling. The problems are aerodynamic in origin and arise when the air flow into the sampling volume is either faster or slower than the air passing around the probe. In the more usual case where the air flow into the probe is faster than the air flowing

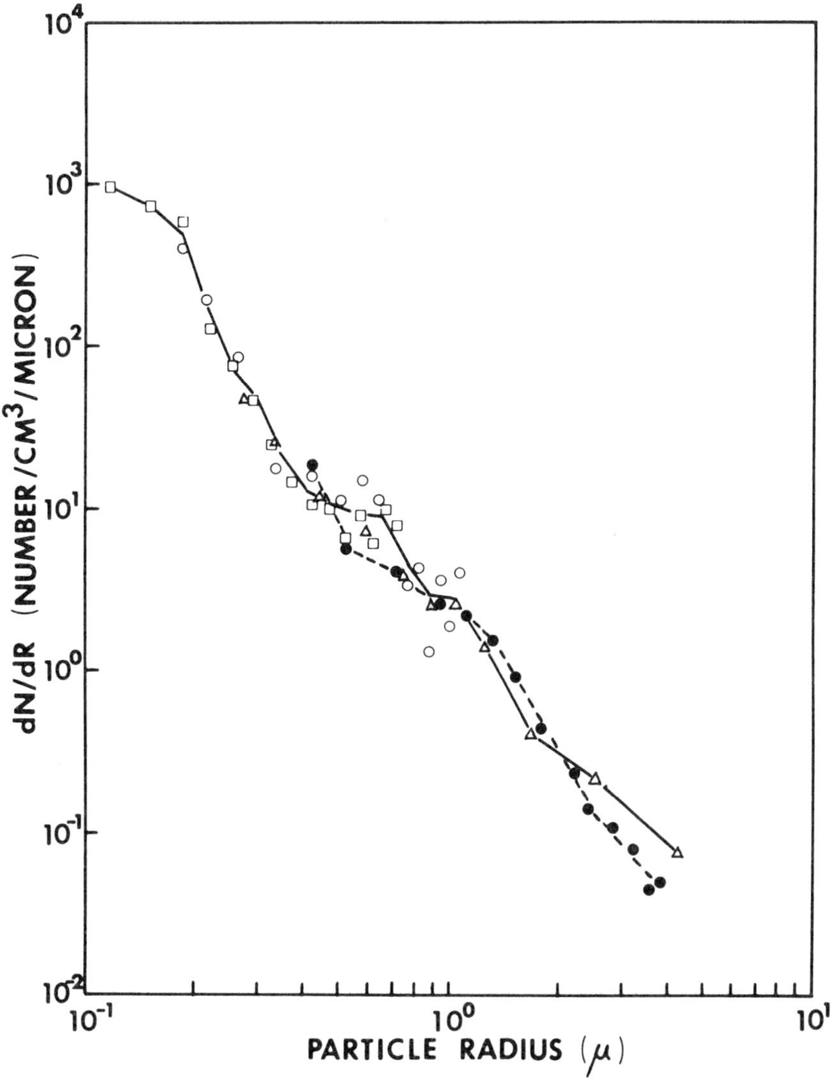

Figure 3. Comparison of the ASSP and the ASAS in the measurement of a clear air, continental aerosol size spectrum at NRL, Washington, D.C. □, ○, Δ data from three ranges of the ASAS, courtesy of Dr. Gary L. Trusty, Optical Radiation Branch NRL. ° data from the 0.5–8.5 μm range of the author's ASSP. The solid and broken lines are drawn to highlight the trends in the data from the ASAS and ASSP, respectively.

outside, particles in the vicinity of the probe entry may be rapidly accelerated either from rest or from some original trajectory. Particles with sufficiently large momentum may slip across streamlines and be diverted away from the sensitive volume of the probe in some cases. Such conditions would have the effect of biasing the count against large particles. To compensate for such effects PMS makes use of a funnel-like entrance horn, or accelerator, about 40 cm long. The sample air first passes through this horn so that the particles can accelerate gradually before reaching the sensitive volume.

In a typical stationary application then, one would assume that the ASSP would be properly set up with the horn affixed to the inlet and a suction pump attached to the exhaust side of the probe by means of a suitable length of tubing. The pump capacity and the tubing diameter would probably be chosen so that air could be pulled through the probe at a speed of several tens of knots. This would be desirable in order to minimize fluctuations due to variable, ambient winds, to ensure that the particle rate is well within the range of the time responses that are normally set for aircraft speeds, and to provide sufficient sampling volume to accumulate a statistically significant number of counts in the sparsely populated channels.

Unfortunately, such reasoning will lead to additional adverse effects which increase with the aspiration rate. We have found that forced aspiration of the probe with air velocities of 80 kt (about 45 cfm), for example, causes a factor of two decrease in the indicated number density in all size channels. In addition, use of the accelerator horn is counterproductive at these velocities. Its use results in an additional depression of the indicated number density as well as a distortion of the shape of the size spectrum.

The first problem is illustrated in Figure 4 where there are shown two successive measurements of the droplet spectrum in a maritime fog with the ASSP. The upper curve was obtained using the steady, 10 kt ambient wind to aspirate the probe isokinetically. This indicated spectrum is presumably correct except for the drop off in the first four size channels, as described in Item 5. The lower curve (which also appears in Figure 2) is the indicated size spectrum when the probe is suction aspirated at 80 kt with the accelerator horn in place. For a given particle size the number densities obtained by suction aspiration at 80 kt are in error by a factor of two or more compared to the values obtained isokinetically. We have found this overall decrease in the indicated number density to occur not only in fogs but also when measuring clear air aerosols at similar, high aspiration rates.

An unanswered question at the moment is whether these errors are actually due to anisokinetic sampling effects or whether they occur for some other reason at high aspiration rates, even under isokinetic, airborne conditions.

Figure 5 shows the appearance of a fog droplet spectrum when measured with and without the accelerator horn. With the horn in use additional counts are indeed picked up in the larger size channels but this gain is numerically offset by a much greater loss of counts in the lower channels. Also the apparent mode in the size distribution is shifted from 10 μm up to 20 μm. Keeping in mind that both curves are already depressed by about a factor of two, and that there is a false drop off in counts in the lower size channels for both cases, one is left wondering what the real size spectrum looks like.

At the moment it is up to the user to anticipate these problems himself. The author is not aware of any factory literature which describes proper aspiration techniques and the associated problems.

A final concern that is not listed in Table 1 has to do with calibration of the instrument. The ASSP can be used for the size measurement of any particles which scatter light with intensities within the range of the instrument, provided that a calibration curve can be established. For volatile particles such as water

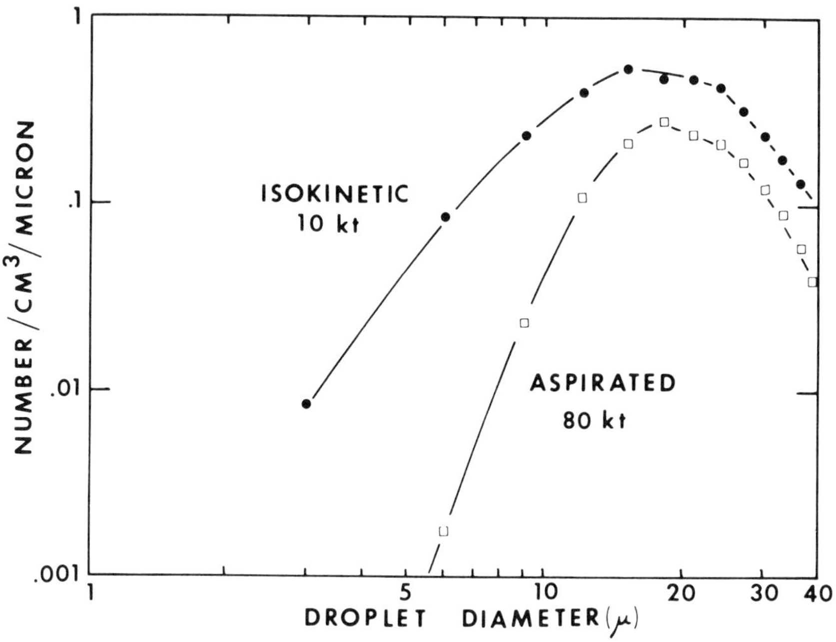

Figure 4. Example of the artificial depression in the indicated number density when the ASSP is suction aspirated at 80 kt (~45 cfm). Data is from a maritime fog, using the inlet accelerator horn and the 3-45 μm of the ASSP.

droplets it is very difficult to produce calibrated droplets of known size and remain certain that they have not partially evaporated (or grown) before reaching the sensitive volume of the probe. For this reason the ASSP and other optically scattering instruments are generally "calibrated" by measuring the scattered light signal from micron size, monodisperse glass or latex spheres and then inferring what the probe response would be if the spheres were water instead. This is done by assuming that any spherical particle in the appropriate size range scatters light in the sensitive volume according to Mie theory. Theoretical Mie scattering curves for particles having an index of refraction, n, appropriate to water and to the glass or latex spheres are then compared.

In particular, the principal procedure followed by PMS to calibrate the ASSP is to select the 2–30 μm (diameter) range setting and to aspirate into the probe glass beads with a narrow size distribution sharply peaked at about 13 μm, as is determined independently with an optical microscope. The overall gain of the signal amplifying electronics is then adjusted until the peak of the glass bead distribution falls mainly in channel 6 as seen on the readout/display. As is illustrated in Figure 6, this adjustment is equivalent to shifting the scattering curves vertically

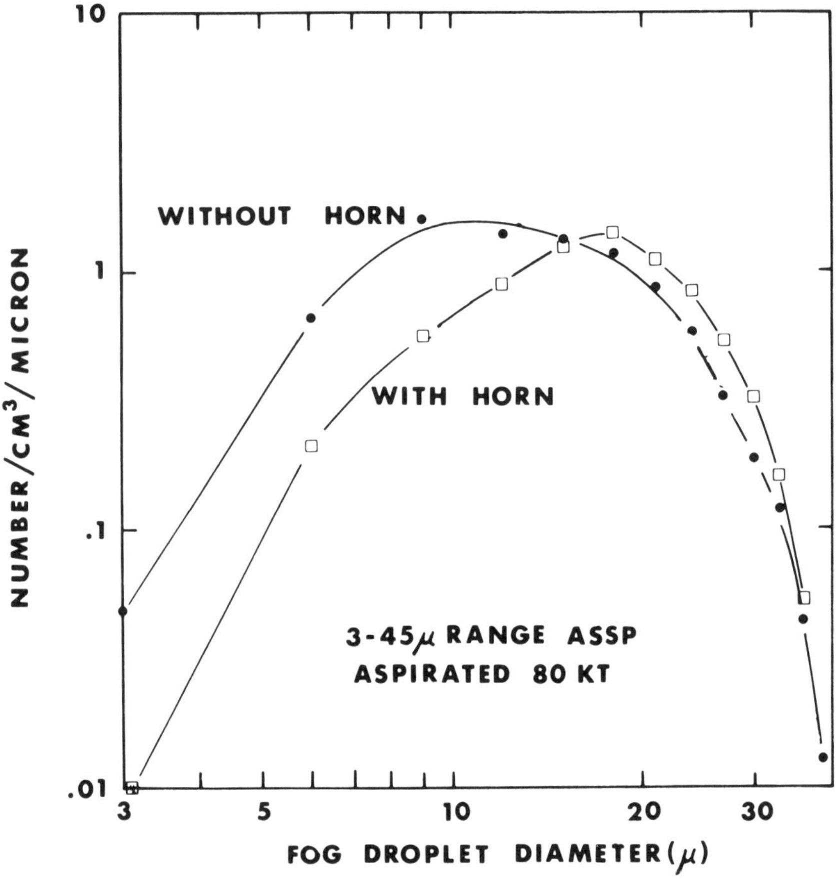

Figure 5. Example of the effect of the accelerator (sample conditioning) horn on the indicated size spectrum of fog droplets.

until the signal voltage corresponding to 13 μm glass beads falls into the lower half of channel 6 in the grid overlay representing the pulse height analyzer. By comparing the two Mie curves one can see that 15 μm diameter water droplets should then fall into the middle of channel 8, as desired. This procedure on the 2–30 μm range and a similar one on the 1–15 μm range, constitute the main calibration for all ranges of the probe. The switch settings for the other ranges, nominally covering the sizes 0.5–8.5 μm, 1–15 μm, and 3–45 μm diameters, correspond to fixed multiples or fractions of the gain established in the primary procedure just described. In other words, changing ranges corresponds to shifting the scattering signal curves vertically by fixed amounts such that for each range setting a different portion of the Mie curves contributes to a given size channel.

Difficulties in interpretation arise when one contemplates the effects of the oscillatory portions of the Mie curves, especially when the ASSP is tuned to the 0.5–8.5 μm range as shown in Figure 7. If the actual response curve of the ASSP

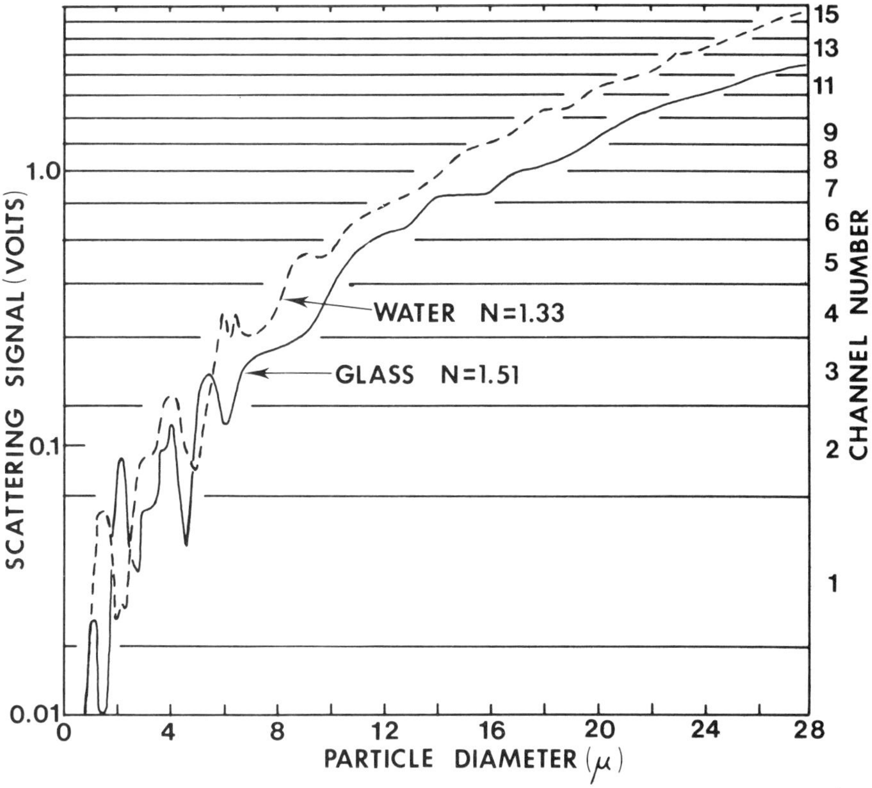

Figure 6. Scattering signal curves corresponding to the 2–30 μm range of the ASSP. The theoretical Mie scattering curves for water droplets and glass spheres are shown in relation to the fifteen (size) channels of the PHA when the ASSP signal gain is properly set for the 2–30 μm range for water droplets.

followed the oscillations in the Mie curves, many of the size channels would be multiple valued in particle size and the size resolution of the probe would be poor in this range. Fortunately, for various reasons including multiple phase wavefronts in the illuminating laser beam, according to PMS, the oscillatory nature of the Mie scattering is suppressed in the actual instrument and the true response function is something like the smooth curve shown in Figure 7. That the actual response curve is smooth and lacks even small inflections can be demonstrated by plotting natural fog droplet size spectra, as measured on the 0.5–8.5 μm range of the ASSP, in the form dn_i/dr_i vs. r_i, where dn_i is the particle count in the *i*th size channel of width dr_i, and r_i is the median particle radius for the *i*th channel. Assuming that the natural fog droplet spectrum is a smooth distribution without significant fine structure, only a smooth response curve like the one shown in Figure 7 will result in a plot of dn_i/dr_i vs. r_i that is free from spurious peaks or dips. Such artifacts will occur in the data plots, however, when inflections on the smooth response curve are postulated in any attempt to make it resemble the theoretical Mie curve.

The main question is, which smooth curve in the vicinity of the theoretical curve represents the true instrument response? The curve shown is essentially the

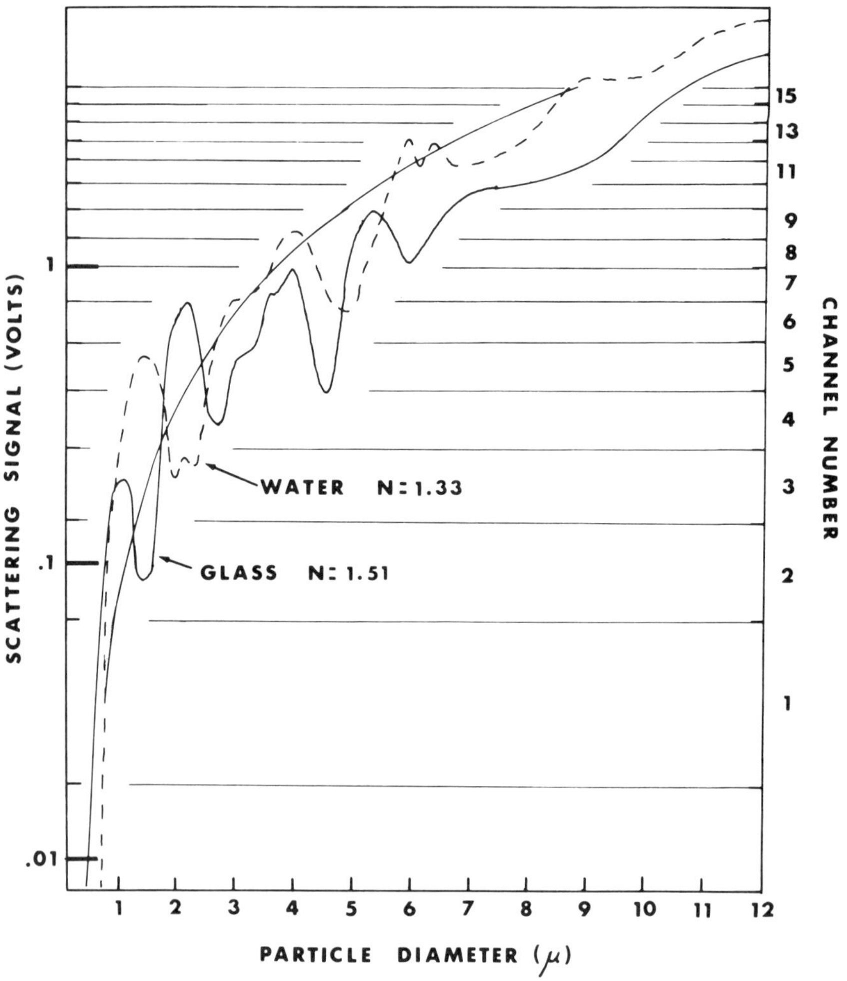

Figure 7. Scattering signal curves corresponding to the 0.5–8.5 μm range of the ASSP. The theoretical Mie scattering curves for water droplets and glass spheres are shown in relation to the fifteen (size) channels of the PHA when the ASSP signal gain is properly set for the 0.5–8.5 μm range for water droplets. The smooth, solid curve is the response curve that the ASSP apparently actually follows for water droplets, as determined empirically by PMS.

calibration curve given in the ASSP manual, except that we have modified the original curve slightly to force closer agreement with the Mie theory below 1 μm. Without any more information than is available so far, a number of other curves could just as well be chosen. However, most of the justifiable, alternate curves will result in a shift of the calibration by no more than plus or minus one size channel. This amount of uncertainty is within the advertised accuracy of the probe anyway.

It is probably of little use to attempt a refinement in the accuracy of calibration for water droplets in this size range by using monodisperse latex spheres since the true instrument response for them will be just as uncertain as it is for water droplets. If one believes that the true response is equal to the theoretical response for sizes below about 1 μm diameter where the Mie curve is monotonically decreasing with particle size, then perhaps monodisperse latex spheres near 0.5 μm could establish the threshold for channel 1 more accurately.

The newer ASAS is reportedly insensitive to index of refraction so that a range of monodisperse latex spheres can be used to calibrate the ASAS, for all optically nonabsorbing particles at least. Thus the ASAS could be used to calibrate the ASSP if they both sample the same aerosol. The agreement shown in Figure 3 indicates that our modified version of the factory authorized calibration for the ASSP is reasonably well chosen, according to the ASAS data. The original factory calibration would pull the datum points for the first two size channels significantly below the lowest values registered on the ASAS for the same two sizes.

A comment is in order on the reliability of field calibrations done by the user for routinely checking the performance of the ASSP. The procedure recommended by PMS (but not described in the ASSP manual) is to insert special plugs with axially aligned, 0.5 cm (3/16 inch) i.d., laboratory-type, gas hose fittings into the intake and exhaust side of the probe sampling aperture. A moderate amount of glass calibration beads is then spread out on a piece of paper and the beads are sampled by "vacuuming" them into the intake hose fitting when a suction pump is attached to the exhaust fitting. Figure 8 shows the observed spread in channel counts for a typical run of three consecutive calibration tests. Also shown is the distribution that would be expected to show up on the display, based on an independent measurement of the bead sizes using an optical microscope with a Porton graticule. It is seen that the measured spectrum is artificially broadened either as a result of the sampling procedure or due to poor resolving power in the instrument itself. The "error" bars also illustrate that there is a fair amount of irreproducibility from one test to another. In the three tests represented in Figure 8, the mode of the distribution was spread over both channels 5 and 6 in the first test, fell in channel 5 alone in the second test, and was spread over channels 4 and 5 in the third test. With this much variation the actual calibration for water droplets could run anywhere from the advertised range of 2–30 μm to a possible span of 3–37 μm. The accuracy claimed by the manufacturer for any size measurement is ± 1 size channel or $\pm 10\%$, whichever is greater. This means that the span for a 2–30 μm range can actually be something like 4–33 μm and still be

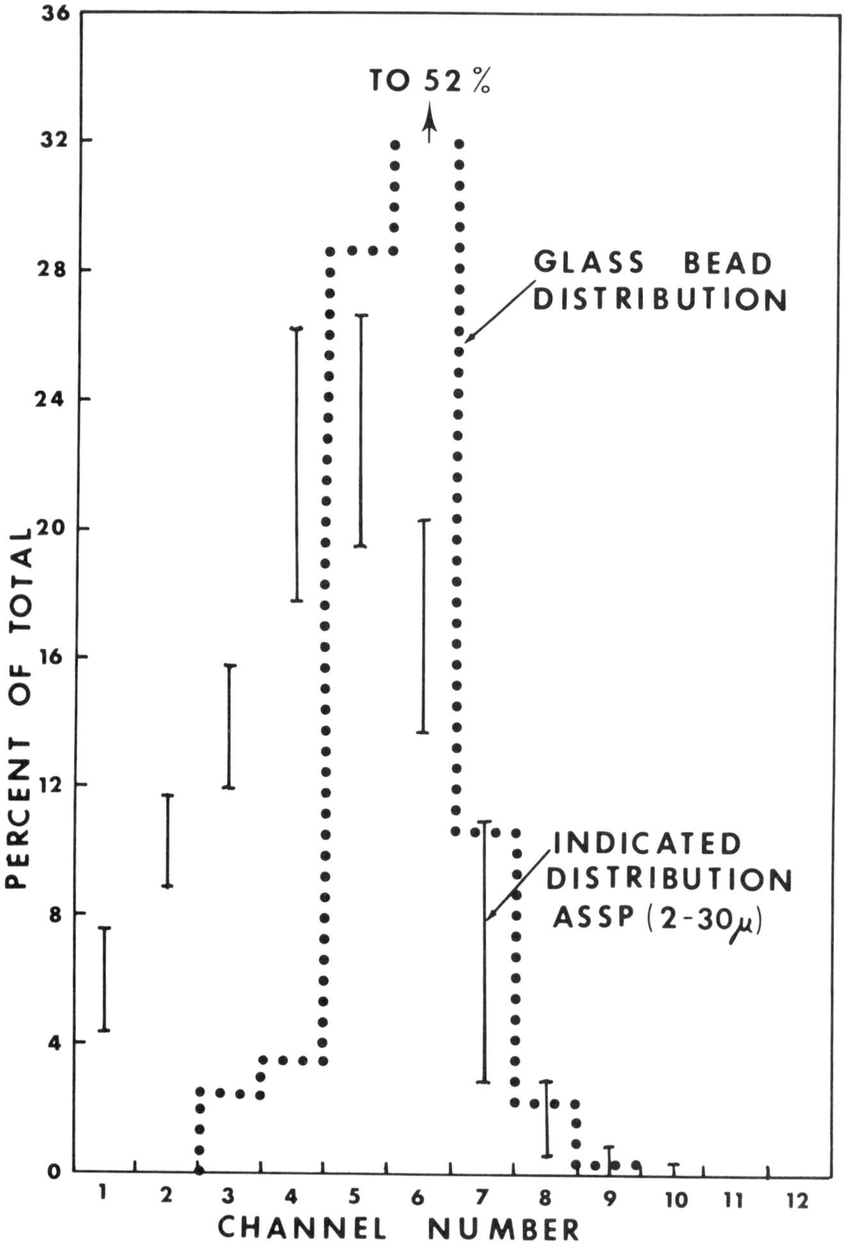

Figure 8. Size distribution of glass beads used for calibration of the ASSP. The dotted histogram gives the expected particle count in the various size channels based on a piecewise linear approximation to the theoretical Mie curve for glass beads in the 2–30 μm range and on an independent size measurement using an optical microscope. The ''error'' bars represent the range of particle counts observed in the ASSP size channels during three typical, consecutive calibration checks in the field.

within the advertised limits of accuracy. The uncertainty in the glass bead field calibrations thus limits the usefulness of these tests to checking for major losses of calibration. Our experience indicates that if the mode of the indicated glass bead distribution falls within ± 2 channels of the correct channel, it is best to assume that the advertised calibration still applies unless the ASSP can be calibrated against some other instrument.

In summary, we reiterate that even though the PMS design is neat and sophisticated, generally trouble free from outright electronic failures, and apparently simple to use, one should be alert to the possible existence of subtle sources of error in the instrument itself. Of equal importance is the requirement that the probes be deployed and aspirated properly in order to avoid subtle but serious sampling errors. It is unfortunate that written guidelines from the manufacturer are lacking on this subject.

In order to help guard against unrecognized errors resulting from instrumental effects or from improper sampling techniques we recommend that alternate instruments such as transmissometers or other particle sizing devices be used simultaneously for comparison whenever possible, especially in untried situations.

Although the operational problems described in this article are specifically concerned with the ASSP, the precautions could just as well be applied to any other newly marketed instrument until it has seen several years of service in the field in a variety of operating conditions.

Note Added in Proof.—Later versions of the ASSP and other models of light-scattering probes since introduced by PMS are equipped with a more sophisticated design for the edge rejection circuitry. The new design determines the particle transit time by measuring the pulse width of the scattering signal at half height rather than at the threshold of the first size channel. The intention is to eliminate the bias against smaller particles that evidently was inherent in the original design. At this writing the author has not had an opportunity to personally evaluate the performance of any of the PMS probes equipped with the newer design of edge rejection circuitry.

AEROSOL SIZE AND SHAPE DETERMINATION USING A LASER LIGHT SCATTERING SPECTROMETER

J. ALLEN AND R. B. HUSAR

Air Pollution Research Laboratory, Washington University, St. Louis, Mo.

AND E. S. MACIAS

Department of Chemistry, Washington University, St. Louis, Mo.

Aerosol particles in the size range from about 0.1 μm to 3 μm can be studied using light scattering techniques. These particles are small enough to be respirable but still large enough to be the principal contributors to atmospheric turbidity through their ability to scatter light. Mie (1908) theory provides the set of formulae relating a particle's light scattering cross section to its geometric size and refractive index (n). Thus if n is known, it is possible to measure aerosol sizes by passing the particles through a light beam, one at a time, and recording the intensity of scattered radiation from each. Such a method has the advantage of making the measurements *in situ* and in real time, while most impaction and filtration methods require subsequent laboratory analysis.

The objective of this work was to develop an instrument which is sensitive to very small particles and is able to give information regarding particle shapes. As particle sizes become small compared to the wavelength of light, the scattering cross section falls very much faster than does the geometrical cross section. In the small-size limit, the scattering cross section is proportional to the fourth power of the geometrical area, or the sixth power of the diameter. This necessitates intense illumination and sensitive detectors, yet the very steepness of the scattering vs. size function facilitates excellent size resolution. In addition not all aerosol particles are spherical; some are crystalline or irregularly shaped. It is possible to extract some shape information from the scattering pattern which permits one at least to discriminate between spherical and non-spherical aerosols.

Supported by the U.S. Environmental Protection Agency, Environmental Sciences Lab., Grant No. R803115.

APPARATUS

The apparatus consists of a low-power helium-neon laser operating with a remote mirror in order to exploit the high intensity light within the actual standing wave cavity. The aerosol to be analyzed is hydrodynamically focused into a narrow stream which intersects the light beam. The scattered light is collected by a lens and photomultiplier on each side of the beam as shown in Figure 1. Thus, whenever a particle crosses the laser beam, a pulse appears at the output of each detector. These pulses are processed electronically to give one signal equal to the sum and another equal to the difference. The sum gives the effective optical size of the particle, while the difference, called the asymmetric scattering gives information related to shape.

The Laser

A laser was used because of its high intensity and well-defined parallel beam. If constructed as an emitting laser, the plasma tube used in this work would have a nominal output of 2 milliwatts, but with a totally reflecting mirror 7 cm from the

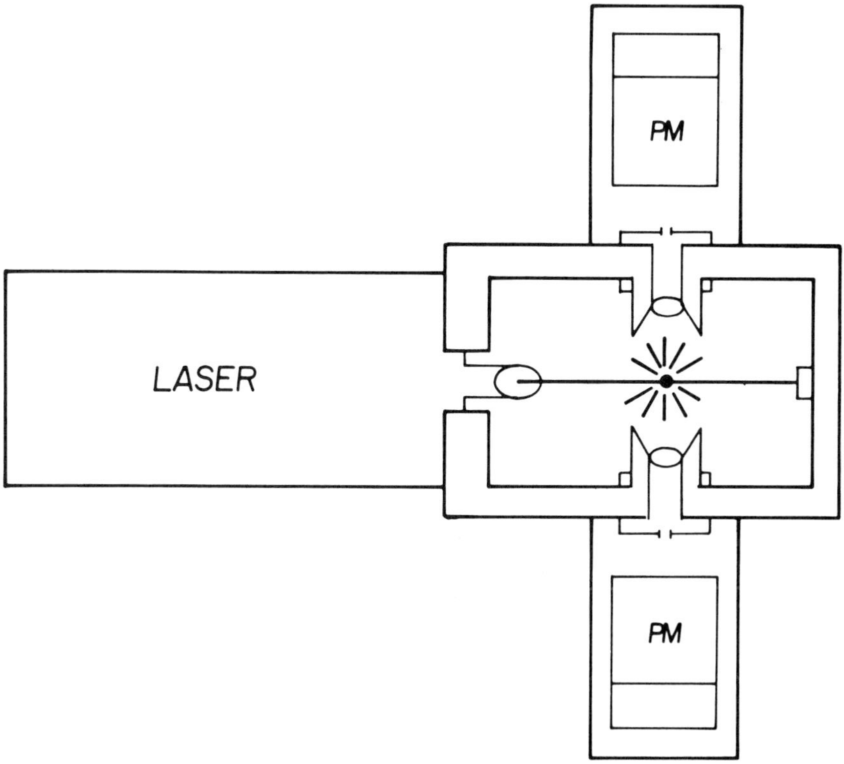

Figure 1. A schematic diagram of the light scattering apparatus. The aerosol particles (moving into the page) cross the standing wave laser beam and scatter light which is collected by the lenses and detected by the two photomultipliers.

Brewster window at the tube end, the intensity within the cavity is at least 20 times that. In the TEM_{00} mode, the beam has a Gaussian profile with a 2 σ diameter of about 0.5 mm, and thus a central intensity of nearly 50 watts per cm^2.

Maintaining the mirror in critical angular alignment requires a very rigid structure. A more serious problem is the cavity's lengthwise dimensional stability, because every time the mirror spacing (cavity length) changes by 1/2 wavelength, the laser must change to another longitudinal mode. In the course of that transition, a phenomenon known as mode-hopping occurs during which the laser oscillates randomly between the two adjacent modes, with the result that its light output is noisy and erratic.

The frame of the present instrument was made of cherry wood which combines

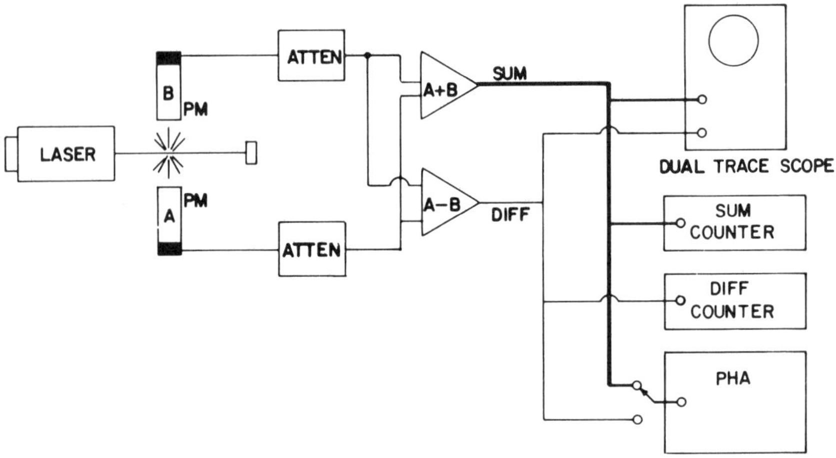

Figure 2. Signal processing. The attenuators correct for differences in photomultiplier sensitivities. The signals are then sent to two operational amplifiers. The upper adds the two signals to produce the sum pulse, while the lower, operating as a differential amplifier, yields a difference pulse whenever the A and B signals are unequal.

excellent mechanical and working properties with a low thermal expansion in the direction parallel to the grain (less than 1/10 that of aluminum). This small change is easily compensated by a short length of brass. The structure has so far proved to be entirely satisfactory, offering excellent stability, stiffness and inherent vibration damping.

The Injector

In order to achieve a pulse height resolution of 10%, the aerosol stream must cross the beam within a region where the intensity is at least 90% of its central value. This limits the stream diameter to 0.1 mm for a 0.5 mm diameter laser beam. This order of focusing has been achieved by means of an aerodynamically focused nozzle. The aerosol passes through a series of telescopically fitted pieces of hypodermic tubing, the last being 0.25 mm I.D. The aerosol then passes through a tapered nozzle where the accelerating flow of filtered sheath air reduces its diameter to the required fineness. The particles cross the laser beam at about 10 meters/sec, yielding pulse durations of 50 microseconds. With flow rate of 0.042 cc/sec, an aerosol density of 48,000 particles per cc yields a 1% probability of two particles being in the beam at one time. Normal ambient aerosols have on the order of 1000 particles/cc in the countable size range of the instrument (0.18 to 1.1 μm).

The Photodetectors

The axes of the laser, the injector, and the optical system are all mutually perpendicular. A lens of 13 mm focal length and f 2.5 aperture is on each side of the interaction zone. A diaphragm stop is at the conjugate focal point of each lens and behind this is a 1.5 inch, end-window photomultiplier with a type S-20 cathode for good response to the red (6328 Å) He-Ne laser light. In the housing of each photomultiplier is an op-amp follower to provide low-impedance output for transmission to the signal processing amplifiers.

Signal Processing

The signal from each phototube is attenuated to adjust for differences in tube sensitivity. The adjusted signals are now fed to two operational amplifiers. One of these is connected as a summing amplifier so that the pulses from the right and left tubes are added. The other operational amplifier operates in its differential mode in order to subtract the two pulses. If a particle scatters light symmetrically there will be an output pulse in the sum channel, but complete cancellation will result in no output in the difference channel. On the other hand, if a particle scatters asymmetrically, the cancellation will be imperfect and output pulses will appear in both the sum and the difference channels. The two signals are sent to a dual-trace oscilloscope and to two electronic counters. Either signal can be sent to a logarithmic amplifier (to accommodate the more than 300:1 range in pulse heights between 0.18 and 1.1 μm particles) connected to a multichannel analyzer. An audio amplifier and loudspeaker permit listening to the pulses to aid in making adjustments or estimating counting rates. Additional electronics includes a single-channel analyzer and a linear gate which permit the sum pulses to be gated by the difference pulses or vice versa. In this way, the spectrum of only those sum pulses accompanied by difference pulses, or the spectrum of difference pulses associated with a restricted range of sum pulse heights can be displayed.

PERFORMANCE AND RESULTS

Small Particle Sensitivity

The instrument is capable of clearly resolving the signal produced by 0.18 μm polystyrene latex (PSL) above the background noise. The signal-to-noise ratio can undoubtedly be improved by appropriately limiting the bandwidth of the signal amplifiers, as well as by improvements in the optics, such as faster lenses and better control of stray light. The upper size limit of 1.1 μm results solely from the restricted dynamic range of the logarithmic amplifier ahead of the multichannel analyzer.

Shape Discrimination

Information on particle shape is obtained by counting the total numbers of sum and difference pulses obtained during a time interval for a variety of aerosols. The counter discriminators are set just above the background noise levels. The

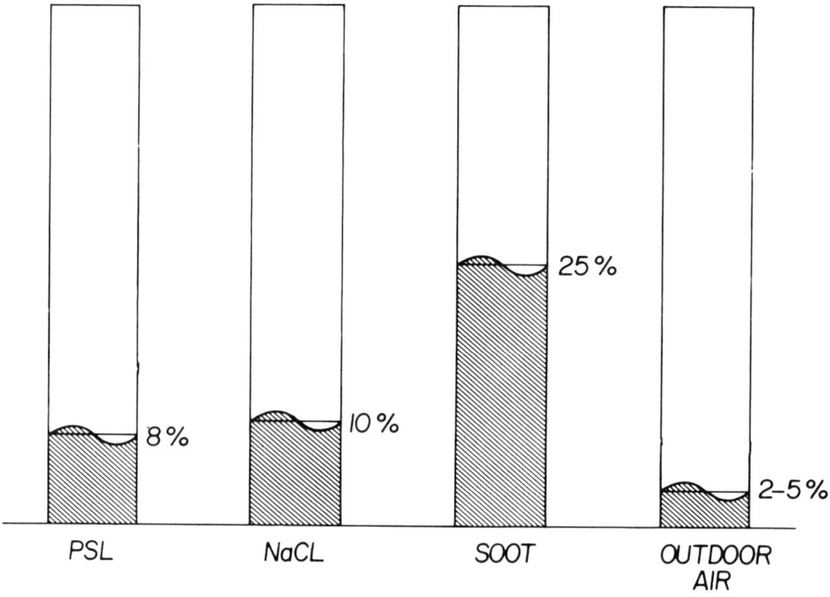

Figure 3. Asymmetric scattering probabilities for three laboratory aerosols and atmospheric air. The wavy lines suggest the spread in the data.

ratio of difference pulses to sum pulses counted represents the probability that a particle shows asymmetric scattering. Figure 3 represents an average derived from many such runs. The wavy line at the top of each bar suggests the approximate scatter in the data. It is clear that the soot, which is known to consist of highly irregular, chainlike aggregates (Williamson, 1973) shows the most asymmetric scattering (25%). Sodium chloride, made up of cubic crystals, shows less asymmetric scattering (10%), while polystyrene-latex spheres show less still (8%).

The relatively large amount of asymmetric scattering for PSL particles is due to doublets (or multiplets) consisting of two (or more) still stuck together while being counted. This was indicated by the increase in the asymmetric counting ratio when a thicker suspension was used in the nebulizer which generated the aerosol, thereby increasing the probability that a water droplet would contain more than one sphere and thereby produce a multiplet upon evaporating.

This was shown more conclusively using the multichannel analyzer which allows the resolution of the singlet, doublet, and triplet peaks of 0.365 μm PSL. A single-channel analyzer in the sum pulse line was used to trigger a gate in the difference line. The sum counter was tapped into the SCA output while the difference counter read the pulses at the gate output as shown in Figure 4. Thus, the sum counter would see only those pulses whose heights fell within the SCA window, and the difference counter would see only those difference pulses

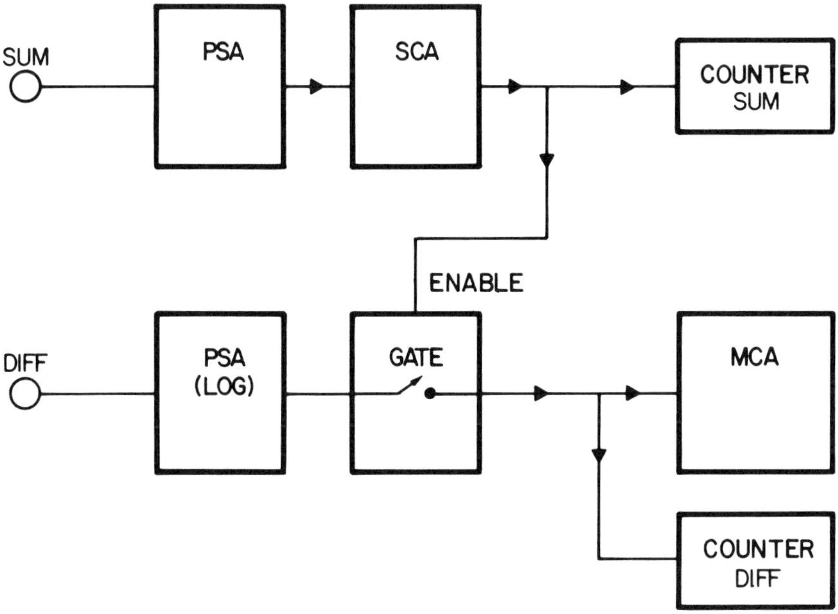

Figure 4. The addition of a single-channel analyzer, a linear gate, and another pulse shaping amplifier makes it possible to select and count only those sum and difference pulses associated with a restricted range or "window" of effective particle sizes.

coincident with the selected sum pulses. Three runs made with the SCA window set respectively at the singlet, doublet, and triplet peaks yielded the following asymmetric counting ratios also shown in Figure 5; singlet-1.25%, doublet-30%, triplet-34%. This confirms that any asymmetric scattering from PSL spheres is due to multiplets whose collective shape is non-spherical.

Dry sodium chloride shows a sizeable (10%) asymmetric scattering probability. However, when the aerosol was aged in a humidity chamber, the asymmetric count ratio varied considerably as a function of humidity, but invariably fell to near zero at relative humidities greater than 75% as shown in Figure 6. This is undoubtedly due to deliquescence of the salt crystals to form spherical solution droplets (Friedlander, 1976).

All inland atmospheric aerosols which have been sampled so far have had asymmetric counting ratios between 2 and 5%. This low value has two important consequences: first, the assumption that all the particles are spherical will not introduce statistically serious errors into the size spectrograms of common atmospheric aerosols obtained by optical counters. On the other hand, the appearance of a high asymmetric counting ratio in an environmental sample would indicate the presence of what is ordinarily a comparatively rare particulate species. It would also serve as a warning that in analyzing particle sizes, it might not be safe to assume sphericity.

One further question is whether there exists any correlation between particle size and asymmetric scattering for heterodisperse aerosols. To determine this, the

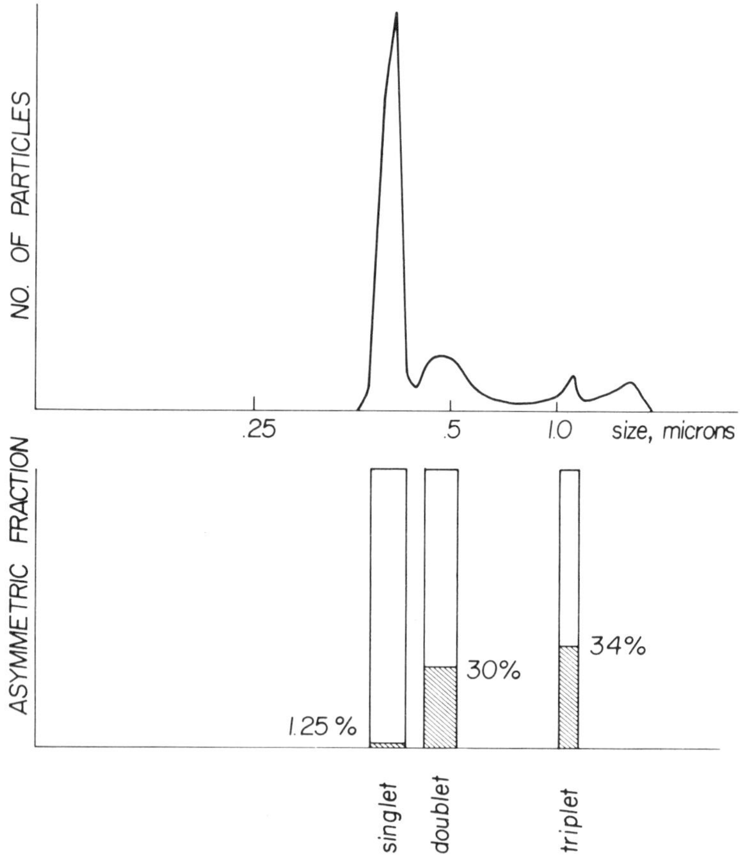

Figure 5. Proof that multiplets cause whatever asymmetric scattering is observed from polystyrene-latex. Above is the multichannel analyzer spectrum of 0.365 micron PSL. The bar graphs below indicate the asymmetric scattering fraction obtained when the apparatus in Figure 4 restricted counting to each of 3 narrow windows in the size spectrum. The widths of the bars correspond to the widths of the windows. The vertical scale of the spectrum has been exaggerated to the right of the singlet peak in order to show the multiplet peaks more clearly.

sum signal was sent to the multichannel analyzer but its input was gated by the difference channel. This gives the size spectrum of only those particles which exhibit asymmetric scattering. The resulting spectrum for NaCl showed a sharp drop below 0.3 μm and almost no particles below 0.25 μm appeared in the gated spectrum, even though these smaller sizes were very abundant in the ungated spectrum. Outdoor air samples (collected near our St. Louis, Mo., laboratory) showed similar behavior as did an aerosol generated by atomizing sea water. Soot also showed a higher asymmetric scattering probability associated with larger particles, but unlike the other aerosols did not show a clear cutoff. Even particles as small as 0.19 μm (PSL equivalent) showed a significant asymmetric scattering probability. It should be noted, of course, that soot, unlike the other

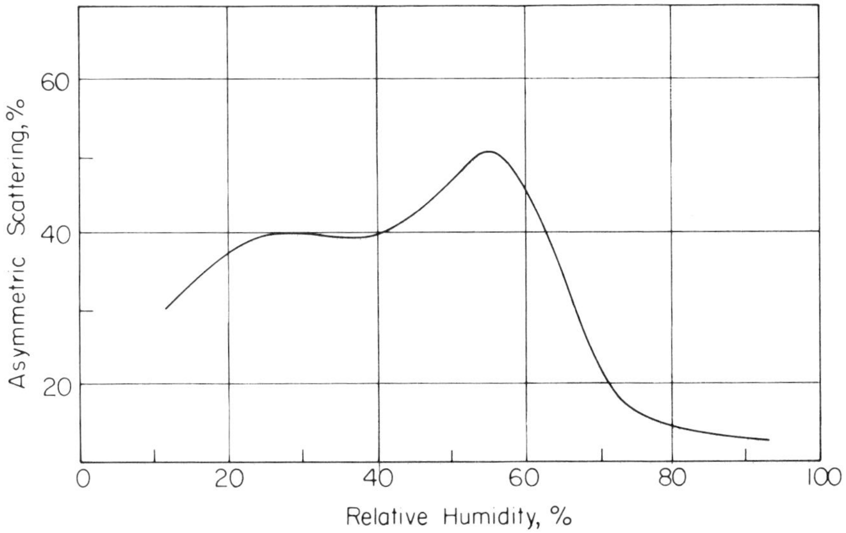

Figure 6. Asymmetric scattering of NaCl as a function of relative humidity. As the humidity exceeds the deliquescence point of NaCl (about 78%) the salt changes from a crystal to a spherical solution droplet, and the asymmetric scattering decreases rapidly to a very small value.

aerosols, absorbs a sizeable amount of the incident light rather than scattering it conservatively. Thus, a soot particle measured by scattered light is probably much larger than its scattering intensity would indicate. It may be significant that 0.3 μm, the minimum size for efficient asymmetric scattering, is about one half the wavelength of the incident light.

FUTURE PLANS

From this work it is apparent that the optical size spectrometer described in this work has the potential to become a useful tool for aerosol analysis. Used in conjunction with other devices, it may prove valuable in determining the composition and origin of air pollutants without having to make conventional chemical analyses.

REFERENCES

Friedlander, S. K., 1977: *Smoke, Dust and Haze*. John Wiley and Sons.
Mie, G., 1908: *Ann. Physik, 25,* 377.
Williamson, S. J., 1973: *Fundamentals of Air Pollution,* Addison-Wesley, 245.

COUNTING EFFICIENCIES OF THREE SINGLE PARTICLE AEROSOL COUNTERS

DAVID RIMBERG*

Aerosols Control Corp., Montvale, N.J.

One of the most frequently used methods to detect and measure aerosol size and concentration is the light scattering, single particle counter. This technique is automatic, non-destructive, continuous, and yields instantaneous results. In addition, for certain situations, light scattering calculations can easily be made using Mie theory. The major limitations of the present instrumental methods are minimum detectable size, inaccuracy in measurement of number concentration, and poor size resolution. Also, the responses of the light scattering aerosol counters are heavily dependent on particle surface irregularities and refractive index. Although the limitations seem to outweigh the advantages for using light scattering counters, industry often depends on these instruments to measure airborne particulate contamination because the particle size and count data are generated in real time.

DESIGN CRITERIA FOR SINGLE PARTICLE COUNTERS

The principles of operation in all light scattering aerosol counters are the same. Aerosol particles in the micrometer (10^{-4}cm; denoted by μm) size range are allowed to flow single file in a narrow stream through a beam of light. Hence, each time a particle traverses the light beam a pulse of scattered light is produced. The pulse of scattered light is collected and sensed. The collection is usually done by a system of lenses and reflecting mirrors, which focus the scattered light onto a photoelectric sensor, usually a photomultiplier tube. The photomultiplier tube senses the light and converts it to an electronic signal. The number of light pulses received by the photomultiplier tube per unit of time for a given aerosol

*Present Address: North American Pemco, Inc., P.O. Box 655, Bardonia, N.Y. 10954.
This research was supported by The Health and Safety Laboratory, ERDA, New York.

flow rate determines the number concentration (particles/volume), and the intensity of the scattered light pulse indicates the particle size.

The principles described have been used for the last 25 years in the design of automatic light scattering, single particle counters. A survey of the work in the field from 1954 to 1970 is given in Gucker and Rose (1954), O'Konski and Doyle (1955), Fisher et al. (1955), O'Konski et al. (1959), Thomas et al. (1961), Zinky (1962), Sinclair (1967), Lavery and Leavitt (1967), Martens and Keller (1968), Moretz et al. (1970) and Grafendorfer (1970). In recent years, improvements in optical, sampling and electronic systems have been made and several counters have become available commercially.

PROBLEMS WITH PRESENT COMMERCIAL COUNTING SYSTEMS

Numerous publications (Whitby and Vomela, 1967; Quenzel, 1969; Rimberg and Thomas, 1970a,b; Rimberg and Keafer, 1970; Jaenicke, 1972; Pinnick and Hoffmann, 1973; Preining, 1968; Graedel, 1974; Nikiforova and Sedunov, 1972) have described some of the problems associated with the present commercial counters. A complete list of the problems has been compiled and is discussed below.

Light Source

In commercial counters the sampled aerosol traverses the image of a tungsten lamp filament. The tungsten filament has been used because it is sturdy, inexpensive, does not require auxiliary power supplies, and is available in numerous sizes, voltages and configurations. However, the filament gets very hot, cannot be modulated rapidly, and its output cannot be varied over an appreciable range without affecting spectral output. High infrared radiation emitted by the lamp accounts for excessive heat, varied spectral output, and eventual breakdown of the lamp.

Lens System

In these counters the light is focused by a complicated system of lenses which often introduce optical distortion and stray light, thereby preventing the high signal to noise ratio necessary for accurate scanning of particles, especially in the submicron range.

Poor Size Resolution

Particle size data have been obtained using commercial counters for both laboratory and field aerosols. The size distributions obtained by these instruments are inaccurate as shown by several experimental investigations (Whitby and Vomela,

1967; Rimberg and Thomas, 1970a; Jaenicke, 1972; Graedel, 1974; Nikiforova and Sedunov, 1972). Poor size resolution for monodisperse aerosols also has been shown (Rimberg and Thomas, 1970b).

Inefficient Counting

Every particle greater than 0.3 μm diameter should, according to most manufacturers' specifications, produce a pulse of scattered light which is subsequently collected and recorded. However, comparison of aerosol concentrations made by optical counters to those obtained from microscopic counting on membrane filters has indicated large discrepancies. Experiments have shown that as high as 50% of 0.5 μm particles are not counted by the optical counter (Rimberg and Keafer, 1970).

Contamination of Optical Chamber

Most aerosol sampling systems carry the particles across a gap between two small diameter tubes. This allows the focused light beam to interact with the particle. Very often, due to local eddy currents and turbulence, particles stray from the aerosol stream and find their way into the optical chamber, thereby contaminating the optics. (The turbulence and the eddies result from the sudden expansion and contraction of the streamlines which occur as the aerosol flow traverses the gap between sampling tubes). This contamination results in significant reduction of the signal to noise ratio necessary for accurate counting and sizing.

Multi-Valued Response

The angular orientation of the light beam, aerosol flow stream and phototube is critical to the response of a system. In several commercial instruments the situation arises where the response of the instrument is the same for two different size particles. Theoretical and experimental response curves have been obtained (Quenzel, 1969; Hodkinson and Greenfield, 1965; Cooke and Kerker, 1975) verifying this situation. Hence, if a counter is being used to sample aerosols whose size can produce a multi-valued response, incorrect particle size measurements result.

Index of Refraction

The amount of light scattered from a particle is dependent on the refractive index of the particle. The standard calibration procedure for present aerosol counters is to use monodisperse polystyrene latex microspheres of real refractive index 1.60. If particles of other indices are used in the counter, incorrect size determinations result, since calibration was performed with latex (Quenzel, 1969; Hodkinson and Greenfield, 1965; Cooke and Kerker, 1975).

Coincidence

In certain situations aerosol counters must be used to measure highly concentrated aerosols. High concentrations produce a condition where more than one particle at a time may appear in the sensing volume. This effect is commonly referred to as coincidence. Hence, the number concentration of particles observed by the counter must be corrected to indicate the actual number concentration present in the aerosol stream (Jaenicke, 1972).

Considering the aforementioned problems and difficulties of presently available commercial counters, and with the advent of the laser, a second generation of aerosol counters has appeared (Jacobi et al., 1968; Neitzel, 1968; Cooper and Byers, 1970; Gebhart et al., 1970; Wyatt and Phillips, 1972; Schehl et al., 1973; Gravatt, 1973; Gucker et al., 1973) which attempt to overcome many of these shortcomings. Similarly, the present investigation has also attempted to design and evaluate a new laser aerosol counter. The first objective of this study was to compare commercial particle counting efficiencies to the new aerosol counter.

DESIGN OF A NEW AEROSOL COUNTER

In previously developed particle counters the scattered light from flowing aerosols is detected either directly at a fixed range of angles by a photomultiplier tube (PMT), or indirectly by a light collecting mirror which reflects the scattered light to a PMT. Regardless of the method used to collect the scattered light from an aerosol, it is obvious that the more light detected the greater the PMT response and therefore the more accurate the particle size determination and number concentration measurement. Hence, to collect the maximum amount of scattered light, an ellipsoidal mirror should be used. The unique property of this type of mirror is that all the light scattered from one of its foci is reflected back to the other. Therefore, if a narrow beam of light is allowed to intersect an aerosol flow stream at one focal point, and a photodetector placed at the other, the maximum amount of scattered light is collected and measured. Therefore, a system employing these ingredients was used in the development of a new aerosol counter. A description of the [1] optical, [2] sampling, and [3] electronic systems associated with the new counter follows.

[*1*] *Optical System*

Ellipsoidal Collector Mirror.—A prototype system of the ellipsoidal light collector is shown in Figure 1. A laser beam enters the ellipsoidal collector near one end and passes through the focus at the opposite end. The aerosol particles enter through a sampling tube and pass through the focus in a direction approximately at right angles to the laser beam. A portion of the incident light is scattered in all directions while the remaining light passes into the light trap where the beam intensity is monitored. The scattered light pulses are reflected by the ellipsoidal mirror to the PMT at the other focus where the light is converted to electrical pulses. The duration of the pulses depends on the transit time the aerosol particle spends in the beam.

Laser Light Source.—To eliminate the many problems of focusing the light from a tungsten filament lamp, laser light was employed. At present, the most suitable type of laser for a light source is He-Ne. A wide variety of intensities are available and relatively inexpensive compared to other lasers. They operate continuously and have a primary output wavelength of 0.6328 μm.

The illuminating light source used for the prototype optical system was, according to the manufacturer's specifications, 1 milliwatt power output, spatial coherence with a TEM_{00} uniphase wavefront, and temporal coherence with three axial modes at 450 megahertz. Divergence of the beam was 0.8 milliradians with a beam diameter of 1.4 mm at the aperture.

Light Trap.—To prevent the laser beam which exits through the mirror from re-entering the optical chamber, a light trap was installed. This light trap was oriented so that it was impossible for any extraneous reflected laser light to find its way back into the optical chamber.

Beam Intensity Monitor.—A plane mirror was mounted in the light trap which reflected the laser onto a Si-photodiode. The photodiode was connected to a D.C. millimeter which detected any changes in the output of the laser. The Si-photodiode used was a PIN 10 metal semiconductor with an anti-reflective coating.

[2] *Aerosol Sampling System*

The design criteria for an optimum aerosol sampling system are: minimum aerosol loss in sampling tubes from mechanisms such as gravity settling or diffusion, prevention of particles from diverging out of the aerosol stream into

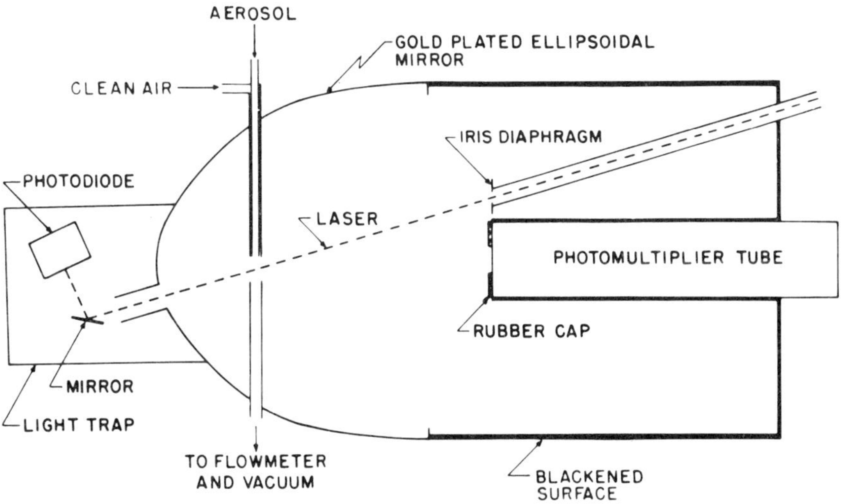

Figure 1. Prototype system of an ellipsoidal light collector.

the optical sensing volume, and minimization of the size of the optical sensing volume. Each of these criteria was considered in detail and the following designs were employed.

Particle Loss in Sampling Tubes.—The loss of aerosol in sampling tubes from gravity settling can easily be eliminated by allowing the sampling tube axis to be parallel to earth's gravitational field. Losses from diffusion which are prevalent for sub-micron particles cannot be easily eliminated. However, for particles greater than 0.1 μm losses due to diffusion can be neglected. An approximation to the loss of aerosol from diffusion can be obtained from theoretical calculations (Thomas, 1970), assuming that the size of the aerosol and sampling tube diameter is known.

Sampling Tube Design and Operation.—The simplest way of preventing particles traversing the light beam from being diverted into the optical chamber because of turbulence or eddy diffusion is to surround the aerosol flow stream with a sheath of clean air, thereby keeping the path of the aerosol straight.

To establish the operating flow rates of the sheath air and aerosol in the prototype optical system, a duplicate sampling system was constructed and installed in a Millikan type apparatus. This facilitated visual observation of the flow, by using a highly concentrated aerosol of cigarette smoke. The smoke was observed as it traversed the gap between tubes. The smoke flow and clean air sheath flow were regulated to provide smooth streamline flow. Photographs of these phenomena were taken, and shown in Figure 2. The sampling tubes used were hypodermic needles. The diameter of the outer sheath air needle and the aerosol exit needle was 1.80 mm and the inner aerosol sampling needle 0.68 mm.

Optical Sensing Volume.—The light scattering or sensing volume is defined as the volume in which the laser beam and aerosol stream intersect. A small scattering volume is highly desirable since it reduces the possibility of more than one particle at a time intersecting the laser beam (i.e., low coincidence). For the prototype system, an approximation to the sensing volume is obtained by multiplying the laser beam diameter, 1.4 mm, by the cross sectional area of the aerosol flow stream, 0.369 mm^2, obtaining 0.512 mm^3.

If uniform spatial distribution of the particles is assumed, the maximum allowable particle concentration free of coincidences is the reciprocal of the sensing volume of 1.95×10^3 particles/cm^3. However, because of random spatial distribution, the number concentration must be kept much lower. The actual concentration of an aerosol can be calculated from the relation

$$N_o = N_a e^{-VN_o} = N_a\left[1 - VN_o + \frac{(VN_o)^2}{2!} - \frac{(VN_o)^3}{3!} + \ldots\right] = N_a(1-VN_o)$$

where V is the sensing volume, N_0 the observed number of particles per unit volume, and N_a the actual concentration of particles per unit volume. For a 3% coincidence loss, $VN_0 = 0.03$, and $N_0 = 0.03/0.512$ mm^3 = 58.5 particles/cm^3. Since $N_0/N_a = 0.97 = (1 - VN_0)$, then $N_a = N_0/(0.97) = 58.5/(.97) = 60.0$

particles/cm^3. Hence, for a random spatial distribution, a number concentration of 60 particles/cm^3 or less must be present if a coincidence of 3% is not to be exceeded.

[3] *Electrical System*

Photomultiplier Tube.—The light photons scattered by the aerosol particles at the focus are reflected back to the PMT located at the other focus. The cathode on the PMT face converts only a small fraction of the incident photons into photoelectrons. By definition, the quantity which indicates the number of photoelectrons emitted from the cathode per incident photon of energy hc/λ is called the photocathode quantum efficiency. The quantum efficiency available for a wavelength of 0.6328 μm is approximately 6%.

The cathode selected for the wavelength under consideration had a stoichiometry of Na_2KSbCs and is commonly referred to as a S_{20} or trialkali cathode. The PMT has a 2″ diameter flat faced, end window, 44 mm cathode

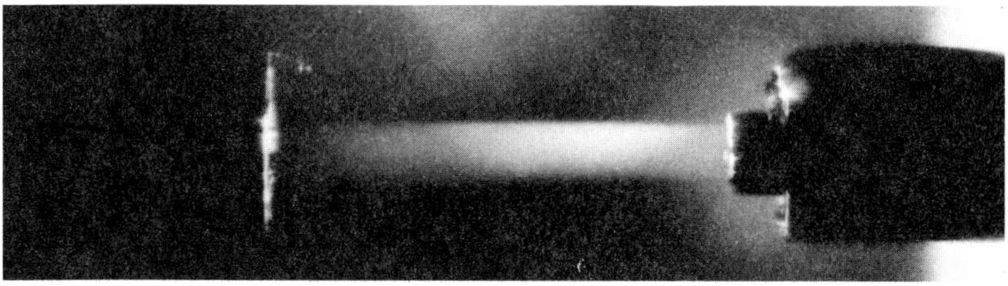

Figure 2. Streamline flow of aerosol across sampling tubes.

surface with 11 venetial blind dynodes having CsSb secondary emitting surfaces. The high red response, coupled with very low dark current, and a gain of 2×10^6 made this tube suitable for the prototype system.

To avoid effects from environmental magnetic and electrostatic fields a mu-metal shield surrounding the tube was provided. To eliminate the light scattering from air molecules, other than at the focus, from entering the PMT a ¼″ rubber aperture was placed over the PMT thereby allowing light only from the other focus to enter the tube.

Preamplifier, Amplifier, Discriminator.—The output from the PMT was passed into a Nuclear Data ND-520 PAD (Palatene, Ill.) module which contained a charge sensitive preamplifier, linear amplifier, and integral discriminator. A block diagram of the module is shown in Figure 3. Referring to Figure 3, the charge sensitive preamplifier accepts the negative charge from dynode eleven. This charge is converted to a voltage signal obtained by integrating the input across a capacitor. The signal from the preamplifier is coupled directly to the first of two stages in the linear amplifier. The amplified signal of the first stage is passed into the second stage and further amplified. The output of the second stage is a positive clipped voltage pulse which, if desired, can be passed to an analog to digital converter or single channel analyzer. Because of its availability, the output from the linear amplifier was coupled to an integral discriminator circuit. The integral discriminator provides an output pulse of fixed magnitude (10 volts rectangular) and duration whenever an input pulse is supplied by the discriminator control. The output of the discriminator is coupled to an Ortec Model 431 and 430 Timer-Scaler (Oak Ridge, Tenn.). The purpose of the scaler is to count the number of pulses per unit time passed to it from the discriminator output.

The value of the discriminator settings was critical since it determines if a pulse of light from the aerosol is to be counted or not. If the threshold bias is too low, electronic noise, dark current, stray light and aerosol will be counted. On

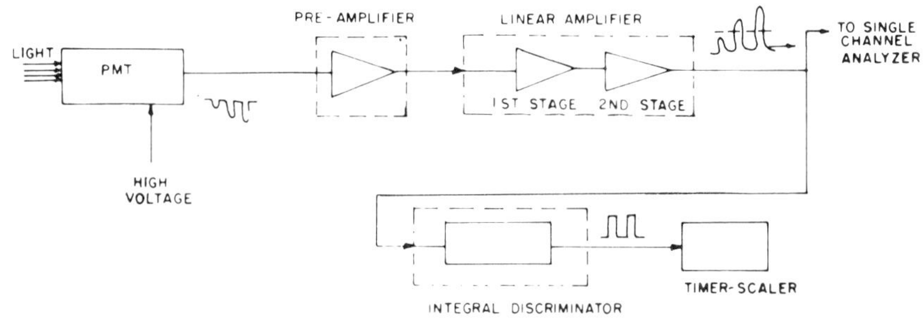

Figure 3. Block diagram of electronic counting system.

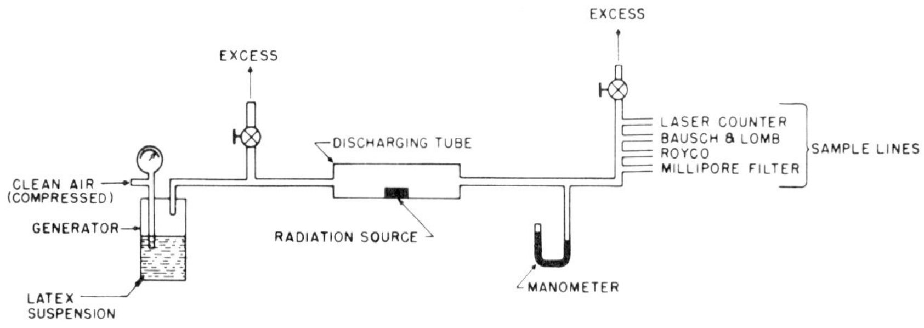

Figure 4. Aerosol flow system.

the other hand, if set too high, the signal will not be large enough to overcome threshold bias and therefore no counts will be recorded. Hence, trial discriminator settings were made to determine when only signal from aerosol was being detected and nothing else.

EXPERIMENTAL PROCEDURE

Aerosol Flow System

The aerosol flow system used to test the response of several particle counters is shown in Figure 4. A diluted latex suspension was atomized and passed through a tube about three feet long. This length allowed enough time for temperature equilibrium and drying. The aerosol as generated contained electronic charge which could result in extraneous loss of aerosol in the flow system. Hence, the aerosol was passed through a discharging tube containing a 20 mCi strontium-90 source. A flow rate less than 20 l/min was maintained in order to ensure charge equilibrium of the aerosol (Thomas and Rimberg, 1967). From the discharging tube, the apparatus was arranged so that the aerosol could be sampled by four different detectors: the laser counter, Bausch and Lomb, Royco, and Millipore filter.

A slight positive pressure of 1 cm of H_2O was maintained in the aerosol flow system. The positive pressure was used to prevent the possibility of any ambient aerosol from leaking into the system.

Aerosol Generation

Monodisperse polystyrene and polyvinyltoluene latex microspheres were atomized and used to test the response of the various particle counters. The latex suspensions were diluted with doubly distilled, de-ionized water and atomized

TABLE I. Characteristics of Low Latexes

Type Latex	Part. Dia. (μm)	Stand. Dev. (μm)	No. Measurements	Run No.	Solids (%)
Polyvinyltoluene	1.947	Not Available	Not Available	866-38	10
Polystyrene	1.099	0.0059	106	LS-1028-E	30
Polystyrene	0.81	Not Available	Not Available	LS-1114-B	30
Polystyrene	0.500	0.0027	130	LS-1029-E	10

with clean compressed air at 15 psig using a Laskin-type nozzle (Echols and Young, 1963) immersed in an 8-inch diameter by 9 1/2 inch high stainless steel beaker.

The characteristics of the latex suspensions supplied by Dow are shown in Table I.

The latex suspensions received from Dow contain an emulsifier (stabilizer) to prevent the microspheres from aggregating in solution. In atomizing the suspension, the emulsifier may be carried along with the latex particles, and affect optical measurements. It has previously been found (Rimberg and Thomas, 1970a) that only the 1.947 μm polyvinyltoluene emulsifier affects light scattering measurements. Therefore, the emulsifier was removed from the 1.947 μm particles from the suspension. The precipitated 1.947 μm particles were redispersed in distilled water and the centrifuging process repeated. This procedure was performed a total of five times. Finally, the precipitate was diluted with doubly distilled-deionized water and then atomized for testing purposes.

Aerosol Detection and Measurement

The test aerosols were simultaneously counted by three light scattering instruments and a membrane filter.

Laser Counter

The optical, aerosol flow, and electrical systems for this counter have been described. The aerosol flow rate for the experiment was 365 cm^3/min with an optical sensing volume of 0.512 mm^3. The aerosol was collimated by a sheath of clean air. Light pulses from the particles traversing the laser were received by the PMT and recorded by a scaler using a 200 sec. time interval.

Bausch and Lomb 40-1

This particle counter (Martens and Keller, 1968) uses a parabolic mirror to collect the light scattered from particles passing through the sensing volume. The mirror in turn passes the light to a PMT. The parabolic mirror collects the scattered light in the near forward direction from 23.75° to 53.75°. The sample flow rate is 170 cm^3/min and is collimated by a sheath of clean air. The optical sensing volume is 0.5 mm^3. Maximum number concentration is one million particles per cubic foot (35 per cm^3) with 3% coincidence. Particle counts of sizes equal to and greater than a selected size are read from a digital display. Seven size ranges are provided, 0.3, 0.5, 1.0, 2.0, 3.0, 5.0, 10.0 μm. The particle size range chosen for number concentration measurements was 0.3 μm and greater with a counting period of 100 sec. To determine the number concentration in each size range the difference in counts between successive size selector settings was obtained. Theoretical light scattering calculations have also been made for the Bausch and Lomb optical system (Quenzel, 1969; Hodkinson and Greenfield, 1965; Cooke and Kerker, 1975). It was shown using these calculations that the response of the Bausch and Lomb is less dependent on the aerosol index of refraction compared to the Royco. However, multivalued responses are present in the Bausch and Lomb.

Royco 220

This particle counter (Zinky, 1962; Cooke and Kerker, 1975; Lieberman and Allen, 1969) is designed to receive scattered light directly by a PMT from aerosols at a 90° angle from the incident light. It can monitor the particle size and concentration up to one million particles per cubic foot (35 per cm^3). Two size ranges can be monitored simultaneously and displayed on two, five-digit ''Nixie'' indicators. Sample flow rate through the system is 0.1 cubic foot per minute, and the optical sensing volume is 2.63 mm^3. Corrections for coincidence (electronic and particle) have been obtained empirically by the manufacturer (i.e., a counting rate of approximately 75,000 particles/0.1 cfm results in a 10% coincidence).

The Royco particle size ranges chosen for the experiment were 0.3–0.5, 0.5–1.0, 1.0–2.0, 2.0–3.0 μm, with a counting period of one minute. To monitor the four size ranges, two replaceable amplifier attenuation boards each containing two size ranges were purchased. In the experiments, two size ranges were obtained simultaneously, and the other two immediately afterwards by exchanging the attenuation boards.

Theoretical light scattering calculations have been made for this optical system by Hodkinson and Greenfield (1965) and extended by Quenzel (1969). The conclusions drawn from these calculations are that in a right angle scattering system response from an aerosol of known refractive index is relatively linear as a function of particle size. However, for an atmospheric aerosol with refractive indices of water m = 1.33–0i at one extreme, to carbon with m = 1.95–0.66i at the other, the response is heavily dependent on refractive index.

Millipore Filter

The number concentrations for the test aerosols were obtained by sampling with type GS Millipore filter paper and counting the particles microscopically (American Optical "Phase Star": microscope, oil immersion × 97 objective, × 15 eyepiece). The filters were made transparent with 1.510 refractive index oil. The number concentration of particles in the aerosol was calculated from the number

TABLE II. Aerosol Characteristics

Particle Diameter (μm)	Dilution (%)	Number Concentration (particles/cm^3)	Singles (%)
0.500	0.01	39	97
0.81	0.0227	32	94
1.099	0.0625	24	95
1.947	2.0	11	97

of particles per unit filter area viewed under the microscope, sampling flow rate per unit filter area, and sampling time. Also, the percentages of single latex particles were determined by counting the numbers of singles and clusters (doublets, triplets, etc.) per unit of filter area. A 25 mm diameter filter was chosen for the experiment.

EXPERIMENTAL DATA

Aerosol Characteristics

The latex suspensions (except 1.947 μm) were diluted by known volumes of doubly distilled-deionized water and then atomized at 15 p.s.i. This pressure allowed the aerosol concentration to be low enough so that the coincidence error would be minimum. The volume dilutions used in the experiment are shown in Table II with their respective initial number concentrations and percentage single particles obtained from millipore sampling.

Aerosol number concentration measurements were made simultaneously by the Royco, Bausch and Lomb, and laser counters. The PMT of the laser counter was operated at 1100, 1200, and 1300V. Many discriminator settings were used for each voltage in order to determine the range of settings which counted every particle. The PMT voltages chosen were large enough to produce a signal whose amplitude was suitable for the counting electronics but not large enough to produce a signal from PMT noise. A millipore sample was taken during each voltage setting from which number concentration measurements were also made.

Number Concentration Measurements

Laser Counter.—Pulses of light scattered by the various monodisperse aerosols were counted by the scaler for each discriminator setting, thereby determining the aerosol number concentration. Since Millipore concentrations were

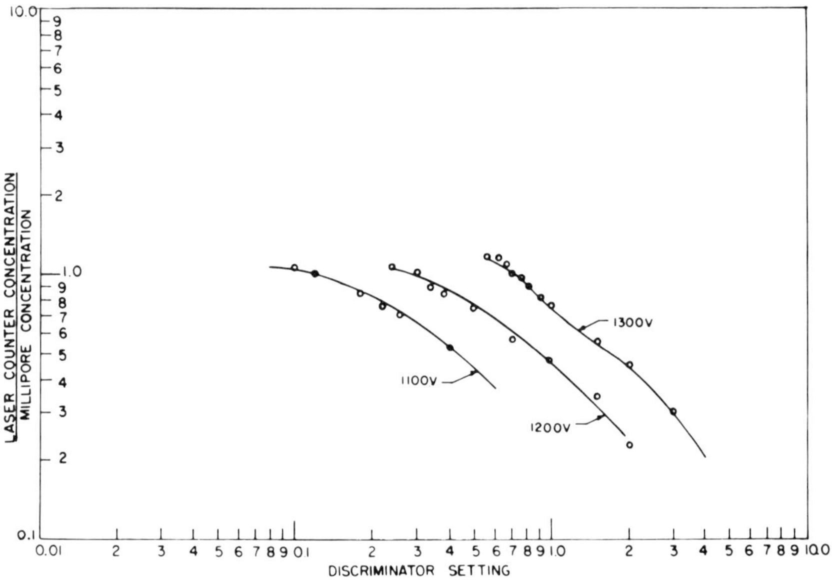

Figure 5. Laser counter efficiency for 0.500 μm particles as a function of discriminator setting.

also taken, comparison of these two measurements determine the efficiency of the laser counter. A value of "one" for the ratio of the laser counter concentration to the Millipore concentration indicated that every particle that passed through the laser optical sensing volume was detected and counted. As an example, the range of discriminator settings for a 0.500 μm latex is shown in Figure 5 (similar data for the 0.81, 1.099 and 1.947 μm were obtained and for convenience will not be shown). Ratios less than one showed that signal from some of the aerosol was not large enough to overcome the discriminator threshold bias.

Royco; Bausch and Lomb Counters.—The number concentrations obtained by the commercial counters can also be compared to Millipore concentrations and are shown in Figure 6. It appears that the Royco and Bausch and Lomb counters are quite accurate for determining number concentrations of aerosols 0.81 μm or larger. However, for smaller aerosols, less than 0.81 μm, the commercial counters detect only a fraction of the aerosol which traverses the optical sensing volume.

Millipore Samples.—The number concentrations obtained from the Millipore filter samples of the different aerosols are shown in Table III and correspond to the designated PMT voltages in the laser counter. Sampling time was at least 50 min. per sample, and was increased according to the suspected decrease in aerosol concentration with time. After a designated sampling time the filters were mounted on glass slides, rendered transparent, and observed in the optical microscope. Many fields were counted and a minimum of 1,000 particles per sample

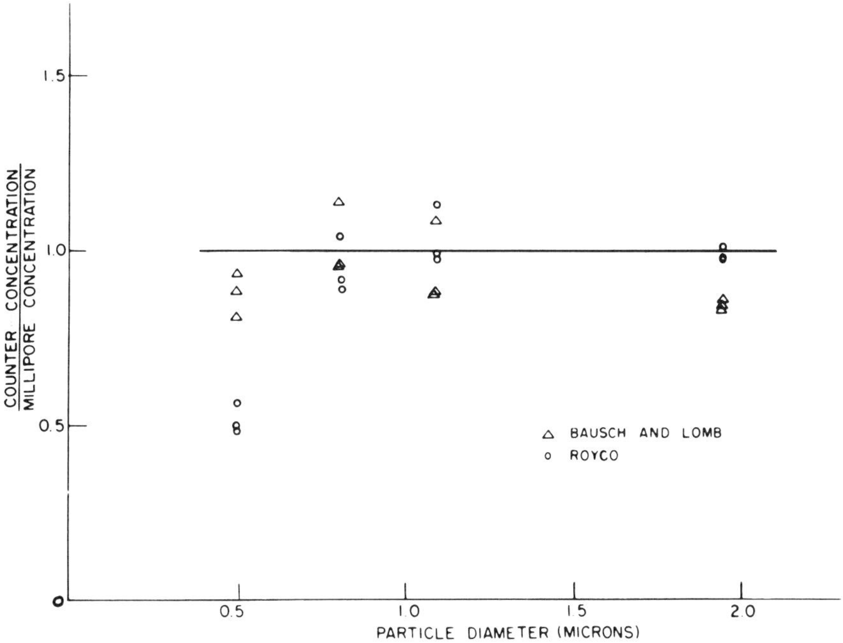

Figure 6. Counting efficiency for two commercial counters.

observed. The percentage of single particles obtained in the aerosols was always greater than 90.0%.

Particle Size Measurements

As the number concentration measurements were being made, particle size determinations were also obtained. It would appear at the outset that since the latex as supplied by Dow is highly monodisperse, an aerosol containing these particles could be easily sized. The commercial counters classify particle size according to the amplitude of the light pulses received from the particles as measured with a pulse height analyzer. Since each particle is the same size, it seems reasonable

TABLE III. Millipore Filter Number Concentration Measurements

Particle Diameter (μm)	PMT Voltage (volts)	Number Concentration (particles/cm^3)	Singles (%)	Sampling Flow Rate (cm^3/min)
1.947	1100	10.8	96.5	911
	1200	10.5	94.9	
	1300	10.8	96.9	
1.099	1100	23.5	95.0	600
	1200	31.5	93.5	
	1300	31.9	93.8	
0.81	1100	31.7	94.2	620
	1200	29.4	96.2	
	1300	17.8	90.2	
0.500	1100	39.3	97.0	620
	1200	35.7	97.0	
	1300	25.5	94.7	

that each light pulse from a monosized aerosol would have the same approximate amplitude. However, this was not the case. Due to the unavailability of a multichannel pulse height analyzer for the laser counter, particle size measurements could not be made although it would be a simple matter to do so.

Royco; Bausch and Lomb Counters.—Table III indicates the experimental results for the two commercial counters on particle sizes of 1.947, 1.099, 0.81, 0.500 μm. If we examine the data for the 0.81 μm aerosol, all the particles should appear in the 0.5–1.0 μm channel. However, the table shows large percentages of aerosol in all channels. This data could be construed as polydisperse aerosol. However, it has been shown (Rimberg and Thomas, 1970a) that the aerosols are quite monodisperse.

The explanation of this discrepancy can be found in the optical system of the counters. If the illumination of the optical sensing volume is not uniform and

TABLE IV. Percent Size Discrimination of Monodisperse Aerosols Recorded by the Royco and Bausch and Lomb Counters

Aerosol Diameter (μm)	Counter	Size Range (μm) 0.3–0.5	0.5–1.0	1.0–2.0	2.0–up
1.947[a]	Royco	19.5	13.5	36.2	30.9
1.947[a]	B & L	0	0	70.6	29.1
1.099	Royco	14.4	24.0	45.0	16.6
1.099	B & L	2.4	15.2	80.4	2.0
0.81	Royco	20.9	44.2	32.1	2.8
0.81	B & L	7.6	39.4	52.8	0.2
0.500	Royco	51.9	33.8	3.0	0.3
0.500	B & L	49.2	39.2	1.2	0.4

a. Centrifuged 5 times.

homogeneous, monodisperse particles at different points in the sensitive volume can generate scattered light pulses of different intensities. Also, the photocathode may not be uniformly sensitive. If so, electrical pulses can be produced which vary in intensity, even if the scattered light from each particle is identical. Hence, a monodisperse aerosol can produce a pulse spectrum, which appears as a heterodisperse aerosol distribution.

EXPERIMENTAL CONCLUSIONS

The response of a new aerosol counter to latex aerosols 0.500, 0.81, 1.099, and 1.947 μm, has been measured to determine its counting efficiency. Two commercial light scattering counters, the Royco 220 and Bausch and Lomb 40-1, were also evaluated for comparison purposes. The new counter, showed its ability to measure with 100% efficiency, the number of particles per unit volume of 0.500 μm and greater, whereas the two commercial counters indicated their counting inefficiencies at this size. The commercial counters also showed a wide spread in size distribution on monodisperse aerosols.

REFERENCES

Cooke, D. D. and M. Kerker, 1975: Response calculations for light scattering aerosol particle counters. *Appl. Opt., 14,* 734–739.

Cooper, D. W. and R. L. Byers, 1970: Laser light scattering from laboratory aerosols. *J. Air Pollution Control Assoc., 20,* 43–47.

Echols, W. H. and J. A. Young, 1963: Studies of portable air operated aerosol generators. Naval Res. Lab. Rep., NRL-5929.

Fisher, M. A., S. Katz, A. Lieberman and N. E. Alexander, 1955: The aerosoloscope: An instrument for the automatic counting and sizing of aerosol particles. *Proceedings of 3rd National Air Pollution Symp.,* Pasadena, Calif. Apr. 20th.

Gebhart, J. B., J. Bol, W. Heinze and W. Letschert, 1970: A particle-size spectrometer for aerosols utilizing low-angle scattering of the particles in a laser beam. *Staub-Reinhalt. Luft, 30,* 5–14.

Graedel, T. E., 1974: Channel width determination and electronic pulse processing losses in optical particle counters. *J. Aerosol Sci., 5,* 125–131.

Grafendorfer, I. W., 1970: A classifying optical particle counter with variable adjustments of scattering angle. Report of Aerosol Research at the First Physics Institute of the University of Vienna, May.

Gravatt, C. C., 1973: Real time measurement of the size distribution of particulate matter in air by a light scattering method. *J. Air Pollution Control Assoc., 23,* 1035.

Gucker, F. T. and D. G. Rose, 1954: A photoelectronic instrument for counting and sizing aerosol particles. *Brit. J. Appl. Phys. Suppl., 3,* 138–143.

———, J. Tuma, H. M. Lin, G. M. Huang, S. C. Ems and T. R. Marshall, 1973: Rapid measurement of light scattering diagrams from single particles in an aerosol stream and determination of latex particle size. *J. Aerosol Sci., 4,* 389–404.

Hodkinson, J. R. and J. R. Greenfield, 1965: Response calculations for light-scattering aerosol counters. *Appl. Opt., 4,* 1463–1474.

Jacobi, W., J. Eichler and N. Stolterfoht, 1968: Particle size spectrometry of aerosols by light scattering in a laser beam. *Staub-Reinhalt. Luft, 28,* 15–22.

Jaenicke, R., 1972: The optical particle counter: Cross-sensitivity and coincidence. *J. Aerosol Sci., 3,* 95–111.

Lavery, A. L. and W. Z. Leavitt, 1967: Improved atmospheric particle analyzer. NASA Tech. Brief, 67-10231, July.

Lieberman, A. and R. J. Allen, 1969: Theoretical and experimental light scattering data for a near forward system. *Proc. 8th Annual Meeting Amer. Assoc. Contamination Control,* May.

Martens, A. E. and J. D. Keller, 1968: An instrument for sizing and counting airborne particles. *J. Amer. Industrial Hygiene Assoc., 29,* 257.

Moretz, W. J., V. D. Withstandley and G. W. Anderson, 1970: A portable counter and size analyzer for airborne dust. *RSI, 41,* 978.

Neitzel, W. E., 1968: "Real Time" monitor uses laser beam. *Contamination control, VII,* 12.

———, 1969: A high-volume, real-time aerosol monitor. SC-DR-69-56, Sandia Laboratories, Albuquerque, N.M. 18pp., June.

Nikiforova, N. K. and Y. S. Sedunov, 1972: An investigation of errors arising in photoelectric counters—aerosol particle analyzer. *J. Aerosol Sci., 3,* 441–453.

O'Konski, C. T. and G. T. Doyle, 1955: Light scattering studies in aerosols with a new counter-photometer. *Anal. Chem., 27,* 694–701.

———, M. D. Bitron and W. I. Higuchi, 1959: Light scattering instrumentation for particle size distribution measurement. Special Technical Publ. No. 234, ASTM, Philadelphia, 180–205.

Pinnick, R. G. and D. J. Hoffmann, 1973: Efficiency of light-scattering aerosol particle counters. *Appl. Opt., 12,* 2593–2597.

Preining, O., 1968: *Staub-Reinhalt. Luft, 28,* 29.

Quenzel, H., 1969: Influence of refractive index on the accuracy of size determination of aerosol particles with light scattering aerosol counters. *Appl. Opt., 8,* 165–169.

Rimberg, D. and D. Keafer, 1970: Accuracy of measuring aerosol concentration with particle counters. *J. Colloid and Interface Sci., 33,* 628.

——— and J. W. Thomas, 1970a: Comparison of particle size of latex aerosols by optical and gravity settling methods. *J. Colloid and Interface Sci., 32,* 101–105.

——— and ———, 1970b: Response of an optical counter to monodisperse aerosols. *Atmos. Environ., 4,* 681–688.

Schehl, R., S. Ergun and A. Headrick, 1973: Size spectrometry of aerosols using light scattering from the cavity of a gas laser. *RSI, 44,* 1193–1196.

Sinclair, D., 1967: A new photometer for aerosol particle size analysis. *J. Air Pollution Control Assoc., 17,* 105–108.

Thomas, A. L., A. N. Bird, R. H. Collins and P. C. Rice, 1961: A portable photometer and particle size analyzer. *J. Instr. Soc. of Amer., 8,* 52–56.

Thomas, J. W., 1970: Mass transfer of aerosols with axial diffusion in circular tubes. U.S. Atomic Energy Comm., HASL-238, Health and Safety Laboratory, October.

——— and D. Rimberg, 1967: A simple method for measuring the average charge on a monodisperse aerosol. *Staub-Reinhalt. Luft, 27,* 18–22.

Whitby, K. T. and R. A. Vomela, 1967: Response of single particle optical counters to nonideal particles. *Environ. Sci. Technol., 1,* 801–814.

Wyatt, P. and D. T. Phillips, 1972: A new instrument for the study of individual particles. *J. Colloid and Interface Sci., 39,* 5–14.

Zinky, W. R., 1962: The aerosol particle counter. *J. Air Pollution Control Assoc., 12,* 578–583.

QUESTIONS

Rimberg: I want to propose at this meeting that some type of standard method be reported which everybody could recommend or reference, for instance through an ASTM committee or any other professional society. Then everybody would know what to do as far as generating polystyrene latex, or methylene blue, or DOP. We could then leave the calibration and move on to other things. Let's have a method for particular limitations and proceed with the business of measuring these aerosols in the real environment, not necessarily the laboratory environment.

Lepper: I'm on the ASTM committee on air sampling. We welcome you on the committee because there can only be less benefactors on the committee than there are users. We need you. So we are doing this work on calibration procedure.

Whitby: All these instruments were originally developed for clean room counting, and it's clear that they are not really suitable for the research applications that we and other people have described. The manufacturers have done a fairly good job but I still feel that plain reliability in the basic hardware is far more important than whether you have an elliptical mirror. William Clark's data shows what we found with all (particulate or elliptical) counters and Climet too; that here are two base lines' inability to shape the thing from one place to another, difficulties in getting replacement parts in a hurry for field work, and signal levels that are so low you have to build amplifiers to get them into counting data position systems. All these sorts of things are more important than the various differences in the optical systems.

If the instrument's basic reliability and good calibration procedures are developed, then some of the theoretical advantages of optical systems might be useful. So far we have not been able to take advantage of these and we have tried hard to do it in the field. In the laboratory it's a different matter. There one can keep a $100,000.00 piece of machinery and have the electronics engineer check it all the time. You can't do this in the field and there is where the problem is.

I was struck today that in the presentations we've seen very little optical data from these instruments. The reality of using them is quite different than what can be obtained with calibration aerosols. The manufacturers so far have done a good job on first and second generation instruments, and as we get into the third generation I hope that some of these practical things will improve so that our Climet will finally be stable.

There is a great difference from instrument to instrument in repeatability and stability. I think we have worked with seven or eight 220's, four Bausch & Lomb's, and three or four 218's. We find that in the optical system, such as the Climet's (and we've only worked with one of those but I have indicated that we have chosen a very stable one, as are the Bausch & Lomb and the Royco 220), the forward scattering units, because of slight differences in manufacture, have no two calibration curves exactly alike. Therefore, they have to be calibrated individually. I think if we took all of the calibrations (sometimes a dozen on the same instrument) over seven or eight instruments and put them all on one graph then they would probably all agree. So the forward scattering instruments are much more touchy and the manufacturers' tolerances are obviously not enough to make one instrument identical to another one. This is the only advantage for going away from the forward scattering units. Nonforward scattering instruments are definitely superior in other respects. On the whole, I think the optical counters are pretty good machines.

PART III

Electrical Aerosol Analyzer

ELECTRICAL AEROSOL ANALYZER: HISTORY, PRINCIPLE, AND DATA REDUCTION

BENJAMIN Y. H. LIU, DAVID Y. H. PUI, AND ABDE KAPADIA

Mechanical Engineering Department, University of Minnesota, Minneapolis

Over the years, many techniques have been developed for measuring aerosols. Only a few have survived the test of time and come into more general use. In order for a technique to be more generally adopted it must compete favorably with alternative techniques in terms of range of applicability, measurement speed, accuracy, or cost. It is unlikely that a given technique will be more generally adopted if it does not excel in one or more of these respects.

The electrical aerosol analyzer (EAA) was initially developed as a means for measuring aerosol size distribution *in situ* in the range below that of the light scattering optical counter, i.e., below about 0.3 μm in particle diameter. At the time of its initial development, no such *in situ* measuring device was available for small particles below 0.3 μm. Although the electron microscope could be used for measuring particles in this range, the technique is tedious, time-consuming, and not suited for repetitive measurements. Such repetitive measurements are needed, for instance in order to follow the changing size distribution of aerosols, or for atmospheric monitoring in pollution and air quality studies. In addition, the electron microscope technique cannot be applied to volatile particles, which would evaporate in the high-vacuum environment of the microscope. For these and other applications, an automatic, *in situ* measuring device such as the EAA is needed.

The research was supported by the Energy Research and Development Administration under Contract E (11-1)-1248.

HISTORY OF THE ELECTRICAL AEROSOL ANALYZER

The development of the EAA began at the Particle Technology Laboratory, University of Minnesota, in 1963. K. T. Whitby, together with his students, principally C. Peterson, G. Sem, W. E. Clark, and R. B. Husar, pioneered the development and the application of the instrument (Whitby and Clark, 1966; Whitby et al., 1966; Clark, 1966; Clark and Whitby, 1967; Whitby and Husar, 1970a, b, c). The first operational laboratory instrument, described in detail in the paper by Whitby and Clark (1966), is shown in Figure 1. This device contained all the essential elements of the instrument known today including the

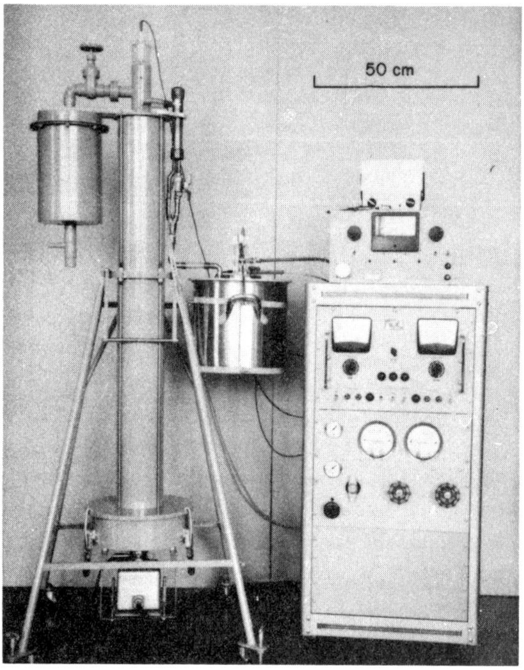

Figure 1. The University of Minnesota prototype electrical aerosol analyzer.

diffusion charger, the mobility analyzer, and the electrometer current sensor. The instrument was subsequently developed into a commercial device by TSI, Inc. under license from the University of Minnesota. The commercial instrument, the TSI Model 3000, was introduced in 1967.

The U of M prototype and the commercial TSI instruments were used in a number of aerosol studies, the most notable of which was the 1969 Pasadena Smog Study (see Hidy, 1972), a cooperative research project involving the University of Minnesota, the California Institute of Technology, the University of Washington, and the California State Department of Health. The purpose of the project was to study the physico-chemical properties of aerosol in the Los Angeles area. The study led to the finding that the smog aerosol size distribution in Los Angeles is bimodal (Whitby et al., 1972). The bimodal nature of atmospheric aerosol size distributions has since been confirmed by a number of independent studies, including measurements made at various localities and by such independent methods as the impactor. The Pasadena study was the first major study in which the usefulness of the EAA was demonstrated. It was concluded that the EAA, in spite of its limitations, was capable of giving valuable data not obtainable by other means.

Although the early EAA worked well in the Pasadena and other studies, it had one severe limitation: the instrument was rather large and heavy. The TSI Model 3000 weighed nearly 182 kg and had a volume of 0.9 m^3. It was realized that unless the size and the weight of the instrument could be substantially reduced, the usefulness of the device would be severely limited. For instance, it would be totally impractical to use the instrument on board small aircraft to perform airborne measurements, which were then becoming important in air pollution studies. It was decided that a second-generation instrument should be developed to substantially reduce the size and weight of the instrument.

That the instrument's size and weight could indeed be substantially reduced was apparent from the aerosol charging and precipitation studies then being conducted at the Particle Technology Laboratory by B. Y. H. Liu and his students. It was realized that the sonic jet diffusion charger used in the early EAA, which required a source of compressed air, could be replaced by a diffusion charger of a simpler design that did not require such a compressed air source. In addition, the mobility analyzer tube, which was 16 cm in diameter and 150 cm long in the original EAA, could be substantially reduced in size without affecting its performance. Further, miniaturized electronic components, including high voltage power supplies and electrometer operational amplifiers, were becoming available and could be used to substantially reduce the overall size, weight, complexity, and cost of the instrument.

The development of this second-generation EAA was undertaken primarily by B. Y. H. Liu with help from his students, principally D. Y. H. Pui. The result of this development is the laboratory prototype instrument described by Liu et al. (1974). The instrument was completed in the summer of 1971 and was used in the ACHEX (Aerosol Characterization Experiments) Program in California to perform several specialized experiments. The laboratory prototype instrument performed well in these field trials, and the decision was made to further develop the instrument into a commercial device. The primary responsibility for the commercial development was undertaken by G. Sem at TSI, Inc. with help from Liu, Pui, and others. The result is the TSI Model 3030 EAA, pictured in Figure 2. This instrument has a volume of 0.05 m^3 and a weight of 24 kg. This represents roughly an eight-fold reduction in weight and a factor of 18 reduction in total volume compared to the predecessor instrument. This was accomplished by the careful design and optimization of all electronic and mechanical parts, including such principal components as the charger, the mobility analyzer, and the electrometer current sensor. The operating principle of this most recent device, which was first introduced by TSI in 1973, is described below.

OPERATING PRINCIPLE

The operating principle of the EAA can be explained with the help of the theoretical curves shown in Figure 3. In this figure, the mean electrical mobility, Z_p, of a charged aerosol is shown as a function of particle size for several different values of the charging parameter, nt, where n (ions/cm^3) is the concentration of unipolar ions to which the particles have been exposed, and t (sec) is the charging time. These curves apply to the case of "diffusion charging," i.e., particle charging by ion diffusion. This is in contrast to the case of "field charging", which occurs as a result of an externally applied electric field.

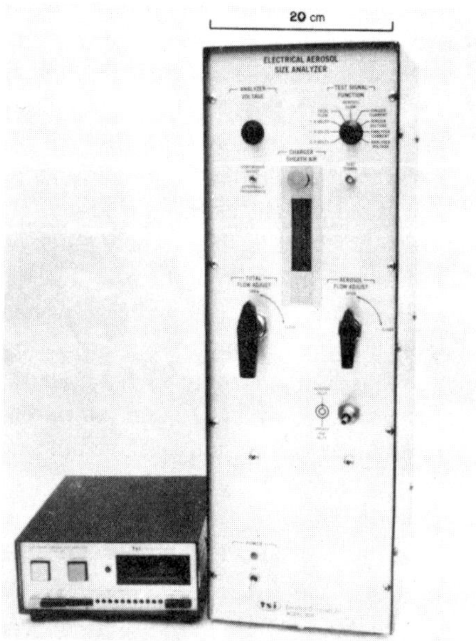

Figure 2. The TSI Model 3030 electrical aerosol analyzer.

As Figure 3 shows, in diffusion charging, the mean electric mobility of the aerosol is a monotonically decreasing function of particle size for particles smaller than about 1 μm in diameter. The size of a diffusion-charged aerosol can thus be determined from mobility measurement using this functional relationship between particle size and particle mobility. This is the basis on which the EAA operates.

Figure 4 is a schematic diagram of the EAA showing how the measurement is accomplished. The aerosol to be measured is first sampled into the aerosol charger, where it is exposed to unipolar positive ions and becomes electrically charged. The charged aerosol then enters the mobility analyzer which functions as a low-pass filter to precipitate the high mobility particles while allowing the low mobility particles to pass through. The "cut-off" mobility of the analyzer is determined by the applied voltage on the analyzer. By varying this voltage and

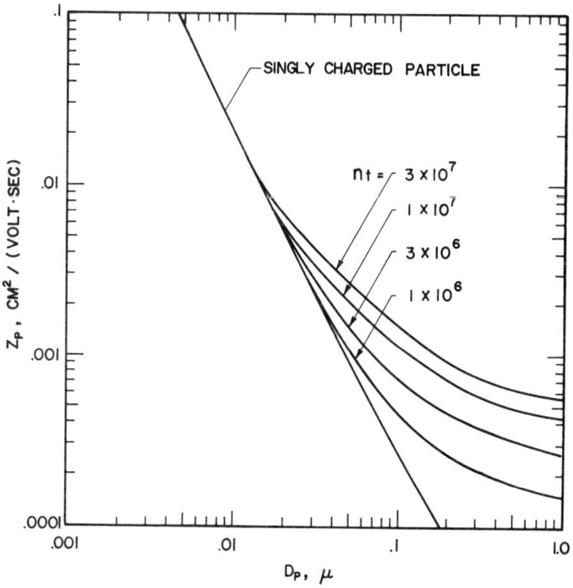

Figure 3. Theoretical relationship between mean electric mobility and size for aerosols in unipolar diffusion charging.

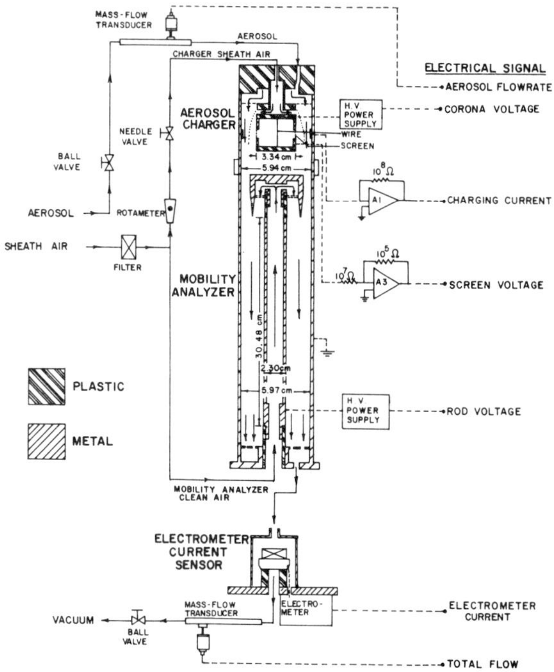

Figure 4. Schematic diagram of the electrical aerosol analyzer.

measuring the corresponding current carried by the charged particles with the electrometer current sensor, a voltage-current curve is generated which can be further analyzed to yield the desired particle size distribution curve.

Diffusion Charger

Figure 5 is a schematic diagram showing the aerosol charger in more detail. The charger is composed of two concentric metal cylinders with a fine, 25 μm diameter tungsten wire placed along the axis. A section of the inner cylinder is open and covered with a metal screen to permit the positive ions, which are produced by a high voltage corona discharge from the wire, to pass through. Charging of aerosol particles occurs in the annular space between the cylinders. The flow of ions through the screen is obtained by maintaining the inner cylinder at a positive voltage relative to the outer cylinder.

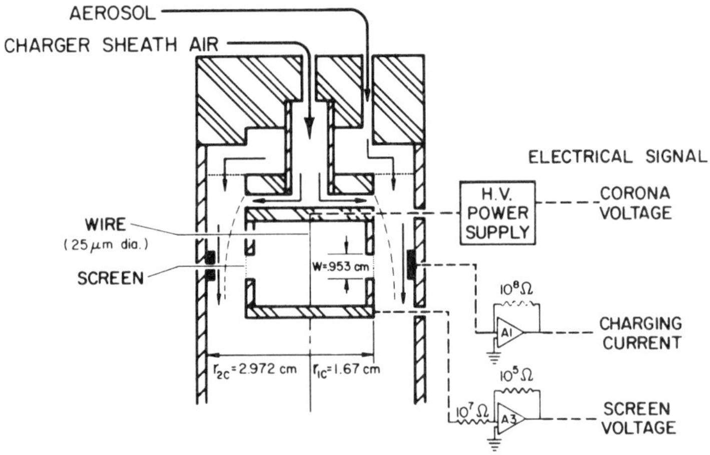

Figure 5. Schematic diagram of the unipolar charger.

The residence time of the particles in the annular charging region is equal to the volume of the charging region, $\pi(r_{2c}^2 - r_{1c}^2)$ w (cm^3), divided by the total air flow, q_{tc} (cm^3/sec) through the charger, or

$$t = \frac{\pi(r_{2c}^2 - r_{1c}^2)w}{q_{tc}} \tag{1}$$

where $r_{1c} = 1.670$ cm and $r_{2c} = 2.972$ cm are the inner and outer radii of the annular region, and w = 0.953 cm is the width of the screen opening. The total flow through the charger, q_{tc}, consists of an aerosol flow of 4ℓpm and a clean, sheath air flow of 1ℓpm. The clean sheath air is used here to displace the aerosol stream away from the screen and to prevent particles from entering the high intensity corona region within the inner cylinder. This is necessary in order to obtain uniform charging conditions, and to prevent particle loss by precipitation in the corona discharge region. With a total flow of 5ℓpm through the charger, the calculated residence time of the particles is 0.217 sec.

To determine the operating conditions of the charger, let us consider the motion of ions in the annular charging region. In this region, a radial electric field is established due to the potential difference between the cylinders. If we ignore the space charge due to the ions, this electric field, E, at a point r cm from the axis, can be obtained from the Gauss law,

$$E = \frac{V_s}{\ell n(r_{2c}/r_{1c})} \frac{1}{r} \quad (V/cm) \tag{2}$$

where V_s is the voltage on the inner cylinder (the screen voltage), the outer cylinder being kept near ground potential. This electric field, when acting upon the positive ions with a mobility Z_i (cm^2/V-sec), would cause the ions to move with a drift velocity of $E\,Z_i$ (cm/sec) toward the outer cylinder. This gives rise to a current of

$$I_c = EZ_i n 2\pi rwe \ (A) \tag{3}$$

where $e = 1.602 \times 10^{-19}$ (C) is the elementary unit of charge. Combining Eqs. (2)and (3) and solving for the concentration of ions, n (ions/cm^3) in the charging region, we have

$$n = \frac{I_c\, \ell n(r_{2c}/r_{1c})}{V_s Z_i 2\pi we}. \tag{4}$$

Eq. (4) allows the ion concentration n in the charging region to be calculated from the charging current, I_c, the screen voltage, V_s, and the dimensions of the charger.

To measure the charging current, I_c, a section of the outer cylinder is isolated from ground and connected to the input of operational amplifier (A1) as shown. A $10^8\,\Omega$ resistor is used as a feedback resistor to allow the few nA of charging current to be measured with a digital voltmeter. The operating condition of the charger is stabilized by comparing this voltage with a stabilized DC reference. The resulting difference is used as an error signal to control the high voltage applied to the corona wire. By adjusting the DC reference voltage, the charging current can also be adjusted. Similarly, the screen voltage is measured by operational amplifier (A3). To obtain the necessary screen voltage, the inner cylinder is connected to ground through a $10^7\,\Omega$ screen resistor, R_s. The current flow, I_t, through this resistor causes the screen potential to rise to the level, $V_s = I_t R_s$, where I_t is the total corona current collected by the inner cylinder.

The operating characteristics of this cylindrical aerosol charger have been determined by Liu and Pui (1975); the results are shown in Figures 6 and 7. In

Figure 6, the charging current is shown as a function of the screen voltage. Similarly, in Figure 7, the ion concentration, n, and the corresponding nt product are shown. The nt product shown is for the standard charger flow of 5 ℓpm and a particle charging time of 0.217 sec. It is seen that in order to obtain an nt product of 1×10^7 (ions/cm^3) (sec) as recommended for the operation of the instrument, the screen voltage should be set to 70 volts and the charging current, I_c, to 7.6 nA. However, individual instruments do differ, and the values of I_c and V_s

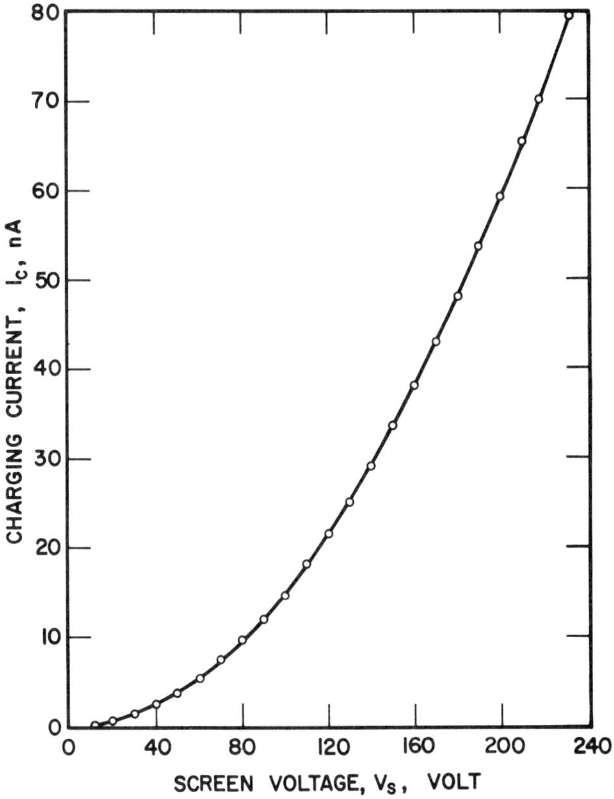

Figure 6. Relationship between screen voltage and charging current.

needed to obtain the same nt product of 10^7 may differ somewhat (±10%) among the individual instruments.

In calculating the ion concentration and the nt product, we have assumed a positive ion mobility of 1.4 cm²/V-sec. Further, in deriving Eq. (4) we have ignored the space charge effects. Thus, the n and nt values given above are nominal values, which are used here for the purpose of specifying the operating conditions of the charger. In addition, there is a weak electric field (approximately 55 V/cm) in the charging region, which may cause the particle charge to differ somewhat from the case of pure diffusion charging. Thus, the charging characteristics of the charger must be determined experimentally. This is discussed in the following article.

Mobility Analyzer

The mobility analyzer used in the EAA is in the form of a cylindrical condenser, with two concentric metal cylinders forming the electrodes of the condenser. The outer cylinder is grounded while a negative high voltage is applied to the inner electrode. The aerosol stream is introduced into the condenser near the outer

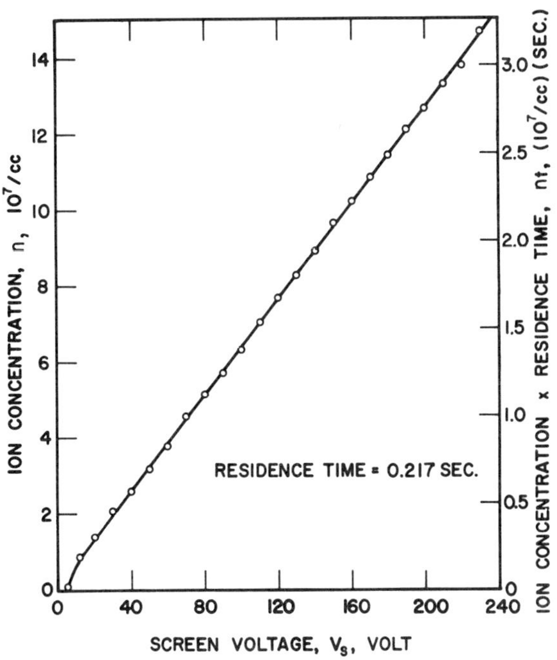

Figure 7. Ion concentration and nt product for the diffusion charger.

cylinder, while a particle-free, clean air sheath is introduced around the inner electrode. With a negative, high voltage applied to this inner electrode (the collector rod), the positively charged particles are attracted and are deflected through the clean air stream. It can be shown (Wahi and Liu, 1971) that for a particle with a mobility Z_p entering the instrument at a point adjacent to the outer cylinder, the particle would be precipitated on the lower end of the collector rod at an applied collector rod voltage of

$$V_{max} = \frac{q_{tm} \ell n(r_{2m}/r_{1m})}{2\pi L Z_p}. \quad (5)$$

Similarly, if the particle is carried by the aerosol stream into the instrument at the mid-point of the aerosol stream, the precipitation voltage would be

$$V = \frac{(q_{tm} - \frac{1}{2}q_{am})\ell n(r_{2m}/r_{1m})}{2\pi L Z_p}. \quad (6)$$

Finally, if the particle enters the instrument at the interface between the aerosol and clean air streams, the precipitating voltage would be

$$V_{min} - \frac{(q_{tm} - q_{am})\ell n(r_{2m}/r_{1m})}{2\pi L Z_p}. \quad (7)$$

In the above equations, q_{am} (cm^3/sec) and q_{tm} (cm^3/sec) are the aerosol and total mobility analyzer flows, respectively, r_{1m} (1.151 cm) and r_{2m} (2.985 cm) are the inner and outer radii of the annular precipitating space of the analyzer, and L (30.48 cm) is the length of the collector rod. If the aerosol stream entering the mobility analyzer has a uniform electric mobility, the precipitating voltage will spread over a range of values from V_{min} to V_{max} due to the finite width of the stream. This spread can be made small by making the aerosol flow rate, q_{am}, small in comparison with the total flow, q_{tm}. In the EAA, $q_{am} = 5\ \ell$pm and $q_{tm} = 50\ \ell$pm. Under these conditions the precipitating voltage will cover a range spreading over ±5% from the mean. This represents the ultimate resolution limit of the instrument for measuring particle mobility.

It should be noted that the derivation of Eqs. (5), (6), and (7) does not depend on the velocity of the fluid in the mobility analyzer being uniform. The equations will apply as long as the flow is laminar. The actual velocity profile that exists in the mobility analyzer is immaterial—it can be flat, fully developed, or developing. However, there is a small error in the equations due to the neglect of the "fringing" of the electric field near the ends of the collector rod. In making the derivation, the assumption is made that the electric field is everywhere radial. Near the upper and lower ends of the collector rod, the electric field necessarily diverges from a completely radial direction due to the end effect. This error is quite small in the actual mobility analyzer since the collector rod length ($L = 30.48$ cm) has been made large in comparison with the gap width ($r_{2m} - r_{1m} = 1.834$ cm).

Electrometer Current Sensor

The current sensor shown in Figure 3 is used to measure the charged particles penetrating through the mobility analyzer. The sensor consists of an absolute filter placed inside a Faraday cage, and an electrometer operational amplifier connecting to this filter. As the charged particles are removed by filtration, the charge is collected on the filter, thereby producing a current flow in the electrometer. The sensitivity of the current sensor for current detection is limited by the noise current in the electrometer, which is in the vicinity of 1×10^{-15} A.

DATA REDUCTION

As it is explained in the next paper in this series, the measurement range of the EAA has been divided into equal geometrical intervals of four per decade of particle size. Each instrument channel thus covers a particle size range of $10^{1/4}$ to 1, or 1.78 to 1. The channel boundaries have been chosen to be 0.0032, 0.0056, 0.01, 0.0178, 0.032, 0.056, 0.1, 0.178, 0.32, 0.56, and 1.0 μm.

To measure the size distribution of an aerosol with the EAA, the flow rates are first set to their respective standard values: 4 ℓpm for the aerosol flow, 1 ℓpm for the charger sheath air flow, and 50 ℓpm for the total flow of the instrument. The collector rod voltage is then stepped through the sequence of preselected values corresponding to the particle sizes at the channel boundaries. The electrometer currents, I_0, I_1, ... I_{11}, are then recorded. Here, I_0 refers to the electrometer current when the collector rod voltage is zero, I_1, the current when the collector rod is at the first, 0.0032 μm, voltage step, etc. These "integral currents" are then converted to the differential currents, ΔI's,

$$\begin{aligned} \Delta I_1 &= I_1 - I_2 \\ \Delta I_2 &= I_2 - I_3 \\ &\cdot \\ &\cdot \\ &\cdot \\ \Delta I_{10} &= I_{10} - I_{11} \end{aligned} \tag{8}$$

before the data are further processed. The fundamental problem in data reduction is to convert these ΔI's into the corresponding ΔN's of the number concentration of particles in the respective size intervals.

Data Reduction Using Instrument Sensitivities for Monodisperse Aerosols

The sensitivity of an instrument may be defined as the output of the instrument per unit input. In the case of the EAA, the output is the ΔI's and the input, ΔN's, the number concentration of the aerosol. If the instrument has an ideal behavior so that particles of a given size will produce a response in a given channel only, and no response in the neighboring channels, we can define an instrument sensitivity as

$$S_{N,i} = \frac{\Delta I_i}{\Delta N_i}\ , \ \text{pA/(particles/cc)} \tag{9}$$

where the subscript, N, indicates a number sensitivity and ΔN_i is the number of particles (per unit aerosol volume) in the channel. Similarly, we can define a surface area sensitivity, $S_{S,i}$, and a volume sensitivity, $S_{V,i}$, as follows,

$$S_{S,i} = \frac{\Delta I_i}{\Delta S_i},\ pA/(\mu m^2/cc) \tag{10}$$

$$S_{V,i} = \frac{\Delta I_i}{\Delta V_i},\ pA/(\mu m^3/cc) \tag{11}$$

where

$$\Delta S_i = \Delta N_i(\pi D_{p,i}^2) \tag{12}$$

$$\Delta V_i = \Delta N_i(\frac{1}{6}\pi D_{p,i}^3) \tag{13}$$

are the particle surface area and volume, both per unit aerosol volume. For such an ideal instrument, data reduction consists of simply dividing the ΔI's by the corresponding number, surface or volume sensitivities,

$$\Delta N_i = \Delta I_i/S_{N,i} \tag{14}$$

$$\Delta S_i = \Delta I_i/S_{S,i} \tag{15}$$

$$\Delta V_i = \Delta I_i/S_{V,i}. \tag{16}$$

In the actual instrument, monodisperse aerosols of a given size will produce a response not only in the nominal channel of the instrument for that size, but also in the neighboring channels. Table II of the following article illustrates this behavior. This table shows how the current is distributed among the various channels of the instrument when the instrument is used to measure monodisperse aerosols of various sizes located at the midpoint of each channel.

In the conventional data reduction procedure recommended by Liu and Pui (1974), this nonideal behavior of the instrument is ignored. The instrument sensitivity,

$$\hat{S}_{N,i} = \frac{I_0}{\Delta N_i},\quad \hat{S}_{S,i} = \frac{I_0}{\Delta S_i},\quad \hat{S}_{V,i} = \frac{I_0}{\Delta V_i}, \tag{17}$$

is obtained by dividing the total current, I_0 indicated by the electrometer when the collector rod voltage is set to zero, by the input (monodisperse) aerosol concentrations, ΔN_i, ΔS_i, or ΔV_i. These sensitivities are then used in place of the true instrument sensitivities in Eqs. (14)–(16) to calculate the number, surface area, or volume concentrations. This procedure, though not theoretically correct, is by far the most commonly used because of its simplicity and ability to give what appear to be reasonable results in the majority of cases. Table I lists the inverse of the instrument sensitivities, i.e., $1/\hat{S}_{N,i}$, $1/\hat{S}_{S,i}$, $1/\hat{S}_{V,i}$, based on the midpoint of the individual channels. In obtaining the inverse instrument sensitivities, we have fitted the following two equations to the original data points of Liu and Pui by regression:

$$\hat{S}_{N,i} = 2.351 \times 10^{12} D_{pi}^{6.262} \qquad \text{for } D_{pi} < 0.0125\ \mu m$$

$$\hat{S}_{N,i} = 426.4 \times D_{pi}^{1.156} \qquad \text{for } 0.0125\ \mu m \leqslant D_{pi} < 2.05\ \mu m \qquad (18)$$

where $\hat{S}_{N,i}$ is in units of pA/(10^6 particles/cm^3) and D_{pi} is in units of μm. The values given in Table I are calculated by the use of Eq. (18) rather than the original data of Liu and Pui.

Data Reduction Using Apparent Instrument Sensitivities for Polydisperse, Log-Normally Distributed Aerosols

In most cases, the EAA is used to measure aerosols whose size distribution is close to, or can be approximated by, a log-normal distribution. The data reduction procedure for such cases can be developed by consideration of the EAA response to polydisperse, log-normally distributed aerosols.

For purposes of developing the procedure, let us suppose that the EAA is used to measure an aerosol having a log-normal size distribution. Let the instrument indicate a ΔI_i in the ith channel. If ΔN_i is the number of particles (per unit aerosol volume) in the channel, we can define an *apparent* instrument sensitivity for the channel by,

$$\tilde{S}_{N,i} = \frac{\Delta I_i}{\Delta N_i} \qquad (19)$$

from which we have

$$\Delta N_i = \frac{\Delta I_i}{\tilde{S}_{N,i}}\,. \qquad (20)$$

Eq. (20) shows that if the apparent instrument sensitivity for the polydisperse aerosol is known, the correct aerosol concentration, ΔN_i, can be obtained by dividing the ΔI_i by the corresponding apparent sensitivity, $\tilde{S}_{N,i}$.

In actual practice, the $\tilde{S}_{N,i}$'s are not known a priori and must first be estimated before the data reduction can proceed.

To facilitate data reduction using the apparent instrument sensitivities, we have prepared Tables II, III, and IV for polydisperse, log-normally distributed aerosols having various mean sizes and standard deviations. The tables contain both the apparent number sensitivities defined by Eq. (19) and the surface area and volume sensitivities defined by

$$\tilde{S}_{S,i} = \frac{\Delta I_i}{\Delta S_i} \qquad (21)$$

$$\tilde{S}_{V,i} = \frac{\Delta I_i}{\Delta V_i} \qquad (22)$$

where the ΔI's were calculated from the equation

$$\Delta I_i = \sum_k \Delta N_k \hat{S}_{N,k} P_{i,k}\,. \qquad (23)$$

In Eq. (23), ΔN_k is the number concentration of particles in the kth size interval, $\hat{S}_{N,k}$ is the number sensitivity defined by Eq. (17) for monodisperse aerosols in the kth size interval, and $P_{i,k}$ is the fraction of current appearing in the ith channel due to particles in the kth size interval. The symbol $\sum_k$ indicates summation over all size intervals.

Although Eq. (23) can be evaluated by means of the basic response matrix of the instrument obtained by calibration (see Table II in the following article), the size interval used in this basic response matrix is too wide. To improve the accuracy of the apparent instrument sensitivities calculated by Eq. (19), we have expanded the original response matrix, which contains four equal geometrical size intervals per decade of particle size, into a more detailed matrix containing 16 equal geometrical size intervals per decade of particle size. This expanded matrix is obtained from the original matrix by interpolation and is shown in Table

TABLE I. Inverse of Instrument Sensitivities for Monodisperse Aerosols

Chnl. No.	Chnl. Bd. (μm)	Chnl. Mdpt. (μm)	$\hat{S}_{N,i}$[a] pA/(10^6 part./cm^3)	$1/\hat{S}_{N,i}$ part./(cm^3 − pA)	$1/\hat{S}_{S,i}$ μm^2/(cm^3 − pA)	$1/\hat{S}_{V,i}$ μm^3/(cm^3 − pA)
	0.0032					
1		0.0042	.00308	3.25(8)	1.80(4)	12.6
	0.0056					
2		0.0075	.116	8.61(6)	1.52(3)	1.90
	0.0100					
3		0.0133	2.85	3.50(5)	1.95(2)	4.32(−1)
	0.0178					
4		0.0237	5.57	1.79(5)	3.17(2)	1.25
	0.0316					
5		0.0422	10.87	9.19(4)	5.14(2)	3.62
	0.0562					
6		0.0750	21.18	4.72(4)	8.34(2)	10.42
	0.1000					
7		0.1330	41.14	2.43(4)	1.35(3)	2.99(1)
	0.1780					
8		0.2370	80.38	1.24(4)	2.20(3)	8.64(1)
	0.3160					
9		0.4220	156.8	6.37(3)	3.57(3)	2.51(2)
	0.5620					
10		0.7500	305.5	3.27(3)	5.78(3)	7.23(2)
	1.0000					

a. Based on value at the geometrical midpoint of each channel.

V. The apparent instrument sensitivities shown in Tables II, III, and IV were obtained by means of this expanded response matrix.

In using the apparent instrument sensitivities for data reduction, the number sensitivities shown in Table II should be used when the data are to be converted to a number distribution. Similarly, when the surface area or the volume distributions are desired, the corresponding surface area or volume sensitivities shown in Tables III and IV should be used.

To illustrate the data reduction procedure using these apparent instrument sensitivities, let us consider the data shown in Table VI obtained by the EAA on a polydisperse aerosol produced by a pneumatic atomizer of the Collison type. Using the instrument sensitivities for monodisperse aerosols given in Table I, we first obtain the ΔN's in Column 6 using the conventional data reduction procedure. The quantity $\Delta N/\Delta \log D_p$ is then calculated as shown in Column 7. The resulting distribution is plotted in Figure 8. Using the procedure described below

Table 2a.

INVERSE NUMBER SENSITIVITY (NUM./CC/PA) 5/77

GEOMETRIC STANDARD DEVIATION = 1.25

CHNL NO	CHNL BD MICRON	CHNL MDPT MICRON	MEAN(NMD) 0.004	0.006	0.008	0.010	0.020	0.040	0.060
	0.0022								
1		0.0042	2.17E 08	9.99E 07	7.24E 07	6.29E 07	5.39E 07	--------	--------
	0.0056								
2		0.0075	2.55E 07	1.02E 07	4.80E 06	3.03E 06	1.63E 06	1.45E 06	--------
	0.0100								
3		0.0133	1.21E 06	9.40E 05	6.98E 05	5.31E 05	2.60E 05	2.61E 04	2.78E 03
	0.0178								
4		0.0237	--------	2.29E 05	2.37E 05	2.35E 05	1.98E 05	4.07E 04	2.02E 03
	0.0316								
5		0.0422	--------	--------	1.36E 05	1.37E 05	1.57E 05	2.17E 05	8.58E 04
	0.0562								
6		0.0750	--------	--------	--------	--------	3.82E 02	2.99E 04	7.67E 04
	0.1000								
7		0.1330	--------	--------	--------	--------	--------	6.04E 01	5.19E 03
	0.1780								
8		0.2270	--------	--------	--------	--------	--------	--------	5.29E 00
	0.3160								
9		0.4220	--------	--------	--------	--------	--------	--------	--------
	0.5620								
10		0.7500	--------	--------	--------	--------	--------	--------	--------
	1.0000								

CHNL NO	CHNL BD MICRON	CHNL MDPT MICRON	MEAN(NMD) 0.080	0.100	0.200	0.400	0.600	0.800	1.000
	0.0032								
1		0.0042	--------	--------	--------	--------	--------	--------	--------
	0.0056								
2		0.0075	--------	--------	--------	--------	--------	--------	--------
	0.0100								
3		0.0133	--------	--------	--------	--------	--------	--------	--------
	0.0178								
4		0.0237	9.67E 01	6.52E 00	--------	--------	--------	--------	--------
	0.0316								
5		0.0422	2.56E 04	7.59E 03	4.96E 01	--------	--------	--------	--------
	0.0562								
6		0.0750	5.99E 04	3.44E 04	6.50E 02	9.84E-01	--------	--------	--------
	0.1000								
7		0.1330	2.50E 04	3.90E 04	1.36E 04	7.80E 01	6.66E-01	--------	--------
	0.1780								
8		0.2270	2.97E 02	2.87E 03	2.43E 04	3.66E 03	8.38E 01	1.16E 00	1.46E-02
	0.3160								
9		0.4220	1.20E-02	1.93E 00	2.37E 03	1.45E 04	5.44E 03	8.11E 02	7.42E 01
	0.5620								
10		0.7500	--------	--------	1.18E 00	2.56E 03	1.00E 04	8.85E 02	4.27E 03
	1.0000								

Table 2b.

INVERSE NUMBER SENSITIVITY (NUM./CC/PA) 5/77

GEOMETRIC STANDARD DEVIATION = 1.50

CHNL. NO.	CHNL. BD. MICRON	CHNL. MDPT. MICRON	MEAN(NMD) 0.004	0.006	0.008	0.010	0.020	0.040	0.060
	0.0032								
1		0.0042	1.79E 08	1.34E 08	1.12E 08	9.89E 07	7.51E 07	6.40E 07	--------
	0.0056								
2		0.0075	9.66E 06	6.40E 06	4.87E 06	4.03E 06	2.57E 06	1.98E 06	1.80E 06
	0.0100								
3		0.0133	6.54E 05	5.45E 05	4.76E 05	4.29E 05	3.13E 05	2.05E 05	1.39E 05
	0.0178								
4		0.0237	2.29E 05	2.23E 05	2.16E 05	2.08E 05	1.64E 05	7.64E 04	3.24E 04
	0.0316								
5		0.0422	1.51E 05	1.58E 05	1.66E 05	1.73E 05	1.95E 05	1.65E 05	1.11E 05
	0.0562								
6		0.0750	--------	1.91E 03	3.75E 03	6.17E 03	2.44E 04	5.62E 04	5.80E 04
	0.1000								
7		0.1330	--------	--------	1.81E 01	4.72E 01	8.65E 02	1.02E 04	2.37E 04
	0.1780								
8		0.2370	--------	--------	--------	--------	3.67E 01	1.22E 03	5.53E 03
	0.3160								
9		0.4220	--------	--------	--------	--------	--------	1.20E 01	2.40E 02
	0.5620								
10		0.7500	--------	--------	--------	--------	--------	--------	1.49E 00
	1.0000								

CHNL. NO.	CHNL. BD. MICRON	CHNL. MDPT. MICRON	MEAN(NMD) 0.080	0.100	0.200	0.400	0.600	0.800	1.000
	0.0032								
1		0.0042	--------	--------	--------	--------	--------	--------	--------
	0.0056								
2		0.0075	1.71E 06	1.66E 06	--------	--------	--------	--------	--------
	0.0100								
3		0.0133	9.73E 04	7.09E 04	2.31E 04	--------	--------	--------	--------
	0.0178								
4		0.0237	1.48E 04	7.43E 03	6.43E 02	4.27E 01	--------	--------	--------
	0.0316								
5		0.0422	7.48E 04	5.23E 04	1.42E 04	3.13E 03	1.20E 03	--------	--------
	0.0562								
6		0.0750	4.79E 04	3.71E 04	1.08E 04	2.06E 03	6.90E 02	3.03E 02	1.56E 02
	0.1000								
7		0.1330	3.08E 04	3.20E 04	1.83E 04	4.28E 03	1.28E 03	4.88E 02	2.20E 02
	0.1780								
8		0.2370	1.12E 04	1.56E 04	1.85E 04	6.53E 03	1.65E 03	4.01E 02	9.95E 01
	0.3160								
9		0.4220	9.72E 02	2.06E 03	8.57E 03	1.01E 04	5.33E 03	2.21E 03	8.63E 02
	0.5620								
10		0.7500	9.05E 00	3.40E 01	9.84E 02	5.86E 03	7.28E 03	5.78E 03	3.95E 03
	1.0000								

Table 2c.

INVERSE NUMBER SENSITIVITY (NUM./CC/PA) 5/77

GEOMETRIC STANDARD DEVIATION = 1.75

CHNL. NO.	CHNL. BD. MICRON	CHNL. MDPT. MICRON	MEAN(NMD) 0.004	0.006	0.008	0.010	0.020	0.040	0.060
	0.0032								
1		0.0042	1.70E 08	1.45E 08	1.31E 08	1.21E 08	9.80E 07	8.32E 07	7.70E 07
	0.0056								
2		0.0075	7.08E 06	5.68E 06	4.91E 06	4.41E 06	3.29E 06	2.63E 06	2.37E 06
	0.0100								
3		0.0133	5.11E 05	4.60E 05	4.26E 05	4.02E 05	3.34E 05	2.72E 05	2.34E 05
	0.0178								
4		0.0237	2.11E 05	2.01E 05	1.92E 05	1.83E 05	1.44E 05	9.03E 04	5.96E 04
	0.0316								
5		0.0422	1.76E 05	1.83E 05	1.87E 05	1.89E 05	1.83E 05	1.49E 05	1.19E 05
	0.0562								
6		0.0750	1.11E 04	1.72E 04	2.29E 04	2.81E 04	4.59E 04	5.41E 04	4.98E 04
	0.1000								
7		0.1330	3.36E 02	8.33E 02	1.56E 03	2.50E 03	9.25E 03	2.19E 04	2.74E 04
	0.1780								
8		0.2370	--------	1.09E 02	2.51E 02	4.69E 02	2.67E 03	9.21E 03	1.40E 04
	0.3160								
9		0.4220	--------	--------	--------	6.29E 00	1.40E 02	1.44E 03	3.27E 03
	0.5620								
10		0.7500	--------	--------	--------	--------	7.12E 00	6.99E 01	2.37E 02
	1.0000								

CHNL. NO.	CHNL. BD. MICRON	CHNL. MDPT. MICRON	MEAN(NMD) 0.080	0.100	0.200	0.400	0.600	0.800	1.000
	0.0032								
1		0.0042	7.35E 07	7.12E 07	--------	--------	--------	--------	--------
	0.0056								
2		0.0075	2.22E 06	2.13E 06	1.91E 06	--------	--------	--------	--------
	0.0100								
3		0.0133	2.08E 05	1.87E 05	1.25E 05	7.54E 04	5.42E 04	--------	--------
	0.0178								
4		0.0237	4.16E 04	3.04E 04	9.86E 03	2.71E 03	1.21E 03	6.74E 02	4.25E 02
	0.0316								
5		0.0422	9.74E 04	8.17E 04	4.34E 04	2.11E 04	1.34E 04	9.61E 03	7.38E 03
	0.0562								
6		0.0750	4.35E 04	3.77E 04	2.04E 04	9.18E 03	5.44E 03	3.68E 03	2.70E 03
	0.1000								
7		0.1330	2.86E 04	2.79E 04	1.97E 04	9.84E 03	5.71E 03	3.68E 03	2.56E 03
	0.1780								
8		0.2270	1.64E 04	1.75E 04	1.51E 04	7.30E 03	3.34E 03	1.56E 03	7.58E 02
	0.3160								
9		0.4220	4.91E 03	6.27E 03	9.51F 03	7.78E 03	4.74E 03	2.75E 03	1.62E 03
	0.5620								
10		0.7500	5.21E 02	9.05E 02	3.36E 03	5.96E 03	5.69E 03	4.69E 03	3.74E 03
	1.0000								

Table 2d.

INVERSE NUMBER SENSITIVITY (NUM./CC/PA) 5/77

GEOMETRIC STANDARD DEVIATION = 2.00

CHNL. NO.	CHNL BD. MICRON	CHNL. MDPT. MICRON	MEAN(NMD) 0.004	0.006	0.008	0.010	0.020	0.040	0.060
	0.0032								
1		0.0042	1.67E 08	1.50E 08	1.40E 08	1.33E 08	1.14E 08	9.92E 07	9.25E 07
	0.0056								
2		0.0075	6.25E 06	5.42E 06	4.92E 06	4.58E 06	3.73E 06	3.12E 06	2.85E 06
	0.0100								
3		0.0133	4.60E 05	4.28E 05	4.06E 05	3.91E 05	3.44E 05	3.01E 05	2.76E 05
	0.0178								
4		0.0237	1.94E 05	1.83E 05	1.74E 05	1.65E 05	1.34E 05	9.66E 04	7.47E 04
	0.0316								
5		0.0422	1.87E 05	1.88E 05	1.87E 05	1.85E 05	1.70E 05	1.42E 05	1.22E 05
	0.0562								
6		0.0750	2.74E 04	3.42E 04	3.91E 04	4.27E 04	5.06E 04	5.02E 04	4.59E 04
	0.1000								
7		0.1330	3.16E 03	5.37E 03	7.60E 03	9.75E 03	1.81E 04	2.53E 04	2.69E 04
	0.1780								
8		0.2370	9.72E 02	1.91E 03	2.98E 03	4.08E 03	8.96E 03	1.42E 04	1.61E 04
	0.3160								
9		0.4220	3.43E 01	1.13E 02	2.51E 02	4.48E 02	1.90E 03	4.56E 03	6.28E 03
	0.5620								
10		0.7500	--------	--------	2.70E 01	4.40E 01	1.87E 02	6.78E 02	1.30E 03
	1.0000								

CHNL. NO.	CHNL. BD. MICRON	CHNL. MDPT. MICRON	MEAN(NMD) 0.080	0.100	0.200	0.400	0.600	0.800	1.000
	0.0032								
1		0.0042	8.84E 07	8.56E 07	7.82E 07	--------	--------	--------	--------
	0.0056								
2		0.0075	2.69E 06	2.58E 06	2.31E 06	2.11E 06	2.01E 06	--------	--------
	0.0100								
3		0.0133	2.58E 05	2.44E 05	2.00E 05	1.57E 05	1.33E 05	1.18E 05	1.06E 05
	0.0178								
4		0.0237	6.04E 04	5.03E 04	2.62E 04	1.23E 04	7.58E 03	5.32E 03	4.02E 03
	0.0316								
5		0.0422	1.08E 05	9.66E 04	6.58E 04	4.24E 04	3.22E 04	2.63E 04	2.24E 04
	0.0562								
6		0.0750	4.16E 04	3.78E 04	2.59E 04	1.60E 04	1.16E 04	9.20E 03	7.61E 03
	0.1000								
7		0.1330	2.66E 04	2.57E 04	2.02E 04	1.33E 04	9.69E 03	7.53E 03	6.10E 03
	0.1780								
8		0.2370	1.66E 04	1.64E 04	1.31E 04	7.45E 03	4.36E 03	2.66E 03	1.69E 03
	0.3160								
9		0.4220	7.40E 03	8.11E 03	8.78E 03	6.47E 03	4.35E 03	2.98E 03	2.10E 03
	0.5620								
10		0.7500	1.94E 03	2.55E 03	4.61E 03	5.43E 03	4.91E 03	4.23E 03	3.63E 03
	1.0000								

Table 2e.

INVERSE NUMBER SENSITIVITY (NUM./CC/PA) 5/77

GEOMETRIC STANDARD DEVIATION = 2.25

CHNL. NO.	CHNL BD MICRON	CHNL MDPT. MICRON	MEAN(NMD) 0.004	0.006	0.008	0.010	0.020	0.040	0.060
1	0.0032	0.0042	1.66E 08	1.53E 08	1.45E 08	1.39E 08	1.24E 08	1.11E 08	1.04E 08
2	0.0056	0.0075	5.87E 06	5.29E 06	4.93E 06	4.67E 06	4.00E 06	3.47E 06	3.22E 06
3	0.0100	0.0133	4.34E 05	4.11E 05	3.96E 05	3.84E 05	3.50E 05	3.17E 05	2.98E 05
4	0.0178	0.0237	1.79E 05	1.69E 05	1.60E 05	1.53E 05	1.28E 05	1.00E 05	8.34E 04
5	0.0316	0.0422	1.86E 05	1.83E 05	1.80E 05	1.76E 05	1.60E 05	1.38E 05	1.24E 05
6	0.0562	0.0750	3.89E 04	4.35E 04	4.63E 04	4.80E 04	5.03E 04	4.75E 04	4.38E 04
7	0.1000	0.1330	8.65E 03	1.18E 04	1.44E 04	1.65E 04	2.24E 04	2.57E 04	2.59E 04
8	0.1780	0.2370	4.20E 03	6.13E 03	7.73E 03	9.05E 03	1.30E 04	1.56E 04	1.59E 04
9	0.3160	0.4220	5.73E 02	1.14E 03	1.72E 03	2.28E 03	4.42E 03	6.64E 03	7.67E 03
10	0.5620	0.7500	8.25E 01	1.54E 02	2.35E 02	3.24E 02	8.18E 02	1.79E 03	2.60E 03
	1.0000								

CHNL. NO.	CHNL BD MICRON	CHNL MDPT. MICRON	MEAN(NMD) 0.080	0.100	0.200	0.400	0.600	0.800	1.000
1	0.0032	0.0042	1.00E 08	9.75E 07	8.96E 07	8.32E 07	--------	--------	--------
2	0.0056	0.0075	3.06E 06	2.95E 06	2.66E 06	2.43E 06	2.32E 06	2.25E 06	2.20E 06
3	0.0100	0.0133	2.85E 05	2.74E 05	2.42E 05	2.09E 05	1.90E 05	1.77E 05	1.67E 05
4	0.0178	0.0237	7.20E 04	6.27E 04	4.13E 04	2.49E 04	1.81E 04	1.42E 04	1.18E 04
5	0.0316	0.0422	1.13E 05	1.05E 05	8.04E 04	5.95E 04	4.92E 04	4.28E 04	3.83E 04
6	0.0562	0.0750	4.06E 04	3.78E 04	2.90E 04	2.08E 04	1.68E 04	1.43E 04	1.25E 04
7	0.1000	0.1330	2.53E 04	2.45E 04	2.04E 04	1.53E 04	1.24E 04	1.05E 04	9.13E 03
8	0.1780	0.2370	1.56E 04	1.50E 04	1.18E 04	7.44E 03	4.96E 03	3.47E 03	2.53E 03
9	0.3160	0.4220	8.18E 03	8.40E 03	7.87E 03	5.70E 03	4.12E 03	3.09E 03	2.40E 03
10	0.5620	0.7500	3.24E 03	3.74E 03	4.93E 03	4.99E 03	4.49E 03	3.99E 03	3.56E 03
	1.0000								

Table 2f.

INVERSE NUMBER SENSITIVITY (NUM. /CC/PA) 5/77

GEOMETRIC STANDARD DEVIATION = 2.50

CHNL. NO.	CHNL. BD. MICRON	CHNL. MDPT. MICRON	MEAN(NMD) 0.004	0.006	0.008	0.010	0.020	0.040	0.060
	0.0032								
1		0.0042	1.65E 08	1.55E 08	1.48E 08	1.44E 08	1.30E 08	1.19E 08	1.13E 08
	0.0056								
2		0.0075	5.66E 06	5.22E 06	4.93E 06	4.73E 06	4.17E 06	3.72E 06	3.49E 06
	0.0100								
3		0.0133	4.19E 05	4.02E 05	3.90E 05	3.81E 05	3.53E 05	3.27E 05	3.12E 05
	0.0178								
4		0.0237	1.68E 05	1.59E 05	1.51E 05	1.45E 05	1.25E 05	1.02E 05	8.88E 04
	0.0316								
5		0.0422	1.81E 05	1.77E 05	1.72E 05	1.69E 05	1.54E 05	1.36E 05	1.25E 05
	0.0562								
6		0.0750	4.48E 04	4.73E 04	4.86E 04	4.92E 04	4.90E 04	4.56E 04	4.25E 04
	0.1000								
7		0.1330	1.40E 04	1.69E 04	1.90E 04	2.05E 04	2.40E 04	2.54E 04	2.51E 04
	0.1780								
8		0.2370	8.14E 03	1.01E 04	1.14E 04	1.24E 04	1.47E 04	1.55E 04	1.52E 04
	0.3160								
9		0.4220	2.09E 03	3.02E 03	3.77E 03	4.36E 03	6.17E 03	7.58E 03	8.05E 03
	0.5620								
10		0.7500	3.58E 02	5.55E 02	7.47E 02	9.30E 02	1.73E 03	2.84E 03	3.55E 03
	1.0000								

CHNL. NO.	CHNL. BD. MICRON	CHNL. MDPT. MICRON	MEAN(NMD) 0.080	0.100	0.200	0.400	0.600	0.800	1.000
	0.0032								
1		0.0042	1.09E 08	1.07E 08	9.89E 07	9.24E 07	8.90E 07	8.68E 07	--------
	0.0056								
2		0.0075	3.34E 06	3.24E 06	2.95E 06	2.72E 06	2.60E 06	2.52E 06	2.47E 06
	0.0100								
3		0.0133	3.01E 05	2.93E 05	2.67E 05	2.41E 05	2.26E 05	2.15E 05	2.07E 05
	0.0178								
4		0.0237	7.96E 04	7.26E 04	5.29E 04	3.68E 04	2.91E 04	2.45E 04	2.13E 04
	0.0316								
5		0.0422	1.16E 05	1.10E 05	9.00E 04	7.19E 04	6.24E 04	5.63E 04	5.18E 04
	0.0562								
6		0.0750	4.00E 04	3.78E 04	3.09E 04	2.42E 04	2.06E 04	1.83E 04	1.66E 04
	0.1000								
7		0.1330	2.44E 04	2.37E 04	2.05E 04	1.65E 04	1.42E 04	1.26E 04	1.14E 04
	0.1780								
8		0.2370	1.45E 04	1.39E 04	1.10E 04	7.39E 03	5.35E 03	4.06E 03	3.19E 03
	0.3160								
9		0.4220	8.16E 03	8.10E 03	7.12E 03	5.21E 03	3.97E 03	3.16E 03	2.59E 03
	0.5620								
10		0.7500	4.02E 03	4.34E 03	4.91E 03	4.68E 03	4.24E 03	3.84E 03	3.52E 03
	1.0000								

Table 2g.

INVERSE NUMBER SENSITIVITY (NUM./CC/PA) 5/77

GEOMETRIC STANDARD DEVIATION = 2.75

CHNL. NO.	CHNL. BD. MICRON	CHNL. MDPT. MICRON	MEAN(NMD) 0.004	0.006	0.008	0.010	0.020	0.040	0.060
	0.0032								
1		0.0042	1.64E 08	1.56E 08	1.51E 08	1.47E 08	1.35E 08	1.25E 08	1.20E 08
	0.0056								
2		0.0075	5.52E 06	5.17E 06	4.94E 06	4.77E 06	4.30E 06	3.89E 06	3.69E 06
	0.0100								
3		0.0133	4.10E 05	3.96E 05	3.86E 05	3.78E 05	3.55E 05	3.33E 05	3.21E 05
	0.0178								
4		0.0237	1.60E 05	1.52E 05	1.45E 05	1.40E 05	1.22E 05	1.03E 05	9.24E 04
	0.0316								
5		0.0422	1.76E 05	1.71E 05	1.67E 05	1.63E 05	1.50E 05	1.35E 05	1.25E 05
	0.0562								
6		0.0750	4.73E 04	4.85E 04	4.89E 04	4.90E 04	4.77E 04	4.43E 04	4.17E 04
	0.1000								
7		0.1330	1.79E 04	2.01E 04	2.16E 04	2.25E 04	2.46E 04	2.50E 04	2.45E 04
	0.1780								
8		0.2370	1.11E 04	1.25E 04	1.34E 04	1.40E 04	1.51E 04	1.50E 04	1.43E 04
	0.3160								
9		0.4220	3.83E 03	4.74E 03	5.38E 03	5.85E 03	7.12E 03	7.87E 03	7.96E 03
	0.5620								
10		0.7500	8.51E 02	1.17E 03	1.45E 03	1.69E 03	2.58E 03	3.57E 03	4.09E 03
	1.0000								

CHNL. NO.	CHNL. BD. MICRON	CHNL. MDPT. MICRON	MEAN(NMD) 0.080	0.100	0.200	0.400	0.600	0.800	1.000
	0.0032								
1		0.0042	1.16E 08	1.14E 08	1.06E 08	1.00E 08	9.66E 07	9.44E 07	9.27E 07
	0.0056								
2		0.0075	3.55E 06	3.46E 06	3.18E 06	2.95E 06	2.83E 06	2.76E 06	2.70E 06
	0.0100								
3		0.0133	3.12E 05	3.05E 05	2.84E 05	2.62E 05	2.50E 05	2.41E 05	2.34E 05
	0.0178								
4		0.0237	8.47E 04	7.88E 04	6.16E 04	4.65E 04	3.89E 04	3.41E 04	3.06E 04
	0.0316								
5		0.0422	1.18E 05	1.13E 05	9.65E 04	8.09E 04	7.24E 04	6.67E 04	6.25E 04
	0.0562								
6		0.0750	3.96E 04	3.78E 04	3.21E 04	2.65E 04	2.34E 04	2.13E 04	1.98E 04
	0.1000								
7		0.1330	2.38E 04	2.32E 04	2.05E 04	1.73E 04	1.54E 04	1.40E 04	1.30E 04
	0.1780								
8		0.2370	1.37E 04	1.30E 04	1.04E 04	7.34E 03	5.60E 03	4.48E 03	3.69E 03
	0.3160								
9		0.4220	7.85E 03	7.66E 03	6.54E 03	4.88E 03	3.87E 03	3.20E 03	2.72E 03
	0.5620								
10		0.7500	4.39E 03	4.58E 03	4.79E 03	4.46E 03	4.07E 03	3.75E 03	3.49E 03
	1.0000								

Table 3a.

INVERSE SURFACE SENSITIVITY <MICRON2/CC/PA> 5/77

GEOMETRIC STANDARD DEVIATION = 1. 25

CHNL. NO.	CHNL. BD. MICRON	CHNL. MDPT. MICRON	MEAN<SMD> 0. 004	0. 006	0. 008	0. 010	0. 020	0. 040	0. 060
	0. 0032								
1		0. 0042	1. 46E 04	8. 55E 03	6. 83E 03	6. 21E 03	5. 71E 03	--------	--------
	0. 0056								
2		0. 0075	3. 27E 03	1. 85E 03	1. 10E 03	7. 97E 02	5. 18E 02	4. 91E 02	--------
	0. 0100								
3		0. 0133	3. 84E 02	3. 37E 02	2. 88E 02	2. 52E 02	2. 17E 02	4. 30E 01	5. 10E 00
	0. 0178								
4		0. 0237	--------	2. 27E 02	2. 34E 02	2. 43E 02	3. 02E 02	1. 68E 02	1. 49E 01
	0. 0316								
5		0. 0422	--------	--------	3. 96E 02	4. 06E 02	5. 08E 02	1. 13E 03	8. 64E 02
	0. 0562								
6		0. 0750	--------	--------	--------	--------	1. 63E 00	2. 02E 02	1. 01E 03
	0. 1000								
7		0. 1330	--------	--------	--------	--------	--------	5. 07E-01	6. 99E 01
	0. 1780								
8		0. 2370	--------	--------	--------	--------	--------	--------	9. 69E-02
	0. 3160								
9		0. 4220	--------	--------	--------	--------	--------	--------	--------
	0. 5620								
10		0. 7500	--------	--------	--------	--------	--------	--------	--------
	1. 0000								

CHNL. NO.	CHNL. BD. MICRON	CHNL. MDPT. MICRON	MEAN<SMD> 0. 080	0. 100	0. 200	0. 400	0. 600	0. 800	1. 000
	0. 0032								
1		0. 0042	--------	--------	--------	--------	--------	--------	--------
	0. 0056								
2		0. 0075	--------	--------	--------	--------	--------	--------	--------
	0. 0100								
3		0. 0133	--------	--------	--------	--------	--------	--------	--------
	0. 0178								
4		0. 0237	9. 49E-01	7. 34E-02	--------	--------	--------	--------	--------
	0. 0316								
5		0. 0422	3. 50E 02	1. 26E 02	1. 16E 00	--------	--------	--------	--------
	0. 0562								
6		0. 0750	1. 24E 03	1. 00E 03	4. 20E 01	9. 33E-02	--------	--------	--------
	0. 1000								
7		0. 1330	6. 26E 02	1. 43E 03	1. 44E 03	2. 15E 01	2. 41E-01	--------	--------
	0. 1780								
8		0. 2370	8. 27E 00	1. 19E 02	3. 45E 03	1. 65E 03	8. 10E 01	1. 91E 00	3. 85E-02
	0. 3160								
9		0. 4220	2. 25E-04	8. 12E-02	3. 96E 02	7. 52E 03	5. 86E 03	1. 55E 03	2. 25E 02
	0. 5620								
10		0. 7500	--------	--------	2. 00E-01	1. 46E 03	1. 19E 04	1. 82E 04	1. 40E 04
	1. 0000								

Table 3b.

INVERSE SURFACE SENSITIVITY (MICRON2/CC/PA) 5/77

GEOMETRIC STANDARD DEVIATION = 1.50

CHNL. NO.	CHNL. BD. MICRON	CHNL. MDPT. MICRON	MEAN(SMD) 0.004	0.006	0.008	0.010	0.020	0.040	0.060
	0.0032								
1		0.0042	1.30E 04	1.06E 04	9.34E 03	8.57E 03	7.04E 03	6.31E 03	6.07E 03
	0.0056								
2		0.0075	1.83E 03	1.38E 03	1.14E 03	9.92E 02	7.15E 02	5.92E 02	5.54E 02
	0.0100								
3		0.0133	2.83E 02	2.61E 02	2.47E 02	2.39E 02	2.21E 02	1.94E 02	1.59E 02
	0.0178								
4		0.0237	2.53E 02	2.64E 02	2.73E 02	2.80E 02	2.99E 02	2.43E 02	1.52E 02
	0.0316								
5		0.0422	4.75E 02	5.13E 02	5.51E 02	5.89E 02	7.75E 02	9.89E 02	9.33E 02
	0.0562								
6		0.0750	--------	8.56E 00	1.80E 01	3.12E 01	1.54E 02	5.72E 02	8.99E 02
	0.1000								
7		0.1330	--------	--------	1.38E-01	3.67E-01	7.54E 00	1.33E 02	5.14E 02
	0.1780								
8		0.2370	--------	--------	--------	--------	5.81E-01	2.78E 01	1.95E 02
	0.3160								
9		0.4220	--------	--------	--------	--------	--------	2.30E-01	7.62E 00
	0.5620								
10		0.7500	--------	--------	--------	--------	--------	--------	1.76E-01
	1.0000								

CHNL. NO.	CHNL. BD. MICRON	CHNL. MDPT. MICRON	MEAN(SMD) 0.080	0.100	0.200	0.400	0.600	0.800	1.000
	0.0032								
1		0.0042	--------	--------	--------	--------	--------	--------	--------
	0.0056								
2		0.0075	5.36E 02	5.25E 02	--------	--------	--------	--------	--------
	0.0100								
3		0.0133	1.26E 02	1.00E 02	3.87E 01	--------	--------	--------	--------
	0.0178								
4		0.0237	8.80E 01	5.16E 01	6.19E 00	4.86E-01	--------	--------	--------
	0.0316								
5		0.0422	7.72E 02	6.18E 02	2.27E 02	6.05E 01	2.52E 01	--------	--------
	0.0562								
6		0.0750	1.00E 03	9.67E 02	4.81E 02	1.27E 02	4.83E 01	2.29E 01	1.24E 01
	0.1000								
7		0.1330	9.97E 02	1.38E 03	1.63E 03	7.16E 02	2.86E 02	1.27E 02	6.30E 01
	0.1780								
8		0.2370	5.94E 02	1.15E 03	3.09E 03	2.56E 03	1.21E 03	4.81E 02	1.82E 02
	0.3160								
9		0.4220	6.51E 01	2.25E 02	2.10E 03	5.76E 03	5.62E 03	3.86E 03	2.25E 03
	0.5620								
10		0.7500	1.19E 00	5.00E 00	2.75E 02	4.25E 03	9.89E 03	1.25E 04	1.21E 04
	1.0000								

Table 3c.

INVERSE SURFACE SENSITIVITY (MICRON2/CC/PA) 5/77

GEOMETRIC STANDARD DEVIATION = 1.75

CHNL. NO.	CHNL. BD. MICRON	CHNL. MDPT. MICRON	MEAN(SMD) 0.004	0.006	0.008	0.010	0.020	0.040	0.060
	0.0032								
1		0.0042	1.26E 04	1.13E 04	1.05E 04	9.92E 03	8.53E 03	7.58E 03	7.18E 03
	0.0056								
2		0.0075	1.49E 03	1.28E 03	1.15E 03	1.06E 03	8.59E 02	7.28E 02	6.75E 02
	0.0100								
3		0.0133	2.56E 02	2.46E 02	2.40E 02	2.36E 02	2.25E 02	2.13E 02	2.02E 02
	0.0178								
4		0.0237	2.75E 02	2.83E 02	2.87E 02	2.90E 02	2.90E 02	2.57E 02	2.15E 02
	0.0316								
5		0.0422	6.14E 02	6.66E 02	7.09E 02	7.46E 02	8.70E 02	9.44E 02	9.18E 02
	0.0562								
6		0.0750	5.94E 01	9.93E 01	1.41E 02	1.84E 02	3.94E 02	7.00E 02	8.52E 02
	0.1000								
7		0.1330	2.73E 00	7.07E 00	1.39E 01	2.33E 01	1.12E 02	4.38E 02	8.05E 02
	0.1780								
8		0.2370	--------	1.75E 00	4.27E 00	8.47E 00	6.50E 01	3.85E 02	8.74E 02
	0.3160								
9		0.4220	--------	--------	--------	1.12E-01	2.99E 00	6.78E 01	3.03E 02
	0.5620								
10		0.7500	--------	--------	--------	--------	8.53E-01	9.95E 00	3.97E 01
	1.0000								

CHNL. NO.	CHNL. BD. MICRON	CHNL. MDPT. MICRON	MEAN(SMD) 0.080	0.100	0.200	0.400	0.600	0.800	1.000
	0.0032								
1		0.0042	6.94E 03	6.79E 03	6.41E 03	--------	--------	--------	--------
	0.0056								
2		0.0075	6.45E 02	6.25E 02	5.78E 02	--------	--------	--------	--------
	0.0100								
3		0.0133	1.94E 02	1.85E 02	1.48E 02	1.05E 02	8.10E 01	--------	--------
	0.0178								
4		0.0237	1.78E 02	1.48E 02	6.68E 01	2.31E 01	1.13E 01	6.66E 00	4.35E 00
	0.0316								
5		0.0422	8.60E 02	7.96E 02	5.44E 02	3.16E 02	2.19E 02	1.65E 02	1.31E 02
	0.0562								
6		0.0750	9.06E 02	9.08E 02	7.23E 02	4.31E 02	2.89E 02	2.11E 02	1.63E 02
	0.1000								
7		0.1330	1.10E 03	1.31E 03	1.56E 03	1.22E 03	8.87E 02	6.61E 02	5.08E 02
	0.1780								
8		0.2370	1.37E 03	1.80E 03	2.83E 03	2.66E 03	1.97E 03	1.38E 03	9.40E 02
	0.3160								
9		0.4220	6.77E 02	1.10E 03	3.06E 03	5.00E 03	5.07E 03	4.37E 03	3.51E 03
	0.5620								
10		0.7500	1.02E 02	2.05E 02	1.40E 03	5.44E 03	8.64E 03	1.02E 04	1.04E 04
	1.0000								

Table 3d.

INVERSE SURFACE SENSITIVITY (MICRON2/CC/PA) 5/77

GEOMETRIC STANDARD DEVIATION = 2.00

CHNL. NO.	CHNL. BD. MICRON	CHNL. MDPT. MICRON	MEAN(SMD) 0.004	0.006	0.008	0.010	0.020	0.040	0.060
	0.0032								
1		0.0042	1.25E 04	1.16E 04	1.10E 04	1.06E 04	9.49E 03	8.61E 03	8.19E 03
	0.0056								
2		0.0075	1.37E 03	1.24E 03	1.15E 03	1.10E 03	9.43E 02	8.28E 02	7.75E 02
	0.0100								
3		0.0133	2.47E 02	2.41E 02	2.37E 02	2.34E 02	2.27E 02	2.19E 02	2.14E 02
	0.0178								
4		0.0237	2.85E 02	2.89E 02	2.90E 02	2.90E 02	2.84E 02	2.61E 02	2.33E 02
	0.0316								
5		0.0422	7.21E 02	7.66E 02	7.99E 02	8.25E 02	8.95E 02	9.24E 02	9.03E 02
	0.0562								
6		0.0750	1.79E 02	2.44E 02	3.01E 02	3.52E 02	5.43E 02	7.42E 02	8.23E 02
	0.1000								
7		0.1330	2.99E 01	5.54E 01	8.51E 01	1.18E 02	3.04E 02	6.66E 02	9.41E 02
	0.1780								
8		0.2370	1.86E 01	4.12E 01	7.13E 01	1.08E 02	3.53E 02	9.26E 02	1.42E 03
	0.3160								
9		0.4220	6.33E-01	2.25E 00	5.50E 00	1.10E 01	8.52E 01	4.71E 02	9.87E 02
	0.5620								
10		0.7500	--------	--------	3.41E 00	5.76E 00	2.85E 01	1.32E 02	3.03E 02
	1.0000								

CHNL. NO.	CHNL. BD. MICRON	CHNL. MDPT. MICRON	MEAN(SMD) 0.080	0.100	0.200	0.400	0.600	0.800	1.000
	0.0022								
1		0.0042	7.93E 03	7.74E 03	7.26E 03	6.88E 03	--------	--------	--------
	0.0056								
2		0.0075	7.43E 02	7.20E 02	6.63E 02	6.20E 02	6.01E 02	5.88E 02	--------
	0.0100								
3		0.0133	2.10E 02	2.06E 02	1.90E 02	1.69E 02	1.54E 02	1.43E 02	1.33E 02
	0.0178								
4		0.0237	2.17E 02	1.98E 02	1.36E 02	8.00E 01	5.51E 01	4.14E 01	3.27E 01
	0.0316								
5		0.0422	8.80E 02	8.48E 02	7.06E 02	5.35E 02	4.40E 02	3.78E 02	3.34E 02
	0.0562								
6		0.0750	8.62E 02	8.71E 02	7.99E 02	6.26E 02	5.11E 02	4.33E 02	3.77E 02
	0.1000								
7		0.1330	1.13E 03	1.26E 03	1.48E 03	1.38E 03	1.20E 03	1.05E 03	9.30E 02
	0.1780								
8		0.2370	1.80E 03	2.08E 03	2.69E 03	2.63E 03	2.26E 03	1.89E 03	1.57E 03
	0.3160								
9		0.4220	1.49E 03	1.93E 03	3.46E 03	4.62E 03	4.70E 03	4.38E 03	3.93E 03
	0.5620								
10		0.7500	5.45E 02	8.28E 02	2.59E 03	5.90E 03	7.96E 03	9.00E 03	9.39E 03
	1.0000								

Table 3e.

INVERSE SURFACE SENSITIVITY (MICRON2/CC/PA) 5/77

GEOMETRIC STANDARD DEVIATION = 2.25

CHNL. NO.	CHNL. BD MICRON	CHNL. MDPT. MICRON	MEAN(SMD) 0.004	0.006	0.008	0.010	0.020	0.040	0.060
	0.0032								
1		0.0042	1.24E 04	1.17E 04	1.13E 04	1.10E 04	1.01E 04	9.33E 03	8.94E 03
	0.0056								
2		0.0075	1.31E 03	1.22E 03	1.16E 03	1.11E 03	9.93E 02	8.95E 02	8.47E 02
	0.0100								
3		0.0133	2.42E 02	2.38E 02	2.36E 02	2.34E 02	2.28E 02	2.22E 02	2.19E 02
	0.0178								
4		0.0237	2.88E 02	2.89E 02	2.89E 02	2.88E 02	2.81E 02	2.64E 02	2.48E 02
	0.0316								
5		0.0422	7.87E 02	8.20E 02	8.43E 02	8.60E 02	9.01E 02	9.13E 02	9.02E 02
	0.0562								
6		0.0750	3.02E 02	3.72E 02	4.27E 02	4.73E 02	6.23E 02	7.58E 02	8.14E 02
	0.1000								
7		0.1330	9.99E 01	1.54E 02	2.07E 02	2.58E 02	4.84E 02	8.04E 02	1.01E 03
	0.1780								
8		0.2370	1.10E 02	1.87E 02	2.69E 02	3.52E 02	7.46E 02	1.34E 03	1.74E 03
	0.3160								
9		0.4220	1.42E 01	3.51E 01	6.54E 01	1.04E 02	3.88E 02	1.06E 03	1.64E 03
	0.5620								
10		0.7500	1.12E 01	2.23E 01	3.62E 01	5.25E 01	1.62E 02	4.66E 02	8.30E 02
	1.0000								

CHNL. NO.	CHNL. BD MICRON	CHNL. MDPT. MICRON	MEAN(SMD) 0.080	0.100	0.200	0.400	0.600	0.800	1.000
	0.0032								
1		0.0042	8.69E 03	8.51E 03	8.00E 03	7.59E 03	7.38E 03	7.25E 03	--------
	0.0056								
2		0.0075	8.17E 02	7.95E 02	7.26E 02	6.89E 02	6.66E 02	6.52E 02	6.41E 02
	0.0100								
3		0.0133	2.16E 02	2.13E 02	2.05E 02	1.94E 02	1.86E 02	1.79E 02	1.74E 02
	0.0178								
4		0.0237	2.24E 02	2.22E 02	1.79E 02	1.32E 02	1.06E 02	8.98E 01	7.80E 01
	0.0316								
5		0.0422	8.85E 02	8.66E 02	7.80E 02	6.64E 02	5.91E 02	5.38E 02	4.99E 02
	0.0562								
6		0.0750	8.39E 02	8.48E 02	8.20E 02	7.20E 02	6.42E 02	5.83E 02	5.37E 02
	0.1000								
7		0.1330	1.14E 03	1.24E 03	1.42E 03	1.41E 03	1.33E 03	1.24E 03	1.16E 03
	0.1780								
8		0.2370	2.01E 03	2.20E 03	2.60E 03	2.58E 03	2.37E 03	2.13E 03	1.91E 03
	0.3160								
9		0.4220	2.11E 03	2.49E 03	3.63E 03	4.40E 03	4.47E 03	4.31E 03	4.05E 03
	0.5620								
10		0.7500	1.22E 03	1.62E 03	3.48E 03	6.10E 03	7.56E 03	8.35E 03	8.73E 03
	1.0000								

Table 3f.

INVERSE SURFACE SENSITIVITY <MICRON2/CC/PA> 5/77

GEOMETRIC STANDARD DEVIATION = 2.50

CHNL. NO.	CHNL. BD. MICRON	CHNL. MDPT. MICRON	MEAN(SMD) 0.004	0.006	0.008	0.010	0.020	0.040	0.060
	0.0032								
1		0.0042	1.23E 04	1.18E 04	1.15E 04	1.12E 04	1.05E 04	9.83E 03	9.48E 03
	0.0056								
2		0.0075	1.28E 03	1.20E 03	1.16E 03	1.12E 03	1.03E 03	9.42E 02	8.99E 02
	0.0100								
3		0.0133	2.40E 02	2.37E 02	2.35E 02	2.33E 02	2.29E 02	2.24E 02	2.21E 02
	0.0178								
4		0.0237	2.88E 02	2.88E 02	2.87E 02	2.86E 02	2.79E 02	2.65E 02	2.53E 02
	0.0316								
5		0.0422	8.25E 02	8.50E 02	8.66E 02	8.77E 02	9.02E 02	9.07E 02	8.98E 02
	0.0562								
6		0.0750	4.02E 02	4.66E 02	5.14E 02	5.52E 02	6.68E 02	7.65E 02	8.06E 02
	0.1000								
7		0.1330	1.98E 02	2.72E 02	3.37E 02	3.96E 02	6.20E 02	8.89E 02	1.05E 03
	0.1780								
8		0.2370	2.90E 02	4.24E 02	5.46E 02	6.57E 02	1.09E 03	1.62E 03	1.92E 03
	0.3160								
9		0.4220	8.44E 01	1.63E 02	2.53E 02	3.49E 02	8.31E 02	1.60E 03	2.13E 03
	0.5620								
10		0.7500	5.78E 01	9.78E 01	1.41E 02	1.87E 02	4.32E 02	9.45E 02	1.44E 03
	1.0000								

CHNL. NO.	CHNL. BD. MICRON	CHNL. MDPT. MICRON	MEAN(SMD) 0.080	0.100	0.200	0.400	0.600	0.800	1.000
	0.0032								
1		0.0042	9.25E 03	9.08E 03	8.60E 03	8.18E 03	7.97E 03	7.83E 03	7.72E 03
	0.0056								
2		0.0075	8.71E 02	8.51E 02	7.95E 02	7.48E 02	7.24E 02	7.08E 02	6.97E 02
	0.0100								
3		0.0133	2.19E 02	2.17E 02	2.12E 02	2.05E 02	2.00E 02	1.96E 02	1.93E 02
	0.0178								
4		0.0237	2.43E 02	2.35E 02	2.03E 02	1.68E 02	1.46E 02	1.31E 02	1.19E 02
	0.0316								
5		0.0422	8.86E 02	8.74E 02	8.16E 02	7.36E 02	6.82E 02	6.42E 02	6.10E 02
	0.0562								
6		0.0750	8.25E 02	8.33E 02	8.24E 02	7.65E 02	7.13E 02	6.71E 02	6.37E 02
	0.1000								
7		0.1330	1.15E 03	1.22E 03	1.37E 03	1.41E 03	1.37E 03	1.32E 03	1.27E 03
	0.1780								
8		0.2370	2.12E 03	2.26E 03	2.54E 03	2.54E 03	2.40E 03	2.25E 03	2.10E 03
	0.3160								
9		0.4220	2.53E 03	2.84E 03	3.71E 03	4.26E 03	4.32E 03	4.23E 03	4.07E 03
	0.5620								
10		0.7500	1.91E 03	2.34E 03	4.09E 03	6.19E 03	7.31E 03	7.94E 03	8.29E 03
	1.0000								

Table 3g.

INVERSE SURFACE SENSITIVITY <MICRON2/CC/PA> 5/77

GEOMETRIC STANDARD DEVIATION = 2.75

CHNL NO.	CHNL BD MICRON	CHNL MDPT MICRON	MEAN(SMD) 0.004	0.006	0.008	0.010	0.020	0.040	0.060
	0.0032								
1	0.0056	0.0042	1.23E 04	1.19E 04	1.16E 04	1.14E 04	1.08E 04	1.02E 04	9.88E 03
2	0.0100	0.0075	1.26E 03	1.20E 03	1.16E 03	1.13E 03	1.05E 03	9.75E 02	9.37E 02
3	0.0178	0.0133	2.38E 02	2.36E 02	2.34E 02	2.33E 02	2.29E 02	2.25E 02	2.23E 02
4	0.0316	0.0237	2.88E 02	2.87E 02	2.85E 02	2.84E 02	2.77E 02	2.65E 02	2.56E 02
5	0.0562	0.0422	8.48E 02	8.66E 02	8.77E 02	8.85E 02	9.01E 02	9.03E 02	8.96E 02
6	0.1000	0.0750	4.76E 02	5.32E 02	5.72E 02	6.03E 02	6.94E 02	7.69E 02	8.00E 02
7	0.1780	0.1330	3.02E 02	3.85E 02	4.54E 02	5.12E 02	7.18E 02	9.44E 02	1.07E 03
8	0.3160	0.2370	5.17E 02	6.83E 02	8.22E 02	9.40E 02	1.35E 03	1.80E 03	2.04E 03
9	0.5620	0.4220	2.49E 02	4.03E 02	5.52E 02	6.93E 02	1.27E 03	2.01E 03	2.48E 03
10	1.0000	0.7500	1.64E 02	2.49E 02	3.32E 02	4.14E 02	7.97E 02	1.46E 03	2.03E 03

CHNL NO.	CHNL BD MICRON	CHNL MDPT MICRON	MEAN(SMD) 0.080	0.100	0.200	0.400	0.600	0.800	1.000
	0.0032								
1	0.0056	0.0042	9.67E 03	9.51E 03	9.07E 03	8.67E 03	8.45E 03	8.31E 03	8.21E 03
2	0.0100	0.0075	9.12E 02	8.93E 02	8.41E 02	7.95E 02	7.71E 02	7.56E 02	7.44E 02
3	0.0178	0.0133	2.21E 02	2.20E 02	2.16E 02	2.10E 02	2.07E 02	2.04E 02	2.02E 02
4	0.0316	0.0237	2.49E 02	2.42E 02	2.18E 02	1.91E 02	1.73E 02	1.61E 02	1.51E 02
5	0.0562	0.0422	8.87E 02	8.77E 02	8.36E 02	7.78E 02	7.38E 02	7.07E 02	6.82E 02
6	0.1000	0.0750	8.15E 02	8.23E 02	8.23E 02	7.87E 02	7.52E 02	7.22E 02	6.97E 02
7	0.1780	0.1330	1.15E 03	1.21E 03	1.34E 03	1.39E 03	1.38E 03	1.35E 03	1.33E 03
8	0.3160	0.2370	2.19E 03	2.29E 03	2.50E 03	2.51E 03	2.41E 03	2.31E 03	2.20E 03
9	0.5620	0.4220	2.81E 03	3.06E 03	3.75E 03	4.16E 03	4.22E 03	4.16E 03	4.06E 03
10	1.0000	0.7500	2.51E 03	2.94E 03	4.51E 03	6.24E 03	7.14E 03	7.66E 03	7.97E 03

Table 4a.

INVERSE VOLUME SENSITIVITY (MICRON3/CC/PA) 5/77

GEOMETRIC STANDARD DEVIATION = 1.25

CHNL. NO.	CHNL. BD. MICRON	CHNL. MDPT. MICRON	MEAN(VMD) 0.004	0.006	0.008	0.010	0.020	0.040	0.060
	0.0032								
1		0.0042	1.10E 01	7.32E 00	6.17E 00	5.77E 00	5.52E 00	--------	--------
	0.0056								
2		0.0075	3.40E 00	2.27E 00	1.53E 00	1.19E 00	8.65E-01	8.46E-01	--------
	0.0100								
3		0.0133	6.38E-01	5.90E-01	5.38E-01	5.05E-01	5.71E-01	1.60E-01	2.04E-02
	0.0178								
4		0.0237	--------	6.56E-01	6.87E-01	7.27E-01	1.07E 00	9.52E-01	1.14E-01
	0.0316								
5		0.0422	--------	--------	2.01E 00	2.07E 00	2.69E 00	7.27E 00	7.76E 00
	0.0562								
6		0.0750	--------	--------	--------	--------	9.77E-03	1.51E 00	1.02E 01
	0.1000								
7		0.1330	--------	--------	--------	--------	--------	4.30E-03	7.10E-01
	0.1780								
8		0.2370	--------	--------	--------	--------	--------	--------	1.18E-03
	0.3160								
9		0.4220	--------	--------	--------	--------	--------	--------	--------
	0.5620								
10		0.7500	--------	--------	--------	--------	--------	--------	--------
	1.0000								

CHNL. NO.	CHNL. BD. MICRON	CHNL. MDPT. MICRON	MEAN(VMD) 0.080	0.100	0.200	0.400	0.600	0.800	1.000
	0.0032								
1		0.0042	--------	--------	--------	--------	--------	--------	--------
	0.0056								
2		0.0075	--------	--------	--------	--------	--------	--------	--------
	0.0100								
3		0.0133	--------	--------	--------	--------	--------	--------	--------
	0.0178								
4		0.0237	8.57E-03	7.18E-04	--------	--------	--------	--------	--------
	0.0316								
5		0.0422	3.72E 00	1.48E 00	1.64E-02	--------	--------	--------	--------
	0.0562								
6		0.0750	1.58E 01	1.52E 01	9.74E-01	2.67E-03	--------	--------	--------
	0.1000								
7		0.1330	8.64E 00	2.43E 01	4.18E 01	1.02E 00	1.34E-02	--------	--------
	0.1780								
8		0.2370	1.22E-01	2.11E 00	1.16E 02	9.75E 01	6.98E 00	2.15E-01	5.40E-03
	0.3160								
9		0.4220	2.84E-06	1.26E-03	1.45E 01	4.79E 02	5.32E 02	1.87E 02	3.41E 31
	0.5620								
10		0.7500	--------	--------	7.35E-03	9.74E 01	1.14E 03	2.29E 03	2.19E 33
	1.0000								

Table 4b.

INVERSE VOLUME SENSITIVITY (MICRON3/CC/PA) 5/77

GEOMETRIC STANDARD DEVIATION = 1.50

CHNL NO.	CHNL BD. MICRON	CHNL MDPT. MICRON	MEAN(VMD) 0.004	0.006	0.008	0.010	0.020	0.040	0.060
	0.0032								
1		0.0042	1.01E 01	8.67E 00	7.87E 00	7.36E 00	6.23E 00	5.84E 00	5.69E 00
	0.0056								
2		0.0075	2.26E 00	1.83E 00	1.58E 00	1.42E 00	1.10E 00	9.53E-01	9.07E-01
	0.0100								
3		0.0133	5.37E-01	5.19E-01	5.12E-01	5.11E-01	5.29E-01	5.32E-01	4.77E-01
	0.0178								
4		0.0237	7.78E-01	8.32E-01	8.85E-01	9.34E-01	1.12E 00	1.15E 00	8.67E-01
	0.0316								
5		0.0422	2.49E 00	2.72E 00	2.95E 00	3.19E 00	4.43E 00	6.56E 00	7.24E 00
	0.0562								
6		0.0750	--------	5.23E-02	1.14E-01	2.03E-01	1.10E 00	4.88E 00	9.23E 00
	0.1000								
7		0.1330	--------	--------	1.12E-03	3.01E-03	6.43E-02	1.28E 00	6.01E 00
	0.1780								
8		0.2370	--------	--------	--------	--------	6.66E-03	3.62E-01	2.98E 00
	0.3160								
9		0.4220	--------	--------	--------	--------	--------	2.88E-03	1.04E-01
	0.5620								
10		0.7500	--------	--------	--------	--------	--------	--------	5.52E-03
	1.0000								

CHNL NO.	CHNL BD. MICRON	CHNL MDPT. MICRON	MEAN(VMD) 0.080	0.100	0.200	0.400	0.600	0.800	1.000
	0.0032								
1		0.0042	--------	--------	--------	--------	--------	--------	--------
	0.0056								
2		0.0075	8.86E-01	8.74E-01	--------	--------	--------	--------	--------
	0.0100								
3		0.0133	4.06E-01	3.38E-01	1.46E-01	--------	--------	--------	--------
	0.0178								
4		0.0237	5.77E-01	3.71E-01	5.46E-02	4.77E-03	--------	--------	--------
	0.0316								
5		0.0422	6.75E 00	5.86E 00	2.57E 00	7.68E-01	3.35E-01	--------	--------
	0.0562								
6		0.0750	1.20E 01	1.30E 01	8.80E 00	2.82E 00	1.16E 00	5.73E-01	3.20E-01
	0.1000								
7		0.1330	1.42E 01	2.30E 01	4.06E 01	2.48E 01	1.17E 01	5.71E 00	3.01E 00
	0.1780								
8		0.2370	1.08E 01	2.45E 01	1.04E 02	1.27E 02	7.96E 01	3.99E 01	1.84E 01
	0.3160								
9		0.4220	1.13E 00	5.34E 00	8.62E 01	3.47E 02	4.46E 02	3.86E 02	2.74E 02
	0.5620								
10		0.7500	3.86E-02	1.69E-01	1.19E 01	2.84E 02	8.92E 02	1.42E 03	1.64E 03
	1.0000								

Table 4c.

INVERSE VOLUME SENSITIVITY (MICRON3/CC/PA)

GEOMETRIC STANDARD DEVIATION = 1.75

CHNL NO	CHNL BD MICRON	CHNL MDPT MICRON	MEAN(VMD) 0.004	0.006	0.008	0.010	0.020	0.040	0.060
	0.0032								
1	0.0056	0.0042	9.88E 00	9.10E 00	8.60E 00	8.24E 00	7.24E 00	6.70E 00	6.43E 00
2	0.0100	0.0075	1.94E 00	1.73E 00	1.60E 00	1.50E 00	1.27E 00	1.12E 00	1.05E 00
3	0.0178	0.0133	5.18E-01	5.14E-01	5.14F-01	5.15E-01	5.23E-01	5.23E-01	5.22E-01
4	0.0316	0.0237	9.03E-01	9.54E-01	9.94E-01	1.03E 00	1.12E 00	1.14E 00	1.06E 00
5	0.0562	0.0422	3.35E 00	3.68E 00	3.98E 00	4.24E 00	5.25E 00	6.36E 00	6.77E 00
6	0.1000	0.0750	3.95E-01	6.80E-01	9.92E-01	1.32E 00	3.11E 00	6.47E 00	8.92E 00
7	0.1780	0.1330	2.27E-02	5.97E-02	1.19E-01	2.02E-01	1.04E 00	4.82E 00	1.04E 01
8	0.3160	0.2370	--------	2.00E-02	5.00E-02	1.01E-01	8.52E-01	6.04E 00	1.63E 01
9	0.5620	0.4220	--------	--------	--------	1.39E-03	3.80E-02	1.00E 00	5.80E 00
10	1.0000	0.7500	--------	--------	--------	--------	2.68E-02	3.31E-01	1.39E 00

CHNL NO	CHNL BD MICRON	CHNL MDPT MICRON	MEAN(VMD) 0.080	0.100	0.200	0.400	0.600	0.800	1.000
	0.0032								
1	0.0056	0.0042	6.27E 00	6.17E 00	5.92E 00	--------	--------	--------	--------
2	0.0100	0.0075	1.02E 00	9.93E-01	9.37E-01	--------	--------	--------	--------
3	0.0178	0.0133	5.25E-01	5.15E-01	4.52E-01	3.49E-01	2.82E-01	--------	--------
4	0.0316	0.0237	9.54E-01	8.47E-01	4.66E-01	1.86E-01	9.76E-02	5.95E-02	3.98E-02
5	0.0562	0.0422	6.80E 00	6.64E 00	5.29E 00	3.44E 00	2.50E 00	1.95E 00	1.58E 00
6	0.1000	0.0750	1.05E 01	1.13E 01	1.13E 01	8.04E 00	5.84E 00	4.46E 00	3.55E 00
7	0.1780	0.1330	1.62E 01	2.15E 01	3.49E 01	3.48E 01	2.88E 01	2.33E 01	1.90E 01
8	0.3160	0.2370	2.94E 01	4.32E 01	9.47E 01	1.20E 02	1.08E 02	8.85E 01	6.96E 01
9	0.5620	0.4220	1.67E 01	3.28E 01	1.34E 02	2.95E 02	3.67E 02	3.76E 02	3.50E 02
10	1.0000	0.7500	3.76E 00	7.96E 00	6.80E 01	3.71E 02	7.46E 02	1.05E 03	1.24E 03

Table 4d.

INVERSE VOLUME SENSITIVITY (MICRON3/CC/PA) 5/77

GEOMETRIC STANDARD DEVIATION = 2.00

CHNL NO.	CHNL BD MICRON	CHNL MDPT MICRON	MEAN(VMD) 0.004	0.006	0.008	0.010	0.020	0.040	0.060
	0.0022								
1		0.0042	9.80E 00	9.28E 00	8.93E 00	8.68E 00	7.97E 00	7.40E 00	7.12E 00
	0.0056								
2		0.0075	1.82E 00	1.69E 00	1.60E 00	1.54E 00	1.37E 00	1.24E 00	1.17E 00
	0.0100								
3		0.0133	5.16E-01	5.15E-01	5.16E-01	5.17E-01	5.22E-01	5.28E-01	5.30E-01
	0.0178								
4		0.0237	9.78E-01	1.02E 00	1.04E 00	1.06E 00	1.12E 00	1.13E 00	1.10E 00
	0.0316								
5		0.0422	4.06E 00	4.39E 00	4.66E 00	4.87E 00	5.60E 00	6.28E 00	6.55E 00
	0.0562								
6		0.0750	1.27E 00	1.79E 00	2.28E 00	2.73E 00	4.61E 00	7.17E 00	8.77E 00
	0.1000								
7		0.1330	2.60E-01	4.94E-01	7.75E-01	1.10E 00	3.12E 00	8.02E 00	1.29E 01
	0.1780								
8		0.2370	2.26E-01	5.17E-01	9.23E-01	1.44E 00	5.33E 00	1.70E 01	3.02E 01
	0.3160								
9		0.4220	--------	2.82E-02	7.00E-02	1.41E-01	1.23E 00	9.15E 00	2.48E 01
	0.5620								
10		0.7500	--------	--------	1.09E-01	1.86E-01	9.66E-01	4.83E 00	1.20E 01
	1.0000								

CHNL NO.	CHNL BD MICRON	CHNL MDPT MICRON	MEAN(VMD) 0.080	0.100	0.200	0.400	0.600	0.800	1.000
	0.0022								
1		0.0042	6.94E 00	6.81E 00	6.49E 00	6.23E 00	--------	--------	--------
	0.0056								
2		0.0075	1.14E 00	1.11E 00	1.04E 00	9.88E-01	9.64E-01	9.49E-01	--------
	0.0100								
3		0.0133	5.31E-01	5.30E-01	5.19E-01	4.89E-01	4.62E-01	4.39E-01	4.19E-01
	0.0178								
4		0.0237	1.05E 00	1.01E 00	7.97E-01	5.39E-01	3.99E-01	3.13E-01	2.56E-01
	0.0316								
5		0.0422	6.64E 00	6.63E 00	6.19E 00	5.20E 00	4.49E 00	3.98E 00	3.60E 00
	0.0562								
6		0.0750	9.79E 00	1.04E 01	1.13E 01	1.03E 01	9.06E 00	8.04E 00	7.23E 00
	0.1000								
7		0.1330	1.71E 01	2.07E 01	3.07E 01	3.48E 01	3.37E 01	3.16E 01	2.94E 01
	0.1780								
8		0.2370	4.27E 01	5.38E 01	8.93E 01	1.11E 02	1.11E 02	1.04E 02	9.51E 01
	0.3160								
9		0.4220	4.46E 01	6.55E 01	1.58E 02	2.69E 02	3.22E 02	3.42E 02	3.43E 02
	0.5620								
10		0.7500	2.24E 01	3.58E 01	1.37E 02	4.10E 02	6.67E 02	8.69E 02	1.02E 03
	1.0000								

Table 4e.

INVERSE VOLUME SENSITIVITY (MICRON3/CC/PA) 5/77

GEOMETRIC STANDARD DEVIATION = 2.25

CHNL. NO.	CHNL. BD. MICRON	CHNL. MDPT. MICRON	MEAN(VMD) 0.004	0.006	0.008	0.010	0.020	0.040	0.060
	0.0032								
1		0.0042	9.76E 00	9.37E 00	9.11E 00	8.92E 00	8.36E 00	7.87E 00	7.62E 00
	0.0056								
2		0.0075	1.77E 00	1.67E 00	1.61E 00	1.56E 00	1.43E 00	1.32E 00	1.26E 00
	0.0100								
3		0.0133	5.16E-01	5.16E-01	5.17E-01	5.17E-01	5.21E-01	5.25E-01	5.28E-01
	0.0178								
4		0.0237	1.02E 00	1.05E 00	1.07E 00	1.08E 00	1.11E 00	1.12E 00	1.10E 00
	0.0316								
5		0.0422	4.57E 00	4.85E 00	5.06E 00	5.23E 00	5.76E 00	6.23E 00	6.43E 00
	0.0562								
6		0.0750	2.28E 00	2.90E 00	3.42E 00	3.88E 00	5.56E 00	7.53E 00	8.68E 00
	0.1000								
7		0.1330	9.15E-01	1.45E 00	2.01E 00	2.57E 00	5.36E 00	1.02E 01	1.43E 01
	0.1780								
8		0.2370	1.45E 00	2.59E 00	3.88E 00	5.26E 00	1.28E 01	2.77E 01	4.04E 01
	0.3160								
9		0.4220	1.83E-01	4.68E-01	9.05E-01	1.50E 00	6.82E 00	2.57E 01	4.85E 01
	0.5620								
10		0.7500	--------	--------	1.23E 00	1.81E 00	5.90E 00	1.85E 01	3.52E 01
	1.0000								

CHNL. NO.	CHNL. BD. MICRON	CHNL. MDPT. MICRON	MEAN(VMD) 0.080	0.100	0.200	0.400	0.600	0.800	1.000
	0.0032								
1		0.0042	7.45E 00	7.33E 00	6.99E 00	6.71E 00	6.57E 00	6.48E 00	--------
	0.0056								
2		0.0075	1.22E 00	1.20E 00	1.13E 00	1.07E 00	1.04E 00	1.03E 00	1.01E 00
	0.0100								
3		0.0133	5.29E-01	5.29E-01	5.28E-01	5.21E-01	5.12E-01	5.04E-01	4.96E-01
	0.0178								
4		0.0237	1.08E 00	1.06E 00	9.46E-01	7.81E-01	6.70E-01	5.90E-01	5.29E-01
	0.0316								
5		0.0422	6.52E 00	6.55E 00	6.43E 00	5.95E 00	5.54E 00	5.21E 00	4.93E 00
	0.0562								
6		0.0750	9.43E 00	9.94E 00	1.10E 01	1.09E 01	1.04E 01	9.81E 00	9.32E 00
	0.1000								
7		0.1330	1.76E 01	2.02E 01	2.79E 01	3.28E 01	3.37E 01	3.35E 01	3.28E 01
	0.1780								
8		0.2370	5.09E 01	5.96E 01	8.61E 01	1.04E 02	1.07E 02	1.06E 02	1.03E 02
	0.3160								
9		0.4220	7.10E 01	9.18E 01	1.70E 02	2.54E 02	2.96E 02	3.16E 02	3.24E 02
	0.5620								
10		0.7500	5.46E 01	7.60E 01	1.95E 02	4.30E 02	6.22E 02	7.70E 02	8.82E 02
	1.0000								

Table 4f.

INVERSE VOLUME SENSITIVITY (MICRON3/CC/PA) 5/77

GEOMETRIC STANDARD DEVIATION = 2.50

CHNL NO	CHNL BD MICRON	CHNL MDPT MICRON	MEAN(VMD) 0.004	0.006	0.008	0.010	0.020	0.040	0.060
	0.0022								
1		0.0042	9.73E 00	9.42E 00	9.22E 00	9.07E 00	8.61E 00	8.19E 00	7.97E 00
	0.0056								
2		0.0075	1.73E 00	1.66E 00	1.61E 00	1.57E 00	1.46E 00	1.37E 00	1.32E 00
	0.0100								
3		0.0133	5.16E-01	5.17E-01	5.17E-01	5.18E-01	5.21E-01	5.24E-01	5.26E-01
	0.0178								
4		0.0237	1.04E 00	1.06E 00	1.08E 00	1.09E 00	1.11E 00	1.11E 00	1.11E 00
	0.0316								
5		0.0422	4.90E 00	5.14E 00	5.31E 00	5.45E 00	5.85E 00	6.20E 00	6.36E 00
	0.0562								
6		0.0750	3.19E 00	3.81E 00	4.31E 00	4.73E 00	6.17E 00	7.74E 00	8.63E 00
	0.1000								
7		0.1330	1.91E 00	2.72E 00	3.48E 00	4.20E 00	7.28E 00	1.19E 01	1.52E 01
	0.1780								
8		0.2370	4.18E 00	6.46E 00	8.69E 00	1.09E 01	2.07E 01	3.61E 01	4.74E 01
	0.3160								
9		0.4220	1.18E 00	2.42E 00	4.00E 00	5.84E 00	1.76E 01	4.47E 01	7.00E 01
	0.5620								
10		0.7500	--------	3.44E 00	5.07E 00	6.82E 00	1.69E 01	4.03E 01	6.56E 01
	1.0000								

CHNL NO.	CHNL BD MICRON	CHNL MDPT MICRON	MEAN(VMD) 0.080	0.100	0.200	0.400	0.600	0.800	1.000
	0.0022								
1		0.0042	7.82E 00	7.71E 00	7.39E 00	7.12E 00	6.97E 00	6.87E 00	6.80E 00
	0.0056								
2		0.0075	1.29E 00	1.27E 00	1.20E 00	1.14F 00	1.11E 00	1.10E 00	1.08E 00
	0.0100								
3		0.0133	5.27E-01	5.28E-01	5.29E-01	5.27E-01	5.25E-01	5.22E-01	5.19E-01
	0.0178								
4		0.0237	1.09E 00	1.08E 00	1.01E 00	9.09E-01	8.34E-01	7.75E-01	7.28E-01
	0.0316								
5		0.0422	6.44E 00	6.48E 00	6.48E 00	6.25E 00	6.02E 00	5.82E 00	5.64E 00
	0.0562								
6		0.0750	9.22E 00	9.63E 00	1.06E 01	1.09E 01	1.07E 01	1.05E 01	1.02E 01
	0.1000								
7		0.1330	1.78E 01	1.99E 01	2.61E 01	3.08E 01	3.25E 01	3.30E 01	3.31E 01
	0.1780								
8		0.2370	5.62E 01	6.32E 01	8.40E 01	9.92E 01	1.04E 02	1.05E 02	1.04E 02
	0.3160								
9		0.4220	9.20E 01	1.11E 02	1.77E 02	2.45E 02	2.78E 02	2.97E 02	3.07E 02
	0.5620								
10		0.7500	9.15E 01	1.17E 02	2.40E 02	4.42E 02	5.94E 02	7.09E 02	7.98E 02
	1.0000								

Table 4g.

INVERSE VOLUME SENSITIVITY (MICRON3/CC/PA) 5/77

GEOMETRIC STANDARD DEVIATION = 2.75

CHNL. NO.	CHNL BD MICRON	CHNL MDPT. MICRON	MEAN(VMD) 0.004	0.006	0.008	0.010	0.020	0.040	0.060
	0.0032								
1		0.0042	9.72E 00	9.47E 00	9.29E 00	9.16E 00	8.78E 00	8.42E 00	8.22E 00
	0.0056								
2		0.0075	1.71E 00	1.65E 00	1.61E 00	1.58E 00	1.49E 00	1.41E 00	1.36E 00
	0.0100								
3		0.0133	5.17E-01	5.17E-01	5.18E-01	5.18E-01	5.20E-01	5.23E-01	5.24E-01
	0.0178								
4		0.0237	1.06E 00	1.07E 00	1.08E 00	1.09E 00	1.11E 00	1.11E 00	1.11E 00
	0.0316								
5		0.0422	5.14E 00	5.33E 00	5.47E 00	5.58E 00	5.90E 00	6.18E 00	6.31E 00
	0.0562								
6		0.0750	3.91E 00	4.51E 00	4.97E 00	5.34E 00	6.58E 00	7.87E 00	8.60E 00
	0.1000								
7		0.1330	2.06E 00	4.06E 00	4.93E 00	5.72E 00	8.82E 00	1.30E 01	1.58E 01
	0.1780								
8		0.2370	8.11E 00	1.14E 01	1.43E 01	1.70E 01	2.78E 01	4.24E 01	5.23E 01
	0.3160								
9		0.4220	3.87E 00	6.84E 00	1.01E 01	1.35E 01	3.11E 01	6.25E 01	8.77E 01
	0.5620								
10		0.7500	--------	9.22E 00	1.26E 01	1.60E 01	3.30E 01	6.62E 01	9.74E 01
	1.0000								

CHNL. NO.	CHNL BD MICRON	CHNL MDPT. MICRON	MEAN(VMD) 0.080	0.100	0.200	0.400	0.600	0.800	1.000
	0.0032								
1		0.0042	8.09E 00	7.99E 00	7.70E 00	7.44E 00	7.30E 00	7.20E 00	7.12E 00
	0.0056								
2		0.0075	1.34E 00	1.31E 00	1.25E 00	1.20E 00	1.17E 00	1.15E 00	1.14E 00
	0.0100								
3		0.0133	5.25E-01	5.26E-01	5.28E-01	5.28E-01	5.28E-01	5.27E-01	5.25E-01
	0.0178								
4		0.0237	1.10E 00	1.09E 00	1.05E 00	9.78E-01	9.26E-01	8.85E-01	8.51E-01
	0.0316								
5		0.0422	6.38E 00	6.42E 00	6.47E 00	6.37E 00	6.24E 00	6.12E 00	6.01E 00
	0.0562								
6		0.0750	9.08E 00	9.42E 00	1.03E 01	1.07E 01	1.08E 01	1.07E 01	1.06E 01
	0.1000								
7		0.1330	1.80E 01	1.97E 01	2.49E 01	2.92E 01	3.10E 01	3.19E 01	3.23E 01
	0.1780								
8		0.2370	5.97E 01	6.55E 01	8.25E 01	9.57E 01	1.00E 02	1.02E 02	1.03E 02
	0.3160								
9		0.4220	1.08E 02	1.25E 02	1.82E 02	2.38E 02	2.66E 02	2.83E 02	2.93E 02
	0.5620								
10		0.7500	1.27E 02	1.55E 02	2.74E 02	4.50E 02	5.75E 02	6.69E 02	7.43E 02
	1.0000								

TABLE V. Expanded Matrix Obtained from the Basic Response Matrix by Interpolation

Chnl. No.	Chnl. Mdpt. (μm)	Chnl. Bd. (μm)	Col.–rod volt. (V)	.0037 / .0013	.0042 / .0032	.0049 / .0078	.0056 / .0191	.0065 / .0469	.0075 / .116	.0087 / .286	.0100 / .703	.0116 / 1.78	.0133 / 2.85	.0154 / 3.38	.0178 / 4.00	.0205 / 4.71	.0237 / 5.57	.0274 / 6.59
		Particle Diameter (μm) / Sensitivity, pA/(10^6 part./cm^3)																
		0.0032	19															
1	0.0042			1	1	1	0.50	0	0	0	0	0	0	0	0	0	0	0
		0.0056	59															
2	0.0075			0	0	0	0.50	1	1	1	0.50	0	0	0	0	0	0	0
		0.0100	186															
3	0.0133			0	0	0	0	0	0	0	0.50	1	1	1	0.48	0.05	0.08	0
		0.0178	588															
4	0.0237			0	0	0	0	0	0	0	0	0	0	0	0.52	0.95	0.92	1
		0.0316	1870															
5	0.0422			0	0	0	0	0	0	0	0	0	0	0	0	0	0	0
		0.0562	2600															
6	0.075			0	0	0	0	0	0	0	0	0	0	0	0	0	0	0
		0.100	4440															
7	0.133			0	0	0	0	0	0	0	0	0	0	0	0	0	0	0
		0.178	6600															
8	0.237			0	0	0	0	0	0	0	0	0	0	0	0	0	0	0
		0.316	8380															
9	0.422			0	0	0	0	0	0	0	0	0	0	0	0	0	0	0
		0.562	9600															
10	0.750			0	0	0	0	0	0	0	0	0	0	0	0	0	0	0
		1.000	10600															
			(>10600)	0	0	0	0	0	0	0	0	0	0	0	0	0	0	0

	Particle Diameter (μm)			.0316	.0365	.0422	.0487	.0562	.0649	.0750	.0866	.1000	.1160	.1330	.1540	.1780	.2050
	Sensitivity, pA/(10^6 part./cm^3)			7.78	9.19	10.9	12.8	15.2	17.9	21.2	25.0	29.6	35.1	41.1	48.3	57.7	67.9
Chnl. No.	Chnl. Mdpt. (μm)	Chnl. Bd. (μm)	Col.-rod Volt. (V)														
		0.0032	19														
1	0.0042			0	0	0	0	0	0	0	0	0	0	0	0	0	0
		0.0056	59														
2	0.0075			0	0	0	0	0	0	0	0	0	0	0	0	0	0
		0.0100	186														
3	0.0133			0	0	0	0	0	0	0	0	0	0	0	0	0	0
		0.0178	588														
4	0.0237			0.51	0.44	0.49	0.15	0.145	0.035	0	0	0	0	0	0	0	0
		0.0316	1870														
5	0.0422			0.48	0.52	0.07	0.48	0.355	0.345	0.14	0.05	0	0	0	0	0	0
		0.0562	2600														
6	0.075			0.01	0.04	0.44	0.195	0.325	0.515	0.65	0.61	0.51	0.45	0.40	0.13	0.03	0
		0.100	4440														
7	0.133			0	0	0	0.175	0.175	0.085	0.17	0.31	0.405	0.48	0.51	0.67	0.47	0.35
		0.178	6600														
8	0.237			0	0	0	0	0	0.02	0.02	0.03	0.085	0.07	0.09	0.17	0.38	0.46
		0.316	8380														
9	0.422			0	0	0	0	0	0	0.02	0	0	0	0	0.03	0.08	0.14
		0.562	9600														
10	0.750			0	0	0	0	0	0	0	0	0	0	0	0	0.04	0.04
		1.000	10600														
			(>10600)	0	0	0	0	0	0	0	0	0	0	0	0	0	0.01

Table V—*continued*

	Particle Diameter (μm) / Sensitivity, pA/(10^6 part./cm^3)			.2370 / 80.4	.2740 / 95.1	.3160 / 112.2	.3650 / 132.6	.4220 / 156.9	.4870 / 185.2	.5620 / 218.7	.6490 / 258.4	.7500 / 305.5	.8660 / 360.9	1.000 / 426.4	1.160 / 506.4	1.330 / 593.4	1.540 / 703.3	1.780 / 831.9
Chnl. No.	Chnl. Mdpt. (μm)	Chnl. Bd. (μm)	Col.-rod Volt. (V)															
		0.0032	19															
1	0.0042			0	0	0	0	0	0	0	0	0	0	0	0	0	0	0
		0.0056	59															
2	0.0075			0	0	0	0	0	0	0	0	0	0	0	0	0	0	0
		0.0100	186															
3	0.0133			0	0	0	0	0	0	0	0	0	0	0	0	0	0	0
		0.0178	588															
4	0.0237			0	0	0	0	0	0	0	0	0	0	0	0	0	0	0
		0.0316	1870															
5	0.0422			0	0	0	0	0	0	0	0	0	0	0	0	0	0	0
		0.0562	2600															
6	0.075			0	0	0	0	0	0	0	0	0	0	0	0	0	0	0
		0.100	4440															
7	0.133			0.22	0.15	0.08	0.02	0	0	0	0	0	0	0	0	0	0	0
		0.178	6600															
8	0.237			0.48	0.45	0.42	0.32	0.31	0.20	0.095	0.13	0.11	0.07	0.04	0.04	0.04	0.04	0.04
		0.316	8380															
9	0.422			0.20	0.25	0.32	0.38	0.36	0.39	0.305	0.27	0.28	0.23	0.20	0.20	0.20	0.20	0.20
		0.562	9600															
10	0.750			0.06	0.09	0.11	0.18	0.18	0.21	0.26	0.25	0.25	0.25	0.26	0.26	0.26	0.26	0.26
		1.000	10600															
			(>10600)	0.04	0.06	0.07	0.10	0.15	0.20	0.34	0.35	0.36	0.45	0.50	0.50	0.50	0.50	0.50

TABLE VI. Sample Calculation for Obtaining Number Distributions[a]

Chnl. No.	Chnl. Bd. (μm)	Chnl. Mdpt. (μm)	ΔI_i pA	$1/S_{N,i}$ part./cm^3-pA	$\Delta\dot{N}_i$ part./cm^3	$\Delta\dot{N}_i/\Delta\log D_p$[b] part./cm^3	$1/\tilde{S}_{N,i}$[c] part./cm^3-pA	$\Delta\tilde{N}_i$ part./cm^3	$\Delta\tilde{N}_i/\Delta\log D_p$[b] part./cm^3
	0.0032								
1		0.0042	—	3.25(8)	—	—	7.51(7)	—	—
	0.0056								
2		0.0075	—	8.61(6)	—	—	2.57(6)	—	—
	0.0100								
3		0.0133	0.07	3.50(5)	2.45(4)	9.8 (4)	3.13(5)	2.19(4)	8.76(4)
	0.0178								
4		0.0237	0.8	1.79(5)	1.43(5)	5.73(5)	1.64(5)	1.31(5)	5.25(5)
	0.0316								
5		0.0422	0.46	9.19(4)	4.23(4)	1.69(5)	1.95(5)	8.97(4)	3.59(5)
	0.056								
6		0.075	0.73	4.72(4)	3.45(4)	1.38(5)	2.44(4)	1.78(4)	7.12(4)
	0.100								
7		0.133	0.32	2.43(4)	7.78(3)	3.1 (4)	8.65(2)	2.77(2)	1.11(3)
	0.178								
8		0.237	0.068	1.24(4)	8.43(2)	3.38(3)	3.67(1)	2.5(−1)	1.00(1)
	0.316								
9		0.422	0.015	6.37(3)	9.56(1)	3.82(2)	—	—	—
	0.562								
10		0.750	0.006	3.27(3)	1.96(1)	7.84(1)	—	—	—
	1.000				$\Sigma\Delta\dot{N}_i$ = 2.53(5)			$\Sigma\Delta\tilde{N}_i$ = 2.61(5)	

a. Aerosol produced by an inverted Collison atomizer with 0.01% (by wt.) NaCl solution.
b. $\Delta\dot{N}_i/\Delta\log D_p = 4 \times \Delta\dot{N}_i$; $\Delta\tilde{N}_i/\Delta\log D_p = 4 \times \Delta\tilde{N}_i$.
c. Sensitivities for geometrical mean number diameter of 0.02 μm and σ_g of 1.5.

we estimate the geometrical mean number diameter and σ_g of the aerosol to be 0.024 μm and 1.32, respectively, for this particular distribution. Using these values as guides, we go to Table II and select the instrument sensitivities for a geometrical mean number diameter of 0.02 μm and a σ_g of 1.5 and list the values of $1/\tilde{S}_{N,i}$ in column 8 of Table VI. These are then used to calculate the ΔN and $\Delta N/\Delta \log D_p$ values in Columns 9 and 10. The resulting distributions are plotted in Figure 8 for comparison with the previously obtained distribution. By a similar procedure, the volume distributions are calculated as in Table VII. The results are plotted in Figure 9.

To estimate the size distribution parameters needed for selecting the apparent instrument sensitivities for polydisperse aerosols, we make use of the fact that for a log-normal distribution, the peak of the distribution occurs at $\bar{D}_p$, where $\bar{D}_p$ is the geometrical mean diameter of the particles. Further, the ratio of the probability density functions at D_{p1} and $\bar{D}_p$ is 0.61 when

$$D_{p1} = \bar{D}_p/\sigma_g \tag{24}$$

and, the ratio of the probability density functions at D_{p2} and $\bar{D}_p$ is also 0.61 when

$$D_{p2} = \bar{D}_p \cdot \sigma_g \cdot \tag{25}$$

Thus, for a log-normal distribution,

$$\sigma_g = \sqrt{D_{P2}/\,D_{P1}} \tag{26}$$

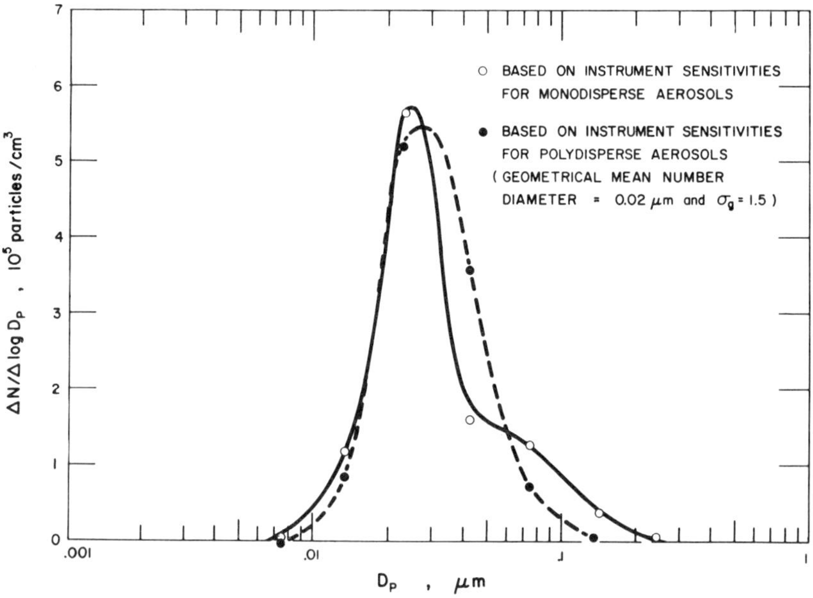

Figure 8. Atomizer aerosol number distributions calculated by means of instrument sensitivities for monodisperse and polydisperse aerosols.

TABLE VII. Sample Calculation for Obtaining Volume Distributions[a]

Chnl. No.	Chnl. Bd. (μm)	Chnl. Mdpt. (μm)	ΔI_i pA	$1/\hat{S}_{V,i}$ $\mu m^3/cm^3$-pA	$\Delta \hat{V}_i$ $\mu m^3/cm^3$	$\Delta \hat{V}_i/\Delta \log D_p$[b] $\mu m^3/cm^3$	$1/\tilde{S}_{V,i}$[c] $\mu m^3/cm^3$-pA	$\Delta \tilde{V}_i$ $\mu m^3/cm^3$	$\Delta \tilde{V}_i/\Delta \log D_p$[b] $\mu m^3/cm^3$
	0.0032								
1		0.0042	—	12.6	—	—	6.81	—	—
	0.0056								
2		0.0075	—	1.9	—	—	1.11	—	—
	0.0100								
3		0.0133	0.07	4.32(−1)	3.02(−2)	1.21(−1)	5.30(−1)	3.7 (−2)	1.48(−1)
	0.0178								
4		0.0237	0.8	1.25	1.00	4.00	1.01	8.08(−1)	3.23
	0.0316								
5		0.0422	0.46	3.62	1.67	6.67	6.63	3.05	1.22(1)
	0.056								
6		0.075	0.73	10.42	7.61	3.04(1)	1.04(1)	7.59	3.04(1)
	0.100								
7		0.133	0.32	2.99(0)	9.57	3.83(1)	2.07(1)	6.62	2.65(1)
	0.178								
8		0.237	0.068	8.64(1)	5.88	2.35(1)	5.38(1)	3.66	1.46(1)
	0.316								
9		0.422	0.015	2.51(2)	3.77	1.51(1)	6.55(1)	9.83(−1)	3.93
	0.562								
10		0.750	0.006	7.23(2)	4.34	1.74(1)	3.58(1)	2.15(−1)	8.59(−1)
	1.000				$\Sigma \Delta \hat{V}_i = 33.87$			$\Sigma \Delta \tilde{V}_i = 22.96$	

a. Aerosol produced by an inverted Collison atomizer with 0.01% (by wt.) NaCl solution.
b. $\Delta \hat{V}_i/\Delta \log D_p = 4 \times \Delta \hat{V}_i$; $\Delta \tilde{V}_i/\Delta \log D_p = 4 \times \Delta \tilde{V}_i$.
c. Sensitivities for geometrical mean volume diameter of 0.1 μm and σ_g of 2.0.

where D_{p1} and D_{p2} are the sizes for which probability density function is 61% of the peak value. Thus, to determine the σ_g of an aerosol, we draw a horizontal line through the size distribution curve at the 61% level and determine D_{p1} and D_{p2} as shown in Figure 10. The σ_g can then be calculated by means of Eq. (26). This is

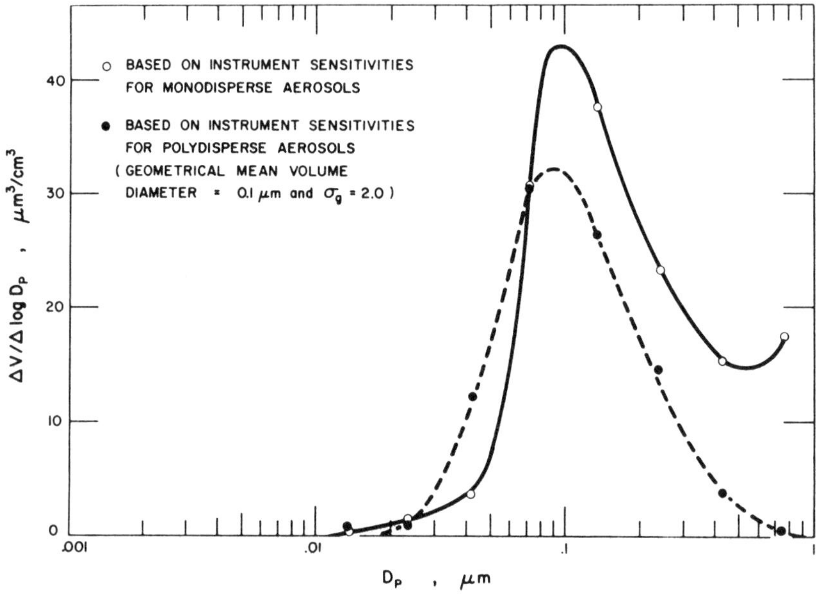

Figure 9. Atomizer aerosol volume distributions calculated by means of instrument sensitivities for monodisperse and polydisperse aerosols.

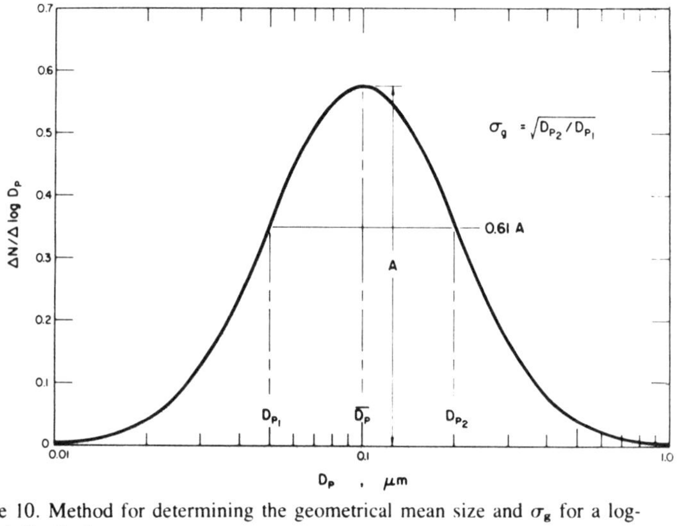

Figure 10. Method for determining the geometrical mean size and σ_g for a log-normal distribution.

the method by which the σ_g is determined for the size distribution shown in Figure 8. In that distribution, the size at the peak of the distribution is simply taken as the geometrical mean size, $\bar{D}_p$, for the distribution.

REFERENCES

Clark, W. E., 1966: The concentration and size distribution of atmospheric aerosols. M.S. Thesis, Univ. of Minn.

——— and K. T. Whitby, 1967: Concentration and size distribution measurements of atmospheric aerosols and a test of the theory of self-preserving size distributions. *J. Atmos. Sci.*, *24*, 677–687.

Hidy, G. M., 1972: *Aerosols and Atmospheric Chemistry*. New York, Academic Press.

Liu, B. Y. H. and D. Y. H. Pui, 1975: On the performance of the electrical aerosol analyzer. *J. Aerosol Sci.*, *6*, 249–264.

———, K. T. Whitby and D. Y. H. Pui, 1974: A portable electrical aerosol analyzer for size distribution measurement of submicron aerosols. *J. Air Pollution Control Assoc.*, *24*, 1067–1072.

Wahi, B. N. and B. Y. H. Liu, 1971: The mobility of polystyrene latex particles in the transition and the free molecular regimes. *J. Colloid Interface Sci.*, *37*, 374–381.

Whitby, K. T. and W. E. Clark, 1966: Electric aerosol particle counting and size distribution measuring system for the 0.015 to 1.0 μm size range. *Tellus*, *18*, 573–586.

——— and R. B. Husar, 1970a: The Minnesota aerosol analyzing system. Report of the International Cloud and Ice Nuclei Workshop. Fort Collins, Colorado, 19–29.

——— and ———, 1970b: Generation of artificial aerosols. Report of the International Cloud and Ice Nuclei Workshop, Fort Collins, Colorado, 29–32.

——— and ———, 1970c: The particle size spectra of atmospheric aerosols measured on August 10, 11, and 12 during the IWCIN. Report of the International Cloud and Ice Nuclei Workshop, Fort Collins, Colorado, 84–91.

———, ———, and B. Y. H. Liu, 1972: The aerosol size distribution of Los Angeles smog. *J. Colloid Interface Sci.*, *39*, 177–204.

———, B. Y. H. Liu, W. E. Clark, R. A. Vomela, and H. H. S. Yu, 1966: Generation and decay of small ions. Progress report to USPHS on Grant No. AP00136-05, Particle Technology Laboratory, Dept. of Mech. Eng., Univ. of Minn.

ELECTRICAL AEROSOL ANALYZER: CALIBRATION AND PERFORMANCE

David Y. H. Pui and Benjamin Y. H. Liu

Mechanical Engineering Department, University of Minnesota, Minneapolis

The purpose of a calibration study is to determine the instrument response empirically and in sufficient detail so that measurement made with the device can be interpreted. In order to calibrate an instrument, it is necessary to produce a known input to apply to the instrument and to determine the corresponding instrument output. In the case of the Electrical Aerosol Analyzer (EAA) the input needed is in the form of monodisperse aerosols of a known size and concentration.

Two techniques were used to generate the monodisperse aerosols needed in the calibration. Figure 1 shows the size range covered by each technique, the σ_g of the aerosol produced, the aerosol material and the estimated accuracies in particle size and concentration.

GENERATION OF MONODISPERSE TEST AEROSOLS

Aerosol Generation by the Electrical Classification Method

The principal method used for generating the monodisperse aerosols is the electrostatic classification method described previously by Liu and Pui (1974a). This is a technique whereby a monodisperse aerosol is generated by mobility classification using particles that are singly charged.

For particles carrying a single unit of charge, the electrical mobility is, according to aerosol theory,

$$Z_{p_1} = \frac{eC}{3\pi\mu D_p(300)} \text{cm}^2/\text{V-sec} \tag{1}$$

The research is supported by the Energy Research and Development Administration under Contract E(11-1)-1248.

where e = 4.8×10^{-10} esu is the elementary unit of charge, μ (poise) is the gas viscosity, D_p (cm) is the particle diameter, and C (dimensionless) is the slip correction. The factor 300 in the denominator converts the mobility in electrostatic unit to the common unit of cm^2/V-sec. This relationship between particle size and particle mobility for singly charged particles is shown in Figure 2.

To effect particle separation by means of Eq. (1), the apparatus shown in Figure 3 is used. The system consists of a Collison atomizer to produce the polydisperse aerosols needed for classification, a diffusion dryer to evaporate the aqueous solution droplets, a Krypton-85 neutralizer to reduce the particle charge to a sufficiently low level in preparation for electrostatic classification, and finally, a differential mobility analyzer to effect actual particle separation by

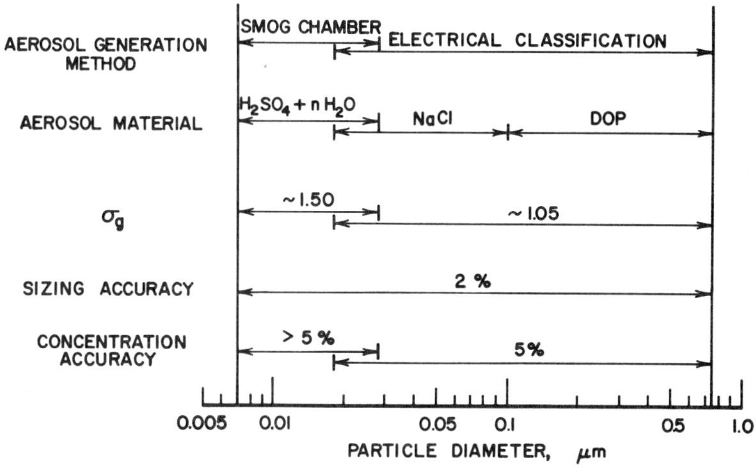

Figure 1. Aerosol generation techniques and the characteristics of aerosols used in calibrating the EAA.

mobility classification. In addition, an electrometer current sensor is used to determine the concentration of the monodisperse aerosols produced.

The differential mobility analyzer shown in Figure 3 is similar to the EAA mobility analyzer described in the preceding paper. The only difference is that in the differential mobility analyzer, there is a circumferential slit near the lower end of the collector rod through which a small air flow can be extracted. This air flow contains the particles that have been deflected to the vicinity of the slit by the electric field. The mean electric mobility of these particles is given by

$$Z_p = \frac{q_0 - \frac{1}{2}(q_s + q_a)\ell n(r_2/r_1)}{2\pi VL} \text{ cm}^2/\text{V-sec} \qquad (2)$$

where r_1 and r_2 (cm) are the inner and outer radii of the annular precipitating space in the mobility analyzer, L is the length of the collector rod between the

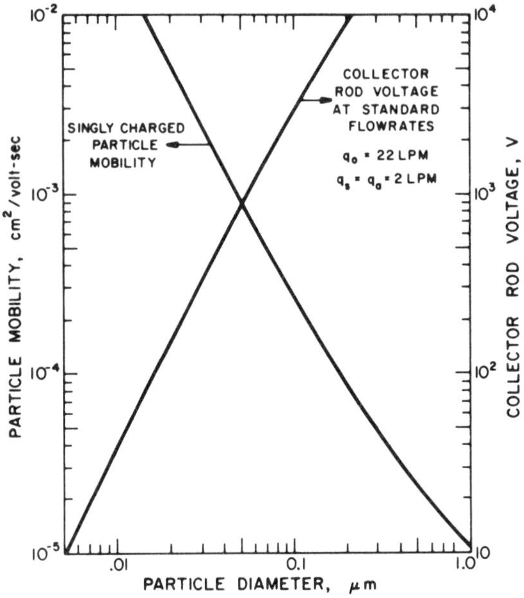

Figure 2. Relationship between particle size and particle mobility for singly charged particles and the relationship between particle size and collector rod voltage for the differential mobility analyzer.

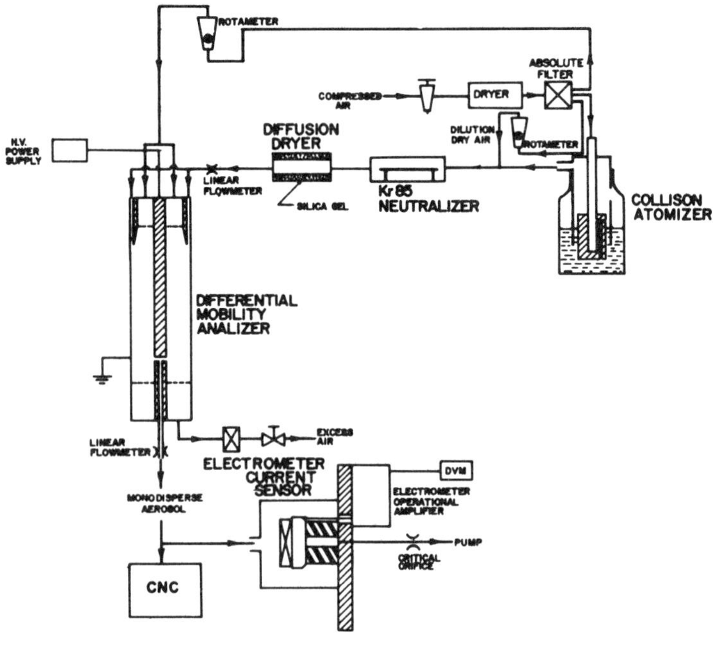

Figure 3. Apparatus for generating monodisperse aerosols by mobility classification of singly charged particles (particle size 0.02–0.1 μm).

aerosol inlet and the exit slit, V (Volts) is the applied voltage on the collector rod, q_o (cm^3/sec) is the total flow in the mobility analyzer, q_a (cm^3/sec) is the aerosol flow at the inlet to the instrument, and q_s (cm^3/sec) is the aerosol flow through the slit. Eq. (2) shows that the particle mobility can be calculated when the dimensions of the device, the flow rates, and the voltage on the collector rod are known. By means of Eq. (1), the particle size can then be calculated. The relationship between particle size and collector rod voltage is shown in Figure 2 for the apparatus used in the calibration and for $q_o = 22$ ℓpm and $q_a = q_s = 2$ ℓpm.

In order that the particles extracted by the differential mobility analyzer be singly charged, the particles must be sufficiently small and also, the average charge on the particles must be low. The low level of charge on the particles is obtained in the present case by exposing the aerosol to a source of ionizing radiation using the "Krypton-85 neutralizer" shown in Figure 3. This is done in order that the particles may be brought to a state of charge equilibrium with bipolar ions. At the equilibrium state, the distribution of charges on the particles is given by the Boltzmann's law (see Liu and Pui, 1974b),

$$\frac{N_{n_p}}{N_0} = \exp\left(-n_p{}^2 e^2/2a\,kT\right), \quad (3)$$

where N_{n_p} and N_0 are the relative number of particles carrying n_p and 0 units of charge, a (cm) is the particle radius, k is the Boltzmann's constant, and T (K) is the absolute temperature. Table I shows the predicted charge distribution according to Boltzmann's law for particles of various sizes. For instance, for a particle diameter of 0.1 μm, ~43% of the particles are electrically neutral, ~48% of the particles are singly charged, while ~9% of the particles are multiply charged. For a particle diameter of 0.06 μm, only ~2.5% of the particles are multiply charged, the remainder being either singly charged or electrically neutral.

When the differential mobility analyzer shown in Figure 3 is supplied with a polydisperse aerosol that has been passed through a Kr-85 neutralizer, it will serve to reject all particles except those within a narrow mobility range. If the particles are sufficiently small and in Boltzmann equilibrium, the particles will be either electrically neutral or singly charged. Under such conditions, the differential mobility analyzer will reject the neutral particles while allowing the singly charged particles of the desired size to pass through. By sampling these singly charged particles into the electrometer current sensor shown in Figure 3 and measuring the electrometer current, I (A), where

$$I = q_e\, e\, N, \quad (4)$$

the particle concentration, N (particles/cm^3), can be determined. In Eq. (4), q_e (cm^3/sec) is the volumetric aerosol flow rate into the sensor and $e = 1.602 \times 10^{-19}$ (C) is the elementary unit of charge.

This method of generating monodisperse aerosols worked well for small particles below 0.1 μm in diameter and was used to generate monodisperse NaCl particles in the 0.02 to 0.1 μm range, as shown in Figure 1. The size and concentration of the aerosol were measured to an estimated accuracy of ±2% and ±5%, respectively, using Eqs. (2) and (4). However, when the method was used to generate larger particles above 0.1μm, the aerosol was no longer monodisperse due to the interfering effect of the multiply charged particles.

The source of these interfering, multiply charged particles can be seen by referring to Figure 4. In Figure 4, we have plotted the size distribution curve for the polydisperse aerosol produced by the Collison atomizer. Let us suppose that the differential mobility analyzer is set to extract singly charged particles of 0.2 μm. Let us further suppose that the collector rod voltage is negative, so that only positively charged particles are extracted. According to Table I, for a 0.2 μm aerosol in Boltzmann equilibrium, 22.6% of the particles would carry one positive unit of charge. This is represented by the cross-hatched area under the curve at 0.2 μm. Since doubly charged particles of 0.32 μm diameter have the same

TABLE I. Distribution of Charges on Aerosol Particles According to Boltzmann's Law

	Percent of Particles Carrying n_p Elementary Charge Units								
D_p (μm)	$n_p = -4$	−3	−2	−1	0	+1	+2	+3	+4
0.01				0.34	99.32	0.34			
0.02				5.23	89.53	5.23			
0.04			0.23	16.22	67.10	16.22	0.23		
0.06		0.01	1.25	21.30	54.88	21.30	1.25	0.01	
0.08		0.08	2.78	23.37	47.53	23.37	2.78	0.08	
0.10		0.26	4.39	24.09	42.52	24.09	4.39	0.26	
0.20	0.32	2.33	9.66	22.63	30.06	22.63	9.66	2.33	0.32
0.40	2.19	5.92	12.05	18.44	21.26	18.44	12.05	5.92	2.19
0.60	3.82	7.41	11.89	15.79	17.36	15.79	11.89	7.41	3.82
0.80	4.83	7.94	11.32	14.00	15.03	14.00	11.32	7.94	4.83
1.00	5.42	8.06	10.71	12.70	13.45	12.70	10.71	8.06	5.42

electrical mobility as the singly charged particles of 0.2 μm, these particles will also be extracted. This is represented in Figure 4 by the cross-hatched area under the curve at 0.32 μm.

It is clear from Figure 4 that in order to reduce the concentration of the multiply charged particles, the aerosol supplied to the differential mobility analyzer must be made more monodisperse. In Figure 5, we show the use of a condensation generator to produce aerosols that are moderately monodisperse. In this device, a Collison atomizer is also used to produce a polydisperse DOP (di-octyl phthalate) aerosol initially, but the aerosol is made uniform by vaporization and condensation. The result is a DOP aerosol having a σ_g of 1.2 to 1.4. When this aerosol is supplied to the differential mobility analyzer, a monodisperse, singly charged aerosol can be extracted for particles as large as 1 μm with a negligible concentration of multiply charged particles. This device was used to produce the monodisperse calibrating aerosols in the 0.1 to 1 μm range, as shown in Figure 1.

In generating the DOP aerosol by the above technique, the DOP was first dissolved in isopropyl alcohol and then sprayed by the Collison atomizer. Since the alcohol remained in the vapor state, there was a high concentration of alcohol vapor in the aerosol. This vapor, if not removed, would cause significant changes in the mobility of the ions, thus causing changed operating characteristics for the charger, and possibly changed charging characteristics. In order to remedy the situation, a counter-current packed tower scrubber was added to the system as shown in Figure 5, to remove the alcohol vapor. The particular scrubber used was made of a vertical glass column, 5 cm in diameter by 65 cm high, packed with Rachig rings, which are short segments of glass tubes, each measuring 8

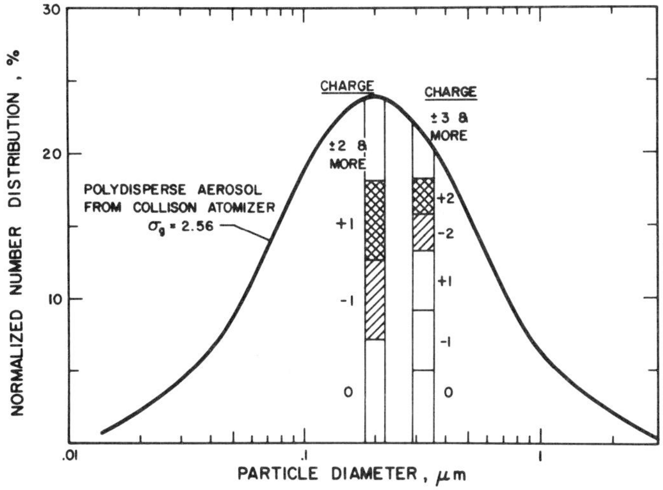

Figure 4. Diagram showing the source of multiply charged particles in aerosols generated by the mobility classification method.

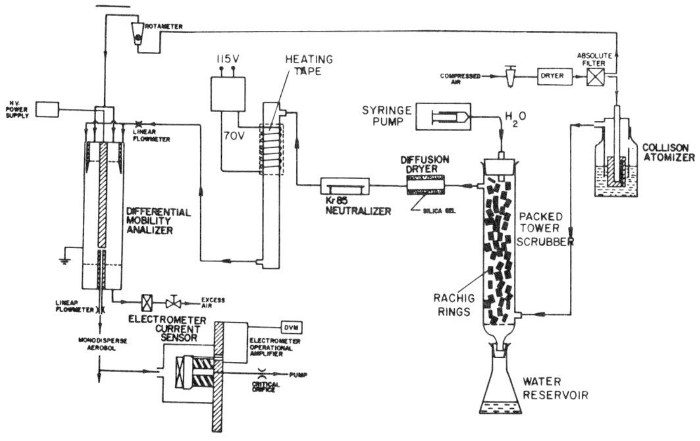

Figure 5. Apparatus for generating monodisperse aerosols by mobility classification of singly charged particles (particle size 0.1–1 μm).

mm in diameter and 8 mm in length. Aerosol was passed upward through the tower while water was supplied to the tower at the top and flowed down over the surface of the glass packings. Alcohol vapor was removed by absorption into water. It was found that at an aerosol flow of 3 ℓpm, a water flow of 12 cc/min was adequate to reduce the alcohol vapor to a negligible level. Since the aerosol stream leaving the scrubber was saturated with water vapor, it was passed through a silica-gel diffusion dryer to remove the water vapor before being introduced into the Kr-85 neutralizer and the differential mobility analyzer.

By the use of the above techniques, monodisperse aerosols were generated in

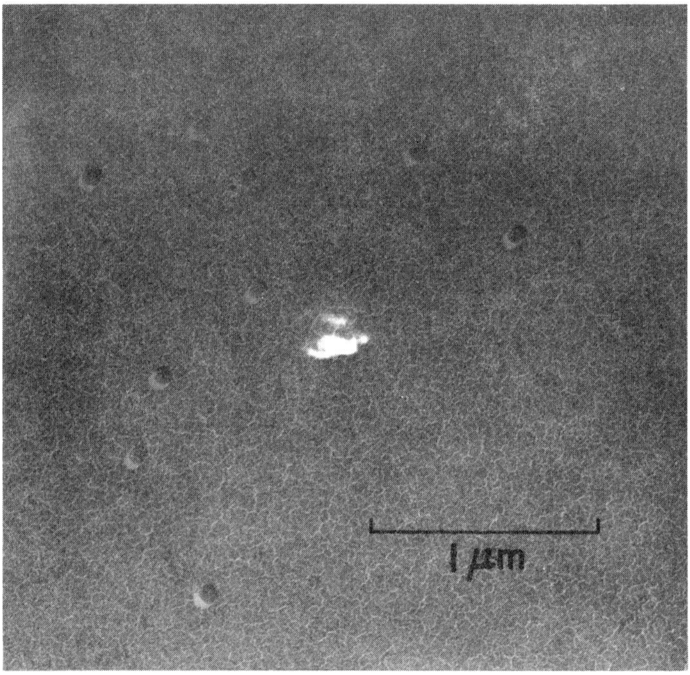

Figure 6. Electron microscope picture of the 0.05 μm NaCl particles generated by mobility classification technique.

the 0.02 to 1 μm diameter range. The size and the concentration of the aerosols were determined to an accuracy of ±2% and ±5%, respectively, by means of Eqs. (2) and (4). These values were checked by comparison with an independent electron microscope technique and good agreement was obtained. Figure 6 is an electron microscope picture showing some of the monodisperse NaCl particles generated by the technique. The particles are seen to be very monodisperse.

Smog Chamber Aerosol

For calibrating the instrument below 0.02 μm, aerosols were generated in a smog chamber by means of the photo-oxidation of SO_2. The smog chamber, a 18 m^3 volume Teflon bag, was irradiated by ultraviolet light from 72 daylight fluorescent lamps of 40 watts each. The SO_2 gas was injected into the bag by a syringe and was allowed to diffuse uniformly throughout the particle-free air in the bag. The lights were then turned on for periods ranging from 2 to 30 minutes, producing a high concentration of small particles in the bag. By allowing the particles to coagulate, the size of the particles could be increased.

The size of the smog chamber aerosol was measured by exposing it to unipolar ions and measuring the mobility of the particles. Since the particles were sufficiently small, they were singly charged. Their size could thus be measured accurately by means of the EAA without calibration. However, the concentration of the particles must be established independently, and this was accomplished as follows.

A General Electric condensation nuclei counter was first calibrated in the 0.02 to 0.1 μm range by means of monodisperse aerosols generated by the mobility classification method described above. The counter was then used to measure the concentration of aerosols below 0.02 μm. There is some uncertainty in the aerosol concentration thus measured, since the response characteristics of the GE counter are not known for particles smaller than about 0.02 μm. However, no alternative technique is available at the present time for determining the absolute concentration of aerosols in this ultrafine particle size range. Consequently, the final absolute calibration of the instrument in this ultrafine particle size range must await further development of absolute aerosol standards.

PERFORMANCE OF THE EAA

The performance of the EAA was evaluated by means of monodisperse aerosols generated by the techniques described. For each monodisperse aerosol generated, the instrument sensitivity

$$\hat{S} = \frac{I_0}{N} \tag{5}$$

was first determined by dividing the electrometer current, I_0, obtained by setting the collector rod voltage to zero, by the input aerosol concentration, N. Next the collector rod voltage was varied, and the corresponding electrometer current was measured. This led to a series of voltage-current curves which provided data for the operating characteristics of the aerosol charger and the performance of the EAA.

Sensitivity of the EAA as an Aerosol Detector

The sensitivity of the EAA for detecting monodisperse aerosols is described by the quantity, $\hat{S}$, defined by Eq. (5). The sensitivity is shown in Figure 7 for two operating conditions of the aerosol charger: nt = 7×10^6 and 1×10^7 (ions/cm^3)(sec). The early instrument produced by TSI, Inc., was set to a charger nt of 7×10^6, whereas 1×10^7 is now adopted as the standard. These charger nt's were obtained with charging currents of 3.7 nA and 7.6 nA, respectively. The corresponding screen voltages were 50V and 70V for the instrument (Serial No. 13) used in this calibration.

It is interesting to note in Figure 7 that the sensitivity of the instrument is a strong function of particle size. At the lower particle size limit of the instrument, the sensitivity is on the order of 10^{-2} pA/(10^6 particles per cm^3); at the upper size limit the sensitivity is about 400 pA/(10^6 particles per cm^3). Since the minimum current measurable by the electrometer is approximately 0.002 pA

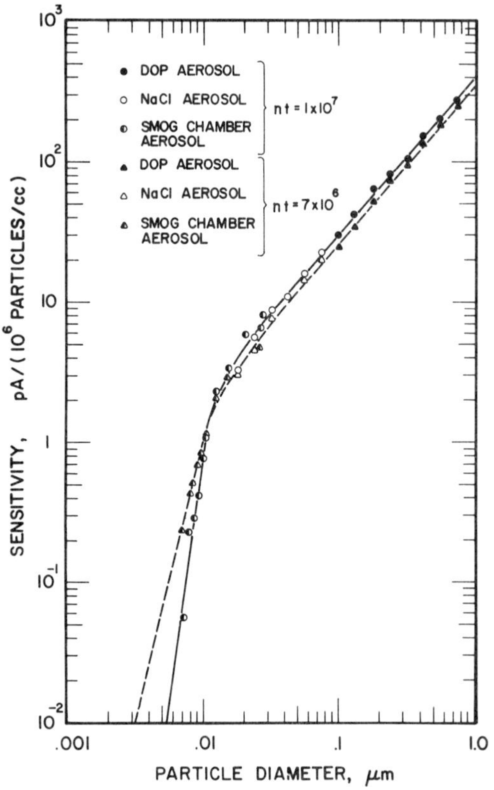

Figure 7. Sensitivity of the EAA for monodisperse aerosols.

(noise current ~ 0.001 pA), the above sensitivities translate to a minimum detectable aerosol concentration of 0.2×10^6 particles per cm^3 at the lower particle size limit of the instrument to 5 particles per cm^3 at the upper limit of 1 μm.

The sensitivity of the instrument can be related to the more fundamental instrument characteristics as follows. If N is the concentration of the monodisperse aerosols being measured by the instrument and $n_p e$ is the average charge per particle, then the electrometer should indicate a theoretical current of

$$I_{th} = q_{ac}\, n_p\, e\, N, \tag{6}$$

where q_{ac} is the aerosol flow rate into the charger. In the actual case, not all the particles are charged, and some particles are lost in the charger, in the mobility analyzer, and in the connecting tubings between the aerosol inlet and electrometer current sensor. If we define an empirical factor, f, such that

$$f = I_o/I_{th} \tag{7}$$

we have, by Eq. (6),

$$f = \frac{\hat{S}}{q_{ac} n_p e}, \tag{8}$$

where $\hat{S}$ is the instrument sensitivity defined by means of Eq. (5). Eq. (8) relates the empirical ''loss and incomplete charging'' factor, f, to the measured instrument sensitivity, $\hat{S}$ defined by Eq. (5).

The actual calculation of f from experimental data requires not only the instrument sensitivity, $\hat{S}$, but also the average charge on the particles, n_p e. Since the true average is not easily measured, we have substituted the median charge based on the median particle mobility, measured by the mobility analyzer (see Figure 11 below) in calculating this ''loss and incomplete charging factor'', f. The result is shown in Figure 8. It is interesting to note that the factor f is essentially 1.0 for large particles, indicating nearly complete charging of the particles and low particle loss. However, the factor drops sharply below a particle size of 0.02 μm due to the incomplete charging of the small particles. The drop in instrument sensitivity below 0.02 μm can be attributed primarily to this incomplete charging effect.

Voltage-Current Curves and Charging Characteristics

The relationship between collector rod voltage and electrometer current is shown in Figure 9 for a charger nt of 1×10^7 (ions/cm^3) (sec). The electrometer current has been made dimensionless in these figures by dividing the current by the maximum electrometer current, I_0. The curves are therefore in the form of dimensionless penetration curves showing the fraction of current collected by the electrometer for various applied voltages on the collector rod. The curves are related to the mobility distribution of the charged particles and, consequently, the distribution of charges on the particles.

There are a number of interesting features to be noted in these voltage-current curves. In Figure 9, we see that the curve for the 0.018 μm aerosol has a large negative slope. This results from the fact that the particles are monodisperse and singly charged, and have a narrow mobility distribution. The steepness of the curve also shows the good resolution of the mobility analyzer. This shows that the mobility analyzer is operating ideally and that there are no extraneous effects such as fluid turbulence affecting the performance of the instrument.

For the two smaller particle sizes shown in Figure 9, viz. 0.012 and 0.015 μm, the slopes of the voltage-current curves are less steep. This can be attributed to the polydispersity of the smog chamber aerosol. For these particles, the charging remains uniform and the particles are singly charged.

For particles in the intermediate size range of 0.024 to 0.075 μm, the voltage-current curves shown in Figure 9 become more complicated. There are apparent discontinuities in the curves which can be attributed to the multiple charging of the particles and the discrete nature of electronic charges. For instance, for the 0.032 μm aerosol, the voltage-current curve shows a discontinuity at a collector rod voltage of 1,500 volts. To the left of the discontinuity, the decrease in

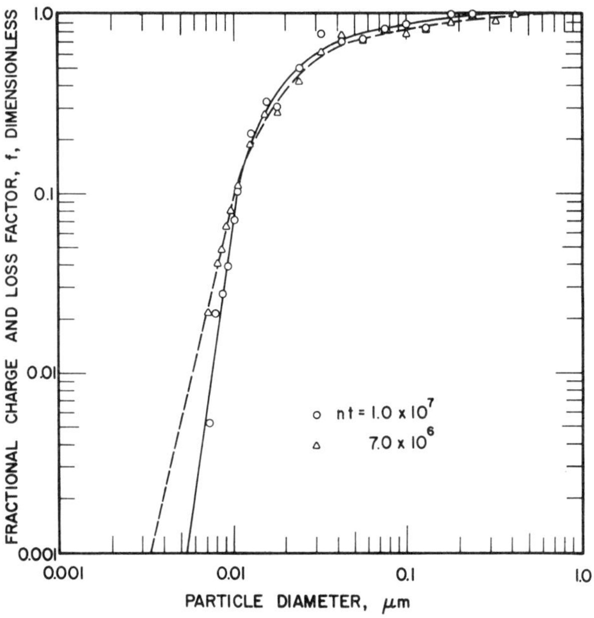

Figure 8. The empirical "particle loss and incomplete charging factor" f for the EAA.

electrometer current with increasing collector rod voltage is caused by the precipitation of the doubly charged particles; to the right of the curve, the decrease can be attributed to the singly charged particles. That these indeed correspond to the precipitation of singly and doubly charged particles can be seen from the fact that the median collector voltage needed to precipitate the particles on the two sides of the curve are 1,000 and 2,000 volts, respectively, which are in exact ratios of 1 to 2, as one would expect for the precipitation of doubly and singly charged particles. Further, we see that the precipitation of the doubly charged particles results in a decrease in electrometer current from the 100% to the 70% level, and the precipitation of the singly charged particles caused the remaining current drop from 70% to zero. Since each doubly charged particle carries twice the charge of a singly charged particle, these figures show that the ratio of the singly and doubly charged particles is 70 to (30/2), or 70 to 15, in the present case. Thus, for this particular aerosol, 70/(70 + 15), or 82.4% of the particles are singly charged, while the remaining 17.6% are doubly charged.

The remaining curves in Figure 9 are for particles in the 0.1 to 0.75 μm diameter range. For these particles, the particle size is sufficiently large and the charges on the particles sufficiently numerous so that the voltage-current curves again become continuous.

Channel Boundaries and Response Matrix

In order to establish the appropriate channel boundaries for the EAA, the relationship between the particle size and the collector rod voltage must be known. This relationship is shown in Figure 10 for the median collector rod voltage, i.e.,

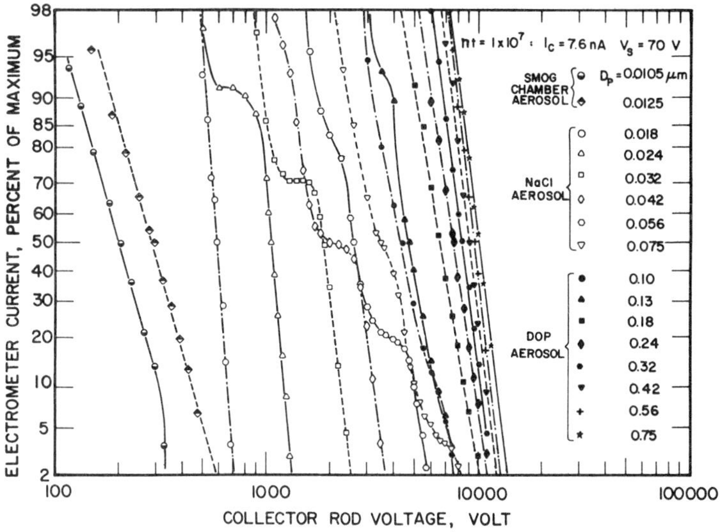

Figure 9. EAA voltage-current curves for nt = 1 × 10^7 (ions/cc) (sec).

the voltage needed to reduce the electrometer current to 50% of its maximum value, as a function of particle size. Included in this figure are the data for the charger nt of 7×10^6, 1×10^7, and 1.9×10^7 (ions/cm^3) (sec). Also included is the theoretical relationship used by the manufacturer in setting the collector rod voltage values for the first production models of the EAA.

In choosing to set the appropriate channel boundaries for the EAA, it was decided that equal geometrical intervals should be used. It was further decided that equal geometrical intervals of four per decade of particle size were appropriate for the instrument. Thus, the adjacent channel boundaries are in constant ratios of $10^{1/4}$ to 1, or 1.78 to 1. The individual channel boundaries selected are: 0.0032, 0.0056, 0.01, 0.0178, 0.032, 0.056, 0.1, 0.178, 0.32, 0.56, and 1.0 μm respectively.

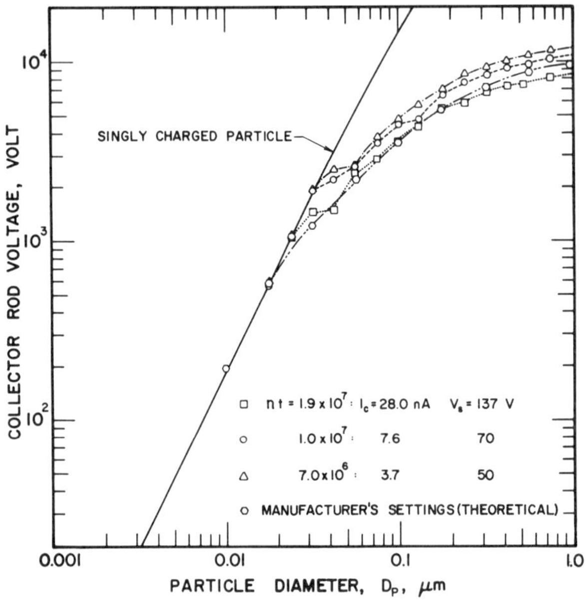

Figure 10. Collector rod voltage as a function of particle size.

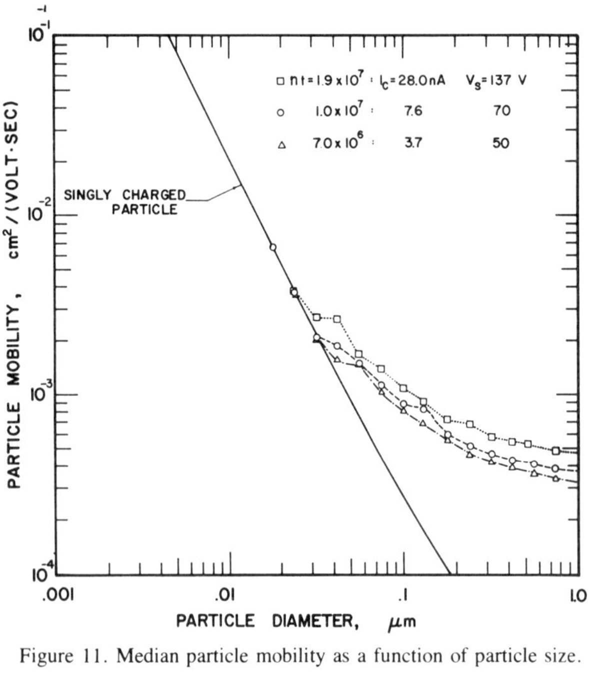

Figure 11. Median particle mobility as a function of particle size.

TABLE II. Response Matrix of the EAA for nt = 1×10^7 (ions/cc)(sec)

Channel	Channel Midpoint (μm)	Channel Boundary (μm)	Particle Diameter (μm) / Sensitivity pA/(10^6 part./cc) / Collector-rod Voltage (V)	0.0042	0.0075	0.013	0.024	0.042	0.075	0.13	0.24	0.42	0.75
			Sensitivity pA/(10^6 part./cc)		0.105	2.40	6.00	11.5	22.5	41.5	81.0	150	285
		0.0032	19										
1	0.0042				0	0	0	0	0	0	0	0	0
		0.0056	59										
2	0.0075				1	0	0	0	0	0	0	0	0
		0.0100	186										
3	0.0133				0	1	0.08	0	0	0	0	0	0
		0.0178	588										
4	0.0237				0	0	0.92	0.49	0	0	0	0	0
		0.0316	1,870										
5	0.0422				0	0	0	0.07	0.14	0	0	0	0
		0.0562	2,600										
6	0.075				0	0	0	0.44	0.65	0.42	0	0	0
		0.100	4,440										
7	0.133				0	0	0	0	0.17	0.49	0.22	0	0
		0.178	6,600										
8	0.237				0	0	0	0	0.02	0.09	0.48	0.31	0.11
		0.316	8,380										
9	0.422				0	0	0	0	0.02	0	0.20	0.36	0.28
		0.562	9,600										
10	0.750				0	0	0	0	0	0	0.06	0.18	0.25
		1.000	10,600										
			(>10,600)								(0.04)	(0.15)	(0.36)

With the above choice of instrument channels, the performance of the instrument can then be described. This can best be done with the "response matrix" shown in Table II. In this table, Columns 1, 2, and 3 give, respectively, the channel number, the geometrical midpoint of the channel, and the channel boundaries. In Column 4, the collector rod voltage corresponding to the particle size at the channel boundaries are shown. The remainder of the columns in the table give the fractional distribution of electrometer current in various channels of the instrument for the monodisperse particle sizes shown. For instance, in Column 8 we see that, for a particle size of 0.024 μm, 8% of the drop in electrometer current occurs in Channel 3, while 92% of the drop occurs in Channel 4. For an ideal instrument, all the current drop should occur in one channel, viz. in the appropriate size channel to which the particles belong. In such a case, the response matrix should be an identity matrix with diagonal elements only, i.e., $[\delta_{ij}]$, where $\delta_{ij} = 1$ when $i = j$, and $\delta_{ij} = 0$ when $i \neq j$. However, Table II shows that the response of the EAA is not ideal, except for particles smaller than 0.013 μm. For larger particles, the distribution of currents among different instrument channels generally occurs over a 2 to 5 channel range. This nonideal behavior of the instrument has led to data reduction difficulties which can be minimized by means of procedures described in the preceding paper.

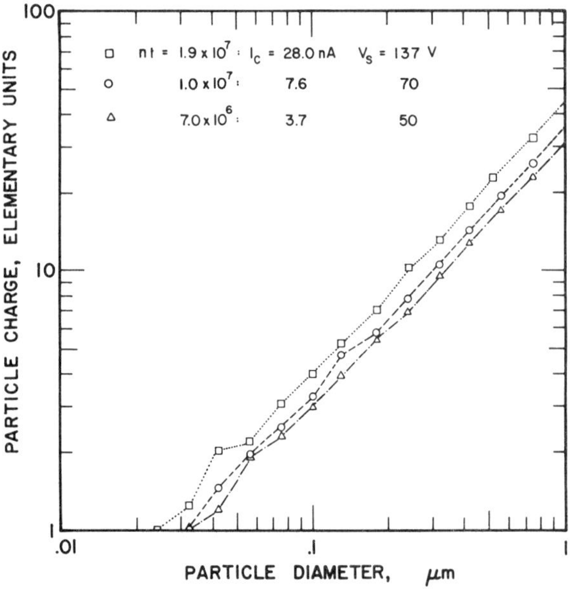

Figure 12. Median particle charge as a function of particle size.

Particle Mobility and Charge

The relationship between the median mobility of the particles and particle size can be derived from the collector rod voltage vs. particle size curve shown in

Figure 10. Knowing the mobility of the particles, the particle charge can then be determined. Figure 11 shows the relationship between particle mobility and particle size thus determined, whereas in Figure 12, the relationship between particle charge and particle size is shown. The result in Figure 11 shows that in unipolar diffusion charging, the median particle mobility is indeed a monotonically decreasing function of particle size in the submicron size range as predicted by diffusion charging theory. Thus, the "diffusion charging-mobility analysis" principle can be used as the principle for size distribution measurement of submicron aerosols.

REFERENCES

Liu, B. Y. H. and D. Y. H. Pui, 1974a: A submicron aerosol standard and the primary, absolute calibration of the Condensation Nuclei Counter. *J. Colloid and Interface Sci.*, *47*, 155–171.

——— and ———, 1974b: Equilibrium bipolar charge distribution of aerosols. *J. Colloid and Interface Sci.*, *49*, 305–312.

ELECTRICAL AEROSOL ANALYZER: OPERATION, MAINTENANCE, AND APPLICATION

GILMORE J. SEM

TSI Incorporated, St. Paul, Minn.

Theoretical aspects of the EAA have been discussed by Liu, Pui, and Kapadia (1976), Liu and Pui (1975), Liu, Whitby, and Pui (1974), Pui (1976), and Sem (1975a). Primary calibration of the EAA has been discussed by Pui and Liu (1976), Liu and Pui (1975), and Pui (1976). The application of the EAA, including conditioning systems for aerosol dilution, drying, neutralization, damping, generation, and batching, has been discussed by Sem (1975c).

The purpose of this paper is to discuss certain operation characteristics, maintenance procedures, personal experiences, and applications of the electrical aerosol analyzer (EAA). Also included is a bibliography of all known references covering the theory, calibration, operation, application, and data of the EAA.

FACTORY CALIBRATION

A newly assembled EAA, shown in Figure 1, is first electronically checked and adjusted. All amplifier gains are set. Voltage dividers are checked and adjusted. Power supplies are checked. Ionizer current, voltage and analyzer voltage are adjusted. Step voltages, step timing, electrometer gain, zeroing, noise level, and performance are checked. The performance of all electrical systems as a unit is then verified.

Each entire flow system is then leak tested by plugging both aerosol and sheath air inlets with a Swagelok fitting and pulling a vacuum of about 60 cm of Hg gauge. If all o-rings are properly installed and lubricated with high vacuum grease, the vacuum should decrease by no more than 1 mm Hg per minute. If a leak is present, we first isolate the leak to either the charger, mobility analyzer, electrometer, or connecting tubing. Isopropyl alcohol is then squirted onto the suspected joint, and if the leak is there, the alcohol is quickly drawn into the joint. We then perform proper corrective action.

All three flowmeters are calibrated before installation in the EAA. The charger sheath air rotameter is calibrated at the 1 ℓpm operational flow rate, the aerosol thermal flowmeter is calibrated over the 1–10 ℓpm range and the total air thermal flowmeter is calibrated over the 6–60 ℓpm range. The air flow calibrations are rechecked at the proper operating flow rate after leak testing is completed. Aerosol and total air flow are checked by connecting a standard thermal flowmeter to the aerosol and total air inlets, respectively, on the front panel of the EAA. Charger sheath air is checked by connecting the standard thermal flowmeter to the inlet of the rotameter. The standard thermal flowmeter consists of a standard TSI Model 2011 flowmeter, calibrated from 0.6–60 ℓpm, with a Mine

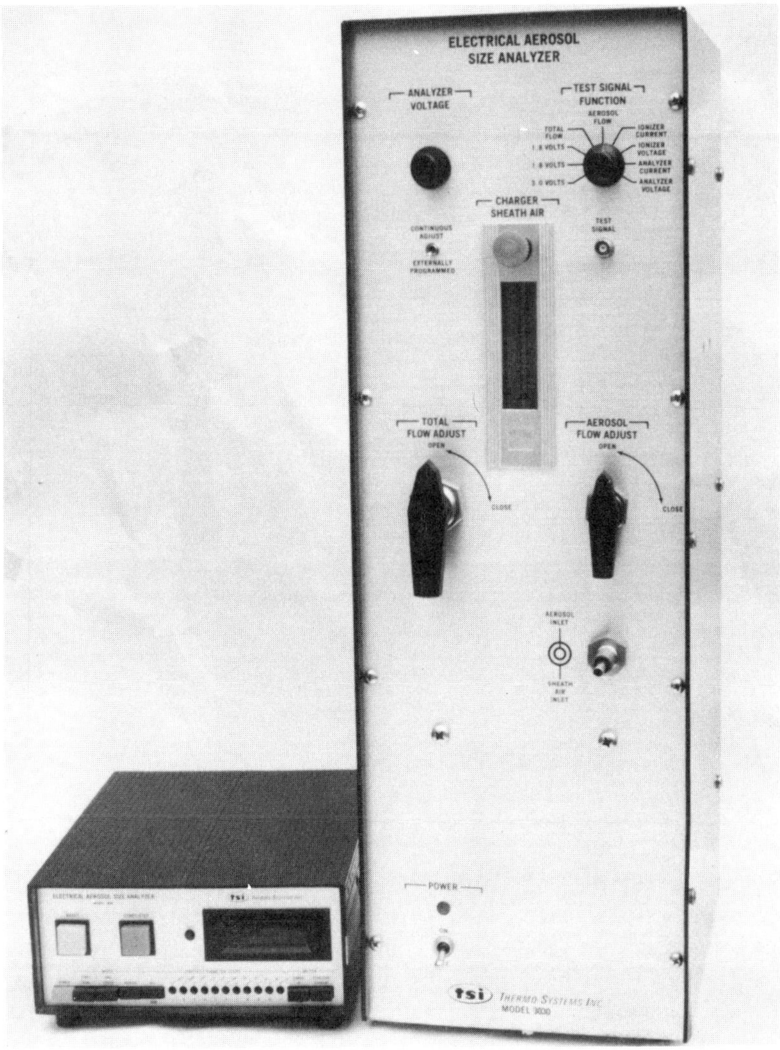

Figure 1. Electrical aerosol analyzer model 3030.

Safety Appliances (MSA) filter holder No. 78006 with filter cartridge MSA No. 95302 attached to the inlet of the flowmeter. While the calibration of the aerosol flowmeter will change as it becomes contaminated by particles, the standard flowmeter calibration remains constant.

The final factory checkout and calibration phase is a comparison of the new EAA with a factory standard EAA. The factory standard EAA has been compared directly with the EAA which was calibrated by Liu and Pui (1975) and has been found to have nearly identical performance. When calibrating a new EAA,

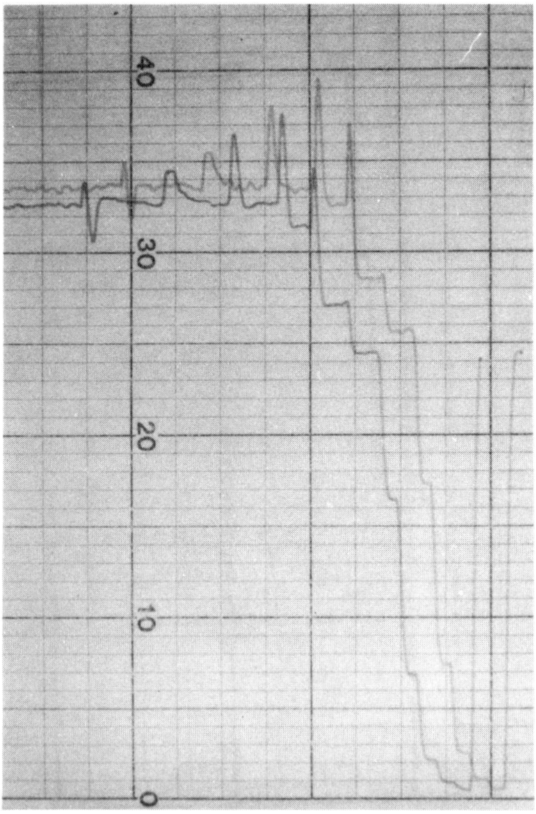

Figure 2. Two-pen strip chart recording of a new EAA and the factory standard EAA.

a plastic bag with about 4 m^3 capacity is filled with room air containing primarily tobacco and several other smokes. The new EAA and the standard EAA then draw both aerosol and sheath air (50 ℓpm) from the bag simultaneously. Analyzer current signals are recorded on a 2-pen strip chart recorder. Figure 2 shows a typical acceptable recording of the standard and test EAA. Nearly all EAAs measure identical analyzer current levels within $\pm 5\%$ on all steps before leaving the factory. The maximum allowable deviation from the standard on any single analyzer current step is $\pm 10\%$. These same standards also apply to repaired and re-conditioned instruments as they leave the factory. The ΔI values from 0.03–0.3 μm also match within ±10% on all channels. ΔI values in the first 4 and last 2 channels sometimes deviate by as much as ±25% when leaving the factory.

MAINTENANCE AND CLEANING

Other than routine daily operational checks of the ionizer current, ionizer voltage, and flow rates, the primary maintenance procedure is periodic cleaning of the aerosol flowmeter sensor, charger assembly, mobility analyzer tube and rod, and electrometer filter holder.

Cleaning the Aerosol Flowmeter

Since the sensing element of the aerosol flowmeter is located in the aerosol flow stream it must be cleaned periodically. If it is not cleaned, the response of the flowmeter will drift until the calibration is no longer valid. For aerosols which cause maximum analyzer currents in the 1–2 × 10^{-12} ampere range, clean the aerosol flowmeter sensor every 200–300 hours of operation. For aerosols which cause maximum analyzer currents in the 10–20 × 10^{-12} ampere range, clean the sensor every 20–30 hours of operation. If these rules of thumb are followed, the aerosol flowmeter calibration will generally remain valid within about ± 5%.

To clean the aerosol flowmeter sensor, remove the sensor from the Venturi tube by loosening 2 screws. The smaller appendage (about 1 mm dia × 1 cm long) is the flow sensor and the larger appendage (about 2 mm dia × 1 cm long) is the temperature compensation winding. Wash both carefully with clean, reagent-grade isopropyl alcohol and a small, soft, artist's paint brush. Do not touch the 2 sensors with your hands and do not bend them or damage them in any way. When they are thoroughly clean and air dried, reassemble the flowmeter. A thorough cleaning will restore the sensor response to within about ± 2% of the initial calibration. If a high quality standard flowmeter is available, the calibration of the aerosol flowmeter should be checked after reassembly. A rotameter will cause significant restriction of the aerosol flow and cannot be used for accurate calibration.

Cleaning the Charger Assembly

As the EAA operates, the corona wire becomes slowly coated with a white, granular powder. After several hundred hours of operation in most atmospheres, the ion current and ionizer screen voltage (corresponding EAA panel functions are ionizer current and ionizer voltage) are no longer within the proper ratio to maintain an nt value of 10^7 sec per cc where n is the ion concentration in the charging region and t is charging time. When this happens, either the ionizer current and ionizer voltage must be adjusted to the proper ratio or the corona wire must be cleaned.

Since t is maintained constant at t = 0.217 sec by charged aerosol flow rate and charger geometry, $nt = 10^7$ corresponds to $n = 4.61 \times 10^7$ ions/cc. From Eq. (4) of Liu, Pui, and Kapadia (1976), the ratio I_c/V_s then becomes:

$$\frac{I_c}{V_s} = 1.07 \times 10^{-10} \text{ amperes/volt} \tag{1}$$

where I_c is the ion current and V_s is the ionizer screen voltage. As described by Liu and Pui (1975), the readout signal for I_c is a voltage V_{Ic} corresponding directly in volts to $I_c \times 10^8$ amps. The readout signal for V_s is a voltage $V_{Vs} = (0.01)V_s$. The readout voltage ratio then becomes

$$\frac{V_{Ic}}{V_{Vs}} = 1.07. \tag{2}$$

When an EAA leaves the factory, V_{Ic} is adjusted until the ratio is $1.07 \pm 5\%$. Normally, this ratio occurs with V_{Ic} in the 0.65–0.80 range.

As the EAA operates, V_{Ic} is maintained constant by a feedback control system. V_{Vs} usually decreases as the corona wire becomes contaminated. When the ratio drops out of tolerance, V_{Ic} may be decreased by any value greater than 0.50 volts which will restore the I_c/V_s ratio to the proper value. This is possible because a decrease in I_c will result in a proportionally smaller decrease in V_s. I_c is adjusted by turning potentiometer 3 as shown in the instruction manual.

If the value of I_c/V_s cannot be adjusted within the limits described above, the charger subassembly SA0337, shown in the instruction manual, must be removed and the corona wire cleaned by dipping SA0337 into a detergent and water solution for a few minutes. Then rinse with clean water, rinse quickly with clean alcohol, and dry the assembly in air. Do not soak the assembly in a solvent for more than a few seconds or the acrylic will be damaged. Be careful to avoid breaking the wire. Since the wire can be easily broken, a spare charger assembly should be kept available. Also, wipe the inside surface of the charger tube assembly with a soft cloth soaked with clean reagent-grade isopropyl alcohol. Do not damage the 3 electrical contact points on the ion current collector foil inside the tube. Reassemble, making sure the system is leak-free.

Cleaning the Mobility Analyzer

If the analyzer current averages $1 - 2 \times 10^{-12}$ amperes, the central collector rod of the EAA needs cleaning after about 500–1000 hrs of operation. The mobility analyzer tube needs cleaning after about 1000–2000 hrs of operation at the same aerosol concentration.

Wipe the central rod and the inside of the mobility analyzer tube with a soft cloth soaked with clean alcohol. During this process, be very careful not to damage the edge of the cone, the central rod, the inside of the tube, or the centering pins on the cone assembly. A tiny scratch, barely visible, is sufficient to cause severe malfunctioning of the EAA.

The concentricity of the tube with the central rod is very critical so do not damage or dent the tube, the centering edge of the base plate, or the centering pins at the top of the central rod assembly. When dry, reapply vacuum grease to the o-ring at the bottom of the tube and reassemble, eliminating all air leaks.

Cleaning the Electrometer Assembly

A ''noisy'' analyzer current signal may indicate that the electrometer assembly needs cleaning. The electrometer assembly should be cleaned every 500 hrs of operation when operated at analyzer currents in the 1 - 2×10^{-12} ampere range or more often if ''noisy'' output signals appear.

To clean the electrometer assembly, first unsolder the electrical lead connecting the aerosol filter with its holder and remove the filter cartridge. Wash the inside and outside diameters of the Teflon insulator EO2017 (see instruction manual) with clean, reagent grade isopropyl alcohol. Also wash the bottom end of the insulator which is a part of SA0350, but do not remove it from the mounting plate. Do not allow alcohol inside the electronics box on the top side of the mounting plate. When the washed parts are completely air dried, replace the aerosol filter with a new cartridge MSA No. 95302 and solder the electrical lead on SA0350 to the filter rim. Reassemble the system into the EAA making sure no air leaks remain.

COMMON CAUSES OF INSTRUMENT FAILURE

In general, EAAs have experienced few failures. The most serious failure which has occurred on several instruments is breakage of the corona wire in the charger assembly. Since the corona wire is not field replaceable, the time lost in sending the assembly back to the factory for repair suggests that each user keep a spare charger assembly for fast repairs.

On early EAAs, the flow straightener cylinder in the charger assembly did not contain a set screw, but rather depended on a tight fit to hold the charger wire assembly in place. Occasionally, especially during transport or shipment, the flow straightener cylinder on the older units will slip down far enough to prevent electrical contact to the corona wire or the charger screen. This problem usually first becomes apparent at low or zero values of ionizer current and/or ionizer voltage, and is repaired by pushing the flow straightener cylinder further up on the center white plastic tube.

Low, zero, or erratic ionizer current and/or ionizer voltage may signal that the corona wire is broken. When looking into the acrylic at the bottom end of the charger wire assembly SA0337, the wire should be visible along the entire centerline of the assembly. If it is broken, one end will usually not match up with the other end or no wire will be visible. The wire is not field repairable and must be returned to the factory for replacement. A spare assembly, consisting of

SA0337, EO2031, EO2030, and connecting hardware, is recommended for all EAA users to prevent excessive delay in a measurement program.

Most other failures are associated with random electrical component failure. In these cases, the field technician should maintain telephone contact with a factory technician to speed field repairs.

SELDOM USED OPERATIONAL CONTROLS

Electrometer Range Switching

When EAAs are shipped from the factory, they are set for analyzer current sensitivity of 1 volt = 10^{-12} ampere. This range suits most applications, including most ambient air and smog chamber measurements.

However, high concentrations found near freeways, in stack and exhaust measuring systems, etc., can yield analyzer currents higher than 10×10^{-12} amperes which is the upper limit on the standard range. A toggle switch located on the underside of the electrometer mounting plate allows the operator to switch to the 1 volt = 10×10^{-12} ampere range so that analyzer currents as high as 100×10^{-12} amperes may be sensed. The charger becomes overloaded at about 30×10^{-12} amperes so avoid higher concentrations. On old EAAs, the electrometer range switch is located on the printed circuit board in the flow cabinet. Electronic noise levels are reduced by moving the electrometer range switch to the mounting plate.

Continuous Analyzer Voltage Adjustment

A toggle switch and variable adjustment on the front panel of the flow cabinet permits the setting of any positive voltage up to about + 11,000 volts D.C. on the central rod of the mobility analyzer. This is useful for calibration with monodisperse aerosols. Any arbitrary particle size steps may be chosen manually.

Be sure the analyzer voltage toggle switch is in the proper position when operating in the standard 11 step mode.

Custom-Designed Particle Size Steps

The 11 particle size steps and the length of time for each step are programmed on a single read-only-memory integrated circuit chip which is located in the control cabinet of the EAA on a plug-in socket. The operator may order any values of analyzer voltage (particle size) and any reasonable values of step time to fit his special need. For example, many smog chamber aerosols do not grow as large as 0.1 μm during an experiment so all 11 steps may be chosen to fit within the applicable range. Some aerosols, e.g., some stack emissions, have few particles below 0.05 μm. Custom-designed program chips are available from the factory.

The most common size increments are logarithmically equal, often with 4 or 8 steps per size decade, to simplify the data reduction.

nt Adjustment

The resolution of the EAA can often be enhanced by changing the nt value of the charger. Liu and Pui (1975) have calibrated the EAA for 2 values of nt, Pui (1976) has obtained charger data for other nt values. Using Eq. (4) of Lui, Pui, and Kapadia (1976), the correct values of ionizer current and ionizer voltage can be calculated for a range of n values. The value of t for the EAA is fixed by the charged aerosol flow rate and the charger geometry. Thus, the operator may choose the best charging condition for his aerosol.

To adjust the nt value, turn potentiometer 3 on the printed circuit board in the flow cabinet of the EAA until the proper values of ionizer current and ionizer voltage appear. Note that ionizer current and ionizer voltage cannot be adjusted independently.

Sheath Air Composition

There is some controversy at this time about the preferred air composition in the corona region around the corona wire in the charger. The air in the corona region is sheath air so the operator may choose the proper air composition for his experiment. For example, when measuring sulfuric acid droplets or aerosol containing SO_2, the operator may want to use sheath air which is free of sulfur-containing air. Be sure that the sheath air is nearly the same temperature and humidity as the aerosol.

Switching Off Unused Size Steps

Two switches on the printed circuit board in the control cabinet of the EAA permit the operator to skip any number of sizing steps from either the small or large particle size range. For example, the first 2 steps may be switched off for ambient air monitoring because the data below 0.01 μm is usually not useful for such aerosol. The last step may be switched out if an optical particle counter covers the size range above 0.5 μm. The switch positions are numbered. To adjust the starting and ending step, turn the first switch to the desired first step and the other switch to the desired last step.

"NEGATIVE" CONCENTRATION MEASUREMENTS BELOW 0.01 μm

When measuring polydisperse aerosols with constant size distribution, the analyzer current signal may occasionally increase slightly from step 1 to step 2. Thus, the EAA is indicating "negative" aerosol concentration in the size range below 0.005 μm, an obviously incorrect measurement. Newer EAA's with serial numbers of 91 and above rarely display "negative" concentrations below 0.005 μm and do not display "negative" concentrations above 0.005 μm unless the aerosol concentration in a given channel is increasing during the measurement. Older EAA's, with serial numbers 90 and below, may often display "negative" concentrations between steps 2 and 3 and occasionally between steps 3 and 4.

This problem is caused by surface charge on the Teflon insulator at the bottom of the mobility analyzer central rod. The insulator obtains considerable surface charge by electret and/or piezoelectric processes while the analyzer voltage is in the high voltage (large particle size) positions and when the insulator is mechanically stressed. When the insulator is nearly hidden from direct line-of-sight to the aerosol, as in newer EAA's, or when the surface charge is allowed to drain off the insulator in older EAA's, the "negative" measurements disappear.

With older EAA's, several methods exist to obtain data in the first 2 channels, i.e., below 0.01 μm particle diameter. One method is to keep the EAA in step 0 or step 1 for about 1 hour between size measurements. A second, usually better, method is to avoid the high analyzer voltage steps (i.e., large particle sizes) on each cycle. This may be done by setting the upper size limit switch to step 6 or 7 and operating in the automatic continuous or one cycle mode or by manually resetting the EAA after reaching step 6 or 7 (0.05–0.1μm). The second method is generally applicable to smog chamber studies. A third method is factory update.

The operator must recognize the limitations of all EAA's, especially the older ones, in the size range below 0.01 μm. Assuming the use of restricted operating procedures as suggested in the previous paragraph, the measurements below 0.005 μm are, at best, semiquantitative. Some conclusions based on relative measurements in this range are certainly valid. However, Liu and Pui (1975) did not assign any calibration factor for calculating particle concentration below 0.005 μm and the calibration factor for 0.005–0.01 is only tentative. Some smog chamber operators report good agreement between EAA measurements below 0.01 μm and predictions based on independent theories. However, accurate measurements of aerosols below 0.01 μm are difficult by any technique. Until accurate calibration methods exist in this range, EAA measurements below 0.01 μm should be considered semiquantitative.

USING A STRIP CHART RECORDER

A strip chart recorder is probably the least expensive and most accurate method of recording the analyzer current output data from the EAA. Strip chart recorders are not suitable for recording large numbers of size distributions (several hundred or more) or for routine recording with no operator present. However, even in cases where a minicomputer or other type of data aquisition system is necessary, a few strip chart recordings of the data can often lead to better methods of data recording and analysis.

A primary requirement of strip chart recording is that the operator must choose the proper chart voltage range, choose the optimum chart speed, and zero the chart recorder.

Figure 2 shows the factory calibration data for an EAA. If we wanted to calculate the size distribution of this relatively "well behaved" aerosol, the first step is to choose an analyzer current value (or output voltage value) which represents each size step of the EAA. The chart recorder allows the operator to use his best judgment in choosing an "average" representative value. Most digital recorders would record a voltage which is averaged over some time period much less than 1 sec. Thus, the digital value may be erroneously recorded at a

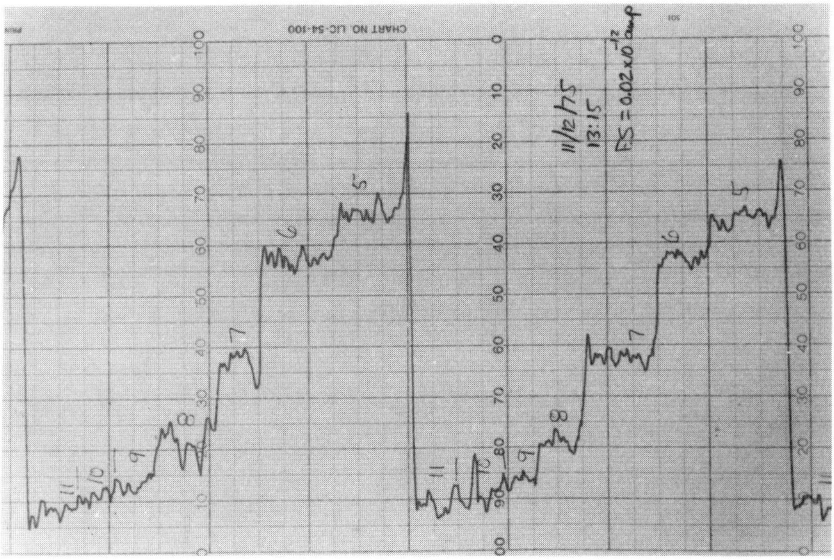

Figure 3. Strip chart recording of an EAA downstream of a baghouse filter on a coal-fired boiler with full scale = 0.020×10^{-12} amperes.

value considerably higher or lower than the proper average value. Some mini-computer systems are programmed to record an average of multiple values over a time span of 1 or several seconds.

Figure 3 shows a strip chart recording of emissions from a coal-fired boiler downstream of a bag-house filter particulate collector. The data includes only steps 5–11 because very few particles were found below 0.03 μm. Notice that the concentration level is very low, with a maximum analyzer current of about 0.013×10^{-12} amperes. The signal is very "noisy" and could be recorded digitally only with great difficulty. However, the strip chart recorder allowed repeatable measurements because the operator could estimate the average value for each size step even in the presence of large fluctuations.

Figure 4 shows a strip chart recording of an aerosol which is increasing in concentration during the EAA measurements. The strip chart recording allows the operator to correct for the changing concentration by drawing a line between given steps of consecutive size runs and using this new line as the base line for analyzer current. By modifying the measured analyzer current in this way, he may make accurate measurements even in the presence of increasing or decreasing aerosol concentration. This technique assumes that the aerosol concentration is changing at a constant rate in all size ranges, an assumption which may not be true in dynamic systems where the size distribution is changing rapidly such as in photochemical smog chambers.

APPLICATIONS

Ambient Atmospheric Aerosols

It was for the measurement of atmospheric aerosol size distributions that the electrical mobility technique was first developed into a practical instrument. The primary application of the old Model 3000 analyzer was the measurement of

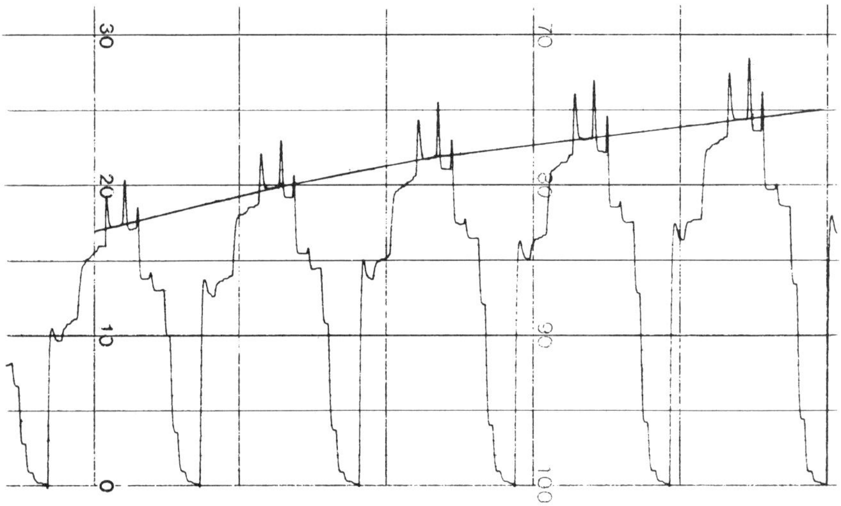

Figure 4. Strip chart recording of an EAA showing continuously increasing aerosol concentration with a new base line connecting successive levels of step 3.

urban and non-urban atmospheric aerosol. The EAA is now being used in several atmospheric aerosol measurement programs.

Until about 1970, very little was known about submicron atmospheric aerosols. Recent measurements of both submicron and supermicron aerosols, e.g., Willeke, Whitby, Clark, and Marple (1974), have shown that at least 2 distinct modes exist in the volume weightings of size distributions of urban aerosols. The submicron mode consists almost entirely of particles generated by combustion sources or formed by the photochemical conversion of gaseous materials into aerosols. Submicron particles usually are removed from the atmosphere by collision with and adhesion to larger objects such as raindrops, snowflakes, buildings, foliage, etc. The supermicron mode consists primarily of mineral dusts which become airborne by wind action, sometimes combined with such activities as agriculture, ground transportation, and stack emissions. Supermicron particles usually settle from the air by gravitational force. A change in the level of one mode does not seem to affect the other in a predictable pattern. A third mode sometimes exists in the 0.01 μm range caused by fresh combustion or fresh photochemical aerosols which have not yet "grown" by coagulation and condensation into the more stable 0.1–1.0 μm range.

Much remains to be learned about atmospheric aerosols. Nearly all submicron size distribution measurements have been made in California, Denver, St. Louis, Washington State, Minnesota, and Michigan. Although some generalizations can certainly be made from these measurements, some differences in the particle size characteristics have been seen. The characteristics of local sources and local meteorological conditions are certain to affect the size distribution measurements at a given location. Thus, many more measurements are needed to characterize different urban or non-urban locations and to characterize the addition of a single source to the atmospheric mixture. It may be found that size distribution measurement becomes a useful, routine, continuous monitoring tool as more is learned about atmospheric pollution.

This brief summary will not attempt to describe the previous work performed by many investigators using the Model 3000, its several prototypes, and the EAA, but will only reference the published work. Clark and Whitby (1967), Peterson and Paulus (1968), Peterson, Paulus, and Foley (1969), Peterson (1969), and Paulus and Peterson (1969) performed the first extensive field measurements of atmospheric particle size distributions, all in Minnesota urban air. Los Angeles smog aerosol at a single location was measured extensively in late summer 1969 by Whitby et al. (1970), Whitby, Liu, Husar, and Barsic (1972), Whitby, Husar, and Liu (1972), and Husar, Whitby, and Liu (1972). Whitby, Husar, and Rao (1973) reported size distribution measurements of non-urban atmospheric Colorado aerosols. Alkezweeny and Slinn (1972) reported 1971 size distribution measurements at 3 locations in Washington State; one urban site, one inland non-urban site, and one coastal site. Willeke, Whitby, Clark, and Marple (1974), Durham, Wilson, Ellestad, Willeke, and Whitby (1975), Willeke and Whitby (1974) and Willeke and Whitby (1975) reported measurements at two urban sites in Denver in 1971. An extensive field study of aerosols at about a dozen urban and remote sites in California in 1972–1974 is the subject of a series of papers and reports, beginning with Whitby, Clark, Marple, Sverdrup, Sem, Willeke, Liu, and Pui (1975), Sverdrup, Whitby, and Clark (1975), and Hidy et al. (1975a, b). Knutson (1976) compared EAA and diffusion battery measurements while measuring New York City atmospheric aerosol. Results of other atmospheric aerosol size distribution measurement programs include measurements of particles produced by a fleet of automobiles by Wilson, et al. (1977). Groblicki (1976) and Whitby, et al. (1976) and measurements of the accuracy of several analog sensors to measure size distributions by Sverdrup (1977) and Sverdrup and Whitby (1977).

Several of the reports above describe the sampling system in detail, notably Clark and Whitby (1967), Peterson, Paulus, and Foley (1969), Whitby et al. (1970), and Whitby, Liu, Husar, and Barsic (1972). Sem (1974) described the aerosol inlet, transport, and distribution system used by EPA in the 1973 and 1974 St. Louis and Denver programs. This system is very similar to the sampling system used in the mobile laboratory for the 1972 California ACHEX program.

Willeke and Whitby (1975) discuss various methods of presenting and interpreting atmospheric size distributions.

The important considerations when designing a system to deliver submicron atmospheric aerosol to the EAA are:

1. Reduce particle losses, primarily diffusion losses, to a minimum by correct sizing of the sampling tube diameter, tube length, and flow rate. It may be desirable to bring a larger flow rate to the EAA than it needs and then remove the correct flow rate from the larger stream as done by Sem (1974).
2. Prevent condensation on and evaporation from the particles or tube surface which can occur when sampling aerosol from the atmosphere into a cooled or heated room.
3. Provide water escapes, inlet hoods, etc., to prevent rain from interfering with the system.
4. Prevent or remove excessive electrostatic charge which may be present if sampling freshly generated aerosol or which may develop when sampling through long plastic tubes. See Liu and Pui (1974a, b) for an explanation of aerosol charge equilibrium and neutralization. If in doubt, a radioactive Kr-85 charge neutralizer may be used upstream of the analyzer.
5. Prevent excessive aerosol concentration fluctuations during each 2-minute sizing cycle. This can usually be done by providing a simple aerosol damping chamber with about 20 liter capacity (about 20 sec) just upstream of the analyzer in the 50 ℓpm sample line. This will reduce concentration fluctuations which occur within a period of less than about 20 sec. If the aerosol concentration level drifts up or down more than 10% during a 2-minute sizing cycle, a bag sampling system, like the one shown below for auto exhaust sampling, may be needed. This condition could occur if a single source excessively influences the measurements, such as light traffic (one or two cars per min) on a nearby highway or a single combustion point source within 1 km upwind of the sampling site.

Aircraft Systems for Urban and Single Source Plumes

A number of analyzers have been mounted on aircraft and used to measure particle size distributions in both urban plumes and the plumes from a single combustion source. Alkezweeny (1974) described an instrumented Cessna 411 which was used to gather data in the urban plume of St. Louis during 1974. Alkezweeny and Radke (1973) used the Model 3000 in an instrumented B-23 in 1972 to measure AgI particles generated by an airplane flying ahead of the B-23. Liu, Whitby, and Pui (1974) used a prototype EAA on an instrumented Cessna 205 in southern California in 1972. They used (Figure 5) a bag sampling system during the 1972 study and in a St. Louis urban and point source plume study with a commercial EAA in 1973, 1974, and 1975 (Whitby, Cantrell, Husar, Gillani, Anderson, Blumenthal, and Wilson, 1976). Anderson, Blumenthal, and Sem (1977) used the instrumented Cessna 205 and bag system in a Denver urban plume study in 1973. Hobbs, Radke, and Hindman (1974, 1976), Radke, Stith, Hegg, and Hobbs (1978), and Hindman, Hobbs, and Radke (1977) used an instrumented B-23 to extensively study urban and point source plumes. Strand and Varsi (1974) used an aircraft with an EAA in a completely automated sampling and sizing system in 1974 to study high altitude, solid fuel, rocket exhaust plumes. They also used an instrumented helicopter to study the same plumes at low altitudes.

During a typical 2-minute analyzer sizing cycle, the typical airplane has traveled over 5 km and, by that time, is usually in a very different aerosol cloud. Thus, a bag sampling system is usually necessary, whether sampling an urban plume or a single source plume. The bag sampling system in Figure 5 has been

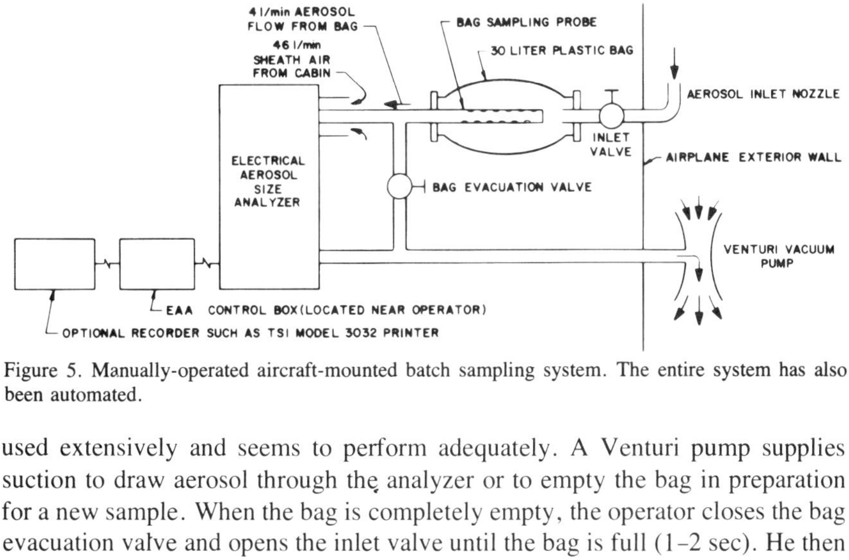

Figure 5. Manually-operated aircraft-mounted batch sampling system. The entire system has also been automated.

used extensively and seems to perform adequately. A Venturi pump supplies suction to draw aerosol through the analyzer or to empty the bag in preparation for a new sample. When the bag is completely empty, the operator closes the bag evacuation valve and opens the inlet valve until the bag is full (1–2 sec). He then closes the inlet valve, adjusts the flow rate through the analyzer, and begins the data cycle. Usually, the sheath air is drawn from the cabin to reduce the bag volume requirement. Data may be recorded automatically on magnetic tape as described by Alkezweeny (1974), may be recorded by a separate paper tape printer, or may be spoken into a cassette tape recorder as the operator watches the data readout on the EAA control box in his lap.

The purpose of single source plume measurements is not only to measure the emissions of the source, but more importantly to identify what happens to the plume as it disperses and to characterize the ultimate contribution of the source to the total air pollution problem. The purpose of urban plume measurements, sometimes called 3-dimensional air pollution mapping, is to better understand the complex aging and dispersion of air pollution from a mixture of sources covering an entire city under a variety of meteorological and source conditions. This information may lead to a more precise air pollution model which can help control personnel to reduce the frequency and intensity of high pollution episodes.

The size of particles in the atmosphere, particularly of submicron particles, is a necessary factor in characterizing what will happen to the particles while airborne and how long the particles are expected to remain airborne. Thus, the EAA data are an important part of the single source and urban plume measurements.

Photochemical Smog Studies

Smog is the result of a photochemically induced reaction which converts certain gaseous components of the atmosphere into aerosol form. The products of the smog reaction appear as submicron particles. A very important parameter in the characterization of the reaction is the conversion rate of gas to aerosol. The EAA can measure the development of the aerosol size distribution in a smog chamber as the reaction progresses through its various stages. The number size distribution can easily be converted to volume size distribution by the assumption of spherical particles and to mass size distribution by the assumption of particle density. Thus, the mass conversion rate can quite accurately be determined.

The EAA and the Model 3000 have been used in a number of smog chamber experiments to measure the gas-aerosol conversion rate and to follow the development of the aerosol distribution with irradiation time. Among those who have reported results are Husar and Whitby (1971, 1973) and Husar, Whitby, and Liu (1972), who irradiated particle free laboratory air with diffuse solar radiation. Clark (1972), and Clark and Whitby (1975) who measured extensively the aerosol resulting from photochemical oxidation of SO_2 in air, and Kocmond et al. (1973, 1975), Kocmond and Yang (1975, 1976), and Kocmond, Kittelson, Yang, Demerjian, and Whitby (1976) studied the formation mechanisms and the composition of photochemical aerosols. Others who have used the EAA for smog chamber analysis are Miller, Levy, Pui, Whitby, and Wilson (1976) and Miller and Levy (1975) who measured photochemical aerosols produced from automobiles using several fuels and several auto operating conditions and McNelis (1974) and McNelis, Ripperton and Wilson (1975) studied the photochemical production of aerosols from ozone and olefin in the presence of SO_2. Clark, Landis, and Harker (1976) and Harker, Richards, and Clark (1977) measured photochemical aerosol production using natural sunlight to irradiate Teflon bags of ambient air near a freeway and elsewhere. Fox et al. (1975) described the equipment they are using for several types of smog chamber experiments in an outdoor irradiation chamber. While Marlow, Reist, and Dwiggins (1976) reported that SO_2 in the sampled aerosol will adversely affect EAA measurements, Liu and Pui (1978) could find no SO_2 effect while trying to repeat the experiment. Liu and Pui suggest that the results of Marlow et al. were influenced by an unstable aerosol source.

The characteristics of several types and sizes of smog chambers have been well described by the above investigators. The cited references also discuss potential problems encountered when conducting smog chamber experiments.

Smoke Stack Emissions

The performance of particulate emissions control devices on submicron particles has not been measured in the past, primarily because submicron particles were only a small fraction of the total mass of particulate emissions. The measurement of the submicron fractional efficiency of control devices has become more important recently.

Sem, Borgos, Whitby and Liu (1972) evaluated basic techniques for measuring aerosol size distributions in coal combustion emissions from power generating plants and concluded that the electrical mobility analyzer technique was the best available technique for measuring particles smaller than 0.5 μm. They suggested that some development remained to be done to adapt present instruments for stack use. Much of that development was done with the development of the EAA.

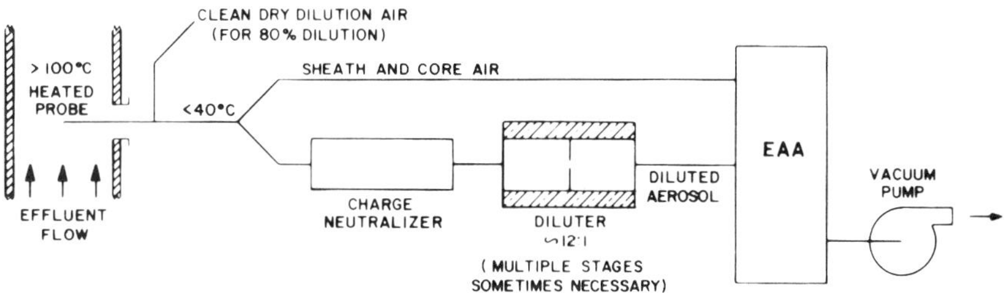

Figure 6. Simple sampling system for high temperature streams with constant aerosol concentration (Schmidt, Geiseke and Allen, 1976).

Several groups are currently adapting the EAA to field measurements of the fractional efficiency of electrostatic precipitators, scrubbers, and fabric filtration systems. Schmidt, Gieseke, and Allen (1976) have used a dilution system like the one in Figure 6 on coal-fired, electric power plant emissions. Concentration fluctuations were not a problem on several sources they measured. Using a dilution system similar to the one shown in Figure 7, McCain, Cushing, and Smith (1974) and Smith, Cushing and McCain (1974) are using an EAA in a number of field emissions measurement projects. Ragland, Pontius and Smith (1976) recently developed an improved sample extraction system using a heated dilution system. Concentration fluctuations are a definite problem in some types of processes. McCain (1975) reports excellent correlation between a diffusion battery and an EAA on emissions from one stack. Sem (1975a, b), Ensor, Hooper, and Scheck (1976), Ensor, Hooper, Scheck, and Carr (1976), and Ensor, Hooper, Carr, and Scheck (1976) discuss the evaluation of a baghouse on a coal-fired boiler. Hersh, Hurley, and Carr (1976) studied the effects of smoke suppressants on the emissions from a utility gas turbine. Carlson and Jackson (1977) evaluated the conversion of SO_2 to particles so the particles can be collected by a fluid bed.

Although dilution sampling as shown in Figures 6 and 7 may be adequate for some process emissions measurements, a dilution system with batch sampling capabilities may be needed for other process emissions where severe particle concentration fluctuations frequently occur. Sem (1975c) suggests the batch sampling system shown in Figure 8 which incorporates a bag system into the

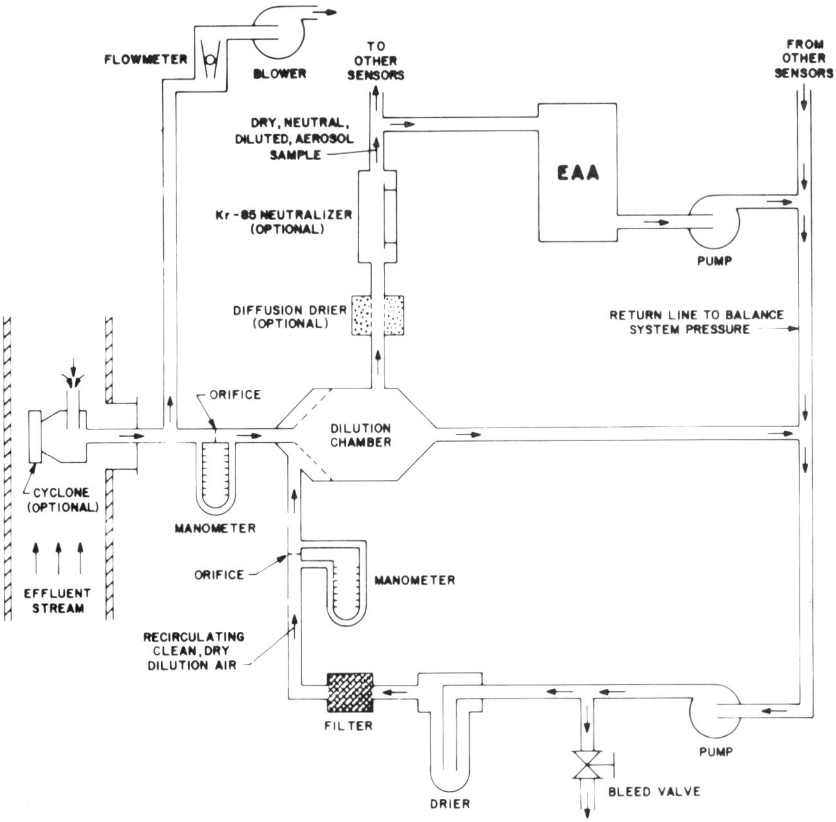

Figure 7. SRI sampling-dilution system for measuring 0.005–2 μm particles from stacks (McCain, Cushing and Smith, 1974).

McCain, Cushing, and Smith (1974) sampling system for measuring stack emissions which fluctuate significantly, e.g., downstream of electrostatic precipitators and baghouses during rapping and shaking, respectively.

Important considerations for a smokestack sampling system for use with the EAA are:

1. Reduce aerosol concentration to a level acceptable to the EAA, usually about 100 $\mu g/m^3$ below 1 μm diameter. This will also reduce coagulation to an acceptable level.

2. Reduce aerosol concentration fluctuations so that the analyzer sees no more than ± 10% change during the 2 minute operating cycle.
3. Reduce aerosol temperature below 50°C and maintain aerosol and sheath air temperature within ± 3°C of each other.
4. Reduce electrostatic charge on the particles to a level near the Boltzmann equilibrium level.
5. Prevent condensation of all vapors on particles and on tubing, instrument, and other surfaces.

Auto and Diesel Exhaust

In many urban areas, a major component of atmospheric aerosol is auto and diesel exhaust emissions. Nearly all of the mass of auto emission particles is contained in submicron particles. A second major contribution by auto emissions to atmospheric pollution is the photochemical transformation of gaseous exhaust products into particulate form (smog). Nearly all photochemically produced ma-

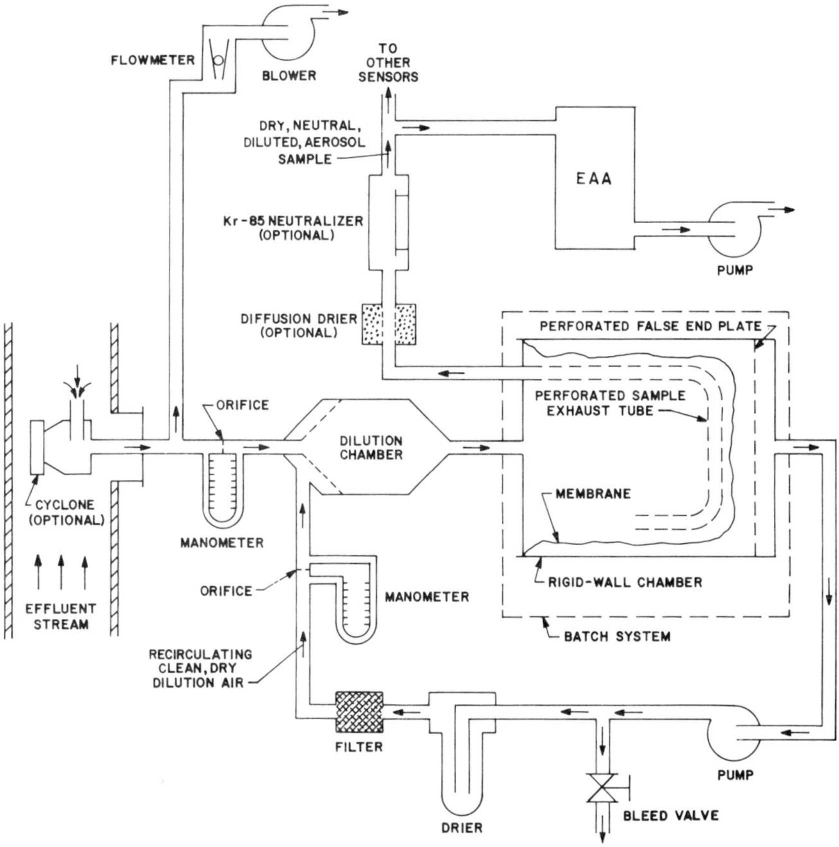

Figure 8. Sampling-dilution system modified with batch sampling system for use on emission streams with fluctuating aerosol concentration and/or size distribution (Sem, 1975).

terial condenses on submicron particles, causing them to grow to about 0.1–1.0 μm. An earlier section of this paper discussed the use of the EAA in smog studies.

Miller, Levy, Pui, Whitby, and Wilson (1976) and Foster et al. (1975) used the EAA to measure the size distribution of primary auto exhaust as a function of fuel and auto operating conditions. They used a sampling system consisting of a high volume dilution tunnel similar to the design described by Habibi (1970) and a sample chamber with additional dilution. The sample chamber doubled as a smog chamber in other experiments. Miller, Trayser, and Joseph (1976) measured sulfuric acid aerosol emissions from vehicles using an EAA. Fissan, Felsper, and Franzen (1976) measured engine emissions.

A major problem with the sampling of auto and diesel emissions is the wide variation of exhaust flow rate when the vehicle is operated in a typical driving cycle. Most instruments and sampling systems require a constant sample flow rate. In recent years, the constant volume sampling (CVS) system has become the accepted method for sampling auto emissions in the U.S. The CVS system mixes enough clean air with the exhaust to maintain a constant total flow rate through the system. Most monitoring instruments sample from the diluted, constant flow rate portion of the system.

Several problems still remain when one attempts to measure the particle size distribution of auto or diesel emissions with the EAA. The primary problems are 1) aerosol concentration fluctuations during a single particle sizing cycle, and 2) condensation of vapors during short periods of the driving cycle.

The EAA requires nearly constant aerosol concentration (within ± 10%) during the particle sizing cycle (1–2 minutes). During extended periods at constant engine speed and operating conditions, this requirement may be met with auto or diesel exhaust. However, under typical urban driving conditions with idling, acceleration, deceleration, constant speed, and other operating modes, aerosol concentration probably varies by more than one order of magnitude. Moreover, these changes may occur within a period of a single second.

During short periods of acceleration or deceleration, the exhaust can become supersaturated with vapors. Thus, for a second or two, condensation occurs on the particles and on any surface which is cooler than the dew point temperature of the supersaturated vapor. When the more normal, unsaturated condition returns, the liquid evaporates from the particles and surfaces. Even during the supersaturated condition, liquid evaporates from the particles when ambient air dilutes the exhaust if the ambient air has a low enough dew point. Thus, the supersaturated condition should normally be avoided in the sampling system, especially if the measuring instrument is adversely affected by condensation as is the EAA.

Both the fluctuating concentration and the condensation problems can be adequately solved by a batch dilution sampling system similar to the one shown in Figure 8. Undiluted aerosol (or CVS constant volume aerosol) is drawn through a probe which is heated to prevent condensation, into a flexible bag which has previously been partially filled with clean, dry air. The sampling rate is controlled by the amount of vacuum drawn in the chamber surrounding the bag, or, if the stream being sampled is pressurized, by the ambient upstream pressure. After the appropriate amount of aerosol, measured by a totalizing aerosol flowmeter, has been sampled into the bag, the inlet valve is closed and the bag is filled with clean air. The measured incoming aerosol and the total bag volume determine the dilution ratio, which may range from no dilution to a factor as high as 10,000. The analyzer may begin its particle sizing cycle just after the bag is filled and the vacuum is released from the chamber around the bag. If the bag volume is sufficient, the entire 50 ℓpm needed by the analyzer may be drawn from the bag. If the bag volume is limited, as little as 4 ℓpm may be used by the EAA with the 46 ℓpm needed for sheath and core air drawn from the room. One must be sure the aerosol and sheath air temperatures do not differ by more than about 3°C. As the diluted aerosol is drawn from the bag, the bag collapses, maintaining constant aerosol concentration and pressure. When the analyzer has completed its cycle, the bypass valve is opened drawing all the remaining aerosol from the bag in preparation for the next sample.

A second technique for measuring auto and diesel exhaust is to measure single or multiple plumes in the atmosphere produced by one or several vehicles, as was done during part of the 1972 ACHEX study reported by Whitby et al. (1975). Their sampling was done alongside a freeway in Los Angeles. Depending on wind direction, the background or fresh freeway aerosol was sampled. Wilson et al. (1977), Whitby et al. (1976), and Groblicki (1976) measured the aerosol produced by automobiles traveling on a test track in Detroit during a 1975 sulfate dispersion experiment. A 1974–1975 study reported by Clark, Landis, and Harker (1976) had separate sampling facilities located on both sides of a freeway in Los Angeles. In this way, the same ''parcel'' of air was sampled before and after the addition of the freeway aerosol. In these studies, a bag sampling system was not used. If the aerosol emitted to the atmosphere by a single vehicle is of interest, a bag sampling system, probably without dilution, would be required.

Jet Engine Exhaust

The exhaust of a jet engine contains high concentrations of submicron particles. This was verified previously in unreported measurements made by Anderson, Blumenthal, and Sem (1977) with an instrumented aircraft. As the instrumented aircraft was taking off, a bag sample was obtained from the visible plume about 2 minutes after a commercial jet had used the same runway. The resulting distribution is shown in Figure 9. The jet plume distribution obtained may not represent a single jet exhaust because earlier airport traffic and the surrounding pollution interfered, and dilution and aging of the aerosol changed its size distribution from the initial condition.

Although the plume measurement technique may be sufficient for some jet engine exhaust studies, a batch sampling system similar to the system described in Figure 8 may yield more detailed information about the emissions during various modes of operation of the engine. With this system, the investigator could obtain experimental information for the evaluation of methods of reducing exhaust emissions. One could also study changes in the size distribution as the aerosol ages.

Dehne et al. (1975) have constructed a sampling system for measuring jet engine exhaust with an EAA. No details of the system have been released.

Combustion Aerosol

Combustion processes, such as internal combustion engines and the production of heat for electrical power generation, industrial and home heating, and many other uses, are the primary sources of submicron atmospheric aerosols. Many

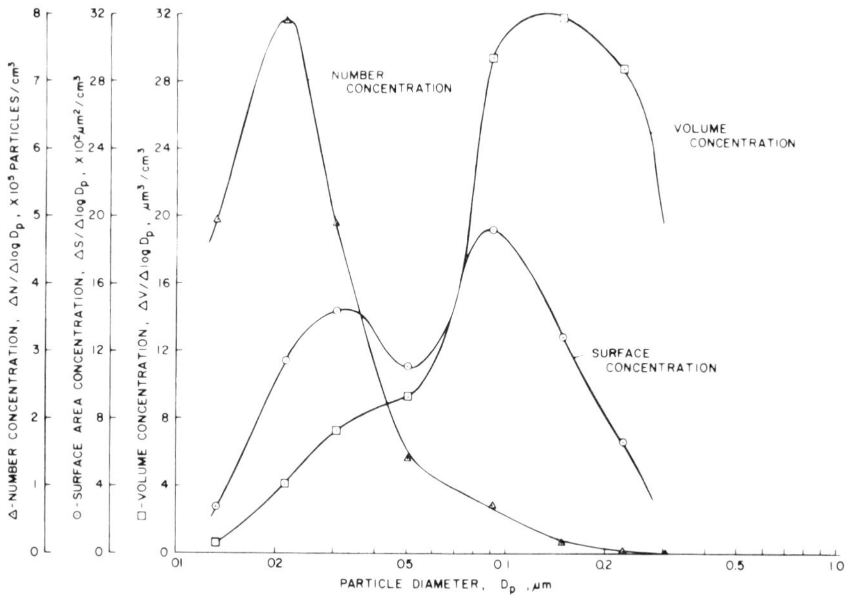

Figure 9. EAA size distribution of atmospheric jet aircraft plume sampled at runway two minutes after takeoff.

home and industrial fire detectors operate by detecting submicron particles in the 0.01 μm range. The often-deadly smoke produced by the burning of clothing, furniture fabrics, aircraft seat cushions, plastics, and other common materials consists primarily of very small submicron particles.

The initial aerosol produced by nearly all combustion processes contains particles almost entirely within the 0.002–0.05 μm range. These particles then grow, by condensation and coagulation, into larger particles at a growth rate determined by such factors as particle size and concentration. Measurement of the characteristics of these particles, especially size and concentration, and the determination of the growth rate and its characteristics, are important for a better understanding of the process of formation of such particles. This information may form a basis for the development of combustion processes which produce fewer particles and less of the resulting air pollution.

The first reported use of the Model 3000 analyzer for measurement of combustion aerosol was by Vomela and Whitby (1967) who burned a benzene-kerosene mixture in a cotton-wick lamp. Later, Husar (1971), in work also described by Whitby, Liu, Husar, and Barsic, (1972), reported the measurement of propane torch and cigarette aerosols using a Model 3000 analyzer. Hanson (1974) measured propane torch aerosol with a Model 3000. Fissan, Kittelson, and Whitby (1972) and Barsic (1977) measured propane-air aerosols while controlling the oxidation ratio, fuel flow rate, initial particle concentration of the air, and the humidity of the aerosol. Boffa and Pfender (1973) and Boffa, Mazza, and Rosso (1975) measured aerosols produced by a transpiration-cooled anode operated in a high intensity electric arc. Mulholland, Lee, and Baum (1976) studied the coagulation of several combustion smokes. McMahon and Ryan (1976) and Benner et al.

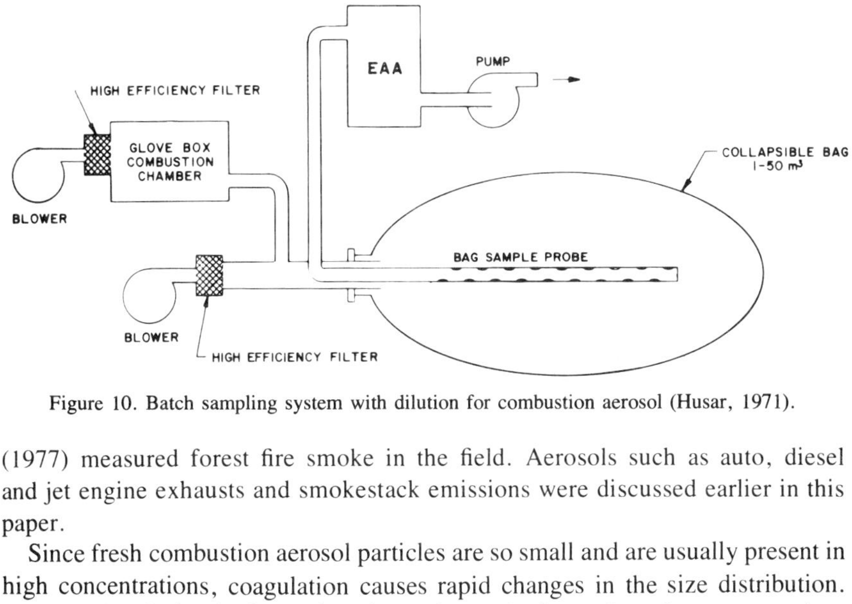

Figure 10. Batch sampling system with dilution for combustion aerosol (Husar, 1971).

(1977) measured forest fire smoke in the field. Aerosols such as auto, diesel and jet engine exhausts and smokestack emissions were discussed earlier in this paper.

Since fresh combustion aerosol particles are so small and are usually present in high concentrations, coagulation causes rapid changes in the size distribution. Thus, rapid dilution by large air volumes is required to reduce the concentration, thereby slowing the coagulation rate to a reasonable level.

Some combustion processes are largely uncontrolled resulting in large variations in particle concentration. For this reason, it is usually necessary to use a bag sampling system to provide the EAA with a relatively stable concentration during its sampling cycle. The bag system can also provide enough aerosol for repeated size distribution measurements for the study of the coagulation process.

Figure 10, adapted from Husar (1971) and also used by Hanson (1974), shows a convenient bag sampling system with high dilution capability. It can easily be adapted to a large variety of combustion processes by placing the combustion device in the glove box chamber. Other sampling systems are described by the other investigators cited above. All contain dilution capability and most provide for batch sampling. The details of these systems are described in the references and will not be repeated here.

Fractional Efficiency of Filters and Precipitators

As it becomes important to remove submicron particles from the air to which we are exposed, the fractional efficiency of fabric filters and electrostatic precipitators must be improved and measured. The EAA can measure the submicron fractional efficiency of moderate efficiency devices such as the furnace filters and electrostatic precipitators used in homes. As shown in Figure 11, the investigator installs the filter or precipitator in a suitable test duct or wind tunnel and seeds the upstream end of the test duct with the desired aerosol. With an analyzer, he then measures the particle concentration upstream and downstream of the filter. The

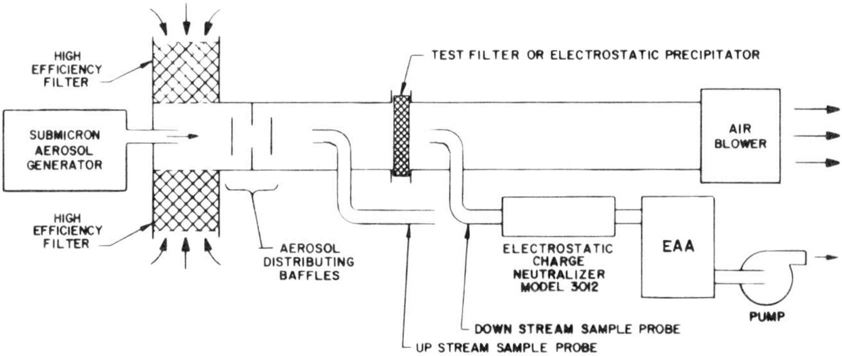

Figure 11. Aerosol test duct system for submicron fractional efficiency measurement of filters and electrostatic precipitators with moderate (less than 98%) efficiency.

particle number concentration ratio of particles downstream to particles upstream within any size range is the fractional penetration $P_{n,i}$ of the size range i. The number concentration fractional efficiency of that size range $E_{n,i}$ is equal to $(1 - P_{n,i})$. The volume concentration fractional efficiency can easily be calculated from $E_{n,i}$. By assuming that particle density is constant, the volume concentration fractional efficiency is equal to the mass concentration fractional efficiency.

It is important to use an electrostatic charge neutralizer ahead of the EAA, especially when measuring freshly generated aerosol and when measuring downstream of an electrostatic precipitator. The aerosol concentration must remain constant, within ± 10%, during both the upstream and downstream sizing cycles. The particle concentration level must be within a range where the EAA is not overloaded upstream of the filter and is able to detect the aerosol downstream of the filter. If the concentration range is chosen correctly and the ratio of upstream to downstream concentration does not exceed 100, the EAA will measure both upstream and downstream accurately.

The upper concentration limit for the EAA depends on the size distribution. If the distribution corresponds with a "typical" urban atmospheric aerosol, the EAA can accurately measure aerosols with a total mass concentration above 1,000 micrograms per cubic meter. For devices with efficiencies approaching 99%, dilution of the upstream aerosol just before it enters the EAA may be necessary to prevent overloading of the EAA while maintaining a detectable concentration downstream of the device as is described above for smokestack control device evaluation.

Powders and Liquid Suspensions

The particle size distribution of a powder, suspension, or slurry can be an important factor in characterizing or predicting certain aspects of a process or its resulting product. For example, the size of the particles may determine a chemical reaction rate or the viscosity of a slurry. The submicron particle size distribution of carbon black used in tire manufacturing is an important factor in determining the long term wear characteristics of the resulting tire. In some such cases. careful control of particle size is necessary for the economical production of a quality product.

In some cases, it is possible to aerosolize the particles in such a way that the aerosol particle size distribution is similar to the original powder, slurry, or suspension particle size distribution. In the case of carbon black and silicon dioxide, this technique has been demonstrated to be highly sensitive to small changes in particle size distribution in the 0.02–0.5 μm range.

The key to this technique is the quality of the aerosolization process. One acceptable aerosol generator is a constant output air-jet atomizer such as TSI Model 3076.

The complete aerosol generation procedure is as follows. The original powder or slurry is first dispersed into a suitable liquid suspension. The concentration of particles in the suspension must be dilute enough so that the chance of 2 or more particles in each atomized droplet is slight, but concentrated enough so that the resulting aerosol is detectable by the EAA. The ratio of suspension liquid to particles should be about 10,000:1 to meet these two requirements. The particles must be well dispersed. This can be done by using a suitable dispersing agent and a high energy ultrasonic probe (50–200 watt output at about 20 kHz; e.g., Branson Sonifier available from Heat Systems Co., Plainview, N.Y.) for several minutes. It may be necessary to cool the suspension during sonification by placing the suspension vessel in a cool bath. Then place the suspension in the open reservoir of the freshly cleaned atomizer, continue sonification, and begin the sizing run immediately.

Figure 12 shows the equipment setup for this application. The atomized aerosol passes through a drying chamber where the aerosol is mixed with about 55 ℓpm of clean dry air and through an electrostatic neutralizer to prevent excessive particle charge. The drying chamber should be at least 10 or 15 ℓ in capacity to damp out rapid aerosol concentration fluctuations. The aerosol is then fed to the EAA with any excess aerosol bypassed and disposed.

It is very important to prevent any contamination of the suspension and the aerosol. One must be extremely careful when cleaning the parts of the system in contact with the suspension. The suspending liquid must be as pure as possible. To illustrate the importance of suspension purity, assume we have a suspension with 40 ppm solid impurities, a typical value for "clean" distilled water or reagent grade alcohol. Every 1 μm droplet exiting the atomizer will have 40 ppm of solids in it, in addition to a possible particle of interest. Droplets without the particle of interest will dry up leaving only the solid impurity residue. The volume of the resulting solid particle (assuming it is spherical) will be 0.00004 times the volume of the 1 μm droplet. Thus, the diameter of the resulting solid residue particle is 0.034 μm. If the particles of primary interest are 0.05–0.1 μm, one must be careful not to "see" the residue particles as part of the primary size distribution (Whitby and Liu, 1968).

Thus, use only fresh, clean suspension liquid. Even distilled water will dissolve enough glass from a glass storage jar in one month to result in a few ppm solid impurities. Clean the hardware in contact with the suspension carefully. Do not delay the experiment after sonifying the suspension. Continue sonification during the run. Run only short, 5–10 minute, sizing runs. If the atomizer operates longer, more residue particles from the container will appear. Finally, be sure the aerosol is completely dried and neutralized before it reaches the EAA. Neutralize after drying for most effective neutralization.

Laboratory Aerosols

Many applications exist for the EAA in addition to the above. A research scientist or engineer may wish to characterize the aerosol produced by his specially-built aerosol generator. He may be interested in the size distribution of aerosol produced by an electric motor, a hot bearing, a soldering iron, or a welder. He may be interested in characterizing aerosols in the home such as smoke from a toaster, from a pop corn popper, from frying hamburgers, from vacuuming a carpet, or from smoking a cigarette. He may be interested in the difference between indoor and outdoor aerosols. He may be interested in the difference between aerosols which enter and leave the human respiratory system. He may wish to characterize or monitor changes in his laboratory room aerosol which he is using as a "standard" for testing a new air cleaner. He may wish to measure the size distribution of a new type of room deodorant produced by an aerosol spray can. He may wish to characterize oil mist aerosol from a compressor and study the effects of mist reduction techniques as Feldman (1972) did with a Model 3000 analyzer. He may wish to study humidity effects on aerosols as did Hanson (1974). He may wish to measure the residue size distribution of an atomizer as did Whitby and Liu (1968).

Countless such applications exist. This brief survey cannot describe all of them. It is hoped that the applications which are described will help the person who is planning his own particle size measurement program.

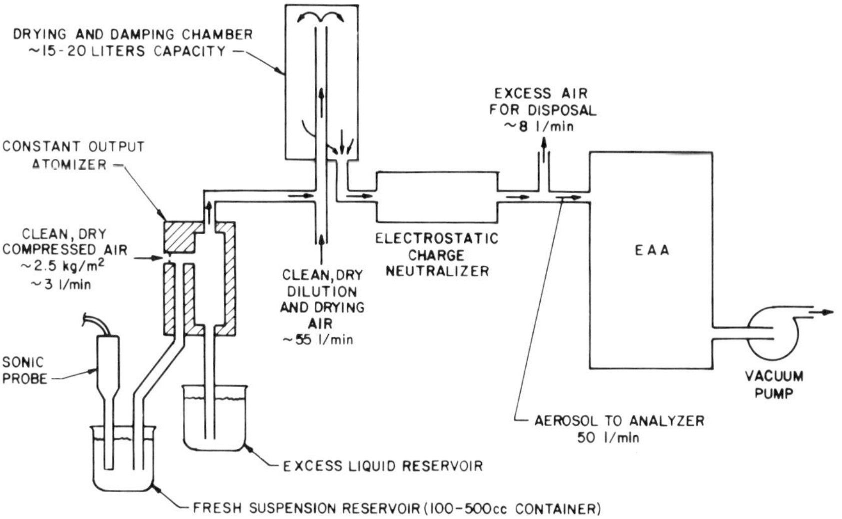

Figure 12. Aerosol generation and sampling system for particle size measurement of liquid suspensions using an air-jet atomizer.

Listed below are all known references discussing EAA theory, design, calibration, operation, and data. References before 1965 generally discuss electrostatic aerosol charging theory and/or its early application to aerosol size measurement. References after 1965 generally refer more specifically to the EAA introduced in 1973 or its predecessor, the Model 3000, introduced in 1966.

REFERENCES

Alkezweeny, A. J., 1974: Airborne laboratory for air pollution study. Battelle Pacific Northwest Laboratory Report No. BNWL-1850 PT3 UC-11, prepared for the U.S. Atomic Energy Commission under Contract AT(45-1):1830, April.

——— and L. F. Radke, 1973: Airborne measurements of the size distribution of AgI particles. Battelle Pacific Northwest Laboratory Report No. BNWL-1751 PT1 UC-53, prepared for the U.S. Atomic Energy Commission under Contract (AT45-1):1830, April.

——— and W. G. N. Slinn, 1972: Aerosol particle size distributions in the northwest. Battelle Pacific Northwest Laboratory Report No. BNWL-1651 PT1 UC-53, prepared for the U.S. Atomic Energy Commission under Contract AT(45-1):1830, December.

Anderson, J. A., D. L. Blumenthal, and G. J. Sem, 1977: Characterization of Denver's urban plume using an instrumented aircraft. Proceedings of the Denver air pollution study-1973, Volume II, EPA Report No. EPA-600/9-77-001, J. L. Durham, Project Officer.

Barsic, N. J., 1977: Size distributions and concentrations of fine particles produced by propane-air combustion in a controlled humidity environment. Ph.D. Dissertation, University of Minnesota.

Benner, W. H., P. Urone, C. K. McMahon, and P. Ryan, 1977: Photochemical potential of forest fire smoke. Presented at the 70th Annual Meeting of the Air Pollution Control Association, Toronto, Ontario, June 20–24, Paper No. 77-5.3.

Boffa, C. V. and E. Pfender, 1973: Controlled generation of monodisperse aerosols in the submicron range. *Aerosol Sci., 4,* 103.

———, A. Mazza and D. M. Rosso, 1975: Generation of monodisperse submicron aerosols by ablation from transpiration-cooled porous matrices. *Fine Particles. Aerosol Generation, Measurement, Sampling, and Analysis,* B. Y. H. Liu, Ed., Academic Press, 111–127.

Carlson, T. E. and M. L. Jackson, 1977: Conversion of SO_2 to particulate for fluidized bed collection. Proceedings of the Fourth International Clean Air Congress, Tokyo, Japan, May 16–20, pp. 739–742.

Clark, W. E., 1972: Measurement of aerosols produced by the photochemical oxidation of SO_2 in air. Ph.D. Dissertation, University of Minnesota.

——— and K. T. Whitby, 1967: Concentration and size distribution measurements of atmospheric aerosols and a test of the theory of self-preserving size distributions. *J. Atmos. Sci., 24,* 677.

——— and ———, 1975: Measurements of aerosols produced by photochemical oxidation of SO_2 in air. *J. Colloid and Interface Sci., 51.*

———, D. A. Landis and A. B. Harker, 1976: Measurements of the photochemical production of aerosols in ambient air near a freeway for a range of SO_2 concentrations. *Atmos. Environ., 10,* 637.

Dehne, H. et al., 1975: Personal communication. Aerotherm Division, Acurex Corp., Mountain View, Calif., April.

Durham, J. L., W. E. Wilson, T. G. Ellestad, K. Willeke and K. T. Whitby, 1975: Comparison of volume and mass distributions for Denver aerosols. *Atmos. Environ., 9,* 717.

Ensor, D. S., R. G. Hooper, and R. W. Scheck, 1976: Determination of the fractional efficiency, opacity characteristics, engineering and economic aspects of a fabric filter operating on a utility boiler. Report No. EPRI FP-297, Palo Alto, Calif., Robert Carr, Proj. Manager.

———, ———, R. C. Carr, and R. W. Scheck, 1976: Evaluation of a fabric filter on a spreader stocker utility boiler. Presented at the 69th Annual Meeting of the Air Pollution Control Association, Portland, Oregon, June 27–July 1, Paper No. 76-27.6.

———, ———, R. W. Scheck and R. C. Carr, 1976: Performance and engineering evaluation of the Nucla baghouse. Presented at the Symposium on Particulate Control in Energy Processes, San Francisco, California, May 11–13.

Feldman, P. L., 1972: Characterization of an aerosol formed by nucleation in a flow system. Ph.D. Dissertation, Pennsylvania State University, University Park, Pennsylvania, September.

Fissan, H. J., C. Felsper, and H. Franzen, 1976: Bestimmung der grossenverteilung von aerosolen aus verbrennungsvorgangen mit einem elektrostatischen aerosol analysator. Proceedings of 4th Tagung der Gesellschaft fur Aerosolforschung, Bad Soden/Ts., W. Germany.

———, D. B. Kittelson and K. T. Whitby, 1972: Measurement of aerosols produced by a propane-air flame in a controlled environment. Report PTL-190, Particle Technology Laboratory, Mechanical Engineering Department, Minneapolis, Minnesota 55455. Prepared for EPA under Research Grant No. R-801175, August.

Foster, J. F., D. A. Trayser, E. R. Blosser, F. A. Creswick, D. F. Miller, A. Levy, R. I. Mitchell and J. H. Oxley, 1975: Chemical and physical characterization of automotive exhaust particulate matter in the atmosphere. Final Summary Report to the Coordinating Research Council (CAPE-19-70), Prepared by Battelle-Columbus Laboratories, August.

Fox, D. L., J. E. Sickles, M. R. Kuhlman, P. C. Reist and W. E. Wilson, Jr., 1975: Design and operating parameters for a large ambient aerosol chamber. *J. Air Pollution Control Assoc., 25,* 1049.

Groblicki, P. J., 1976: General Motors sulfate dispersion experiment: aerosol sizing measurements. General Motors Research Publication No. GMR-2127, EV #28; G.M. Technical Center, Warren, Mich.

Habibi, K., 1970: Characterization of particulate lead in vehicle exhaust—experimental technique. *Environ. Sci. Technol., 4,* 239.

Hanson, H. A., 1974: The deliquescent properties of aerosols. Ph.D. Dissertation, University of Minnesota.

Harker, A. B., L. W. Richards, and W. E. Clark, 1977: The effect of atmospheric SO_2 photochemistry upon observed nitrate concentration in aerosols. *Atmos. Environ., 11,* 87.

Hersh, S., J. F. Hurley, and R. C. Carr, 1976: The effects of smoke suppressant additives on the particulate emissions from a utility gas turbine. Presented at the 69th Annual Meeting of the Air Pollution Control Association, Portland, Oregon, June 27–July 1, Paper No. 76-8.1.

Hidy, G. M., et al., 1975a: Characterization of aerosols in California (ACHEX). Final report of ARB Contract No. 358, Air Resources Board, State of California, Sacramento, California.

———, et al., 1975b: Summary of the California aerosol characterization experiment. *J. Air Pollution Control Assoc., 25,* 1106.

Hindmann, E. E., II, P. V. Hobbs and L. F. Radke, 1977: Airborne investigations of aerosol particles from a paper mill. *J. Air Pollution Control Assoc., 27,* 224.
Hobbs, P. V., L. F. Radke and E. E. Hindmann, II, 1974: Preliminary results from a new airborne aerosol sampling system. *Trans. Amer. Geophys. Union, 56,* 1127.
———, ——— and ———, 1976: An integrated airborne particle measuring facility and its preliminary use in atmospheric aerosol studies. *Aerosol Sci., 7,* 195.
Husar, R. B., 1971: Coagulation of Knudsen aerosols. Ph.D. Dissertation, University of Minnesota.
——— and K. T. Whitby, 1971: The growth rate and size spectrum of photochemical aerosols. Presented at the Conference on Science in the Control of Smog, California Institute of Technology, Pasadena, California, November 15–16.
——— and ———, 1973: Growth mechanisms and size spectra of photochemical aerosols. *Environ. Sci. Technol., 7,* 241.
———, ——— and B. Y. H. Liu, 1972: Physical mechanisms governing the dynamics of Los Angeles smog aerosol. *J. Colloid and Interface Sci., 39,* 211.
Knutson, E. O., 1976: Comparative measurements of New York City Aitkin particles with a diffusion battery and an electrical aerosol analyzer. HASL Technical Memorandum 76-13, December 29, Health and Safety Laboratory, Energy Research and Development Administration, New York, N.Y.
Kocmond, W. C., D. B. Kittelson, J. Y. Yang and K. L. Demerjian, 1973: Determination of the formation mechanisms and composition of photochemical aerosols. Report No. EPA-650/3-73-002, prepared by Calspan Corporation for EPA under Contract No. 68-02-0557, M. Bufalini, Project Officer, December.
———, ———, ——— and ———, 1975: Study of aerosol formation in photochemical air pollution. Report No. EPA-650/3-75-007, prepared by Calspan Corporation for EPA and Coordinating Research Council, August.
——— and J. Y. Yang, 1975: A smog chamber study of SO_2 photo-oxidation rates and aerosol formation mechanism. Calspan Report No. NA5365-M-3, prepared for EPA, October, 1975.
——— and ———, 1976: Sulfur dioxide photo-oxidation rates and aerosol formation mechanisms, a smog chamber study. EPA Report No. EPA-600/3-76-090, B. Dimitriades, Project Officer.
———, ———, ———, ——— and K. T. Whitby, 1976: Aerosol formation in simple photochemical systems. *Fate of Pollutants,* I. H. Suffet, Ed., Wiley.
Liu, B. Y. H. and D. Y. H. Pui, 1974a: Electrical neutralization of aerosols. *Aerosol Sci., 5,* 465.
——— and ———, 1974b: Equilibrium bipolar charge distribution of aerosols. *J. Colloid and Interface Sci., 49,* 305.
——— and ———, 1975: On the performance of the Electrical Aerosol Analyzer. *Aerosol Sci. 6,* 249.
——— and ———, 1978: Comments on the paper aspects of the performance of the electrical aerosol analyzer under nonideal conditions; by W. H. Marlow, P. C. Reist, and G. A. Dwiggins. Submitted to *Aerosol Sci.*
———, ——— and A. Kapadia, 1976: Electrical Aerosol Analyzer: history, principle, and data reduction. Presented at the Aerosol Measurement Workshop, University of Florida, Gainesville, Florida, March 24–26.
———, K. T. Whitby and D. Y. H. Pui, 1974: A portable electrical analyzer for size distribution measurement of submicron aerosols. *J. Air Pollution Control Assoc., 24,* 1067.
Marlow, W. H., P. C. Reist, and G. A. Dwiggins, 1976: Aspects of the performance of the electrical aerosol analyzer under nonideal conditions. *Aerosol Sci., 7,* 457.
McCain, J. D., 1975: Evaluation of centrifield scrubber. Report EPA-650/2-74-129-a, prepared by Southern Research Institute for EPA Contract No. 68-02-1480, June.
———, K. M. Cushing and W. B. Smith, 1974: Methods of determining particulate mass and size properties: laboratory and field measurements. *J. Air Pollution Control Assoc., 24,* 1172.
McMahon, C. K. and P. W. Ryan, 1976: Some chemical and physical characteristics of emissions from forest fires. Presented at the 69th Annual Meeting, Air Pollution Control Association, Portland, Oregon, June 27–July 1, Paper No. 76-2.3.
McNelis, D. N., 1974: Aerosol formation from gas phase reactions of ozone and olefin in the presence of sulfur dioxide. Report No. EPA-650/4-74-034, W. E. Wilson, Project Officer.
McNelis, D. N., L. A. Ripperton, and W. E. Wilson, 1975: Formation mechanism and growth characterization of aerosol formed in the propylene-ozone-sulfur dioxide thermal reaction. Submitted for publication.
Miller, D. F., and A. Levy, 1975: Exhaust hydrocarbon relationships with photochemical aerosol

formation. Paper 75-16.3, presented at the 68th Annual Meeting, Air Pollution Control Association, Boston, Massachusetts, June 15–20.

———, D. A. Trayser and D. W. Joseph, 1976: Size characterization of sulfuric acid aerosol emissions. Presented at the SAE Automotive Engineering Congress, Detroit, Michigan, February.

Miller, D. F., A. Levy, D. Y. H. Pui, K. T. Whitby, and W. E. Wilson, 1976: Combustion and photochemical aerosols attributable to automobiles. *J. Air Pollution Control Assoc.*, *26*, 576.

Mulholland, G., T. Lee and H. Baum, 1976: The effect of the initial size distribution on the coagulation of aerosols. NBS, Gaithersburg, Maryland. Submitted for publication.

Paulus, H. J. and C. M. Peterson, 1969: Urban aerosol: count size related to meteorologic data. School of Public Health, University of Minnesota, Minneapolis, Minnesota 55455. Final Report, Research Grant AP 00453-03, November.

Peterson, C. M., 1969: Aerosol and meteorological sampling of an urban atmosphere to determine particle dispersion characteristics. Ph.D. Dissertation, University of Minnesota.

——— and H. J. Paulus, 1968: Continuous monitoring of aerosols over the 0.001 to 10 micron spectrum. *J. Amer. Indust. Hygiene Assoc.*, *29*, 111.

———, ——— and G. H. Foley, 1969: The number-size distribution of atmospheric particles during temperature inversions. *J. Air Pollution Control Assoc.*, *19*, 795.

Pui, D. Y. H., 1976: Experimental study of diffusion charging of aerosols. Ph.D. Dissertation, University of Minnesota.

——— and B. Y. H. Liu, 1976: Electrical Aerosol Analyzer: calibration and performance. Presented at the Aerosol Measurement Workshop, University of Florida, Gainesville, Florida, March 24–26.

Radke, L. F., J. L. Stith, D. A. Hegg, and P. V. Hobbs, 1978: Airborne studies of particles and gases from forest fires. *J. Air Pollution Control Assoc.*, *28*, 30.

Ragland, J. W., D. H. Pontius and W. B. Smith, 1976: Personal communication. Southern Research Institute, Birmingham, Alabama 35205, March.

Schmidt, E. W., J. A. Geiseke and J. M. Allen, 1976: Characterization of fine particulate emissions from a coal-fired power plant. Battelle Columbus Laboratories, Columbus, Ohio. To be submitted to *Atmos. Environ.*

Sem, G. J., 1974: Design of the EPA semi-trailer mobile air pollution laboratory. Report EPA-650/4-74-017, Prepared by Thermo-Systems Inc. for EPA under Contract No. 68-02-0654, J. L. Durham, Project Officer, September.

———, 1975a: Submicron particle size measurement of stack emissions using the electrical mobility technique. Presented at the Workshop on Sampling, Analysis and Monitoring of Stack Emissions, Dallas, Texas, October 2–3. Sponsored by Electric Power Research Institute, Palo Alto, California.

———, 1975b: Submicron particle sizing experience on a smoke stack using the Electrical Aerosol Size Analyzer. Presented at the Seminar on In-Stack Particle Sizing for Particulate Control Device Evaluations, EPA, Research Triangle Park, North Carolina, December 3–4.

———, 1975c: Design and application of an electrical size analyzer for submicron aerosol particles. *Analysis Instrumentation,* Vol. 13, Proceedings of the 21st Annual ISA Analysis Instrumentation Symposium, Instrument Society of America, 400 Stanwix St., Pittsburgh, Pennsylvania 15222.

———, J. A. Borgos, K. T. Whitby and B. Y. H. Liu, 1972: State-of-the-art: 1971, Instrumentation for measurement of particulate emission from combustion sources, Volume III: Particle size. Prepared by Thermo-Systems Inc. under EPA Contract CPA 70-23, John Burckle, Project Officer.

Smith, W. B., K. M. Cushing and J. D. McCain, 1974: Particulate sizing techniques for control device evaluation. Report No. EPA-650/2-74-102, prepared by Southern Research Institute for EPA under Contract No. 68-02-0273, July.

Strand, L. and J. Varsi, 1974: Personal communication. Jet Propulsion Laboratory, California Institute of Technology, 4800 Oak Grove Drive, Pasadena, California 91103.

Sverdrup, G. M., K. T. Whitby and W. E. Clark, 1975: Characterization of California Aerosols—II. Aerosol size distribution measurements in the Mojave Desert. *Atmos. Environ.*, *9*, 483.

——— and ———, 1977: Determination of submicron atmospheric aerosol size distributions by use of continuous analog sensors. *Environ. Sci. Technol.*, *11*, 1171.

———, 1977: Parametric measurement of submicron atmospheric aerosol size distributions. Ph.D. Dissertation, University of Minnesota.

Vomela, R. A. and K. T. Whitby, 1967: The charging and mobility of chain aggregate smoke particles. *J. Colloid and Interface Sci.*, *25*, 568.

Whitby, K. T. and B. Y. H. Liu, 1968: Polystyrene aerosols—electrical charge and residue size distribution. *Atmos. Environ.*, *2*, 103.

——— (Principal Investigator), et al., 1970: Aerosol measurement in Los Angeles smog. Publication PTL-141, Particle Technology Laboratory, Mechanical Engineering Department, University of Minnesota. Report prepared for National Air Pollution Control Administration under Research Grant AP-00839-01.

———, B. Y. H. Liu, R. B. Husar and N. J. Barsic, 1972: The Minnesota aerosol analyzing system used in the Los Angeles smog project. *J. Colloid and Interface Sci.*, *39*, 136.

———, R. B. Husar, and B. Y. H. Liu, 1972: The aerosol size distribution of Los Angeles smog. *J. Colloid and Interface Sci.*, *39*, 177.

———, ——— and A. K. Rao, 1973: Size distribution of non-urban Colorado aerosols. Publication No. PTL-173, Particle Technology Laboratory, Mechanical Engineering Department, University of Minnesota, Minneapolis, Minnesota.

———, W. E. Clark, V. A. Marple, G. M. Sverdrup, G. J. Sem, K. Willeke, B. Y. H. Liu and D. Y. H. Pui, 1975: Characterization of California aerosols - I. Size distribution of freeway aerosols. *Atmos. Environ.*, *9*, 463.

———, B. K. Cantrell, R. B. Husar, N. V. Gillani, J. A. Anderson, D. L. Blumenthal and W. E. Wilson, Jr., 1976; Aerosol formation in a coal-fired power plant plume. Submitted to *Science*.

———, D. B. Kittelson, B. K. Cantrell, N. J. Barsic, D. F. Dolan, L. D. Tarvestad, D. J. Nieken, J. L. Wolf, and J. R. Wood, 1976: Aerosol size distributions and concentrations measured during the General Motors proving grounds sulfate study. Contained in EPA Report No. EPA-600/3-76-035, R. K. Stevens, et al., editors.

Willeke, K. and K. T. Whitby, 1974: Physical characteristics of Denver-area aerosols. Paper 74-262, Presented at the 67th Annual Meeting, Air Pollution Control Association, Denver, Colorado, June 9-13.

———, ———, W. E. Clark and V. A. Marple, 1974: Size distributions of Denver aerosols - a comparison of two sites. *Atmos. Environ.*, *8*, 609.

——— and ———, 1975: Atmospheric aerosols: size distribution interpretation. *J. Air Pollution Control Assoc.*, *25*, 529.

Wilson, W. E., L. L. Spiller, T. G. Ellestad, P. J. Lamothe, T. G. Dzubay, R. K. Stevens, E. S. Macias, R. A. Fletcher, J. D. Husar, R. B. Husar, K. T. Whitby, D. B. Kittelson, and B. K. Cantrell, 1977: General Motors sulfate dispersion experiment: summary of EPA measurements. *J. Air Pollution Control Assoc.*, *27*, 46.

The following articles discuss certain aspects of the equipment used in the primary calibration of the EAA:

Knutson, E. O., 1972: The distribution of electric charge among the particles of an artificially charged aerosol. Ph.D. Dissertation, University of Minnesota.

——— and K. T. Whitby, 1975a: Aerosol classification by electric mobility: apparatus, theory, and applications. *Aerosol Sci.*, *6*, 443.

——— and ———, 1975b: Accurate measurement of aerosol electric mobility moments. *Aerosol Sci.*, *6*, 453.

Liu, B. Y. H., 1974: Laboratory generation of particulates with emphasis on submicron aerosols. *J. Air Pollution Control Assoc.*, *24*, 1170.

——— and D. Y. H. Pui, 1974: A submicron aerosol standard and the primary absolute calibration of the Condensation Nuclei Counter. *J. Colloid and Interface Sci.*, *47*, 155.

———, ———, A. W. Hogan and T. A. Rich, 1975: Calibration of the Pollak counter with monodisperse aerosols. *J. Appl. Meteor.*, *14*, 46.

The following articles discuss early designs of the EAA and its charger:

Hewitt, G. W., 1957: The charging of small particles for electrostatic precipitation. *Trans. Amer. Inst. Electrical Engr.*, *76*, 300.

Whitby, K. T., R. C. Jordan and C. M. Peterson, 1964: Development of a particle counter system and development of a technique for studying the charge on an evaporating drop. Publication No. PTL-80, Particle Technology Laboratory, Mechanical Engineering Department, University of Minnesota, Minneapolis, Minnesota 55455. Report to U.S. Public Health Service, June.

——— and W. E. Clark, 1966: Electrical aerosol particle counting and size distribution measuring system for the 0.015 to 1 micron size range. *Tellus*, *18*, 573.

Several important historical references on the electrical measurement of aerosol particles include:

Millikan, R. A., 1910: A new modification of the cloud method of determining the elementary electrical charge and the most probable value of that charge. *Philosophical Mag. and J. of Science,* Series 6, *19,* 1.

Thomson, J. J., 1898: On the charge of electricity carried by the ions produced by Roentgen rays. *Philosophical Mag. and J. of Science,* Series 5, *46,* 528.

Wilson, H. A., 1903: A determination of the charge on the ions produced in air by Roentgen rays. *Philosophical Mag. and J. of Science,* Series 6, *5,* 429.

QUESTIONS

Q: I would like to ask two questions. First, can you tell us the minimum number of particles required for each size channel in order for an aerosol to be detected? The second question is, if you are using the EAA to monitor aerosols outside where the relative humidity is high and the particles in the air are largely solution droplets, does the instrument measure the size of the particles that are actually floating around in the air or do they get dried down to their dry size before they are measured by the instrument?

Liu: The first question is very difficult and the answer depends upon the instrument sensitivity for the particular size range of interest. The EAA is limited primarily by the noise of the electrometer, about 1.0×10^{-15} amperes. If you divide that number by the instrument sensitivity for each channel, you get a minimum concentration that you can detect. As I remember, at the upper instrument limit of 1 μm diameter it is about 5 particles per cc and at the lower instrument limit of about .005 μm diameter it is about 200,000 particles per cc. As far as the effect of humidity on particles, what you measure is the size of the particle as it exists at the instrument. If you have a water droplet at the point of measurement, you would measure the water droplet. If you have a dry particle at the point of measurement, you would measure the dry particle.

Q: Are the particles broken down by drying them out as they are drawn from the outside into the EAA?

Liu: Well, that would depend upon the temperature history of the sampling process. If you start off with a wet condition and if the temperature is higher in the room containing the EAA than it is outside, you would dry out the particle. On the other hand, if the temperature of the room is lower than the outside, as in summertime, then you may cause some condensation on the particle. So, it depends entirely on the sampling conditions.

Q: If you want to determine a size distribution using the electrical aerosol analyzer, is it necessary that all the particles be charged with only one elementary charge or at least have a charge distribution which is narrow compared to the size distribution of the aerosol?

Liu: You want the charge distribution to be as narrow as possible, but you see in work that Dr. Pui presented that you do have a spread of the charges over several elementary units. This is why data reduction procedure becomes very important. You have the same problem with the diffusion battery. You get the penetration as a function of particle size and you somehow have to invert those readings to get the particle size distribution. That is why the data reduction procedures are a very important process. I don't think we have a complete answer on the data reduction procedure yet, but I think we are getting there. At least it's much better now than it was a year ago.

Q: In reference to the article which appeared in *Journal of Applied Meteorology* entitled "Calibration of the Pollak Counter with Monodisperse Aerosols," by Liu, Pui, Hogan and Rich, did you compare the Nolan-Pollak counter with the electrical aerosol analyzer? Also, in your comparison of the two instruments, you measured higher concentrations with your analyzer than indicated by the Nolan-Pollak counter. Do you have an explanation for this discrepancy?

Liu: I don't think we compared the electrical aerosol analyzer with the Nolan-Pollak counter; what we compared was a differential mobility analyzer with the counter. We call the differential mobility analyzer the Minnesota Stripper, a name coined by David Sinclair. It is a device which puts a single charge on a particle. With a single charge on a particle you can measure the concentration with a high degree of accuracy because all you need to do is measure the current flow. You can then calculate the concentration within about 5% accuracy. So our comparison was based on the Nolan-Pollak counter and the differential mobility analyzer, or Minnesota Stripper as we call it. I really don't know the reason why it differs. The discrepancy on larger particles and lower concentrations could be partly our fault and could be partly the Nolan-Pollak calibration. I think one problem with the Nolan-Pollak counter is that when you get down to really low concentrations, statistical fluctuations become significant. There were also problems of not being able to get very low concentrations. If we get under 200 particles per cc it becomes very difficult to filter out the error completely. We were very happy to get agreement within 17% at the upper end. You know Nolan and Pollak did their work twenty some years ago and they counted individual particles with an Aitken pocket counter and a dilution bridge. We were overjoyed that we were using completely different methods and were able to get agreement within 18%.

Hogan: Every experiment that I have ever done seems to be limited to an accuracy of about ± 15% when you try to move aerosols around. I think once you get down to that degree of accuracy the problem lies with the connections you make, the pumps you use, and everything else. You really cease to calibrate the instruments when you get to that level and you are really calibrating your complete experimental system.

Q: What is an explanation for the consistently higher counts measured by the differential mobility analyzer?

Liu: That could be a systematic error in the case of the Nolan-Pollak counter and dilution bridge, which divided the concentration by a factor of two repeatedly. I think the cumulative error would be different at high concentrations than at low concentrations so that the disagreement between the counters could be a reflection of the cumulative errors in the dilution procedure, but I'm not sure. I still think we are not quite at a point where we can talk about five to ten percent agreements in aerosol measurements, although I, personally, am somewhat biased. I feel the differential mobility analyzer method is capable of giving you 5% absolute accuracy in concentration.

THE DIFFUSION CHARGING MOBILITY ANALYSIS HYPOTHESIS REVISITED

W. H. MARLOW

Department of Applied Science, Brookhaven National Laboratory, Upton, N.Y.

It is an established fact that in the atmosphere, free charges (electrons or ionized molecules or atoms) cluster available atoms and molecules about themselves in sequences of reactions that generally take well under 10 msec following the appearance of the unclustered charged species. Further, these "cluster ions" are extremely sensitive to the trace gases of the atmosphere so that their masses and mobilities are determined by their gas-phase environment. In all cases these ions exhibit a spectrum of physical characteristics rather than single values characterizing one ion. Thus, a change in atmospheric trace gases will shift the mass and mobility spectrum rather than cause the disappearance of a single species and the emergence of another, different, single species of ion. It follows that any process, such as aerosol charging, which is dependent in detail upon cluster ion masses and mobilities is implicitly also dependent upon atmospheric trace gas composition.

Aerosol charging itself is similarly a process whose application must be viewed with circumspection. In fact, it was my experiences of calculating time dependent, bipolar, polydisperse aerosol charging (Marlow and Brock, 1975) and examining available data for the atmosphere which caused me to question the diffusion charging mobility analysis hypothesis three years ago. In those studies, I assumed that the positive and negative atmospheric cluster ions were of two different mobilities with one species of ion of each sign present. Under conditions of constant ion generation rate, I learned that simply changing the total number of particles without altering the number ratios among particles of different sizes actually would change the sign of the overall charge acquired by the smallest particles due to competition from the larger particles. Since larger particles are much more effective sinks for ions than smaller particles, the more mobile ions tend to attach more rapidly to larger particles leaving a preponderance of less mobile ions which become the dominant charging species for the smaller particles. From this point of view, removal of the more mobile ions forced an essential change in the charge character of the smallest aerosol particles.

With this general information as background, I will describe one interpretation of the results presented by Prof. Reist (1976). Because the ambiguities he has presented occurred for SO_2 on the 0.05 to 0.2 ppm range with no effect observed with NO_2 in the 0.1 to 0.8 ppm range, I am inclined to attribute these observations either to the addition of SO_2 to the NO_x-laden air from the charger corona or to the apparently greater tendency of SO_2 to participate in clustering with a positive ion compared to NO_2 (Kadlecek, 1974). Such a conclusion must be considered tentative because of the uncertainty in present knowledge of atmospheric NO_2- and SO_2-ion reactions. If the cause is related to the clustering of SO_2 in the presence of H_2O, positive ions in the aerosol charger and possibly NO_x should then shift the ion mobility spectrum toward lower mobilities (Bricard et al., 1972; Castleman et al., 1971). This may have several consequences; one would be to lower the ion current flowing across the aerosol stream. This has been noted by Liu and Pui (1975) in the case of alcohol as an accidental contaminant. A second effect would be differential charging of different size aerosol particles. Since unipolar ion capture by particles is a nonequilibrium process for the charging method under consideration here, it is critically dependent upon *both* particle size and ion mobility. If SO_2 removes the most mobile small ions which are necessarily most responsible for charging the smallest particles in this nonequilibrium process (as opposed to the steady state achieved in the bipolar situation just described), then only the larger particles stand any appreciable chance of becoming properly charged. Since these less-mobile small ions now are more numerous due to increased clustering, the larger particles will become more charged at the expense of the charging of the small ones. This picture would seem to be borne out by data (Figure 1). Because the smallest particles constitute only a small part of the number spectrum in this case, they do not remove a significant number of small ions from the vicinity of the larger particles. However, the addition of SO_2 does lower the mobility of the cluster ions which are the most mobile, and therefore most likely to charge the small particles, without noticeably increasing the large particles' charge.

Whether or not the hypothesized mechanism of variability in the cluster ion mobility spectrum is valid will be difficult to decide. There is some evidence for it, but the field of clustering is one which has considerably more questions than answers. The plausibility of this mechanism may be adduced from computational studies of time-dependent unipolar aerosol charging of species polydisperse in both aerosol and ion compositions which I am currently undertaking.

In addition, more rigorous experiments need to be done to determine whether the observed effect of SO_2 and water vapor is associated only with the 1 ℓpm flow of charger sheath air, or with the 4 ℓpm aerosol air as well.

If my picture of the relevant processes is correct, then it has the most serious consequences for aerosol sizing by all means which rely upon gas-phase electrical charging. The diffusion-charging mobility analysis hypothesis becomes irrelevant because the *implicit* assumptions of independence of the charging process from effects of aerosol and cluster ion polydispersities and variabilities therein is not valid. This is not to say the hypothesis is wrong; it says that charging depends upon uncontrollable variables. Any calibration of the instrument's performance

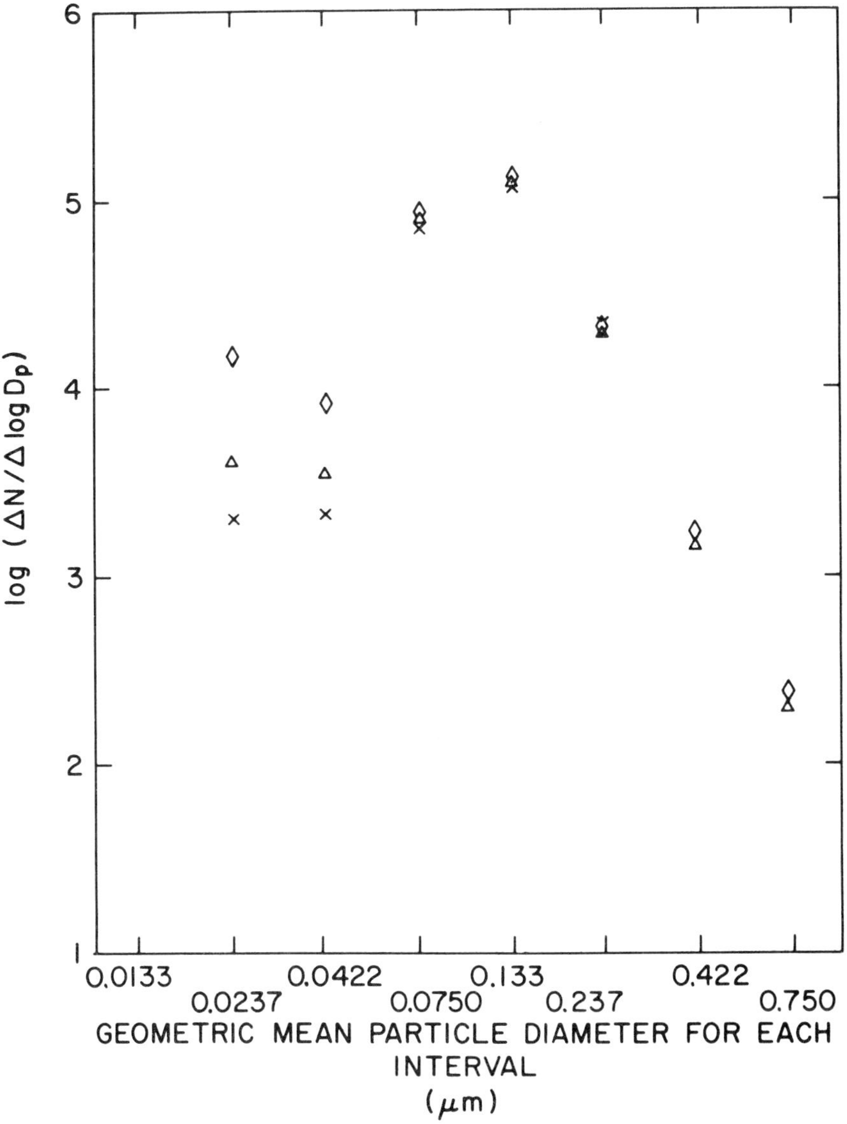

Figure 1. Comparison among measurements for an aerosol lacking large numbers of very small particles. ◇ and Δ are breathing air containing sheath air samples taken prior to and following a .17 ppm SO_2 containing sample (X on figure).

will be of little value under nonlaboratory conditions due to the indeterminate nature and concentrations of atmospheric trace gases that are an inseparable part of the aerosol.

REFERENCES

Bricard, J., M. Cabane, G. Madelaine and D. Vigla, 1972: Formation and properties of neutral ultrafine particles and small ions conditioned by gaseous impurities of the air. *J. Colloid and Interface Sci., 39,* 42.

Castleman, A. W., I. N. Tang and H. R. Munkelwitz, 1971: Clustering of sulfur dioxide and water vapor about oxonium and nitric acid ions. *Science, 173,* 1025.

Kadlecek, J. A., 1974: Publication Number 263. Atmospheric Sciences Research Center, State University of New York at Albany.

Liu, B. Y. H. and D. Y. H. Pui, 1975: On the performance of the Electrical Aerosol Analyzer. *J. Aerosol Sci., 6,* 244–264.

Marlow, W. H. and J. R. Brock, 1975: Calculations of bipolar aerosol charging. *J. Colloid and Interface Sci., 51,* 23–31.

Reist, P., 1976: Discussion, presented at Aerosol Measurement Workshop, University of Florida, Gainesville, Florida, March 24–26.

QUESTIONS

Liu: I would like to put forth an alternative hypothesis as to why the data presented by Dr. Reist appeared the way it did. We know that small particles are extremely unstable, especially those formed by photochemical reactions. Any-time you change the atmosphere around a particle, the particle itself can change; it can evaporate or be changed by additional material condensing on it. I think the data that you have presented could be explained equally well by an actual change in the particle itself rather than by a change in charging conditions. I think that the refractive index of the particles, and the mobility of the ions could account for perhaps 20 or 30% of the observed differences, but not much more than that. My feeling is that the aerosol particles could have changed in size and this is actually what was measured.

Sem: I'd also like to comment to anyone who is looking into this any further that the sheath air is a sheath air not only for the clean air core in the mobility analyzer, but also for the charger. You can change the sheath air at will, essen-tially, to make whatever composition you want.

THE REDUCTION OF DATA FROM THE ELECTRICAL AEROSOL ANALYZER

L. Willard Richards

Air Monitoring Center, Rockwell International, Newbury Park, Calif.

One of the most useful instruments for determining aerosol size distributions in the 0.01 to 0.5 μm diameter size range is the electrical aerosol analyzer (EAA). The design of the model of the EAA currently available commercially has been described by Liu, Whitby, and Pui (1974). This model has been calibrated with a series of monodisperse aerosols (Liu and Pui, 1975), and a data reduction procedure and constants based on this calibration have been recommended by these authors, and will be referred to here as the Liu-Pui data reduction procedure (LPDRP). From the time the LPDRP was available until recent months, most users of the EAA have employed these data reduction constants.

In this article, the same experimental data of Liu and Pui are used to derive a more accurate, but more complicated data reduction procedure, referred to here as the revised data reduction procedure (RDRP). It is further shown that the LPDRP and RDRP can give results for the aerosol concentration in some instrument size ranges which differ by a factor of two or three. These differences arise mainly from the assumption in the LPDRP that the signal (current difference) in each instrument channel is proportional to the aerosol concentration in the nominal size range corresponding to that channel. The fact that a monodisperse aerosol can cause an instrument response in two to five different channels shows that this simplification is not always valid. Changes in the concentration of aerosol in the nominal size range of one instrument channel can cause changes in the instrument response in neighboring channels. This effect becomes more important as the aerosol size increases.

Another factor which causes the apparent instrument calibration constants to vary is that the instrument response to different size aerosols all within the nominal range of one instrument channel can be rather different. In most chan-

Supported in part by California Air Resources Board Contract Nos. ARB3-986 and ARB4-335.

nels, the instrument sensitivity varies by about a factor of two, so aerosol particles near the large particle boundary of the channel will carry about twice the current at a given concentration compared to particles near the small particle size boundary. These problems can be overcome by using more closely spaced aerosol size increments in the data reduction procedure.

The Liu and Pui calibration data for the EAA were obtained for a geometric progression of sizes of monodisperse aerosols, with a spacing of eight sizes per factor of 10. This spacing is close enough that it is possible to interpolate to estimate accurately the response of the instrument to any size of monodisperse aerosol. Since the EAA has a linear response to variations in the aerosol concen-

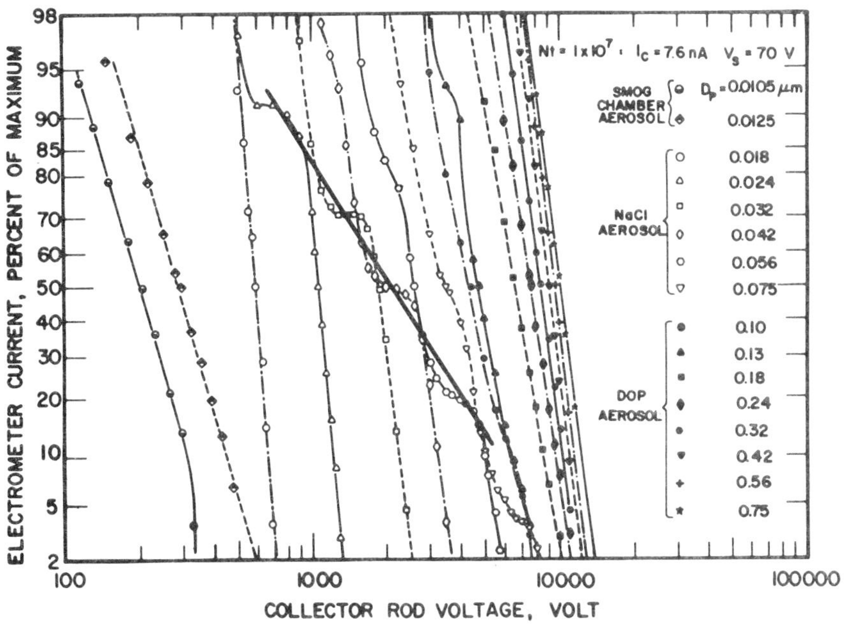

Figure 1. Voltage-current curves for charging conditions with Nt of 1 × 10^7 (ions/cc)(sec) from Liu and Pui (1975). The locus of inflection points has been added.

tration or to combinations of different aerosols, it is possible to calculate the response of the EAA to any aerosol size distribution by representing that distribution by a series of closely spaced, monodisperse aerosols, determining the instrument response to each monodisperse aerosol, and then summing these responses. The following text describes how the interpolation was carried out to obtain a closely spaced set of calibration constants, how these constants are used, and some results from their application to experimental data.

Figure 1 shows the calibration data of Liu and Pui (1975) for the aerosol charging conditions of Nt = 1 × 10^7 (ions/ cc) (sec). For each size aerosol, as the collector rod voltage increases, the fraction of the aerosol which migrates to the collector rod increases, hence the fraction which passes through the drift tube to the electrometer decreases. As shown in the figure, the smaller aerosols are collected at the lower voltages, and successively larger particles are collected at successively higher voltages. In part, the fact that the aerosol of a given size is collected over a range of voltages is because the aerosol is introduced into the drift tube in a zone with a range of distances from the collector rod.

The discrete nature of electrical charge also has an important effect on the shape of the curves. For particles 0.018 μm in diameter and smaller, essentially all particles carry one unit of charge. Slightly larger particles may have either one or two units of charge, and there is a distinct inflection in the curves at the voltage where essentially all of the doubly charged particles have been collected and the singly charged particles pass through uncollected. Triply charged particles begin to show in the data at 0.042 μm diameter, and account for 15 or 20% of the electrometer current with 0.056 μm particles. The larger particles carry more units of charge, and the factors which limit the resolution of the instrument smear out the inflections caused by the discrete quantities of charge.

The interpolation between the curves in Figure 1 is made possible by the observation that the shape of the curves near the inflection from one to two units of charge is the same for all curves. In addition, the inflection points are nearly equally spaced on a straight line, which has been added to the figure. For the interpolation, the curves can be translated without rotation in such a way that the inflection point remains on the locus of inflections. The particle size is approximately linearly related to the position of the inflection point on this locus. This procedure permits constructing the calibration curve for any monodisperse particle size between 0.018 μm and 0.75 μm. There is an irregularity in the experimental data at 0.13 μm, which was largely ignored in the interpolations reported below.

In normal operation, the collector rod voltage is advanced stepwise through a series of preset voltages, and the electrometer current read at each voltage. Liu and Pui determined the voltages to be used by:

1. Selecting a geometric progression of aerosol sizes with four sizes per factor of 10.
2. Determining the collector rod voltage for each size which reduces the electrometer current by half.

The interpolation procedure described above permits constructing a continuous curve for the voltage which reduces the electrometer current by half at each particle size, and this is shown in Figure 2. The points are from the data in Figure 1, and the points which were used to determine the collector rod voltage steps associated with the LPDRP are represented by solid circles.

It can be seen that the transition from one to two units of charge on a particle causes an inflection at 0.042 μm, and from two to three units an inflection at 0.075 μm. As a result, the uniform progression of particle sizes produces a non-uniform progression of voltages. It is interesting to note that the voltage progression would have been much more uniform if the particles' size progression corresponding to the open circles had been used. If new instrument operating voltages and calibration constants are developed for the LPDRP in the future, the use of a smoothed curve (such as that suggested by the open circles) should be considered.

On first glance at Figure 2, it appears that the EAA responds equally to several different particle sizes near 0.042 μm, in a manner similar to the response of optical particle counters to somewhat larger monodisperse aerosols. In fact, examination of Figure 1 shows that no two curves for different size aerosols are alike, so the EAA can always distinguish between monodisperse aerosols of different size.

Liu and Pui (1975) used the curves in Figure 1 to determine the calibration constants for the LPDRP. The same procedure was used with interpolated curves to determine the calibration constants reported in Table I for a geometric progression of particle sizes with 24 increments per factor of 10 as given by the equation

$$D_i = 10^{(i-61)/24} \ \mu m, \ i = 1,61. \tag{1}$$

The values of F_{ij} in this table are the fraction of the maximum electrometer current due to particles of size i associated with instrument channel j. The fraction of the electrometer current associated with an instrument channel is the difference between the fraction of the current at the two voltages at the channel boundaries as read from the (interpolated) curves in Figure 1.

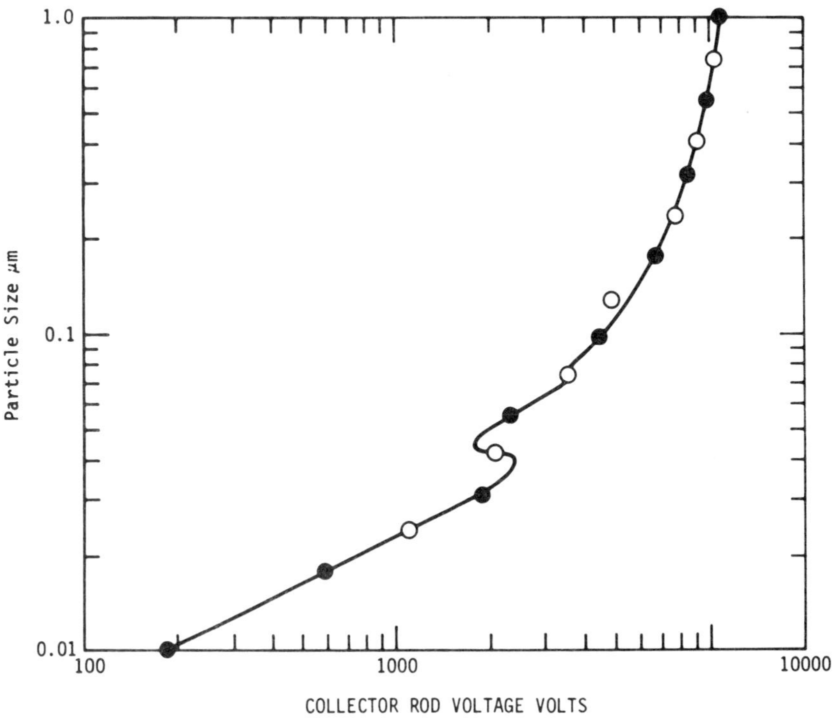

Figure 2. The collector rod voltage for which the electrometer current is half of its maximum value.

TABLE I. Calibration Constants for the Revised Data Reduction Procedure

Particle Size Subscript i	Particle Diameter D_i (μm)	Sensitivity $\frac{pA}{10^6\ particles/cm^3}$	Fraction of the Electrometer Current F_{ij} in Each Instrument Channel j										
			1	2	3	4	5	6	7	8	9	10	>10
1	0.0032	↑	1.	—	—	—	—	—	—	—	—	—	—
2	0.0035		1.	—	—	—	—	—	—	—	—	—	—
3	0.0038		1.	—	—	—	—	—	—	—	—	—	—
4	0.0042		1.	—	—	—	—	—	—	—	—	—	—
5	0.0046		1.	—	—	—	—	—	—	—	—	—	—
6	0.0051		.95	.05	—	—	—	—	—	—	—	—	—
7	0.0056	See Eq. (2)	.50	.50	—	—	—	—	—	—	—	—	—
8	0.0062		.05	.95	—	—	—	—	—	—	—	—	—
9	0.0068		—	1.	—	—	—	—	—	—	—	—	—
10	0.0075		—	1.	—	—	—	—	—	—	—	—	—
11	0.0083		—	1.	—	—	—	—	—	—	—	—	—
12	0.0091		—	.95	.05	—	—	—	—	—	—	—	—
13	0.010	↓	—	.50	.50	—	—	—	—	—	—	—	—
14	0.011	1.3	—	—	.95	—	—	—	—	—	—	—	—
15	0.012	1.7	—	—	1.	—	—	—	—	—	—	—	—
16	0.013	2.0	—	—	1.	—	—	—	—	—	—	—	—
17	0.014	2.4	—	—	1.	—	—	—	—	—	—	—	—
18	0.016	3.0	—	—	.95	.05	—	—	—	—	—	—	—
19	0.018	3.9	—	—	.50	.50	—	—	—	—	—	—	—
20	0.019	4.8	—	—	.05	.95	—	—	—	—	—	—	—
21	0.021	5.6	—	—	.07	.93	—	—	—	—	—	—	—
22	0.025	6.3	—	—	.09	.91	—	—	—	—	—	—	—
23	0.026	7.0	—	—	—	.97	.03	—	—	—	—	—	—
24	0.029	8.0	—	—	—	.84	.16	—	—	—	—	—	—
25	0.032	↑	—	—	—	.51	.49	—	—	—	—	—	—
26	0.035		—	—	—	.37	.56	.07	—	—	—	—	—
27	0.038		—	—	—	.44	.27	.29	—	—	—	—	—

28	0.042	See Eq. (3)	—	—	—	.48	.08	.44	—	—	—	—	—
29	0.046		—	—	—	.45	.16	.39	—	—	—	—	—
30	0.051		—	—	—	.20	.47	.27	.06	—	—	—	—
31	0.056		—	—	—	.14	.36	.33	.17	—	—	—	—
32	0.062		—	—	—	.09	.20	.53	.18	—	—	—	—
33	0.068		—	—	—	—	.28	.60	.09	.03	—	—	—
34	0.075		—	—	—	—	.14	.58	.24	.04	—	—	—
35	0.083		—	—	—	—	.08	.69	.19	.04	—	—	—
36	0.091		—	—	—	—	.02	.58	.32	.08	—	—	—
37	0.100		—	—	—	—	—	.57	.36	.07	—	—	—
38	0.110		—	—	—	—	—	.55	.37	.08	—	—	—
39	0.12		—	—	—	—	—	.46	.46	.08	—	—	—
40	0.13		—	—	—	—	—	.42	.50	.08	—	—	—
41	0.14		—	—	—	—	—	.14	.71	.12	.03	—	—
42	0.16		—	—	—	—	—	.10	.60	.24	.06	—	—
43	0.18		—	—	—	—	—	.03	.47	.39	.08	.03	—
44	0.195		—	—	—	—	—	—	.40	.42	.13	.05	—
45	0.21		—	—	—	—	—	—	.31	.46	.16	.05	.02
46	0.24		—	—	—	—	—	—	.22	.48	.21	.06	.03
47	0.26		—	—	—	—	—	—	.17	.50	.23	.07	.03
48	0.29		—	—	—	—	—	—	.11	.48	.27	.09	.05
49	0.32		—	—	—	—	—	—	.08	.42	.32	.11	.07
50	0.35		—	—	—	—	—	—	.05	.40	.35	.12	.08
51	0.38		—	—	—	—	—	—	.03	.34	.36	.15	.12
52	0.42		—	—	—	—	—	—	—	.30	.36	.21	.13
53	0.46		—	—	—	—	—	—	—	.26	.37	.20	.17
54	0.51		—	—	—	—	—	—	—	.22	.35	.24	.19
55	0.56		—	—	—	—	—	—	—	.18	.36	.25	.21
56	0.62		—	—	—	—	—	—	—	.14	.35	.25	.26
57	0.68		—	—	—	—	—	—	—	.13	.31	.25	.31
58	0.75		—	—	—	—	—	—	—	.12	.29	.24	.35
59	0.83		—	—	—	—	—	—	—	.08	.25	.24	.43
60	0.91		—	—	—	—	—	—	—	.08	.20	.24	.48
61	1.00		—	—	—	—	—	—	—	.07	.15	.26	.52

The calibrations of Liu and Pui (1975) also include the determination for each size aerosol of the instrument sensitivity S_i, i.e., the electrometer current divided by the aerosol concentration at low enough collector rod voltages that the aerosol passes through the drift tube and is collected on the electrometer filter (except for aerosol losses in the instrument, which are largely accounted for by this calibration). The data for aerosol particle sizes between 0.01 μm and 0.03 μm are given in Table I. For particles smaller than 0.01 μm, values of S_i are given by

$$\log_{10} S_i = 13.975 + 7.033 \log_{10} D_i, \tag{2}$$

and for particles larger than 0.03 μm by

$$\log_{10} S_i = 2.618 + 1.113 \log_{10} D_i \tag{3}$$

where S_i is in units of pA per 10^6 particles per cc and D_i is in units of μm.

In a hypothetical case in which a number concentration N_i particles/cc of each size D_i is supplied to the EAA, the observed current difference for each instrument channel would be

$$\Delta I_j = \sum_{i=1}^{61} N_i S_i F_{ij} \times 10^{-6}. \tag{4}$$

The factor of 10^{-6} arises because S_i is given in units of pA per 10^6 particles per cc.

The particle size distribution which most nearly gives the same instrument response as an observed response can be found by:

1. Adopting a criterion for goodness of fit, so various fits can be ranked in order of decreasing quality.
2. Assuming a trial size distribution N_i.
3. Calculating the instrument response from Eq. (4).
4. Determining the goodness of fit.
5. Adjusting the trial size distribution so as to improve the fit, and
6. Continuing with steps 3 through 5 until the goodness of fit parameter is minimized.

In the work reported below, the problems at hand were adequately solved by using a visual observation of a plot of the data in step 1, and having the computer operator choose new parameters in step 5. However, it should be straightforward to automate the entire procedure.

The above procedure is simplified if some functional form is assumed for the size distribution. For atmospheric aerosols, three size classes appear to be important, and the experimental data are usually adequately represented by a trimodal log-normal distribution. Only two of the aerosol size classes fall within the range of EAA, so a bimodal, log-normal distribution is usually the mathematical form of choice in reducing EAA data for atmospheric aerosols, and many aerosols from other sources.

To represent a log-normal distribution containing a concentration of N_t particles per cc by a series of monodisperse aerosols of the sizes given by Eq. (1), the concentration of each size N_i is given by the relation

$$N_i = N_t \frac{(\ell n\ 10)/24}{\sqrt{2\pi}\ \sigma} \exp - \frac{(\ell n\ D_i - \ell n\ D_g)}{2\sigma^2}, \quad (5)$$

where D_g is the geometric mean diameter and σ is the standard deviation. A bimodal, log-normal distribution can be represented by the sum of two such relations. Therefore, there are six adjustable parameters for a bimodal log-normal distribution, N_t, D_g and σ for each mode.

The first example of the use of this procedure shows data for an aerosol obtained from a flow reactor (Lipeles, 1976). In this case, a functional form closely approximating the aerosol size distribution was chosen, and the instrument response to this size distribution was calculated. Then, this calculated instrument response was used to determine new calibration constants for the LPDRP, and the data given by the dotted line segments in Figure 3 were obtained. The solid line segments show the results obtained by the LPDRP. The results in this figure were the first obtained by the methods presented here, and encouraged the further use of this method.

Another result is shown in Figure 4. These data were obtained immediately downwind from a major freeway in Los Angeles. (The EAA was operated with charging conditions of $Nt = 7 \times 10^6$ (ions/cc) (sec), so calibration constants derived for this charging condition were used in the reduction of these data.) The points representing the experimental data were obtained from the observed instrument response using the LPDRP. The curve represents the best-fit, log-normal distribution, and is made up of straight line segments joining the values of N_i when presented in this aerosol surface vs aerosol size representation. The points representing the calculated instrument response were obtained from the current differences calculated for each instrument channel from the N_i represented in the curve. The LPDRP was used to convert the calculated instrument response into the points in the figure. When the calculated and observed points in the figure are equal, the calculated and observed instrument current differences are equal. Thus, the transformation used for the observed and calculated points in Figure 4 is one way of weighting the differences between the observations and calculations when searching for the best fit.

In the fitting procedure, values of D_g and σ for each mode were selected by hand, and the values of N_t for each mode were selected to obtain the best least-squares fit of the data points as represented in Figure 4. Thus, the data were transformed to the $\Delta S/\Delta$ log D representation before the sum of the square of the errors was determined. Needless to say, the choice of a goodness of fit criterion was arbitrary, and other choices could have been used.

The results in Figure 4 show that it was possible to obtain close agreement between the observed and calculated instrument response, (hence the criterion used to determine the best fit is not of great importance). The quality of this fit is not strong evidence for the bimodal, log-normal nature of aerosol distributions,

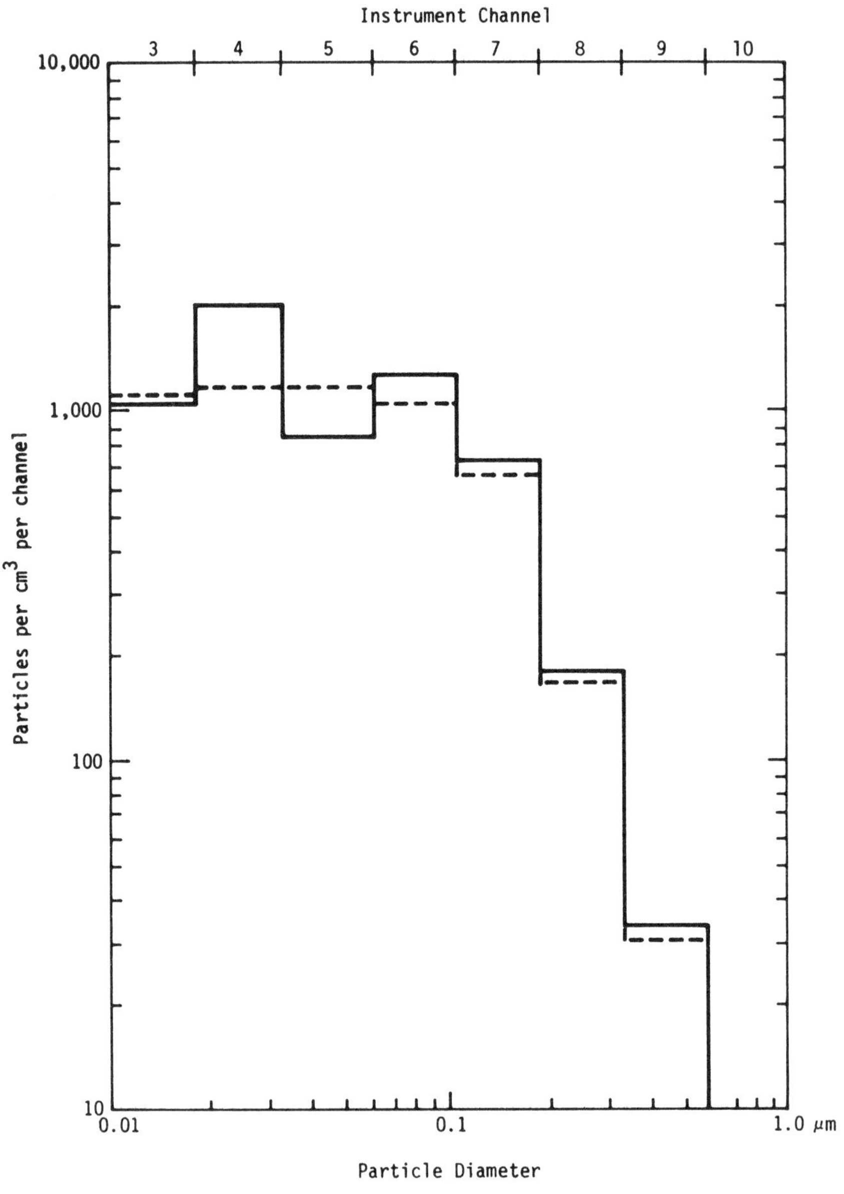

Figure 3. Number concentration of a flow-reactor produced aerosol measured with Nt = 1×10^7 cm^{-3} sec. The solid line contains results from the LPDRP and the dotted line contains results from the RDRP.

because six parameters were adjusted in the fit to nine non-zero data points. The best evidence for the bimodal nature of the aerosol was the observation that the smaller particle size mode (nuclei mode) was directly due to the freeway traffic, and the larger particle size mode (accumulation mode) was not importantly influenced by the traffic. However, the quality of the fit does show that the bimodal, log-normal distribution is adequate to represent the data.

It is apparent that there is some difference between the results obtained by the LPDRP (shown by the points) and the RDRP (shown by the curve). For one thing, the values of D_g for each mode from the RDRP are smaller than from the LPDRP.

Another comparison between the two data reduction procedures is shown in

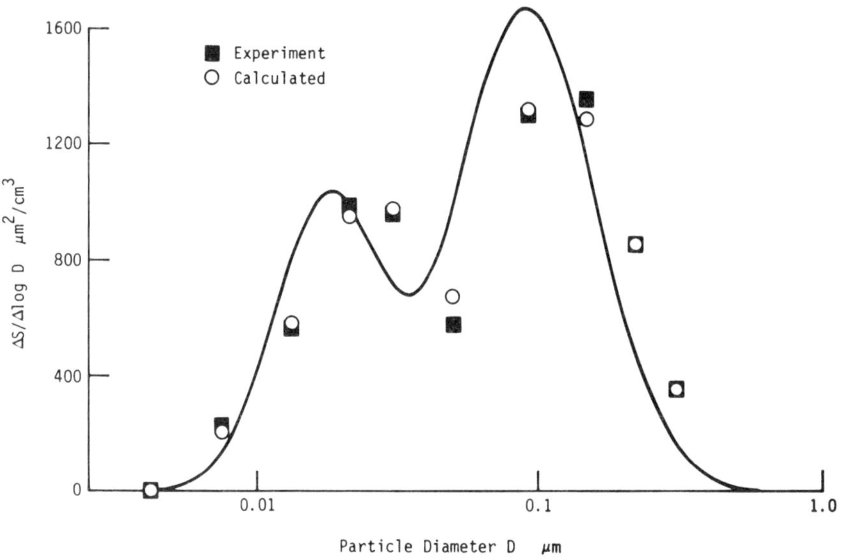

Figure 4. Particle size distribution data obtained near a freeway with Nt = 7×10^6 (ions/cm^3)(sec).

Table II. Two different log-normal aerosol size distributions were assumed, one appropriate for accumulation mode aerosols and the other for nuclei mode aerosols. The instrument response for an arbitrary concentration of each aerosol was calculated. Then the concentration of aerosol particles within the size range of each nominal instrument channel specified by the LPDRP was divided by the calculated current difference for that channel to obtain calculated values for the instrument calibration constants. The calibration constants reported by Liu and Pui were then divided by the calculated constants to obtain the ratios given in Table II.

The data for Channels 7 and larger for nuclei mode aerosols are not reported for two reasons:

1. In this size range, the nuclei mode aerosol concentration falls off extremely

rapidly with increasing particle size, with the result that the calculated calibration constants are orders of magnitude larger than those of the LPDRP, and

2. There is no practical use for the calculated-constants because no actual aerosol would fall off as rapidly in concentration in this size range as does the assumed size distribution.

The data for Channel 6 are enclosed in parentheses because the above comments are partly relevant to this result.

There are two important things to note in Table II. The first is that the calibration constants calculated here for the accumulation mode and nuclei mode size distributions are as much as a factor of three different from those recommended

TABLE II. Ratios of Calibration Constants
Nt = 1 × 10^7 (ions/cc) (sec)
(Liu, Pui Constants/Constants Calculated Here)

	Assumed Size Distributions			
	Accumulation Mode		Nuclei Mode	
	D_g	σ	D_g	σ
Number	0.061	0.67	0.013	0.42
Surface	0.150		0.018	
Volume	0.235		0.022	

Instrument Channel	Nominal Aerosol Size (μm)	Ratios of Calibration Constants	
2	0.0075	3.85	3.04
3	0.0133	1.36	0.90
4	0.0237	3.10	0.94
5	0.0422	0.80	0.66
6	0.0750	1.28	(6.7)
7	0.133	1.04	
8	0.237	0.96	Too
9	0.422	1.36	Few
10	0.750	3.70	Particles

for use in the LPDRP. The second is that the calibration constants calculated here depend on the aerosol size distribution. This is especially notable in instrument Channel 4 for the size distributions in Table II, but such differences can occur in other instrument channels for other size distributions.

In summary, it is believed that the existing data reduction procedures for the EAA contain approximations which have a significant effect on the derived results, but that more elaborate data reduction procedures are easily derived which remove these problems, and allow experimentalists to obtain data from the EAA which exploit its full potential. It should be possible to make computer routines generally available to provide easy access to these data reduction procedures.

REFERENCES

Lipeles, M., 1976: The formation of aerosols in a photochemical flow reactor. Final Report to CAPA-8 Committee of the Coordinating Research Council, SC5012.12FRD, Air Monitoring Center, Rockwell International.

Liu, B. Y. H., K. T. Whitby, and D. Y. H. Pui, 1974: A portable electrical analyzer for size distribution measurement of submicron aerosols. *J. Air Pollution Control Assoc.*, *24*, 1067–1072.

——— and D. Y. H. Pui, 1975: On the performance of the electrical aerosol analyzer. *J. Aerosol Sci.*, *6*, 249–264.

QUESTIONS

Q: Could it be that a unimodal distribution with some special median size and standard deviation could result in the same ΔI values being measured as a bimodal distribution?

Richards: I feel comfortable saying in this case (Figure 4) that the particle size distribution is bimodal rather than unimodal in this surface representation. On the question of uniqueness; six adjustable parameters are used to fit about nine data points, so obtaining a fit does not really demonstrate that this is a bimodal, log-normal, aerosol size distribution. There are many parameters to adjust in this fit, and the results obtained will, of course, depend on the criterion used to measure the goodness of fit. In aerosol work, many different measures of the goodness of fit could be used, and there will never be agreement as to which one is best. Each person will have his own application, and, for his own reasons emphasize certain aspects of the aerosol size distribution in determining the goodness of fit. In this work, we did not worry much about the problems of the uniqueness of the fit, or demonstrating from the fit that the aerosol size distribution was actually bimodal. We knew from other information that the size distribution was bimodal, and the results obtained were adequate for our purposes at the time.

Whitby: We just fit about 900 size distributions. Later I'm going to show a couple of those and say a little bit about their fitting. I doubt that any of our group or any of the people who have been associated with our group have much doubt left about the fact that these modes are real. We've cross-checked them with optical counters, we've taken the modal distributions and computed the response of nephelometers and nuclei counters and, within the accuracy of these instruments, they all agree. So there have been enough cross-checks to show that these modal characteristics that the electrical aerosol analyzer measures in the atmosphere are real. Whether the σ_g of the mode is 1.5 or 1.7, I won't argue. Whether the mean size is up or down 20% or so, I won't argue about that. Whether the volume is plus or minus 30%, I won't argue about that. But the modal characteristics and the general shapes I think are quite well established.

A COMPARISON OF THE SIMULTANEOUS RESPONSE OF FOUR ELECTRICAL AEROSOL ANALYZERS

William E. Clark* and Edward L. Avol**

Air Monitoring Center, Rockwell International, Newbury Park, Calif.

The Rockwell International Air Monitoring Center is involved in an EPA sponsored program to install aerosol concentration and size distribution monitoring instruments into several EPA Community Health Air Monitoring Program (CHAMP) field stations. As part of this program, four Thermo-Systems, Inc. Model 3030 Electrical Aerosol Analyzers (EAA) will be utilized to measure the concentration and size distribution of ambient aerosol particles from 0.0056 to 1.0 μm diam. In order to gain an understanding of the relative accuracy with which concentration and size distribution measurements could be made with the Electrical Aerosol Analyzers, it was of interest to determine how closely measurements made with several instruments compared with one another. A study was therefore conducted to compare the simultaneous response of four EAA's to the same aerosol.

The calibration of the EAA depends upon the product of the ion concentration, N, and residence time, t, of particles in the charging section of the analyzer. This parameter, referred to as the Nt product, can be varied by the instrument operator. Experiments were performed to determine how the response of the analyzer was affected by varying the Nt product between the extremes of the tolerance range considered acceptable by the instrument manufacturer.

EQUIPMENT AND PROCEDURE

The EAA measures the number concentration of aerosol particles in specific size increments, in-situ, by drawing a continuous sample of aerosol into a steady flow system, charging the aerosol particles to an electrical mobility which is a function

*Present address: Environmental Engineering Department, California Polytech State University, San Luis Obispo, Calif. 93407.

**Present address: Rancho Los Amigos Hospital, 7601 E. Imperial Highway, Downey, Calif. 90242.

of particle size and then measuring the mobility distribution of the aerosol particles. A detailed description of the instrument has been presented by Liu (1974, 1975). To determine a size distribution, the instrument is stepped through eleven preset voltages and the electrometer current for each voltage is recorded. The difference in current, ΔI, between adjacent voltage steps is then calculated and this value is multiplied by a calibration constant to determine the particle number concentration, particles per cc, for each size increment. From the number concentration data, the surface area and volume distributions may also be calculated by multiplying the respective number concentration values by the particle surface area, πD_p^2, and particle volume, $1/6\ \pi D_p^3$, for each size increment, where Dp refers to the geometric mean diameter of the size increment.

The charge acquired by particles passing through the diffusion charging section of the EAA depends upon the product of the ion concentration, N, and the residence time, t, of particles in the charging region. This Nt product, in turn, depends upon the geometry of the diffusion charger, the aerosol flow rate, and

TABLE I. EAA Differential Current Values for Cigarette Smoke Aerosol (Run 13)

EAA Channel	Diameter (μ)	ΔI Values-Pico amps				
		No. 80	No. 9	No. 77	No. 81	Average
3	0.0133	—	—	—	—	—
4	0.0237	0.730	0.650	0.740	0.650	0.693
5	0.0422	0.880	0.820	0.790	0.840	0.833
6	0.0750	1.30	1.64	1.41	1.61	1.49
7	0.133	1.30	1.56	1.39	1.46	1.43
8	0.237	0.928	1.09	0.948	1.17	1.03
9	0.422	0.318	0.396	0.236	0.448	0.350
10	0.750	0.115	0.138	0.081	0.163	0.124

the ratio of two parameters used to monitor the operation of the charger: the ionizer current, and the ionizer voltage. Since the geometry of the charging section and the aerosol flow rate are constant from instrument to instrument, the Nt product can be made equivalent between instruments by adjusting the ionizer current to ionizer voltage ratio. The EAA instruction manual indicates that this ratio should be adjusted to a value of $1.1\times10^{-10} \pm 0.16\times10^{-10}$ amps per volt. This value for the ionizer current to voltage ratio corresponds to an Nt product of 1×10^{7} ions-sec/cc and a tolerance range of about ± 14.5%.

The EAA's used in this study included three recently purchased instruments with serial numbers 77, 80 and 81, and an older instrument with serial number 9, which was converted to operate with an Nt product of 1×10^{7} ions-sec/cc rather than with a value of 7×10^{6} as originally delivered. The four analyzers were thoroughly checked and adjusted to make sure that all of the operating parameters such as power supply voltages, D/A converter zero and full scale settings, and preset center rod voltage values were within their respective tolerance limits. In addition, immediately preceding each comparative experiment, the aerosol and total flow rates for each instrument were precisely adjusted to 4 ℓpm and 50 ℓpm, respectively, and the ionizer current to voltage ratio was adjusted to a value of 1.1×10^{-10} amps per volt with a tolerance range of $\pm$ 1.0%.

A 300 ft^3 Teflon bag was used to contain the aerosols simultaneously sampled by the four analyzers. The sampling lines leading from the Teflon bag to each instrument were of equal length. Two types of combustion aerosols were used in this study: cigarette smoke, and aerosol produced by the burning of a mixture of

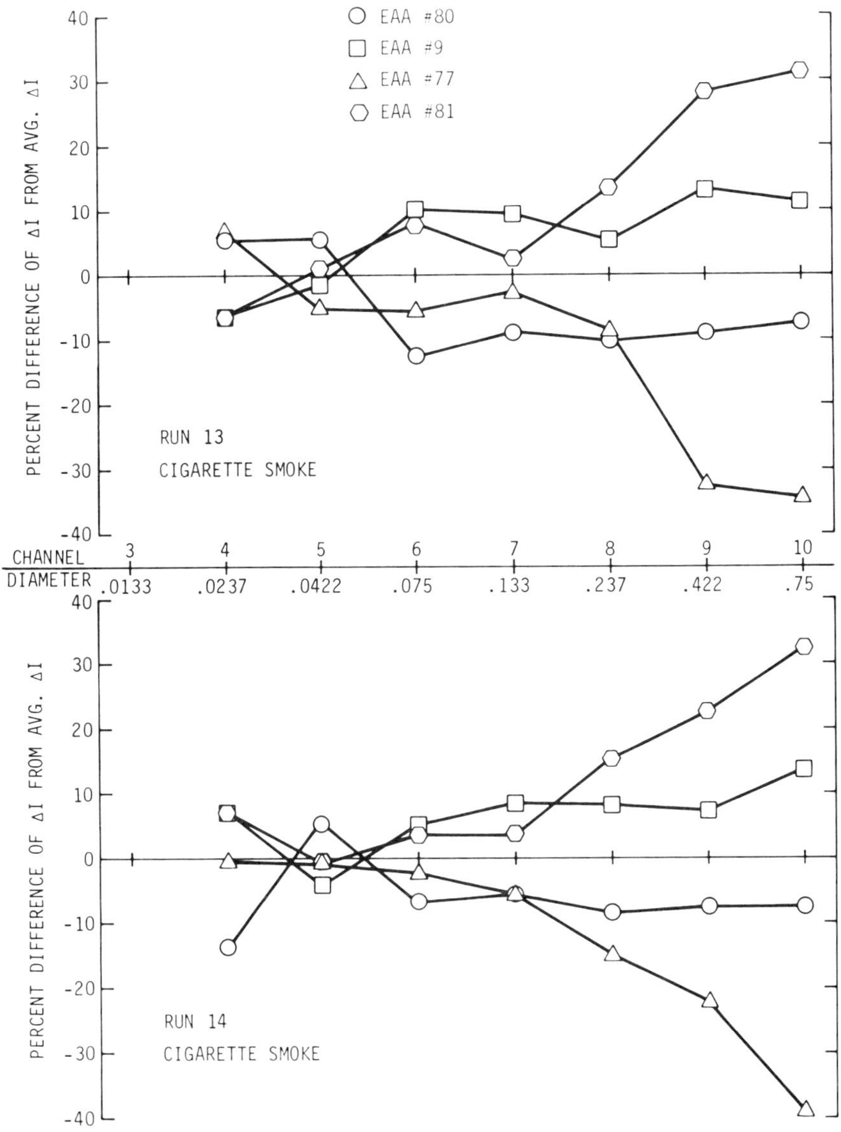

Figure 1. The relative response of four electrical aerosol analyzers to cigarette smoke aerosol.

natural gas and air. These aerosols were introduced to the bag through the blower inlet as the bag was filling, and allowed to age for 10 to 20 minutes before the size distribution measurements were made by simultaneously stepping the four analyzers through their preset voltage increments.

RESULTS AND DISCUSSION

Table I shows the results of a typical comparative response experiment with cigarette smoke aerosol. For each EAA particle size increment, as designated by a channel number and the geometric mean diameter of the size increment, are tabulated the differential current values, ΔI, measured for each of the four analyzers. Also shown for each size increment is the average value of the four ΔI

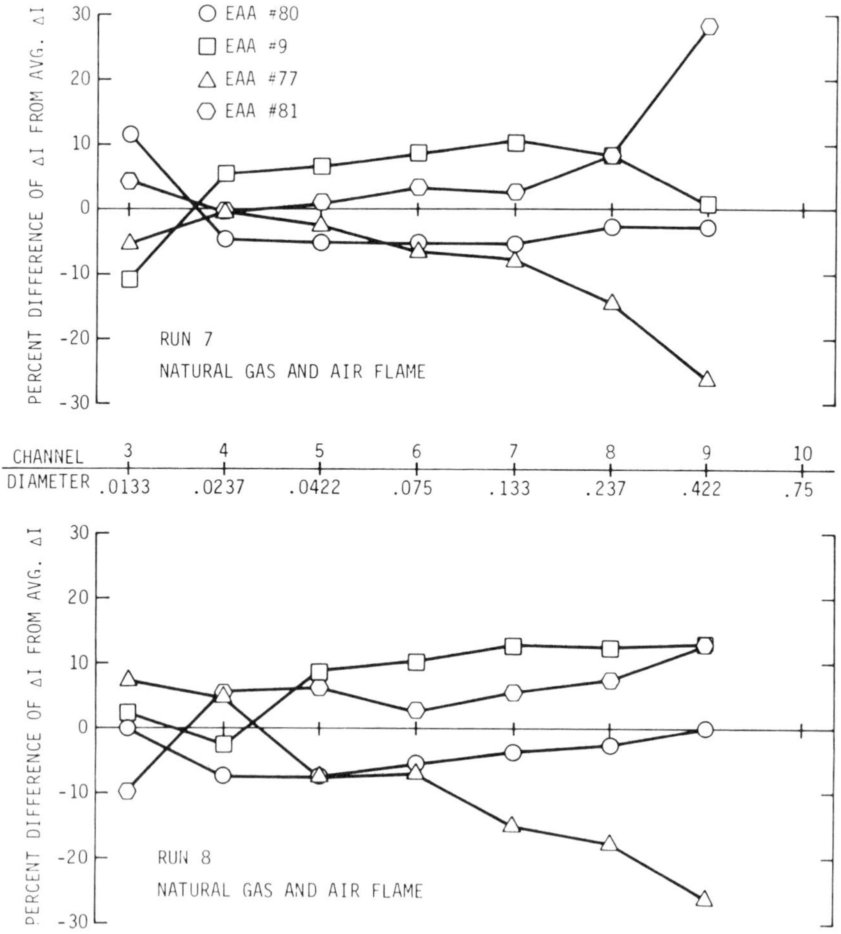

Figure 2. The relative response of four electrical aerosol analyzers to an aerosol produced by a natural gas and air flame.

measurements. For the cigarette smoke aerosols accurate comparative measurements could be made from Channel 4, 0.0237 μm, to the upper limit of the EAA, 0.750 μm. The aerosol produced by burning a mixture of natural gas and air produced a particle size distribution which was shifted more towards smaller sizes, and comparative measurements could be made from Channel 3, 0.0133 μm, to Channel 9, 0.422 μm.

To compare the response of the four analyzers, the percent differences in the measured differential current values, ΔI, from the average differential current value, AVG. ΔI, were calculated for each instrument and plotted as a function of

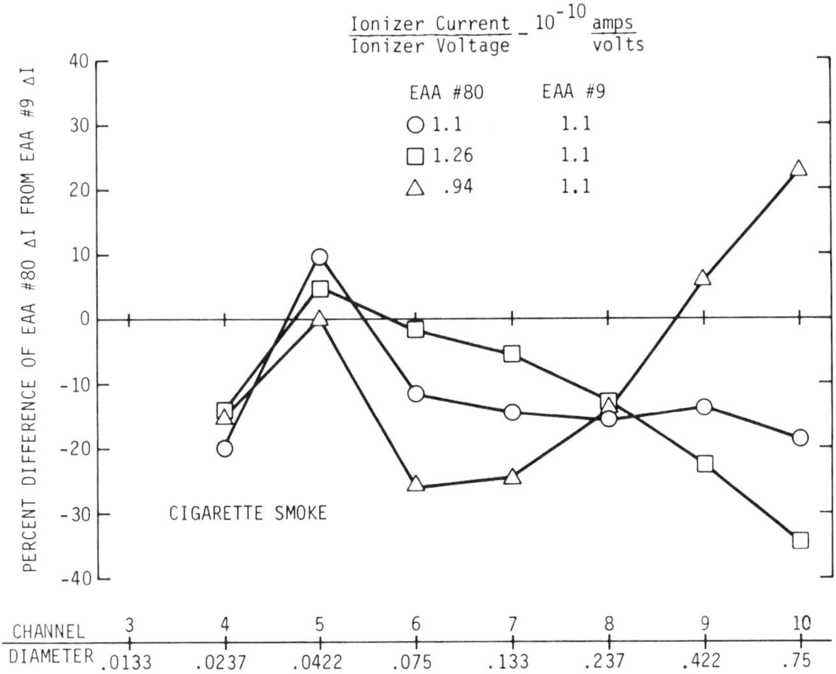

Figure 3. The effect of changing the ionizer current to ionizer voltage ratio on the electrical aerosol analyzer response for cigarette smoke aerosol.

particle size. Figures 1 and 2 show relative response curves for two runs with cigarette smoke aerosol and two runs with aerosols produced by a natural gas and air burner. These four experimental runs were typical of several experimental runs using the two aerosol types. The results of the comparative analysis can be summarized in the following way: with few exceptions, the four EAA's agreed to within ± 10% of the average ΔI value up through Channel 7, 0.133μm. For Channels 8 through 10 the agreement with the average ΔI value was within ± 15% for 0.237 μm particles, ± 30% for 0.422 μm particles, and ± 40% for 0.750 μm particles. The response of each new EAA is compared to the response of a factory standard EAA before shipment (Sem, 1976). From 0.03 to 0.3μm, the ΔI values have been found to match within ± 10% on all channels. However, in the first 4 and last 2 channels, the ΔI values have been found to deviate by as much as ± 25%. If it is assumed that the average ΔI values measured in the present study are comparable to the response of the factory standard instrument, the deviations in ΔI measured in this study are somewhat higher than those observed by Thermo-Systems for the last 3 channels.

The effect of varying the Nt product between the extremes of the tolerance range suggested by the instrument manufacturer, ± 14.5% was determined by simultaneously measuring the size distribution of cigarette smoke aerosols with two analyzers. One analyzer was adjusted to an ionizer current to voltage ratio value of 1.1×10^{-10} while the other analyzer was adjusted to the extremes of the tolerance range suggested for this ratio. The results of this experiment are shown in Figure 3. The percent difference in ΔI for EAA No. 80 as compared to EAA No. 9 is plotted as a function of particle size for various ionizer current to voltage ratios. Varying the current to voltage ratio between the tolerance limits resulted in skewing the measured particle size distribution toward smaller particle sizes when the ratio was at the upper limit and towards larger particle sizes when the ratio was at the lower limit. The percent difference in ΔI for EAA No. 80 from ΔI for EAA No. 9 changed from 10 to 15% for Channels 6 and 7 and from 10 to 40% for Channels 9 and 10. The results of this experiment suggest that the Nt product should be adjusted to within a considerably narrower tolerance range than that suggested by the manufacturer if reproducible results are to be achieved with a single instrument or if comparable results are to be achieved with more than one instrument.

REFERENCES

Liu, B. Y. H., K. T. Whitby and D. Y. H. Pui, 1974: A portable electrical analyzer for size distribution measurements of submicron aerosols. *J. Air Pollution Control Assoc., 24,* 1067–1072.

——— and D. Y. H. Pui, 1975: On the performance of the Electrical Aerosol Analyzer, submitted to *Aerosol Science*.

Sem, G. J., 1976: Electrical Aerosol Analyzer: operation, maintenance, and application. Presented at the Aerosol Measurement Workshop, University of Florida, Gainesville, Florida, March 24–26.

QUESTIONS

Clark: I would like to know how I can make the four analyzers agree better. Also, if the four analyzers agreed well with your standard instrument at the factory, why didn't they agree better with each other at our laboratory?

Sem: Well, I think the analyzers did agree well up to about Channel 7 or 8. In our factory check we use ambient laboratory aerosol and we don't measure much current in the last two channels. So, I really can't say how the ΔI values compare in our factory check for the last two channels.

Whitby: I think you should keep in mind that one of the basic characteristics of the analyzer is the diminishing slope of the mobility vs. size curve. Furthermore, current levels normally encountered at the upper end of the instrument approach the noise level, and so for large particles one is running out of both sensitivity and resolution. When I look at the numbers, I don't really see any way of improving this situation on a commercial basis. I think that by a vast amount of work, one might track down what some of these differences are, but I seriously doubt at this point that the effort is worth it. This is one of the reasons why we have always felt that one should run an optical counter starting at about 0.4 of a micron. Then the overlap between the two instruments is useful and an interesting check, but I have no doubts at all that the optical counter is the best way to measure the size range from about 0.4 μm to 1.0 μm.

Sem: If I had a well calibrated optical particle counter and a correctly operating electrical analyzer, I would believe the optical particle counter in the size range corresponding to the last channel of the analyzer, the one just below 1 μm. In the other ranges, personally, I would tend to believe the electrical analyzers. I essentially verify what Dr. Whitby has said.

THE ROLE OF THE ELECTRICAL AEROSOL ANALYZER IN PHOTOCHEMICAL AEROSOL RESEARCH

W. C. KOCMOND AND J. Y. YANG

Calspan Corporation, Environmental & Energy Systems Department, Buffalo, N.Y.

AND D. B. KITTELSON

Mechanical Engineering Department, University of Minnesota, Minneapolis

Even though major advances in the knowledge of gas-phase reactions involved in the formation of photochemical smog have been gained in recent years, the important chemical and physical characteristics governing gas to particle conversions remain largely unresolved. Most laboratory reaction chambers in the past have been designed more specifically for investigating homogeneous gas-phase processes. The 600 m^3 Calspan chamber which represents the largest photochemical reaction vessel in the United States is a more suitable facility for simulating aerosol formation processes in the atmosphere.

It is the purpose of this discussion to review smog chamber aerosol research done at Calspan and, in so doing, illustrate some of the possible uses of the electrical aerosol analyzer in this area of research. Much of the data reported here was generated jointly by Calspan and the University of Minnesota in cooperative studies performed for the Environmental Protection Agency and the Coordinating Research Council.[1]

EXPERIMENTAL FACILITIES

The smog chamber used at Calspan (Kocmond et al., 1973) consists of a cylindrical chamber 30 ft (9.14 m) in diameter and 30 ft (9.14 m) high enclosing a volume of 20,800 ft^3 ($590m^3$). The 1.25 cm thick chamber walls are coated with a specially-formulated fluoroepoxy-type urethane which has surface energy and reactivity properties comparable to those of FEP Teflon. Illumination within the chamber is provided by 28.6 kw of fluorescent blacklight and sun lamps installed inside 24 lighting modules and arranged in eight vertical channels attached to the wall of the chamber. Each lighting module contains two 40-watt, sun lamps,

1. EPA Contract Nos. 68-02-0557, 68-01-1231; CRC Contract No. CAPA-8-71.

eight 85-watt, high output black lamps, and two 215-watt, specially produced black lamps. The lighting modules are covered with 0.5 cm Pyrex glass and are sealed from the chamber working volume. Measured light intensity using the $k_d[NO_2]$ method reported by Steadman and Niki (1973) is $k_d \sim 0.35\ min^{-1}$.

Chamber air is purified through a recirculation system consisting of a series of absolute and activated charcoal filters. Nearly all gaseous contaminants and particulate matter can be removed from the chamber air in about four to five hours of filtration. Filtered air generally contains no measurable particles, less than 0.1 ppm NO_x, 0.2 ppm non-methane HC, and no measurable SO_2 or ozone.

Instrumentation used to monitor aerosol behavior and reactant concentrations within the chamber includes a Bendix Model 8002 chemiluminescent ozone analyzer, Model 8101-B nitrogen oxides analyzer, Model 8300 total sulfur analyzer, and the Model 8201 reactive hydrocarbon analyzer; a Hewlett-Packard 5750 gas chromatograph; a Meteorology Research, Inc., Integrating Nephelometer; a Gardner Associates small particle detector; a General Electric condensation nucleus counter; and a Royco Optical Particle Detector. More complete descriptions of the chamber and analytical instrumentation facilities are given in an earlier Calspan report (Kocmond et al., 1973). Measurements of aerosol behavior were made with the aforementioned particle detectors and also a Thermo-Systems Model 3030 Electrical Aerosol Analyzer (EAA). A description of the commercial prototype is given in Liu and Pui (1975).

AEROSOL BEHAVIOR IN PHOTOCHEMICAL SYSTEMS

The simplest of the photochemical systems studied with the EAA was the SO_2 + clean air system. Clark (1972, 1975) was perhaps the first investigator to show how measurements of aerosol growth could be used to determine SO_2 photooxidation rates for this system. In addition, it was found that the details of aerosol growth could be examined through measurements of particle number, surface and volume concentrations. The general mechanisms governing aerosol behavior when SO_2 is irradiated with light are briefly described below.

The SO_2 + Clean Air System

In most photochemical aerosol studies, the usual test procedure is first to purify the sample air and then to introduce known concentrations of pollutants into the chamber. The sample is then irradiated while observations are made of aerosol formation, particle size distribution, and gaseous behavior. Achieving adequate air purity prior to an experiment is essential since even trace amounts of gaseous contaminants can lead to unwanted chemical reactions and the formation of substantial aerosol.

The production of aerosol from SO_2 photooxidation is generally attributable to chemical conversion of SO_2 to SO_3 followed by rapid reactions with atmospheric constituents, such as H_2O and NH_3, to form condensable products. Homogeneous photooxidation of SO_2 by irradiation under clean atmospheric conditions produces sulfuric acid aerosol via the following sequence of reactions (Castleman et al., 1975):

(1) Oxidation of SO_2 to SO_3
(2) Reaction of SO_3 with water to form H_2SO_4
(3) Clustering of H_2SO_4 and water molecules to form pre-nucleation embryos

A number of different reaction schemes of SO_2 oxidation by free radical and active intermediates have been suggested to account for the observed rates of SO_2 photooxidation. A review of the current understanding of the chemical kinetic schemes involved in the oxidation process is given by Calvert and McQuigg (1975). A review of the general mechanisms of the physical behavior of aerosol formation, growth and decay in the SO_2-clean air system can be found in Clark (1972) and Kocmond et al. (1975, 1976).

Briefly, however, three main mechanisms govern aerosol behavior when SO_2 is irradiated by sunlight: nucleation, condensation, and coagulation. Initially, after irradiation begins, there is a short induction period (usually a few seconds to a few minutes, depending on SO_2 concentration and light intensity) followed by rapid homogeneous nucleation to form tiny particles in the supersaturated vapor mixture. Nuclei initially formed are of the order of 15 Å in diameter, too small to be detected by condensation nuclei counters (CNC) or the EAA.

Once nucleation begins, another path for gas to particle conversion is made available, i.e., condensation. Condensation tends to reduce the supersaturation (SS) and thus the nucleation rate. At the same time, the high concentration of particles leads to an increase in the rate of coagulation. Both of these factors cause the number of particles to increase less rapidly, reach a maximum, and finally decrease. This behavior is illustrated by the plot of particle number against time in the insert of Figure 1. This figure also shows a three-dimensional view of particle number and size as a function of time for a typical SO_2 + clean air irradiation (0.80 ppm SO_2, 40% RH).

The total aerosol surface area, $S(\mu m^2/cc)$, Figure 2, behaves in a somewhat different manner. As the aerosol grows, new surface is produced by nucleation and by condensation on existing particles. At the same time, surface is lost by coagulation of particles. After a period of time, a balance is reached between particle surface losses and gains and an equilibrium surface concentration is reached. The three-dimensional view of surface concentration vs. time more clearly illustrates the contributions of the different particle sizes to the total surface area produced.

Aerosol volume, on the other hand, is produced by both condensation and nucleation but is unchanged by coagulation. Hence, there is steady growth in volume as a function of time, as shown by the plot and the three-dimensional view of aerosol volume in Figure 3. As long as the processes of nucleation and condensation continue, the volume of aerosol also rises. If the stationary state approximation is applied to the condensing and nucleating species, the rate of production of aerosol should be directly proportional to the rate of production of these species and thus the rate of photooxidation of SO_2.

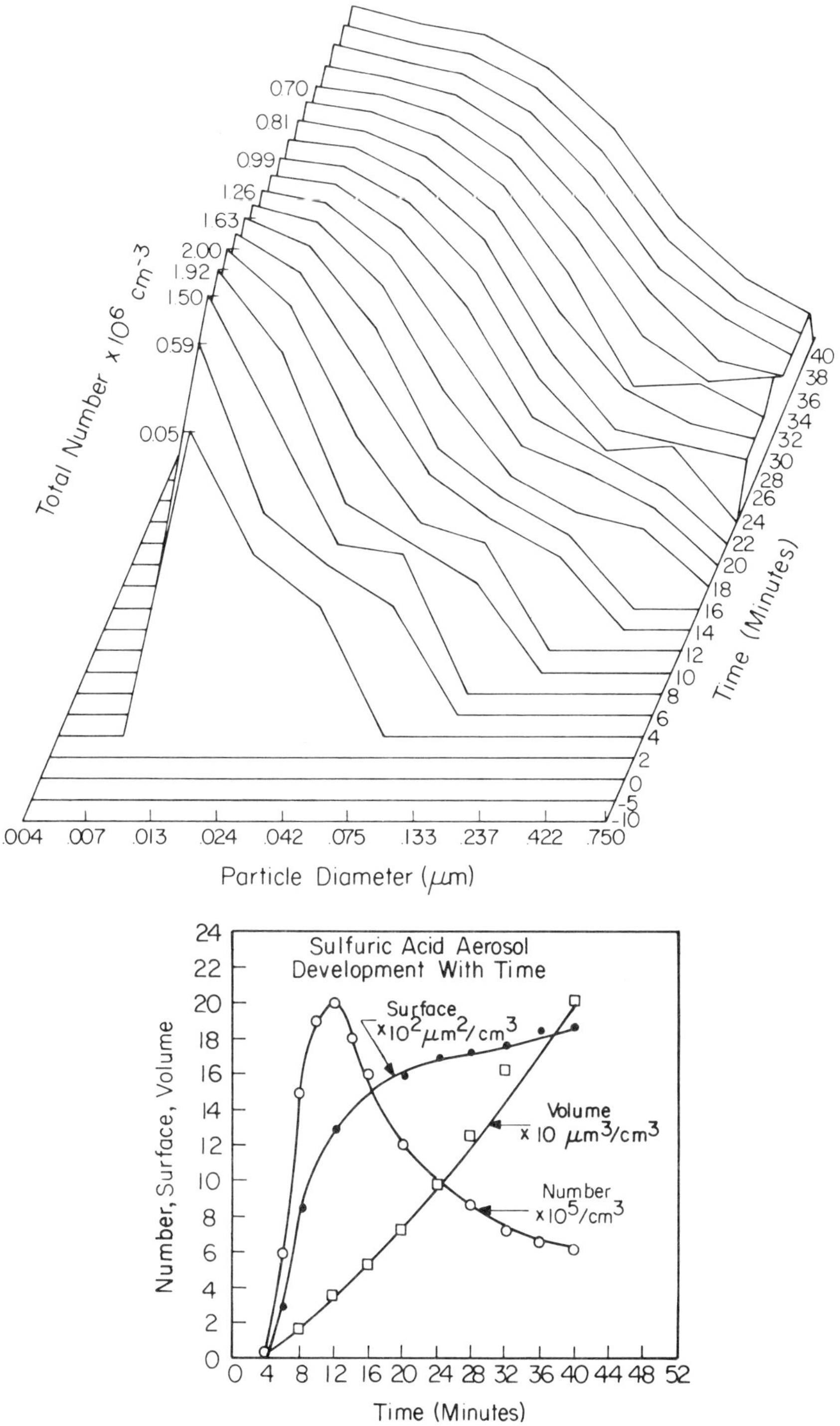

Figure 1. Aerosol development with time (number concentration).

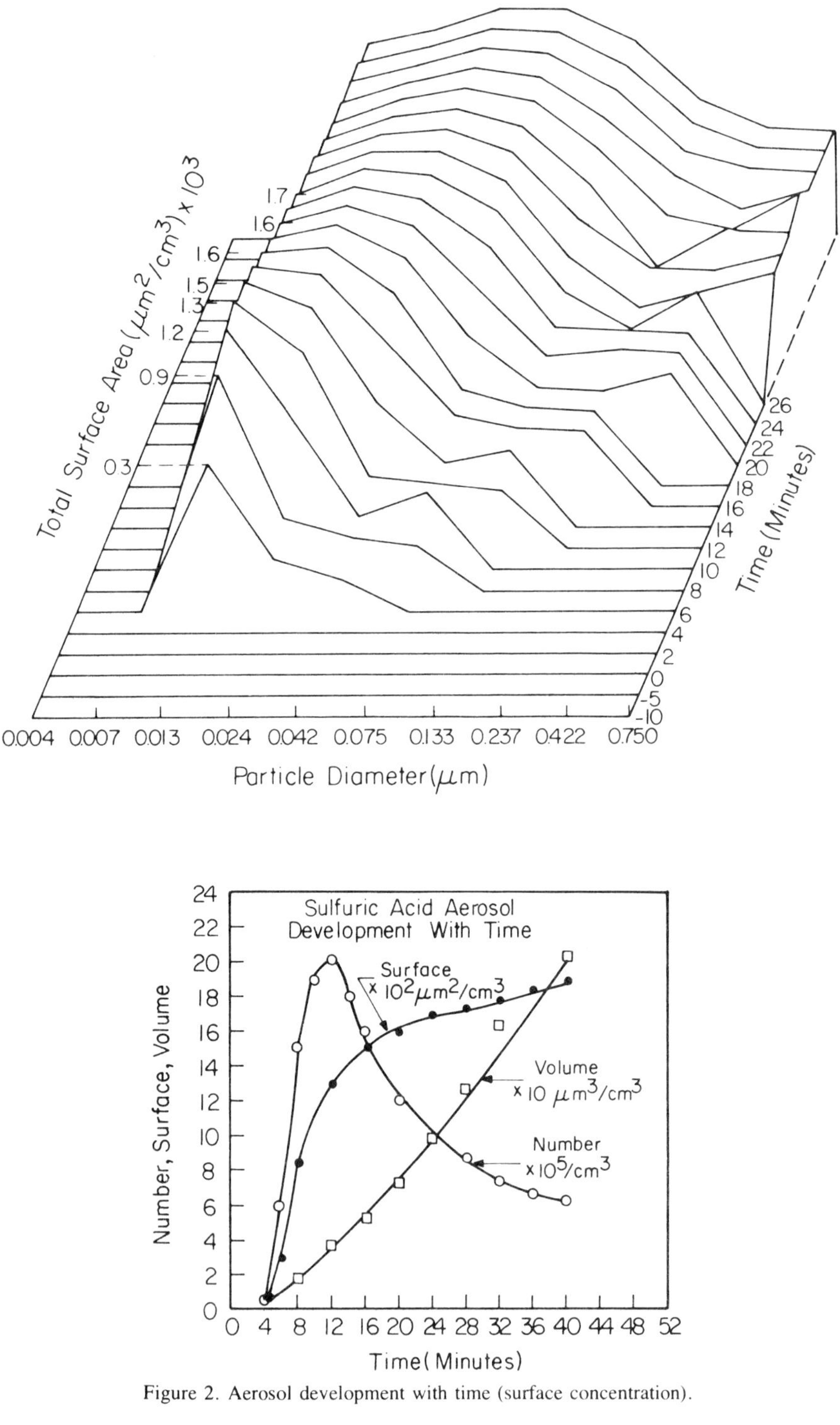

Figure 2. Aerosol development with time (surface concentration).

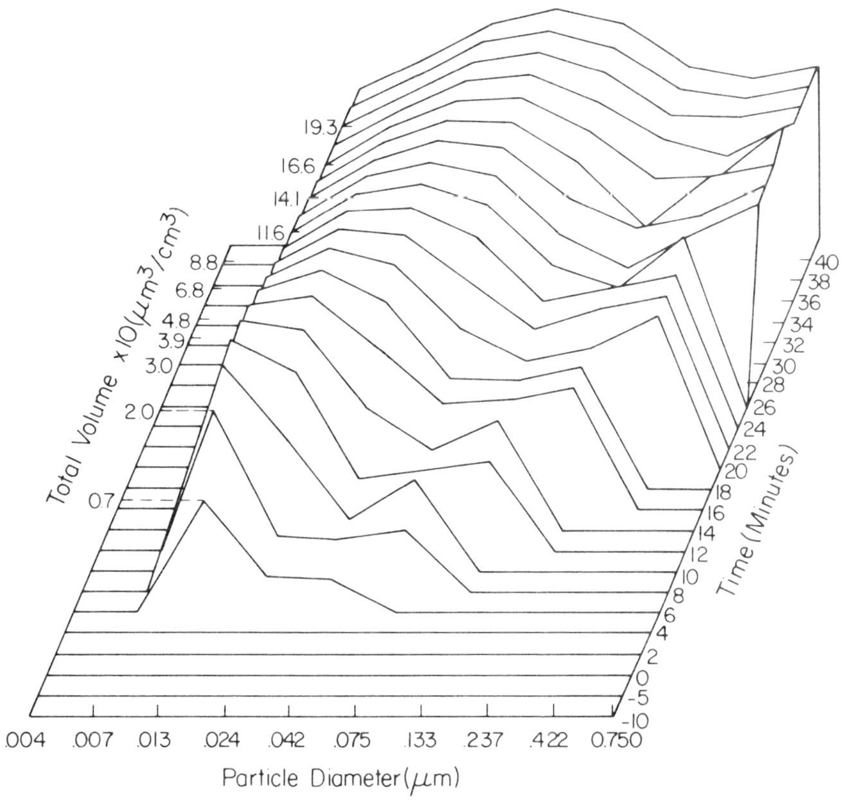

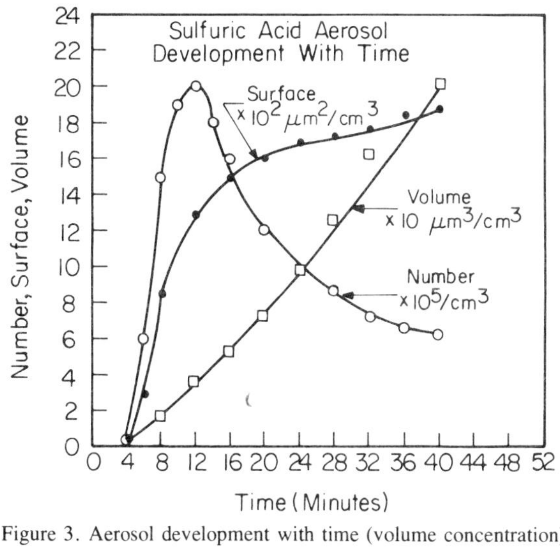

Figure 3. Aerosol development with time (volume concentration).

Determination of SO_2 Photooxidation Rate from EAA Data

In general, it has been found that the rate of SO_2 photooxidation in the SO_2 + clean air system approaches a constant value and that a plot of volume against time yields a straight line (Clark, 1972, 1975). The reasoning is that the rate of oxidation of SO_2 to SO_3 is equal to the rate of removal of SO_3 to form sulfuric acid droplets. The rate of production of H_2SO_4 aerosol, corrected for molecular weight change and water concentration, therefore equals the rate of photooxidation of SO_2 which is constant during the linear growth phase of the experiments. Under these conditions, the slope of the straight line volume growth curve can be related directly to the rate of photooxidation of SO_2 according to:

$$-\frac{d[SO_2]}{dt} = \frac{dv}{dt} \times \rho \times \underline{P} \times \frac{MW_1}{MW_2} \qquad (1)$$

where ρ is the density of the H_2SO_4 droplet, $\underline{P}$ is the weight fraction of H_2SO_4 in the drop, MW_1 is the molecular weight of SO_2, and MW_2 is the molecular weight of H_2SO_4. The quantities ρ and $\underline{P}$ can be determined from data given by Bray (1970) assuming that water vapor in the gas phase is in equilibrium with water in the aerosol droplets.

The above equation is true if one assumes that (1) all of the SO_3 formed combines with water to form H_2SO_4 droplets; (2) the sulfuric acid droplets are in equilibrium with the water vapor in the gas phase; and (3) the droplets are represented by a pure H_2SO_4 solution. For a clean, contamination-free system, these assumptions appear valid; however, in the presence of even small amounts of particulates, reactive hydrocarbons or nitrogen oxide species, large deviations from linearity can be expected with attendant accelerated aerosol growth. In addition to homogeneous gas-phase reactions involving the reactive intermediates generated from trace contaminants, heterogeneous paths may also contribute to accelerated SO_2 oxidation and aerosol growth. Thus, it has been found from careful examination of aerosol data in SO_2 + clean air experiments that frequently there is a slight upward curvature in the volume growth during the later stages of the experiment. This can be seen from the data shown in Figure 3. The upward curvature suggests a role of trace contaminants in the air which contribute to aerosol production. According to Calvert and McQuigg (1975), contributions to SO_2 photooxidation by HO_2-SO_2 and OH-SO_2 reactions would reach maximum values after about 30 minutes in the slightly polluted air. As the SO_2 + clean air irradiation proceeds, reactive intermediates build up within the chamber (depending on the background levels of NO_x and contamination of chamber surfaces) and the photooxidation rate of SO_2 increases. Three rates of SO_2 oxidation can therefore be computed from EAA data—one during the initial 30 minutes of an experiment, one during the maximum aerosol growth period and an average value which encompasses the full length of the experiment.

For the large Calspan chamber, average SO_2 photooxidation rates of about 1.0% hr^{-1} are typical. Variations in the SO_2 photooxidation rate have not been found to be tied to any obvious experimental variable, such as SO_2 concentration, relative humidity or chamber size. On the other hand, the history of chamber conditioning arising from preceding experiments seems to have an influence on the results in that somewhat high rates are usually observed after first completing experiments using auto exhaust or reactive hydrocarbons in the presence of NO_x. Another positive indicator of a reactive SO_2 system is the presence of a dark

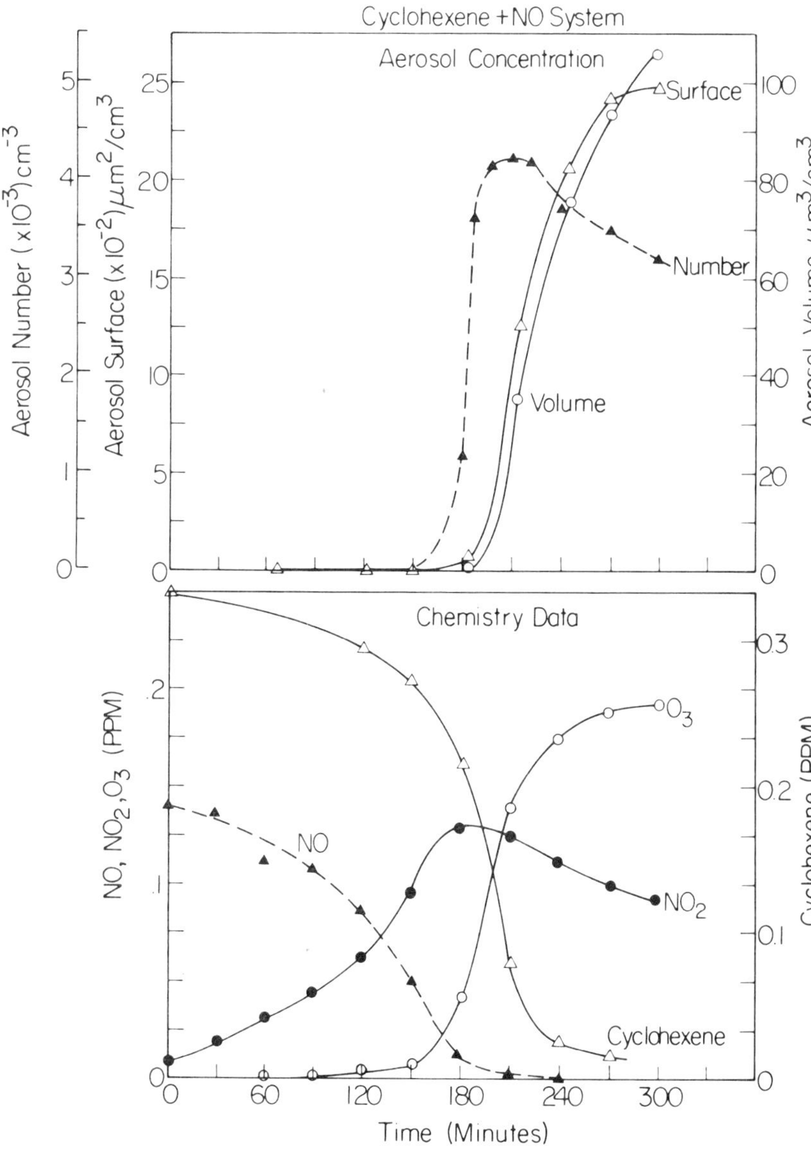

Figure 4. Aerosol development and chemistry data for a cyclohexene + NO system.

reaction after admitting SO_2 into the chamber. In both the Calspan and the University of Minnesota smog chambers, a conditioning effect is noted such that lower rates are observed after repeated SO_2 irradiations.

Hydrocarbon-Containing Systems

The HC + NO system represents another type of photochemical system which behaves quite differently from the simple SO_2 + clean air experiments. Each hydrocarbon + NO experiment can be divided into two phases. In the first phase, NO is converted to NO_2, and some oxidation of the hydrocarbon occurs. Ozone (O_3) starts to appear near the end of this phase as NO concentrations become very low. The second phase begins as soon as the initial NO has been oxidized out of the system and NO_2 reaches its maximum concentration. During this phase [NO] remains low and [NO_2] gradually decreases as NO_2 is converted to high oxides, acids, peroxyacyl nitrates, and other nitrogen compounds; ozone grows rapidly and approaches a maximum; and aerosol formation takes place.

A typical example of this two-phase behavior is shown in Figure 4 in which aerosol and chemistry data from a cyclohexene + NO experiment is given. It may be seen from the data that the NO disappears and NO_2 maximizes in about 180 minutes. This time is considered as the duration of the first phase. The plot also shows that by the end of this phase some ozone has started to appear and the hydrocarbon concentration has started to decrease rapidly.

Early in the second phase aerosol growth begins and proceeds rapidly. The same physical mechanisms control aerosol formation in this system as in the SO_2 + clean air system, i.e., nucleation, condensation, and coagulation. Note the sharp rise in aerosol concentration at this point and also the corresponding increase in the surface and volume concentration of aerosol. For all of the hydrocarbon + NO systems studied (toluene, hexene, m-xylene and cyclohexene) the number of particles formed was less than in the SO_2 + clean air case; but in every case, except for hexene + NO, the volume was always larger, implying the production of fewer but larger particles. A more thorough discussion of aerosol behavior in SO_2 + clean air, HC + NO_x, and HC + NO_x + SO_2 systems can be found in Kocmond et al. (1976).

The chemical behavior of the system in the second growth phase is also shown in Figure 4. Production of ozone proceeds rapidly and ozone concentration maximizes at about the same time that the aerosol volume curve reaches its plateau; NO_2 decreases continuously while hexene is oxidized out of the system.

Hydrocarbon + NO_x + SO_2 Systems

An additional degree of complexity is introduced when SO_2 is added to the HC + NO_x system. In terms of the chemical species which are monitored, the experiments with and without SO_2 are virtually identical. The aerosol behavior, however, is much different when SO_2 is added. During the first phase of the experiment, aerosol growth occurs in a manner similar to that of a simple SO_2 + clean air experiment; i.e., an essentially linear curve of volume against time. As the second phase of the experiment begins, the rate of volume production increases markedly. During this phase the growth curves are more like the hydrocarbon + NO experiments, except that the aerosol number concentrations are higher. In terms of qualitative behavior, a hydrocarbon + NO + SO_2 experiment behaves almost as though it resulted from a linear combination of the hydrocarbon + NO system and the SO_2 + clean air system. Quantitatively, however, more aerosol is usually formed in both growth phases than would be predicted by a simple linear combination of the hydrocarbon + NO system and the SO_2 + clean air system.

The large changes in aerosol behavior produced by the addition of SO_2 to the cyclohexene + NO_x system are shown in Figure 5. During the first phase of

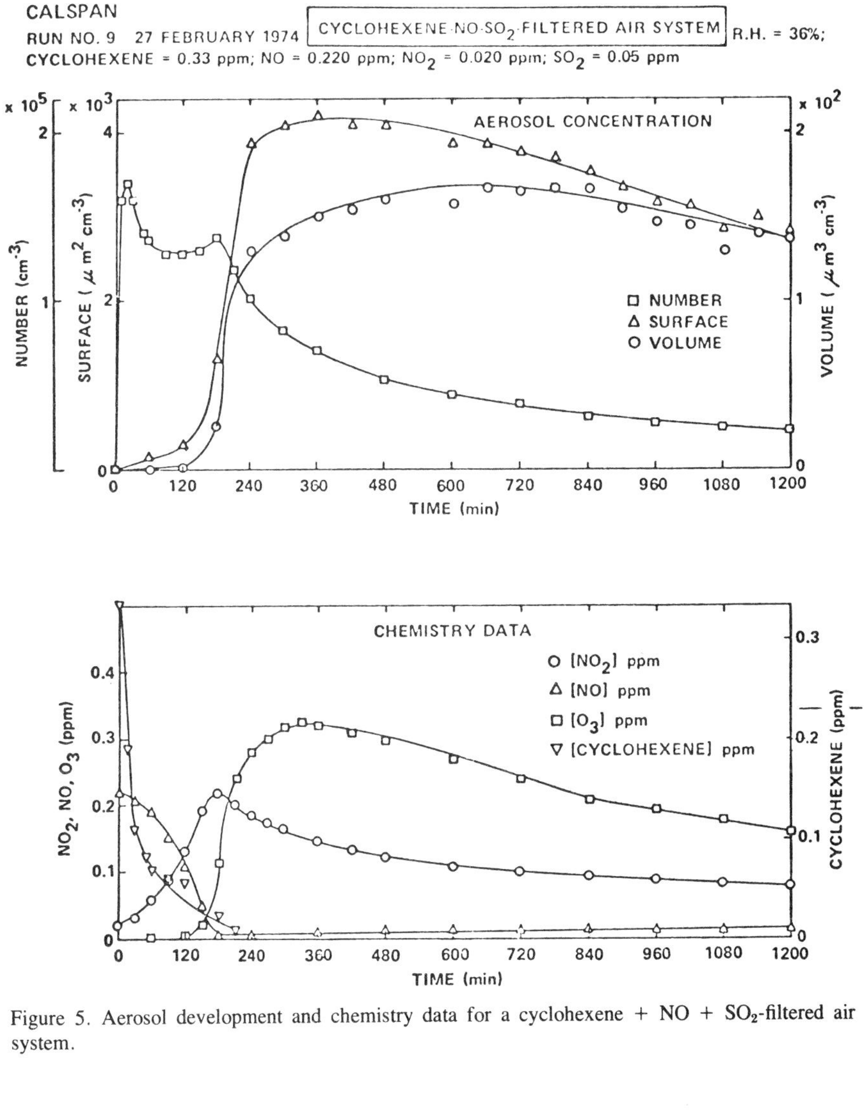

Figure 5. Aerosol development and chemistry data for a cyclohexene + NO + SO_2-filtered air system.

growth, aerosol production occurred as though a slightly contaminated SO_2 + clean air system was being irradiated.

In terms of volume production, the second phase growth is similar to the cyclohexene + NO system in the absence of SO_2. The differences that are observed are not considered significant because of the very rapid particle growth beyond the upper particle size limit of the EAA during this phase. The changes in surface and number concentrations associated with the addition of SO_2, however, are very significant. Both number and maximum surface concentration are much larger for the system with added SO_2. Since the total aerosol volumes produced in all cyclohexene cases are about the same, the additional surface area can only be accounted for by a dramatic decrease in particle size. This effect can be seen in Figure 6, in which mean surface diameter has been plotted as a function of time for several HC + NO and HC + NO + SO_2 systems studied. Data for a typical SO_2 irradiation are also shown in the figure. The almost explosive growth in the cyclohexene + NO case occurs at about the time of complete oxidation of NO and appearance of appreciable ozone. In the case shown, particle diameters of nearly 0.5 μm were produced in less than an hour from the time of initial nucleation. With SO_2 in the system, the particle sizes are much smaller. Under these conditions, initial nuclei formation results from SO_2 oxidation followed by additional condensation on existing particles during the second phase of growth. Although the particles are smaller, the number concentration is higher.

The m-xylene + NO system behaves like the cyclohexene + NO system but to a lesser extent. The addition of SO_2 results in the same effect; initial particle formation and growth at a rate similar to that for SO_2 alone, followed by accelerated growth during the second phase.

By comparison, the much less reactive hexene + NO + SO_2 system was observed to follow essentially the SO_2 particle growth curve for the duration of the experiment.

The implications of these data relative to the production of light scattering aerosol can be seen in Figure 7. In the figure the light scattering function b(SCAT) is plotted as a function of time for several HC + NO, HC + NO + SO_2 systems in the Calspan chamber. The production of very large particles in the cyclohexene + NO case results in the rapid formation of light scattering aerosol. Although the particle diameters are much larger than in the comparable cyclohexene + NO + SO_2 experiment, b(SCAT) is nearly the same because fewer particles are produced. The addition of SO_2 results in smaller but more numerous particles whose effect on light scattering is about the same.

The addition of SO_2 to the m-xylene + NO case can also be seen to produce substantial light scattering aerosol during the second phase of aerosol growth. For several hours there is no effect; but once NO oxidation is complete and second stage growth begins, large visibility losses are observed. The fact that the concentration of particles is much smaller in the m-xylene + NO case (compared to m-xylene + NO + SO_2) accounts for the lack of appreciable light scattering aerosol.

By contrast, the hexene system, even in the presence of SO_2, did not produce

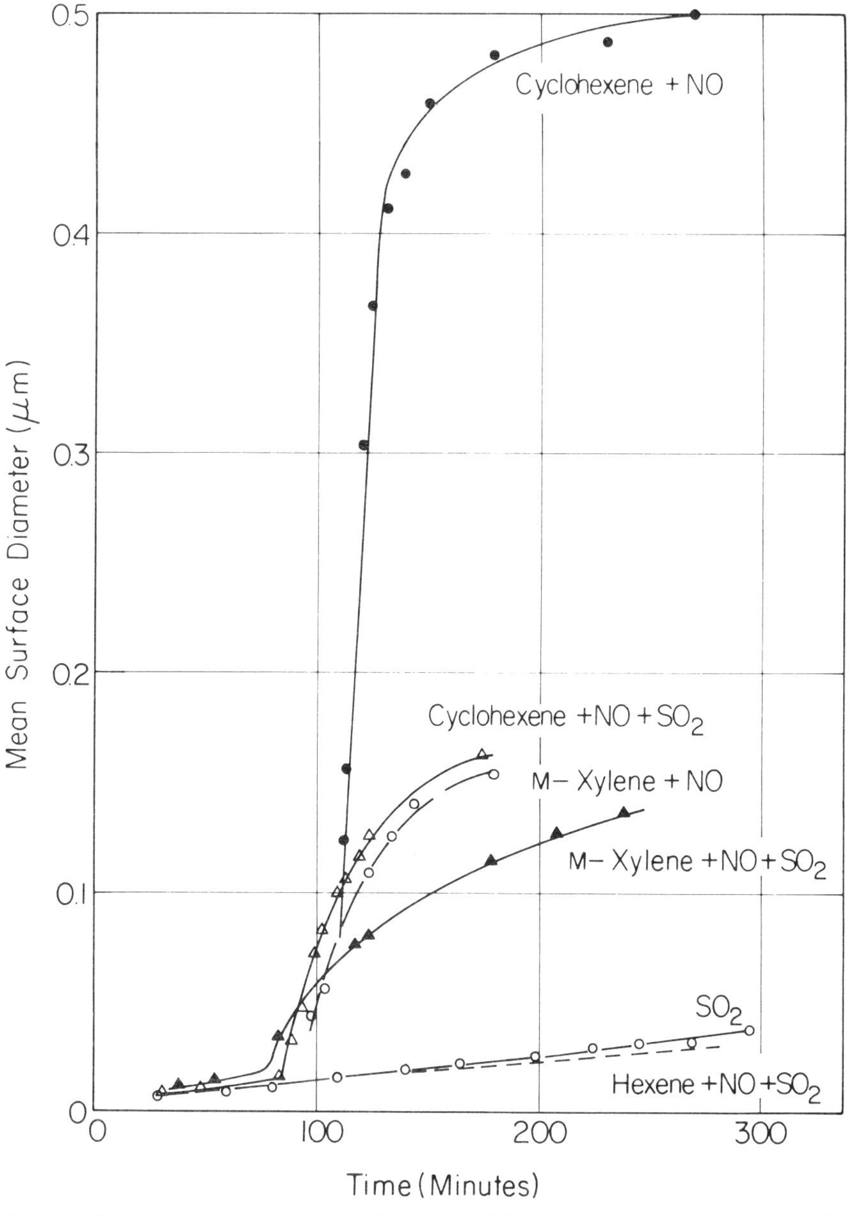

Figure 6. Mean surface diameter vs. time for several HC + NO, HC + NO + SO_2, and SO_2 experiments.

significant light scattering aerosol or visibility losses. Finally, the SO_2 alone, while producing very large numbers of particles, did not produce significant light scattering aerosol over the duration of these tests.

CONCLUSIONS

From data generated with the EAA, the following conclusions can be made relative to the photochemical systems studied.

(1) Average SO_2 photooxidation rates of about 1.0% hr^{-1} are typically ob-

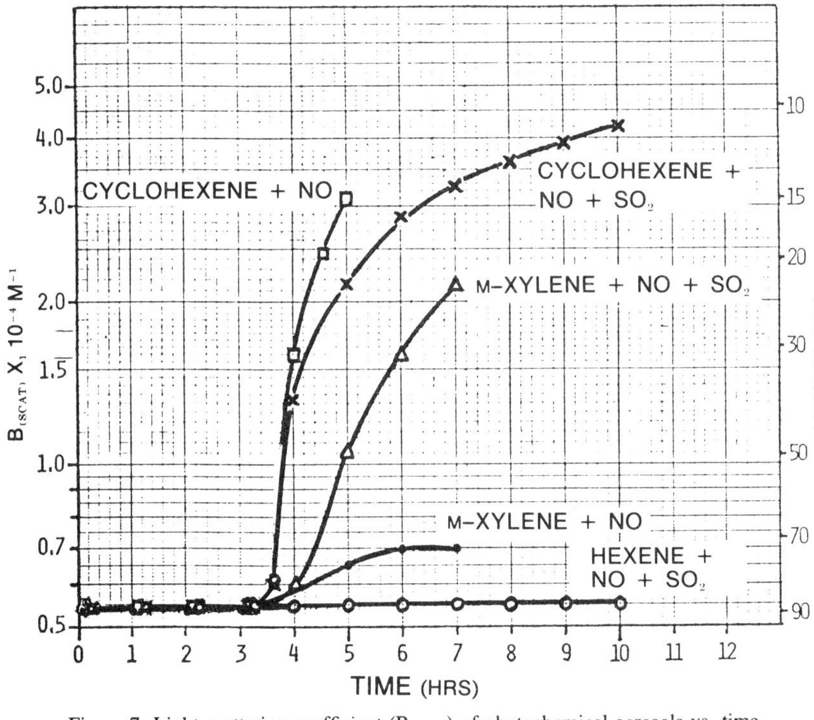

Figure 7. Light scattering coefficient (B_{SCAT}) of photochemical aerosols vs. time.

served in the Calspan chamber for a light intensity of about 80% noonday sun.

(2) It is possible to characterize system reactivity in terms of aerosol behavior. The most important variables were found to be maximum number concentration, equilibrium surface concentration, and volumetric growth rate. These aerosol measures of reactivity have been found to correlate well with other conventional parameters, such as time to $[NO_2]_{max}$ and $[O_3]_{max}$.

(3) Each HC + NO experiment can be divided into two phases. In the first phase, NO is converted to NO_2 and some oxidation of hydrocarbon occurs. Ozone starts to appear near the end of this period. The second phase, accompanied by substantial aerosol formation, begins as soon as NO is oxidized out of the system and NO_2 reaches its maximum; ozone grows rapidly and approaches a maximum.

(4) The addition of SO_2 to the HC + NO system was generally found to exert a synergistic effect on aerosol formation. Possible synergistic effects in the cyclohexene system were masked by the explosive growth of aerosol with and without the addition of SO_2.

(5) The addition of SO_2 to the HC + NO system produces a dramatic decrease in the mean particle diameter. This results from the initial formation of very high concentrations of nuclei during the initial stages of the experiment. During the second stage of aerosol growth, condensation proceeds on the existing particles rather than forming fewer but larger particles typical of the HC + NO system.

REFERENCES

Bray, W. H., 1970: Water vapor pressure control at aqueous solutions of sulfuric acid. *J. Materials, 5,* 233–248.

Calvert, J. G., and R. D. McQuigg, 1975: *Int. J. Chem. Kinetics,* Symp. No. 1, 113.

Castleman, A. W., Jr., R. E. Davis, H. R. Munkelwitz, I. N. Tang, and W. P. Wood, 1975: *Int. J. Chem. Kinetics,* Symp. No. 1, 629.

Clark, W. E., 1972: Measurement of aerosol produced by the photooxidation of SO_2 in air, Ph.D. Dissertation, University of Minnesota.

——— and K. T. Whitby, 1975: Measurement of aerosols produced by the photochemical oxidation of SO_2 in air. *J. Colloid Interface Sci., 51,* 477.

Kocmond, W. C., D. B. Kittelson, J. Y. Yang, and K. L. Demerjian, 1973: Determination of the formation mechanisms and composition of photochemical aerosols. First Annual Summary Report, Calspan Report No. NA5365-M-1, Calspan Corporation, Buffalo, New York 14221.

———, ———, ———, and ———, 1975: Study of aerosol formation in photochemical air pollution. Calspan Report No. NA5365-M-2, Calspan Corporation, Buffalo, New York 14221.

———, ———, ———, ———, and K. T. Whitby, 1976: Aerosol Formation in Simple Photochemical Systems. Accepted for publication in *Advances in Env. Sci. and Tech.,* Vol. 9, Part I and II. New York, Wiley-Interscience.

Liu, B. Y. H. and D. Y. H. Pui, 1975: On the performance of the electrical aerosol analyzer. Accepted for publication in *J. Aerosol Sci.*

Steadman, D. H. and H. Niki, 1973: Photolysis of NO_2 in air as measurement method for light intensity. *Environ. Sci. Technol., 7,* 735.

QUESTIONS

Q: Several speakers have mentioned that under certain circumstances they have been able to measure particles in the 0.0032 μm region with the EAA. Has anyone seen such particles other than when there is sulfur dioxide present?

Kocmond: We have not seen particles in that size range in systems which do not contain SO_2.

Marlow: I put forth the speculation that very small particles are forming because of the presence of SO_2. Is there SO_2 oxidation in the charger?

Whitby: In the earlier TSI Model 3000 Electrical Aerosol Analyzer, such as the one used by Dr. Clark in his thesis work, there were extremely small particles generated in the charger section when SO_2 was present. However, we have seen no evidence of particle formation in the TSI Model 3030 analyzer. Dr. Kittelson and Dr. Kocmond have worked with a number of chemical systems and, from all evidence, the small particles which were measured were formed in the smog chamber. I would guess that our uncertainty in the concentration is about a factor of five or ten and the uncertainty in the size is about a factor of two at the very low end of the EAA. However, we are quite certain that we are detecting these small particles.

Q: I worked briefly with an aerosol generator which produced very small aerosol particles, the platinum generator at HASL, and it was not possible to measure particles with the EAA below Channel 4.

Whitby: One time in Detroit we had four or five analyzers side by side, and another time we had three. We found out in comparing them that in one instrument we could see the very small particles very nicely and in the other two we couldn't. The instruments are not all alike in their behavior in the first two channels.

Marlow: In the initial runs, and even beyond that, of the experimental data which I presented, I ran two instruments side by side because I thought it would be a very easy way to get simultaneous data. However, the systematic uncertainties were so great that I had to go to switching back and forth with the same instrument setup because I just couldn't compare the measurements made with two instruments.

Liu: There are occasions I know where people have observed the 0.0032 μm particles in a high temperature laboratory. They were producing particles in a plasma jet by vaporizing carbon and were able to saturate the EAA with 0.0032 μm particles.

Marlow: In that process you are going to end up having charged particles that are charged by a process that has absolutely nothing to do with your device. In a plasma situation you are going to have unipolar charged particles at practically any size due to any number of different processes such as thermoionic emission and so on.

MEASUREMENT OF ULTRAFINE PARTICLE SIZE DISTRIBUTIONS IN INDUSTRIAL FLUE GASES USING AN ELECTRICAL AEROSOL ANALYZER

JAMES W. RAGLAND

Southern Research Institute, Birmingham, Ala.

In conjunction with the environmental resources work at Southern Research Institute (SRI), a Thermo-Systems, Inc. Model 3030 Electrical Aerosol Analyzer (EAA) is used to measure ultrafine particle size distributions in industrial flue gases. A major part of this task involves the extraction, conditioning, and dilution of the sample gas to concentrations and temperatures within the design limitations of the Model 3030. The following paragraphs briefly describe the sample extraction-dilution process, the minor hardware modifications necessary to make the Model 3030 more adaptable to a rugged field environment, and the special data acquisition procedures.

SAMPLE EXTRACTION-DILUTION SYSTEM

When possible, instack sampling is preferred because it eliminates many condensation and sample loss problems which occur when probes are used for extractive sampling. Unfortunately, the existing ultrafine sizing techniques (diffusional mobility analysis and electrical mobility analysis) do not lend themselves to instack sampling. Particulate concentrations are extremely high in an industrial flue and vary by orders of magnitude from one industrial process to another, and from the inlet side of a control device to the outlet of the same device. Temperature, pressure, moisture content, and the physical properties of the particulate also vary widely from one industrial process to another. Because of this complexity and the limited useful concentration range for these sizing techniques (e.g., due to coagulation losses) as shown by Haberl and Fusco (1970) extensive sample dilution and conditioning is required to obtain information on ultrafine particles in an industrial gas stream.

Figure 1 is a block diagram of the sample extraction-dilution system developed

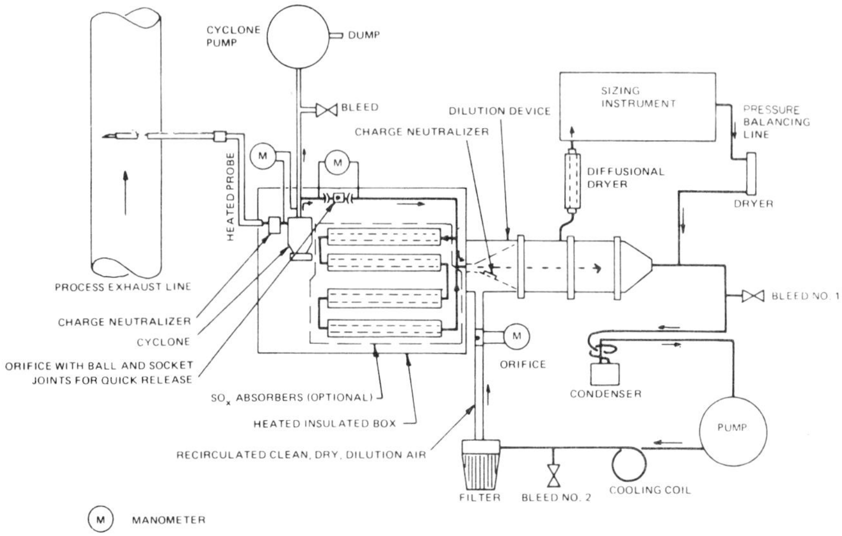

Figure 1. Sample extraction-dilution system (SEDS).

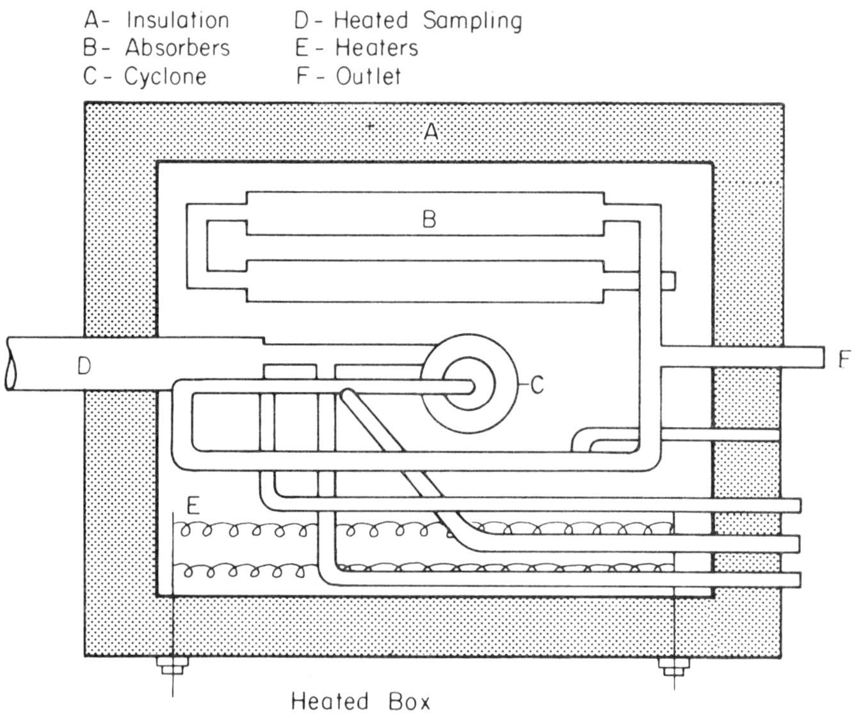

Figure 2. Hot box, top view.

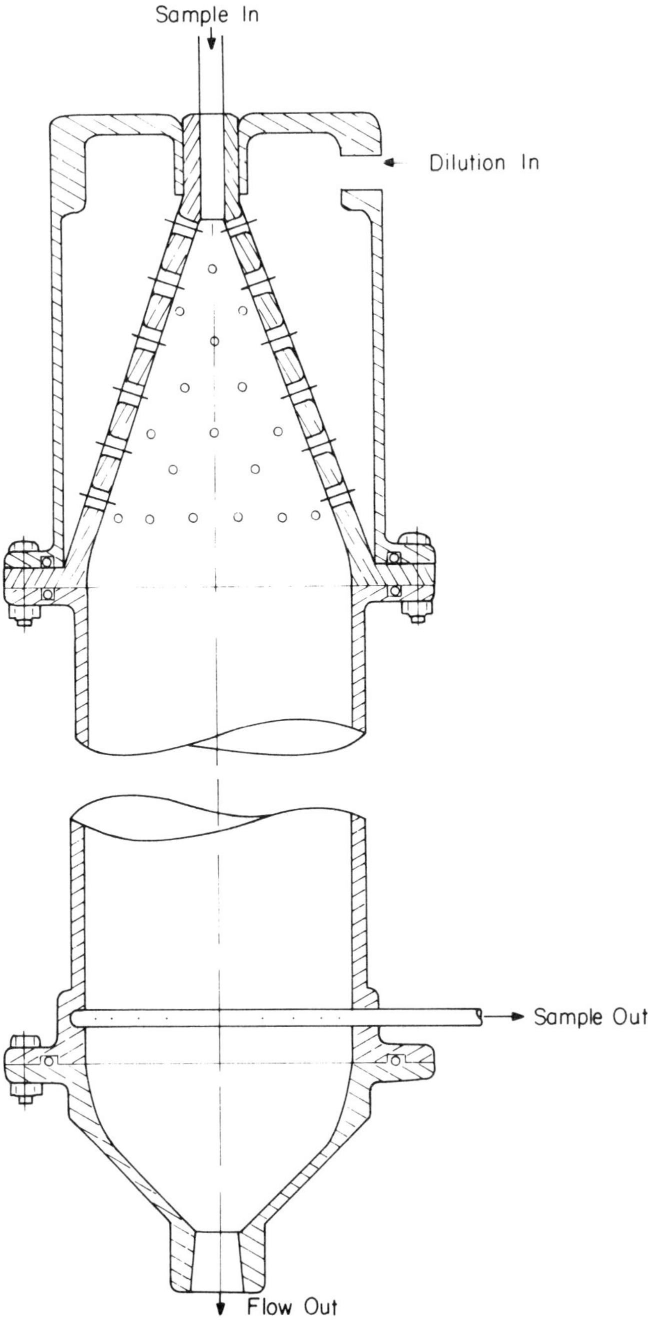

Figure 3. Diluter, cutaway view.

for this purpose by SRI. A 0.5 ACFM sample flow is removed from the process exhaust stream and is pulled through a rigid probe, a flexible connector hose, and a cyclone into a "T" where the flow splits. The excess flow (needed to maintain a constant 0.5 ACFM flow through the cyclone) is dumped to ambient and the desired sample flow (0.003 to 0.5 ACFM) goes into the diluter via a calibrated orifice and an optional sulfur oxide absorber bank. The cyclone, orifice, and sulfur oxide absorber bank are housed in a heated box (Figure 2) so that all components of the system except the diluter can be maintained at 400°F to prevent condensation. Pressure taps for the cyclone and the orifice allow continuous monitoring of the cyclone flow rate and the orifice flow rate by reading the pressure drop across the respective component.

The Charge Neutralizer shown will be added at a future date. Currently, charge neutralization is done in the cone of the diluter by two, 500 μc, Polonium-210 strips, mounted as shown. The sample gas then enters the dilution chamber at the apex of a perforated cone into which clean, dry air is pumped

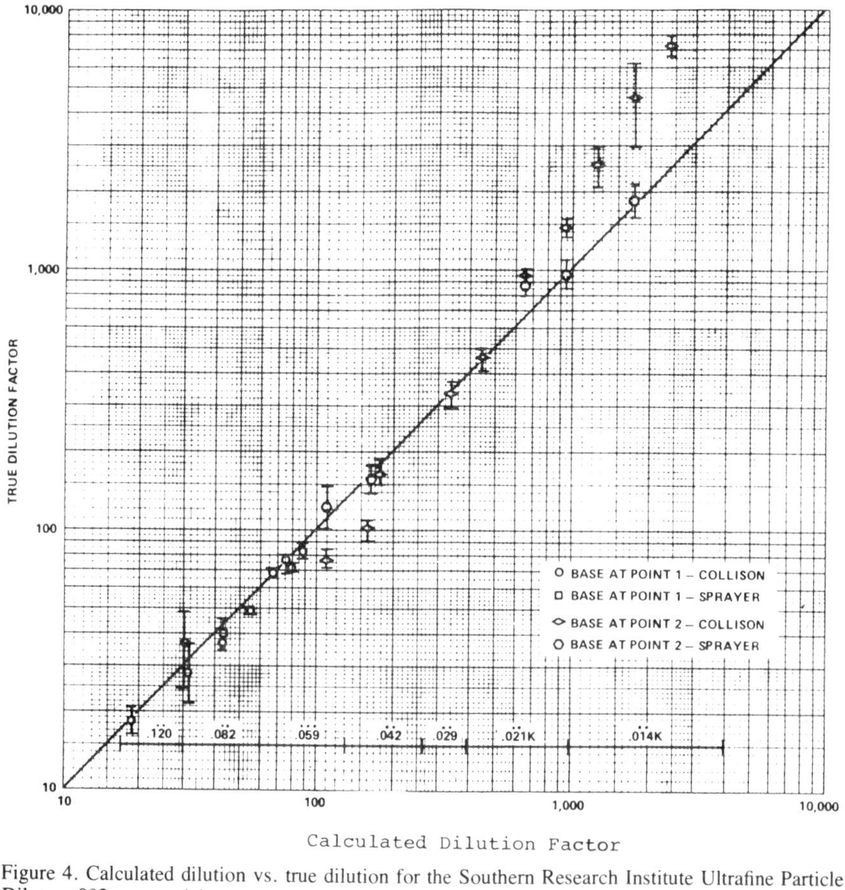

Figure 4. Calculated dilution vs. true dilution for the Southern Research Institute Ultrafine Particle Diluter, .092 μm particles.

through the perforations, creating a highly turbulent mixing zone. Calibration data for the SRI Diluter (Figure 3) is shown in Figures 4 and 5. At a downstream point after adequate mixing has occurred, the diluted sample is extracted and conveyed to the sizing instrument. This diluted sample passes through a diffusional dryer where any remaining moisture is removed (the major drying action is accomplished by using dilution air which has been passed through an ice bath condenser). The exhaust instrument lines are returned to the diluter to maintain a minimal pressure drop across the instruments. Further drying of this recycled air is accomplished by passing this instrument exhaust gas through an absorption drier.

Changing the sample air flow and the dilution air flow allows one to change the dilution ratio (the ratio of the total flow to the sample flow). Sample air and dilution air flow rates are controlled by two bleed valves on the dilution air pump, one upstream of the pump (#1) and one downstream (#2). Manipulation of these valves changes the internal pressure of the diluter which, in turn, sets the sampling rate. As the pressure in the diluter is reduced, the sample flow rate is increased.

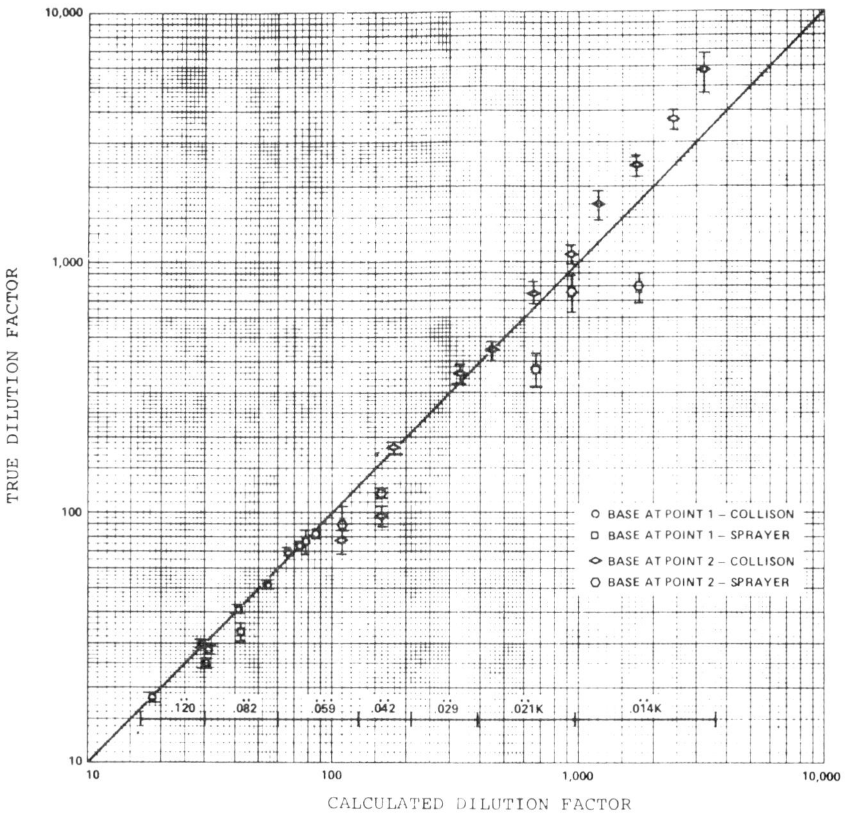

Figure 5. Calculated dilution vs. true dilution for the Southern Research Institute Ultrafine Particle Diluter, .15 μm particles.

In practice, the operation of these valves changes the dilution air flow only slightly (approximately ± 10%) for a many fold change in sample flow (from negative sample flow, or backflush condition, to the maximum (design) flow value attainable with any particular sample metering orifice). Foust (1960) and Spink (1957) described flow metering orifices (head flow meters).

McCain (1974, 1975) reported some problems with the former sample conditioning system: growth of particles when high concentrations of SO_3 (SO_3/H_2SO_4) are present, pluggage of orifices, single point sampling limitations, and diffusional losses in sampling lines. The new hot box configuration includes optional SO_x absorber chambers (partial removal of SO_x gases by diffusion to an absorber

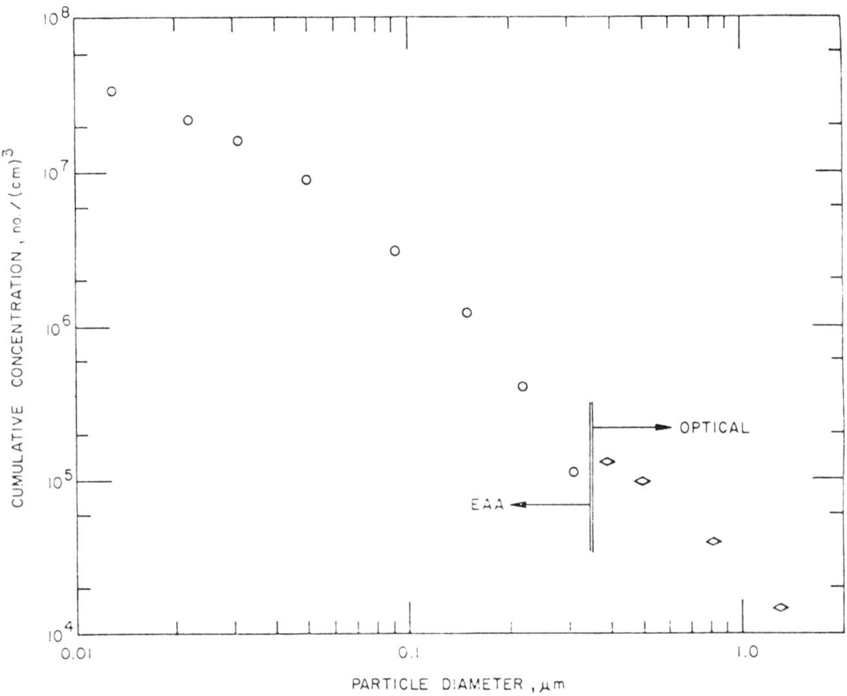

Figure 6. Inlet size distribution obtained at a coal fired power boiler.

medium), rapid replacement for plugged orifices, quick positive determination of partial pluggage, full traverse sampling capability, and decreased sample line losses (by using higher flow rates in the system). The hot box configuration also increases the total time available for data acquisition by decreasing the time needed to change orifices or dilution ratios (2 minutes compared to about 30 minutes).

The new hot box configuration appears to represent a significant engineering improvement to the sample extraction and dilution system by allowing more rapid and convenient data acquisition. If the SO_x removal system is successful a severe problem which has introduced large uncertainties in the ultrafine particle size measurements will also have been solved. Lead oxide, activated charcoal, copper, barium oxide and calcium appear to be promising reagents for this purpose.

Subsequent testing has shown activated charcoal to be a successful SO_x absorber reagent.

THERMO-SYSTEMS INC. MODEL 3030

Field testing to date, using the EAA, has revealed three items worthy of note:

1. The automatic scan mode should not be used because of intermittent concentration fluctuations. A simple remedy is to display the instrument output on a chart recorder and manually cycle channels when it is obvious that a good average (one to two minutes) has been obtained.

2. The instrument contains two calibrated mass flow meters for metering flow rates, however it should be noted that the output of a mass flow meter is not actual volume flow (ALPM). Correct data reduction from existing calibration data depends on obtaining the proper residence time in the charger section and the proper velocity and residence time of the particles in the collection region, both of which are determined by the actual volumetric flow rate (ALPM).

3. The charger section of the instrument utilizes spring loaded electrical contacts to connect to an element held in place by a friction fit. This connection can be vibrated loose in shipment and/or in a field environment. A set screw installed on the Flow Straightener Cylinder seems to have cured this problem (Drawing No. SA0336). Current production models incorporate this modification.

Additional minor hardware modifications include moving the electrometer gain switch to the front panel and replacing the terminal block connectors with the push-on, quick-disconnect circular connectors.

To date the EAA has been used in nine separate field tests. On six of those it was the primary sizing instrument for ultrafine particles and on three it served as secondary instrumentation. The industrial processes tested included an asphalt plant, a clinker cooler for a Portland cement plant, a paper mill kraft recovery boiler, one oil fired and four coal-fired power boilers. The EAA performed with varying degrees of success, the main difficulty being susceptibility to a high vibration environment as described earlier. Figure 6 shows typical field data obtained using the EAA.

REFERENCES

Foust, Alan S., et al., 1960: *Principles of Unit Operations*, John Wiley & Sons, p. 406.

Haberl, J. B. and S. J. Fusco, 1970: Condensation Nuclei Counters: theory and principles of operation. General Electric Technical Information Series, No. 70-POD 12.

McCain, J. D., K. M. Cushing and W. B. Smith, 1974: Methods for determining particulate mass and size properties: laboratory and field measurements. *J. Air Pollution Control Assoc., 12,* 1172–1176.

———, J. P. Gooch and W. B. Smith, 1975: Results of field measurements of industrial particulate sources and electrostatic precipitator performance. *J. Air Pollution Control Assoc., 2,* 117–121.

Spink, L. K. and D. M. Considine, 1957: *Process Instruments and Controls Handbook,* McGraw-Hill, pp. 4–6.

COMPARISON OF ELECTRON MICROSCOPY AND THE ELECTRICAL AEROSOL SIZE ANALYZER FOR DETERMINATION OF SIZE DISTRIBUTION OF A SUBMICRONIC SALT AEROSOL

R. F. PHALEN, A. T. HO, AND J. L. KENOYER

Department of Community and Environmental Medicine, University of California, Irvine

The aerosol parameters of interest in our controlled inhalation toxicity studies include physical form(s), chemical composition, size distribution, airborne mass concentration and the time stability of these aerosol parameters during the course of an inhalation exposure. In some instances other aerosol parameters, such as distribution of electrical charge, specific surface of the particles and rates of dissolution in aqueous media are additionally important to interpreting the inhalation toxicity. As a means of simplification here we will concentrate only on size distributions of aerosols. Further, we will use a sodium chloride aerosol whose geometric diameter is well below 1 μm and will make the useful assumption that our aerosol is describable by the log-normal distribution function.

In order to obtain a reasonably valid description of size distribution, one should: a) obtain a representative sample; b) have a means of classifying the aerosol into a sufficient number of size intervals; c) understand what aerosol property is being measured (e.g., aerodynamic, light scatter, geometric diameters, etc.); and d) verify the result by an independent method.

The major objective of this paper is to describe and evaluate some of our experiences with the Electrical Aerosol Size Analyzer (EASA: Thermo-Systems, Inc. Model 3030, Serial No. 054) in an aerosol inhalation toxicology research program. This instrument is of interest to us because: a) it provides information on airborne particles in the submicron size range; b) it has several size channels so that size distributions are obtainable; c) it has rapid readout of data so that information on particles can be obtained during exposures, and d) it provides a measurement of electrical mobility, which presumably can be related to geometrical diameter (Liu and Pui, 1975). The EASA was evaluated by carefully prepar-

This research was supported by the California Air Resources Board under Contract Number ARB 5-725.

ing submicron-size aerosols by nebulization, and sampling simultaneously with combinations of instruments including electrostatic (Morrow and Mercer, 1964) and thermal (Lauterbach et al., 1952) precipitators, and the Electrical Aerosol Size Analyzer (Liu and Pui, 1975). Additionally, the expected aerosol size distribution was calculated from the atomizer droplet size distribution. This droplet size distribution was calculated using information supplied by the manufacturer and verified by us by electron microscopy of dried fluorescein particles generated by the atomizer. Comparisons of size-distribution data obtained using each of the sizing instruments were made in order to shed light on the accuracy of each instrument. From the outset electron microscopy was relied upon for obtaining accurate size distributions. Our experience has been that electron microscopy is indispensable as a method for evaluating other sizing instruments when solid, stable, dry aerosols larger than about 0.04 μm in diameter are being sampled.

METHOD

During this study, aerosol was generated by atomizing a 10% (by weight) solution of the salt, NaCl. The generation and sampling instruments are described in Table I. Each sampler was used to sample both clean and aerosol-laden air. Sampling rates and other operating parameters were maintained at the optimal values for each instrument. Flowrates, regulated by rotameters, were calibrated using spirometers. Aerosols were generated by a compressed-air atomizer[1] brought to charge equilibrium by passage through a 2mCi ^{85}Kr ion source, diluted with filtered air and blown into a clean 1 m^3 volume stainless steel chamber.[2] The overall set up for generation and sampling is shown in Figure 1. The protocol for generating and sampling aerosol used during this investigation involved establishing a flow of clean air (90 ℓpm) through the chamber, turning on the aerosol generator, allowing 30 minutes for equilibration, and sampling while aerosol was continuously generated. The concentration of aerosol present in the chamber was approximately 4 mg/m^3 during the sampling period. Aerosol was sampled from the center of the chamber either directly into each sampler, or through a large bore (⅜ in diameter) minimum length sampling line (75 cm of aluminum tubing). All sampling instruments except the EASA were inside the chamber and therefore did not require sampling lines. The aerosol chamber was operated at the temperature and relative humidity of the surrounding laboratory room (about 23°C and 40–50%, respectively). Sheath air for the EASA was drawn from the laboratory room.

The atomizer was operated with filtered compressed air at 45 psig, and a flow rate of about 9 ℓpm. Clean dried dilution air was provided at a flow rate of about 15 ℓpm to the freshly nebulized droplets.

Samples for electron microscopy were collected on carbon-coated grids. Care was taken in E.M. sizing: a) to avoid artifacts associated with inadequate cleanliness, and poor electrical and thermal conductivity of grid surfaces; b) to deter-

1. Collison type, Model 7330, Sierra Instruments, Inc., St. Paul, Minn.
2. University of Rochester type (Leach et al., 1958).

TABLE I. Major Apparatus Used in Study

Instrument	Manufacturer	Description	Comments
Electrical Aerosol Size Analyzer (Model 3030) (Liu and Pui, 1975)	TSI (St. Paul, Minn.)	Charges & collects particles 1.0 to less than 0.01 μm diam.	Channel cutoff diameters supplied by manufacturer
Electrostatic Precipitator (Morrow and Mercer, 1964)	ARIES (Albuquerque, N.M.)	Point-to-plane sampler, charges and collects particles on electron microscope grid	Collection reported to be relatively representative down to about 0.05 μm, unknown below that diameter
Particle-Size Analyzer (Model TGZ3)	Carl Zeiss, Inc. (W. Germany)	Sizes images from photographs into 48 channels	Measurement of circles of known diameter indicated that the channel intervals were properly calibrated by the manufacturer
Fluid Atomizer (Model 7330)	Sierra Instruments Inc. (St. Paul, Minn.)	Collison-type compressed air nebulizer with impactor for larger droplets, and ^{85}Kr ion source	Droplet distribution derived for nebulizer by nebulization of fluorescein and sodium chloride aerosols from high concentration solutions. Sizing determined from E.M. samples
Thermal Precipitator (Lauterbach et al. 1952)	Built after design of Lauterbach	Oscillating type; hot wire above electron microscope grid	Collection reported to be 99.9+% efficient for all particles below 1 micrometer

mine accurately magnifications on E.M. photographs; c) to insure sufficient representative sampling from each grid; and d) to obtain high resolution photographs such that particles in the 0.01 μm size range could be seen. Magnifications of 28 thousand to 36 thousand on prints were obtained using a Zeiss Model 9-S2 electron microscope.

RESULTS AND DISCUSSION

Each sizing instrument appeared to give reproducible results and each was able to provide relatively unambiguous information for estimating a count median diameter and geometric standard deviation. The assumption of log-normality appeared to be valid for the NaCl aerosol.

Agreement between instruments is shown in Figure 2. The reasonable agreement of the thermal and electrostatic sample with the expected size distribution supported our confidence in electron microscopy. Each of these 3 estimates of the count median diameter and geometric standard deviation were within 12% of the mean results of 0.093 and 1.86μm, respectively. Electron micrographs obtained for each precipitator are shown in Figure 3. The EASA indicated the presence of large numbers of particles (greater than 50% of the total number) in the size range between 0.01 and 0.04 μm (channels 3, 4 and 5). Channels 1 and 2 of the EASA

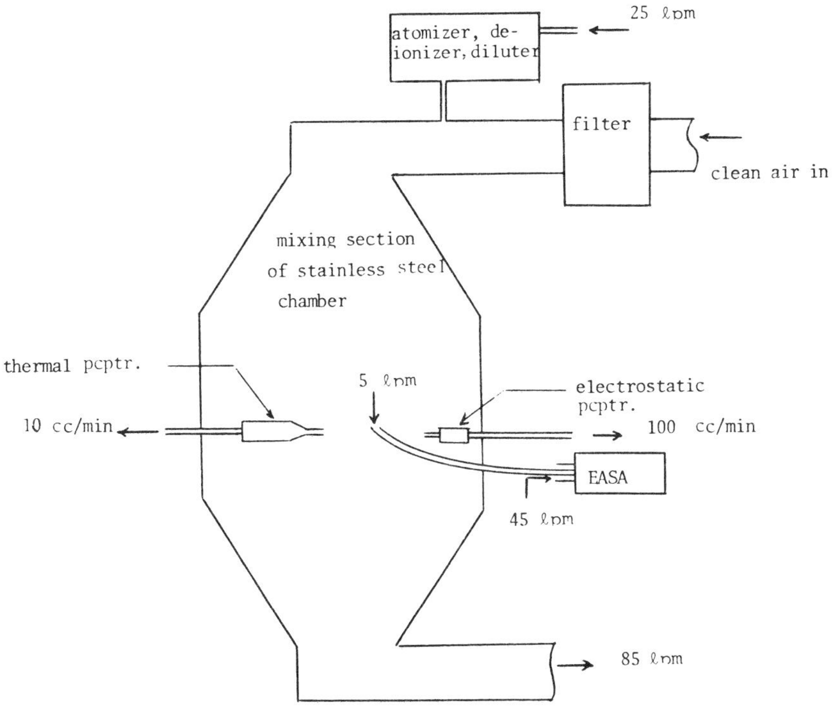

Figure 1. Apparatus used for generation and sampling of NaCl aerosol.

were ignored as recommended by the manufacturer. Predictions from the expected nebulizer size distribution and analysis of electrostatic and thermal E.M. samples indicated a much smaller number of particles in this size range.

New constants for calculating the number of particles in each channel of the EASA were used to determine size distributions from our data. These new constants led to slightly larger estimates of the count median diameter but did not appear to significantly alter the outcome of the study.

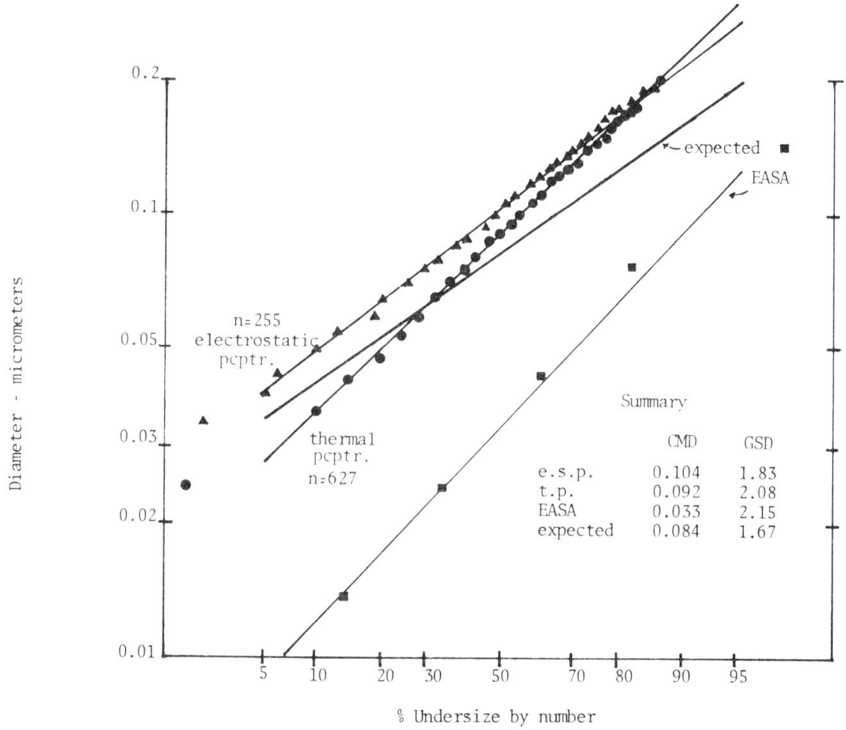

Figure 2. Size distributions for NaCl aerosol, generated from a 10% solution with a Collison atomizer, as obtained with various sampling instruments. The "expected" line was drawn using information supplied by the atomizer manufacturer and verified to be approximately correct in a separate study (n = # particles).

The thermal precipitator appeared somewhat more efficient in collecting the smaller particles than was the electrostatic precipitator, but both instruments provided nearly identical size distributions when the same aerosol was sampled. The discrepancy between the size distributions obtained by electron microscopy and by the EASA is the main subject of the remaining discussion.

In runs with clean air in the chamber, both electron microscopy and the EASA failed to see a significant number of particles in the size range below 0.04 μm. If these particles were present during nebulization they were not expected and were not collected by either the thermal or electrostatic precipitators. Since the collection efficiencies of these precipitators are not known in the very small size range, it is possible that the size distributions given by the EASA are correct. Additional experimental studies are required to confirm or deny this possibility.

Four possible explanations for this discrepancy should be considered. First, the EASA may have collected charges not associated with solid particles (e.g., clusters of charged gaseous molecules, as suggested by Marlow, 1976) and thus led to errors in the particle size distributions. Second, the use of a correction factor for each channel of data in the EASA is strictly valid only if the aerosol distribution is known within the channel. If the actual aerosol distribution is not used to determine this factor, multiplication by the correction factor may over- or

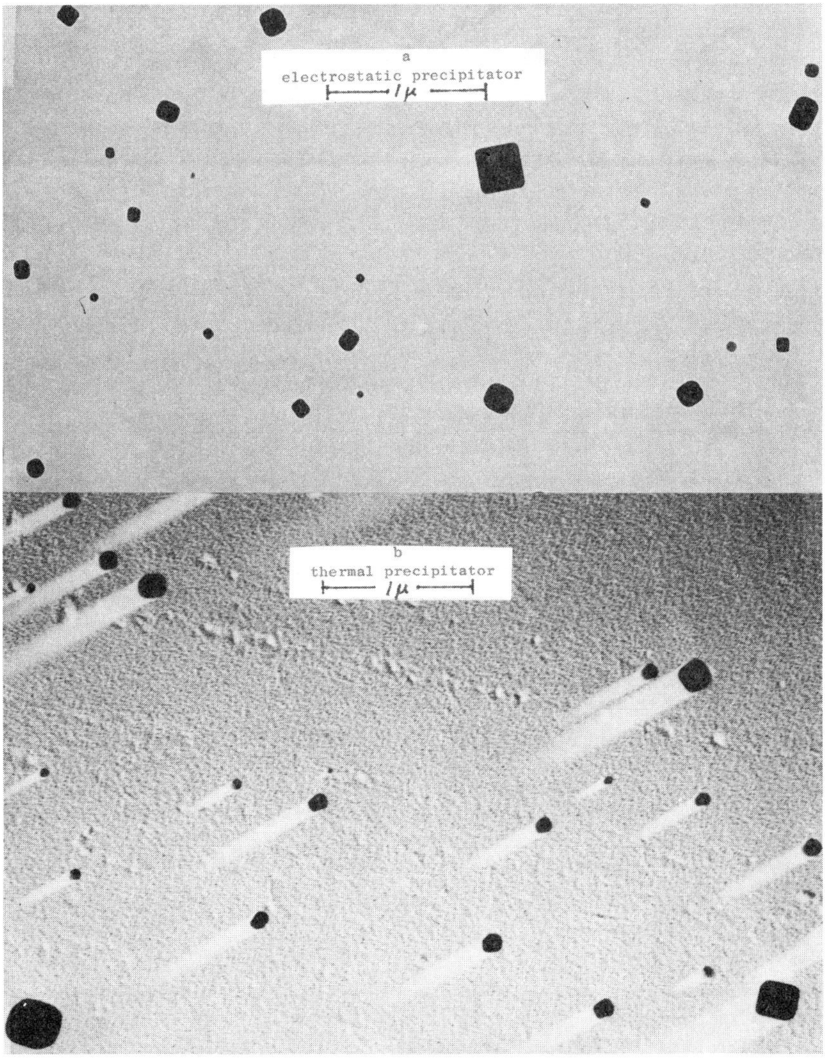

Figure 3. Electron micrographs of NaCl aerosol sampled with (a) electrostatic precipitator and (b) thermal precipitator.

under-estimate the number of particles in that channel. A third explanation is that the collection efficiency of the EASA is dependent upon composition of the aerosol particle. Thus, charging and collection efficiencies would have to be known for the specific aerosol that is being sampled. A fourth possibility is that the nebulizer produced many unexpected small particles and both precipitators had large losses in this small diameter size range. Until the size distributions as determined by the EASA are verified by independent means, the results using this instrument for size distributions should be viewed with discretion. In general, however, the EASA has been very useful in our laboratory since it does provide reproducible data and gives us information on the time and size stability of aerosols in the submicron size range.

The major conclusions of this study are: a) the EASA gave consistent, reproducible data on successive runs; b) above 0.04 μm diameter the electron microscopy and EASA size distributions are quite similar; c) there is an unexplained difference between electron microscopy results and EASA results as to the number of particles present in the 0.01 to 0.04 μm diameter range.

Acknowledgments.—Though the authors are responsible for any shortcomings associated with this paper, we wish to acknowledge the assistance of several persons. The use of the thermal precipitator in addition to our electrostatic precipitator was suggested by Werner Stöber of the Fraunhofer-Gesellschaft Institut für Aerobiologie, Germany. Marvin Tillery of Los Alamos Laboratories loaned the thermal precipitator to us. Nancy Kain and Kathy Miller of the University of California, Irvine, assisted in preparing the manuscript.

Author's note (added in proof)—Since the paper was delivered orally, the results were checked and confirmed using a thermal precipitator for obtaining an E. M. sample. The thermal precipitator does not have the charging artifacts referred to by Dr. Whitby. Data from both samplers are included in this paper. The discrepancy observed remains unsettled.

REFERENCES

Lauterbach, K. E., et al., 1952: Design of an oscillating thermal precipitator. University of Rochester Report UR-199. Rochester, New York.

Leach, L. J., et al., 1958: A multiple exposure unit designed for chronic inhalation studies. University of Rochester—Atomic Energy Research and Development Report UR-534, Rochester, New York.

Liu, B. Y. H. and D. Y. H. Pui, 1975: On the performance of the electrical aerosol analyzer. *J. Aerosol Sci.*, *6*, 249–264.

Marlow, W., 1976: The diffusion charging mobility analysis hypothesis revisited. Presented at the Aerosol Measurement Workshop, University of Florida, Gainesville, Florida, March 24–26.

Morrow, P. E. and T. T. Mercer, 1964: A point-to-point electrostatic precipitator for particle size sampling. *J. Amer. Industrial Hygiene Assoc.*, *25*, 8–14.

QUESTIONS

Whitby: We worked with the point-to-plane sampler many years ago when it was first developed. Recall from the charging curve that Dr. Liu and Dr. Pui showed that the fraction of particles charged drops off very sharply for smaller particles. Also, there is a large change in mobility with particle size. So the point-to-plane precipitator has very severe biases over a wide range in particle size which depends upon the geometry, operating voltage and sharpness of the needle.

Reist: What, I think, this is pointing out is that the problem with electron microscopy is not in the technique with the microscope or the ability to calibrate the microscope, but in the ability to collect a sample that is representative. We're really stuck with electrostatic precipitators and thermal precipitators and each has their faults.

COMMENTS ON AEROSOL MEASUREMENT

PETER J. GROBLICKI

General Motors Research Laboratories, General Motors Technical Center, Warren, Mich.

I'd like to make some comments about Electrical Aerosol Analyzers (EAA) from a user's viewpoint; someone who is trying to run an experimental program *using* them, rather than an experimental program *on* them.

After resisting buying an EAA for a long time, I finally decided that it would be useful for my Smog Chamber studies of gas to particle conversion. I bought one of the first batch produced. I used it for the same type studies as Warren Kocmond and Dave Kittleson, looking at aerosol formation from systems containing sulfur dioxide and hydrocarbons. I was quite happy with it for several reasons. The University of Minnesota was, at that time, probably the only place in the world able to calibrate the EAA, so I had no standards to compare my EAA against. In addition it ran so well. I would twist the switch on the front panel, and would obtain readings of the check voltages that were close to those on the accompanying calibration sheet. As far as I could tell, the EAA was working fine.

In October, 1975, General Motors ran the Sulfate Dispersion Experiment (SDE) at the GM Proving Grounds in Milford, Michigan. This was a large experiment where 352 catalyst equipped cars were driven on a simulated freeway. We wanted to measure the amount of sulfate that these cars would emit and how it was dispersed from the roadway. I used Electrical Aerosol Analyzers to measure the size distribution of the aerosol contributed by the cars. EPA had also brought the University of Minnesota people there to make similar measurements. I acquired a second EAA at that time. The first comparison of my two analyzers, using laboratory aerosol, showed large differences in the size distribution, especially in the 0.1 to 1.0μm size range. Thermo-Systems advised me to clean the older analyzer. They also said that they expected two analyzers to agree within ten percent on any step. After cleaning, the two EAA's did agree within ten percent when I measured the room aerosol. The ionizer conditions were restored to nearly the original settings by this cleaning.

I then took the EAA's out to the SDE site. Ken Whitby has mentioned that we had a total of four EAA's at the site. To compare the instruments, we all sampled from the same bag of aerosol. For particles from 0.01 to 0.1 μm the volumes ranged from 1.1 to 1.4 cubic microns. For particles from 0.1 to 1.0 μm the volumes ranged from 20 to 26 cubic microns. This spread of about 25 percent seemed tolerable, so we went ahead with the work.

I'm working with about 1200 size distributions taken in the course of the SDE. Some anomalies have been appearing which suggest problems with the EAA's. I do not know if there may have been a malfunctioning instrument in the group, but on the surface, all instruments were operating normally. In spite of the fact that the initial comparisons showed a difference of only 20% between my two instruments, when I compared simultaneous readings of background aerosol, the ratio between my two instruments ranged from about 1.1 to 2.4. This spread appeared in measurements of the total volume of particles between 0.1 and 1.0 μm. This variability in measuring the volume concentration has precluded the simple analysis of the data which we had expected.

I could not take the difference between the downwind and upwind analyzers and get a valid measure of the amount of aerosol contributed by the cars on the roadway. Instead, I had to choose the few times when the upwind analyzer showed that the background remained constant, and compare the readings from the downwind analyzer between time periods when cars were and were not present.

On a day of quite high relative humidity, the ion source voltage dropped on every EAA there. The EAA's still operated and gave size distributions, but I have no idea of their validity. I've not seen any reports on the effect of such a major ion source voltage drop (0.7 v to 0.4 v) on the indicated size distribution. On some days, as we switched from step 9 to step 10, we observed a large negative signal, which barely leveled out at a steady reading before switching to step 11. Then on step 11 a similar negative signal and drift up to a reading occurred. Gil Sem of Thermo-Systems has suggested that this behavior is symptomatic of high voltage arcing or breakdown in the EAA. He suggests checking insulators for breakdown paths and also checking for sharp edges on the analyzer assembly.

It also appeared that on some days, one instrument took one and a half hours to warm up. This can't be proven because I have no guarantee that the background aerosol was stable during that time. However, the behavior of the background aerosol would have to be so strange that I postulate a warmup problem.

I mention for Gilmore Sem's benefit that the data were taken onto strip chart recorders during this study. We have a Hewlett-Packard digitizer system which we use to convert the values from the strip charts into punched paper tapes, which can be entered into our computer program. We used a two pen recorder with one pen operating at 10 or 100 mv full scale, and the other channel recording the same signal on a 10 or 20 volt scale. You can then cover the entire range of voltages that you will encounter, and measure them quite accurately.

In summary, I did obtain the data I set out to obtain. However, the difficulties with the EAA's complicated the job of data analysis, and required the careful validation of every size distribution used. I'm continuing to use the EAA because it's the best means that I have for real time particle size measurements in this size range. However, I'm much more skeptical of the data and try to be alert to any hints of abnormal function. If possible, I try to frequently check our instruments against each other.

QUESTIONS

Q: There are a couple of things you mentioned that indicate something was wrong with at least one of the analyzers; for example, Step 10 and Step 11 going negative. That tends to indicate some kind of corona from the center rod which could be caused by a scratch, bump, or a burr on the rod. That's the way that kind of thing tends to show up, unless it's a bad scratch in which case the current always goes negative. The other thing you mentioned was that the ionizer voltage went down to 0.4 volts. I've never seen that in ambient air. Was that consistent on all the analyzers?

Groblicki: The ionizer voltage was that low on my two analyzers and also the one in the trailer. It was also a very humid day.

Q: If that indicated the actual ionizer conditions, wouldn't that be an NT product considerably lower than normal?

Liu: Yes, there could have been a leakage of an insulator there. But that would have thrown off the electrical systems.

Whitby: I'd just like to point out one thing that Dr. Groblicki didn't mention. Both these analyzers were outside, sitting out in the ambient air. In the early morning the ambient temperature was about 0°C. The one in the trailer and the one in our car were running at about room temperature, so the humidity was a lot lower. We have not seen these kinds of problems with the data we've analyzed. We did see a systematic difference between the analyzer in the car and the one in the trailer and when we correlated the volumes in different size ranges there was a very nice 0.7 slope. We actually used this comparison in the data reduction when we wanted to compare the car which was on the track and the trailer which was off the track. But my feeling right now from having done a lot of work on the data that we got from the Detroit study is that the absolute accuracy of the EAA in terms of volume is certainly no better than 25%. Our two instruments were checked out as well as we could before we sent them there, and still we have a 30% absolute difference in their volume measurements. Now I don't know the reasons for this. There are apparently things going on that we don't understand. At this point, I think that the size distribution measurements and the relative values in channels are better than this, but I think the absolute accuracy of the EAA is probably on the order of 25% with the current calibration procedures and the current hardware.

ELECTRICAL AEROSOL ANALYZER CONSTANTS

K. T. Whitby and B. K. Cantrell

Mechanical Engineering Department, University of Minnesota, Minneapolis

To obtain aerosol particle size distributions from the currents measured with the TSI Model 3030 Electrical Aerosol Analyzer, it is necessary to multiply the ΔI's calculated from the I's by a constant. When the EAA is operated at an nt = 10^7 as it was in the GM study, it has been found that the sensitivities as published by Liu and Pui (1975) are not applicable as they are given in the paper. The published calibration was done with monodisperse aerosols, and the response on heterodisperse aerosols will not be the same. Although various more complex calculation schemes are being developed at Minnesota and by other investigators using the

TABLE I. Electrical Aerosol Analyzer Constants for nt = 10^7 $cm^{-3}sec^{-1}$
$\Delta N_i = (\Delta N/\Delta I)_i \Delta I_i = (\Delta N/\Delta I_i)_i\ (I_{i+1} - I_i)$

		($\Delta N/\Delta I_i$) Particle/pa	
Channel No.	Dp_i-μm	(B) Liu-Pui	(HC) Whitby-Cantrell
1	.0042	—	—
2	.0075	9.52×10^6	1.36×10^6
3	.0133	4.17×10^5	3.16×10^5
4	.0237	1.67×10^5	1.53×10^5
5	.0422	8.70×10^4	1.57×10^5
6	.075	4.44×10^4	3.09×10^4
7	.133	2.41×10^4	2.43×10^4
8	.237	1.23×10^4	1.52×10^4
9	.422	6.67×10^4	5.68×10^3
10	.750	3.51×10^3	1.33×10^3

Note: Constants HC developed for DPGVAN = .04 μm, SGN = 1.7, VAN = 1 $\mu m^3 cm^{-3}$, DPGVAC = 0.35 μm, SGAC = 2 and VAC × 30 $\mu m^3 cm^{-3}$.

EAA, it was decided to develop a single set of constants that would improve the calculated distributions but would not require complex calculations.

Constants "B" were used for the on-line data reduction performed in Milford. Constants "HC" were used for all of the results presented in this report.

To develop these new constants, a new matrix applicable to a distributed aerosol was first derived from the Liu-Pui matrix (1975). Then currents for a model distribution having both a nuclei and accumulation mode were calculated. Next, using the ΔN's from the model distribution and the I's calculated from the matrix, new sensitivities were calculated so that when the new sensitivities were multiplied by the ΔI's, the model ΔN's were obtained.

These new constants were evaluated against a variety of model distributions before use, and it was concluded that the maximum errors in integral parameters such as VT, ST, and NT were less than 15% for the extreme cases observed in the sulfate study. These constants are listed in Table I.

REFERENCE

Liu, B. Y. H. and D. Y. H. Pui, 1975: On the performance of the electrical aerosol analyzer. *J. Aerosol Sci., 6,* 249-264.

PART IV

Condensation Nuclei Counter and Diffusion Battery

AEROSOL DETECTION BY CONDENSATION NUCLEUS COUNTING TECHNIQUES

AUSTIN W. HOGAN

Atmospheric Sciences Research Center, State University of New York at Albany, N.Y.

The condensation nucleus counter serves as an aerosol detector by forcing particles to grow into droplets by suddenly subjecting them to a high water saturation. This is done in practice by bringing the air to be examined to a relative humidity near saturation, and then rapidly cooling the air by expansion to produce supersaturation. Quoting Aitken, "Under these conditions, each dust particle becomes a center for condensation, and grows into an easily visible drop." These drops can then be counted by the optically aided eye, or photographed; or the light extinction of a column of such drops can be measured by a light source–photocell system. Aerosol detectors in the first category are indeed "counters" and are often erroneously called "absolute nucleus counters" as they require only geometric calibration. The latter type are called photoelectric nucleus counters and must be initially calibrated against some reference.

The minimum particle size detectable by a nucleus counter is determined by the saturation ratio achieved. This can be calculated for pure water drops from the Kelvin equation. A simplified form of this equation has been given by Das Gupta and Ghosh (1946) as

$$S = \frac{\text{Vapor Pressure at Radius r}}{\text{Vapor Pressure at Radius } \infty} = \exp\ (1.09 \times 10^{-7}/r)$$

when r is in centimeters.

This constant here was developed for use at laboratory temperatures of 18°C, and it is sufficiently accurate for practical use. This equation expresses the saturation required for water droplet growth; the work of Junge and McLaren (1971) indicates that if particles contain 10% by mass of soluble material they behave like soluble particles. This may, in practice, result in the activation of slightly smaller particles than predicted by theory as in Table I for water drops. Representative saturations required for droplet growth are in Table I.

Kassner and his co-workers have experimented with techniques for rapidly producing and holding desired supersaturations. Ideally the supersaturation would be produced by an adiabatic expansion; in practice there is always some heat exchange with the fog chamber walls. The center of the fog chamber is nearly adiabatic, as Nolan (O'Connor, 1975) experimentally determined that nucleation by positive and negative ions, and homogeneous nucleation occurred very near the expansion ratios calculated from theory.

The minimum particle size detectable by any CN meter remains an open question. Re-examination of Table I shows that the initial supersaturation can be increased from 2 to 3, but that the result would lower the minimum detectable radius by only five (5) Å. A practical answer to the minimum particle size detectable question is that r_{min} is less than 5×10^{-7} cm. In reality, the loss of these particles in the inlet system, and the loss to the other larger sized members of the aerosol population due to the high diffusion coefficient of particles $\sim2\times10^{-7}$ cm radius is probably the limiting factor in small particle detection.

TABLE I.

r, cm	Saturation Required for Growth
1×10^{-7}	2.97
1.5×10^{-7}	2.07
5×10^{-7}	1.24
10^{-6}	1.12
5×10^{-6}	1.02
10^{-5}	1.01
10^{-4}	1.001

HISTORY

"I shall here give an account of a method I have devised and developed for counting the number of dust particles in any sample of air; and the figures which have been obtained will, it is hoped, help to give definiteness to our ideas as to the numbers of the particles floating in the atmosphere, under both natural and artificial conditions.

In the search for a solution of this problem, different ideas suggested themselves as to the manner of counting the particles, but they all depended on the fundamental principle of making them visible by supersaturating with vapour the air in which they were floating. Under these conditions, each dust particle becomes a centre of condensation, and grows into an easily visible drop."

Aitken (1888) began the description of his "dust counter" with this. Since that time, Aitken developed two additional portable counters, and Sholz (1932) built the counter so successfully used by Landsberg (1938) in his Nuclear Climatology. Bradbury and Meuron (1938) published the first description of a photoelectric nucleus counter, i.e., an instrument which photoelectrically sensed the attenuation of a light beam by the droplets formed on the original nuclei rather than actually counting the number of drops. Pollak began the development of a photoelectric counter in Prague at about the same time (Pollak and Fuchs, 1936, 1937, 1938; Pollak and Morgan, 1940), utilizing techniques he had used in measuring atmospheric turbidity. After arriving in Dublin, Pollak collaborated with Morgan in constructing a photoelectric counter in 1943, and later published a calibration (Nolan and Pollak, 1946). Pollak continued his development of the photoelectric counter in collaboration with Nolan, O'Connor (Pollak and O'Connor, 1955), McGreevey, Metnieks (Pollak and Metnieks, 1959, 1960a, b, 1961; Metnieks and Pollak, 1959) and others (Pollak and Murphy, 1953; Pollak and Daly, 1957). A recent review of the history of the development of this photoelectric counter is given by Nolan (1972).

Several photoelectric counters have been commercially manufactured in recent years. Three of these, the Rich (1955) manufactured by Gardner Associates, the Skala (1963) manufactured by General Electric and the Environment/One condensation nuclei monitor comprise the majority of such instruments in use at the present.

THE "ABSOLUTE" NUCLEUS COUNTERS

Aitken and Coulier did not actually count droplets in their earliest experiments but estimated changes in aerosol concentration from the variation in density of the fog formed within a flask. Aitken developed his counting chamber soon after, and it was, along with the Scholz counter, the most common aerosol detector in use until about 1950. Such counting chambers are not frequently used at present because of the great amount of labor required to make a single aerosol determination with one of them. Several such instruments are in use as laboratory standards, and where detection of extremely small numbers of aerosol particles is necessary.

The most recent of these is the Kanter (1970) counter of the Max Planck Institute. This is an automated photographic recording counter, and has been used at sea by Jaenicke, Junge and Kanter (1971), and as a calibration standard. Their (Jaenicke and Kanter, 1976) experiments indicate that absolute counters are not "absolute" in that diffusion losses of particles do occur in filling the chamber, and that pure geometric correction does not precisely predict a fraction of drops lost when the adiabatic expansion is created by venting the cloud chamber to outside air or a ballast volume.

Kassner adapted a Wilson cloud chamber to serve as an "absolute" nucleus counter for use as a reference standard. This counter was used to calibrate the photoelectric counters used in the recent survey of lower stratospheric aerosol.

A relatively small photo recording counter has recently been built at our laboratory for use in polar regions where very low aerosol concentrations are encountered. A replica of this instrument will be used by the National Oceanic and Atmospheric Administration (NOAA) for on-site calibration in its Geophysical Monitoring for Climatic Change program. This instrument is rather compact to conform with Polar logistic requirements, but does not incorporate any new or novel features.

PHOTOELECTRIC CONDENSATION NUCLEUS COUNTERS

This general category constitutes the major fraction of CN meters in use today because of the adaptability of this type instrument to real time data output and automatic recording. All of the commercially available instruments are based on preliminary designs of T. A. Rich and prototypes constructed by W. Winters. These instruments all produce an electrical output proportional to the density of fog in the cloud chamber, and must be initially calibrated, and periodically recalibrated, usually using a Pollak Model 1957 replica as a reference (Pollak and Metnieks, 1959).

The principle of operation is slightly more complicated than that of the absolute counter. Air to be examined is drawn into a cylindrical chamber; it may be humidified en route, or the chamber walls may act as a moisture source. A light source-photo detector system sees "100%" of transmitted light at this time. Expansion of the air and the resultant supersaturation causes a *fixed* quantity of water to be made available for condensation. The number of drops formed is equal to the number of aerosol particles present; the number present determines the radius of the drops formed, as the quantity of water is fixed.

The transmission of light through the chamber can then be approximated by the extinction formula

$$I = I_0 \exp -(K_S \pi r^2 n \ell).$$

This cannot exactly be calculated as the light is not monochromatic, and r may not be exactly monodisperse. The expected performance of a photoelectric nucleus counter can be estimated from that of Pollak's counter if it is assumed that $K_s \pi r^2$ is the same function of n in all counters which form an expansion in a similar manner. The performance of several hypothetical counters of varying tube lengths is shown in Figure 1; the plotted points were obtained from a Gardner counter (Rich, 1955), which has a 28 cm light path.

This approximation is only valid for counters which behave like the Pollak in that the *finite* size of the photo detector diminishes the effective extinction of a portion of the light path near the cell. Equal or greater sensitivity can be obtained with shorter tubes if a narrow beam and a small light sensitive area are used. The curves in Figure 1 show that the Pollak's 60 cm tube length is a good compromise, shorter tubes may be more suitable for urban studies, and longer for geophysical research programs.

The several "continuous" nucleus counters available commercially, employ a different technique to achieve linearity of output. The change in light transmission at some short time after the initiation of expansion is determined, rather than the final value of light transmission at the completion of droplet growth. This approximates a constant value of $K_s \pi r^2$ in the transmission formula, and the variation of I becomes approximately a linear function of n over a wide range of concentrations.

SIZE DETERMINATION WITH CONDENSATION NUCLEUS COUNTERS

The minimum radius of particles which will grow to droplets in a CN meter at several supersaturations has been tabulated in Table I. This supersaturation dependence invites application of the nucleus counter as a size spectrometer by varying the initial supersaturation. Unfortunately, the final drop size (and the final light transmission) is a function of total water condensed which is also a function of the supersaturation. This would require that the instrument be recalibrated at each size if it were to be used as a spectrometer.

Rich (1955) avoided this problem by employing a double chamber expansion

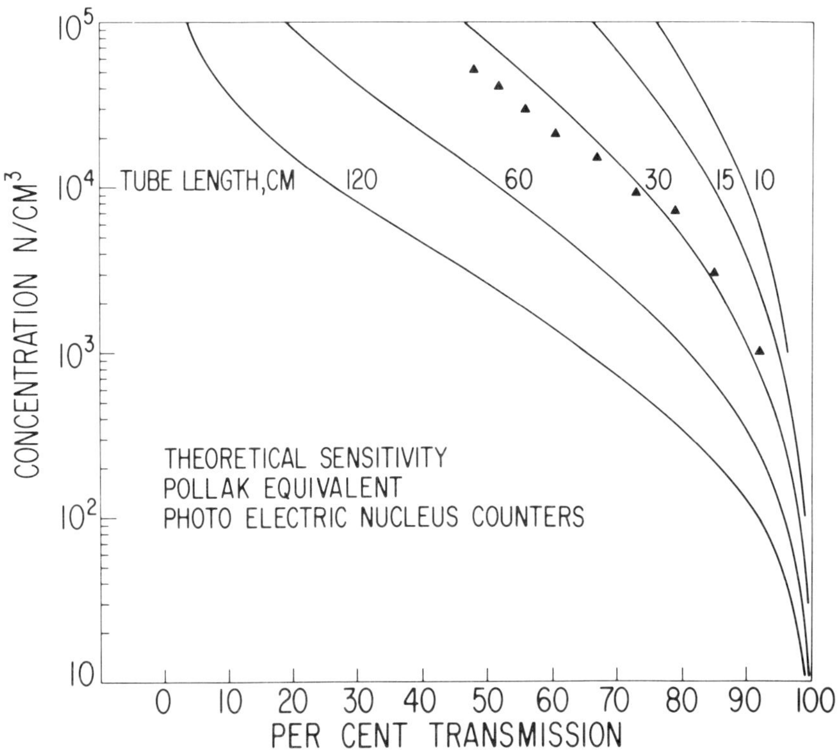

Figure 1. The calculated sensitivity of several photo electric nucleus counters, equivalent to the Pollak Counter. The plotted points were obtained from the calibration of a Rich (1955) Gardner Counter, with 28 cm tube length. Apparently, nearly equal size droplets are formed at a given concentration in both instruments.

volume. The initial supersaturation was achieved by rapid expansion causing particles of a selected size to grow. An additional slow expansion then liberated an equivalent amount of water to the calibrated total expansion. This system is theoretically sound but did not yield to even the most skillful manufacture.

Another technique has been used with some success. The fog tube is placed in a horizontal rather than vertical orientation, and the photo detector output is recorded on a strip chart. An expansion is made in the usual way, and the peak extinction measured. As the fall distance is very short in this direction, the cloud rapidly precipitates, returning the light transmission to its initial value. The fall time, and consequently the drop radius, can be determined from the chart record.

The experiment is then repeated at some other expansion, and the extinction and fall time again measured. The drop radius is determined from Stokes law and, as T, r and ℓ are known, the transmission formula $T = I/I_0$

$$T = I/I_0 = \exp-(K_s \pi r^2 n \ell)$$

can be solved for n

$$n = \frac{-\ell n\ T}{K_s \pi r^2 \ell} .$$

An example of the application of this technique to several aerosols is shown in Figures 2 and 3. The technique apparently works but requires herculean data reduction.

The most reliable sizing techniques utilize the CN meter as a detector only. The aerosol concentration is determined and then some auxiliary device such as a diffusion battery or mobility classifier is inserted ahead of the CN meter. The difference in concentration is then equal to the number of particles with some selected physical property removed by the auxiliary device. This technique allows exploitation of the CN meter's unique ability to detect extremely small changes in concentration.

THE CALIBRATION OF PHOTOELECTRIC NUCLEUS COUNTERS

Nolan and Pollak (1946) first calibrated a cylindrical nucleus counter. Pollak and his many co-workers continued refinement of counters of this geometry in Dublin. The culmination of this work was the Photoelectric Nucleus Counter with Convergent Light Beam, for which Pollak and Metnieks (1960b) derived an "intrinsic" calibration.

All of these calibrations were made by diluting a relatively "stable" aerosol which had been generated by a hot filament and aged several hours in a meteorological balloon "gasometer." The concentration of aerosol in the chamber was determined with an Aitken counter in the earliest experiments. A large, cylindrical cloud chamber with stereo photographic recording (Pollak and

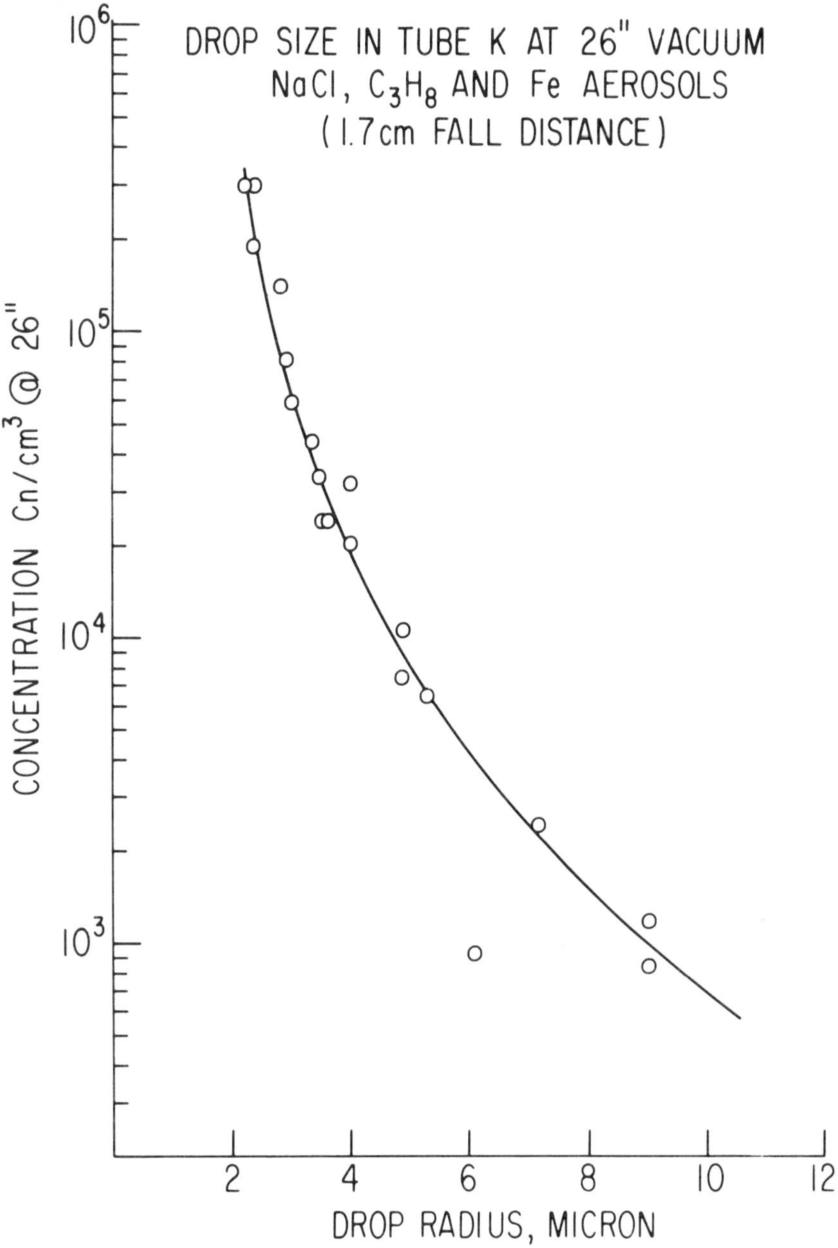

Figure 2. The drop size formed in a cylindrical (1.2 cm radius, 28 cm long) fog tube at several initial aerosol concentrations. The drop size was estimated from the Stokes settling velocity of many drops.

Daly, 1957) was used in the later work. Pollak estimated his overall calibration accuracy to be within 20%.

The Pollak calibration was seriously questioned by Kassner et al. (1968a, b, c, 1970). However, the only attempt at recalibration during the period by Emmanuel and Squires (1969) fell within Pollak's accuracy estimates. Increased interest in submicron aerosols has stimulated several comparisons and recalibrations of nucleus counters during the last several years.

COMPARISON OF SEVERAL POLLAK REPLICAS

A reference laboratory was established at the Atmospheric Sciences Research Center (ASRC) in 1970, to support field and climatological aerosol programs. A Pollak replica, constructed by R. Gussman using the 2.5 cm ceramic fog tube, was designated the "standard" counter. This instrument was #12 in Mr.

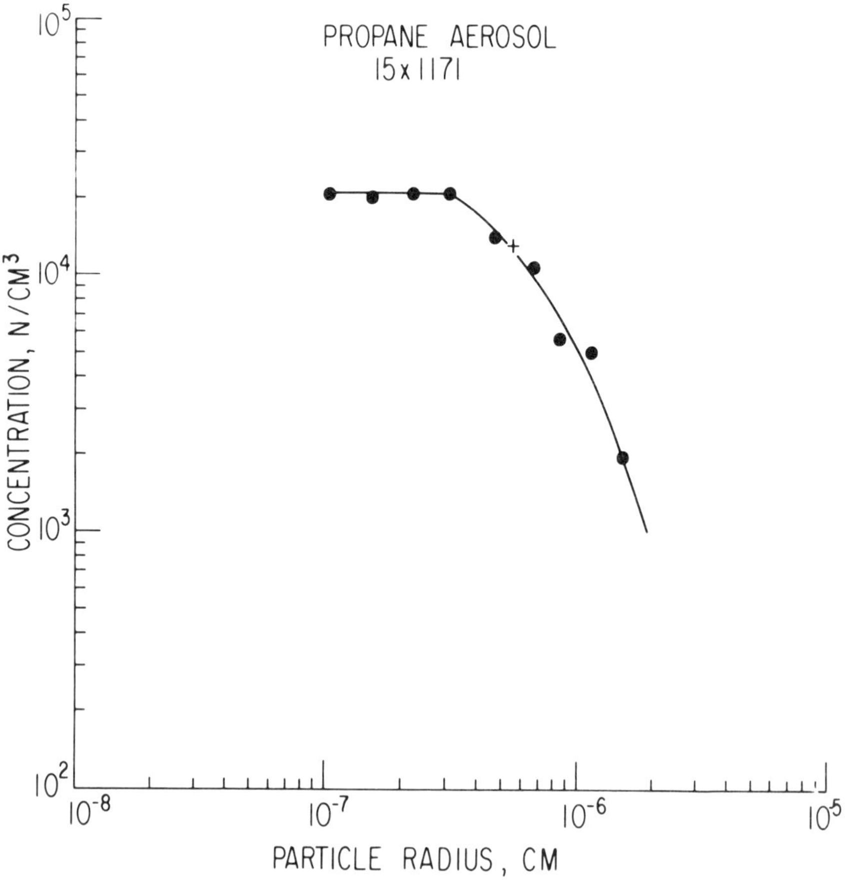

Figure 3. The apparent size distribution of a propane aerosol, determined by the variable expansion method. Diffusion measurements also indicated that most of the particles were less than 3×10^{-7} cm radius.

Gussman's production series which now numbers about 20 instruments. Two additional Gussman replicas were compared with the selected instrument before placing them in the field. Two Dublin Pollaks, constructed under Pollak's supervision by M. Cotter, were retained for laboratory use.

During the last three years three important Pollak replicas have been compared with Gussman Pollak Replica #12 (GP12) in this laboratory: GP10, of the AEC Health and Safety Laboratory (Sinclair, Countess); GP16, now at Amundsen Scott South Pole Station (Hogan, Bodhaine); and a Dublin Pollak by Cotter, used as a reference standard by the General Electric Company and for calibration of the SANDS system (Haberl, Podzimek). A total of 73 comparison points were obtained among these four instruments. The results are tabulated in Table II.

The majority of points lay in the ± 0.5 scale division range; 86% of points

TABLE II. Comparison of Pollak Counters

Deviation, Scale Divisions	−3 to −2	−2 to −1	−1 to 0	0	0 to +1	+1 to +2
	1	6	36	4	24	4

agreed within ± 1 scale division. Considering that all points, and all errors, experimental and sampling, are included here, the agreement is excellent. These experiments verify Pollak's conclusion that properly constructed replicas do indeed possess the "intrinsic" calibration of his prototype.

THE WATER INDEPENDENT CALIBRATION

Although the repeatability among Pollak counters was easily verified, the validity of the initial calibration was not. The development of a stable, monodisperse aerosol generator by Liu, and the electrostatic aerosol analyzer by Whitby et al., presented an opportunity to verify Pollak's calibration by an independent technique. This calibration (Liu, et al., 1975) showed that the Minnesota technique and Pollak's 1959 calibration were within ± 9% in the range of geophysical interest, and within the ± 19% range at very high concentrations using both sodium chloride and incandescent wire aerosols with sizes from 0.01 to 0.12 μm. These sizes had previously been verified by electron microscopy. As the electrostatic technique does not require the condensation of water drops and the calibration proved size independent, this experiment proved not only that Pollak's 1959 calibration was very accurate, but that the photoelectric nucleus counter is indeed an aerosol detector, and not only a "nucleus counter."

COMPARISONS AT THE MAX PLANCK INSTITUTE, MAINZ

A photographic nucleus counter was compared with the Kanter counter and a Scholz replica by Jaenicke and Hogan. The individual observations were within counting error. The overall spread among the several instruments, as referred to the Junge-Kanter automat, was 19%. More recent work by Jaenicke has derived a slight correction to the Kanter's geometry which reduces this discrepancy. The work with this instrument (Jaenicke and Kanter, 1976) has shown the commercial General Electric (Scala, 1963) and Gardner (Rich, 1955) to be in substantial agreement with the Max Planck Institute instruments.

The photographic nucleus counter was compared with GP12 during development; the two commercial counters also derived their calibration from Pollak replicas. These results indirectly verify that the Junge-Kanter Automat and the Pollak calibration are in essential agreement.

COMPARISON AT NCAR

A small workshop was convened by Cadle et al., (1975) at NCAR in 1974 to compare several instruments used to measure the concentration of submicron aerosols in the stratosphere. The ASRC standard (GP12) was transported to NCAR for this experiment. The GE SANDS system is a variation of the GE automat, modified for use at low atmospheric pressures. It has been calibrated at the University of Missouri at Rolla (UMR) under direction of Drs. Podzimek and Kassner.

A unique instrument, using thermal diffusion rather than expansion to produce supersaturation and employing a light scattering detector to count individual droplets, was operated at the session by Rosen.

A comparison of the Wyoming (Hoffman, Rosen) instrument with the reference Pollak is shown in Figure 4; the agreement is very good. The many differences in the technique of creating supersaturation and sensing droplets between these instruments, and the good agreement in the final result indicate that although thermodynamic and optical problems appear theoretically insurmountable, they are, in the practical sense, nearly insignificant.

The comparison of the GE SANDS system with GP12 is given in Figure 5. This provides some cross reference between the UMR (Kassner) instrument and the Pollak, as this GE was calibrated at Rolla. The SANDS system was within 20% of the Pollak except at some lower concentrations where experiment becomes difficult.

The condensation nucleus counter is an accurate and useful aerosol *detector*. An uncertainty of ± 10 to 20% still is present in all calibrations. It appears that this uncertainty is inherent in the sampling and transmission of aerosol concentrations and is not a result of instrument defect. While there are great theoretical, thermodynamic and optical problems to be solved to determine at what real expansion ratio CN meters function, the agreement among the Pollak, the Minnesota, and the Wyoming instruments show these difficulties are minimal in practice. Improvements in the accuracy and performance of aerosol detectors will best be achieved by improving understanding of aerosol mechanics to permit design of more efficient induction systems.

APPLICATION OF CN METERS TO THE STUDY OF ATMOSPHERIC AEROSOLS

A condensation nucleus counter can only provide the value of the total concentration of aerosol particles in some broad size range. This concentration value, in itself, cannot be used to connote any information about the relative size, mass, volume, charge or any other physical property of the aerosol.

Recent field experiments by our laboratory have combined CN meters with a slight modification of the Rich 1966 diffuser-denuder, and light scattering particle detectors to provide additional information about the size distribution of the atmospheric aerosol. This size spectrum is obtained by the CN meter-diffuser-denuder combination in the ''small'' (i.e., $< 10^{-5}$ cm radius) size range, and by a Royco light scattering device in the ''large'' (i.e., $> 2.5\times10^{-5}$ cm radius) size range. A programmable, portable calculator has been adopted to

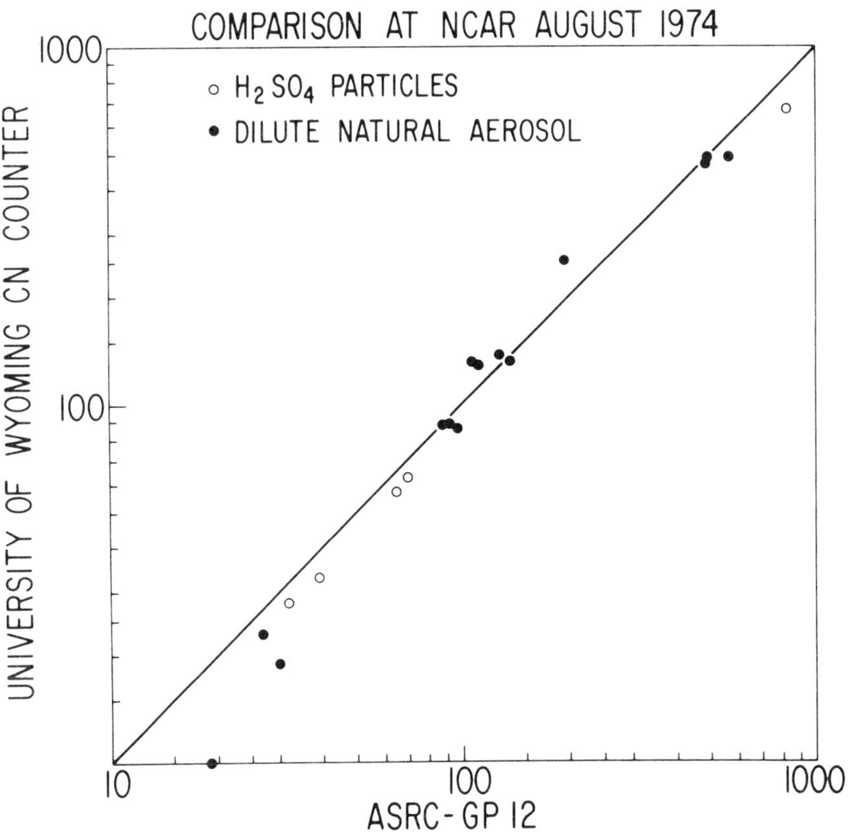

Figure 4. Comparison of the ASRC standard Pollak replica (GP-12) and a thermal diffusion individual particle light scatter counter at NCAR. Although the techniques of applying supersaturation and detecting drops are entirely different, good correspondence is obtained.

handle the large amount of numerical data collected by this technique, and to smooth the transition between techniques by fitting a power law distribution.

This sizing technique is applied in this manner in the field. The light scattering device is started and made to count, the smallest (usually $r = 2.5 \times 10^{-5}$ cm) size range, while the total concentration is measured by the CN meter. The Rich 1966 battery is then inserted ahead of the CN meter, and the number of particles transmitted by the diffusion battery determined; another "large" size class may also be under measurement at this time. Voltage is then applied to the diffuser denuder, and the loss of charged particles determined.

Rich (1966) originally expressed the transmission of this battery in terms of two "equivalent monodisperse radii." The transmission characteristics of the battery in the two modes is shown in Figure 6. Notice that in the diffusion mode, almost all particles less than 10^{-6} cm are removed, but none of those greater than 2×10^{-6} is lost. When voltage is applied, nearly all of the particles large enough to carry charge are taken out; this results in passage of particles $1 < r < 3 \times 10^{-6}$ through a "window" which exists for them due to low diffusion coefficient and low fraction charged. Reasoning from this curve, we can approximate the size of

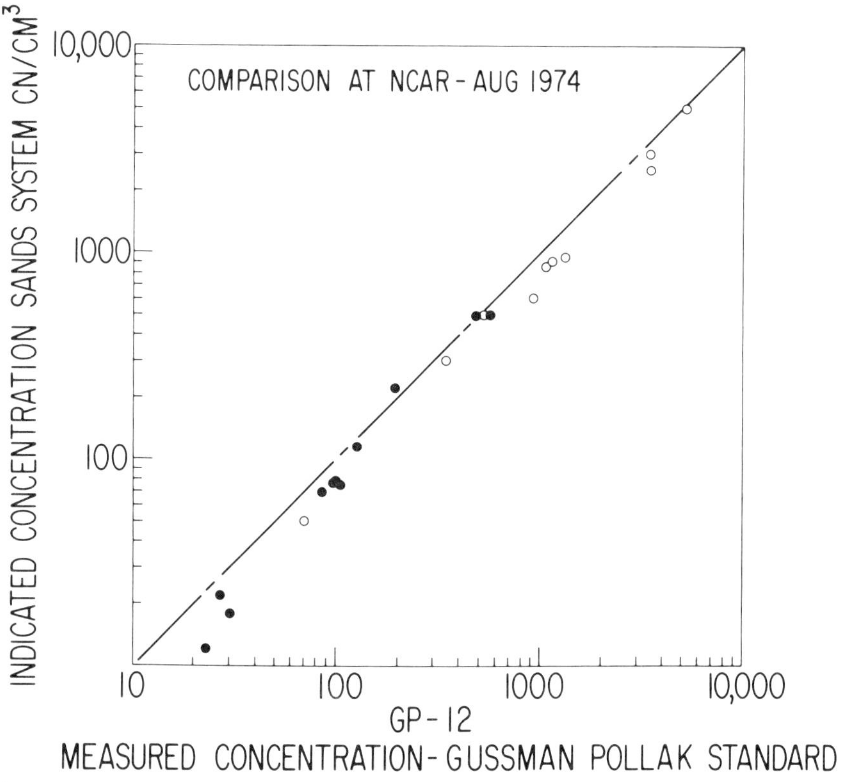

Figure 5. Comparison of the automatic SANDS nucleus counter with GP-12 at NCAR. The SANDS is an analog instrument, and if the slight systematic difference were adjusted out, the correspondence would be exact through most of the range.

the smaller fraction of the aerosol by three size classes, determined by this technique.

Z (all particles) – De (diffuser transmission) = n (small)

De (diffuser transmission) – Ded (denuder transmission) = n (charged, large),

Ded (denuder transmission) = n (passing through midsize window)

These can roughly be placed in these numerical size categories:

Z − De = n ($\bar{r} \approx 10^{-6}$ cm), unless

Z − De ≪ Z, then Z − De = n ($\bar{r} \approx 5\times10^{-7}$ cm)

Ded = n ($\bar{r} \approx 2\times10^{-6}$ cm)

De − Ded = n ($\bar{r} \approx 5\times10^{-6}$ cm)

The mean radii represent the pseudo mean value in the dominant size class. These size classes are then combined with the $>2.5\times10^{-5}$ cm size classes obtained by light scattering to increase many fold the usefulness of field aerosol measurements.

This simplified "size" determination can provide great reasoning insight in the field. These measurements show that the high number concentration of aerosol found near roads and cities is predominantly of small size. These small particles rapidly coagulate with the larger members of the population reducing the *total* number. A measurement made downwind will indicate this smaller number; size measurements made at the two locations will show that these

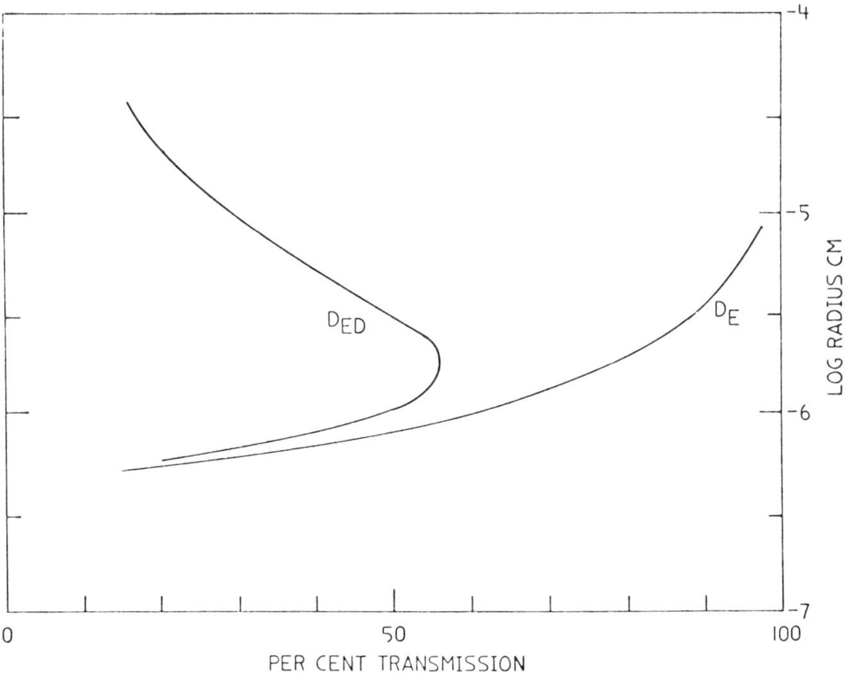

Figure 6. The transmission characteristics of the Rich (1966) diffuser denuder. Most particles less than 10^{-6} cm are removed in the diffusion mode; few particles other than those of $1 < r < 3 \times 10^{-6}$ are transmitted in the combined mode. The difference in transmission between these two modes is due to the number of charged particles of about $r = 5 \times 10^{-6}$ cm radius.

"lost" particles are really attached to the larger members, and that the aerosol has been "conserved."

A few field measurements of these coarse aerosol size parameters have been used to calculate diurnal variation in aerosol concentration, decay of concentration over the seas, and visibility changes in urban areas. Two of these calculations are shown as Figures 7 and 8.

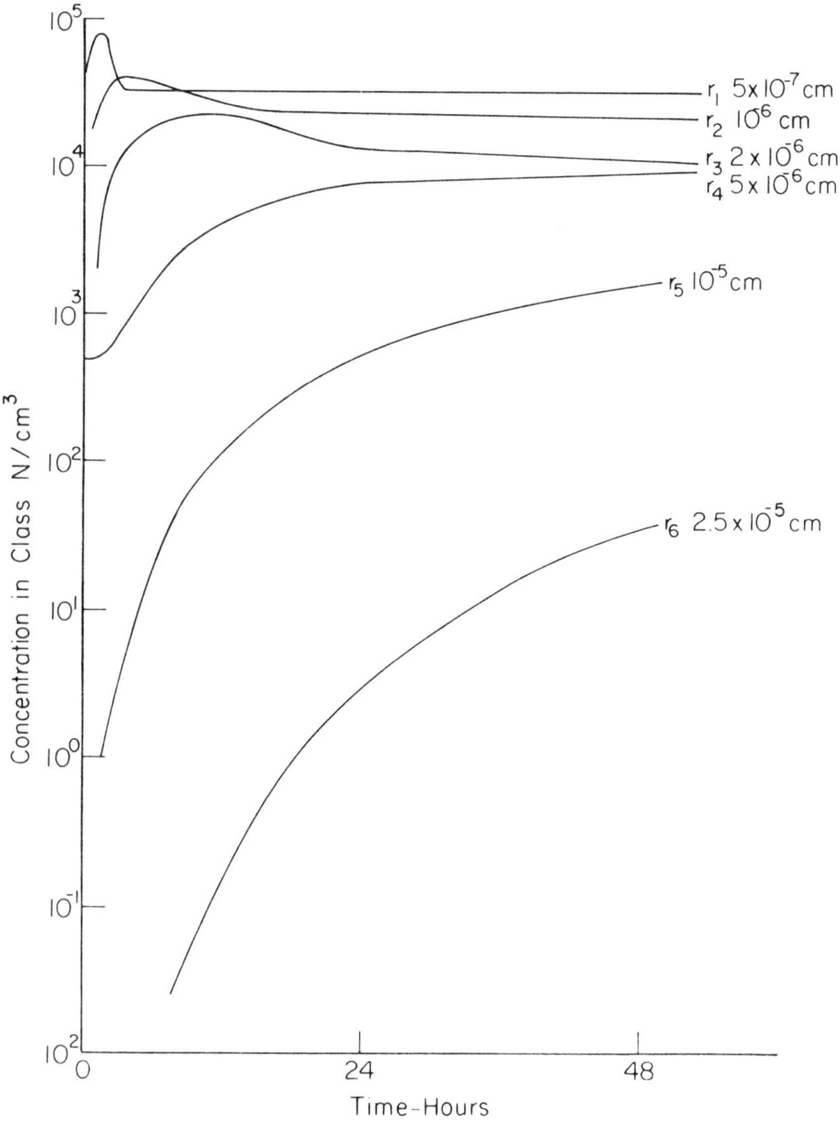

Figure 7. Predicted chronology due to addition of only "small" particles of 5×10^{-7} at a constant rate to a city aerosol. The number concentration rapidly stabilizes and the addition appears in the larger classes through coagulation.

A measurement made just downwind of New York City on an unstable day (just before a thunderstorm) was used to define the "minimum" concentration of particles in several size ranges that would be found there as a starting point for calculation. "Small" particles of radius 5×10^{-7} cm, as would be expected to be generated by "smokeless" cooking gas and automobile fuel combustion, were added at constant rates, simulating what might happen in a more stagnant situation. Figure 7 indicates that the high number concentration of smaller particles rapidly coagulates through the population to appear in the light scattering size range. This is in agreement with observation, as we generally find the *highest*

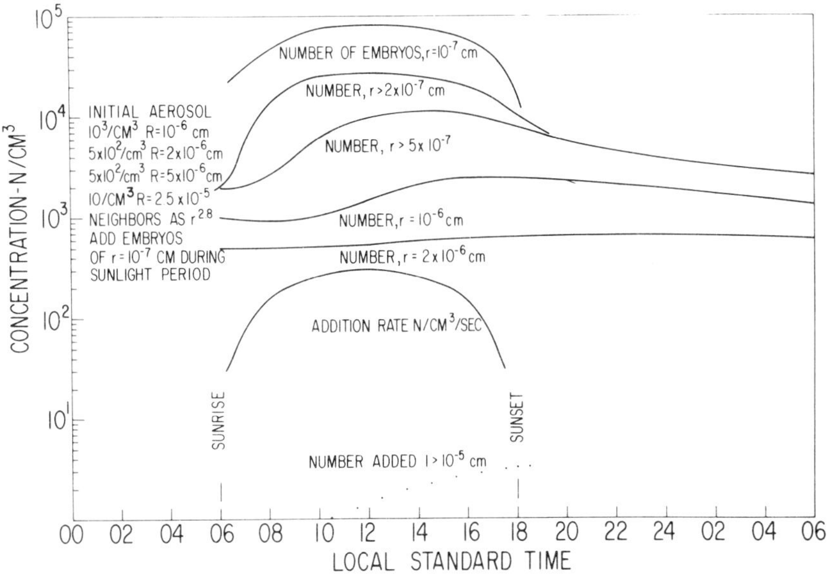

Figure 8. Embryos of sub-detectable size r = 10^{-7} cm are added to a "country" aerosol of initial concentration $2000^n/cm^3$, at a rate proportional to transmission of 4000 A sunlight. The diurnal variation of particles r > 5 × 10^{-7} cm closely approximates that observed in the field. This points out the extreme importance of determining the smallest reliably detectable aerosol size in field situations, and also the important influence of the smaller fraction of the size distribution on the remainder of the aerosol.

number concentration of aerosol at times of fine visibility, and the *lowest* at times of poor visibility.

Some typical early morning rural values of aerosol concentration and size were used as a starting point for the second calculation. "Photochemically" produced aerosol of subdetectable size (r = 10^{-7} cm) was added at a rate proportional to incoming radiation of 4000A on an idealized, 12-hour of sunshine, day. The diurnal variation of aerosol concentration resulting from this photochemical addition is shown in Figure 8. The 2100 hr peak in concentration of 2×10^{-6} cm particles occurs as found experimentally by Lee and Jiusto (1974); the shape of the diurnal variation of total concentration of particles larger than 5×10^{-7} cm is in agreement with that obtained at rural field stations.

These theoretical calculations amplify two important points: concentration measurements may be misleading without concurrent size measurements; and concentration measurements may be a function of minimum detectable size. Any increased knowledge of the mechanics of atmospheric aerosol is dependent on knowing at least the relative fraction of the aerosol in the "small" ($<10^{-6}$ cm), "medium" ($1-3\times10^{-6}$ cm), and "large" ($>3\times10^{-6}$ cm) size ranges of the sublight scattering aerosol. Measurement of the fraction of the aerosol with size less than 5×10^{-7} cm radius may be critical to source studies.

The major unknown, with respect to the CN meter itself, in obtaining such size distributions is the "installed" minimum detectable particle size. Laboratory experiments have shown that particles of $<2\times10^{-7}$ cm radius can be *detected* by CN meters. Whether such 2×10^{-7} cm particles can be induced to enter a CN meter in field installations is an independent concern. Should photochemical aerosol production indeed proceed by the mechanism hypothesized to calculate Figure 7, detection of particles in the 2×10^{-7} cm size range is critical to our understanding of such a mechanism.

CONCLUSIONS AND RECOMMENDATIONS

The Pollak and Metnieks (1960b) calibration of the photoelectric nucleus counter is valid. This calibration is within ± 10 to 20% of that obtained by comparison with several "absolute" nucleus counters, and also within that error band when compared to an independent nondroplet forming aerosol detection technique. The nucleus counters of several laboratories have been found to be within this range of agreement. The limiting factor in these comparisons is most likely sampling and experimental error produced by variation in induction techniques among several instruments. Improvements in calibration accuracy will follow improvements in knowledge of basic aerosol mechanics.

Size estimates should accompany every concentration measurement. The simple diffuser-denuder technique is recommended as a minimum. Development of reliable techniques to determine the fraction of aerosol in the size range smaller than 5×10^{-7} cm would greatly facilitate our understanding of aerosol growth mechanisms.

The important size range of 5×10^{-6} to 2×10^{-5} cm range is presently most difficult to measure. This size range is of extreme importance because it is at the minimum point in the action of nature's aerosol removal processes, and may provide the precursors to visibility obscuring haze. Additional instrument development should be devoted to this range.

REFERENCES

Aitken, J., 1888: On the number of dust particles in the atmosphere. *Proc. Roy. Soc. Edinburgh, 35.*

Bradbury, N. E. and H. J. Meuron, 1938: The diurnal variation of atmospheric condensation nuclei. *Terr. Magn., 43,* 231-240.

Cadle, R. A., G. Langer, J. B. Haberl, A. Hogan, J. M. Rosen, W. A. Sedlacek and J. Wegrzyn, 1975: A comparison of the Langer, Rosen, Nolan-Pollak and SANDS condensation nuclei counters. *J. Appl. Meteor., 14,* 1566-1571.

Das Gupta, N. N. and S. K. Ghosh, 1946: A report on the Wilson Cloud Chamber and its applications in physics. *Rev. Modern Phys., 18,* 223-290.

Emmanuel, M. B. and P. Squires, 1969: The calibration of a photoelectric nucleus counter at low values of nucleus concentration. *Pure Appl. Geophys., 72,* 268-283.

Jaenicke, R., C. E. Junge and H. J. Kanter, 1971: Messungen der aerosol grosverteilung uber der Atlantic. *Meteor. Forsch Ergebneisse, Reihe B #7,* 1-54.

——— and H. J. Kanter, 1976: Direct condensation nuclei counter with automatic photographic recordings, and general problems of "Absolute" counters. *J. Appl. Meteor., 15,* 620-632.

Junge, C. E. and E. McLaren, 1971: Relationship of cloud nuclei spectra to aerosol size distribution and composition. *J. Atmos. Sci., 28,* 382-390.

Kanter, H. J., 1970: Bau und eichung eines automatischen kondensations kernzahlers mit fotografischer registrierung, M. S. Thesis, University Mainz.

Kassner, J. L., Jr., J. C. Carstens and L. B. Allen, 1968a: Analysis of the heat and vapor propagation from the walls of the Nolan, Pollak and Gardner type condensation nucleus counters. *J. Atmos. Sci., 25,* 919-926.

———, ——— and ———, 1968b: The myth concerning the condensation nucleus counters. *J. Rech. Atmos., 3,* 25-31.

———, ———, M. A. Vietti, A. H. Biermann, Paul C. P. Yue, L. B. Allen, M. R. Eastburn, D. D. Hoffman, H. A. Noble and D. L. Packwood, 1968c: Expansion cloud chamber technique for absolute Aitken counting. *J. Rech. Atmos., 3,* 45-51.

———, V. K. Saxena, A. H. Biermann and A. C. Tabelak, 1970: UMR absolute Aitken nuclei counter: Developments since 1967 and operating procedure. *Proc. Second Intern. Workshop on Condensation and Ice Nuclei,* Colorado State University.

Landsberg, H., 1938: Atmospheric condensation nuclei. *Ergeb. Kosm. Phys., 3,* 155-252.

Lee, T. F. D. and J. E. Jiusto, 1974: Diurnal variation in cloud condensation nuclei. *J. Geophys. Res., 79,* 5651-5656.

Liu, B. Y. H., D. Y. H. Pui, A. W. Hogan and T. A. Rich, 1975: Calibration of the Pollak counter with monodisperse aerosols. *J. Appl. Meteor., 14,* 46-51.

Metnieks, A. L. and L. W. Pollak, 1959: Instruction for use of photoelectric condensation nucleus counters. Geophysical Bull. 16, School of Cosmic Physics, Dublin Institute for Advanced Studies.

Nolan, P. J., 1972: The photoelectric nucleus counter. *Proc. Roy. Soc. Dublin, A4,* 161-180.

——— and L. W. Pollak, 1946: The calibration of a photoelectric nucleus counter. *Proc. Roy. Irish Acad., A9,* 9-31.

O'Connor, T. C., 1975: Personal communication.

Pollak, L. W., 1964: Personal communication.

——— and J. Daly, 1957: A condensation nucleus counter with stereo photomicrograph recording. *Geofis. Pura Appl., 36,* 27-34.

——— and F. Fuchs, 1936: Zusätzliche lichtelektrische*β*me einrichtung zum UV-Dosimeter der I. G. Farbenindustrie A. G. *Beitr. Geophys., 47,* 362-368.

——— and ———, 1937: Objektives bioklimatisches UV-Dosimeter mit lichtelektrischer kontrolle der durchleuchtung. *Beitr. Geophys., 50,* 104-118.

——— and ———, 1938: Objektive prufung der quarztestöhrchen des bioklimatischen UV-Dosimeters. *Beitr. Geophys., 53,* 148-154.

——— and A. L. Metnieks, 1959: New calibration of photoelectric nucleus counters. *Geofis. Pura Appl., 43,* 285-301.

——— and ———, 1960a: The influence of pressure and temperature on the counting of condensation nuclei. Part I: The influence of pressure. *Geofis. Pura Appl., 47,* 123-141.

——— and ———, 1960b: Intrinsic calibration of the photoelectric nucleus counter Model 1957 with convergent light beam. Tech. Note 9, Contract AF61 (052)-26, School of Cosmic Physics, Dublin Institute for Advanced Studies.

——— and ———, 1961: The influence of pressure and temperature on the counting of condensation nuclei. Part II: The influence of low temperature. *Geofis. Pura Appl.*, *50*, 2–16.
——— and W. A. Morgan, 1940: On various photoelectric nucleus counters. Unpubl. ms., Irish Meteorological Service.
——— and T. Murphy, 1953: Comparison of Photoelectric nuclei counters. *Geofis. Pura Appl.*, *25*, 44–60.
——— and T. C. O'Connor, 1955: A photoelectric nucleus counter of high precision. *Geofis. Pura Appl.*, *32*, 139–146.
Rich, T. A., 1955: A photoelectric nucleus counter with size discrimination. *Geofis. Pura Appl.*, *31*, 60–65.
———, 1966: Apparatus and method for determining the size of aerosols. *J. Rech. Atmos.*, *2*, 79–86.
Sholz, J., 1932: Vereinfachter Bau eines Kernzahlers. *Meteor. Z.*, *49*, 381–388.
Skala, G. F., 1963: A new instrument for the continuous measurement of condensation nuclei. *Anal. Chem.*, *35*, 702–706.

FINE PARTICLE MEASUREMENT

T. A. Rich*

Atmospheric Sciences Research Center, State University of New York at Albany

The fine particles in an aerosol are arbitrarily taken to be those smaller than 5×10^{-6} cm in "radius." If the aerosol is brought to 100% humidity and then suddenly expanded, the air is cooled and the excess water vapor forms water drops around each particle. While the fine particles are submicroscopic, the water drops can be counted (or photographed) by various means.

There has been some question about formation of drops on hydrophobic particles. There is some evidence that clean oil drops do not serve as condensation centers but that in normal air their surface is contaminated in a short time and water drops can form on them. The comparisons of Lui (1975) between the Pollak CN meter and his electrical device give a comforting reassurance that the two independent methods check one another.

The great wonder in fine particle measurements is not that there are occasional baffling inconsistencies but that there is so much that is predictable even with outrageously oversimplified assumptions. In the early kinetic theory similar assumptions were made and yet many phenomena were rather accurately explained. In the kinetic theory it was reasonable to expect the molecules of a gas to be the same. The assumption of a spherical shape might not be quite right, but in any given test they acted like a sphere of a certain size. In another test a somewhat different size could be found.

The variety of shapes and sizes in natural aerosols is so great that it is necessary to find some kind of an average for calculations as well as a comprehensible description. Photomicrographs show a bewildering variety of crystals, needles, lacy aggregates and occasionally a spherical particle. The literature, more often than not, refers to an average radius without apology, and assumes that the reader realizes that this does not mean the aerosol is spherical but rather that in the circumstances it acts as spheres of that size would. It may not be a good practice, but no better way has been found, and there is a respectable precedent.

Sponsored by the Office of Naval Research Under Contract No. N00014-76-C-0283.

*Deceased.

One of the first steps in any measurement problem is to decide what use will be made of the results. A precision of 1 or 2% is desirable in measuring, say, the loss in connecting tubing even though the accuracy needed may be only 10 or 20%. The number of particles per cc near a high traffic road will show such great variations in time and space that even changes of 50% are without significance. A passion for accuracy per se may be self-defeating or misleading. Accuracy and precision are instrument characteristics, but significance is the result of intelligent use.

The so-called dirty air of a large city usually has less than 10 parts per hundred million of dirt by weight. The fine particles have less than a tenth of this mass and are in lower concentrations by weight than many gaseous pollutants. They have

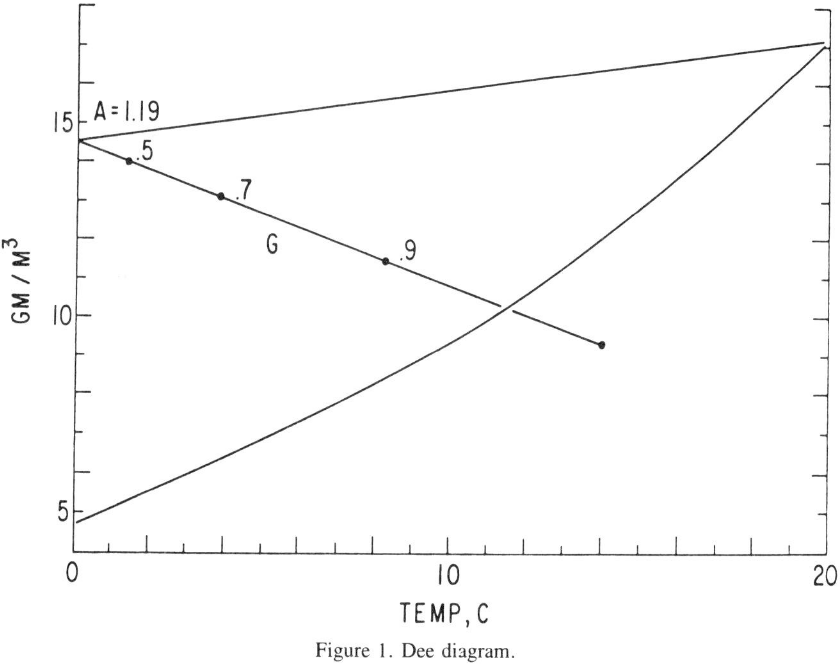

Figure 1. Dee diagram.

little effect on many of the characteristics of air, such as density, viscosity, etc.

A clean concept of the reactions in a fine particle detector is more important for most people than a scientifically impeccable explanation. The following treatment is oversimplified in the details and overdetailed in discussing concepts with the hope of easing the task of the newcomer to the field.

For purposes of illustration, it will be assumed that the ambient temperature is 20°C and the pressure 1000 millibars. The device will be a tube lined with a wet blotter in which the aerosol is confined. A piston is used to expand the volume. An axial light beam will be attenuated by the cloud formed after expansion and a light measuring device is calibrated to indicate particles per cc. Several forms of expansion are in use, but all can be converted to the piston equivalent.

In Figure 1 the Dee (1931) condensation diagram is shown. The grams of water vapor per cubic meter at 100% humidity is plotted against the air temperature going from 4.85 at 0°C to 17.3 at 20°C. The aerosol at $P_1 = 1000$, $T_1 = 20°C$ has a volume V_1. The piston increases the volume to (aV_1) at pressure and temperature P_2, T_2.

$$P_1 V_1^{1 \cdot 4} = P_2 (aV_1)^{1 \cdot 4} \tag{1}$$

$$P_1 V_1 T_2 = P_2 (aV_1) T_1 \tag{2}$$

$$P_1 = a^{1 \cdot 4} P_2 \tag{3}$$

$$T_1 = a^{0 \cdot 4} T_2. \tag{4}$$

If $T_1 = 293°K$ and $T_2 = 273°K = 0°C$, $a = 1.193$

$P_2 = 780$ mb.

$$\text{The water vapor density at } P_2,\ T_2 \text{ is } \rho_a = 14.5 \text{ gm m}^{-3} \tag{5}$$

$$\text{For } T = 0°C,\ H = 1,\ \rho = 4.85 \text{ so the } H = 14.5/4.85 = 3.0 \tag{6}$$

This supersaturation (SS) of 3.0 means that the rate of molecules leaving a water drop at 0°C has been reduced in the proportion of 4.85/17.3 while the bombardment of water vapor has decreased by only 14.5/17.3. The air contains 16% less water vapor than before expansion, not more water vapor as the word supersaturation seems to imply. It may also be noted that the particles are at 20°C initially and see only 84% humidity. The air molecules at 0°C will rapidly cool the particles; at about 17°C the water vapor will start to condense on the larger particles.

For small particles that are or act like water drops, there is an H_c and r_c where condensation starts.

$$H_c = \exp\left(\frac{1.1 \times 10^{-7}}{r_c}\right). \tag{7}$$

TABLE I.

r_c (cm)	10^{-7}	2×10^{-7}	5×10^{-7}	10^{-6}	10^{-5}	10^{-4}
H_c	3	1.7	1.25	1.12	1.01	1.001

For the example given a 10^{-7} particle would have to cool to air temperature (0°C) to start growth, whereas a 2×10^{-7} particle would have to cool to approximately 8°, and 5×10^{-7} to about 13°.

The heat of condensation must be carried away by the air

$$L_V(\Delta\rho_a) = M_{air} \times C_V \times \Delta T_a;\ M_{air} = 1254 \text{ gm/m}^3 \text{ at } 5°C$$

$$C_V = 0.171$$

$$L_V = 575 \text{ cal/gm}$$

$$\frac{\Delta\rho_a}{\Delta T_a} = \frac{1254\times 0.171}{575} = 0.373 \text{ gm m}^{-3}/°\text{C}. \tag{8}$$

A line of this slope starts at 14.5 gm m^{-3} at 0°C to 9.3 gm^{-3} at 14°C. The intersection of this with the H = 1 curve is at 11.4° and the total water condensed = 14.5-10.3 = 4.2 gm m^{-3} because at this point condensation stops.

These figures are for water drops. The particles may behave quite differently—a lacy structure would have capillary channels; the material may be hydroscopic or hydrophobic so there can be little dependence on Eq. (7) as a measurement of size. There is a high likelihood that the device described here would record down to 2×10^{-7} cm radius, but there is going to be some uncertainty at the lowest level of detection.

The Wilson cloud chamber is essentially a condensation nuclei meter, but the condensation nuclei are ions rather than particles. These start to be effective condensation centers when $H>4$. At high values of H so-called spontaneous nucleation occurs. When $H \rightarrow 8$ in clean and (hopefully) ion free air, a fine cloud appears. For fine particles $H \rightarrow 4$ is as high an H as should be used. It will also be noted that in the example used here the temperature fell to 0°C (H = 3). Things are complicated enough without the confusion of vapor pressure over ice vs. vapor pressure over water, so most commercial units use 1.2 as a maximum expansion. The Pollak counter uses a somewhat lower expansion (and his laboratory in Dublin was frequently 17 or 18°C). These are rather cowardly reasons for stopping at 1.2. A better reason is that particles in the range of 10^{-7} and smaller will not last long in natural air, and their losses in the necessary connecting tubing, humidifiers, etc. are high. There is some merit in the argument that these smallest sizes are too ephemeral for field work, although of great importance in following the conversion of gases to bigger particles.

The speed of expansion is a factor as well as the amount of expansion. With a slow expansion, growth starts early on the biggest particles and the condensed vapor reduces the supersaturation to such an extent that the smaller particles never do grow.

The Dee diagram does not give information on the rate of growth of water drops, but it shows vividly that the *initial* growth does not deplete the vapor supply. For the example used here, the total condensable water vapor is 4.2 gm m^{-3} and R_m = maximum size of drops.

$$\frac{4}{3}\pi R_m^3 Z = 4.2\times 10^{-6}$$

$$R_m = .0100/Z^{1/3} \text{ cm}$$

$$R_m = 100/Z^{1/3}\ \mu.$$

It will be a convenience to express the drop radius as a fraction of the maximum radius as G where

$$R = GR_m \text{ cm drop radius.} \tag{10}$$

Here the condensed vapor is $\frac{4}{3}\pi G^3 R_m^3 Z = 4.2\ G^3$ and the remaining vapor 14.5–4.2 G^3.

TABLE II.

G	0.9	0.7	0.5	0.3	0.1
G^3	0.73	0.34	0.125	0.027	0.001
14.5–4.2 G^3	11.4	13.	14.	14.39	14.496

This is shown plotted on Figure 1. It is noteworthy that a drop can grow to 30% of its final size without reducing the water vapor by 1%. It is also a fortunate circumstance, or the speed of expansion would be a far more serious problem.

There are several stages in the growth of fine particles:

1. The particle cools to a point where condensation starts by molecular bombardment of cool air.
2. Molecular deposition of water which warms the particle to a point where heat of net condensation is equal to heat carried away by air.
3. When the drop radius approaches the mean free path length, there is a depletion of water vapor in the vicinity of the drop, as well as a temperature gradient. The mechanism is now ruled by diffusion and thermal conductivity. It can be shown that stages 1 and 2 are fractions of a milli-second and little error will be introduced by assuming the diffusion-thermal conductivity regime throughout the growth period.

$$Q_w = Z_m L_v D(4\pi R)(\rho_a - \rho_d) \text{ cal m}^{-3} \text{ sec}^{-1} \quad (11)$$

$$Q_a = Z_m K(4\pi R)(T_d - T_a) \quad (12)$$

Q_w = heat due to condensation

Z_m = particles/ m^3

L_v = latent heat at constant volume (575)

D = diffusion constant water vapor in air (0.294)

R = drop radius

ρ_a = density of water vapor in air

ρ_d = density of water vapor at drop surface

K = thermal conductivity (5.88×10^{-5})

T_d = temperature of drop surface (°C)

T_a = temperature of the air.

Since $Q_a = Q_w$

$$\frac{\rho_a - \rho_d}{T_d - T_a} = \frac{K}{LD} = 0.348 \qquad (13)$$

K, L, and D are functions of temperature. The values used are for about 5°C for reasons which will become evident later.

In the region of 11°C for H = 1

$$\rho_d = 10 + .7\,(T_d\text{-}11) \equiv 2.3 + 0.7\,T_d \qquad (14)$$

$$(14 \rightarrow 13)\ \rho_a - 2.3 - .7\,T_d = 0.348\,(T_d\text{-}T_a). \qquad (15)$$

This would not apply strictly immediately after expansion, but the behavior of G vs. T_a makes it reasonable to evaluate for $T_a = 0$, $\rho_a = 14.5$, $T_d = 11.64$°C. At the end of condensation cycle $T_d = 11.4°$ from Figure 1, the evidence is that the drop temperature does not change much, if at all. If so, the numerical

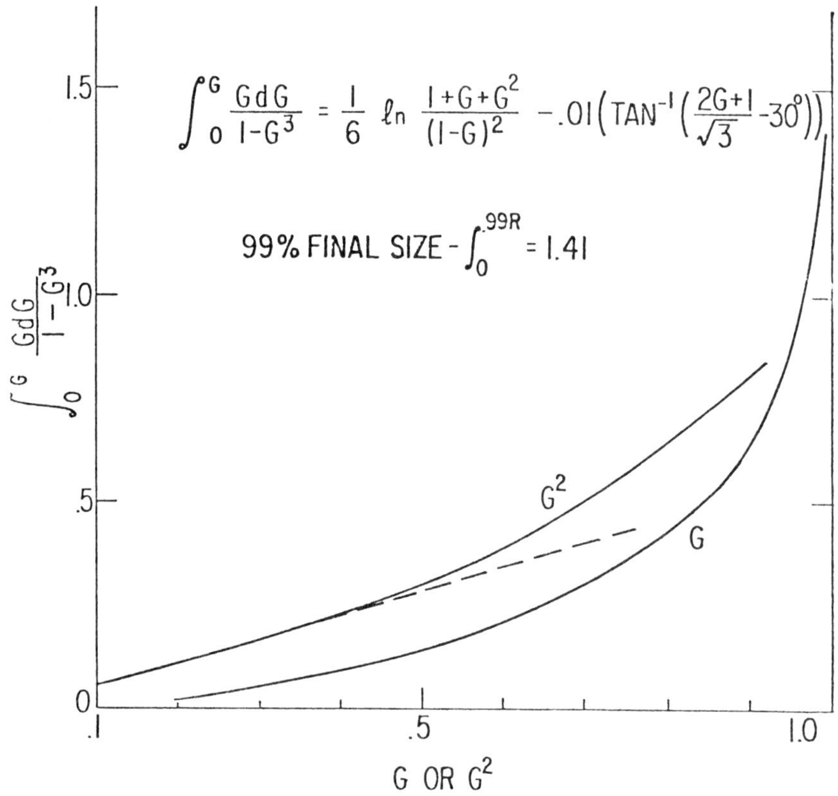

Figure 2. Values of growth integral.

coefficient of 13 should equal Eq. (8): 0.348 vs. 0.373, or a 7% discrepancy. The approximations used here are not that good. The condensation takes place over a range of 0°C to 11.4°C air temperature and the drop temperature over a narrow range around 11.4°C.

The heat released by condensation is at a rate of L dm/dt calories/sec = $L_v Z(4\pi R^2)$ dR/dt and the air carries away $Z(4\pi RK)$ $(T_d\text{-}T_a)$ cal./sec.

$$LRdR = K(T_d - T_a)dt$$

$$T_a = (4/3\pi R^3 Z)/0.373 = 11.3\ R^3 Z$$

$$T_d = 11.4\ \frac{K}{L}(T_d - T_a) = 1.15\ R^3 Z \times 10^{-6} \qquad (16)$$

$$\int_0^r \frac{RdR}{1 - R^3 Z} = 1.15 \times 10^{-6} \int_0^t dt \qquad (17)$$

when $R^3 Z << 1$

$$R^2 = 2.3 \times 10^{-6} t. \qquad (18)$$

If t is say 0.02 secs, $R = 2.1 \times 10^{-4}$ cm and is independent of Z_m if $Z_m << 1$. If $R^3 Z_m = 0.01$, $Z_m = 10^9$ particles/m^3 or 1000/cc when $4/3\,(\pi R^3 Z_m) = 4.2$ gm/m^3 as here, then $R_m{}^3 Z_m = 1$ and Eq. 17 in terms of $G = R/R_m$ becomes

$$R_m{}^2 \int_0^g \frac{GdG}{1 - G^3} = 1.15 \times 10^{-6} \int_0^t dt. \qquad (19)$$

The solution of the integral is:

$$-0.0101\left(\tan^{-1}\frac{(2G+1)}{\sqrt{3}} - 30°\right) = \frac{1}{6}\ell n\frac{1+G+G^2}{(1-G)^2}\ -0.0101 \text{ (see Figure 2).}$$

TABLE III.

G	0.1	0.2	0.4	0.6	0.8	0.9	0.95	0.99
S	0.005	0.019	0.0818	0.198	0.422	0.648	0.88	1.41
G^2	0.01	0.04	0.16	0.36	0.64	0.81		

A substantially linear scale is obtained in the General Electric CN meter by limiting the growth to 0.02 sec. up to 100,000 particles per cc. There are full scale values 100,000, 30,000 10,000, 3,000, and 1,000. Each scale can be adjusted so the departure from linearity from, say, 30,000 to 100,000 is not great. As the concentration increases and the available water is used up, the radius of the drops is proportional to $Z^{-1/3}$, the light scattering area per drop to $Z^{-2/3}$, and the total scattering to $Z^{1/3}$.

In instruments such as the Gardner and Pollak the condensation time is not limited (by design, at least) although the air will warm up from the 11.4°. When there are few particles the time to grow to a given fraction of final size increases and the warming from the 20° ambient may halt the growth sooner than indicated here.

TABLE IV.

Z/cm^3	10^2	10^4	5×10^4	10^5	10^6	
R_m	10	4.64	2.71	2.15	1.0	μ
t	1.22	0.26	0.09	0.056	0.012	sec

SIZE AND SIZE DISTRIBUTION

Size distribution is a higher order of absurdity than size itself. For purposes of calculation or comparison with other aerosols, an imaginary spherical aerosol of 1 or more components is specified that will act like the aerosol under test in certain situations.

The simplest case would be to find an imaginary spherical aerosol that would have the same diffusion loss as the test aerosol under the same set of conditions. The ideal case would be a curve of dZ/dr. In most cases it is impossible to determine more than three components before changes in the aerosol cause uncertainties greater than the accuracy further measurements could give.

Fuchs (1962) calculated the response of families of log normally distributed aerosols when subject to various diffusion losses. With the values of three diffusion losses (one high, one low, one 40–45%), a log normal distribution could be picked out that would most nearly duplicate the test aerosol. This is still only 3 points on a distribution curve, and one cannot place any great dependence on those results for the larger sizes. Megaw et al. (1963) have shown the diffusion method does give good results for the fine particles. The larger particles constitute such a small fraction of the total numbers that no accurate evaluation of their effect can be made with a CN meter, and diffusion losses are small in themselves.

Again, this is not a criticism of getting representations of size distributions; but familiarity with the concept of a radius (as used here and in the literature) should not obscure the fact that aerosol particles are not necessarily spheres.

Another size dependent parameter is the fraction charged when the aerosol is in electrical equilibrium. Essentially this is an application of Boltzmann's law suggested by Fuchs (1964).

$$N_p = N_o \exp\left(-\frac{p^2 e^2}{2rKT}\right) \tag{20}$$

where e = electric charge, K = Boltzmann constant, p = number of electronic charges. The results are shown in Figure 3. For particles larger than 2×10^{-6} cm, the average energy per particle is ½ KT. The capacity of a sphere is numerically equal to its radius. The charge is 4.8×10^{-10} p, so electrical energy = $p^2e^2/2r$.

Below 2×10^{-6}, or what amounts to the same thing, for singly charged particles Boltzmann's law falls off sharply (dotted lines of Figure 3). The assumption that the average energy is ½ KT over the whole range of sizes gives results which seem to give a reasonable check with experiment, or more conservatively, do not create many obvious inconsistencies.

$$F_u = 1\text{-}0.17\,(r\times10^6) \qquad r<2\times10^{-6} \tag{21}$$

$$F_u = 0.95/(r\times10^6)^{1/2} \qquad r<2\times10^{-6} \tag{22}$$

If an aerosol is passed between two plates with an adequate potential difference, only the uncharged particles will penetrate. If such a device is designed to have small diffusion losses, a high voltage will be required which is a serious disadvantage in the field where voltage breakdowns can occur in fog or condensation conditions. For some reason, such a device is called a denuder.

If the plates of a diffusion box are insulated, the device can be used as a diffusion box. Applying a moderate voltage to alternate plates, a combined diffusion-denuder is obtained that will work on low voltage (Rich, 1966). A diffusion box that has 50% transmission for a 10^{-6} cm particle will require a flow for 20 to 30 seconds to reach steady state conditions. When voltage is applied, steady state requires only a few seconds. If, after this sequence, the diffusion box is by-passed to get the unprocessed air to the CN meter, the three readings can be obtained rapidly.

The ratio of diffusion transmission to the direct aerosol and the ratio of the diffusion-denuder to direct aerosol are plotted in Figure 4.

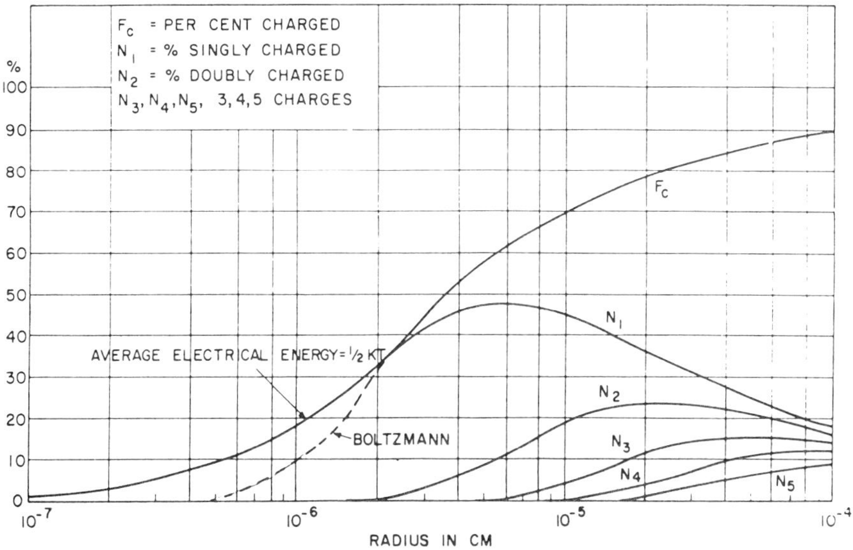

Figure 3. Electrical properties of aerosols.

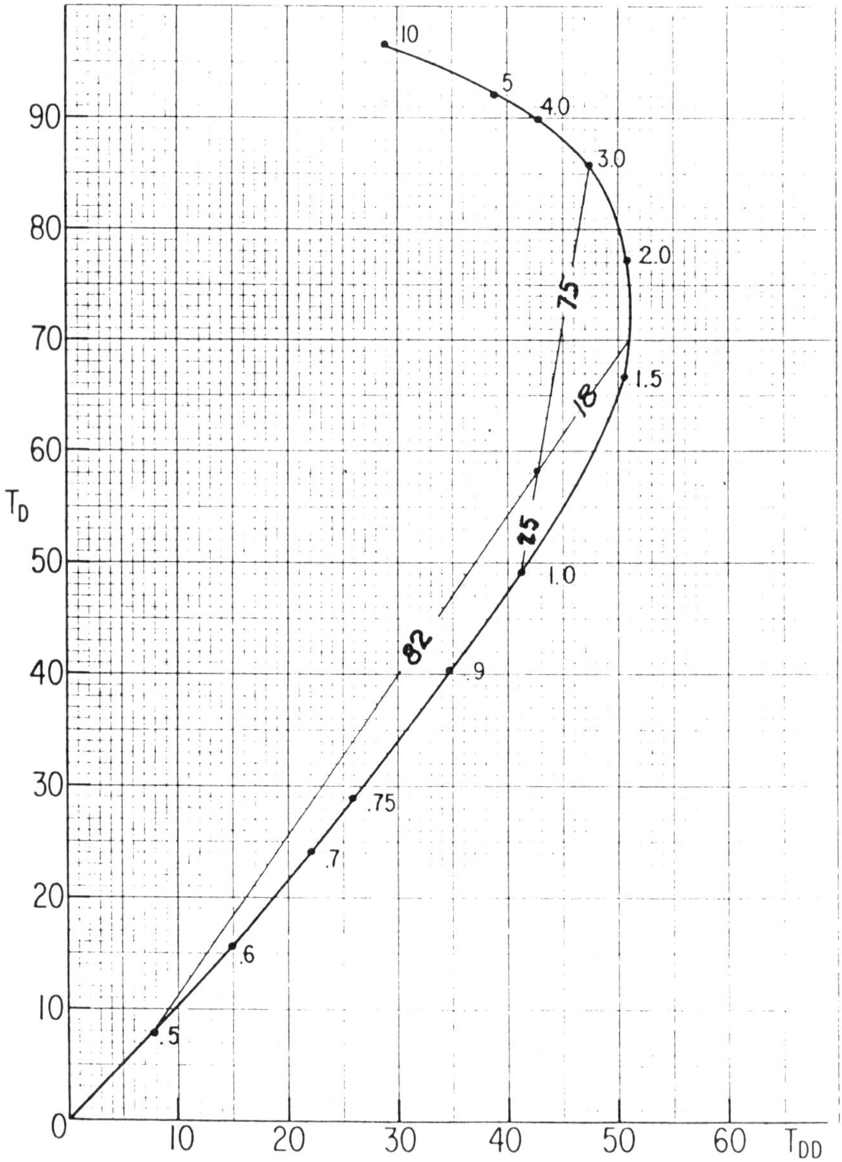

Figure 4. Ratio of diffusion to diffusion-denuder aerosol transmission.

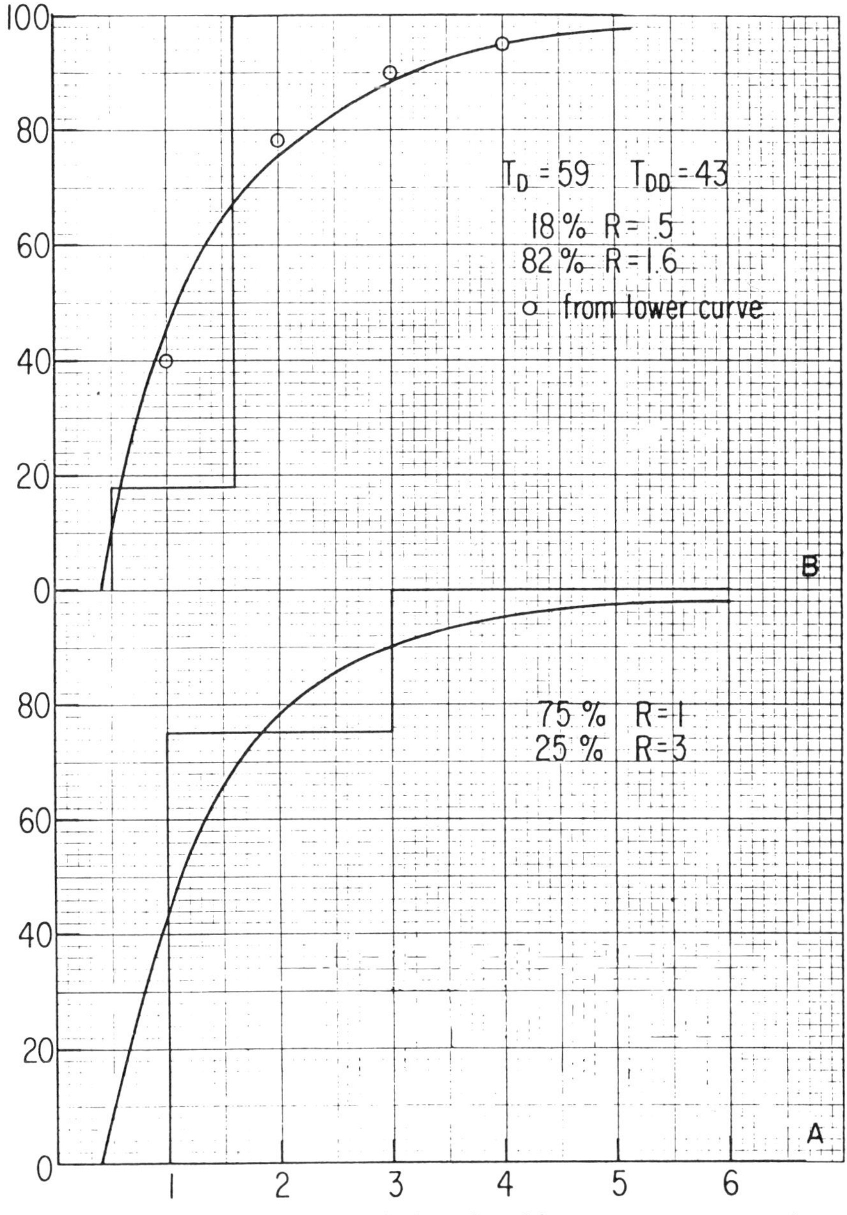

Figure 5. Work sheet. Cumulative distribution estimated from a two-component aerosol.

TABLE V.

r	0.4	0.6	0.8	1.0	1.4	2.0	2.6	3.0	4.0
T_D	0.025	0.181	0.361	0.5	0.666	0.784	0.843	0.868	0.906
T_u	0.932	0.898	0.864	0.83	0.762	0.66	0.589	0.548	0.475
T_{DD}	0.023	0.163	0.313	0.415	0.507	0.517	0.497	0.476	0.43

For example, the point $T_D = 60$, $T_{DD} = 43$ does not represent a monodispersed aerosol. A line connecting 1 and 3×10^{-6} passes close to this point. This could be the result of 75% $R = 10^{-6}$ and 25% $R = 3\times10^{-6}$. On Figure 5A, the solid step is an approximation to the cumulative distribution curve.

A line from 0.5 to 1.6 also passes close to this same point, corresponding to 18% of $R = 0.5\times10^{-6}$ and 82% of $R = 1.6\times10^{-6}$. This is shown in Figure 5B.

In each a curve is drawn to try to get equal areas above and below the step. The curves are similar but obviously not identical. There are many other combinations of 2 components that would give the same T_D and T_{DD} but the cumulative curves should be similar. Again this is not necessarily the true cumulative distribution; it is only a distribution that would act like the real distribution under these conditions.

Fuch's book (1964) contains a wealth of information on diffusion and diffusion losses. The DeMarcus equation is

$$T_o = 0.915 \exp(-1.89\,\mu) + 0.059 \exp(-22.3\mu) + 0.026 \exp(-152\mu) \text{ where}$$

$\mu = D_X/h^2\bar{U}$.

D = diffusion coefficient cm^2/sec

X = length in cm

h = half separation of plates in cm

$\bar{U}$ = average velocity in cm/sec

The diffusion constant is approximately

$$D = \frac{12\times10^{-12}}{r}\left(1 - \frac{10^{-5}}{r}\right).$$

REFERENCES

Dee, P. I., 1931: A condensation diagram. *Proc. Cambridge Phil. Soc.*, p. 93.

Fuchs, N. A. et al., 1962: A determination of particle size in polydisperse aerosols. *Br. J. Appl. Phys.*, *13*, 280.

———, 1964: *Mechanics of Aerosols*. Pergamon Press. 408 pp.

Liu, B. Y. H., 1975: Calibration of Pollak counter with monodisperse aerosols. *J. Appl. Meteor.*, *14*, 46–51.

Megaw et al., 1963: Measurement of diffusion coefficient of homogeneous and other aerosols. *J. R. A.*, *3*, 113–123.

Rich, T. A., 1966: Apparatus and method for measuring the size of aerosols. *J. R. A.*, *79*.

CALIBRATION OF AITKEN NUCLEI COUNTERS DEVIATIONS FROM CLASSICAL PICTURE OF CONDENSATIONAL GROWTH OF NUCLEI

JOSEF PODZIMEK AND JAMES L. KASSNER, JR.

Graduate Center for Cloud Physics Research, University of Missouri-Rolla, Rolla

In this paper we discuss the calibration of nuclei counters and the models describing the growth of embryos in the counters. Mainly, we will discuss Aitken nuclei counters which operate at such a high supersaturation (SS) (greater than 200% related to water surface) that nuclei smaller than 0.1 μm radius will be activated. In this case, the physical-chemical nature of nuclei no longer plays a decisive role in the growth of embryos. On the other hand, the SS should not exceed the threshold for the initiation of the condensational growth of light ions (270%).[1] (This figure is frequently misquoted in the literature because at 350% SS all of the negative ions are active as condensation centers and the positive ions are just beginning to become active (White and Kassner, 1971).) Further, we would like to talk about both absolute nuclei counters (Aitken counter, Scholz counter, University of Missouri at Rolla Absolute Aitken Nuclei counter—UMR-AANC, etc.) and relative counters (Nolan-Pollak counter, General Electric, Environment/One, Gardner counter). In the past absolute nuclei counters were used for the calibration of relative counters.

Absolute nuclei counters like the Aitken counter and the two Scholz counters were characterized by their limited sensitive volume: the Aitken pocket counter in the commercially produced version, 35 cc (the original, 500 cc); the small Scholz counter, 98 cc; and the large Scholz counter, 187 cc. The relatively small sensitive volume influenced the upper and lower limits of nuclei concentrations which could be counted (125 to 200,000 Aitken nuclei (AN) per cc by the Aitken pocket counter, and several tens to 1,000,000 AN per cc by the large Scholz counter). The mean errors in counting nuclei concentration by the Aitken counter were stated to be between 10 and 20% of the concentration. The large Scholz

1. Light (small) ions—ions with the mobility k - 1.0 cm^2/Volt.sec corresponding approximately to a radius r 6.6×10^{-8} cm.

counter was provided with an electrode on which charged nuclei were deposited, and with three different expansion ratios: 1:1.07, 1:1.16, and 1:1.27. The expansion ratio in the smaller counters was approximately 1:1.20.

The operation of the absolute Aitken nuclei counter[2] has evoked discussion on occasion regarding the methodology of the measurements (i.e., Kleinschmidt, 1935; Burckhardt-Flohn, 1939; Grabovskii, 1956). Among the issues raised have been: (1) how long one should wait between individual expansions of the same sample in order to allow the small droplets to fall out and deposit on the counting stage; (2) whether one should add the counts resulting from all the successive expansions of the same sample (i.e., Wigand, 1928; Neuberger, 1935); and (3) how fast should the expansions be made? Further, Ohta (1949) stressed the importance of the time interval between injection of the aerosol into the sensitive volume and the expansion. He claimed that the diffusional losses of very small Aitken nuclei on the walls might strongly influence the measured concentrations. Detailed analyses of the properties of the small Scholz counter were made by Thams and Wieland (1951), and by Podzimek et al. (1961). The mean error in counting was found to be lower than 20% of the mean concentration and was somewhat higher both at very low (about 200 AN per cc) and very high (above 10^5 AN per cc) nuclei concentrations.

The relative Aitken nuclei counters are represented mainly by the photoelectric Nolan-Pollak counter which has been produced with several different modifications since 1945 and described in many articles (Nolan and Pollak, 1948; Metnieks and Pollak, 1959). Because this counter is used currently by many researchers as a secondary standard and because there are several articles published on its operating characteristics (Pollak and Metnieks, 1959; Kassner et al., 1968), we would like to re-emphasize some of the more important research findings.

The Pollak counter is one of the most investigated and best described instruments used in atmospheric sciences today and Pollak and his fellow workers were keenly aware of its design frailties. In reference to the counter air-column diameter (60 cm test volume), Metnieks and Pollak (1959) said: "... If the air-column diameter is 2 cm or less, a non-negligible loss of nuclei takes place in the fog-tube during the 60 sec. between enclosing the sample and releasing the over pressure, especially if the nuclei are small, i.e., their diffusion coefficient large. On the other hand, if the air-column diameter is large, the turbulence in the fog-tube increases and due to processes in the fog (Pollak and Metnieks, 1957) an undesired 'creep' renders it more difficult to determine the correct value of the extinction...."

A simple calculation of particle deposition rate, with r = 10^{-7} cm and $\rho = 1.0$ g/cc in a tube with a radius r = 1.25 cm at very low velocity or in quiescent air, shows that particle losses due to Brownian diffusion might amount to 10% during 60 seconds depending upon the temperature and state of the gaseous medium (Podzimek, 1972). The argument of nuclei wall losses in the Nolan-Pollak

2. This counter is absolute only in the sense that it allows droplets to be observed by a counting technique rather than indirectly by light beam attenuation.

counter has often been used to explain discrepancies between it and absolute counters with larger sensitive volumes such as the UMR-AANC (Saxena et al., 1972). With the exception of an indirect comparison (Podzimek, 1976) there have not been any systematic and conclusive measurements made on this subject.

Several authors have analyzed the effects of the propagation of heat from the walls of the Nolan-Pollak counter. Stressing the combined effects of expansion and heat diffusion from the walls yielded a discrepancy between the SS calculated from the adiabatic law and the actually attained SS (Israel and Nix, 1966; Kassner et al., 1968). Kassner et al., using an infinite cylinder model and a simple definition of the relaxation time of the Nolan-Pollak counter, came to the conclusion that as long as the expansion is quicker than 0.3 seconds there is no danger that the SS decrement can cause a serious error in counting Aitken nuclei with the Nolan-Pollak counter. They claimed that the erroneous interpretation of measurements with fine thermocouples could be responsible for the much lower SS found in the Nolan-Pollak counter by Israel and Nix (1966) and indicated their own experiences with fine wire thermocouples in substantiation of their conclusions.

The most important question is that of the calibration of a relative Aitken nuclei counter against an absolute counter. The Nolan-Pollak Aitken nuclei counter was originally calibrated against the Aitken pocket counter and later against a specially adapted Aitken type counter of 244 cc volume with photographic recording of the cloud density (Pollak and Metnieks, 1959). Other investigators used the small Scholz counter for calibrating purposes. In these calibration reports there were several important points which, in our opinion, should be mentioned. Pollak and Metnieks (1959) found a relatively strong dependence of the measured nuclei concentration on the initial temperature of the Nolan-Pollak counter; on the average, a 1°C increase of initial temperature caused a change in nucleus concentration of up to 10% (Table I). Another important fact was the dependence of the detected nuclei in the photographic Aitken counter upon the state of the graticule. A wet graticule reduced the losses of small drops due to their slower evaporation (Table II). Finally, the "Nachzügler"[3] problem, well known from the absolute counter measurements, apparently had the same importance for relative counters (Table II) as for the absolute ones.

Pollak and Metnieks (1959) in their report on the Aitken type counter stated that "The sum (N) of the average counts per mm^2 from all successive expansions for one sample was used for computing the nucleus concentration N of the sample." We see that the investigators usually stopped the repeated expansions of the same sample after the fourth expansion which probably could cause an error of not more than 1%. However, the number of residual nuclei is dependent upon the mean physical state of the air, the state of dampness of the stage in the sensitive volume, and upon the droplets' fall time (as they must have been deposited to be counted). Higher nuclei concentrations resulted in larger numbers

3. This refers to the question of whether the Aitken type counter should be used to count only the drops falling out after the first expansion, or if it should be used to count the sum from successive expansions of the same sample.

of nuclei being active as condensation centers so that competition for vapor probably limited the size the drops were able to attain and consequently their fall velocity. Another factor relates to the recovery time of the chamber; heat conduction from the walls reheats small chambers so rapidly that droplets experience less growth time. Moreover, the reheating eventually evaporates some droplets, especially those caught in the convection currents and those having to fall from the uppermost parts of the chamber. Thus, the determination of drop concentrations in mid-air as is done in the UMR-AANC counter is by far the more reliable technique. It is questionable whether one can justify adding the counts obtained from successive expansions for Aitken type counters since these in no way

TABLE I. The Influence of Ambient Temperature on Calibration of the Nolan-Pollak Counter

Temperature (°C)	No. 5 $\frac{\text{Photograph}}{\text{Optical Calib.}}$ (%)	No. 6 $\frac{\text{Photograph}}{\text{Optical Calib.}}$ (%)
18.9	124 (6)	111 (6)
20.2	110 (7)	98 (7)
21.0	104 (4)	95 (4)
22.3	82 (3)	79 (3)
23.1	62 (1)	50 (1)

SOURCE: According to Pollak and Metnieks, 1959.
NOTE: The values are averages; the figures in parentheses give the number of sets averaged, each set consisting of 3 to 4 measurements.

TABLE II. Relative Frequency of Droplets per mm² in Various Showers with Wet and Dry Graticule

Shower	1st	2nd	3rd	4th	Total
Wet graticule	805	157	32	6	1000
Dry graticule	640	280	62	18	1000
	575	251	57	17	900

SOURCE: According to Pollak and Metnieks, 1959.

guarantee that the redistribution of nuclei leave one with a number which accurately reflects the concentration of nuclei in the original sample. This also will influence comparisons between the UMR-AANC and all types of relative Aitken nuclei counters in which the original calibrations were based on Aitken type counters employing repeated expansions. Unfortunately there are no systematic measurements on this subject.

Some interesting conclusions can be drawn from an analysis of reheating characteristics of expansion chambers following fast expansions. A detailed, though very simplified, analysis of this subject has been presented by several authors in dealing with the cloud formation in Wilson chambers used in nuclear physics (Williams, 1939; Endt, 1948a, b), and the Nolan-Pollak counter (Kassner et al., 1968). Endt (1948a, b) investigated the incomplete recovery of a

Wilson type chamber (with a height of 2 cm). For an infinitely fast expansion, he calculated the time required for the temperature to rise by an amount δT due to the influx of heat from the walls into the middle of the chamber relative to the temperature drop, ΔT. The time required for the ratio, $\delta T/\Delta T$, to reach a predetermined value constitutes the "sensitive time," τ, of the chamber. The exact magnitude of $\delta T/\Delta T$ is determined by how much the temperature rise, δT, will perturb the measurements at hand; in our case this will depend upon dN/dS for the aerosol being measured.[4] In Figure 1 we plot the sensitive time, τ, as a function $\delta T/\Delta T$ for chambers operating to constant final volume and constant final

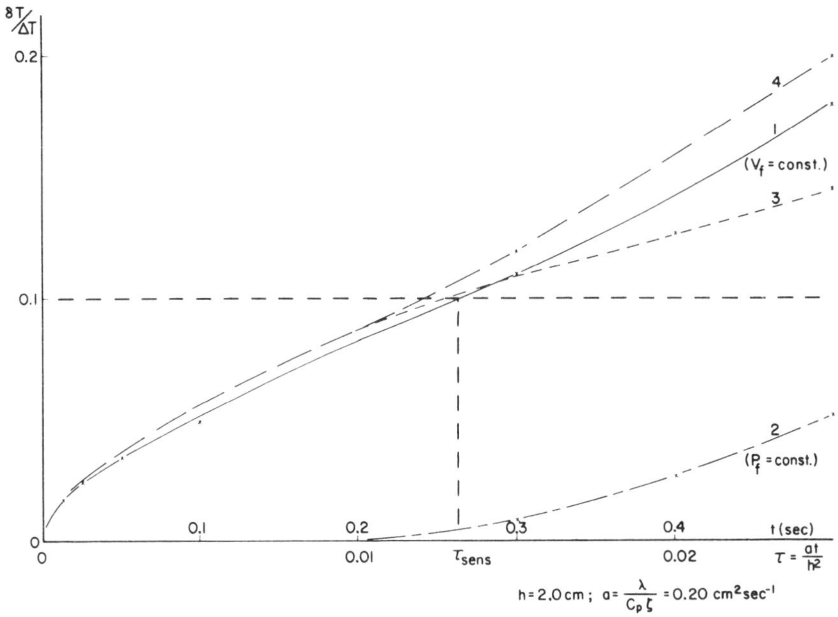

Figure 1. Recovery times for Wilson type chamber (Endt, 1948). 1. Chamber working with constant volume. 2. Chamber working with constant pressure. 3. Williams approximation. 4. van Heerdens approximation.

pressure. It is interesting to note that chambers operating to constant final pressure possess longer sensitive times (the calculations of Endt assume an instantaneous expansion). Chambers operating to constant final pressure warm more slowly because the heated boundary layer of gas near the walls expands in accordance with the gas laws. In the case of chambers which operate to constant final volume, this expansion of gas near the walls compresses the central volume of the chamber causing its temperature to rise. In chambers operating to constant final pressure, such as those which open to the atmosphere, the expansion of the gas near the walls dissipates itself by doing work against the atmosphere and does not contribute to the temperature rise of the gas in the center of the chamber.

4. The change in nuclei concentration N with the change of the supersaturation S.

This would have little or no impact on Aitken counters were it not for the fact that the time required for expansion is finite. Thus, the actual temperature drop achieved by the expansion results from two competing phenomena: (1) the adiabatic cooling part of the expansion, and (2) the reheating due to the walls. In the case of chambers operating to constant final pressure the reheating due to the walls is slightly less efficient.

What does all this mean in terms of the growth rate of droplets? Droplets can grow only as long as SS exists; the slower the rate of thermal recovery of the chamber, the longer the growing time. Aitken-type counters rely upon the deposition of the droplets on a reticule plate as a result of their terminal fall velocity under the influence of gravity. The larger the size the droplets attain the more likely they are to reach the reticule plate before evaporation and to be counted before becoming too small to see.

The rapid-cycling, automatic condensation nucleus counters represent a very different situation. Here the thermal recovery of the chamber becomes totally unimportant because the chamber is mechanically recompressed before natural processes have a chance to take place. Droplet growth reproducibility in these chambers is high for a given drop concentration as a result of the mechanical cycling of the chambers. However, this class of instruments must employ a humidifier to humidify the incoming sample gas to a "standard" relative humidity very close to saturation. It is likely that these humidifiers perform differently depending upon the relative humidity of the air being sampled.

The factors we have discussed which influence Aitken nuclei counting with respect to absolute and relative counters cannot cover all physical processes involved in the calibration of the counters. As an example, we mention some of our experiences in our laboratory comparing the UMR-AANC with the Stratospheric Aitken Nuclei Detection System[5] (SANDS). Both counters have been described elsewhere (Kassner et al., 1968; Haberl, 1975). Several runs were made simultaneously with the Gardner counter, and a detailed description of the arrangement of the experiment (Figure 2) was published one year ago (Wegrzyn and Podzimek, 1975). During the first series of comparisons (September 1974) when we used aged room air, aged residual nuclei from propane, AgI, and NaCl as test aerosols the SANDS instrument showed consistently lower counts than the UMR-AANC, in several cases by more than 30% (Table III). Larger, hygroscopic nuclei, such as NaCl particles, showed higher particle losses in the SANDS (Figure 3). The SANDS counts compared better with the UMR-AANC counts in the concentration range from 300–1000 AN per cc than in the concentration ranges 0–100 and 100–300 AN per cc. In addition to the lower counts of the SANDS, there was also particle loss in the Pressurization Unit which, at ground pressure, amounted to 14.38% for particle concentrations between 3,000 and 10,000 AN per cc and was close to 12% for concentrations between 300 and 1000 AN per cc. Because this calibration was made at UMR just after the calibration of condensation nuclei counters in Boulder and because Haberl in the

5. General Electric Company, Pittsfield, Mass.

GE laboratory originally calibrated the SANDS against a Nolan-Pollak counter, we can conclude that all concentrations communicated by Cadle et al. (1975) should be raised—at least 30 to 40%. We assumed for this statement that the values of the UMR-AANC are closer to reality, bearing in mind the aforementioned results of many investigations.

The second series of comparisons between the SANDS and the UMR-AANC was made in early 1975 when the SANDS was adjusted to the results of the first calibration. Figure 4 for nichrome wire aerosol and Figure 5 for NaCl aerosol show little difference between the two counters. The UMR-AANC counts were

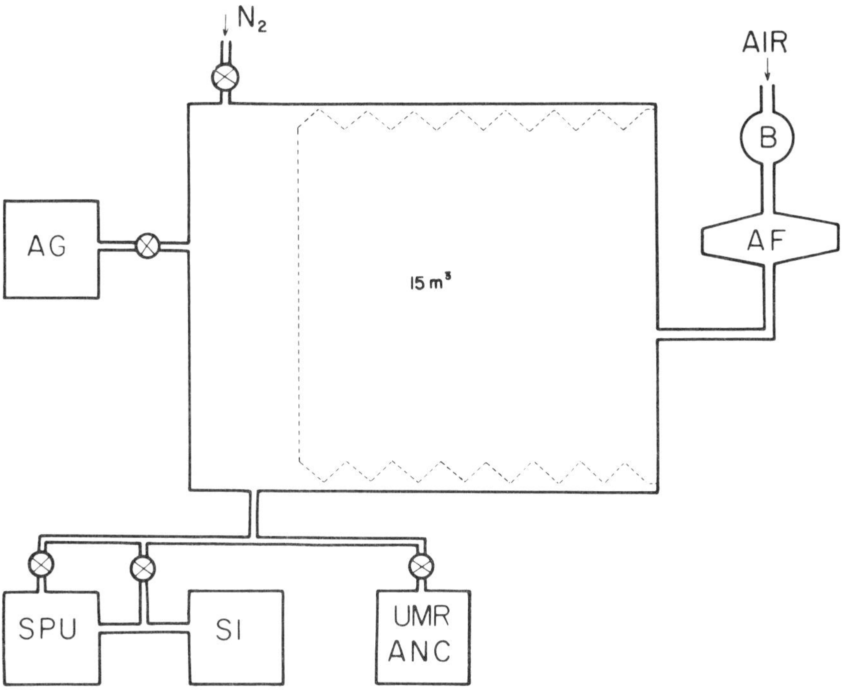

Figure 2. Arrangement of the calibration of the SANDS at the UMR.

still a little higher than the concentrations indicated by the SANDS (Table IV). Both series of the SANDS calibration convinced us strongly of the limited validity of our findings unless both a counter of a similar type and a well defined size distribution are used.

Let us take a moment to discuss some features of nucleation theory applicable to the AANC. The liquid drop theory of homogeneous nucleation provides only a rudimentary analysis of the nucleation process, which in AANC is usually a heterogeneous nucleation process. The classical papers by Volmer and Weber (1926), Farkas (1927), Becker and Döring (1935), Krastanov (1941), Zeldovich (1942) and Frenkel (1946) assumed a quasi-steady-state forward reaction rate

TABLE III. Results of the First Calibration of the SANDS with the UMR-AANC

Type of Aerosol	Concen. Range	No. of Cases	$\bar{Y}$ SANDS	$\bar{X}$ UMR	$\frac{\bar{Y}}{\bar{X}}$ (%)	$r_{x,y}$	b_0	b_1	E.S.E.	$\frac{\text{E.S.E.}}{\bar{Y}}$ (%)
Room air	0–100	9	83.4444	117.9111	70.7690	0.8731	47.4140	0.8448	12.7436	10.8078
	0–300	7	112.4286	163.8571	68.6138	0.8860	−68.2323	2.0643	39.6355	24.1890
		18	195.7778	256.1665	76.4260	0.7730	57.0347	1.0171	15.5684	6.0775
		16	199.4375	299.5625	66.5762	0.9193	−908.4529	6.0571	87.6869	29.2716
Propane	0–100	12	49.7083	71.1333	69.8805	0.9680	−6.4861	1.5615	12.4409	17.4896
		9	55.3333	88.6667	62.4060	0.9495	26.7550	1.1189	10.1477	11.4448
	0–300	4	185.0000	311.0000	59.4835	0.9909	−30.0776	1.8437	60.6955	19.5162
		9	213.3333	317.2222	67.2504	0.9372	38.4299	1.3068	49.9861	15.7574
	0–1000	15	565.8665	684.8665	82.6244	0.9732	124.2644	0.9907	52.0342	7.5977
		9	705.1111	976.5554	72.2039	0.9964	50.1504	1.3138	146.0613	14.9568
A_gI	0–100	7	88.8571	145.0000	61.2808	0.8659	6.1535	1.5626	31.9411	22.0283
	0–300	13	179.4615	255.6923	70.1865	0.9901	25.9352	1.2803	21.1147	8.2579
NaCl	0–100	6	39.5000	72.6667	54.3578	0.9622	−19.7010	2.3384	28.5614	39.3047
	0–300	10	186.4000	317.2000	58.7642	0.9749	26.3367	1.5604	39.4253	12.4292
	0–1000	13	446.4614	651.2307	68.5566	0.7247	−178.2173	1.8578	51.4966	7.9076

through the cluster distribution from embryos containing n molecules to embryos containing n + 1 molecules, and the reverse reaction rate resulting from the evaporation of molecules from the n + 1 embryo. In the Becker and Döring model the rate of nucleation is

$$J = \alpha 4\pi r^{*2} \left(\frac{\bar{v}N_1}{4}\right) ZN_1 \exp\left[-\frac{16\pi\sigma^3}{3KT(n_\ell KT\ell nS)^2}\right]$$

where N_1 is the number of water vapor molecules per cc; α is the probability that the molecule impacting the surface will stick; $4\pi r^{*2}$ is the effective collision cross section of the nucleus with which the impacting water molecule must collide; n_ℓ is the number of molecules per cc in the bulk; r^* is the radius of the critical nucleus; σ is the surface tension of water; $16\pi\sigma^3/3\,(n_\ell KT\ell nS)^2$ is the free energy of formation of the critical embryo; K is the Boltzmann constant; S is the supersaturation ratio; $\bar{v}$ is the average Maxwellian velocity of the molecules; $\bar{v}N_1/4$ is the collision frequency per unit area for vapor molecules; $N_1 = N_0S$ where N_0 is N at saturation vapor pressure; Z is the Zeldovich factor which corrects for deviations from equilibrium; T is the temperature, °K; and $KT\ell nS$ is the molecular heat of condensation (evaporation). Slightly different models were applied by the other authors. Along with the criticism of the thermodynamic basis of homogeneous nucleation theory (i.e., Abraham, 1968; Long, 1969) additional ambiguities arise from the meaning of the collision cross section, surface tension σ and heat of condensation when one considers clusters of less than 100 molecules.

Heterogeneous nucleation theory was considered by many authors to be a formal extension of the homogeneous nucleation model (i.e., Volmer, 1939). Krastanov (1941) and his fellow workers, Frenkel (1946) and Fletcher (1958), developed the more general models of nucleation on insoluble particles introducing correction factors for the more complicated geometry of the condensation on the curved surface of the embryo. Of the few attempts to check the validity of the models for heterogeneous nucleation we might first quote the nucleation studies on ions as the simplest model of nucleation with a relatively simple term for the change in free energy due to the presence of the ion. White and Kassner (1971) deduced a simple model wherein the water molecules in the water surface are preferentially oriented depending upon the sign of the charge on the cluster. The theoretical results were compared with measurements of nucleation rates on (+) and (−) ions in helium and argon atmospheres. In this way, they obtained results which in general show the onset of the nucleation process at slightly lower SS than those predicted by theory; in general, agreement is remarkably good.

A detailed analysis of the kinetics of the adiabatic nucleation of water vapor in a Wilson type chamber with no heat exchange with the environment was made by Bakhanov and Buikov (1967, 1970) and by Bakhanov (1970). They found, in general, a satisfactory agreement with the Becker-Döring formula for the nucleation rate. The nucleation rate study in the UMR-AANC also led to the interesting discovery of the inflection in the nucleation rate curves depending upon the nature of the noncondensible gas (Biermann, 1971). This cannot be predicted by the classical theory and, rather, indicates that some other processes are taking place. Incorporation of gas impurities into the prenucleation clusters could explain this phenomenon. This hypothesis must be necessarily checked in the

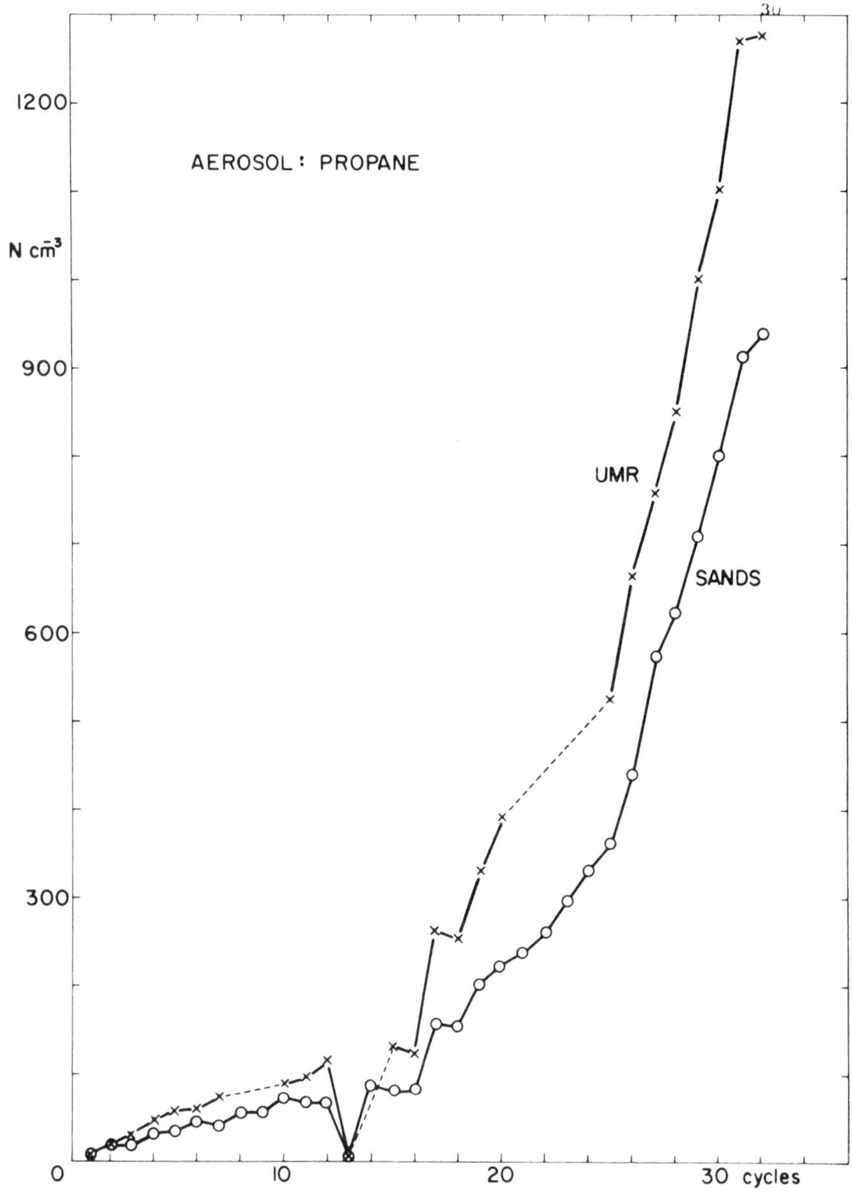

Figure 3. Calibration of the SANDS with propane residual nuclei.

future. Also, the recently established theory of heteromolecular condensation by Stauffer et al. (1973), and Kiang et al. (1973) represents a new field for experimental work.

The nucleation of water droplets in Aitken nuclei counters by soluble submicron particles differs from the previously mentioned processes so that after a very short time the droplets become so large that their growth can be described by a classical, macroscopic, diffusion droplet model. Several articles and monographs have been published on this subject (Fuchs, 1959; Carstens and Kassner, 1968; Carstens et al., 1974, etc.). The diffusion kinetic model seems to be most worthwhile for the application in the Aitken nuclei counters since it covers both

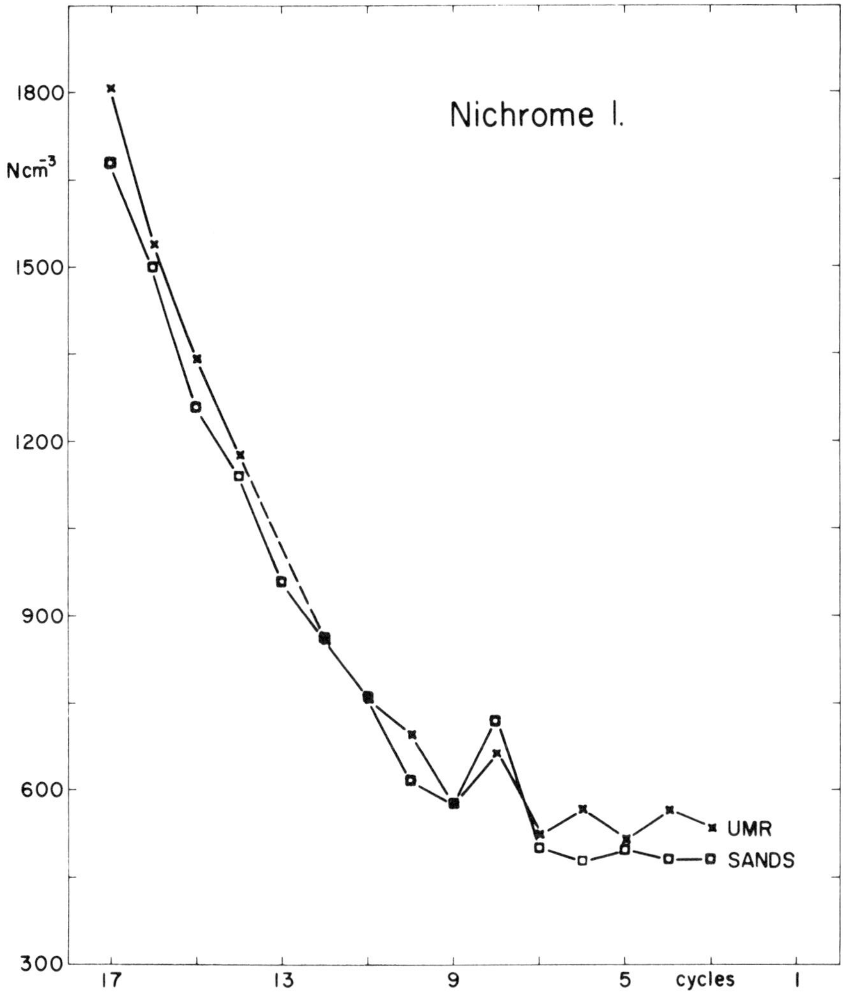

Figure 4. Second calibration of the SANDS with nichrome wire aerosol.

size limits of the growing droplet, including the free molecular flow (Kn >> 1). The applicability of the interpolation formula was discussed by Smirnov (1971). In general, the great merit of this model, which has to be critically judged in the case of extremely fast expansion rates, is its incorporation of the boundary effects on the embryo surface. This happens via the condensation and thermal accommodation coefficients. The values of these constants (although poorly known) form the cornerstone of any confrontation between the theoretical models and experimental measurements of the rate of condensational growth of drops.

Because the paper by Dr. Wagner will discuss this subject further I would like to make some remarks on the measurements recently made by Vietti and Schuster (1973), by Carstens and Carter (1974) and by Wagner (1974). The first two

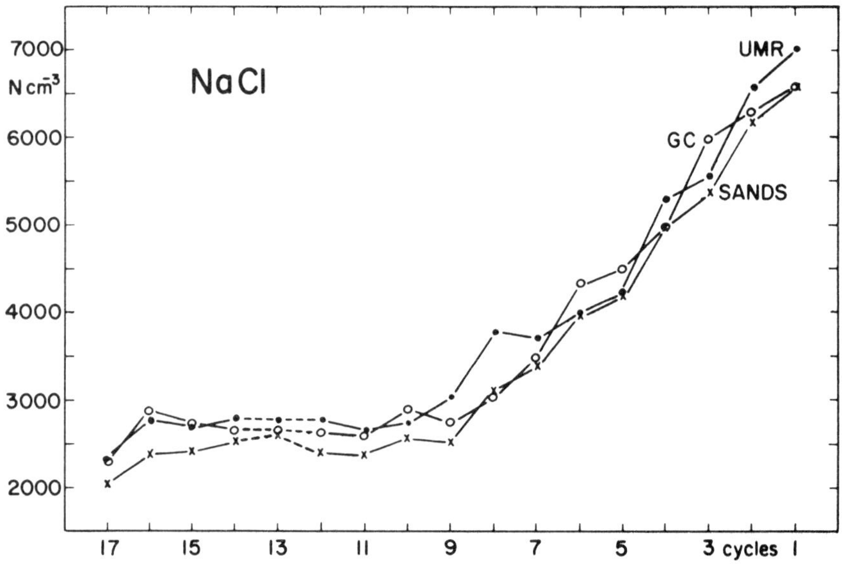

Figure 5. Second calibration of the SANDS with NaCl aerosol. $\bar{d} = 0.13\ \mu m$ and $\sigma/\bar{d} = 0.32$. In the same figure are plotted particle concentrations measured with the Gardner counter (GC).

studies investigated droplet growth in the UMR-AANC and the third investigated the fast expansion chamber with the expansion time equal to five milliseconds. Common to all three studies was the use of light scattering as an indicator of the droplet size and/or of the concentration of drops. Eleven years ago Schuster tried to correct the discrepancies between the observations and the droplet growth model by using a size dependent surface tension (it increases with decreasing radius)

$$\sigma = \sigma_0 \{1 + \exp[-\left(\frac{r}{B}\right)^3]\};$$

$B \simeq 3$ molecule radii.

This technique led to a better fit with observation. However, the question is still open as to whether this approach can be justified while leaving the value of the

TABLE IV. Results of the Second Calibration of the SANDS with the UMR-AANC

Aerosol	Range	No. Cases	$\bar{X}$ UMR	$\bar{Y}$ SANDS	$\frac{\bar{X}}{\bar{Y}}$ (%)	$r_{x,y}$	b_0	b_1	E.S.E.	$\frac{\text{E.S.E.}}{\bar{Y}}$ (%)
Nichrome 1	300–1000	10	627.3	598.0	104.9	0.94760	−115.99	1.13820	76.970	12.071
	1000–3000	4	1467.5	1395.0	105.2	0.99267	90.94	0.88863	164.255	11.775
Sodium	1000–3000	8	2730.6	2417.5	113.0	0.81009	543.60	0.68626	75.628	3.128
chloride	3000–10000	21	6803.0	6619.0	102.8	0.98695	−290.60	1.01567	432.416	6.533
	10000–30000	11	11122.2	10854.5	102.3	0.80902	2495.7	0.75154	320.25	2.950
Nichrome	0–100	2	57.5	48.0	119.8					
	100–300		243.7	231.6	105.2	0.82046	7.99	0.91752	10.667	4.607
	300–1000	3	394.3	380.0	103.8	0.88815	157.44	0.56445	22.240	5.853
Gold	0–100	3	24.67	21.33	115.7					
	100–300	13	260.6	243.2	107.1	0.93234	46.24	0.75584	10.054	4.134

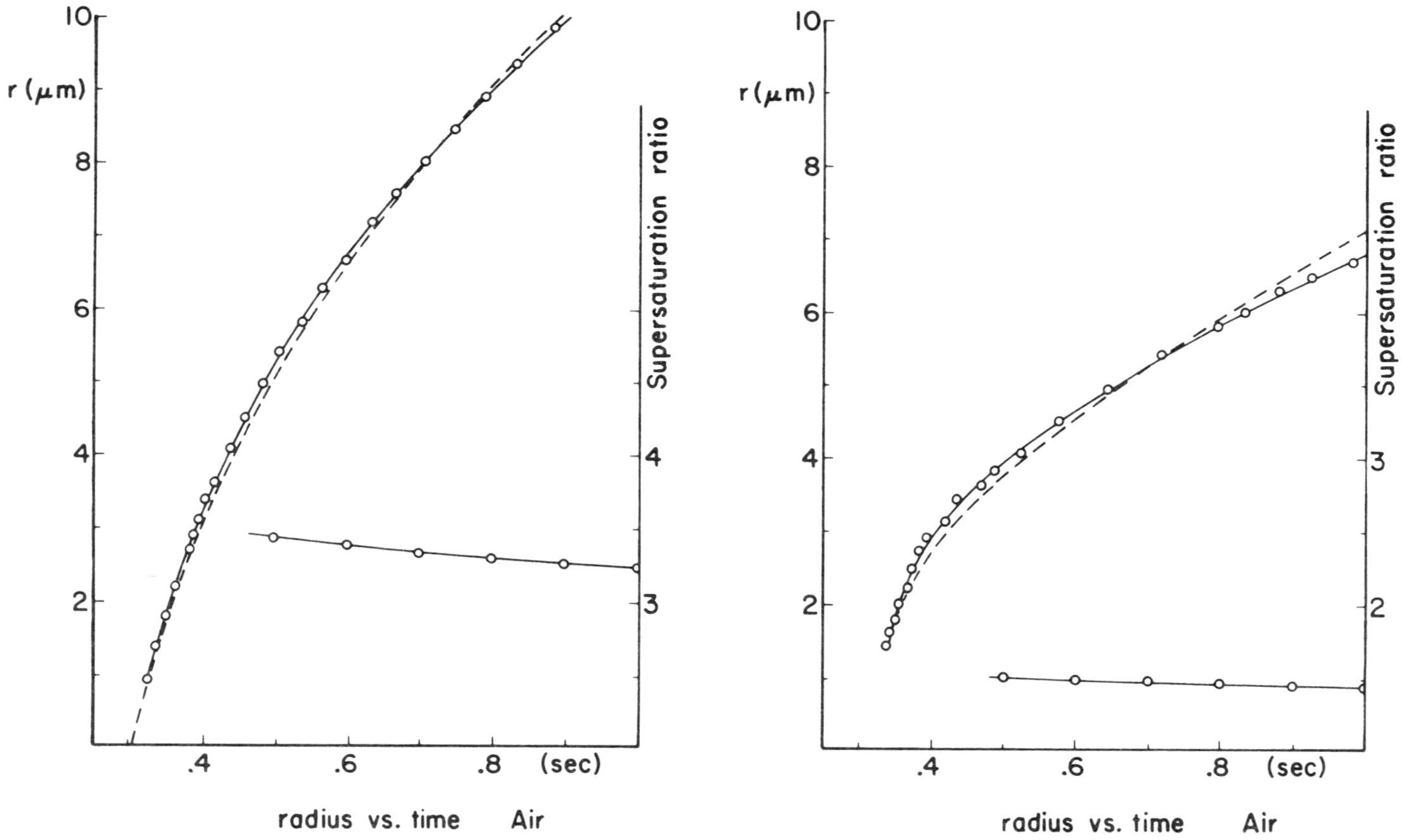

Figure 6. Comparison of droplet growth theory with the measurement (Carstens and Carter, 1974).

condensation coefficient (β) equal to one. Carstens and Carter (1974), on the other hand, showed that for two different expansion rates the observed data are close to the theoretical curves (Figures 6a, b) if one uses for thermal accommodation the coefficient $\alpha = 1$ and $\beta = 0.022$. In Wagner's thesis (1974), the most important finding seems to be the influence of the nuclei concentration upon the rate of growth of the droplets. In inactive and very small nuclei of very high concentrations this leads to considerable differences in nuclei concentrations counted by a counter with a very fast expansion and by a counter with expansion times on the order of 100 milliseconds. Apparently, as in atmospheric cloud condensation, nuclei which activate at small SS consume considerable vapor and reduce the maximum attained SS when slow expansions are employed. For this reason the UMR-AANC employs a sample dilution technique which eliminates vapor depletion effects due to high concentrations of nuclei.

Several remarks on the theoretical models cannot cover the broad field of theoretical and experimental attempts to describe the processes in nuclei counters. We have not mentioned the theoretical models by Reiss (1951) based on statistical mechanical treatment of irreversible condensation or the molecular models describing the formation of small clusters (Plummer and Hale, 1972). There are many other attempts to theoretically describe the processes in counters not mentioned in this contribution. The usefulness of theoretical models in shaping experiments is obvious; they assist in the proper design of counters and give guidance in performing calibrations. The limitations of any theory can be described in our opinion by the words of a great German poet, J. W. Goethe, "Gray is every theory, but green is the living tree."

REFERENCES

Abraham, F. F., 1968: A reexamination of homogeneous nucleation theory: thermodynamic aspects. *J. Atmos. Sci., 25,* 47.

Bakhanov, V. P., 1970: *Koll. Zhur.,* XXXII, 814.

——— and M. V. Buikov, 1967: *Koll. Zhur.,* XXIX, 779.

——— and ———, 1970: *Koll. Zhur.,* XXXII, 654.

Becker, R., and W. Döring, 1935: *Ann. Physik, 24,* 719.

Biermann, A. H., 1971: Homogeneous nucleation of water vapor in inert gas atmospheres, Ph.D. Dissertation, Univ. of Missouri-Rolla.

Burckhardt, H. and H. Flohn, 1939: *Die atmosphärischen Kondensationskerne in ihrer physikalischen und bioklimatologischen Deutung.* Berlin, J. Springer.

Cadle, R. D., G. Langer, J. B. Haberl, A. Hogan, J. M. Rosen, W. A. Sedlacek, and J. Wegrzyn, 1975: A comparison of the Langer, Rosen, Nolan-Pollak and SANDS Condensation Nucleus Counters. *J. Appl. Meteor., 14,* 1566.

Carstens, J. C. and J. M. Carter, 1974: Current meteorological theory of droplet growth by condensation and some comparisons with experiment. Tech. Rep., GCCPR, Univ. of Missouri-Rolla.

——— and J. L. Kassner, Jr., 1968: Some aspects of droplet growth theory applicable to nuclei measurements. *J. Rech. Atmos., 3,* 33.

———, J. Podzimek, and A. Saad, 1974: On the analysis of the condensation growth of a stationary cloud droplet in the vicinity of activation. *J. Atmos. Sci., 31,* 592.

Endt, P. M., 1948a: The recovery of the Wilson cloud chamber working with constant final pressure. *Physica, 14,* 29.

———, 1948b: The recovery of the Wilson cloud chamber working with constant final volume. *Physica, 14,* 97.

Farkas, L., 1927: Speed of nucleus formation in supersaturated vapor. *Phys. Chem.,* Leipzig, A 125, 236.

Fitzgerald, J. W., 1970: *Preprints Conf. Cloud Phys.,* AMS, Ft. Collins, 111.

Fletcher, N. H., 1958: Size effect in heterogeneous nucleation. *J. Chem. Phys., 29,* 572.

Frenkel, J., 1946: *Kinetic theory of liquids,* Dover Publ., New York, 485.

Fuchs, N. A., 1959: *Evaporation and droplet growth in gaseous media,* Pergamon Press.

Fukuta, N. and L. A. Walter, 1970: Kinetics of hydrometeor growth from a vapor-spherical model. *J. Atmos. Sci., 27,* 1160.

Grabovskii, R. I., 1956: *Atmosfernye iadra kondensatsii* (Atmospheric Condensation Nuclei). Leningrad, Gidromet. Izdat.

Haberl, J. B., 1975: Stratospheric Aitken Nuclei Counter. *Rev. Sci. Instrum., 46,* 443.

Israel, H. and N. Nix, 1966: Thermodynamic processes in the condensation nucleus counter. *J. Rech. Atmos., 2,* 185.

Kassner, J. L., Jr., J. C. Carstens and L. B. Allen, 1968: Analysis of the heat and vapor propagation from the walls of the Nolan-Pollak, and Gardner type condensation nucleus counters. *J. Atmos. Sci., 25,* 919.

Kiang, C. S., R. D. Cadle and C. K. Rue, 1973: *Geophys. Res. Lett., 2,* 41.

Kleinschmidt, E., 1935: *Handbuch der meteorologischen Instrumente,* Berlin, J. Springer.

Krastanov, L., 1941: Beitrag zur Theorie der Tropfen und Kristallbildung in der Atmosphäre. *Met. Zeitschr., 58,* 2.

Long, A. B. 1969: A thermodynamic study of homogeneous nucleation. *Rev. of Geophys., 7,* 595.

Metnieks, A. L. and L. W. Pollak, 1959: Instructions for the use of photoelectric condensation nucleus counters. *Geophys. Bull., No. 16,* Dublin, School of Cosmic Physics.

Neuberger, H., 1935: *Met. Zeitschr., 52,* 118.

Nix, N. and N. Fukuta, 1973: Nonsteady-state theory of droplet growth, *J. Chem. Phys., 58,* 1735.

Nolan, P. J. and L. W. Pollak, 1948: *Proc. Roy. Ir. Acad.,* A 51, 9.

Ohta, S., 1949: Investigations on condensation nuclei. *Bull. A.M.S., 30,* 295.

Plummer, P. L. M. and B. N. Hale, 1972: Molecular model for prenucleation water clusters. *J. Chem. Phys., 56,* 4329.

Podzimek, J., 1972: Report on the feasibility study of GE-Aitken Nuclei Counter for stratospheric levels. Rep. to ONR and DOT, GCCPR, Univ. of Missouri-Rolla.

———, A. Hrdinova, and M. Novakova, 1961: Some remarks on the measurement with the small Scholz's condensation nuclei counter (in Czech.), *Met. Zpr., XIV,* 35.

———, and J. L. Kassner, Jr., 1976: Comments on the comparison of condensation nucleus counters. *J. Appl. Meteor., 15,* 1333.

———, W. A. Sedlacek, and J. B. Haberl, 1975: Aircraft measurements of Aitken nuclei in the lower stratosphere. Tech. Rep., GCCPR, University of Missouri-Rolla.

Pollak, L. W. and A. L. Metnieks, 1957: The fog in the photoelectric nucleus counter. *Geofis. Pura e Appl., 31,* 56.

——— and ———, 1959: New calibration of photoelectric nucleus counter. *Geofis. Pura e Appl., 43,* 285.

Reiss, H., 1951: The growth of uniform colloidal dispersions. *J. Chem. Phys., 19,* 482.

Saxena, V. K., D. J. Alofs, and A. C. Tabelak, 1972: A comparative study of Aitken nuclei counters. *J. Rech. Atmos., 6,* 495.

Sedunov, U. S., 1972: Fyzika obrazovania zhidkokapelnoi fazy v atmosfere. Gidrometeoizdat, Leningrad.

Smirnov, V. I., 1971: *Pure Appl. Geophys., 86,* 184.

Stauffer, D., V. A. Mohnen, and C. S. Kiang, 1973: *Aerosol Science, 6,* 461.

Thams, J. C. and W. Wieland, 1951: *Wissensch. Mitt. ETH, 2,* Zurich.

Vietti, M. A. and B. G. Schuster, 1973: Laser scattering measurements of droplet growth in binary mixtures: I. H_2O, and air. *J. Chem. Phys., 58,* 434.

Volmer, M., 1939: *Kinetik der Phasenbildung.* Dresden-Leipzig, Verlag Steinkopff.

——— and A. Weber, 1926: *Phys. Chem., 119,* 227.

Wagner, P., 1974: Untersuchung des Tröpfchenwachstums in einer schnellen Expansionsnebelkammer, Ph.D. Dissertation, University Wien.

Wegrzyn, J. E. and J. Podzimek, 1975: Calibration of the Stratospheric Aitken Nuclei Detection System (SANDS). Tech. Rep., GCCPR, University of Missouri-Rolla.

White, D. R. and J. L. Kassner, Jr., 1971: *Aerosol Science, 2,* 201.
Wigand, A., 1928: *Met. Zeitschr., 63,* 275.
Williams, E. J., 1939: *Proc. Cambr. Phil. Soc., 35,* 512.
Zeldovich, J., 1942: Theory of the formation of a new phase. *Zhur. Teor. Fiz., 12,* 525.

AUTOMATIC ANALYSIS OF SUBMICRON AEROSOLS

DAVID SINCLAIR AND RICHARD J. COUNTESS

U.S. Energy Research and Development Administration, N.Y., N.Y.*

BENJAMIN Y. H. LIU AND DAVID Y. H. PUI

University of Minnesota, Minneapolis

Submicrometer aerosols, or condensation nuclei, in the size range of 0.002 to 0.2 μm diameter are most readily measured for number concentration and size distribution by diffusion batteries and condensation nucleus counters. We at the Health and Safety Laboratory (HASL) have developed a tubular diffusion battery (Sinclair, 1972) of eleven sections and 5000 meters equivalent length. (Equiva-

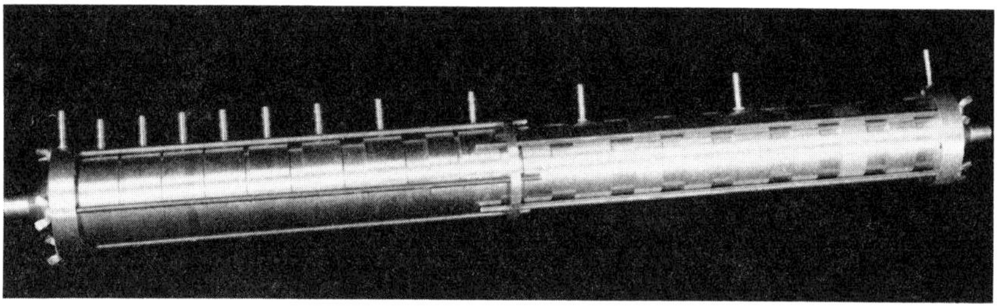

Figure 1. Tubular diffusion battery made of "collimated holes structures." Total length, 61 cm; equivalent length, 5094 meters.

lent length is the number of tubes in parallel times their actual length). Figure 1 shows this battery. Between the twelve sampling ports there is a successively increasing number of "collimated holes structures"[1] of the type shown in Figure 2. Such discs, about 1½ inches diameter and in thickness ⅛ to 1 inch, contain 14,500 holes 0.009 inch diameter of approximately circular cross-section; they make up the tubes of the battery. Tests have shown that these devices behave as if the holes were circular, as they must be if diffusion measurements are to conform with theory.

*Now the U.S. Department of Energy, Environmental Measurements Laboratory.
1. Brunswick Corp., Skokie, Ill.

Equation (1) (Thomas, 1967) describes the molecular diffusion of submicron aerosols flowing at a constant rate through circular tubes, or holes. The fractional penetration of particles through a tube increases greatly with increasing particle size because of the large decrease in diffusion coefficent. Figure 3 shows that the diffusion coefficient decreases about five decades when the particle diameter increases about three decades. Figure 4 shows the percent penetration of particles of diameters from 0.004 to 0.1 μm through the eleven sections of the battery at 4 ℓpm flow rate. As can be seen, only about 1% of the 0.004 μm particles passes through the first battery, while about 15% of the 0.1 μm particles penetrate all eleven batteries.

$$\begin{aligned} f_\alpha = {} & 0.819 \exp(-3.65\,\alpha) \\ & + 0.097 \exp(-22.3\,\alpha) \\ & + 0.032 \exp(-57\,\alpha) \\ & + 0.027 \exp(-123\,\alpha) \\ & + 0.025 \exp(-750\,\alpha) \end{aligned} \tag{1}$$

where

$\alpha = \pi D l/q$

l = length of tube or hole, cm

D = particle diffusion coefficient, cm^2/sec

q = volume flow rate per tube, cc/sec.

Measurement of the number concentration at each of the eleven ports allows calculation of the particle size and size distribution. This measurement can be made with a condensation nucleus counter but requires constant flow rate through the battery. All available nucleus counters operate either by intermittent sampling or varying flow rate, and adiabatic expansion of a humid atmosphere to cause cooling and condensation of water droplets on the nuclei. We at HASL have developed a continuous flow counter (Sinclair and Hoopes, 1975a) which utilizes thermo-electric elements to obtain continuous cooling and ethyl alcohol as the condensate. Figure 5 shows the basic features of this device, while Figure 6 is an end view showing the cooling elements and cooling fins. Additional components required are a 24 volt, 12 amp DC power supply for the cooling elements, and a fan to draw air past the cooling fins along each side of the cold tube. Aerosol is drawn through the alcohol tube and the cold tube in succession at 4 ℓpm. The decrease in light transmission through the cold tube with increase in aerosol concentration may be measured with a microammeter or a millivolt recorder connected to the photocell. The cold tube is usually run at −20°C.

This counter has been calibrated with polydisperse and monodisperse aerosols using the Pollak counter (Pollak and Metnieks, 1959) as a standard to obtain the calibration curve shown in Figure 7. The ordinate shows the concentration of aerosol in number per cm^3, and the abscissa shows the percent light transmission through the cold tube. (The data have been plotted on log-probability paper for convenience.)

Another type of diffusion battery which has recently been developed (Sinclair and Hoopes, 1975b) at HASL utilizes stainless steel screens instead of the tube (hole) clusters. Figure 8 is a photograph of this battery. It contains 55 screens of 635 mesh arranged in ten groups increasing in number of screens from 1 to 10. The calibration and use of this battery will be described later.

AUTOMATED DIFFUSION BATTERIES

The tubular diffusion battery and the screen diffusion battery have each been connected (by one of us, D. S.) to a manifold and a set of timing cams, micro-switches and solenoid valves to allow automatic measurement of aerosol particle

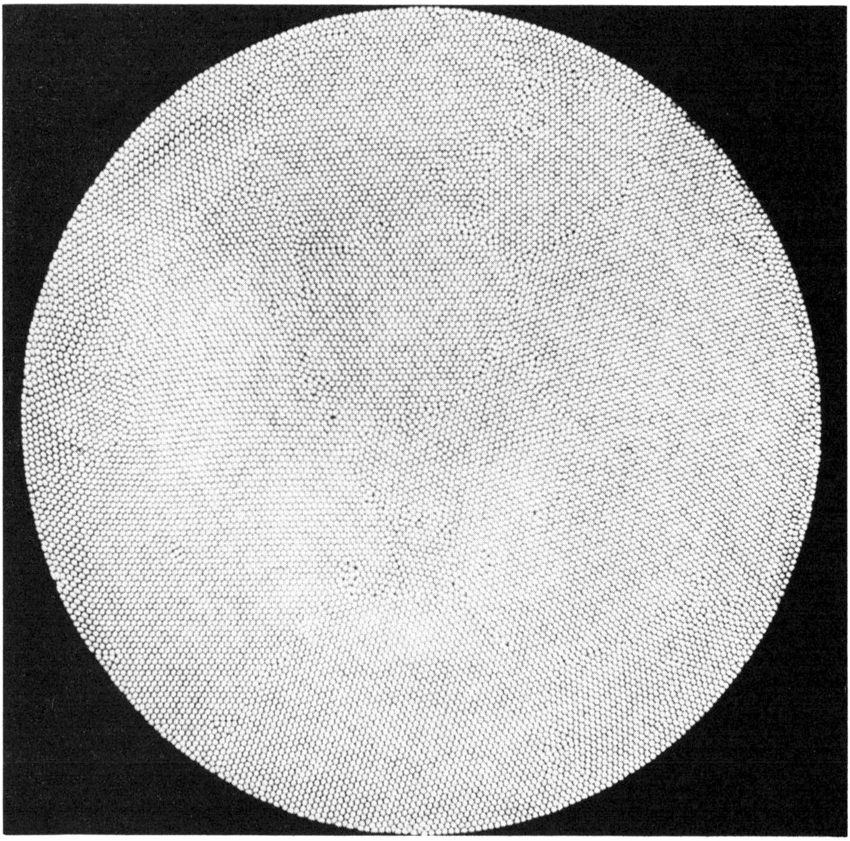

Figure 2. Collimated holes structure, 1 of 18 in diffusion battery. Contains 14,500 holes 0.009 inch diameter.

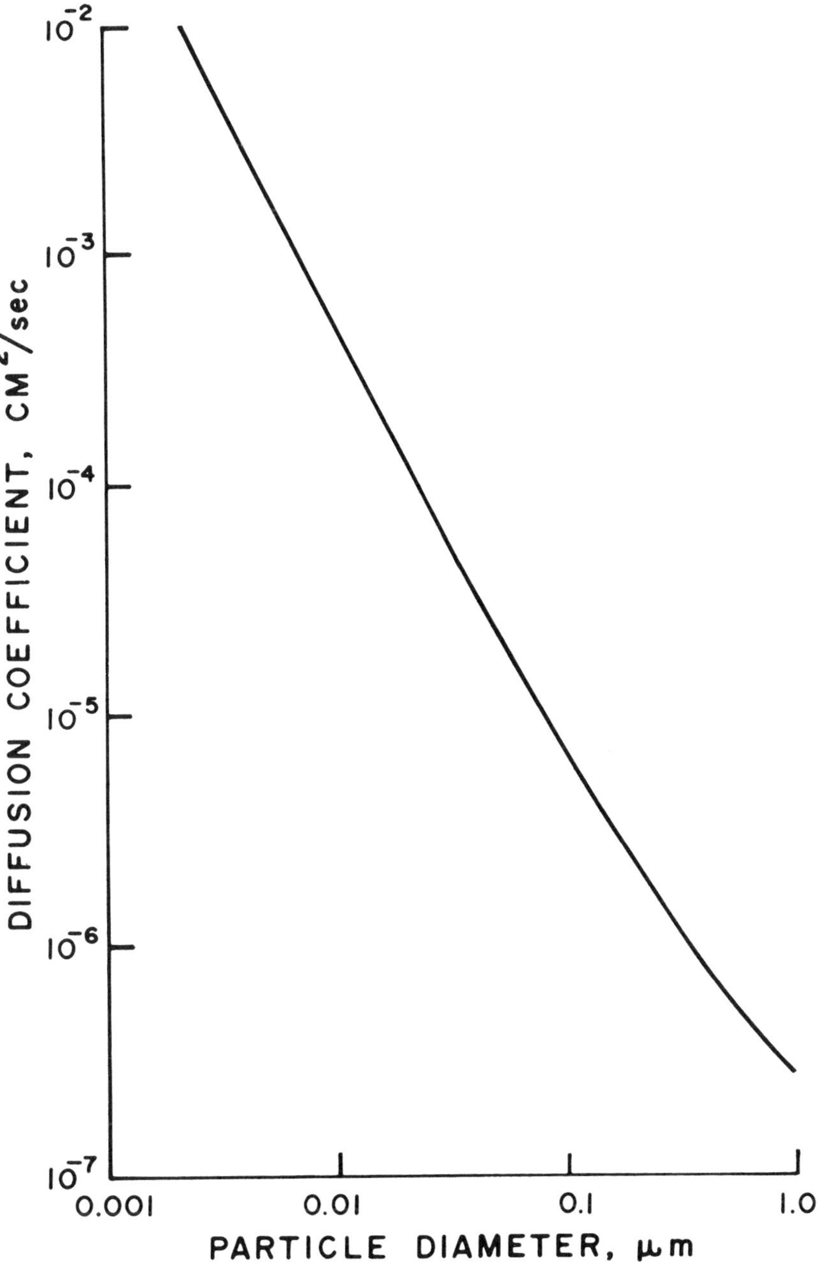

Figure 3. Dependence of diffusion coefficient on particle diameter in submicron range from 0.001 to 1.0 μm.

size distribution. The aerosol may be drawn through the several sections of the battery in sequence by any constant flow rate device such as the continuous flow counter.

Figure 9 shows the assembly for the tubular battery and Figure 10 shows the assembly for the screen battery. The cams, micro-switches and a synchronous

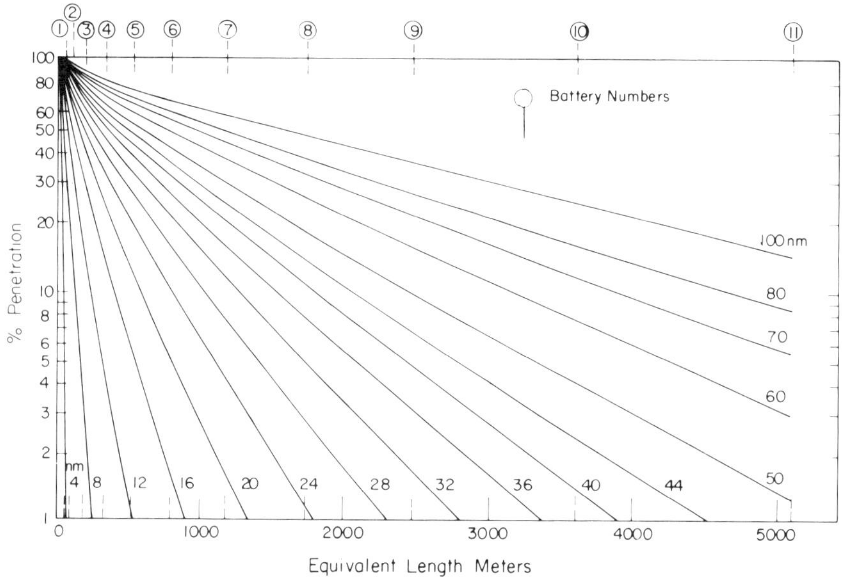

Figure 4. Percent penetration of aerosol particles through the diffusion battery of Figure 1 at 4 ℓpm. Penetration shown as function of particle diameter from 4 to 100 nm.

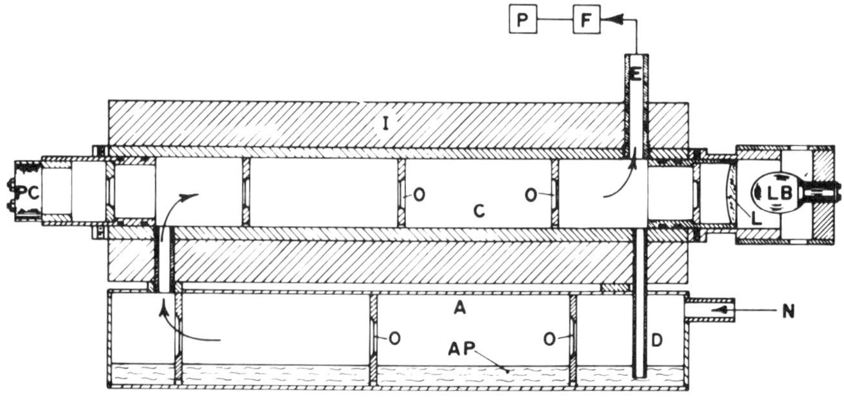

Figure 5. Continuous flow condensation nucleus counter, side view. N = nuclei inlet; A = alcohol tube; AP = alcohol pool; O = orifices; C = cold tube; LB = light bulb; L = lens; PC = photocell; E = exhaust; F = flow meter; P = pump; I = thermal insulation; D = drain.

motor were obtained assembled in a Multi-Cam Recycling Timer.[2] In Figure 9, the aerosol enters the tubular battery at the left rear and the counter is attached to the manifold at the left front. In Figure 10, the aerosol enters the screen battery at the right front and the counter is attached to the manifold at the right rear. The final valve is connected to an ''absolute'' filter to obtain the zero reading of the counter once each cycle. The timer is set to recycle every six minutes with the

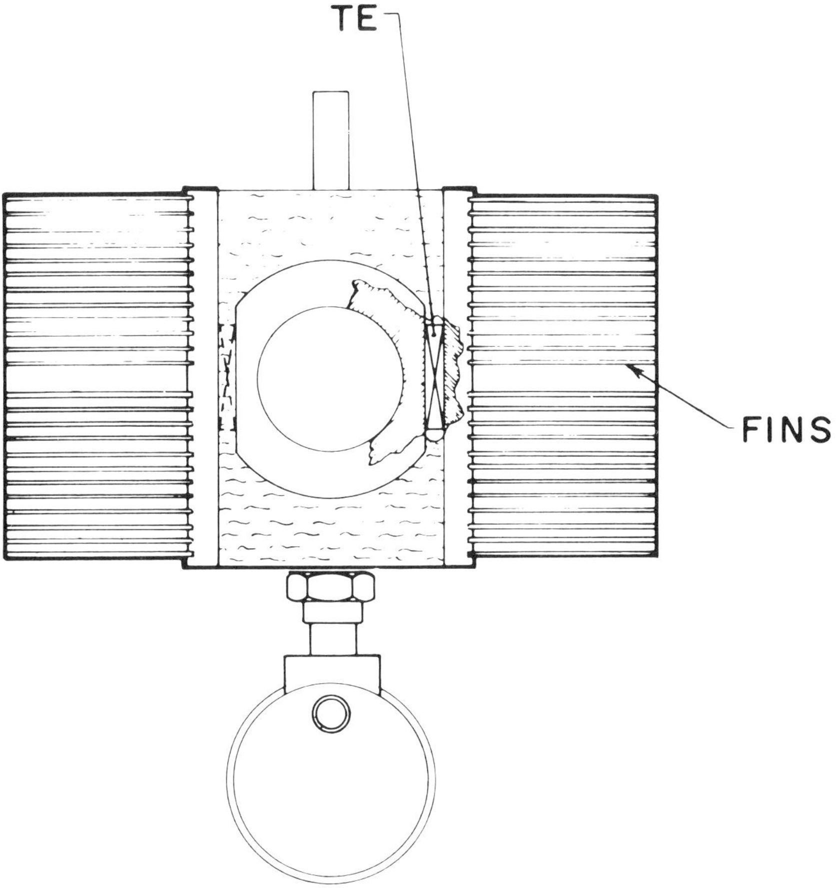

Figure 6. Continuous flow condensation nucleus counter, end view. TE = thermo-electric coolers and cooling fins.

cams adjusted to open each valve in succession for about one-half minute at a time.

Two types of valves were obtained which are being tested for performance and durability. The Valcor[3] valves on the tubular battery are straight-through gate valves with no internal bends and as such should cause minimal loss of aerosol. However, since they were originally intended for liquids they sometimes leak air.

2. Industrial Timer Corp., Parsippany, N.J.
3. Valcor Engineering Corp., Kenilworth, N.J.

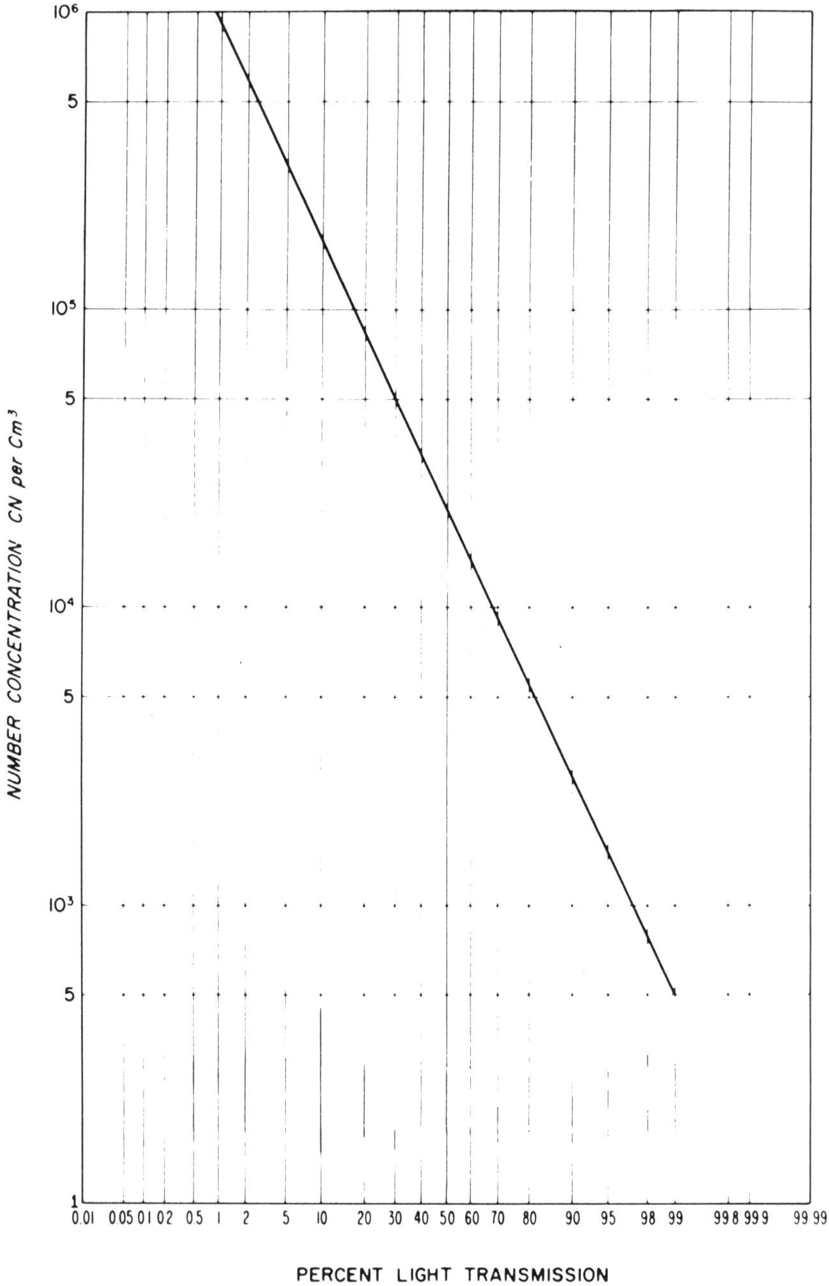

Figure 7. Calibration curve of continuous flow condensation nucleus counter.

Figure 8. Screen diffusion battery contains 55 stainless steel screens of 635 mesh.

This trouble was largely eliminated by placing thin sponge rubber discs behind the Teflon seals. The ASCO[4] valves on the screen battery have soft rubber seals, positively seated by a spring to prevent leakage. However, they contain two right-angle bends. Fortunately, the aerosol losses at 4 ℓpm were observed to be negligible, except for particles a few nm or less in diameter.

4. Automatic Switch Co., Florham Park, N.J.

MEASUREMENTS

In manual operation at the University of Minnesota (Sinclair et al., 1975) theoretical performance of the tubular battery was verified by measurement of monodisperse aerosols. In manual tests at HASL, the screen battery was calibrated with polydisperse aerosols (Sinclair and Hoopes, 1975b).

In recent measurements at the University of Minnesota, the performance of

Figure 9. Tubular diffusion battery mounted for automatic operation and recording, showing assembly of cams, micro-switches and solenoid valves for operation with nucleus counter.

Figure 10. Screen diffusion battery mounted for automatic operation and recording, showing assembly of cams, micro-switches and solenoid valves for operation with nucleus counter.

both automated batteries was tested with monodisperse NaCl and dioctyl phthalate (DOP) aerosols. The aerosols were generated by the Electrical Aerosol Generator described elsewhere (Liu and Pui, 1974). The aerosol was drawn through the battery by either the continuous flow condensation nucleus counter or the Electric Aerosol Detector (Liu and Lee, 1975). The latter instrument can be

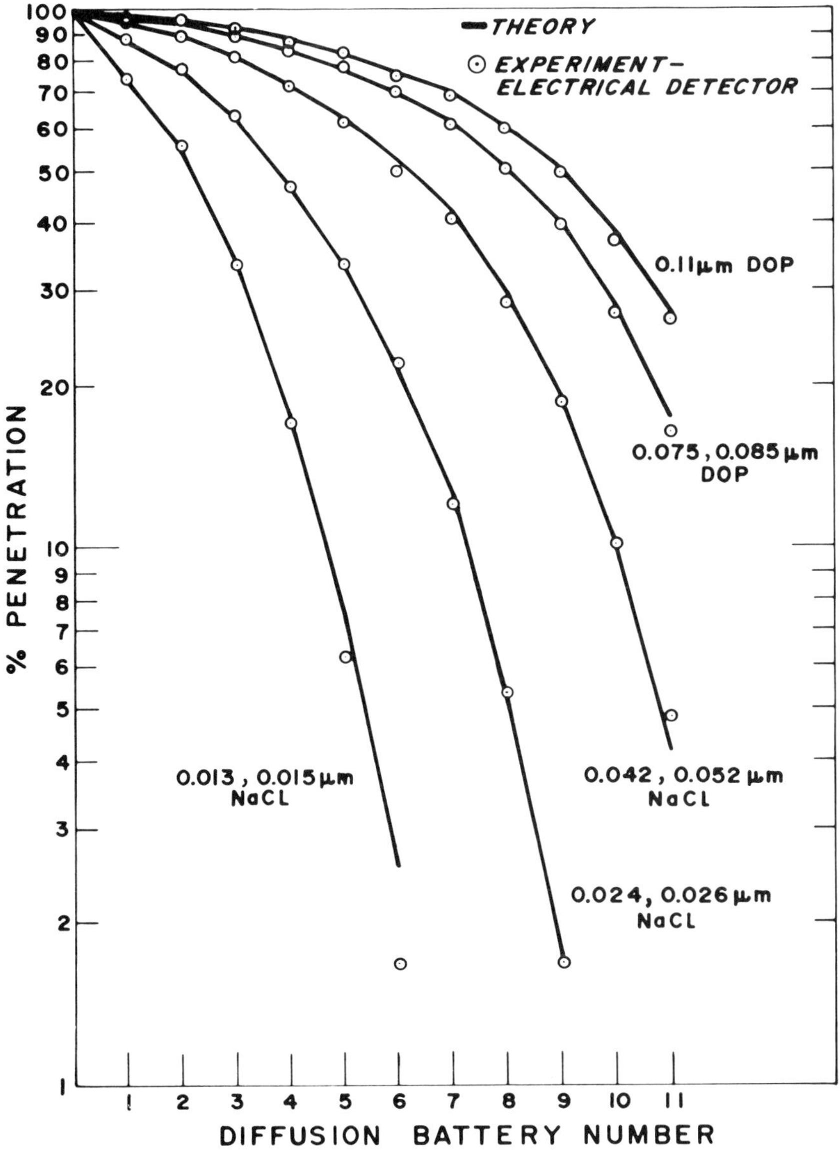

Figure 11. Penetration of monodisperse NaCl and DOP aerosol through tubular diffusion battery measured with Electrical Aerosol Detector at 6 ℓpm. First size values are diameters given by electrical aerosol generator and second values are observed diameters.

used to measure relative aerosol concentrations in diffusion battery applications only when the aerosol is monodisperse. All measurements were recorded on a strip chart. Penetration measurements were made alternately with the tubular battery and the screen battery. These measurements allowed accurate calibration of the screen battery as well as additional performance tests of the tubular battery.

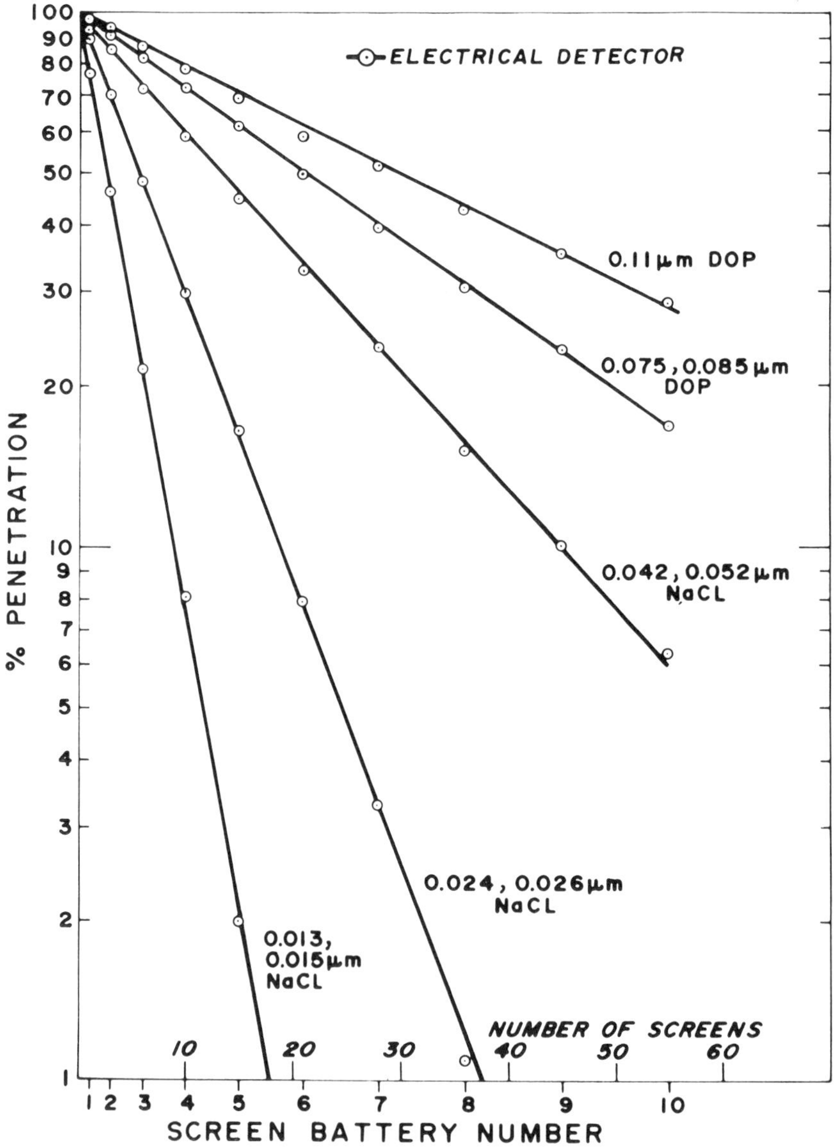

Figure 12. Penetration of monodisperse aerosol of Figure 11 through screen battery at 6 ℓpm.

Figure 11 shows the results obtained with the Electrical Aerosol Detector when measuring NaCl and DOP aerosol with the tubular battery. The ordinate is the percent penetration and the abscissa is the battery number. The circles show the experimental points for aerosols of five different sizes. For four of the resultant penetration curves, two sizes are presented on the graph, the left-hand value

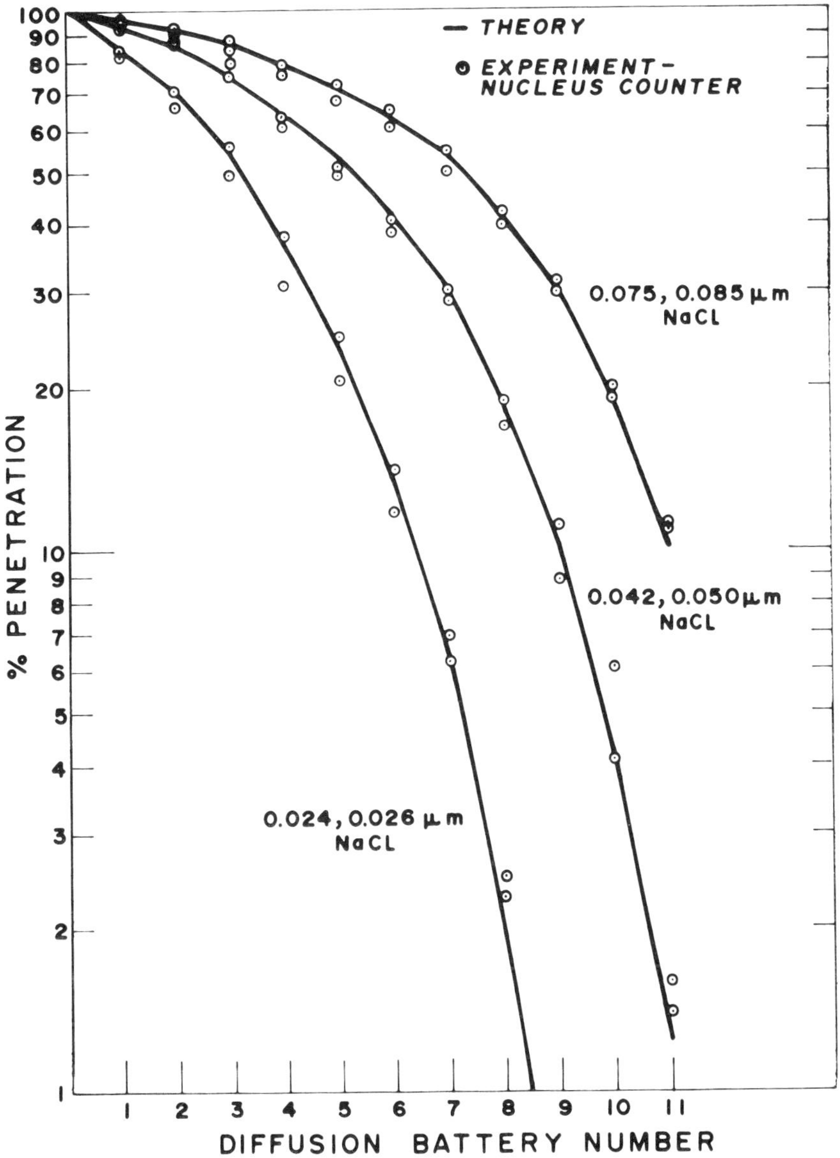

Figure 13. Penetration of monodisperse NaCl and DOP aerosol through tubular diffusion battery measured with continuous flow condensation nucleus counter at 4 ℓpm. First size values are diameters given by electrical aerosol generator and second values are observed diameters.

being the size predetermined for the generator from electrical mobility theory and the right-hand value being the size corresponding to the theoretical curve (solid line) of the diffusion battery that best fits the data. In the case of the largest test aerosol, 0.11 μm, the two values agree almost exactly. Thus, the diameter as measured by the battery is sometimes larger than the diameter calculated from the generator characteristics. The cause of this discrepancy is not understood.

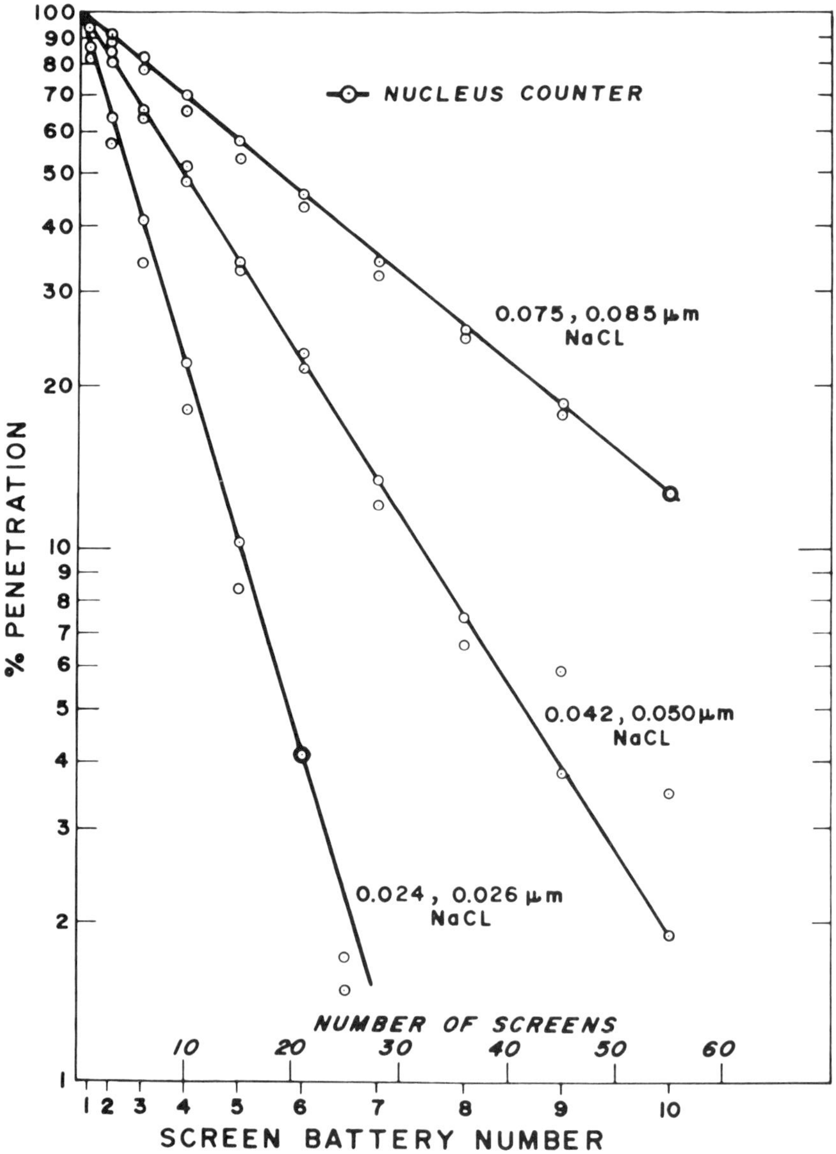

Figure 14. Penetration of monodisperse aerosol of Figure 13 through screen battery at 4 ℓpm.

Similar results are shown by the curves of Figure 12 which are the results of measuring the same aerosols with the screen battery. In these curves the abscissa is the cumulative number of screens. These curves are straight, as they should be for a monodisperse aerosol. As will be described subsequently, such curves serve

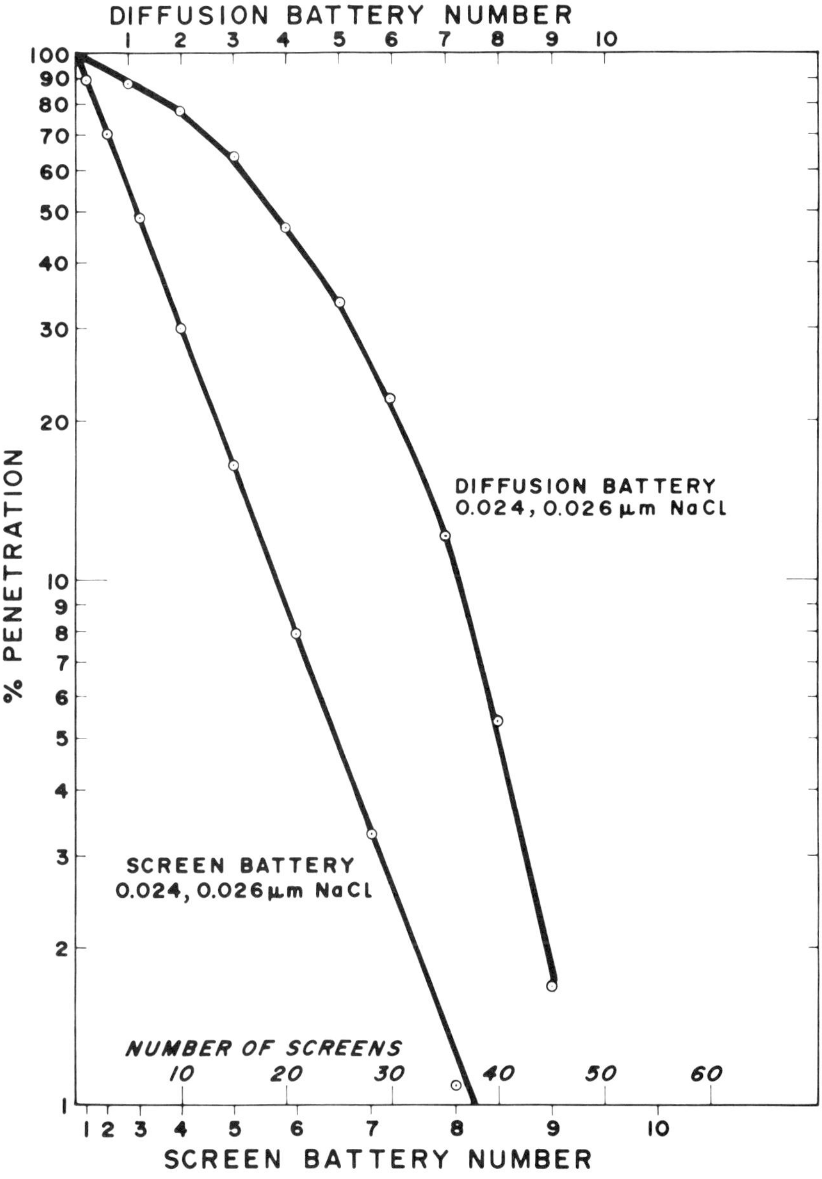

Figure 15. Penetration of 0.024–0.026 μm NaCl aerosol through screen and tubular batteries. This is one method of calibrating the screen battery.

as calibration curves for use with the screen battery to measure polydisperse aerosols.

Figures 13 and 14 show similar results obtained on three sizes with the tubular and screen batteries when using the continuous flow nucleus counter. Again, the measured diameters do not agree exactly with the calculated generator diameters. The pairs of points are caused by another discrepancy: a two-fold, greater number concentration over most of the range given by the generator compared to that measured by the Pollak counter. This resulted in two different calibration curves for the continuous flow counter which converged at each end of the concentration range. However, since diffusion battery measurements are based

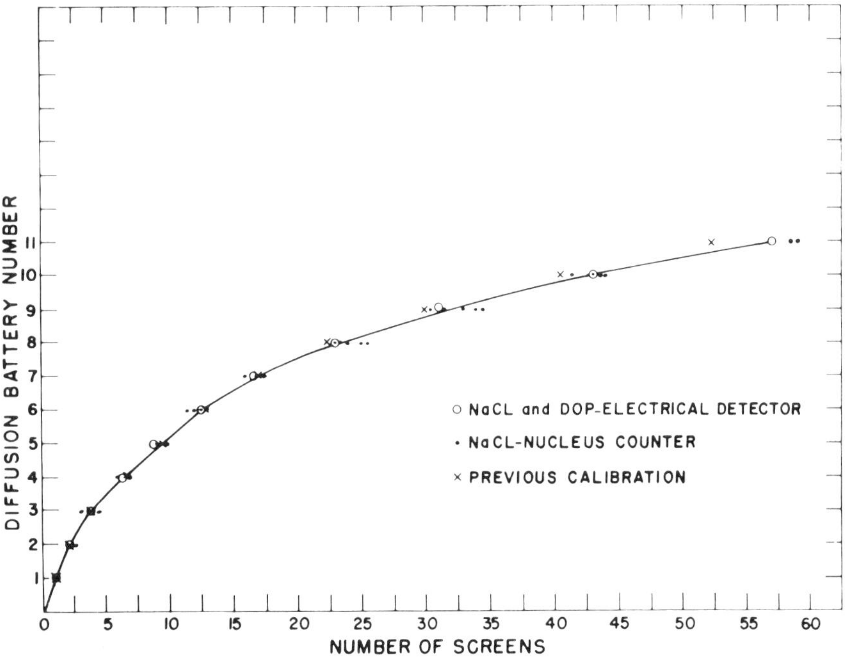

Figure 16. Calibration curve of screen battery for monodisperse (⊙) and polydisperse (×) aerosols.

on relative values, the size measured by the diffusion battery was affected only by the convergence of the two calibration curves. The cause of this two-fold difference in number concentration has yet to be determined.

Figure 15 shows a pair of curves taken from Figures 11 and 12 to show one method of calibrating the screen battery. The abscissa of the screen battery curve having the same percent penetration as a given tubular battery gives the number of screens equivalent to that battery. Figure 16 shows the results of such comparisons obtained with monodisperse aerosols. The results obtained previously (Sinclair and Hoopes, 1975b) with polydisperse aerosols are also included. As described elsewhere (Sinclair and Hoopes, 1975a), this curve can be used to convert the results of measurements with the screen battery to the tubular battery for which theoretical performance curves are available, and thus obtain a size distribution.

However, the screen battery penetration curves for monodisperse aerosols can be used directly with screen battery measurements to determine the size distribution of a polydisperse aerosol. This method of size distribution analysis is the graphical stripping method already described for the tubular battery (Sinclair, 1972; Breslin et al., 1971). In the present case the method consists of subtracting

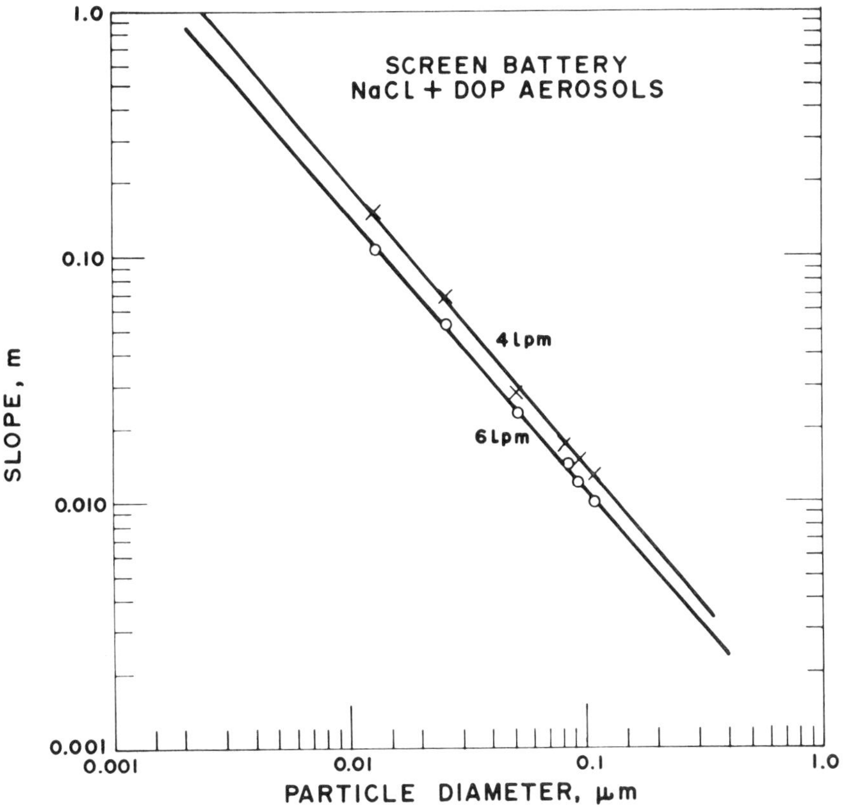

Figure 17. Plot of slope of screen battery penetration curves vs. aerosol particle diameter at flow rates of 4 ℓpm and 6 ℓpm.

calibration curves of the type shown in Figure 12 from the experimental curve for a polydisperse aerosol, beginning with the curve for the largest possible size. This is the size for which the slope of the calibration curve is just less than the final slope at the end of the experimental curve. This procedure gives the percentage of that size, or larger, in the aerosol. After that, a curve for a smaller size is subtracted from the remainder curve, and so on. This method takes some practice, but we have found that the method works quite well for the tubular battery for which the results can be verified by an alternative computer program (Raabe, 1967–68).

For analysis by the graphical stripping method we first plotted screen battery penetration curves for a larger range of sizes than appear in Figure 12. The equation for these curves is log P = − mn + 2, where P is the percent penetration, m is the slope and n is the number of screens. The values of the slope for various particle diameters were obtained from Figure 17, a log-log plot of the

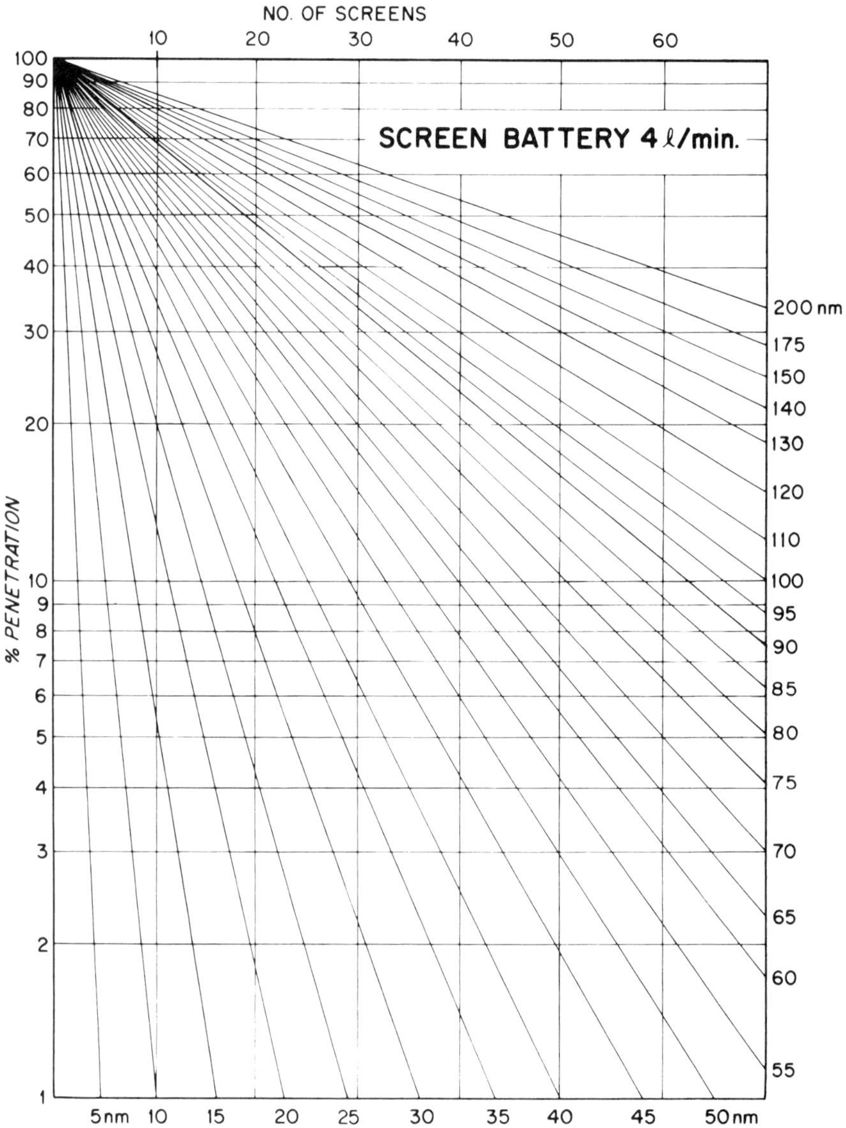

Figure 18. Calculated penetration of monodisperse aerosols through screen battery for 27 diameters from 5 to 200 nm.

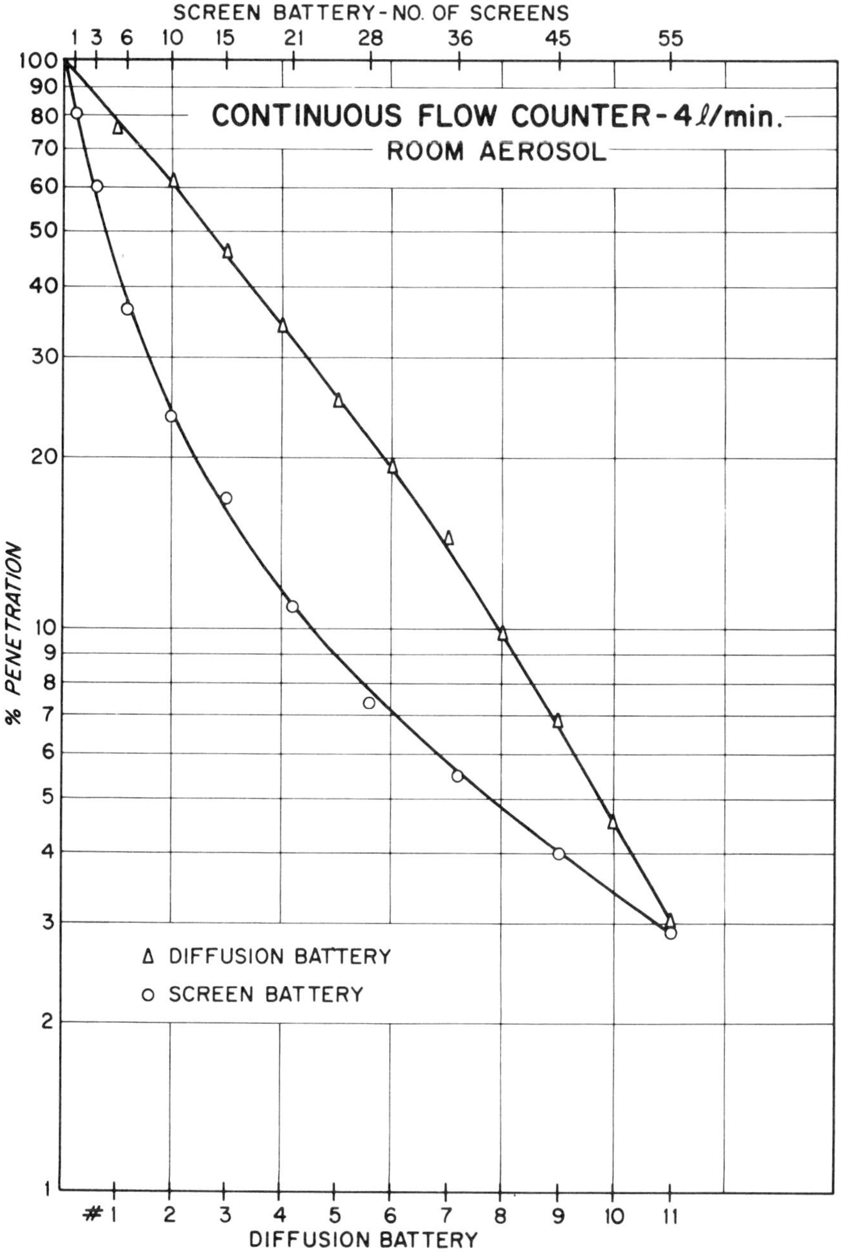

Figure 19. Experimental penetration curves of polydisperse aerosol through tubular and screen batteries measured with continuous flow counter at 4 ℓpm, showing percent penetration vs. tubular battery number (Δ) and number of screens (0).

slope of the screen battery curves vs. particle diameter for two different flow rates and the sizes shown in Table I. By this method, we obtained 27 calibration curves for particle diameters ranging from 0.005 to 0.2 μm as shown in Figure 18.

With the aid of these calibration curves, the graphical stripping method was applied to two sets of data taken at HASL for a polydisperse room aerosol with the tubular diffusion battery and the screen diffusion battery, using the continuous flow counter at a flow rate of 4 ℓpm. The two sets of data are plotted in

TABLE I. Monodisperse Aerosol from Electrical Aerosol Generator

Type of Particle	Generator Diameter (μm)	Measured Diameter (μm)
Electrical Aerosol Detector at 6 ℓ/min		
NaCl	0.013	0.013–0.015
	0.024	0.026
	0.042	0.052
	0.075	0.094
DOP	0.075	0.085
	0.11	0.107
Electrical Aerosol Detector at 4 ℓ/min		
NaCl	0.013	0.014–0.016
	0.024	0.025–0.027
	0.042	0.052
	0.075	0.097
DOP	0.075	0.082
	0.11	0.11–0.12
Continuous Flow Nucleus Counter at 4 ℓ/min		
NaCl	0.024	0.026
	0.042	0.050
	0.075	0.085

Figure 19. The calculated size distribution was identical in each case, namely: number median diameter 0.026 μm, geometric standard deviation 2.92. Our computer program also gave the same values.

CONCLUSIONS

1. The high-efficiency tubular and screen diffusion batteries are useful for measurement of the size distribution of polydisperse aerosols.
2. The continuous flow feature of our condensation nucleus counter makes it suitable for use with high-efficiency diffusion batteries to measure the number concentration and size distribution of polydisperse aerosols.
3. The automated diffusion batteries are suitable for continuous or intermittent measurement and recording of the size distribution and number concentration of ambient or other polydisperse aerosols.

REFERENCES

Breslin, A. J., S. F. Guggenheim, and A. C. George, 1971: Compact high-efficiency diffusion batteries. *Staub-Reinhalt. Luft* (English Trans.), *31,* 1.

Liu, B. Y. H. and D. Y. H. Pui, 1974: A submicron aerosol standard and the primary, absolute calibration of the condensation nuclei counter. *J. Colloid and Interface Sci., 47,* 155.

——— and K. W. Lee, 1975: An aerosol generator of high stability. *J. Amer. Ind. Hygiene Assoc., 36,* 861.

Pollak, L. W. and A. L. Metnieks, 1959: New calibration of photoelectric nucleus counters. *Geof. Pura e Appl., 43,* 285.

Raabe, O. G. 1967–68: Fission product inhalation program annual report 1967–68. Lovelace Foundation Report LF-39, UC-48.

Sinclair, D., 1972: A portable diffusion battery. *J. Amer. Ind. Hygiene Assoc., 33,* 729.

——— and G. S. Hoopes, 1975a: A continuous flow condensation nucleus counter. *J. Aerosol Sci., 6,* 1.

——— and ———, 1975b: A novel form of diffusion battery. *J. Amer. Ind. Hygiene Assoc., 36,* 39.

———, R. J. Countess, B. Y. H. Liu, and D. Y. H. Pui, 1975: Experimental verification of diffusion battery theory. Paper #75-41.4. APCA Annual Meeting, Boston. *J. Air Pollution Control Assoc.,* 1976: *26,* 661.

Thomas, J. W., 1967: Particle loss in sampling conduits. IAEA Symp. Vienna, p. 701.

INSTRUMENTATION FOR REAL-TIME MEASUREMENT OF CONDENSATION NUCLEI

ERNEST A. DEMETRE

Environment/One Corporation, Schenectady, N.Y.

In the evaluation of the particulate constituents in the atmosphere, researchers have been primarily concerned with the measurement of their mass. The concentration, i.e., the number of particles, has not received equal consideration. This has been especially true with respect to the submicron particles, primarily those in the size range smaller than 0.1 μm diameter (Rich, 1961). These small particles, through coagulation in the atmosphere and due to condensation, eventually serve as the sources of many of the microscopic and larger particles which contribute to effects on visibility (Neuberger, 1948). Many effects are the result of very complex gaseous reactions in the atmosphere, therefore an indication of the gaseous contamination that is present. By specifically measuring the number of condensation nuclei produced by a given combustion process, one can study its effect as an air pollution source and the means by which pollutants can be reduced. The availability, within recent years, of instrumentation which has provided field measurements of the concentration of such particles and of their size characteristics, has led to an increasing understanding and growing interest in the importance of such particles.

For the purpose of this discussion, instrumentation capable of real-time detection and measurement of particles smaller than 0.1 μm diameter will be described, particularly for those particles with the ability to act as centers for the condensation of water vapor. These particles are referred to as condensation nuclei. Some examples of the application of condensation nuclei monitors in several areas of research will be noted, e.g., auto emissions, respiratory studies, and air pollution parameters.

CONDENSATION NUCLEI INSTRUMENTATION

The detection and capture of submicron particles (condensation nuclei) is not straightforward. The average size of these invisible atmospheric constituents is in the range of 0.02 μm diameter. This essentially eliminates standard optical techniques for measuring them. Ion chamber-type devices are feasible because of their ability to detect the electrical properties of such particles (Skala, 1966). Another method utilizes the condensation of water droplets around the small particles to increase their diameters and thereby permit detection by optical means. It is this latter principle which has been utilized in the Condensation Nuclei Monitor developed by the Environment/One Corporation.

The Model Rich 100 Condensation Nuclei Monitor operates on the principle of a cloud chamber in which water is condensed upon submicroscopic particles to produce micron sized droplets. A constant air flow sample is periodically diverted through a humidifier into a cloud chamber where a fixed volume expansion occurs, providing a supersaturation of at least 300% which produces a cloud (Das Gupta and Ghosh, 1946). The cloud attenuates a light beam focused on a solid state light sensitive element. The formation of the cloud causes a decrease in light on the light sensitive element which results in a change in the electrical resistance of this element. The electrical signal obtained by the resistance change is amplified and rectified to provide a DC signal proportional to the condensation nuclei concentration. The DC signal is displayed on a panel indicator and is also available for operation of a recorder. After expansion, the cloud chamber is pressurized and flushed out. The total measurement cycle time is approximately once per second.

MAIN ELEMENTS OF THE MODEL RICH 100 CONDENSATION NUCLEI MONITOR (VAN LUIK AND LUDEWIG, 1972)

Vacuum Pump

The Vacuum Pump is used to provide pumping for continuous air sampling and also reduced pressure for Cloud Chamber expansion. A portion of the Vacuum Pump discharge air is used to actuate pneumatic valves which control the flushing of the Cloud Chamber.

Humidifier and Water Supply Circuit

The Humidifier is a chamber comprised of a series of baffles and wetted wicks which humidify the incoming air. The water level in the Humidifier is maintained by means of a thermistor and associated circuitry.

Cloud Chamber Assembly

The Cloud Chamber assembly consists of the Cloud Chamber tube with a light source at one end and a solid state detector at the other end. Flow through the Cloud Chamber is controlled by pneumatic valves; the pneumatic valve actuation is controlled by the Rotary Valve. The fixed volume expansion chamber is used to form the cloud. Expansion is created by decreasing the pressure on the back side of a slack diaphragm. This pressure decrease is also controlled by the Rotary Valve.

Rotary Valve

The Rotary Valve provides the programming to the Condensation Nuclei Monitor. This valve is driven by a 60 rpm motor; the basic cycle time is one (1) second. The Rotary Valve provides control for the pneumatic valves and for expansion in the Cloud Chamber. In addition, a sector on the Rotary Valve shaft controls the solid state switching which is used to detect synchronously the electric signal from the light sensor in the Cloud Chamber.

Amplifier Circuitry

The Amplifier Circuitry consists of a single, plug-in circuit board. This circuitry provides a DC output signal proportional to the concentration of condensation nuclei. Analog signals drive a panel indicator for direct readout in condensation nuclei per cc. The signals are also available for recorder operation. A range switch selects the proper amplifier gain so that the instrument can be read directly on all ranges except the 10M range for which a calibration curve is supplied.

THE MODEL RICH 100 CAPABILITIES

The main discussion has focused on the basic instrument for continuous measurement of condensation nuclei, the Model Rich 100 Condensation Nuclei Monitor, identified by Catalog No. E-1033A-001G1. This instrument measures the concentration of all particles greater than 0.0025 μm diameter at a rate of 1 analysis per second. Concentration of such particles may vary from 300 particles per cc to 10^7 particles per cc. The response time of the instrument is approximately 3.5 seconds with a sample flow rate of 50 cc per second.

The versatility of the Model Rich 100 Condensation Nuclei Monitor permits a broad range of uses including, but not limited to, scientific laboratory studies, atmospheric measurements via vehicular, shipboard and aircraft installation, emission measurements, inhalation toxicology studies and more.

The standard instrument heretofore described can also be supplied with an automatic range changer for remote/unattended operation. With minor modification it has the capability to measure condensation nuclei concentrations down to 50 particles per cc, an increase in differential pressure permitting a response to smaller particles, 0.0016 μm diameter, and a logarithmic output for data bank and similar readout requirements. Two allied instruments, both noncontinuous in operational mode, are an "over the river and through the woods" Portable/Semi-Automatic Condensation Nuclei Monitor and a Portable/Photographic Submicron Particle Counter. Recently delivered to NASA were several Condensation Nuclei Counters with pressurization capabilities for high altitude sampling by commercial jet transports. These units are an integral part of the Global Air Sampling Program (NASA Contract NAS3-19870).

Several ancillary devices can be utilized with the Model Rich 100 Condensation Nuclei Monitor. Atmospheric Gas Detectors for SO_2, NH_3, and Hg, and a Diffusion-Denuder for the determination of characteristic size of submicron particles through their diffusion properties (Rich, 1958) are also available.

SOME APPLICATIONS

There is a broad range of worldwide applications in which the Model Rich 100 Condensation Nuclei Monitor is playing an important part. A few representative examples of the applications of these instruments are:

1. Submicron particle sizing via diffusion battery techniques.
2. Measurement of submicron particle concentration and various gaseous pollutants in an upper atmosphere characterization program.
3. Studies of the emission of submicron particles from automotive engines.
4. Inhalation toxicology studies on primates.
5. Plume detection and tracking by aircraft.
6. Nuclear power plant effluent monitoring.

SUMMARY

The concentration and size distribution of submicron particles (condensation nuclei) with respect to their sources and behavior is not well known. The average or typical concentration worldwide has only recently been considered on a broad scale; this primarily due to the lack of instrumentation suitable for field measurements of condensation nuclei. The increasing availability of such instrumentation is contributing to the growing knowledge and interest in these small particles, their relationship to atmospheric phenomena and their impact upon man and his world.

REFERENCES

Das Gupta, N. N. and S. K. Ghosh, 1946: *Rev. Mod. Phys., 18,* 225–290.

Neuberger, H., 1948: Condensation Nuclei—their significance in atmospheric pollution. *Mech. Engineering Magazine, 70,* 222.

Rich, T. A., 1958: The average size of submicron particles. *Int. J. Air Pollution, 1,* 288–292.

———, 1961: Continuous recorder for condensation nuclei. Fourth Int'l. Symp. on Atmospheric Condensation Nuclei, Heidelburg.

Skala, G. F., 1966: The ion chamber detector as a monitor of thermally produced particulates. *J.R.A., 2,* 189–196.

Van Luik, Jr., F. W. and F. A. Ludewig, Jr., 1972: Measurement of submicron particles and their relation to air pollution. Capital Cities Control Conference, Instrument Society of America.

GENERAL ELECTRIC CONDENSATION NUCLEI COUNTERS

J. Brooks Haberl

General Electric Company, Pittsfield, Mass.

The development and design of condensation nuclei counters within the General Electric Company has received major impetus from the development of military devices designed to detect, rather than measure, sources of airborne condensation nuclei. The first large-scale production of such devices occurred in the late 1950s. This was an aircraft mounted system intended for operation in remote, "clean" areas and the primary emphasis was therefore placed upon high speed sampling and instrument sensitivity. A dark-field, scattered light optical system was employed which is still used in our commercial instruments.

During the Viet Nam War period other instruments of two basic types were produced: high speed, scattered light detection systems for airborne applications, and an attenuated light system for ground-based use where the simplicity of such a system was more important than speed or sensitivity. The experience gained in field operation of the military devices has been extremely useful in adapting these designs to commercial measuring instruments. GE's first commercial instrument was described by Skala (1963), based heavily upon the optical and mechanical elements of the first airborne military instrument of the 1950's. This instrument which I will call the laboratory model (LCNC) has been commercially produced continuously since its development and is today widely used in scientific research in many countries (Figure 1).

A second instrument, which has been produced sporadically since 1969, is a very small, portable, battery operated instrument (PCNC) (Figure 2). The basic characteristics of these two instruments are listed in Table I. The determination of a characteristic as an asset or a liability depends upon the user's particular measurement problem. There have been cases where the small size and portable operation of the PCNC has outweighed its drawbacks in terms of linearity and sensitivity. These cases are, however, greatly in the minority. We have addressed ourselves more strongly to instrument performance, seeking to extend sensitivity, improve reliability and incorporate features of greater benefit to the user.

A recent program with the U.S. Department of Transportation involved the design and deployment of an Aitken Nuclei Counter for airborne measurements to 20 km (65,000 ft) altitude; features of this program are covered in a separate discussion. There were two elements, however, which are highly applicable to commercial instrument improvement. These were the need for much greater

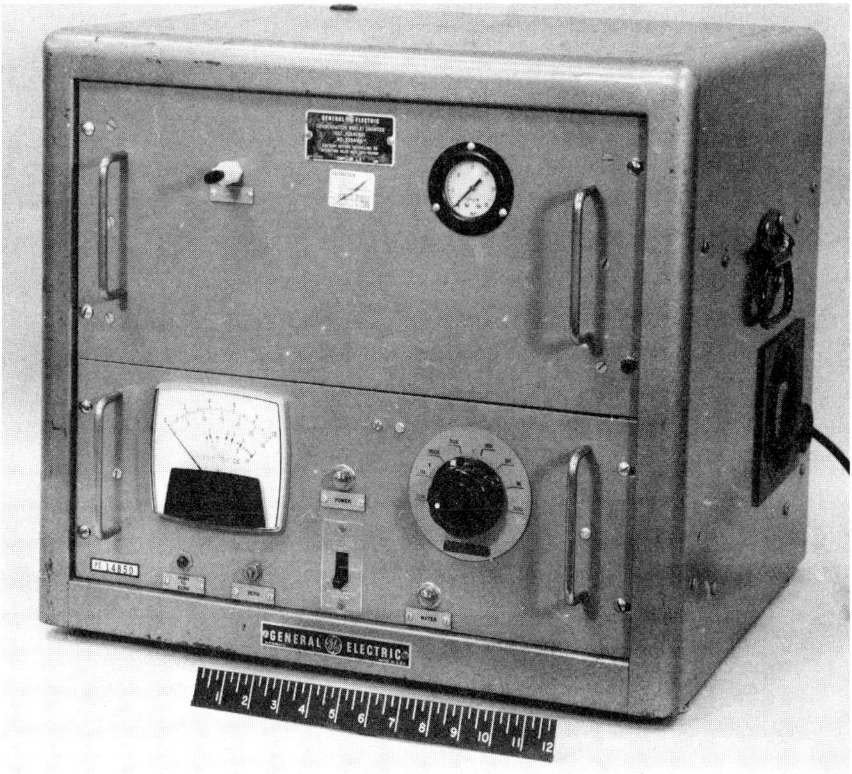

Figure 1. General Electric condensation nuclei counter.

sensitivity, and hands-off operation. The final instrument could reliably measure fewer than 10 particles per cc and could automatically select the proper one of six ranges for high resolution data up to 60,000 particles per cc.

As a result of our experience with military and commercial instruments, the rapid pace of technology in general and a more refined view of the needs of users as we see them, we have recently developed a prototype, new generation instrument designated CNC-2 (Figure 3). The guiding philosophy in the design of this instrument has been to retain the optical and mechanical features of the time-tested laboratory instrument which have proved so reliable over the years, while making major improvements in the electronic portions of the system, extending performance and adding many features more compatible with today's sophisticated technology. This new instrument employs digital readout with automatic ranging although analog outputs are also provided. Binary coded decimal outputs are available at a rear panel connector to allow for digital data recording or auxiliary display. Nuclei count is read directly in a three digit display with automatic decimal point selection, calibrated in units of thousands per cc. The instrument is capable of measuring concentrations from a few particles per cc to 10^6 per cc in 4 linear ranges.

The size and weight have also been reduced to about half that of the LCNC. The new unit weighs 67 pounds and is 20 in. wide, 10 in. high and 19 in. deep. Rear panel connectors are recessed for protection. A front panel function switch allows the digital readout to display the vacuum setting, reading in millimeters of mercury, which is then set by a front panel knob. Humidifier water level control

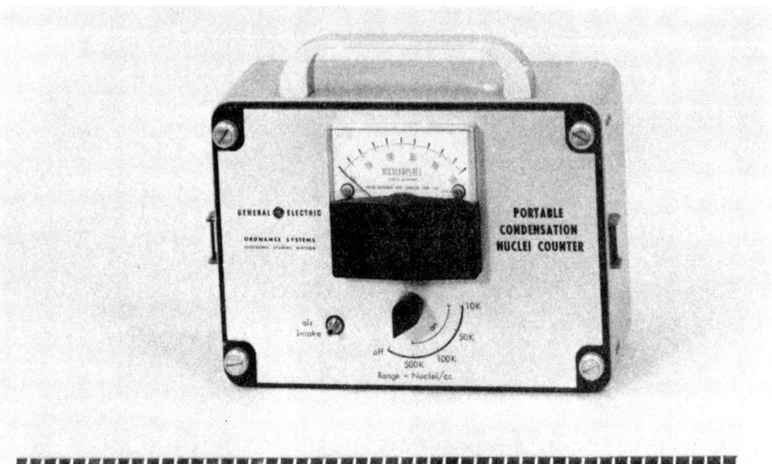

Figure 2. General Electric portable condensation nuclei counter.

TABLE I. Specifications

	LCNC	PCNC
Particle size range	0.001 μm to 0.1 μm diam. (approx.)	
Concentration range	50/cc to 10^7/cc	500/cc to 10^7/cc
Ranges (full scale)		
Linear	300/1000/3000/10000/30000/100000	3000/10000/30000/100000
Compressed	$10^6/10^7$	10^7
Air flow	100 cc/sec	20 cc/sec
Sampling rate	5/sec	2/sec
Power	115 V a.c.	24 and 3 V d.c.
Size	22 × 19 × 20 in.	5 × 7 × 3 in.
Weight	127 pounds	3.5 pounds

is automatic as in the LCNC, admitting distilled water from an auxiliary reservoir when needed. Solid state components are used throughout, including a high gain, very stable, integrated circuit photodetector which has performance characteristics superior to those of the photomultiplier tube previously employed.

In addition to electronic improvements, the new instrument is far easier to service. Physical maintenance is limited to periodic cleaning and relubrication of the rotary valve and cleaning of the humidifier. Access is simple and no mechan-

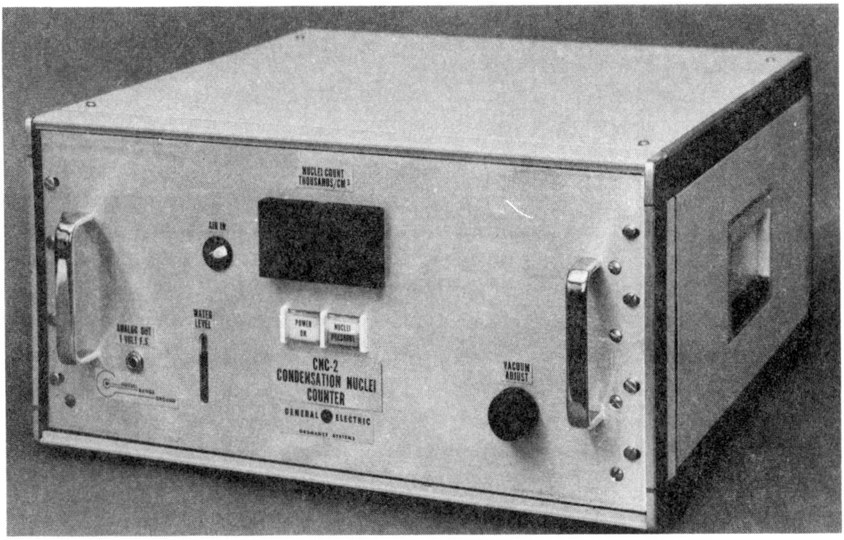

Figure 3. General Electric CNC-2 condensation nuclei counter.

TABLE II. Specifications

Particle size range	0.001 to 0.1 μm diam. (approx.)
Concentration range	<10/cm^3 to 10^6/cm^3
Ranges, automatic (linear), nuclei/cm^3, full scale	1000/10,000/100,000/10^6
Accuracy	± 20%
Sampling rate	1/sec
Air flow	85 cm^3/sec
Time constant	~2sec
Power	115 V a.c., 60 Hz, 3A (can be supplied for 110 or 220 V a.c., 50 Hz)
Outputs	1. 3-digit panel instrument with automatic decimal point, .xxx to xxx. thousands/cm^3 2. Parallel output binary coded decimal, including range indicator, on rear panel, 24-pin connector 3. Front panel jack, analog 0–1.0 V d.c. referred to ground, plus range indication by step voltages in range of 0–1.0 V d.c.
Size	20 in. (51 cm) wide × 10 in. (25.4 cm) high × 19 in. (48.3 cm) deep
Weight	67 lb (30.4 kg)

ical realignment is necessary. The entire unit is contained on a single, slide mounted chassis of standard 19 in. rack width. A summary of CNC-2 characteristics is shown in Table II. The design principles are discussed in Appendix I.

REFERENCE

Skala, G. F., 1963: A new instrument for the continuous measurement of condensation nuclei. *Anal. Chem., 35,* 702.

APPENDIX I—THEORY OF OPERATION, CNC-2

Experiments have shown that the increase of total geometric cross section of water droplets in the cloud chamber of an Aitken Nuclei Counter is essentially a linear function of elapsed time so long as excess water vapor is available. Thus, measurements made while growth is in process will yield a linear relationship between the number of particles (nuclei) and the total geometrical cross section of the droplets formed. The amount of water vapor which can be condensed in a given air sample is a constant which does not depend on the number of droplets produced (assuming no heat exchange through the system boundaries). If all the available water vapor is condensed into droplets, then the total geometrical cross section of the droplets no longer is a linear function of the number, but is instead proportional to the cube root of the number of droplets. Figure 4 shows an idealized representation of these relationships.

In the instrument described by Skala, linearity was obtained to about 10^5 particles per cc by terminating growth at 26 msec and measuring the peak amplitude of the photodetector signal. It can be seen that this is a compromise approach. Low level nuclei concentrations yield smaller signals than they could provide with longer growth allowed, and the detection threshold is therefore higher (in terms of nuclei per cc) than the potential threshold if longer growth times were allowed. Conversely, concentrations above 10^5 per cc cannot be read directly without recourse to special scales or tables.

A more optimized approach would pick the time of measurement to occur during the linear growth phase but allowing as long a growth time as possible for the concentration being measured. The CNC-2 approaches this ideal by measuring the photodetector signal at a number of preselected times in the growth phase. These times are chosen so that the earliest measurement will yield a linear response to the highest concentration, and succeeding measurements (allowing longer growth times) maintain linearity for increasingly smaller nuclei concentrations. The decision logic in this system then becomes a problem of deciding which measurement to use. The approach chosen employs a voltage comparator

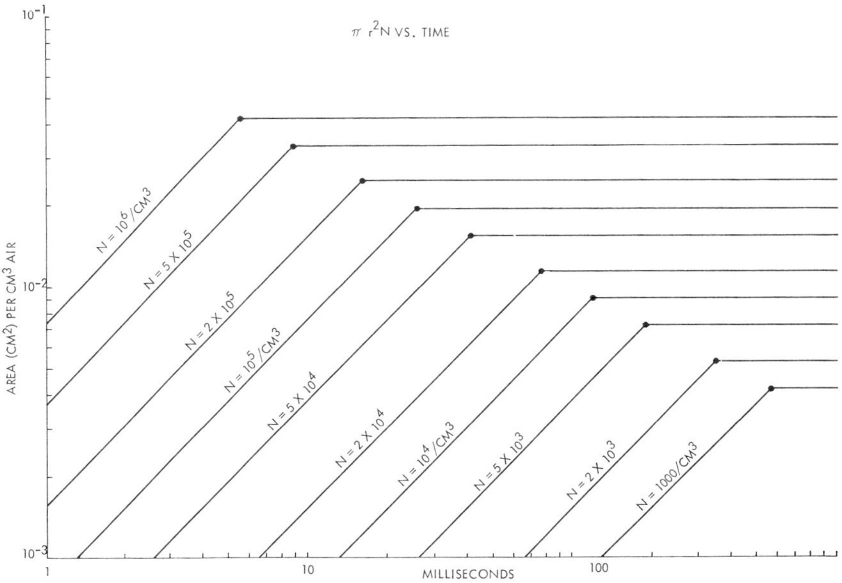

Figure 4. Scattering area vs. time.

which senses when the signal voltage exceeds a predetermined threshold—this event results in preventing any further measurements from being made on that sample. The last (and highest) value measured thus becomes the value to be displayed, and the number of measurements which were allowed indicate the full scale multiplier (or range) for the nuclei count analog.

DROPLET GROWTH KINETICS AND DESIGN PRINCIPLES FOR CONDENSATION NUCLEI COUNTERS

PAUL E. WAGNER

*Clarkson College of Technology, Potsdam, N.Y.**

Condensation nuclei counters (CNC) are very useful instruments for investigation of High Dispersed Aerosols. However, depending on the particular principles of operation, the applicability of CNC's is restricted to certain ranges of nuclei size and number concentration. In many cases, these limitations have not been sufficiently clarified. In order to establish conditions under which the CNC's provide a linear response within well defined limits, the kinetics of droplet growth in the CNC's must be investigated.

THEORETICAL DESCRIPTION OF THE CONDENSATION PROCESS IN CNC's

The growth of *single* droplets was investigated theoretically by many authors (e.g., Maxwell, 1890; Fuchs, 1959; and Brock, 1966). The growth law can be written in the form (Mason, 1971)

$$a\frac{d\,a}{d\,t} = f\,(T, S, \ldots) \tag{1}$$

where a = droplet radius, T = absolute temperature, and S = saturation ratio. The necessary conditions for the validity of Eq. (1) are given (e.g., Wagner and Pohl, 1975). In the case of constant T and S the quadratic growth law is obtained by integration of Eq. (1).

An approximate description of droplet growth in droplet ensembles is based on the so-called cellular model which results in somewhat complicated, nonsteady

Supported in part by the Fonds zur Förderung der Wissenschaftlichen Forschung, Vienna, Austria, and by the United States Public Health Services, Grant 5 RO1 AP 00468-06.

*Present address: I. Physikalisches Institut Universitaet Wien, Strudlhofgasse, 4, A-1090, Vienna, Austria.

state solutions (Reiss and La Mer, 1950). If the mean distance between neighboring droplets is large compared with the droplet radius, another approximation can be applied: "Average" values of the thermodynamic parameters can be defined and the problem reduces to single droplet growth with time-dependent temperature and saturation ratio (Reiss, 1951; Wagner and Pohl, 1975). Based on this model, the time dependence of the droplet radius in a monodisperse droplet

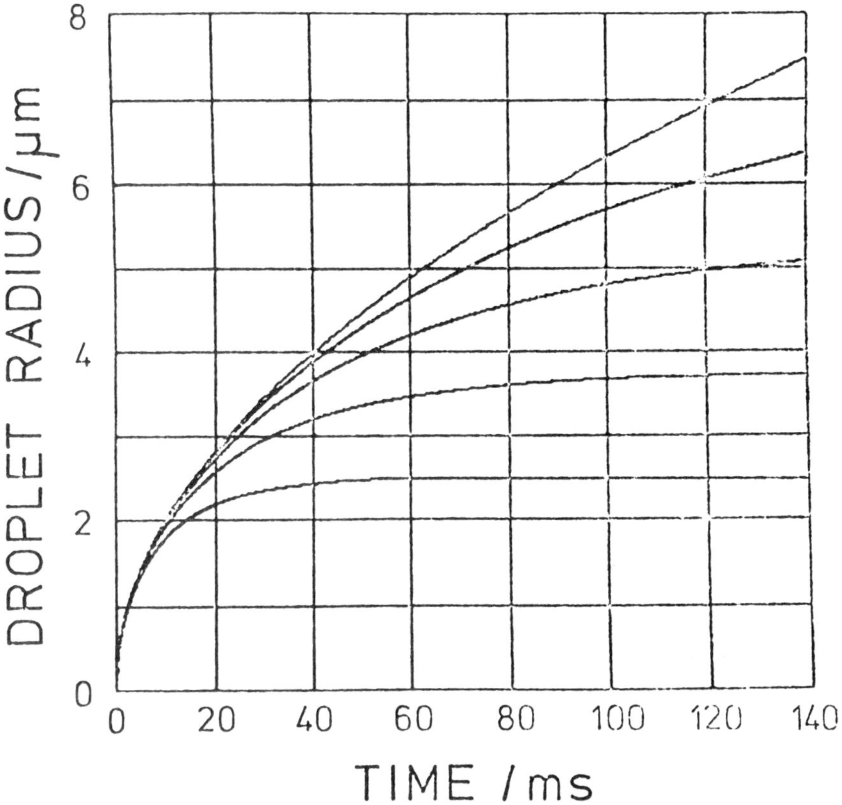

Figure 1. Droplet radius vs. time. Upper curve: Single droplet growth. Lower curves: Droplet concentrations 3.10^3, 10^4, 3.10^4, $10^5/cm^3$.

aerosol at an initial saturation ratio of 400% and various droplet concentrations was computed (Figure 1). It can be seen that droplet growth terminates at certain droplet sizes depending on the number concentration. However, the initial stages of droplet growth are independent from droplet concentration. Computation of the time course of the system temperature shows that a temperature rise of about 15°C occurs during droplet growth. The time dependence of the saturation ratio is shown in Figure 2. As expected, the saturation ratio decreases during droplet growth due to the combined effect of latent heat production and vapor depletion.

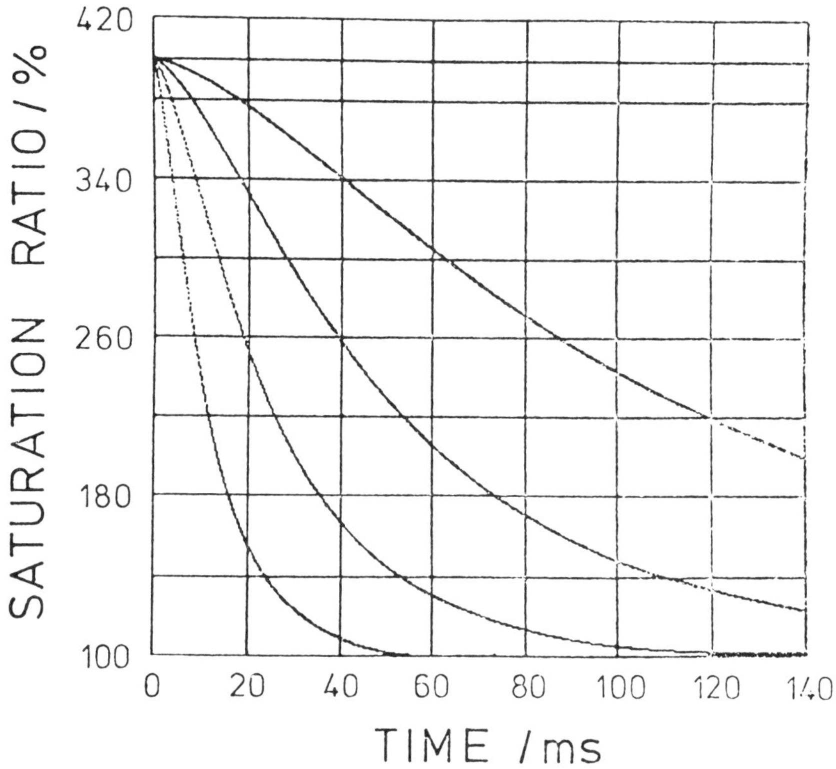

Figure 2. Saturation ratio vs. time, same conditions as in Figure 1.

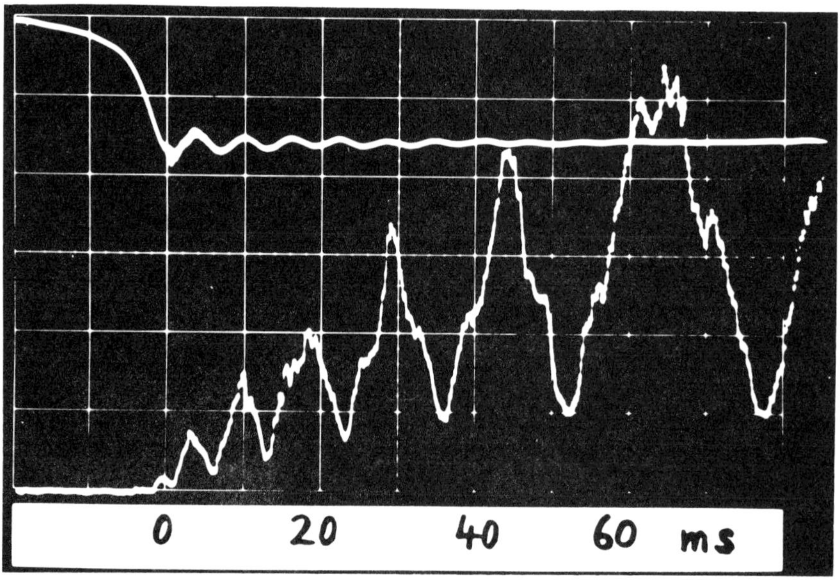

Figure 3. Pressure and scattered intensity at 45° scattering angle during expansion and droplet growth.

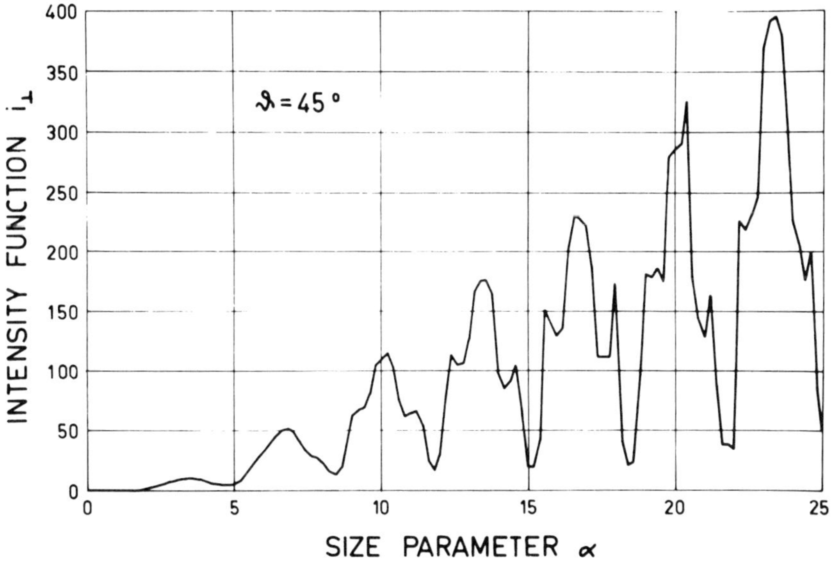

Figure 4. Intensity function vs. size parameter at 45° scattering angle.

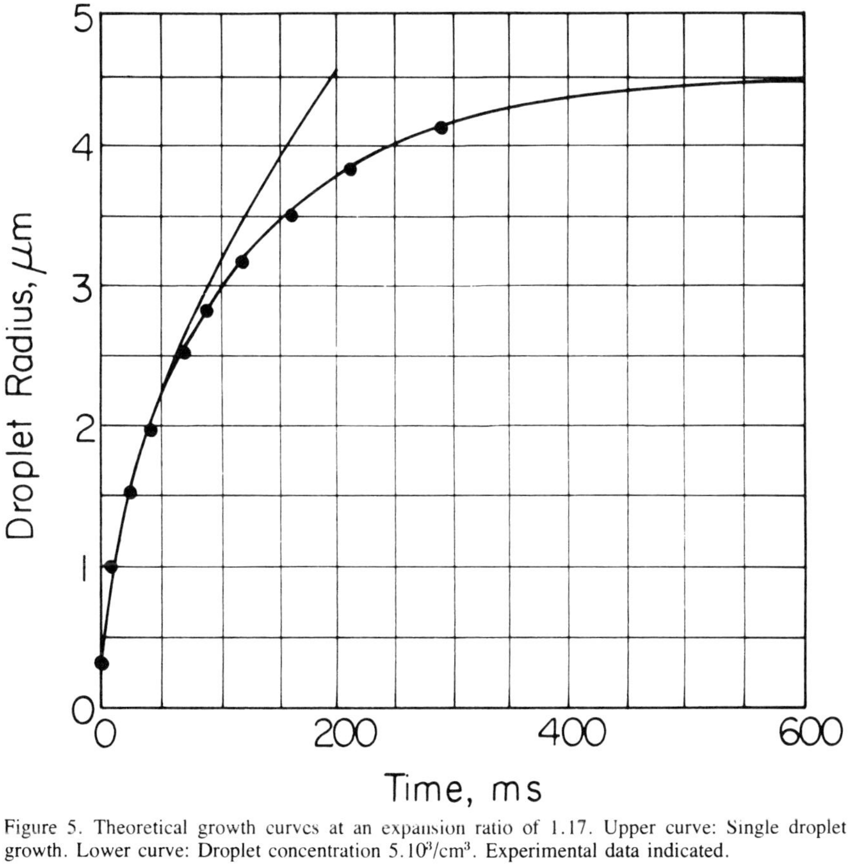

Figure 5. Theoretical growth curves at an expansion ratio of 1.17. Upper curve: Single droplet growth. Lower curve: Droplet concentration $5.10^3/cm^3$. Experimental data indicated.

EXPERIMENTAL TEST

The theory of single droplet growth has been tested by several authors. Early investigations (Owe Berg and George, 1968; and Nix, 1968) showed growth rates several orders of magnitude above the theoretical prediction. However, more recent experiments (Vietti and Schuster, 1973; Carter and Carstens, 1974) seem to confirm the validity of the single droplet growth theory.

The growth process in a monodisperse droplet aerosol was investigated recently (Wagner, 1975). In this experiment humidified nuclei aerosol of defined number concentration was introduced into a cylindrical, pressure-defined, subpressure expansion cloud chamber. The growing droplets were illuminated by a laser beam. Measurement of the scattering intensity as a function of time was performed at different fixed scattering angles. Simultaneously, the gas pressure in the chamber was monitored by means of a fast pressure transducer (Figure 3). Comparison with intensity vs. size curves, as obtained from Mie theory (Kerker, 1969) (Figure 4), shows good agreement with the experimental data (Wagner, 1973). The expansion time of about 5 msec is sufficiently short to provide

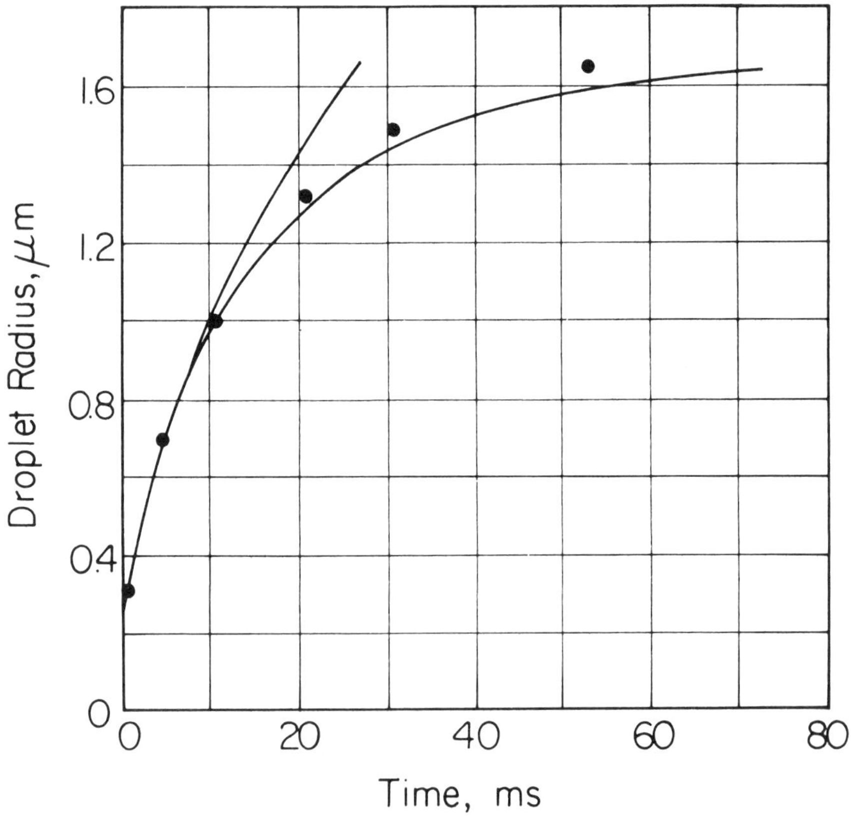

Figure 6. As in Figure 5, droplet concentration $10^5/cm^3$.

approximately dry adiabatic conditions, as can be seen in Figure 3. After establishing a one-to-one correspondence between the experimental and theoretical scattering extrema (Wagner, 1974), the droplet number concentration and size can be obtained from the height and position of the experimental extrema relative to the time axis. Figures 5 and 6 show a comparison of the theoretical growth curves with experimental data at droplet concentrations of 5×10^3 and 10^5/cc, assuming a saturation ratio of 87% within the humidifier. The good agreement proves that the applied theoretical model will yield a valid description of the process occurring in CNC's.

APPLICATION ON CNC

The response of CNC's, based on light extinction or light scattering, can be shown (Wagner and Pohl, 1975) to be approximately proportional to $C\times K_{ext}$ (a), where C denotes the droplet number concentration, and K_{ext} (a) the extinction

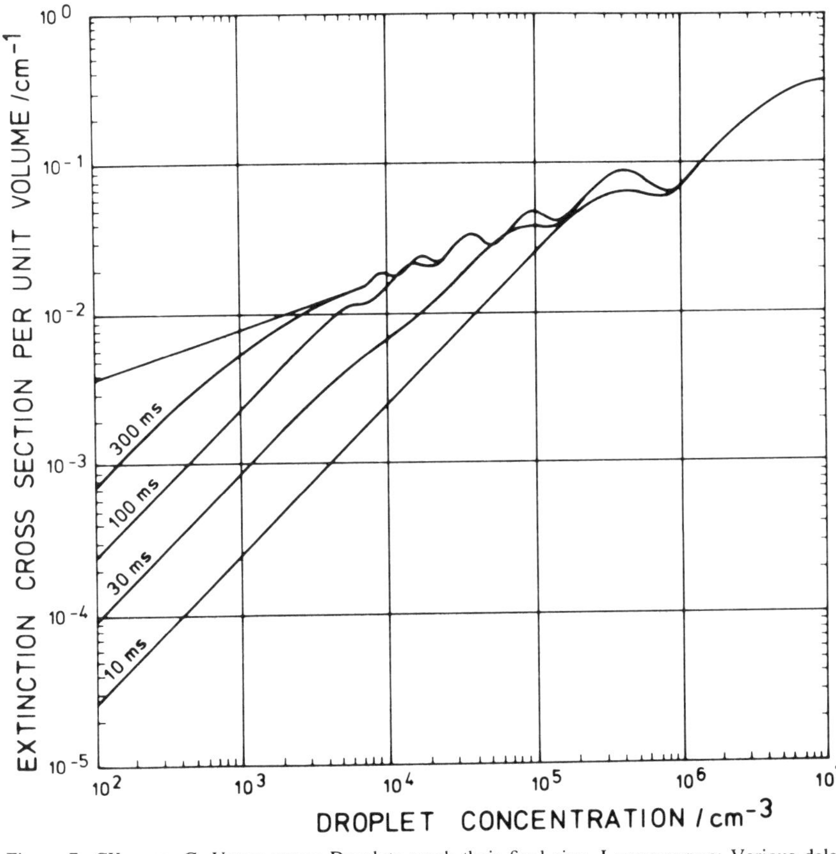

Figure 7. CK_{ext} vs. C. Upper curve: Droplets reach their final size. Lower curves: Various delay times after expansion are chosen.

cross section of a water droplet with radius a. In CNC's of the Pollak-type, the extinction is measured after the droplets have grown to their final size a_{max}, which depends on the droplet concentration C. Figure 7, upper curve, shows $C \times K_{ext}$ (a_{max}) as a function of C, calculated by means of the above droplet growth theory, exhibiting essentially the $C^{1/3}$-dependence observed at Pollak counters. Rich (1961) suggested measuring the scattered intensity at the end of a certain fixed delay-time interval after expansion, rather than waiting for final droplet size. Figure 1 shows that within an appropriate delay-time interval the droplet radius is independent from droplet concentration over a certain concentration range. In this case, K_{ext} (a) is constant and the CNC-response is a linear function of C. The lower curves in Figure 7 show $C \times K_{ext}$ for different delay-

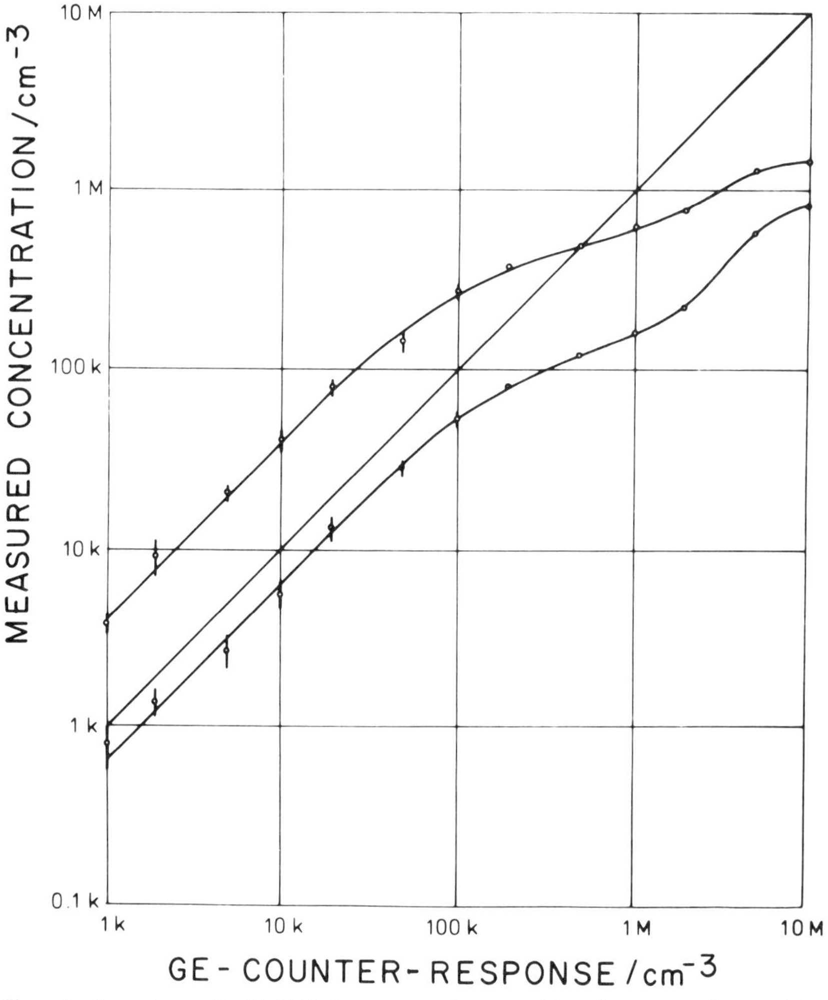

Figure 8. Comparison with GE-CNC. Lower curve: Latex particles. Upper curve: Soot particles.

times after expansion. In the General Electric CNC developed by Rich, the delay-time of 26 msec provides a linear response up to concentrations of about 7×10^4 per cc.

From Figure 2 it can be seen that the supersaturation (SS) decreases considerably during the first 20 msec at high concentrations. The expansion times of many CNC's are on the order of 100 msec, so that the condensation process will be governed by the expansion rate rather than by vapor diffusion and only low SS will occur.

In conclusion, two design-criteria for CNC's are presented:

1. The measurement (light extinction, scattering) should be taken about 10 msec after the end of the expansion to provide a linear response up to concentrations of some 10^5 per cc.
2. Expansion times below about 5 msec are necessary to obtain sufficiently high SS and to avoid counting losses even at high droplet concentrations.

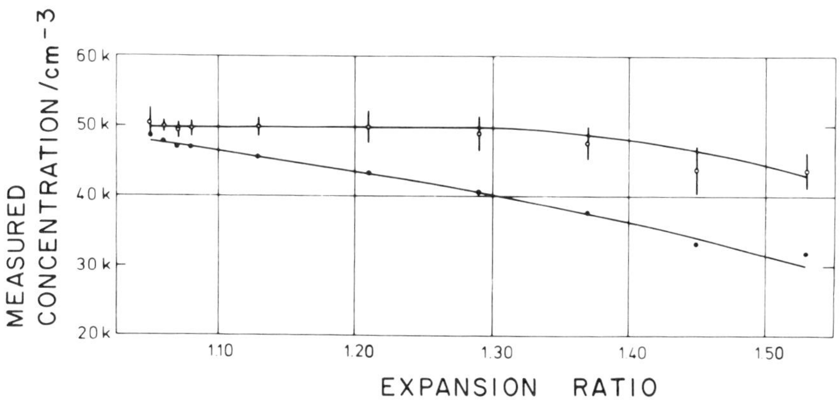

Figure 9. Droplet concentration vs. expansion ratio for Latex particles. Upper curve is corrected for the dilution occurring during expansion.

THE PERFORMANCE OF THE GENERAL ELECTRIC (GE) CNC

The above-mentioned experimental device was compared with a GE-CNC. Two different nuclei aerosols were chosen: spray-dried, monodisperse Latex Aerosol with a mean particle diameter of 0.109 μm; and soot particles obtained from a Benzene-flame. Figure 8, lower curve, shows the calibration curve obtained with Latex Aerosol. Up to about 10^5 particles per cc the concentration values differ by a factor of 0.7; in excellent agreement with measurements by Liu and Pui (1974). However, using the soot particles as condensation nuclei, the readings of the GE-CNC were considerably lower than the measured concentrations. In order to explain these deviations, the droplet concentration was measured as a function of SS at constant nuclei concentrations. The relatively large Latex particles cause droplet nucleation already at low SS (Figure 9) while a higher SS is necessary in order to activate the soot particles (Figure 10). Accordingly, the above-

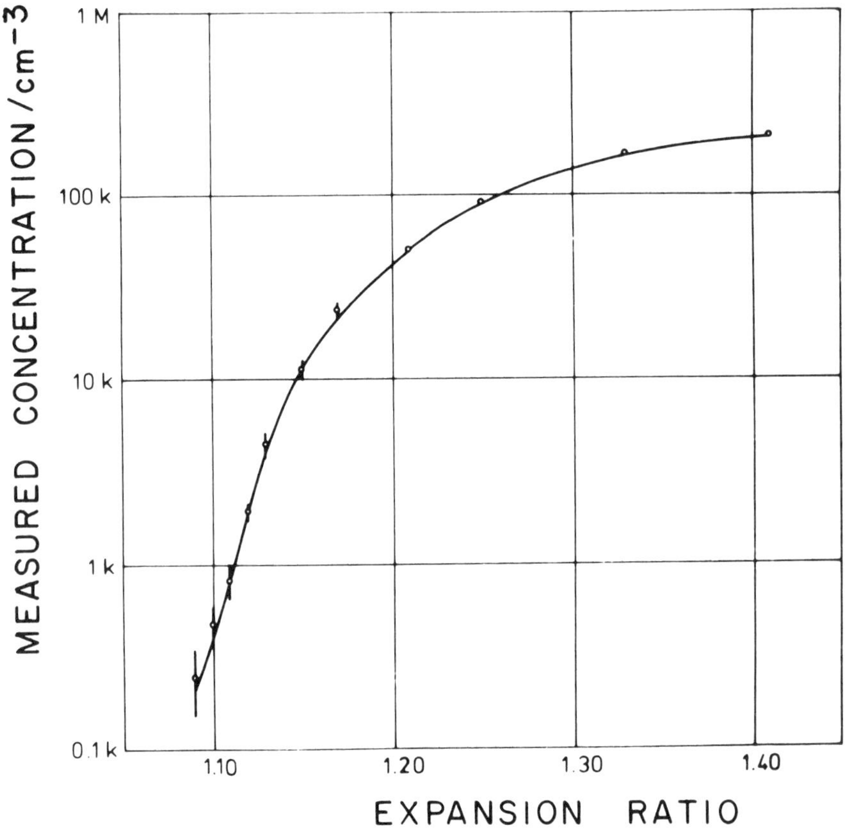

Figure 10. As in Figure 9 for soot particles.

mentioned deviations between measured concentrations and GE-CNC readings could be explained by insufficient humidification and SS in the GE-CNC, causing counting losses in the presence of small condensation nuclei.

REFERENCES

Brock, J. R., 1966: Diffusion to particles in the near free molecular region. *J. Colloid Interface Sci.*, *22*, 513.

Carter, M. J. and J. C. Carstens, 1974: *Trans. A. G. U.*, *55*, 268.

Fuchs, N., 1959: *Evaporation and Droplet Growth in Gaseous Media*. London, Pergamon Press.

Kerker, M., 1969: *The Scattering of Light and Other Electromagnetic Radiation*. New York, Academic Press.

Liu, B. Y. H. and D. Y. H. Pui, 1974: A submicron aerosol standard and the primary, absolute calibration of the Condensation Nuclei Counter. *J. Colloid Interface Sci.*, *47*, 155.

Mason, B. J., 1971: *The Physics of Clouds*. Oxford, Clarendon Press.

Maxwell, J. C., 1890: *Coll. Sci. Pap. Cambridge*, *11*, 625.

Nix, N., 1968: *Kondensation und Verdampfung an Kunstlichen und naturlichen Aerosolen*. Ph.D. dissertation, Techn. Hochschule Aachen.
Owe Berg, T. G. and D. C. George, 1968: *J. Geophys. Res., 73,* 3103.
Reiss, H., 1951: The growth of uniform colloidal dispersions. *J. Chem. Phys., 19,* 482.
——— and V. K. LaMer, 1950: Diffusional boundary value problems involving moving boundaries, connected with the growth of colloidal particles. *J. Chem. Phys., 18,* 1.
Rich, T. A., 1961: *Geof. Pura e Appl., 50,* 46.
Vietti, M. A. and B. G. Schuster, 1973: Laser scattering measurements of droplet growth in binary mixtures—I. H_2O, and air. *J. Chem. Phys., 58,* 434.
Wagner, P., 1973: Optical determination of the size of fast-growing water droplets in an expansion cloud chamber. *J. Colloid Interface Sci., 44,* 181.
———, 1974: Answer to "Note on the optical determination of the size of fast-growing water droplets in an expansion cloud chamber," by Ariel Cohen, *J. Colloid Interface Sci., 48,* 526.
——— and F. G. Pohl, 1975: Interdependence of droplet growth and concentration 1: theory of droplet growth and applications on Condensation Nuclei Counters. *J. Colloid Interface Sci., 53,* 429.
———, 1975: Interdependence of droplet growth and concentration 2: experimental test of droplet growth theory. *J. Colloid Interface Sci., 53,* 439.

QUESTIONS

Q: Should we not inquire about the index of refraction in applying this method in the atmosphere? What happens when the water condenses around the nuclei and the nuclei is still of a comparable mass to the water?

Wagner: If you want to go down to very low times, you will probably end up with very small droplets, and you are right that in this case the nucleus still is not negligible. Then you will have to test your upper limits for a nucleus size as a smaller size. You probably will have to accommodate your instruments to what you want when you want to make measurements of rather large nuclei. You will have to wait a bit longer and take into account that in this case there will be some nonlinearities at high concentrations. If you are sure that the nuclei are very small, then you can go down to smaller times. You end up with smaller droplets, but this isn't a problem because the nuclei on which the water condensed were much smaller than this.

Q: What did you use for a condensation coefficient in your calculations, and would you be able to determine that experimentally because you also had your experimental measurement of the Mie theory to calculate the growth?

Wagner: I hate to answer this now because there is still an open question about the accommodation coefficients. I am well aware of the fact that sticking coefficients down to 0.02 are sometimes used, but my experience goes in the ways that sticking coefficients of unity can be applied and in all my calculations I did this. My explanation for this is that the growth, as I obtained it, is very fast so that the surface of the growing droplets is very new and very clean. In the case where you have a very clean water surface, I think the accommodation coefficient is really unity. If you have growth processes which are allowed to proceed slowly as they are in diffusion chambers from only slightly above 100% saturation, then the droplets' surfaces can be contaminated. This can cause coefficients, sticking coefficients, less than unity. But my answer, I ask you not to quote me right now, is unity.

NCAR AIRBORNE CONDENSATION NUCLEUS COUNTER

G. Langer

National Center for Atmospheric Research, Boulder, Colo.

The development of this condensation nucleus (CN) counter was prompted by several requirements, which could not be satisfied simultaneously by an existing instrument. A counter with a wide range of sensitivity was desired, one that could be operated from the ground (in a polluted atmosphere) up into the stratosphere (where a few nuclei per cc might be encountered). The instrument had to be continuous, with a nonpulsating flow, and simple to operate, i.e., capable of unattended operation. Also, it had to be operable in either a pressurized or unpressurized environment. For the former case, such as on a jet aircraft, provisions had to be made to bring the aerosol sample to cabin conditions without losses. It was also required that the supersaturation (SS) achieved be in the region of meteorological significance, i.e., not over a few percent, which is the maximum encountered in cloud formation. Finally, cost was an important factor.

Figure 1 shows the schematic of the NCAR-CN counter. In its present prototype configuration it weighs about 130 lbs. Basically, the sample air stream of a few cc/min to several hundred cc/min is added to a 3 ℓpm stream of clean air. Another stream of clean, moist air is recirculated continuously through the 7 ℓ cloud chamber at 10 ℓpm. The sample stream is added to this carrier gas as it enters the cloud chamber in the form of a laminar jet. This jet is gradually cooled as the surrounding, cold air is admixed. From the theory of such a mixing process a SS of about 2% is expected (Amelin, 1967). The drops that form on the activated condensation nuclei gradually grow to over 20μm in size as they flow through the cloud chamber. At the bottom of the chamber the whole stream passes through an acoustic counter (Langer, 1972) that counts all the water

This research was performed as part of the National Hail Research Experiment, managed by the National Center for Atmospheric Research and sponsored by the Weather Modification Program, Research Applications Directorate, National Science Foundation.

droplets, which are all over 20μm at that point. This airstream is then filtered and recirculated to provide the air for the sample stream pickup and the moist airstream.

This system is very simple and reliable from a mechanical and electronic point of view. The major questions regarding the operation of the device are related to the loss of drops to the bottom of the chamber, and the problem of establishing the actual value of the SS and verifying that it does not change. With respect to the former, a certain percentage of the drops is lost by deposition onto the

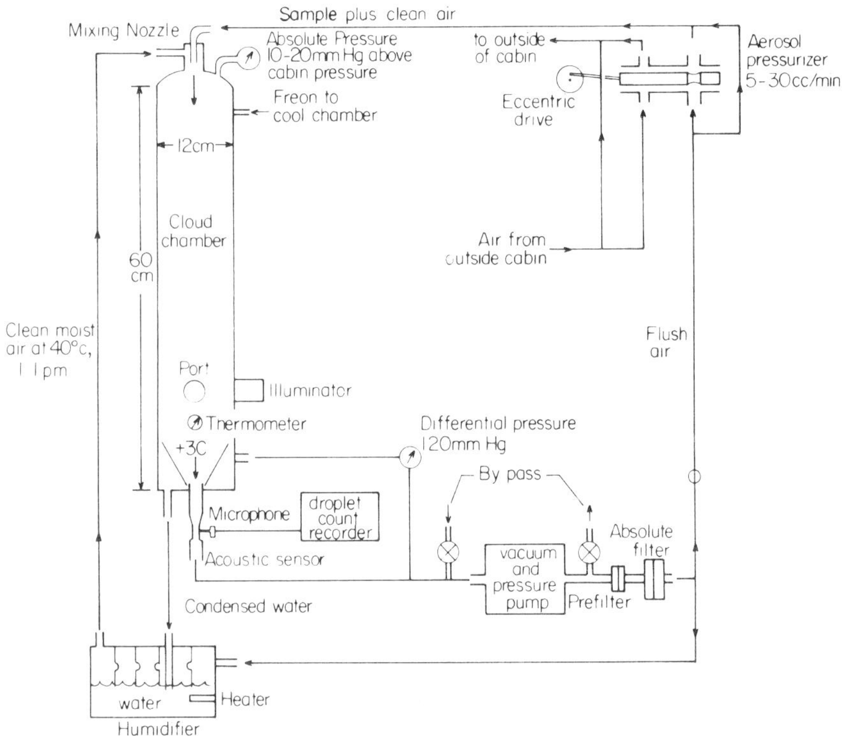

Figure 1. NCAR condensation nucleus counter for operation in a high-altitude, pressurized aircraft.

transition section of the cloud chamber that leads to the acoustic sensor. This loss has been studied with 19 μm ragweed pollen particles as a tracer; i.e., the number of particles entering the cloud chamber was measured with an acoustic counter and compared to the number counted at the exit. The settling loss was found to be constant as long as the chamber pressure remained fixed. This is usually the case because for operation at higher altitudes the unit is pressurized or operated in a pressurized airplane. A major reason for constant-pressure operation is that the acoustic sensor loses sensitivity with reduction in air density; moreover, SS and drop growth are also affected by pressure. The settling loss decreases or increases somewhat with changes in g-loading on an airplane; that is, the force of gravity on the droplets changes momentarily. At present we use a multiplication factor of 4 to correct for the settling loss. We plan to determine this factor accurately by photographing the drops in the chamber to determine their concentration vs. the sensor count.

With respect to the maximum SS achieved in the laminar jet, where the nuclei are activated, the following results are available. The approach has been empirical, comparing the counts from our device against other devices where information on the SS is more readily deduced. During a comparison at NCAR of various condensation nucleus counters (Cadle et al., 1975) the data from the Rosen Counter from the University of Wyoming were of major interest. The other devices operated at very high SS while the Rosen unit worked at 10%, but it used glycol vapor instead of water to produce drops. The Rosen counter gave counts 3.7X as high as the NCAR counter for hygroscopic particles. This is about the same as our factor of 4X for the settling loss. One could conclude that the NCAR-CN counter operates at 10% SS like the Rosen Counter, contrary to our theoretical estimate of 2%. Further study of the intercomparison data shows that nuclei of small size and nonhygroscopic nature were 10X more efficiently detected by the Rosen device. Therefore, the cutoff for the NCAR-CN counter is below 10% SS. Also, Rosen's use of glycol vapor to form the drops by condensation raises some questions as to the validity of the comparison. Work is now in progress on a comparison with a thermal diffusion chamber that uses water vapor and provides very accurate SS data. In addition, we are using monodisperse aerosols generated by the University of Minnesota electrostatic separator. This will provide data for hygroscopic and nonhygroscopic particles of different sizes.

In the meantime, the NCAR-CN counter has been used in the field (Cadle and Langer, 1975) to obtain data on condensation nuclei in the troposphere and lower stratosphere. This work has its origin in research related to various pollutants that may ultimately affect the quality of air in the stratosphere, where pollutants have a very long residence time compared to the troposphere. In an unpressurized aircraft the pump bypass system is adjusted to maintain a pressure of 600 mm Hg. At high altitudes (over 6000 m) an auxiliary pump may be necessary to maintain pressure. To keep the pump motor from overheating due to reduced air density, cooling air from an eyeball (vent) has to be provided. Also, the microphone on the acoustic sensor must be at chamber pressure on both sides of the diaphragm or it will be under such tension that it will not respond. The aerosol pressure pump in Figure 1 meters a known volume of sample regardless of altitude. The device is simple but the piston must fit very snugly to avoid leakage of cabin air in a pressurized aircraft. We have made the pump of Plexiglas, with Teflon piston rings at the end of the piston. High-vacuum, silicone lubricant is applied at the piston rings. The piston requires a break-in period until the generation of spurious particles ceases. The piston usually cycles at a rate of 50/min but can be slowed down to reduce the sample volume. The bore has a volume of 0.3 cc.

The electronic recording system is relatively simple. The acoustic signal produces a 20–40 mv pulse with a signal-to-noise ratio of over 50:1. The count rate is limited by the ringing in the electronic circuit. With a good microphone up to 20,000 counts/min can be handled.

Results from a recent flight test are given below to illustrate the application of the device. The jet aircraft was equipped, in addition to our CN counter, with an Environment/One[1] (E/1) Rich 100 CN counter and an Astro Engineering[2] ozone recorder. The latter obtained its sample from the eyeball air-supply. The E/1 was connected with an open "T" to the exhaust of a two-stage Metal Bellows[3] pump providing an excess flow of pressurized outside air. Figure 2 presents results

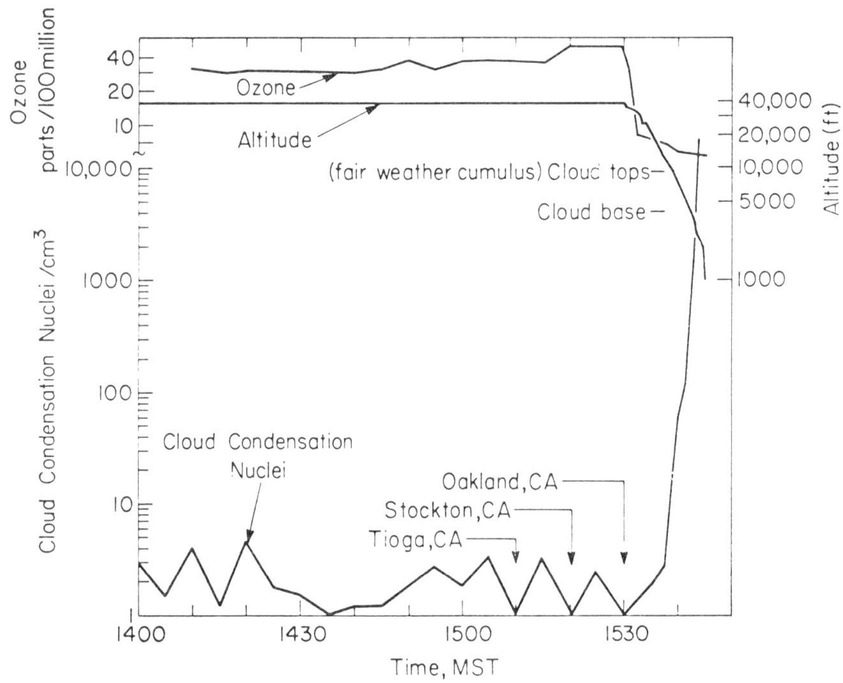

Figure 2. Flight from Denver to McClellan AFB, Ca. Takeoff time: 1317 April 12, 1976.

from part of the flight, mostly in the tropopause and later in the lower stratosphere. Note that no E/1 data are given. On this and previous flights, problems were encountered with spurious counts. It appears now that water enters the cloud chamber and fouls the optics. This apparently happens during turbulence when negative "g" forces may be encountered momentarily and water is lifted out of the humidifier into the cloud chamber suction line. Modifications of the device are now under consideration. Also, one must be very careful not to apply any pressure from the sampling line to the E/1 sample intake; evidently it cannot tolerate a differential pressure.

1. Environment/One Corporation, Schenectady, N.Y.
2. Boulder, Colo.
3. Sharon, Mass.

The counts from the NCAR-CN counter were particularly low for this flight; at other times we had counts up to two orders of magnitude higher. Impactor samples taken during the flight showed no particles over 0.1 μm or so. This supports the low CN counts observed, since our instrument does not count very small particles that require high SS.

REFERENCES

Amelin, A. G., 1967: Theory of fog condensation. Israel Program for Scientific Translations, Jerusalem.

Cadle, R., et al., 1975: A comparison of the Langer, Rosen, Nolan-Pollak, and SANDS condensation nucleus counters. *J. Appl. Met., 14,* 1566–1571.

——— and G. Langer, 1975: Stratospheric Aitken particles near the tropopause. *Geophysical Res. Letters, 2,* 329–332.

Langer, G., 1972: Further evaluation of the acoustical particle counter. *Powder Tech., 6,* 5–8.

TROPOSPHERIC AIRBORNE CONDENSATION NUCLEI SAMPLING: METEOROLOGY RESEARCH, INC., AIR POLLUTION MAPPING PROGRAMS

DAVID S. ENSOR AND J. A. ANDERSON

Meteorology Research, Inc., Altadena, Calif.

The effects of polluting sources on downwind receptors have been a subject of intense study during the last few years. Utilization of airborne sampling has been clearly evident during studies of both local and long range transport and transformation of pollutants from various sources. The instrumentation of airborne platforms has ranged from a single instrument—perhaps a temperature probe—to a full complement of gaseous, aerosol, and meteorological monitors.

The objective of this paper is to describe the use of a Condensation Nuclei Counter (CNC) in airborne sampling. The discussion is limited to an overview of the aircraft system and to a few example cases illustrating the interpretive value of CNC data.

Meteorology Research, Inc. (MRI) has flown instrumented aircraft during tropospheric sampling programs for more than four years. The research projects have been directed mainly towards urban and point source plume characterizations.

Early in the sampling programs it became obvious that no single instrument would provide sufficient information to understand pollutant transformation and transport. The approach was to use a number of carefully selected sensors in the aircraft and ground-based meteorological information. The data then had to be reduced and integrated into a form which could be understood.

AIRCRAFT SAMPLING SYSTEM

One of the aircraft currently used by MRI during environmental sampling programs is shown in Figure 1. The plane is a single engine Cessna Turbo 206 modified with a Roberson STOL conversion, belly pod, and long-range extender tanks. The STOL conversion allows short field takeoffs and landings as well as reduced sampling speeds. The 60–140 mph range of sampling speeds covers those desired to characterize the particular type of source being studied. (The aircraft stalls at about 40 mph.) The belly pod extends the space available for sampling equipment and in the present configuration contains an inverter. The aircraft with a full complement of sampling equipment, pilot, observer, and instrument operator has a flight range of approximately four hours.

The measurement system is similar to that described by Blumenthal and Ensor (1972) and will be summarized only briefly.

The exterior sensors such as temperature, humidity, and turbulence are shown in Figure 1. Five air scoops (three large and two small) are used to provide sample air for the analyzers in the aircraft. These air scoops are opposite the exhaust system on the aircraft, and testing has indicated that the samples are not contaminated by the aircraft's exhaust. Studies by Adams and Koppe (1969) have shown that the effects of propwash and slipstream are minimal on samples obtained in a similar fashion.

Figure 1. Meteorology Research, Inc., Cessna 206 pollution sampling aircraft.

The instrumentation in the cabin is shown in Figure 2. The sensors are summarized as follows:

Aerosol
- Environment/One CNC
- Meteorology Research, Inc. Integrating Nephelometer

Gas
- Theta Sensors, Inc. SO_2
- Monitor Labs, Inc. NO/NO_x
- Rem, Inc. O_3
- Andros, Inc. CO (Not presently used)

Meteorological Parameters
- Temperature
- Humidity (recorded but used only as a check for calculated RH)
- Dew point
- Air speed
- Altitude

Position Parameters
- DME - Distance Measuring Equipment
- VOR - VHF - Omni - Radio

The parameters are all recorded with time on magnetic tape on a Metrodata Model 620 Data Logger. The field tapes are all postprocessed on a computer at MRI.

In Figure 3, a Thermo-Systems Model 3030 Electrical Aerosol Size Analyzer and a Royco Particle Counter are shown mounted for a special experiment.

Figure 2. Aircraft instrumentation in Cessna 206.

EXAMPLE DATA

The concentration of condensation nuclei is a useful indicator of fresh combustion aerosol, while the magnitude of b_{scat} measured with the Integrating Nephelometer is an indicator of aerosol in the 0.1 to 1.0 μm decade of particle size. The concentration of the trace gases also provides information on the age (particularly NO-NO_x) and origin of pollutants. A series of example results (Blumenthal et al., 1974) will be shown to indicate the interrelationships between the various pollutant parameters.

Figure 3. Electrical aerosol size analyzer and Royco particle counter mounted in MRI Cessna 206.

Sounding at Hawthorne Airport, Los Angeles Basin

This sounding (Figure 4) shows the presence of fresh pollutants trapped below the mixing layer as shown by high nuclei concentration, CO, intermediate b_{scat}, and low O_3. There are some aged layers aloft indicated by the correlation of b_{scat} and O_3.

Sounding Over Brackett in the Los Angeles Basin

This sounding (Figure 5) shows concentrated aged pollutants from the previous

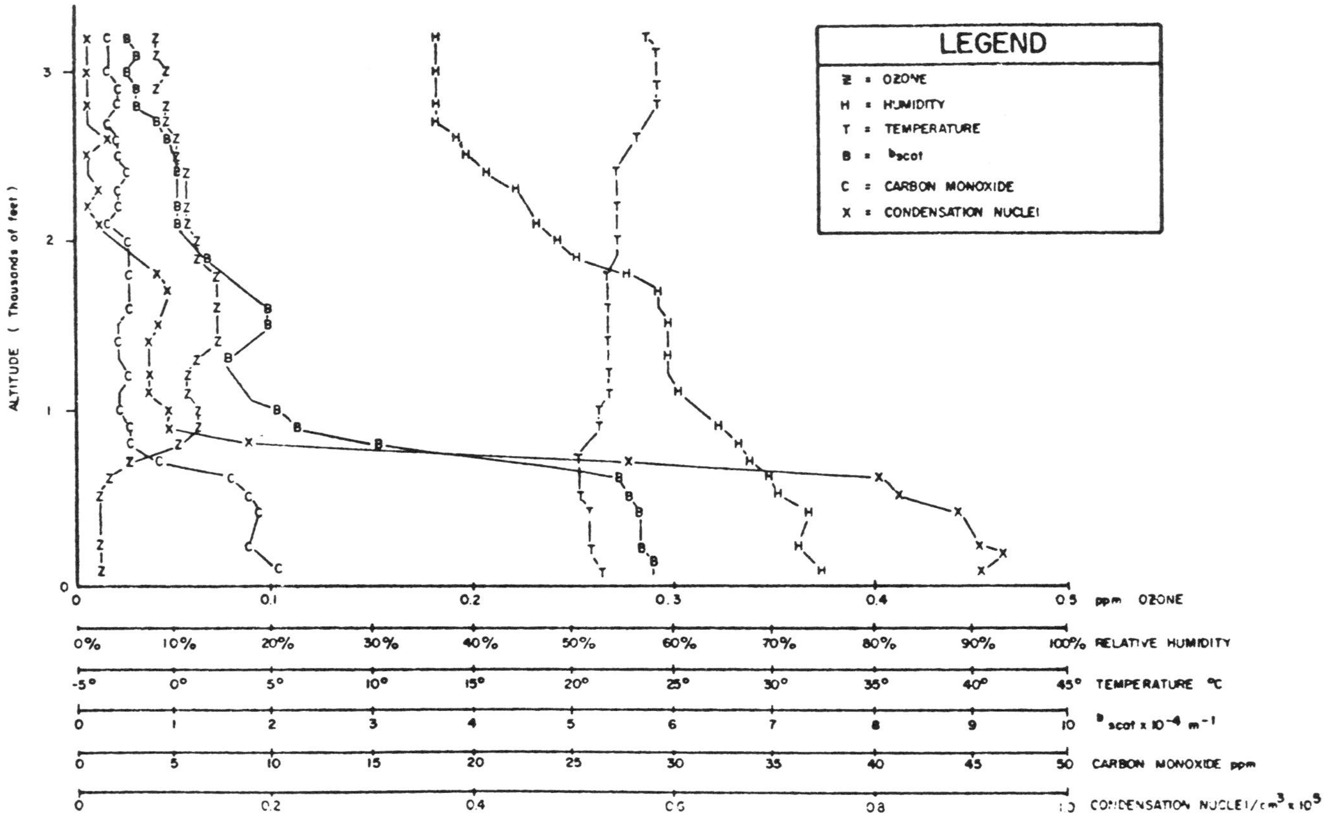

Figure 4. Vertical soundings over Hawthorne Airport (HHR) at 9:15 A.M. PDT, September 20, 1972.

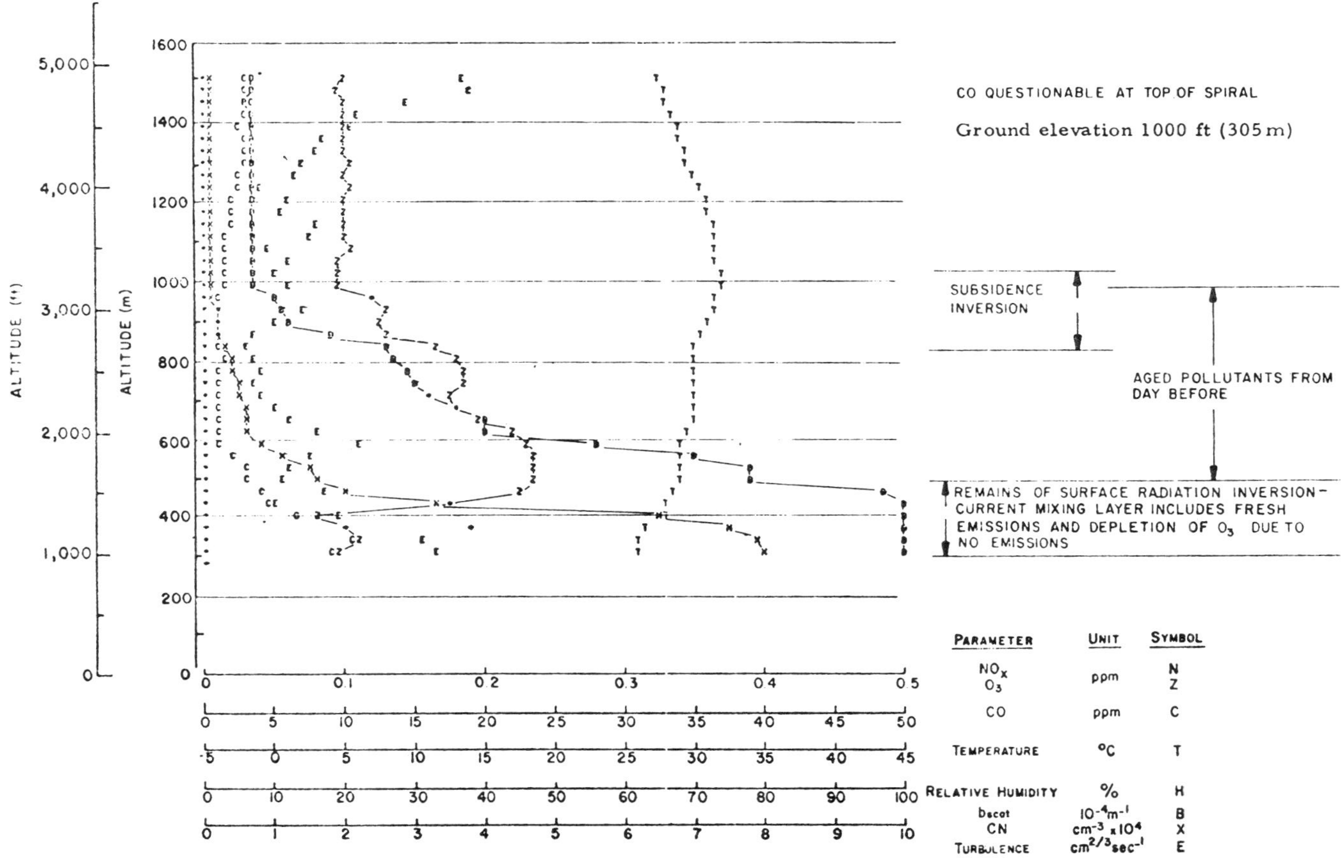

Figure 5. Vertical profile over Brackett (BRA) 0843 PDT, July 25, 1973.

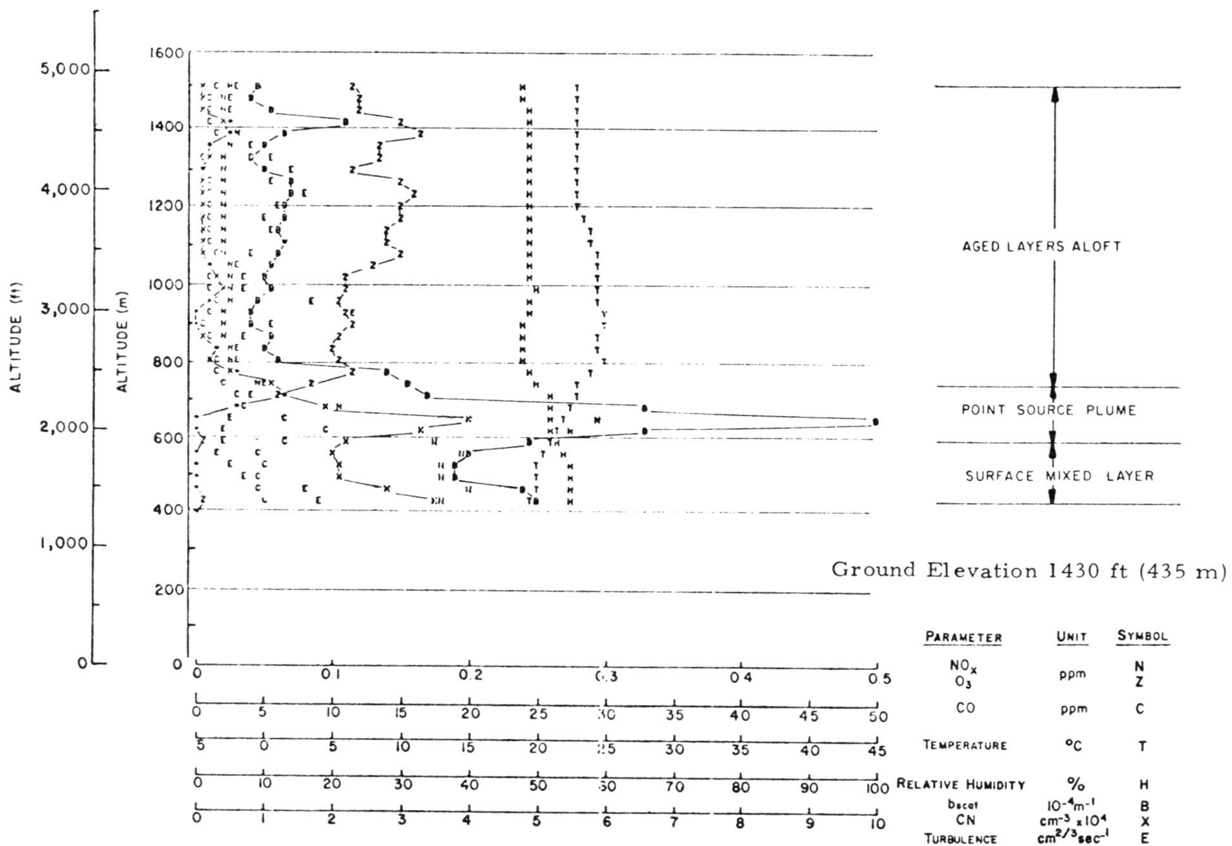

Figure 6. Vertical profile over Rialto (RIA) 0535 PDT, July 27, 1973.

day as well as trapping of pollutants with both a radiation inversion and a subsidence inversion.

Sounding Over Rialto in the Los Angeles Basin

In Figure 6 a more complex situation is shown. The normal pollutant layers are complicated by the presence of a point source plume indicated by high nuclei concentration and b_{scat}.

Oil-Fired Power Plant Plume

In Figure 7 a plume traverse of the Moss Landing Power Plant is shown. The nuclei concentration and NO_x are very high, while b_{scat} is only moderate because

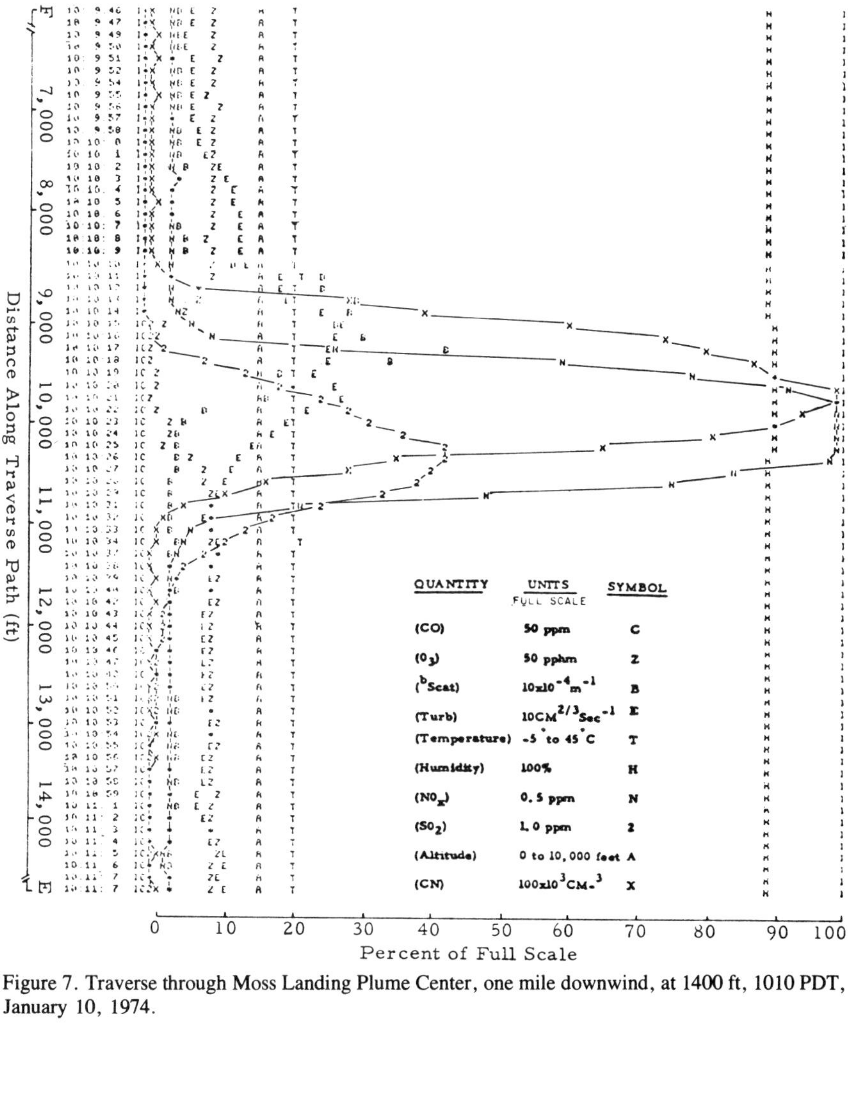

Figure 7. Traverse through Moss Landing Plume Center, one mile downwind, at 1400 ft, 1010 PDT, January 10, 1974.

the plume is only slightly visible. The SO_2 and turbulence show moderate increases. The O_3 drops in the plume because of scavenging by NO. This traverse is typical for an oil-fired plant. It should be noted that different peak locations for the different pollutants are due to the response times of the instruments. Since these data were taken MRI has developed data reduction procedures to correct the data for time lags and "smearing" using the transfer function of the instrument. Additional details of the procedures are contained in a paper by Markowski (1975).

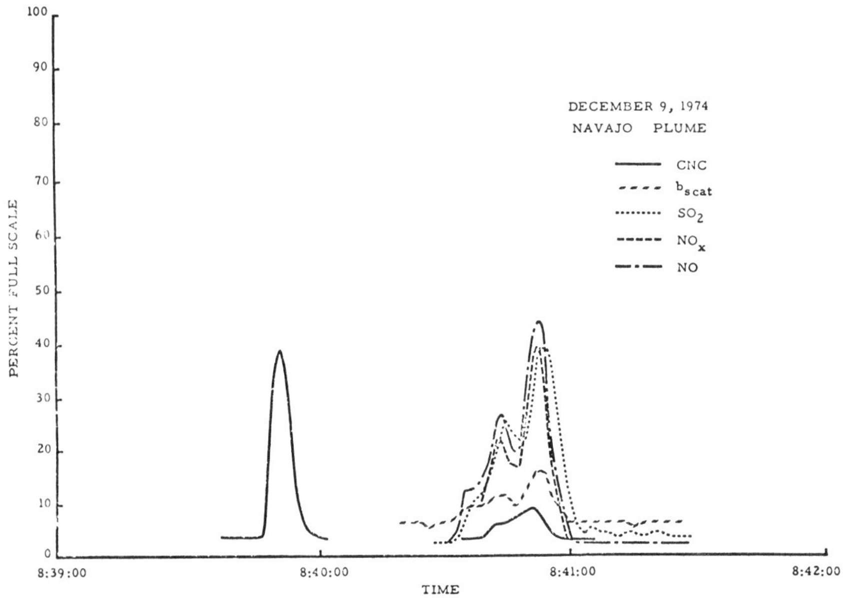

Figure 8. Traverse through Navajo power plant plume.

Coal-Fired Power Plant and the "Phantom CNC Plume"

In Figure 8 a traverse of the Navajo Power Plant plume is shown. The top plume of only condensation nuclei was not related to the plume from the plant. The plant plume has a typical response for a coal-fired plant: moderate levels of b_{scat}, nuclei concentration, and gaseous pollutants. These data have been corrected with the transfer functions of the sensors.

After considerable investigation, it was found that the "phantom CNC plume" was coming from a rock crushing operation near the plant. A 55-gallon oil drum was filled with fuel oil and set afire to dry a conveyor belt.

SUMMARY

The value of using the nuclei concentration to detect fresh combustion aerosol has been illustrated with several examples. However other sensors are required for a complete understanding of pollutant transformation and transport in the atmosphere.

Considerable evidence indicates that the interpretation of a single airborne measurement may (and often does) lead to an incorrect conclusion. Thus an understanding of CNC measurements is greatly increased when supporting data are collected to aid interpretation of the CNC data. In a similar manner, other analyses (e.g., the conversion of NO to NO_2) are aided by the integration of CNC data into the analysis of the data of interest.

REFERENCES

Adams, D. F. and R. K. Koppe, 1969: Instrumenting light aircraft for air pollution research. *J. Air Pollution Control Assoc., 19,* 410–415.

Blumenthal, D. L., and D. S. Ensor, 1972: The use of light aircraft to measure the three-dimensional distribution of air pollutants. Pacific Northwest International Section, Eugene, Oregon, Air Pollution Control Assoc., November 15–17.

———, T. B. Smith, W. H. White, S. L. Marsh, D. S. Ensor, R. B. Husar, P. S. McMurry, S. L. Heisler, and P. Owens, 1974: Three-dimensional pollutant gradient study 1972–1973 program. MRI 74 FR-1262.

Markowski, G. R., 1975: A useful method for transient response analysis and data correction applied to aircraft plume sampling. Air Pollution Control Assoc., Boston, Massachusetts, June 15–20.

EXPERIENCES IN THE DETERMINATION OF PARTICLE SIZE DISTRIBUTIONS IN INDUSTRIAL FLUE GAS EMISSIONS USING DIFFUSIONAL METHODS

JOSEPH D. McCAIN

Southern Research Institute, Birmingham, Ala.

Southern Research Institute has for the past several years been involved in determining particulate concentrations and size distributions for particles in the submicrometer size range for various industrial and governmental organizations. Much of this work has been directed toward characterizing the emissions from particular industrial source types and toward determining the efficiency at which industrial gas cleaning equipment removed particles in the 0.01 to 1.0 μm diameter size range. In making such measurements on industrial sources one encounters a wide range of difficulties, some of which are described in the following discussions.

TABLE I. Typical Flue Gas Sample Conditions

	Range	Typical Value
Temperature	Ambient to 800°C	150°C
Absolute pressure	490–850 mm Hg	750 mm
Pressure differential to ambient	±75 mm Hg	−25 mm
Moisture (vol. percentage)	1–40%	10%
CO_2 (vol. percentage)	0–15%	12%
SO_2 (vol. percentage)	0–1.4%	0.2%
SO_3/H_2SO_4 (vol. percent)	0–0.25%	0.0005%
Particulate mass concentrations	to 20 gram/m^3	Inlets 5 gram/m^3 Outlets .05 gram/m^3
Concentration by number of particles larger than 0.01 μm	to 3 × 10^7/ml	2 × 10^7 inlet 2 × 10^5 outlet

DISCUSSION

Typical sample gas conditions are shown in Table I. At any one source the gas conditions may be any mixture of the ranges shown in the Table and, in fact, these may not represent the extremes that may actually exist. Because of the high particulate concentrations, both by mass and by number, the presence of large quantities of condensible vapors (i.e., water and H_2SO_4), and high concentrations of corrosive gases (H_2SO_4, SO_2 and others), extensive sample conditioning and dilution become mandatory. The required dilutions approach 5000:1 in

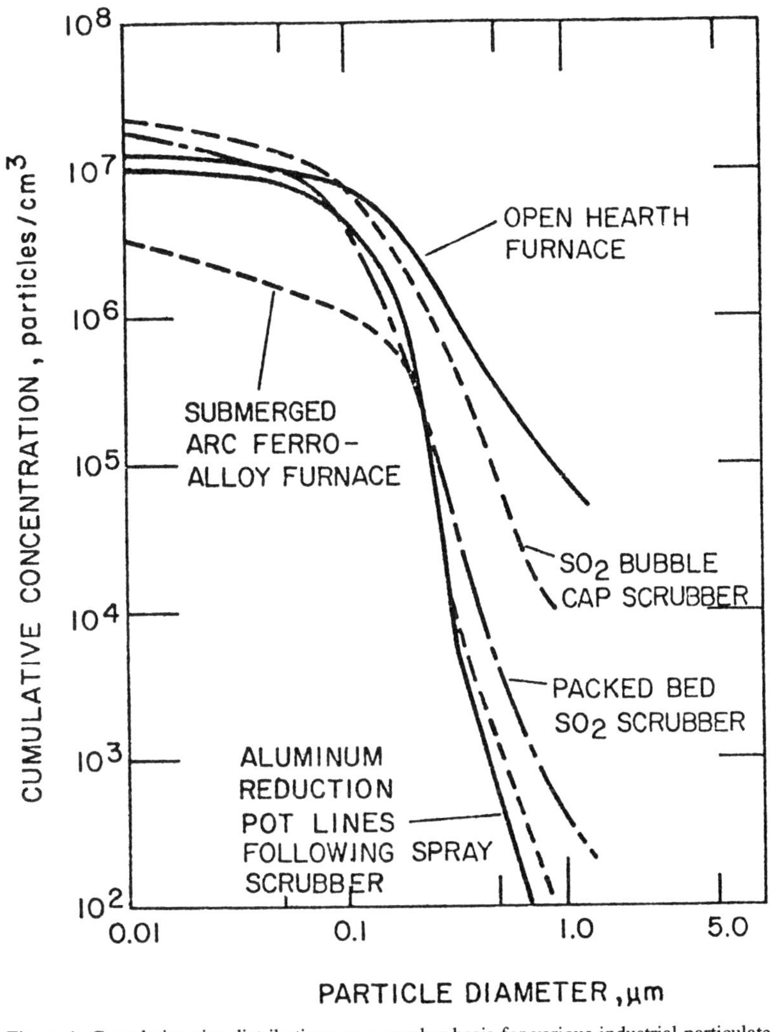

Figure 1. Cumulative size distributions on a number basis for various industrial particulate sources as measured by optical and diffusional methods.

some instances and dilutions by factors of about 500:1 are quite commonly needed. In many instances these dilutions are set by limiting factors in obtaining reliable data from the available detectors, and not by requirements for diluting condensible vapors sufficiently to insure that the sample gas stream to the instrumentation is not saturated. Reliable concentration data for the purposes of

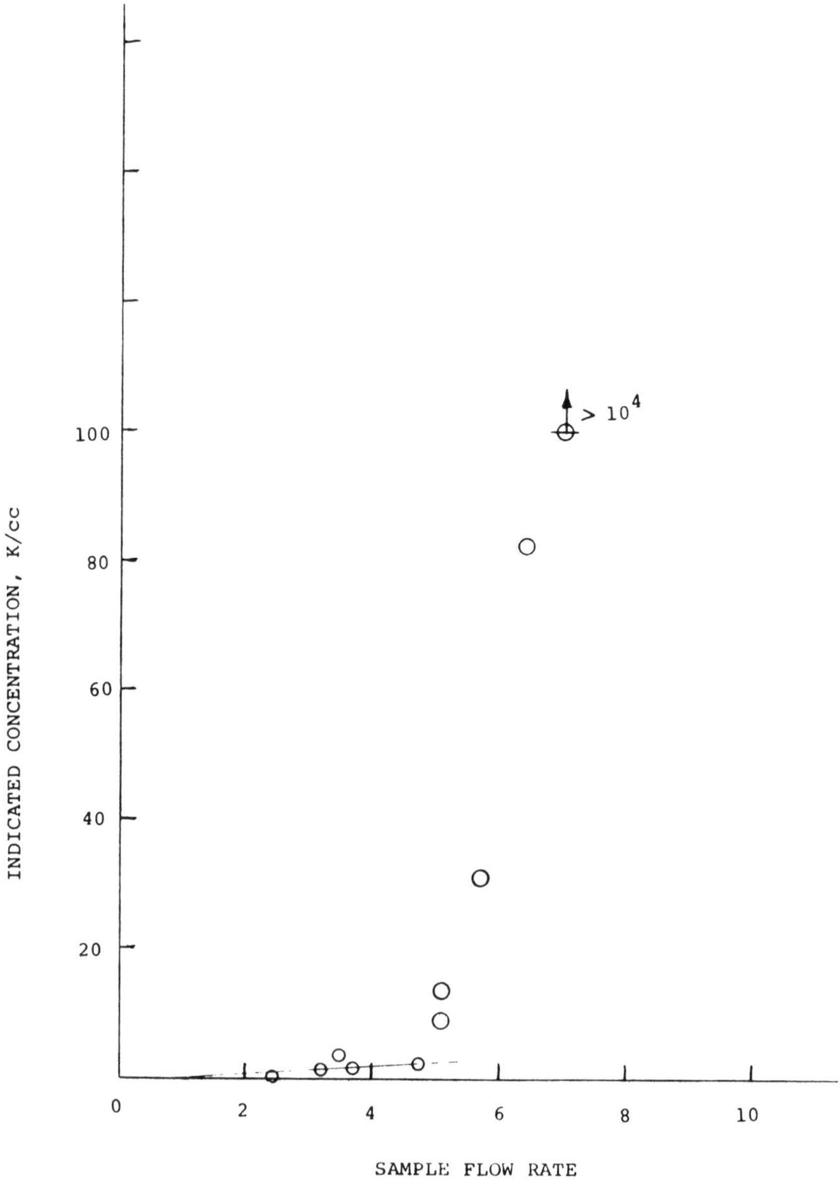

Figure 2. Behavior of diluted sample concentration at onset of sulfuric acid condensation.

diffusional analysis can only be obtained within the linear response concentration limits of the condensation nuclei counters used as detectors. These limits are 10^5 particles per ml for the instrument which we normally use (GE Laboratory Model CNC's and Environment/One Model Rich 100 devices). Actual size distributions obtained at a number of industrial sites are shown in Figure 1. In almost all the cases shown dilutions by at least 100:1 were necessary simply to reduce the concentrations to the linear range limits of the condensation nuclei counters. The dilution system used to date in these measurement programs is that described by Ragland (1976).

In some instances dilution by some minimum value was necessitated in order

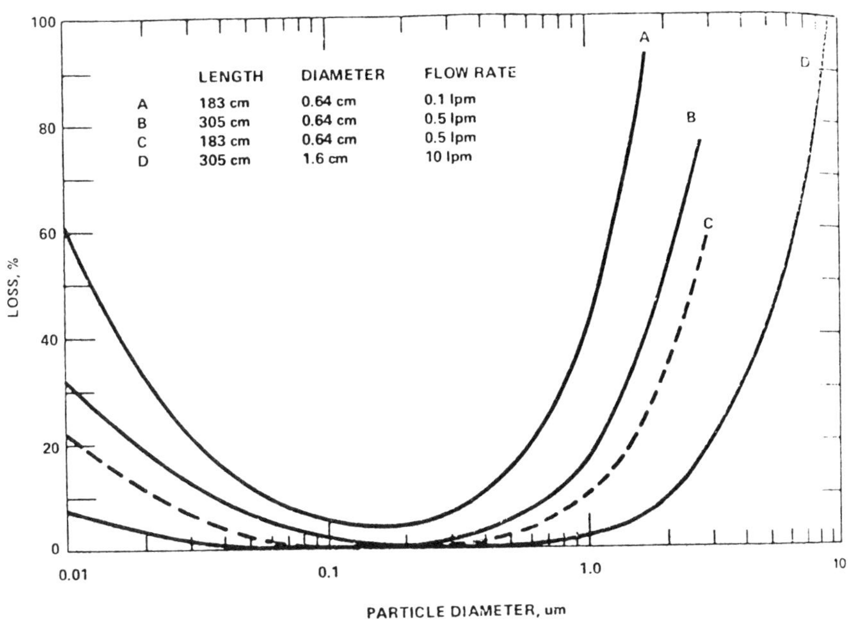

Figure 3. Probe losses due to settling and diffusion for spherical particles having a density of 2.5 g/cm^3, under conditions of laminar flow.

to avoid an apparent condensation phenomenon, which we ascribe to H_2SO_4, within the mixing zone of the dilution system. Figure 2 is a sample data set taken during a test on flue gas from a coal fired boiler which illustrates this phenomenon. As the sample gas flow rate to the diluter was increased, thus reducing the dilution factor (Dilution Factor ~ Constant/Sample Flow Rate in this case), the measured concentrations increased linearly with the sample flow rate, as expected, until some critical value was reached. At this point slight increases in sample flow rates led to very large, nonlinear increases in apparent aerosol concentrations. This same phenomenon has since been reported by other investigators (e.g., Sem, 1976). On a few occasions the onset of this condensation effect occurred at dilutions so high as to preclude obtaining any data at all on the aerosol as it existed in the flue gas stream. The condensation aerosol thus produced has a very small mean size as is evidenced by the fact that a diffusion battery having 50% penetration by 0.016 μm diameter particles passes only a few percent of the condensation aerosol particles.

The use of the diluter adds further difficulties in interpreting data in that sample line losses must be accounted for before one can obtain true concentrations and size distributions. These losses can arise due to diffusion, settling, agglomeration, impaction and electrostatic effects. Estimates of the effect of the first three of these can be made fairly well and Figure 3 shows calculated losses for various probe configurations and flow rates which are typically used in this work. Impaction losses can generally be ignored in these systems; however,

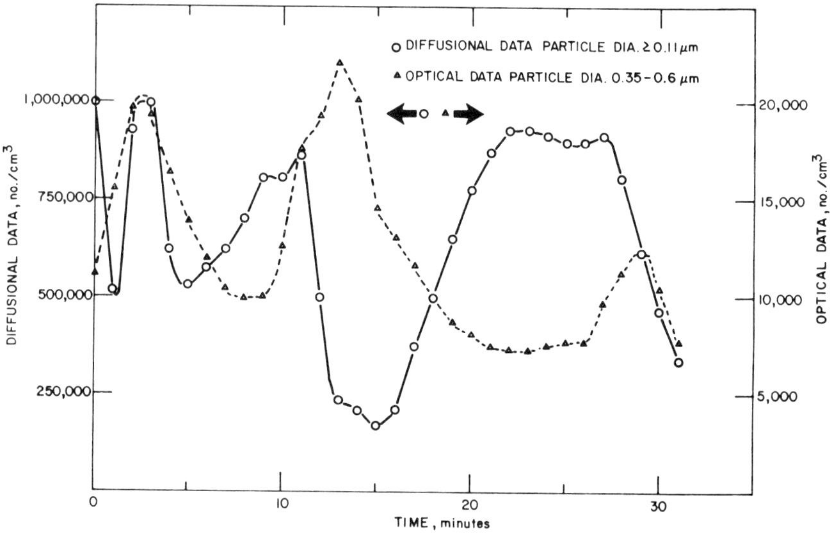

Figure 4. Typical inlet data segment obtained with optical and diffusional methods.

electrostatic effects may introduce serious uncertainties. Unfortunately, electrostatic charge neutralizers suitable for use at probe inlets for the conditions encountered in this type of sampling do not appear to be available.

Data reduction in these measurement programs, which are considered to be first generation attempts, have been carried out using two techniques: graphical stripping as described by Sinclair (1972); and a simplified, effective cut off approach similar to the "D_{50} method" widely used in the analysis of data acquired with cascade impactors. The choice of the data analysis method is generally made on the basis of the quality of the actual data. In many cases the temporal stability of the concentration, and the size distribution of the particulates in the flue gas are so poor that only crude data can be obtained. An example of data from such a source is shown in Figure 4 in which it can be seen that both the size distribution and concentration were subject to rapid, non-repetitive fluc-

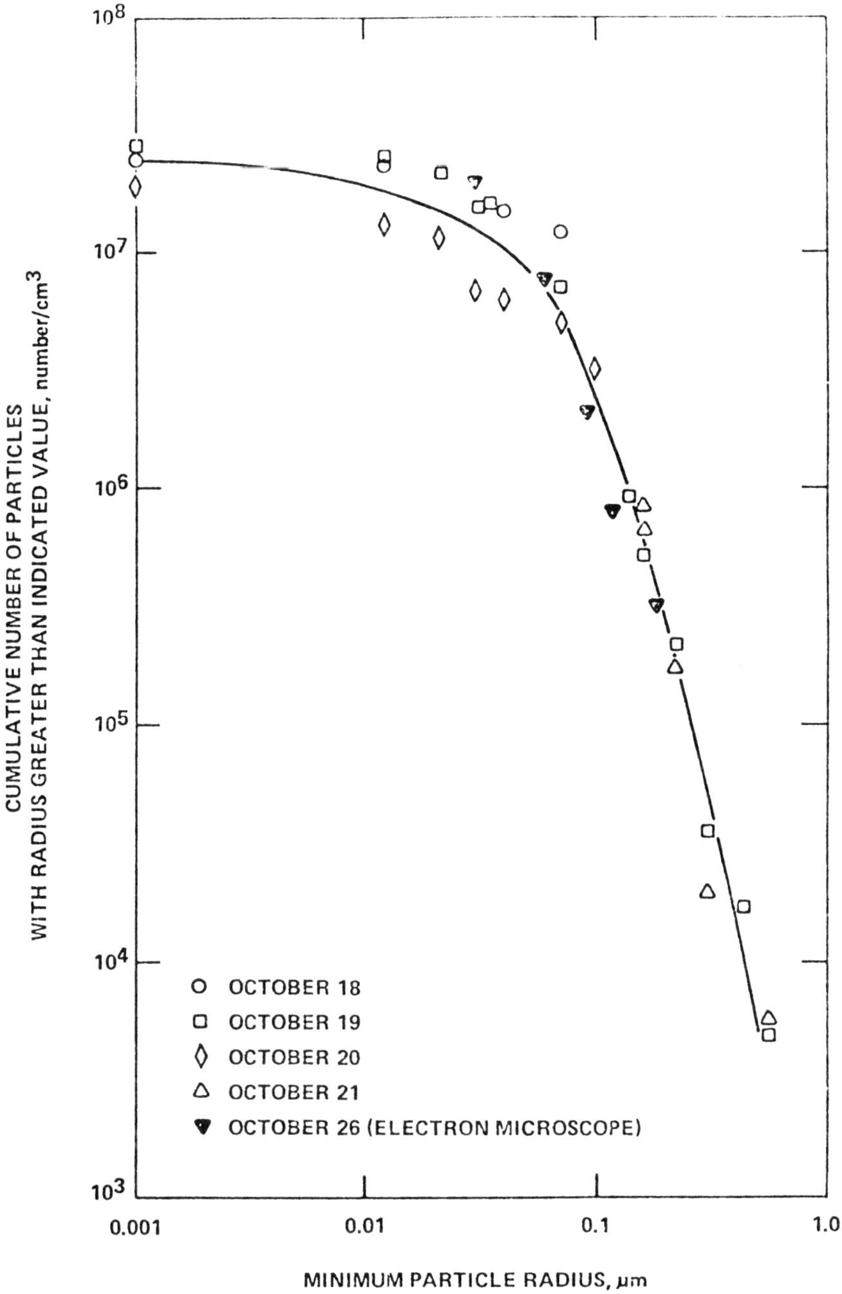

Figure 5. Field test data on stack effluent at a paper mill compared with size distribution measured by electron microscopy.

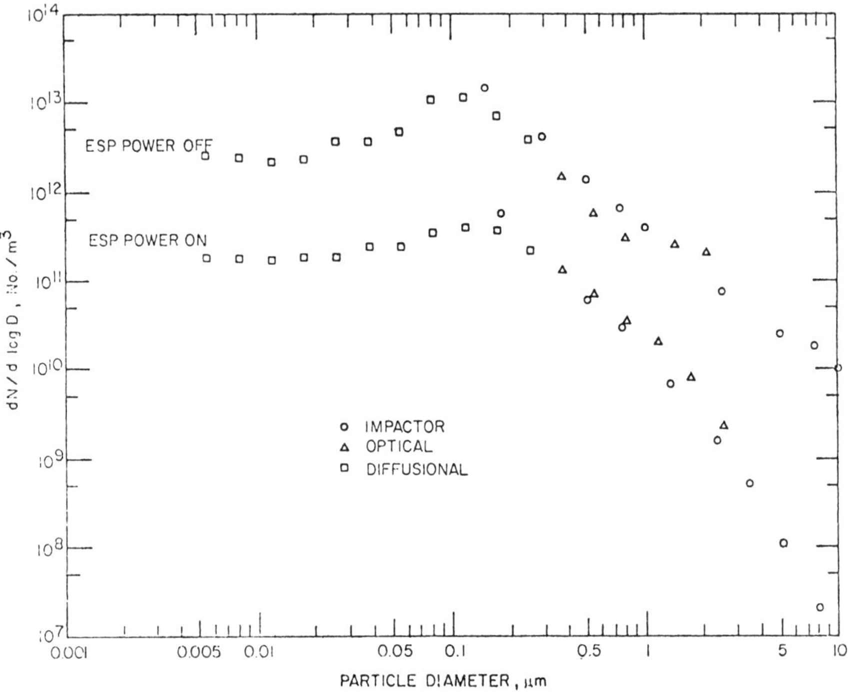

Figure 6. dN/d log D vs. D at inlet and outlet of electrostatic precipitator installed on a coal-fired power boiler.

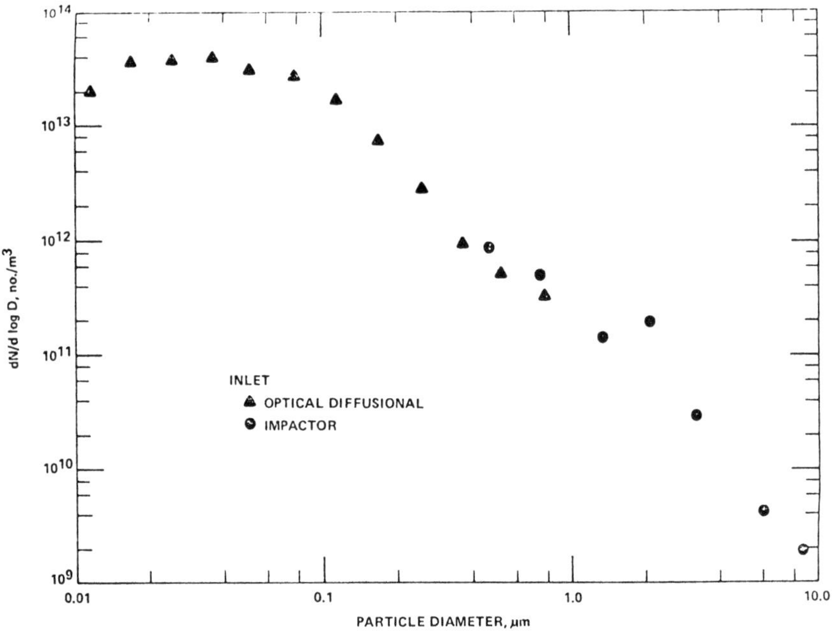

Figure 7. Field test data at the inlet of an electrostatic precipitator at the Meramec Power Plant.

tuations. In such cases order of magnitude estimates of concentration vs. size can be made using the "D_{50}" approach when the graphical stripping inversion technique would break down entirely.

More detailed results from several sampling expeditions are given in Figures 5, 6, 7, and 8. Figure 5 compares results obtained on an SO_2 absorber as determined using diffusional techniques, and a relative size distribution of the same aerosol as determined by electron microscopy counts. Figures 6 and 7 show

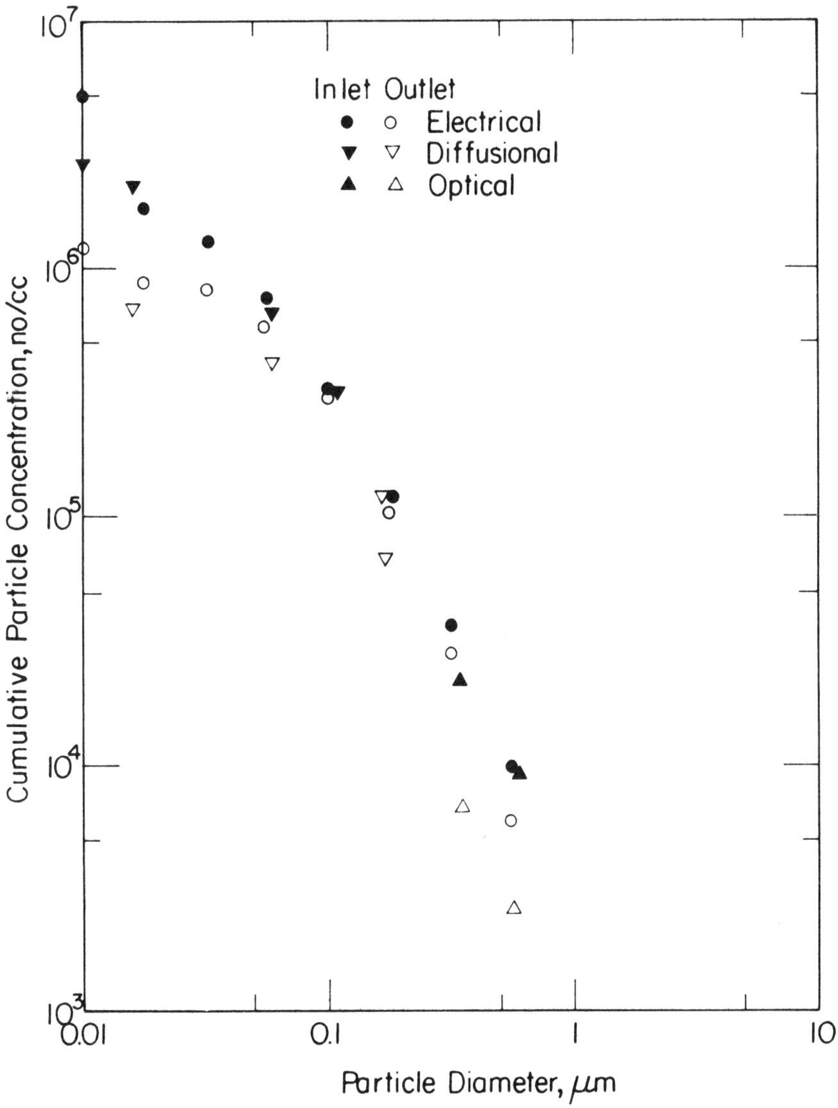

Figure 8. Inlet and outlet size distributions as obtained with optical, diffusional and electrical techniques.

data taken under three sets of circumstances at two different coal fired power plants with comparisons of the results obtained using optical particle counters, cascade impactors, and diffusional techniques. Figure 8 shows the results obtained at an asphalt plant equipped with a low energy wet scrubber using optical, diffusional, and electrical mobility (ThermoSystems Model 3030 Electrical Aerosol Analyzer) techniques.

From our experiences we have concluded that diffusional techniques, especially when used in conjunction with other available measurement methods, can provide useful information on industrial aerosol emissions in the tenth and hundredth micrometer size range.

REFERENCES

Ragland, James W., 1976: Measurement of ultrafine particle size distributions in industrial flue gases using an Electrical Aerosol Analyzer. Aerosol Measurement Workshop, Gainesville, Florida, March 24–26.

Sem, Gilmore J., 1976: Submicron particle size measurement of stack emissions using the electrical mobility technique. *Proc. of the Workshop on Sampling, Analysis, and Monitoring of Stack Emissions,* EPRI SR 41, Special Report, 111–129.

Sinclair, D., 1972: A portable diffusion battery. *J. Amer. Ind. Hygiene Assoc., 33,* 729–735.

THE DIFFUSION PROCESSOR

WILLIAM MARLOW

Brookhaven National Laboratory, Upton, N.Y.

Size-differentiated aerosol sampling of particles below the optical range has not been possible without fatal methodological weaknesses. To my knowledge, the low-pressure impactor is the only technique tried, and is of dubious value for volatile particles. What I shall describe is the next best thing to a clean size separation but which nonetheless appears capable of achieving the desired objective of producing a size-related separation of suboptical particles.

We have heard from Dr. Sinclair about the operation and use of two types of diffusion batteries for determination of size distributions. If we now take such a battery and run the aerosol stream through a filter or any other analytical device rather than through a condensation nucleus counter, we will have sampled the aerosol minus a calculable fraction of the smallest particles. By comparison of this analysis with that of a sample taken directly from the ambient, we may infer the chemical composition of the fraction removed. By operating several such "diffusion processors" with differing removal characteristics in parallel, we can similarly infer the chemical compositions of subfractions of particles in the suboptical range. One such case is illustrated in Figure 1 where curves IIA and IIB are the fractional penetration curves and the dashed curve, A-B, is the distribution according to size of particles removed by the battery between the two filter samples (this example was a Sinclair miniature diffusion battery pumped simultaneously at two ports). By "subtracting" the chemical composition analysis of the second filter from the first, we infer the chemical composition of the number distribution of the original aerosol represented by line A-B. Of course, to know the significance of this distribution, we need a size distribution measurement to indicate exactly how the mass weighting is to be done.

Unlike the conventional diffusion battery where only total numbers of particles are of interest, the diffusion processor must perform in a fairly ideal manner.

Since larger (optical size) particles may quite often contain the greatest part of the atmospheric particulate mass though rarely the particle number, great care must be taken that diffusion is the *only* removal process. This is generally not the case, however. I found there to be on the order of a 5% removal rate per stage of particles in the optical range as the aerosol stream passed through a Collimated Hole Structure battery. For one or two stages in series, this is not significant, but

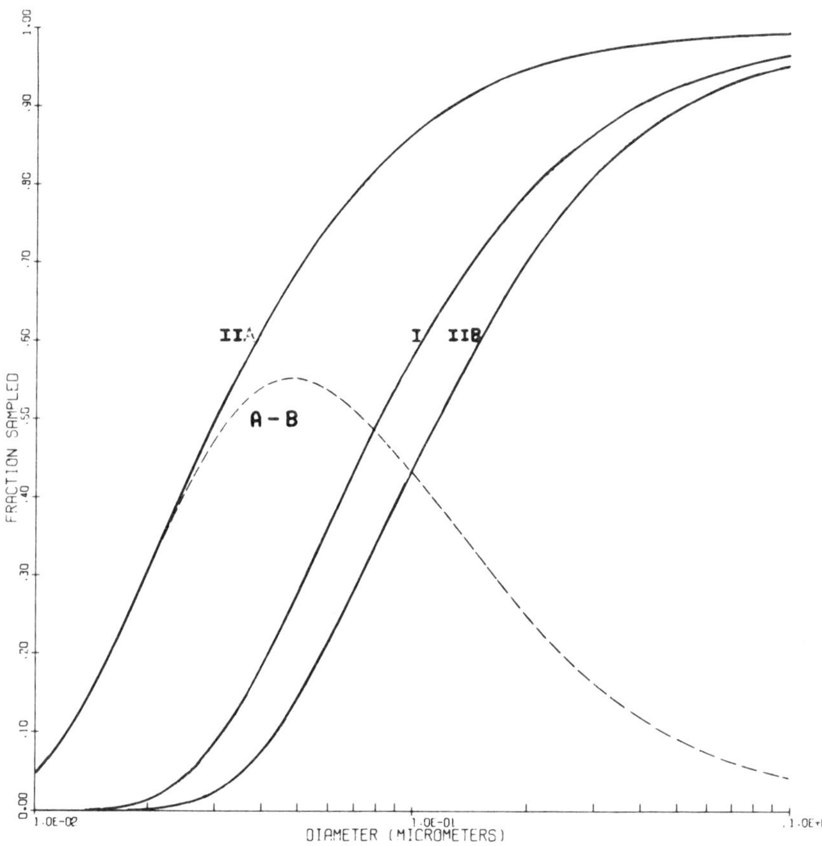

Figure 1. Solid lines represent fractional penetrations through Collimated Hole Structure at two operating conditions. Dashed line represents difference curve (see text).

it can be for numerous such stages. On the other hand, with another battery constructed from fine mesh wire screen, I found up to 80% removal of these particles at the final stage; a figure which clearly eliminates these much less expensive screens as diffusion processors. Since increased air speed through the processor also increased the removal, it is fair to assume impaction and interception to be the dominant processes. To minimize these losses diffusion processors should be constructed of as many parallel diffusion cells as possible and operated at as low a flow rate as feasible.

REFERENCES

Marlow, W. H. and R. L. Tanner, 1976: Diffusion sampling method for ambient aerosol size discrimination with chemical composition determination. *Analyt. Chem.*, *48* (13), 1999.

Tanner, R. L. and W. H. Marlow, 1977: Size discrimination and chemical composition of ambient airborne sulfate particles by diffusion sampling. *Atmos. Environ.*, *11*, 1143.

QUESTIONS

Whitby: What evidence is there that any of these condensation nuclei counters actually go down to 0.002 microns?

Rich: The best evidence I have heard was in Prague in a much bigger hall than this. Somebody brought up the question of how small it would go and one of the foremost authorities in the world, Professor Whitby, popped up and said, "0.002 microns," and there you have it.

Whitby: Well, we just shot down one authority.

Hogan: I don't really know how you define the size of a particle that small; whether you define it in terms of its diffusion coefficient, or whether you catch one and look at it. We have been able to measure particles, or at least sense the presence of particles, that we can't pass through any tubing. We can take a counter of some type, put it very close to a propane flame, and dilute it very rapidly. If the counter itself is very close to that flame or to the diluted flame product, we can detect the presence of a lot of something; but *if* we take any type of diffusion battery or even a couple feet of tubing, we get nothing through. So when you calculate out the theoretical size of the particles or the diffusion coefficient that has to be removed in that amount of time, you find it is down in that neighborhood (0.002). Probably the exact point where we begin to detect particles is the biggest question today. Judging from my practical experience and from some coagulation work that I have done recently, I would say that particles of this size are probably not present in nature except in a couple of rare instances because they coagulate so rapidly with even a thousand particles per cc, ten times that size, that they have a very limited life span. The only place we ever really find anything down here at this bottom end where nothing will get through a diffusion battery is in these rare instances, as I showed at the South Pole where we only had a background of about 50 per cc and suddenly we got a thousand of these real tiny ones to show up. This may happen all the time over the ocean, but with even 500 big ones around we would never find them.

Haberl: We did conduct one experiment using platinum aerosols, in which I acquired samples for both transmission and electron microscope analysis simultaneously. The aerosols in that analysis were almost totally a 0.003 micron particle and certainly the counter was measuring those and those alone.

Podzimek: There is no direct proof of how the particles increase in size, but you have some idea about an image if you are looking at the condensation of ions, too.

Whitby: The condensation of ions really doesn't apply to what we are talking about. The fraction of particles charged below 100 Å (0.01 μm) by Boltzmann's distribution is below 1%. Condensation, of course, occurs on ions of much smaller sizes than uncharged particles. What we are really concerned about is the uncharged particle, and the minimum size in which it exists. The reason I asked this question is that we don't know of any concrete data either, and we try to think of good ways of obtaining it. We also have evidence that the counters often are not going down as small as we have sometimes talked about in the past. One of the points to be made in asking this question is to call for more hard evidence on exactly what the limits are for the specific device. We ought to reference a range more like 20 to 50, or 20 to 60 Å, which is probably the real range in which devices operate.

Sinclair: I think I have a little information on this subject. You saw on one of the earlier slides the calibration curve for the diffusion battery; that was taken at 4 ℓpm and the lowest size indicated there was 0.004 μm. Now we have another set of curves taken at 6 ℓpm which applied to the GE counter referred to by Mr. Haberl which has been around for some time. The lowest curve, the curve corresponding to the lowest size, on that set of calibration curves is around 0.003 or 0.002. That would be about 1% of that size according to theory, and would pass the smallest of our diffusion batteries. I think that is about the best I can say about that. But with this GE counter, we were able to measure the difference between that battery in front of the nuclei counter, and removed. It was a considerable decrease, perhaps about 10% in the concentration as measured by the GE counter through this first diffusion battery. I think that indicates that we were measuring very close to that size. There are sizes that we have found that are too small for this GE counter to measure; they are in the neighborhood of 0.001 or perhaps less. They are radon daughter products unattached after they have been formed in a large box free of condensation nuclei. The radioactive measurements said that there were about 500 particles per cc in this box, but the CN counter, that is the GE counter I have been talking about, was unable to measure anything. So, I think we can say that we get down pretty close to 0.002 or maybe 0.003 but we can't measure 0.001.

Podzimek: I saw in Biermann's thesis that he was not just measuring the charged particles. We can change the supersaturation from the rate of 15 or 20% to 300 and over.

Marlow: I have one or two comments based on my own experiences, and following up what Austin Hogan said. In making field measurements of very fine particles, one is usually calibrating one's own plumbing rather than one's instrument. I think that perhaps we need to view with some skepticism absolute statements about what actually is present in the very, very fine range because of this question of the calibration of one's plumbing. There may very well be atmospheric processes which create very, very fine particles which can coagulate extremely rapidly or grow with each other and form larger particles. That doesn't say that they never exist. I think that with suitable equipment one probably could follow these as, say, a function of radiation or whatever else the cause may be.

Rich: You may remember Professor Pollak always let his aerosols age in the bags he kept them in. He did it because he started at a low reading and the reading increased and finally settled down a little bit, which he never explained very forcefully, but apparently it was the conversion of gases to particles. When you do get gas to particle conversion, as you do in the outdoors, the sizes go through a spectrum from something very small until the instrument sees it. It's a justifiable question as to where that is, but it is still an argument.

Whitby: I want to know about the other end of the problem, the upper range.

This isn't a critical issue but I noticed that the difficult thing is to say that 0.1 is the upper limit. I think that Austin Hogan made an observation which is quite consistent with a lot of size distribution work we are doing now; when you have an aged aerosol, such as you measure from aircraft or in ambient conditions, then what we call the nuclei mode is very often absent or very small. I think what this means is that the number of particles is accounted for by the accumulation mode. Those particles are much larger than 0.1 μm. The mean size of those modes, as seen from the aircraft, were usually 0.3 to 0.4. What this says is that the practical upper limit is determined by number size distribution statistics. When one uses the counter under these aged aerosol conditions, the actual upper limit of what is statistically significant is not 0.1 anymore, but more like 0.7 to 0.6. This is not a very important thing, but it is relevant in the urban area where there are a lot of very small particles.

Podzimek: There is some confusion about the use of the terms Aitken nuclei, and condensation nuclei. I would recommend sticking to what has already been accepted, namely, Aitken nuclei (or condensation nuclei) as those particles are smaller than 0.1 μm. In meteorology we have condensation nuclei which are divided into large nuclei, giant nuclei, and so on. I would really recommend sticking to the original division of nuclei.

Marlow: This is perhaps a terminological question to keep in mind, especially when we start to use condensation nuclei counters using condensing fluids other than water. Are we really talking about condensation nuclei? We are talking about nuclei which can be activated to growth, and I don't think we'd want to pre-empt another field's long-established terminology.

Q: I would like to hear the panel discuss the question of whether all nuclei are condensation nuclei, or whether particles exist that don't serve as condensation sites?

Podzimek: I think we should talk about the activation of nuclei and the critical supersaturation at which nuclei are activated. For this reason, we are using diffusion batteries for the measurement of cloud condensation nuclei, and we are using Aitken nuclei counters which defined nuclei by this very high supersaturation. Due to this, all particles will be activated. In other words, speaking about Aitken nuclei counters, we probably will not bother so much about the nature of nuclei. You know, however, if you are going down or upward you have to think about the size; you have to bear in mind the nature of nuclei, physical and chemical nature. Then it makes sense to measure, as people like Braban and his group in Chicago are doing. They are talking about the activation spectrum of nuclei which is the number of nuclei related to the critical supersaturation at which the nuclei are measured. Those really are condensation nuclei at such conditions, but you have to define them by the activation parameter, supersaturation, and so on. It is very hard to set any point, say 0.05% supersaturation. Nobody can do this.

CALCULATION OF AEROSOL SIZE DISTRIBUTION FROM DIFFUSION BATTERY MEASUREMENTS— A COMPUTER PROGRAM FOR THE GRAPHICAL "STRIPPING" METHOD

DAVID SINCLAIR, DAVID A. CHRISTY, AND KENNETH SNYDER

*Health and Safety Laboratory, U.S. Energy Research and Development Administration,** N.Y., N.Y.

The use of a series of tubular diffusion batteries for measuring sub-micron aerosols has been described in several papers (Sinclair, 1972; Sinclair and Hinchliffe, 1972; Breslin, Guggenheim and George, 1971). Each individual battery in the series consists of "collimated holes structures" containing 14,500 uniform diameter holes (the equivalent of tubes) in parallel. In our method, the penetration of the aerosol through the series of eleven batteries in succession is measured with a continuous flow condensation nucleus counter (Sinclair and Hoopes, 1975). This series of measurements provides eleven fractional penetrations as a function of equivalent length of the battery, from which the aerosol particle size and size distribution can be calculated. (Equivalent length is the actual length of the battery multiplied by the number of tubes in parallel.) The penetration varies greatly with particle size, from about 1% through the first battery for particles of 3 nm diameter to about 40% through all eleven batteries for particles of 200 nm diameter, at a flow rate of 4 ℓ/min.

We have used a manual method called "graphical stripping" which involves plotting the penetration on a logarithmic ordinate and the battery length on a linear abscissa. Straight lines are commonly drawn connecting each pair of points. The slope of the resultant curve decreases in progression from the entrance to the exit battery since penetration increases with increasing particle size. This fact is a key to the use of this method. The method is similar to that used to determine the relative amount of radioisotopes in a mixture from decay or absorption measurements (Cork, 1947). With aerosols it is complicated by the fact that an aerosol is not composed of a limited number of discrete particle sizes. However, experience has shown that a size distribution can be obtained without any assumption as to its form.

*Now the U.S. Department of Energy, Environmental Measurements Laboratory.

COMPUTER PROGRAM

Since the graphical method is very tedious, we have developed a computer program which works well provided ''good'' data have been obtained. The criterion for good data is that the slope of the penetration curve must continuously decrease. If this is not the case the curve will become concave downward at one or more locations. An important sub-routine (CONFIX) of the program finds these concave conditions and corrects for them. It does this by moving one or more data points up or down a few percent. Graphically, this is equivalent to drawing a smooth curve through the data points when some of them are observed to be out of line, as is frequently the case. If many concave conditions occur the data may not be satisfactory. In that case it is advisable to take the average of four or more penetration curves. A common cause of inaccuracy is variation in the concentration of the aerosol while being measured. This usually requires a number of runs to obtain more representative average values.

The method consists of subtracting from the experimental curve theoretical curves for monodisperse sizes, calculated from diffusion theory (Sinclair et al., 1976). These curves are similar in form to the experimental curves except that their curvature is less. The first theoretical curve to be subtracted has a slope for the portion between batteries 10 and 11 somewhat less than that between batteries 10 and 11 for the experimental curve. The slope S between battery 10 and 11 is given by:

$$S(10,11) = \frac{\log p(10) - \log p(11)}{\ell(10) - \ell(11)}. \tag{1}$$

Here p is fractional penetration and ℓ is equivalent length. The preceding batteries having removed 99% or more of the size smaller than about 50 nm, the slope between battery 10 and 11 is determined almost entirely by the larger sizes.

The fraction of aerosol of diameter equal to or greater than d, the size corresponding to the first curve subtracted, is given by:

$$f(d) = \frac{p(e) - R}{p(t)}. \tag{2}$$

Here p(e) is the experimental fractional penetration through battery 11, p(t) is the theoretical fractional penetration of particles of diameter d through battery 11 and R is an arbitrary constant, usually 0.005.

Rewriting equation (2):

$$p(e) - f(d) \cdot p(t) = R \tag{3}$$

shows that R is the remainder left after subtracting particles of diameter d and larger from the aerosol that passed through the whole battery. The remainder cannot be zero since it is recognized that there must be appreciable numbers of smaller sizes in the penetrating aerosol, although they contribute relatively little to the experimental values. Experience has shown that 0.005 is a suitable value for the remainder, which is usually used throughout the computation.

The first subtraction is carried out over the whole penetration curve, using the experimental and theoretical values corresponding to each battery. The result is a reduced experimental curve free of the largest sizes. The above procedure is now repeated, beginning with battery 10 and using the slope between battery 9 and 10 to obtain the second size to subtract. The fraction of the second size is the fraction of aerosol of diameter greater than that size but less than the first size. It

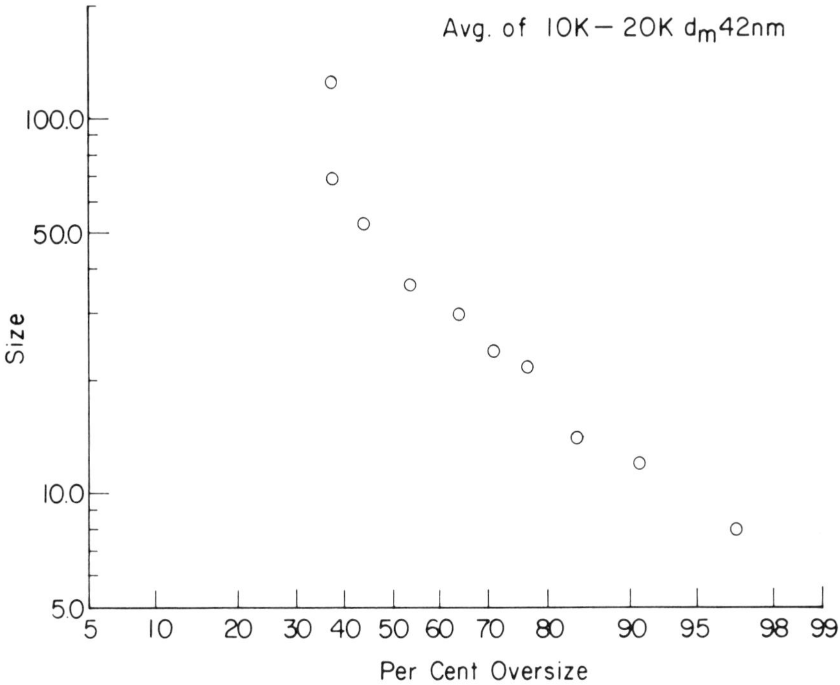

Figure 1. Cumulative size distribution. Average of 18 measurements of concentrations between 10,000 and 20,000 particles per cm³. The median diameter is 42 nm.

is again given by equation (1) in which p(e) is now the reduced experimental value at battery 10 and p(t) is the theoretical penetration at battery 10. This whole process is repeated ten times, once for each interval between batteries. When the 10 fractions so obtained are added together in succession the result is a cumulative size distribution for sizes ranging from about 8 to 150 nm.

The computer performs all the above operations and also looks for concave downward conditions throughout the original and reduced experimental curves. This condition may occur after a series of subtractions and cause a repetition of a previously computed size, but it usually produces only a minor error in the cumulative distribution curve.

Sometimes a negative value occurs during a series of subtractions. This results from a concave condition which may cause a point on the reduced curve to have a lower value than the corresponding point on the theoretical curve. When this happens, the computer starts the computation over again, using 0.01 as the remainder. If a negative value occurs again the data are marked "rejected" but they are available for inspection in the printout.

The computer program (see Appendix) contains the following series of particle diameters: 3,4,6,8,10,12,14,16,18,20,22,24,26,28,30,32,34,36,38,40,42, 44,46,48,50,53,55,60,65,70,75,80,85,90,95,100,110,125,140,165,200 nm.

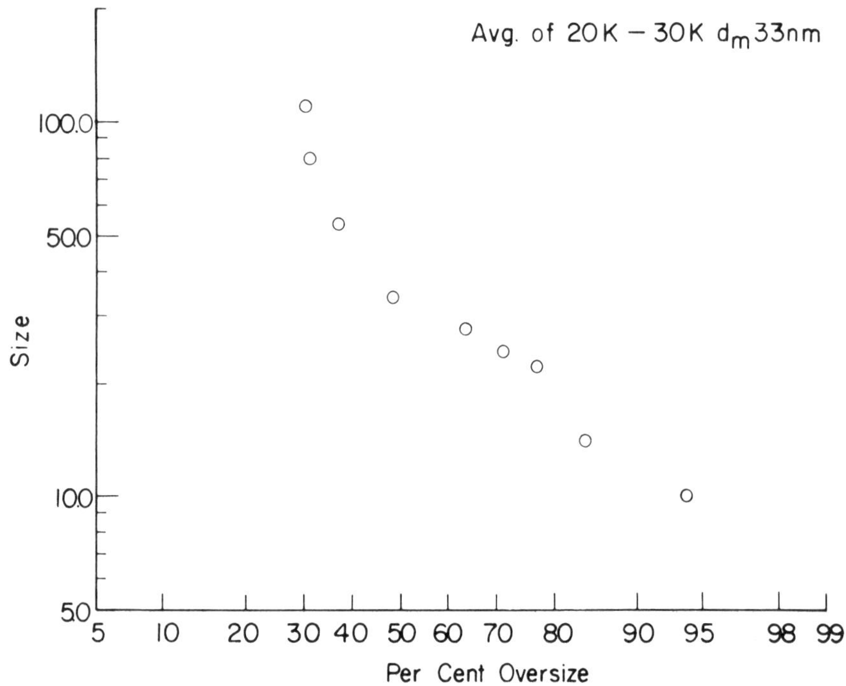

Figure 2. Cumulative size distribution. Average of 32 measurements of concentrations between 20,000 and 30,000 particles per cm^3. The median diameter is 33 nm.

These sizes are selected during computation by means of subroutine FIND. The computation time is about 8 seconds on an IBM 360/30 computer.

The above program was written specifically for tubular diffusion batteries. We also have a diffusion battery made of 635 mesh stainless steel screens (Sinclair and Hoopes, 1975). This device is used to measure submicron aerosol in the same manner as the tubular battery. It had to be calibrated since its performance could not be calculated with sufficient accuracy. This was done with monodisperse aerosols at the University of Minnesota (Sinclair et al., 1978) as reported earlier in this book. The penetration curves for monodisperse aerosols are similar to those of the tubular battery except that they are one-term exponential functions instead of the five-term exponential functions of the tubular batteries. The equations for these curves are presented in the preceding paper

(Sinclair et al., 1978). Their values are contained in the program for the same particle sizes as given above. The procedure for analysis of screen battery measurements is exactly the same as for the tubular battery except that the screen battery has only ten sections.

During the summer of 1976 cooperative measurements were made with New York University and other laboratories of outdoor aerosol in New York City (Knutson, Sinclair and Leaderer, 1977). Figures 1, 2 and 3 show the results of some of the measurements made with the automated tubular diffusion battery and the continuous flow condensation nucleus counter, described in the preceding

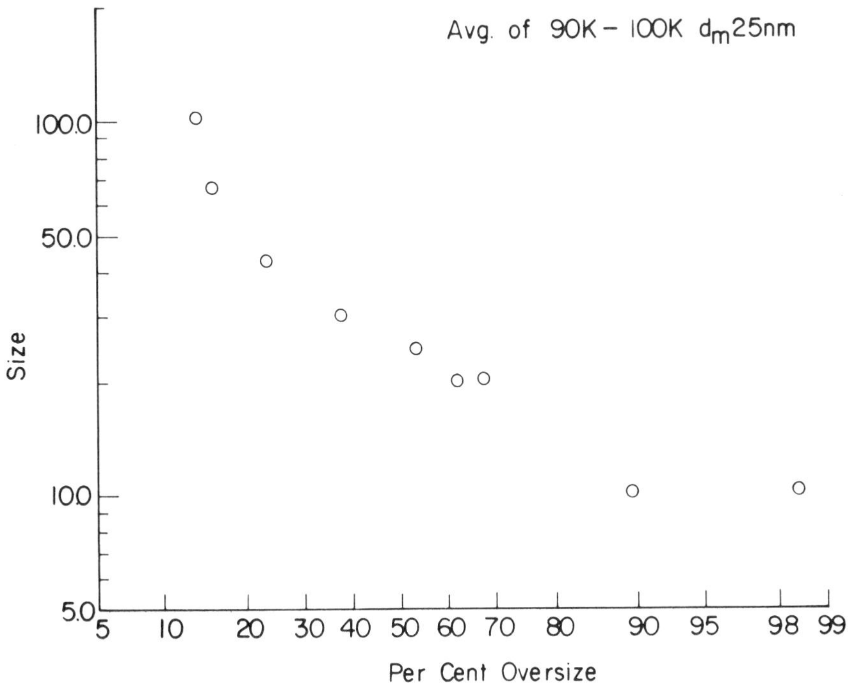

Figure 3. Cumulative size distribution. Average of eight measurements of concentrations between 90,000 and 100,000 particles per cm^3. The median diameter is 25 nm.

paper (Sinclair et al., 1978). These figures show cumulative size distribution curves computed and plotted by the stripping method program. Figure 1 shows the average of 18 measurements of concentrations between 10,000 and 20,000 particles per cm^3. The median diameter is 42 nm. Figure 2 shows the average of 32 measurements of concentrations between 20,000 and 30,000/cm^3. The median diameter is 33 nm. Figure 3 shows the average of eight measurements of concentrations between 90,000 and 100,000/cm^3. The median diameter is 25 nm. It is seen that the particle diameter increases with decreasing concentration. This might be accounted for by coagulation which increases the size and decreases the concentration.

CONCLUSIONS

1. The stripping method, when applied to diffusion battery measurements, yields a cumulative size distribution provided good data are obtained.

2. The computer program performs the computations in a short time (8 seconds) and eliminates the tedium of the graphical stripping method.

3. When the data are unsatisfactory because of too many concave conditions, the program detects and corrects the data in most cases.

4. In the case of fluctuating experimental conditions the average of four or more diffusion battery penetration runs should be used for computation.

REFERENCES

Breslin, A. J., S. F. Guggenheim and A. C. George, 1971: Compact high-efficiency diffusion batteries. *Staub-Reinhalt. Luft, 31,* 1–5.

Cork, J. M., 1947: *Radioactivity and Nuclear Physics.* Van Nostrand, New York.

Knutson, E. O., D. Sinclair and B. Leaderer, 1977: New York summer aerosol study—number concentration and size distribution. Presented at the 1977 Industrial Hygiene Conference, New Orleans.

Knutson, E. O., D. Sinclair and B. Leaderer, 1978: New York summer aerosol study—number concentration and size distribution. Presented at the 1977 Industrial Hygiene Conference, New Orleans, to be published by the New York Academy of Sciences, 1978.

Leaderer et al., 1978: Summary, *JAPCA, 28,* 322.

Sinclair, D., 1972: A portable diffusion battery. *J. Amer. Industrial Hygiene Assoc., 33,* 729–735.

——— and L. Hinchliffe, 1972: Production and measurement of submicron aerosols, II. in *Assessment of Airborne Particles,* Chap. 10, T. T. Mercer, P. E. Morrow, W. Stober, Eds., Charles C. Thomas, Springfield, Ill.

——— and G. S. Hoopes, 1975: A continuous flow condensation nucleus counter. *Aerosol Sci., 6,* 1–7.

——— and ———, 1975: A novel form of diffusion battery. *J. Amer. Industrial Hygiene Assoc., 36,* 39–42.

———, R. J. Countess, B. Y. H. Liu and D. Y. H. Pui, 1976: Experimental verification of diffusion battery theory. *J. Air Poll. Contr. Assoc., 26,* 661–663.

———, ———, ——— and ———. Automatic analysis of submicron aerosols. In *Aerosol Measurement,* Proc. of Aerosol Measurement Workshop, March 24–26, 1976. This volume.

APPENDIX

INSTRUCTIONS FOR JOB STRIP

All commands to the program are in the form of mnemonic codes. They may appear in any order. Commands must be on separate cards. They cannot appear in the title of a data set. The code should be in the first four columns of a command card.

Code	*ACTION TAKEN*
PLOT	Gives a plot of each experimental curve before subtraction.
PPLT	Gives a Log-Probability plot of the sizes found versus their cumulative percentages.
NPLT	No plots.
BOTH	Gives both types of plot.
AVER	Computes to an average of N data sets, where N is an integer from 1 to 9 inclusive appearing in card column 25. The data sets to be averaged must appear immediately after this card. The title of the averaged data set will be that of the last individual data set to be averaged.

Commands affect all data sets following them until another command counteracts it. If no commands are given, the program assumes no plots or averages and output appears only on the printer.

Input of a Data Set

Sample card:
Title (cols. 1–24) ICODE (col. 25) PER CENT PENETRATION (cols. 26–80)
NYU TB 4LPM 76082416 2 176.1360.5443.5931.1621.8615.3010.73 6.81 4.41 2.56 1.43
Cols. 1–24 TITLE—any 24 characters used to identify the data set.
25; ICODE—this value determines the battery type and flow.

ICODE	*BATTERY AND FLOW*
1	Tabular Battery @4 ℓpm
2	Tabular Battery @6 ℓpm
3	Screen Battery @4 ℓpm
4	Screen Battery @6 ℓpm

26–80; Per cent penetration starting with battery 1 in 11F5.2 format.

```
C     THIS PROGRAM COMPUTES CUMULATIVE SIZE DISTRIBUTIONS OF AEROSOLS ACCORDING TO
C     THE GRAPHICAL STRIPPING METHOD AS DEVELOPED BY DR. SINCLAIR.
C     THE ORIGINAL VERSION WAS WRITTEN BY KEN SNYDER.
C     THIS UPDATED AND EXPANDED VERSION WAS WRITTEN BY DAVE CHRISTY.
C
      COMMON/A/OYDATA(11),YDATA(11),NBAT,ITYPE,IFLOW,IBAT,
     & XDIST(11,2),IPLOT,IAVER,NCON,ISIZE
      COMMON/D/HEADR(6)
      DIMENSION BUFF(11)
C
C     IF THIS PROGRAM IS RUN ON A MACHINE WITHOUT A COMPATIBLE PLOTTER, REMOVE
C     THIS CALL, STATEMENT #999, ALL PLOTTING ROUTINES AND REFERRENCES TO THEM.
C
      CALL PLOTS(BUFF,44)
C
 1    IPLOT=0
      IAVER=0
C
C     INPUT THE TITLE, ICODE, AND THE ORIGINAL DATA
C
 5    READ(1,10,END=999)HEADR,ICODE,(OYDATA(I),I=1,11)
C
 10   FORMAT(6A4,I1,11F5.2)
      IFLOW=1
      ITYPE=1
      NCON=0
C
C     TEST FOR A COMMAND
C
      CALL CODE(M)
C
      IF(IAVER.GT.0)CALL AVER(ICODE)
      IF(IAVER.GT.0)GOTO 5
      IF(M.GT.0)GOTO 5
C
C     IFLOW DETERMINES THE FLOW. 1=4LPM, 2=6LPM.
C     ITYPE DETERMINES WHETHER THE DATA IS FOR A TUBULAR BATTERY OR FOR A SCREEN
C     BATTERY. 1=TUBULAR BATTERY, 2=SCREEN BATTERY.
C
      IFLOW=ICODE
      IF(ICODE.LT.3)GOTO 20
      ITYPE=2
      IFLOW=ICODE-2
C
C     NBAT=THE NUMBER OF DATA POINTS
C     OYDATA=ORIGINAL INPUT DATA
C     YDATA=CURRENT DATA SET BEING PROCESSED
C
 20   DO 30N=1,11
      IF(OYDATA(N).LT.0.9)OYDATA(N)=0.0
      YDATA(N)=OYDATA(N)
 30   IF(OYDATA(N).GT.0.0)NBAT=N
 40   CALL CRUNCH
      DO 50L=1,11
 50   YDATA(L)=0.0
      GOTO 5
 999  CALL PLOT(0.0,0.0,999)
      STOP
      END
C
      SUBROUTINE CODE(M)
C
C     THIS ROUTINE CHECKS FOR COMMAND WORDS IN THE TITLE
C
      COMMON/A/OYDATA(11),YDATA(11),NBAT,ITYPE,IFLOW,IBAT,
     & XDIST(11,2),IPLOT,IAVER,NCON,ISIZE
      COMMON/D/HEADR(6)
      DIMENSION WORD(6)
      DATA  WORD/'PLOT','NPLT','PPLT','NPPL','BOTH','AVER'/
      M=0
      IF(HEADR(1).NE.WORD(6))GOTO 5
      IAVER=1
      M=1
      RETURN
 5    DO 10I=1,6
      N=I
      IF(HEADR(1).EQ.WORD(I))GOTO 20
 10   CONTINUE
      RETURN
```

```
 20    M=1
C
C    IPLOT IS A CODE WHICH KEEPS TRACK OF WHAT TYPE OF PLOTS, IF ANY, ARE DESIRED
C
       IPLOT=N
C
       RETURN
       END
C
       SUBROUTINE AVER(ICODE)
C
C    THIS ROUTINE COMPUTES AVERAGES OF DATA SETS
C
       COMMON/A/OYDATA(11),YDATA(11),NBAT,ITYPE,IFLOW,IBAT,
      & XDIST(11,2),IPLOT,IAVER,NCON,ISIZE
       IF(IAVER.GT.1)GOTO 10
       NAVE=ICODE+1
       IAVER=2
       RETURN
 10    DO 20I=1,11
       YDATA(I)=OYDATA(I)+YDATA(I)
       IF(IAVER.EQ.NAVE)YDATA(I)=YDATA(I)/FLOAT(NAVE-1)
       OYDATA(I)=YDATA(I)
 20    CONTINUE
       IF(IAVER.EQ.NAVE)IAVER=-1
       IAVER=IAVER+1
       RETURN
       END
C
       SUBROUTINE CRUNCH
C
C    THIS ROUTINE CALLS MOST OF THE OTHER SUBROUTINES AND PRINTS OUT THE RESULTS
C
       COMMON/A/OYDATA(11),YDATA(11),NBAT,ITYPE,IFLOW,IBAT,
      $ XDIST(11,2),IPLOT,IAVER,NCON,ISIZE
       COMMON/B/THEOR(11,41,2,2),ISIZES(41),IGE1(11,2)
       COMMON/C/PERCEN(11),SIZES(11)
       COMMON/D/HEADR(6)
       DIMENSION SUB(11),ANS(11)
       IERROR=0
       REMAIN=0.5

 1     TPER=0.0
       DO 2K=1,11
       ANS(K)=0.0
 2     SUB(K)=0.0
       DO 5 M=1,11
       PERCEN(M)=0.0
 5     SIZES(M)=0.0
       FLOW=4.0
       IF(IFLOW.EQ.2)FLOW=6.0
       PRINT 10,HEADR,FLOW,REMAIN
 10    FORMAT('1',6A4,//,1X,'FLOW=',F3.1,' LPM',5X,'REMAINDER=',F3.1)
 15    JBAT=NBAT-1
       DO 100I=1,JBAT
       IBAT=NBAT-I+1
       CALL CONFIX
       IF(IPLOT.EQ.0)GOTO 20
       ICALL =I
       IF(IPLOT.EQ.1.OR.IPLOT.EQ.5)CALL GRAPH(YDATA,11,ICALL,ITYPE)
 20    CALL FIND(IS)
C
C    THIS CALCULATES THE PERCENTAGE OF THE SIZE FOUND
C
       PER=(YDATA(IBAT)-REMAIN)/THEOR(IBAT,IS,IFLOW,ITYPE)
C
       IF(PER.LT.0.0)GOTO 200
       ENDPER=TPER+PER
       IF(ENDPER.GT.1.0)GOTO 55
       TPER=TPER+PER
       DO 30N=1,IBAT
       SUB(N)=-THEOR(N,IS,IFLOW,ITYPE)*PER
       ANS(N)=YDATA(N)+SUB(N)
       IF(ANS(N).LE.0.0)GOTO 200
 30    CONTINUE
       ISIZE=ISIZES(IS)
C
C    PERCEN IS AN ARRAY WHICH CONTAINS THE CUMULATIVE PERCENTAGES
C    SIZES IS AN ARRAY WHICH CONATINS THE SIZES CORRESPONDING TO THE PERCENTAGES
```

```
C
      PERCEN(I)=TPER
      SIZES(I)=FLOAT(ISIZE)
C
      PRINT 40,ISIZE,PER,TPER,(YDATA(IA),IA=1,11),(SUB(IB),IB=1,11),
     &(ANS(IC),IC=1,11)
 40   FORMAT(1H0,'SIZE=',I3,4X,'PERCENT=',2PF6.2 ,4X,'TOTAL PERCENT=', F
     &6.2/11X,11(0PF6.2,2X)/11X,11(F6.2,2X)/11X,88('-')/11X,11(F6.2,2X))
      DO 50L=1,NBAT
 50   YDATA(L)=ANS(L)
      IF(I.EQ.JBAT.OR.TPER.GE.1.0)GOTO 55
      GOTO 100
 55   CONTINUE
      IF(IPLOT.EQ.1.OR.IPLOT.EQ.5)CALL GRAPH(YDATA,11,0,1)
      IF(IPLOT.EQ.3.OR.IPLOT.EQ.5)CALL PGRAPH
      PRINT 60,TPER
 60   FORMAT('0','TOTAL PERCENT=',2PF10.6)
      IF(NCON.EQ.0)RETURN
      PRINT 70,NCON
 70   FORMAT('0THERE WERE ',I2,' CONCAVE CONDITIONS.')
      RETURN
 100  CONTINUE
 200  IERROR=IERROR+1
      IF(IERROR.GT.1)GOTO 220
      REMAIN=1.0
      PRINT 210,REMAIN
 210  FORMAT('0','NEGATIVE PERCENT DETECTED'/' REMAINDER RESET TO  ',
     &F4.2)
      IF(IPLOT.EQ.1.OR.IPLOT.EQ.5)CALL GRAPH(YDATA,11,0,1)

      DO 215IP=1,11
 215  YDATA(IP)=OYDATA(IP)
      GOTO 1
 220  PRINT 230
 230  FORMAT('0','SECOND NEGATIVE PERCENT DETECTED'/' DATA SET REJECTED'
     &)
      IF(IPLOT.EQ.1.OR.IPLOT.EQ.5)CALL GRAPH(YDATA,11,0,1)
      RETURN
      END
C
      SUBROUTINE CONFIX
C
C    THIS ROUTINE FINDS AND CORRECTS FOR CONCAVE CONDITIONS
C
      COMMON/A/OYDATA(11),YDATA(11),NBAT,ITYPE,IFLOW,IBAT,
     & XDIST(11,2),IPLOT,IAVER,NCON,ISIZE
      DIMENSION ADIST(11,2)
      DATA ADIST/49.72,94.65,187.46,336.62,527.03,799.57,1172.66,1737.63     0530
     &,2478.28,3599.01,5093.94,1.0,3.0,6.0,10.0,15.0,21.0,28.0,36.0,
     &45.0,55.0,55.0/
      IF(IBAT.LT.3)RETURN
      LBAT=IBAT-2
      DO 100I=1,LBAT
      J=IBAT-I+1
      B1=(ALOG10(YDATA(J-2))-ALOG10(YDATA(J-1)))/XDIST(J-1,ITYPE)
      B2=(ALOG10(YDATA(J-1))-ALOG10(YDATA(J  )))/XDIST(J  ,ITYPE)
C
C    TEST FOR CONCAVE CONDITIONS
C
      IF(B1/B2.GT.1.0001)GOTO 100
C
      NCON=NCON+1
      IF(J.EQ.3)GOTO 50
      YMID=0.0
      IF(I.EQ.1)GOTO 10
C
C    COMPUTING OF NEW ORDINATES FOR POINTS IN ERROR
C
      S1=-(ALOG10(YDATA(J))-ALOG10(YDATA(J+1)))/XDIST(J+1,ITYPE)
      C3=ALOG10(YDATA(J))-S1*ADIST(J,ITYPE)
      YMID=10.**( S1*ADIST(J-1,ITYPE)+C3)
 10   S2=-(ALOG10(YDATA(J-3))-ALOG10(YDATA(J-2)))/XDIST(J-2,ITYPE)
      S3=-(ALOG10(YDATA(J-2))-ALOG10(YDATA(J)))/(ADIST(J,ITYPE)-ADIST(
     &J-2,ITYPE))
      C1=ALOG10(YDATA(J-2))-S3*ADIST(J-2,ITYPE)
      C2=ALOG10(YDATA(J-2))-S2*ADIST(J-2,ITYPE)
      YLOW=10.**( S2*ADIST(J-1,ITYPE)+C2)
      YHIGH=10.**( S3*ADIST(J-1,ITYPE)+C1)
      IF(YMID.GT.YLOW)YLOW=YMID
      YNEW=(YHIGH-YLOW)/2.0+YLOW
```

```
      YDATA(J-1)=YNEW
      GOTO 100
 50   CONTINUE
C
C    THIS SECTION HANDLES RESETTING THE LAST THREE BATTERIES
C
      S=-(ALOG10(YDATA(J-2))-ALOG10(YDATA(J   )))/(ADIST(J,ITYPE)-
     &ADIST(J-2,ITYPE))
      B=ALOG10(YDATA(J   ))-S*ADIST(J   ,ITYPE)
      Y=10.**( S*ADIST(J-1,ITYPE)+B)
      YDATA(J-1)=0.95*Y
 100  CONTINUE
      RETURN
      END
C
      SUBROUTINE FIND(IS)

C
C    THIS ROUTINE FINDS THE SIZE BETWEEN TWO DATA POINTS
C
      COMMON/A/OYDATA(11),YDATA(11),NBAT,ITYPE,IFLOW,IBAT,
     & XDIST(11,2),IPLOT,IAVEP,NCON,ISIZE
      COMMON/B/THEOR(11,41,2,2),ISIZES(41),IGE1(11,2)
      DIMENSION AB(11)
      DATA AB/1.0,3*1.02,2*1.025,5*1.03/
      J=IBAT
C
C    ESLOPE=EXPERIMENTAL SLOPE
C    TSLOPE=THEORETICAL SLOPE
C    IGE1 IS AN ARRAY CONTAINING STARTING POINTS FOR THE SIZE SEARCH. IT IS USED
C    SIMPLY TO REDUCE SEARCH TIME.
C    IS POINTS TO A LOCATION IN AN ARRAY OF SIZES CALLED ISIZES.
C
      ESLOPE=(ALOG10(YDATA(J-1))-ALOG10(YDATA(J)))/XDIST(J,ITYPE)
      IBEGIN=IGE1(J,IFLOW)
      DO 1 I=IBEGIN,41
      TSLOPE=(ALOG10(THEOR(J-1,I,IFLOW,ITYPE))-ALOG10(THEOR(J,I,IFLOW,
     &ITYPE)))/XDIST(J,ITYPE)
      IS=I
      IF(ESLOPE/TSLOPE.GT.1.) GOTO7
 1    CONTINUE
 7    CONTINUE
C
C    THIS CHOOSES THE NEXT SMALLEST SIZE IF THE RATIO OF THE SLOPES IS TOO
C    LARGE.
C
      IF(ESLOPE/TSLOPE.GT.AB(J))IS=IS-1
C
      RETURN
      END
C
      SUBROUTINE PGRAPH
C
C    THIS ROUTINE PLOTS OUT A LOG-PROBABILITY PLOT OF THE RESULTS
C
      COMMON/C/PERCEN(11),SIZES(11)
      COMMON/D/HEADR(6)
      DIMENSION XTIT(5),YTIT(1)
      DATA XTIT/'PER ','CENT',' OVE','RSIZ','E   '/
      DATA YTIT/'SIZE'/
      CALL PLOT(0.0,-11.0,-3)
      CALL PLOT(0.0,1.0,-3)
      XLEN=10.0
      YLEN=8.0
      CALL CAXIS(0.0,0.0,YLEN,-1.0,1.0,3.0,1,YTIT,4,90.0,0)
      CALL PLAXIS(0.0,0.0,XLEN,XTIT,17)
      DO 50I=1,11
      PERCEN(I)=100.0*PERCEN(I)
      IF(PERCEN(I).GT.99.99.OR.PERCEN(I).LE.0.0)GOTO 50
      CALL XGEN(PERCEN(I),X,XLEN)
      IF(SIZES(I).LE.0.0)GOTO 50
      Y=ALOG10(SIZES(I))*(YLEN/3.0)
      CALL SYMBOL(X,Y,.07,11,0.0,-1)
 50   CONTINUE
      X1=XLEN-4.0
      Y1=YLEN-1.0
      CALL SYMBOL(X1,Y1,.14,HEADR,0.0,24)
      XNEW=XLEN+4.0
      CALL PLOT(XNEW,-1.0,-3)
      CALL PLOT(0.0,0.0,3)
```

```
      CALL PLOT(5.0,5.0,3)
      CALL PLOT(0.0,0.0,3)
      RETURN
      END
C
      SUBROUTINE GRAPH(Y,N,ICALL,ITYPE)
C
C   THIS ROUTINE PLOTS OUT THE PENETRATION CURVES BEFORE EACH SUBTRACTION
C
      COMMON/D/HEADR(6)
      DIMENSION Y(11),X(11),AB(11,2)        ,XTIT(2),YTIT(5)
      DATA AB /.028,.048,.098,.168,.264,.4,.591,.872,1.235,0.0,0.0,
     &.0182,.055,.1091,.182,.273,.382,.509,.655,.818,1.0,0.0/
      DATA XTIT/'LENG','TH  '/
      DATA YTIT/'PER ','CENT',' PEN','ETRA','TION'/
      IF(ICALL.GT.1)GOTO 15
      CALL PLOT(0.0,1.0,-3)
      XLEN=-10.0
      ALEN=-XLEN
      YLEN=8.0
      XNEW=-XLEN+4.0
      IF(ICALL.EQ.0)GOTO 100
      XH=-XLEN-4.0
      YH=YLEN-1.5
      CYCL=YLEN/2.0
      CALL CAXIS(0.0,0.0,XLEN,0.5,2.,2.,-9,XTIT,6,0.0,0)
      CALL CAXIS(0.0,0.0,YLEN,-5.,1.0,2., 0,YTIT,20,90.0,4)
      CALL SYMBOL(XH,YH,.14,HEADR,0.0,24)
      DO 10I=1,11
 10   X(I)=-XLEN*AB(I,ITYPE)
 15   CONTINUE
      DO 50J=1,N
      IF(Y(J).LE.0.0)RETURN
      YPT=ALOG10(Y(J))*CYCL
      IF(YPT.LT.0.0.OR.YPT.GT.YLEN)GOTO 50
      IF(X(J).GT.ALEN.OR.X(J).LE.0.0)GOTO 50
      CALL SYMBOL(X(J),YPT,.07,ICALL,0.0,-1)
 50   CONTINUE
      RETURN
 100  CALL PLOT(XNEW,-1.0,-3)
      RETURN
      END
C
      SUBROUTINE CAXIS(X,Y,XLEN,XDIV,ANIT,AINC,NN,ATITLE,NTITLE,THETA1,I
C
C   THIS ROUTINE PLOTS OUT EITHER A REGULAR OR LOG AXIS
C   FOR INSTUCTIONS ON THE USE OF THIS ROUTINE CONTACT DAVE CHRISTY
C          R
     &SKIP)
      DIMENSION ATITLE(1),FACT(10),ALEN(2,100),IPEN(100)
      DATA FACT/0.0,.30103,.47712,.60206,.69897,.77815,.8451,.90309,
     &.95424,1./
      CALL PLOT(0.0,0.0,-3)
      CALL PLOT(X,Y,-3)
      THETA =THETA1/57.29578
      H=.1
      Y1=.16
      IF(XDIV.LT.0.0)GOTO 50
      NPT=IFIX(ABS(XLEN)/XDIV)+1
      IF(XDIV.LT.1.0)H=.07
      N=0
      B=0.0
      A=ANIT
      DO 10IA=1,NPT
      ALEN(1,IA)=B
      ALEN(2,IA)=A
      IPEN(IA)=3
      IF(N.LT.(ISKIP+1))GOTO 5
      N=0
      IPEN(IA)=2
 5    N=N+1
      A=A+AINC
 10   B=B+XDIV
      GOTO 110
 50   NCYC=AINC
      NPT=NCYC*10
      ACYC=ABS(XLEN)/FLOAT(NCYC)
      IF(ACYC.LT.5.0)H=.07
      IS=5
```

```
      IF(ACYC.LT.2.0)IS=10
      N=1
      G=1.0
      DO 100IB=1,NCYC
      DO 90IC=1,10
      ALEN(1,N)=ACYC*FACT(IC)+(ACYC*FLOAT(IB-1))
      IPEN(N)=3
      IF(IC.NE.IS.AND.IC.NE.10)GOTO 90
      ALEN(2,N)=G*ANIT*FLOAT(IC)
      IPEN(N)=2
90    N=N+1
100   G=G*10.0
      ALEN(2,1)=ANIT
110   IPEN(1)=2
      IF(XLEN.LT.0.0)Y1=-.16-H
      IF(NN.EQ.-9)GOTO 160
      DO 150ID=1,NPT
      IF(IPEN(ID).EQ.3)GOTO 150
      X1=ALEN(1,ID)
      CALL ZMAKE(ALEN(2,ID),NN,Z,H)
      X1=X1-Z
      CALL SHIFT(X1,Y1,X2,Y2,THETA)
      CALL NUMBER(X2,Y2,H,ALEN(2,ID),THETA1,NN)
150   CONTINUE
160   N=NPT+1
      IZ=NPT/2
      Y3=.4
      IF(XLEN.LT.0.0)Y3=-.54
      S=(ABS(XLEN)/2.0)-(FLOAT(NTITLE)*.06)
      DO 200IE=1,NPT
      X1=ALEN(1,(N-IE))
      CALL SHIFT(X1,0.0,X2,Y2,THETA)
      IF(IE.EQ.1)CALL PLOT(X2,Y2,3)
      CALL PLOT(X2,Y2,2)
      CALL SHIFT(X1,.03,X5,Y5,THETA)
      CALL PLOT(X5,Y5,3)
      CALL SHIFT(X1,-.03,X6,Y6,THETA)
      CALL PLOT(X6,Y6,2)
      IF(IE.NE.IZ.OR.NTITLE.LT.0)GOTO 200
      CALL SHIFT(S,Y3,X7,Y7,THETA)
      CALL PLOT(X7,Y7,3)
      CALL SYMBOL(X7,Y7,.14,ATITLE,THETA1,NTITLE)
200   CALL PLOT(X2,Y2,3)
      CALL PLOT(-X,-Y,-3)
      RETURN
      END
      SUBROUTINE ZMAKE(A,NN,Z,H)
      V=(6./7.)*H
      Q=(FLOAT(NN))*V
      T=Q+V
      IF(NN.LT.0)T=0.0
      R=3.*V
      IF(A.LT.100.0)R=2.*V
      IF(A.LT.10.0)R=V
      IF(A.LT.0.0)Z=Z+V
      Z=R+T
      Z=Z/2.0
      RETURN
      END
      SUBROUTINE SHIFT(X1,Y1,X2,Y2,THETA)
      H=SQRT(X1**2+Y1**2)
      IF(H.EQ.0.0)GOTO 5
      THETA1=ARSIN(Y1/H)
      GOTO 10
5     THETA1=0.0
10    CONTINUE
      IF(X1.GT.0.0)GOTO 100
      IF(Y1.GT.0.0)GOTO 50
      THETA5=THETA1-((3.0*3.1415926535)/2.0)
      THETA1=THETA1-2.0*THETA5
      GOTO 100
50    CONTINUE
      THETA1=THETA1+(3.1415926535-2.0*THETA1)
100   THETA2=THETA+THETA1
      Y2=H*SIN(THETA2)
      X2=H*COS(THETA2)
      RETURN
      END
C
```

```
      SUBROUTINE PLAXIS(XO,YO,XLEN,XTIT,NTIT)
C
C    THIS ROUTINE PLOTS OUT A PROBABILITY AXIS
C    FOR INSTRUCTIONS ON ITS USE CONTACT DAVE CHRISTY
C
      DIMENSION XTIT(1),PROB(21)
      DATA PROB/.01,.1,.5,1.,2.,5.,10.,20.,30.,40.,50.,60.,70.,80.,90.,
     &95.,98.,99.,99.8,99.9,99.99/
      CALL PLOT(XO,YO,-3)
      XMID=XLEN/2.0
      N=1
      NN=2
      DO 100I=1,21
      CALL XGEN(PROB(I),X,XLEN)
      CALL PLOT(X,0.0,2)
      CALL PLOT(X,.03,3)
      CALL PLOT(X,-.03,2)
      IF(I.GT.1)NN=1
      IF(I.GT.3)NN=-1
      IF(I.GT.18)NN=1
      IF(I.GT.20)NN=2
      CALL ZMAKE(PROB(I),NN,Z,.07)
      X1=X-Z
      CALL NUMBER(X1,-.2,.07,PROB(I),0.0,NN)
      CALL PLOT(X,0.0,3)
      IF(X.LT.XMID.OR.N.EQ.0)GOTO 100
      N=0
      S=XMID-(FLOAT(NTIT)*.06)
      CALL SYMBOL(S,-.44,.14,XTIT,0.0,NTIT)
      CALL PLOT(X,0.0,3)
  100 CONTINUE
      CALL PLOT(-XO,-YO,-3)
      RETURN
      END
      SUBROUTINE XGEN(P,X,XLEN)
      A=-1.0
      PR=P/100.0
      IF(PR.LT.0.5)GOTO 10
      PR=1.0-PR
      A=1.0
   10 T=SQRT(ALOG(1./(PR**2)))
      TERM1=2.515517+.802853*T+.010328*(T**2)
      TERM2=1.0+1.432788*T+.189269*(T**2)+.001308*(T**3)
      X1=T-TERM1/TERM2
      X=(((A*X1)+3.7191243)/3.7191243)*(XLEN/2.0)
      RETURN
      END
C
      BLOCK DATA
      COMMON/A/OYDATA(11),YDATA(11),NBAT,ITYPE,IFLOW,IBAT,
     & XDIST(11,2),IPLOT,IAVER,NCON,ISIZE
      COMMON/B/THEOR(11,41,2,2),ISIZES(41),IGE1(11,2)
      DIMENSION A(451),B(451),C(451),D(451),DUMMY(1804)
      EQUIVALENCE (A(1),DUMMY(1),THEOR(1,1,1,1))
      EQUIVALENCE(B(1),DUMMY(452)),(C(1),DUMMY(903)),(D(1),DUMMY(1354))
      DATA YDATA/11*0.0/
C
C    XDIST CONTAINS THE EQUIVALENT LENGTHS OF THE BATTERIES
C
      DATA XDIST/49.72,44.93,92.81,149.16,190.41,272.54,373.09,564.97,
     G 740.65,1120.73,1494.93,1.0,2.,3.,4.,5.,6.,7.,8.,9.,10.,10./
C
C    IGE1 CONTAINS STARTING VALUES FOR THE SIZE SEARCH
C
      DATA IGE1/1,1,4,5,6,8,10,12,15,19,25,1,1,3,4,5,7,8,10,12,16,20/
C
C    ISIZES CONTAINS THE SIZES USED IN THE PROGRAM
C
      DATA ISIZES/3,4,6,8,10,12,14,16,18, 20,22,24,26,28,30,32,34,36,38,
     X40,42,44,46,48,50,53,55,60,65,70,75,80,85,90,95,100,110,125,140,16
     Z5,200/
C
C    THIS IS THE ARRAY WHICH CONTAINS THE THEORETICAL PENETRATIONS
C
      DATA A/.4757,.0037,9*0.,4.492,.2668,.0009,8*0.,22.39,5.683,.4135,
     A.0069,7*0.,39.45,16.72,3.4858,.316,.0156,.0002,5*0.,51.65,27.997,
     B9.519,1.89,.2548,.0157,5*0.,60.25,37.62,16.75,5.11,1.187,.1599,
     C.011,4*0.,66.499,45.479,23.937,9.481,3.063,.6618,.087,4*0.,
     D71.20,51.87,30.56,14.37,5.768,1.695,.341,.027,3*0.,
     E74.84,57.11,36.48,19.367,9.036,3.283,.8819,.11,3*0. ,
```

```
F77.734,61.44,41.71,24.207,12.612,5.342,1.766,.303,3*0.,
G80.079,65.076,46.3,28.77,16.304,7.751,2.993,.653,.098,2*0.,
H82.017,68.152,50.341,33.019,19.98,10.38,4.518,1.187,.2268,2*0.,
J83.64,70.78,53.9,36.93,23.56,13.15,6.286,1.914,.4408,.048,.0025,
K85.02,73.05,57.06,40.52,26.99,15.97,9.234,2.825,.756,.103,0.,
L86.199,75.03,59.85,43.8,30.26,18.78,10.31,3.904,1.18,.195,.0177,
M87.22,76.75,62.35,46.8,33.33,21.54,12.46,5.127,1.725,.333,.037,
N83.109,78.27,64.58,49.55,36.22,24.23,14.64,6.468,2.376,.52,.069,
O88.88,79.61,66.59,52.06,38.93,26.82,16.84,7.90,3.131,.774,.12,
&89.57,80.8,68.39,54.35,41.46,29.31,19.02,9.41,3.98,1.09,.19,
B90.18,81.867,70.03,56.48,43.83,31.69,21.167,10.96,4.91,1.46,.29,
C90.73,82.82,71.51,58.43,46.04,33.96,23.26,12.55,5.91,1.898,.42,
D91.21,83.678,72.85,60.22,48.11,36.13,25.31,14.16,6.97,2.39,.576,
E91.65,84.45,74.08,61.877,50.046,38.18,27.29,15.77,8.08,2.95,.77,
F92.05,85.16,75.21,63.41,51.86,40.13,29.21,17.375,9.23,3.555,.99,
G92.41,85.81,76.24,64.93,53.56,41.98,31.05,18.97,10.41,4.21,1.257,
H92.89,86.68,77.55,66.78,55.91,44.59,33.7,21.32,12.21,5.268,1.72,
J93.18,87.21,78.49,67.97,57.36,46.22,35.38,22.85,13.43,6.02,2.065,
K93.81,88.36,80.37,70.61,60.63,49.95,39.31,26.55,16.48,8.028,3.075,
L94.35,89.34,81.95,72.87,63.47,53.24,42.85,30.04,19.50,10.16,4.263,
M94.81,90.19,83.31,74.82,65.94,56.16,46.06,33.29,22.44,12.38,5.596,
N95.2,90.93,84.499,76.52,68.11,58.76,48.97,36.32,25.27,14.62,7.043,
O95.55,91.58,85.54,78.02,70.04,61.09,51.6,39.13,27.97,16.85,8.574,
P95.86,92.15,86.47,79.35,71.75,63.18,53.99,41.73,30.53,19.06,10.16,
Q96.13,92.67,87.298,80.54,73.29,65.06,56.18,44.14,32.96,21.2,11.78,
R96.38,93.14,88.04,81.6,74.678,66.77,58.18,46.38,35.26,23.31,13.43,
S96.6,93.559,88.72,82.57,75.93,68.33,60.,48.45,37.42,25.337,15.063,
&         96.99,94.29,89.89,84.26,78.1,71.05,63.2,52.2,41.37,29.16,
A18.28619,
B97.45,95.15,91.32,86.3,80.79,74.38,67.22,56.87,46.49,34.31,22.88,
C97.8,95.82,92.43,87.92,82.92,77.05,70.44,60.73,50.8,38.81,27.099,
D98.24,96.65,93.83,89.99,85.66,80.49,74.61,65.83,56.61,45.08,33.28,
F98.64,97.41,95.16,92.,88.37,83.95,78.84,71.12,62.77,51.98,40.43/
 DATA B/2.646,.0974,.00013,8*0.,
A11.8226,1.684,.037,8*0.,
B34.548,12.982,2.117,.1296,.0038,6*0. ,
C50.728,27.046,8.905,1.683,.212,.0119,5*0. ,
D61.25,38.83,17.788,5.677,1.395,.204,.016,4*0.,
E68.41,48.029,26.499,11.281,3.99,.98,.155,.0086,3*0.,
F73.52,55.18,34.26,17.42,7.71,2.6,.63,.067,3*0.,
G77.33,60.82,40.95,23.47,12.05,4.998,1.61,.26,.0277,2*0.,
H80.26,65.35,46.66,29.14,16.61,7.96,3.11,.69,.106,2*0.,
J82.57,69.04,51.54,34.32,21.15,11.27,5.07,1.4,.28,.025,0.,
K84.445,72.11,55.73,38.99,25.52,14.74,7.37,2.41,.61,2*0. ,
L85.98,74.67,59.34,43.19,29.64,18.24,9.90,3.68,1.09,.17,.015,
M87.27,76.84,62.47,46.95,33.49,21.68,12.57,5.19,1.76,.34,.038,
N88.35,78.69,65.21,50.32,37.05,25.01,15.3,6.89,2.59,.59,.08,
&         89.27,80.28,67.6,53.35,40.34,28.20,18.04,8.72,3.59,.94,.15,
A90.07,81.66,69.71,56.07,43.37,31.22,20.74,10.65,4.72,1.38,.268,
B90.75,82.87,71.53,58.52,46.15,34.08,23.37,12.63,5.96,1.92,.42,
C91.35,83.92,73.24,60.74,48.71,36.76,25.92,14.65,7.31,2.56,.63,
D91.88,84.86,74.725,62.75,51.08,39.29,28.37,16.67,8.72,3.28,.89,
E92.34,85.69,76.06,64.58,53.26,41.66,30.72,18.68,10.19,4.086,1.21,
F92.76,86.45,77.27,66.25,55.27,43.88,32.97,20.66,11.7,4.96,1.58,
G93.14,87.13,78.35,67.78,57.14,45.97,35.12,22.61,13.23,5.898,2.0,
H93.48,87.74,79.36,69.19,58.87,47.93,37.17,24.52,14.78,6.887,2.48,
J93.78,88.31,80.28,70.48,60.47,49.77,39.11,26.37,16.33,7.92,3.02,
K94.069,88.83,81.12,71.68,61.97,51.49,40.97,28.17,17.87,8.99,3.6,
L94.45,89.54,82.26,73.31,64.03,53.91,43.58,30.76,20.15,10.64,4.54,
M94.69,89.96,82.96,74.31,65.296,55.4,45.2,32.42,21.65,11.77,5.22,
N95.2,90.93,84.5,76.52,68.12,58.77,48.98,36.33,25.27,14.62,7.05,
&         95.6,91.7,85.8,76.4,70.55,61.7,52.3,39.9,28.7,17.48,9.02,
A96.02,92.45,86.95,80.04,72.65,64.27,55.27,43.13,31.9,20.3,11.1,
B96.3,93.1,87.9,81.5,74.5,66.55,57.9,46.1,34.95,23.0,13.2,
C96.6,93.6,88.8,82.7,76.1,68.56,60.3,48.77,37.75,25.66,15.3,
D96.89,94.1,89.6,83.83,77.6,70.36,62.41,51.22,40.36,28.16,17.43,
E97.1,94.5,90.3,84.8,78.9,71.98,64.3,53.5,42.77,30.5,19.5,
F97.3,94.9,90.9,85.7,80.04,73.4,66.1,55.56,45.02,32.8,21.5,
G97.5,95.25,91.5,86.5,81.1,74.77,67.68,57.42,47.1,34.9,23.5,
H97.8,95.8,92.45,87.95,82.95,77.1,70.5,60.8,50.86,38.87,27.16,
J98.16,96.5,93.6,89.6,85.2,79.9,73.9,64.97,55.6,44.,32.189,
K98.43,97.0,94.48,90.97,86.98,82.17,76.67,68.39,59.6,48.4,36.64,
L98.76,97.64,95.57,92.64,89.25,85.08,80.23,72.9,64.85,54.38,42.59,
M99.05,98.2,96.58,94.2,91.45,87.96,83.81,77.5,70.4,60.89,50.2/
 DATA C/16.5,.45,9*0.,27.32,2.04,.04,8*0.,44.19,8.63,.74,.03,7*0.,
&55.53,17.13,2.93,.28,.02,6*0.,63.39,25.47,6.49,1.05,.11,.01,5*0.,
& 69.05,32.93,10.84,2.47,.39,.04,5*0.0,
& 73.3,39.39,15.52,4.48,.95,.15,.02,4*0.,
& 76.59,44.93,20.19,6.95,1.83,.37,.06,.01,3*0.,
& 79.21,49.69,24.69,9.72,3.03,.75,.15,.02,3*0.,
```

```
& 81.33,53.79,28.93,12.66,4.5,1.3,.31,.06,.01,2*0.,
& 83.08,57.34,32.88,15.66,6.2,2.04,.56,.13,.02,2*0.,
& 84.55,60.43,36.52,18.66,8.06,2.94,.91,.24,.05,.01,0.,
& 85.79,63.15,39.88,21.61,10.04,4.01,1.37,.4,.1,.02,0.,
& 86.87,65.55,42.97,24.47,12.1,5.2,1.94,.63,.18,.04,0.,
& 87.8,67.68,45.81,27.22,14.2,6.5,2.62,.92,.29,.08,0.,
& 88.61,69.58,48.42,29.85,16.31,7.9,3.39,1.29,.43,.13,0.,
& 89.33,71.29,50.82,32.37,18.41,9.36,4.25,1.72,.62,.2,0.,
& 89.97,72.83,53.04,34.76,20.49,10.87,5.19,2.23,.86,.3,0.,
& 90.54,74.22,55.09,37.02,25.53,12.41,6.19,2.8,1.14,.42,0.,
& 91.06,75.49,56.99,39.1[illegible],24.52,13.98,7.25,3.43,1.47,.58,0.,
& 91.52,76.65,58.75,41.22,26.46,15.55,8.36,4.11,1.85,.76,0.,
&91.94,77.71,60.39,43.15,28.34,17.12,9.5,4.85,2.28,.98,0.,
&92.32,78.68,61.92,44.98,30.17,18.68,10.68,5.64,2.75,1.24,0.,
&92.67,79.59,63.34,46.72,31.93,20.23,11.87,6.46,3.26,1.52,0.C,
&92.99,80.42,64.67,48.36,33.64,21.75,13.08,7.32,3.81,1.84,0.C,
&93.43,81.55,66.51,50.68,36.08,24.,14.91,8.66,4.7,2.38,0.,
&93.69,82.25,67.64,52.12,37.63,25.46,16.13,9.58,5.33,2.78,0.0,
&94.27,83.78,70.19,55.44,41.28,28.97,19.17,11.96,7.03,3.9,0.,
&94.76,85.08,72.39,58.37,44.59,32.28,22.14,14.39,8.87,5.18,0.,
&95.17,86.21,74.31,60.97,47.61,35.38,25.02,16.84,10.79,6.58,0.,
&95.53,87.18,76.,63.3,50.36,38.28,27.79,19.28,12.77,8.09,0.,
&95.84,88.03,77.5,65.39,52.88,40.98,30.44,21.67,14.78,9.67,0.,
&96.11,88.79,78.83,67.27,55.18,43.7,32.96,24.,16.8,11.3,0.,
&96.35,89.46,80.02,69.,57.29,45.84,35.35,26.26,18.8,13.,0.,
&96.57,90.06,81.1,70.53,59.23,48.03,37.62,28.45,20.[illegible]55,0.,
&96.76,90.59,82.07,71.94,61.02,50.08,39.77,30.56,22[illegible].35,0.,
&97.09,91.52,83.76,74.43,64.21,53.78,43.74,34.53,26.[illegible],20.,0.,
&97.48,92.63,85.8,77.47,68.19,58.5,48.93,39.89,31.7,24.56,0.,
&97.78,93.49,87.41,79.91,71.43,62.44,53.36,44.59,36.45,29.12 ,0.,
&98.16,94.57,89.44,83.03,75.66,67.67,59.41,51.2,43.31,35.96,0.,
&98.52,95.62,91.43,86.13,79.94,73.09,65.83,58.42,51.08,43.99,0./
 DATA D/27.23,2.02,.04,8*0.,38.94,5.91,.35,.01,7*0.,
&54.92,16.57,2.74,.25,.01,6*0.,
&64.76,27.16,7.38,1.3,.15,.01,5*0.,
&71.28,36.22,13.12,3.39,.62,.08,.01,4*0.,
&75.87,43.68,19.08,6.32,1.59,.3,.04,.01,3*0.,
&79.26,49.8,24.8,9.79,3.06,.76,.15,.02,3*0.,
&81.86,54.86,30.09,13.51,4.97,1.5,.37,.07,.01,2*0.,
&83.91,59.07,34.9,17.3,7.19,2.51,.74,.18,.04,.01,0.,
&85.56,62.64,39.23,21.02,9.64,3.78,1.27,.36,.09,.02,0.,
&86.92,65.67,43.12,24.61,12.21,5.27,1.97,.64,.18,.05,0.,
&88.06,68.28,46.62,28.03,14.84,6.92,2.84,1.03,.33,.09,0.,
&89.02,70.54,49.77,31.25,17.47,8.69,3.85,1.52,.53,.17,0.,
&89.85,72.53,52.6,34.28,20.07,10.56,4.99,2.12,.81,.28,0.,
&90.56,74.28,55.17,37.12,22.61,12.48,6.23,2.82,1.16,.43,0.,
&91.19,75.83,57.51,39.77,25.08,14.42,7.56,3.62,1.58,.63,0.,
&91.74,77.22,59.63,42.25,16.37,8.96,4.5,2.07,.88,0.,
&92.23,78.47,61.57,44.56,29.74,18.31,10.4,5.45,2.63,1.17,C.,
&92.67,79.59,63.35,46.72,31.94,20.23,11.88,6.46,3.26,1.52,0.,
&93.07,80.61,64.98,48.75,34.04,22.12,13.37,7.53,3.94,1.92,0.,
&93.42,81.54,66.48,50.64,36.04,23.96,14.88,8.64,4.68,2.37,0.,
&93.75,82.39,67.87,52.42,37.95,25.76,16.39,9.78,5.47,2.87,0.,
&94.04,83.16,69.16,54.09,39.78,27.51,17.89,10.94,6.29,3.4,0.,
&94.31,83.88,70.36,55.65,41.52,29.21,19.38,12.13,7.16,3.98,0.,
&94.56,84.54,71.47,57.13,43.18,30.86,20.85,13.33,8.05,4.6,0.,
&94.89,85.44,73.,59.18,45.53,33.24,23.02,15.13,9.44,5.59,0.,
&95.09,85.99,73.94,60.46,47.01,34.75,24.43,16.34,10.39,6.28,0.,
&95.54,87.2,76.04,63.34,50.41,38.33,27.85,19.32,12.81,8.12,0.,
&95.91,88.23,77.84,65.87,53.46,41.61,31.07,22.24,15.28,10.06,0.,
&96.23,89.11,79.41,68.09,56.19,44.62,34.09,25.07,17.74,12.08,C.,
&96.5,89.88,80.78,70.06,58.64,47.37,36.93,27.78,20.17,14.13,0.,
&96.74,90.55,81.99,71.82,60.86,49.9,39.58,30.37,22.55,16.19,C.,
&96.94,91.14,83.06,73.39,62.88,52.23,42.06,32.84,24.86,18.24,0.,
&97.14,91.66,84.02,74.81,64.71,54.37,44.38,35.18,27.1,20.27,0.,
&97.31,92.13,84.88,76.1,66.38,56.35,46.54,37.41,29.25,22.26,0.,
&97.45,92.55,85.66,77.27,67.92,58.18,48.57,39.51,31.33,24.21,0.,
&97.71,93.28,87.01,79.31,70.62,61.45,52.55,43.4,35.23,27.94,C.,
&98.01,94.15,88.64,81.79,73.97,65.57,56.97,48.5,40.48,33.11,C.,
&92.25,94.83,88.92,83.77,76.68,68.95,60.91,52.87,45.08,37.76,0.,
&98.54,95.68,91.54,86.3,80.17,73.39,66.2,58.83,51.53,44.47,0.,
&98.82,96.5,93.12,88.8,83.67,77.92,71.7,65.2,58.58,52.02,0./
 END                                                                4490
```

Sample Data Input for STRIP

```
    TITLE                ICODE                       PENETRATION DATA
        1            2           3          4          5          6          7          8
12345678901234567890123456789012345678901234567890123456789012345678901234567890  Card Column Indicator

NYU TB 4LPM 76082416 2   176.1360.5443.5931.1621.8615.3010.73 6.81 4.41 2.56 1.43 } tubular battery @ 4 lpm.
NYU TB 4LPM 76082406 2   169.0851.2837.2825.4918.3813.3810.04 7.02 5.13 3.46 2.42 } output on printer only.
PLOT
NYU TB 4LPM 76080615 2   182.7070.1956.0244.2134.6127.3521.5516.0012.10 8.39 5.76   as above with cumulative plot.
BOTH                                                                                  next 3 with both plots.
AG AVE.#3 TB 5/5/76      281.2151.5430.3016.3408.7304.7102.5301.3800.7900.3400.00   tubular battery @ 6 lpm.
NYU TB 4LPM 76081916 2   175.3042.8133.3325.6219.0213.48 9.47 6.01 3.94 2.05 1.09     "       "     @ 4 lpm.
RM 2/18/76 SB - 4 LPM    381.6056.5036.7023.0014.5010.0007.4005.5004.0002.90         screen battery  @ 4 lpm.
NPLT                                                                                  next 2 with no plots.
NYU TB 4LPM 76082610 2   180.6066.4153.1641.5332.8225.8220.1214.9110.91 7.44 5.02   tubular battery @ 4 lpm.
UNIMODAL 10NM  TB 4LPM   152.0028.0009.5001.8900.2500.02                              "       "     @ 4 lpm.
PPLT                                                                                  next 3 with probability plot.
UNIMODAL 80NM  TB 4LPM   196.0092.0086.0078.0070.0061.0052.0039.0028.0016.9008.60   tubular battery @ 4 lpm.
BIMODAL 10-80NM  TB 4LPM 174.0060.0048.0040.0035.0030.5026.0019.5014.0008.4504.30     "       "     @ 4 lpm.
AVER                     2                                                          } average of two tubular
AG AVE.#1 TB 5/5/76      168.4350.9532.2519.5611.6106.6403.7301.7600.8100.3700.00   } batteries both at 4 lpm.
AG AVE.#2 TB 5/5/76      180.8851.5628.8415.4507.8503.9702.0601.0400.5600.0000.00   } (only one plot for average)
NPLT                                                                                  no plots for rest of data.
AVER                     3                                                          }
NYU TB 4LPM 76080615 2   182.7070.1956.0244.2134.6127.3521.5516.0012.10 8.39 5.76   } Average of three tubular
NYU TB 4LPM 76082416 2   176.1360.5443.5931.1621.8615.3010.73 6.81 4.41 2.56 1.43   } batteries all at 4 lpm.
NYU TB 4LPM 76082406 2   169.0851.2837.2825.4918.3813.3810.04 7.02 5.13 3.46 2.42   }
```

PART V

Other Methods for Aerosol Measurement

REMOTE SENSING OF ATMOSPHERIC AEROSOLS

Alex E. S. Green

Department of Physics and Astronomy, University of Florida, Gainesville

Remote sensing of atmospheric aerosols can be carried out at various levels of sophistication. The present report describes several techniques which have been used in the University of Florida's atmospheric radiation group and the information which can be gathered by each technique.

SUN RADIOMETERS

The simplest form of remote aerosol sensor is the Volz single wavelength sun photometer (Volz, 1954) (see Figure 1a). This is a very simple photometer which effectively measures the solar irradiance at the ground at 500 nm. The instrumental calibration constant essentially is the extraterrestrial solar irradiance. Then by discounting the attenuation due to Rayleigh scattering by air molecules along the slant path one can establish the so-called turbidity which is related to the vertical aerosol thickness of the atmosphere. This is the simplest overall characterization of the atmospheric aerosol burden.

A two channel sun photometer (Figure 1b) measures the direct solar irradiance at 500 nm and 380 nm (Flowers, 1969). There are many National Oceanic and Atmospheric Administration (NOAA) stations in the country routinely using this instrument. The advantage of two wavelengths is that in addition to giving the optical depth of the atmosphere, it gives some information about the particle size distribution. In particular, if one assumes

$$\tau = \tau_0 (\lambda_0/\lambda)^\alpha, \quad (1)$$

where τ_0 is the optical depth at $\lambda_0 = 500$ nm, then the second measurement yields

Work supported in part by the National Science Foundation, Atmospheric Science Section.

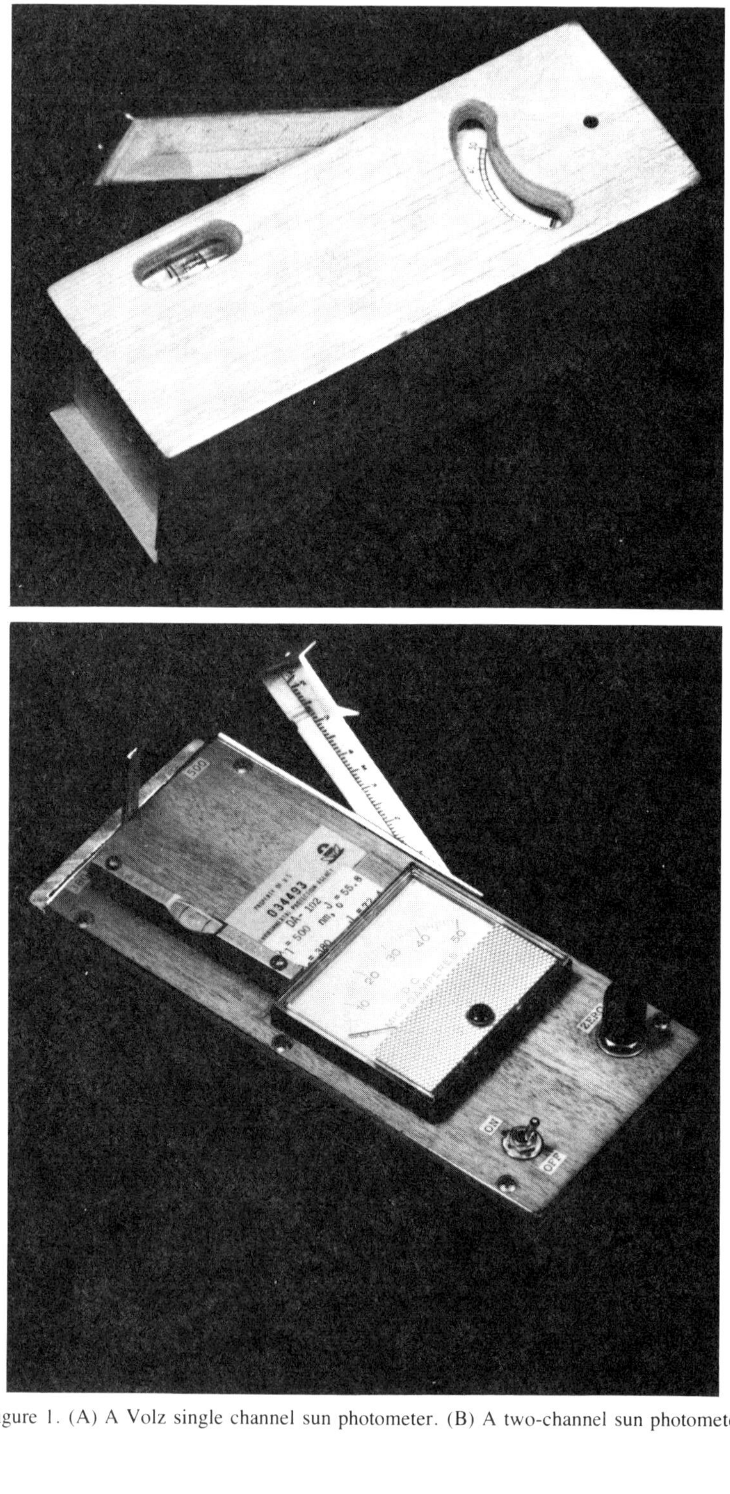

Figure 1. (A) A Volz single channel sun photometer. (B) A two-channel sun photometer.

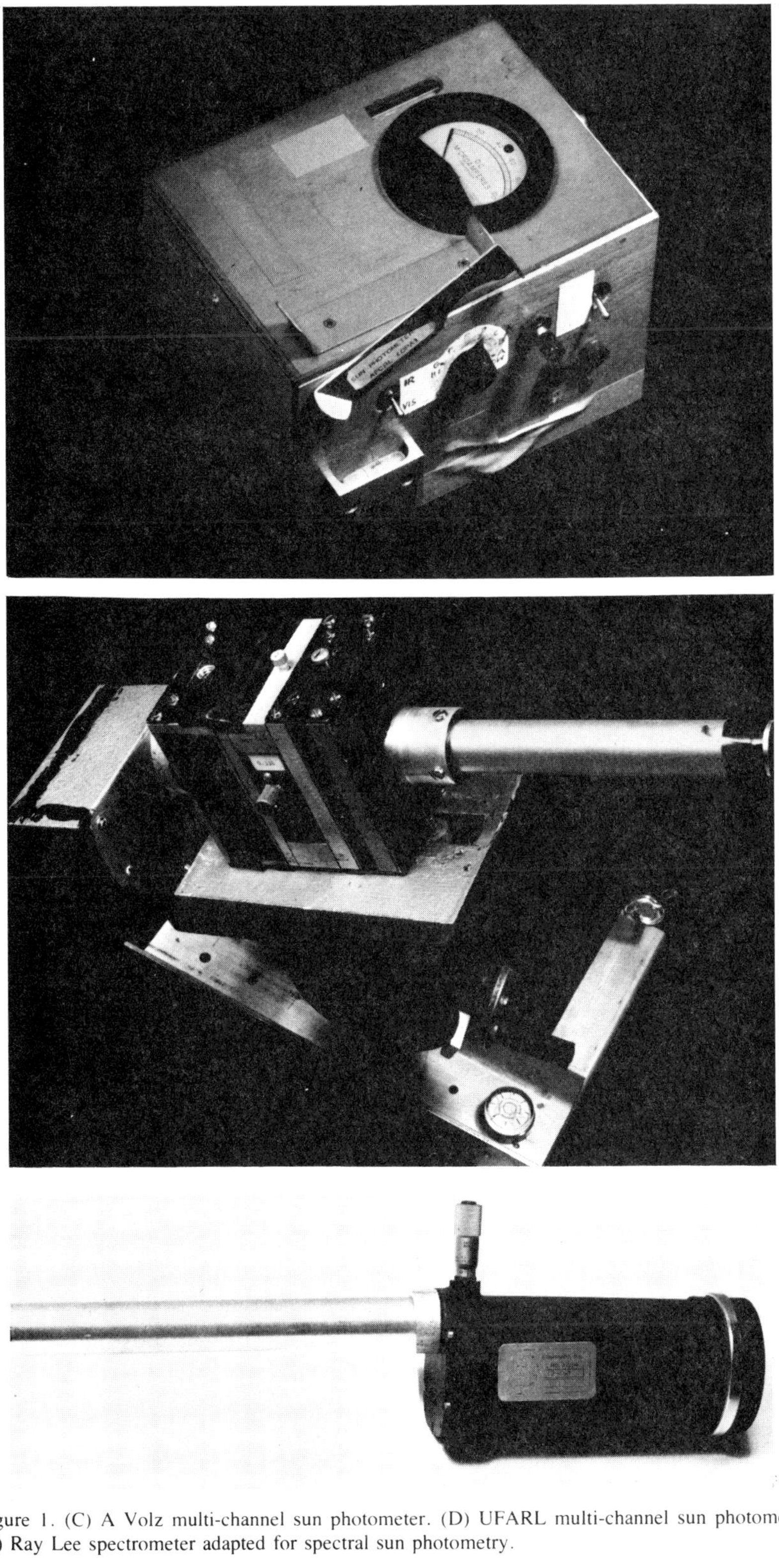

Figure 1. (C) A Volz multi-channel sun photometer. (D) UFARL multi-channel sun photometer. (E) Ray Lee spectrometer adapted for spectral sun photometry.

the power α. This coefficient is quite variable depending upon the size distribution of the atmospheric aerosols. A nominal value usually assumed is $\alpha = 1.3$.

We also have a multi-channeled instrument built for us by Fred Volz (1973) (Figure 1c) which gives the solar irradiance at 6 wavelengths (in nm): 313, 340, 500, 640, 880 and 940 wavelength. We really haven't used it to its full possibilities. Besides giving additional information about the aerosol size distribution, the two longest wave channels give data on the water vapor content.

Finally, we have developed a multi-channel sun photometer (Figure 1d) into which we insert various interference filters. This is used more for research purposes than for monitoring.

Very recently, Ray Lee Instruments, Inc. loaned us a small, compact spectrometer with a 150 mm grating. We have converted it into a spectral sun photometer using a simple photo-resistance detector (Figure 1e). With this detector we can dial any wavelength from about 350 nm to 600 nm. This range could be extended with a better detector.

AUREOLE STUDIES

Measurements of the intensity distribution in the solar aureole, the region of the sky immediately surrounding the sun, provide even more detailed information on atmospheric aerosols (Green et al., 1971–72; Ward et al., 1973; McPeter and Green, 1976). Figure 2 illustrates a remote sensing technique in which we take advantage of the unique properties of forward scattering. In this zone the relative intensity is very sensitive to particle size but is insensitive to irregularities of aerosol particles and the dielectric constant. In this technique, we photograph the sun through a neutral density 4 filter. This attenuates the sun's intensity to about the same level as the sky intensity surrounding it. Then we make comparative densitometry measurements of the ND4 attenuated sun's disc vs. the aureole immediately surrounding it. Using a single scattering model, we may readily invert the aureole intensity distribution to arrive at some reasonable characterization of particle size distribution. We do this at several wavelengths with the aid of filters. More recently we've been making aureole measurements in the middle ultraviolet where one must worry about multiple scattering corrections (Furman et al., 1976).

Our aureole studies take advantage of the information content of skylight in the immediate neighborhood of the sun. If no aerosols were present in the atmosphere the relative intensity as a function of angle would be approximately as shown by curve a in Figure 3. This curve is very flat and as you get very close to the sun's disc it would rise by about 6 orders of magnitude. On the other hand, on a day of moderate clarity the intensity close to the sun would vary as shown by curve b. On a smoggy or hazy day, there would only be a slight variation as one approaches the sun's disc (see curve c). Thus, a dynamic range of 6 orders magnitude of intensity variation is available potentially to extract information as to atmospheric aerosol content and the aerosol size distribution.

Figure 4 gives a densitometric trace of a vertical and a horizontal cut showing how the densities rise (Green et al., 1972) as we approach the sun. Where the filters obscure the sky the trace drops precipitously. At the disc itself the trace comes back to the radiance level of the sun's disc reduced precisely by 4 orders of magnitude. This serves as a standard density point for calibrating the sky radiance.

The analysis by single scattering using analytic models of atmospheric constituents is moderately straightforward. Taking the sun intensity as known we allow for the attenuation along the path to a scattering element; then the Rayleigh

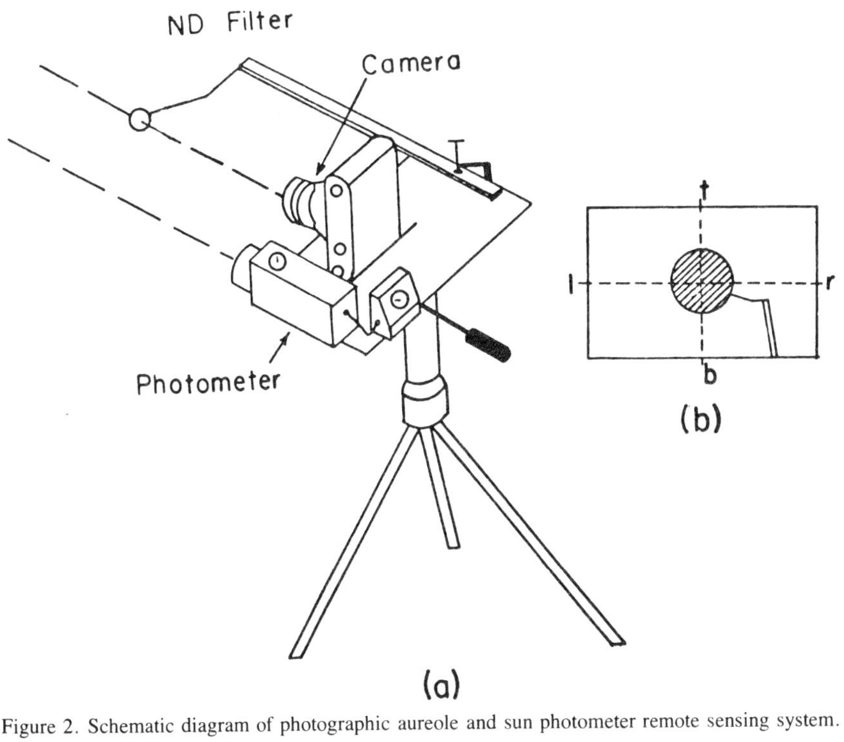

Figure 2. Schematic diagram of photographic aureole and sun photometer remote sensing system.

and aerosol scattering from this element into solid look angle of the instrument and then allow for the attenuation to the detector. Assuming a separable model of the aerosol size and altitude distributions we can relate the intensity at various angles to the parameters in size distribution functions (Green et al., 1971).

Aerosol size distribution inverters go at the problem somewhat differently. The style of inversion technique which we have used so far has been to assume a model of the aerosol size-altitude distribution chosen in the light of all the available prior information. Then we simply seek the values of the parameters which force the results of this model into closest conformity with observation. How well you do in this type of inversion approach depends upon how realistic your model is. If one chooses the model to take advantage of the experience of the point samplers, one starts at quite an advantage. Figure 5 is an example of the type of altitude distribution function which we have used to characterize atmospheric aerosols. These distributions will fit quite well the major features of any of the library of altitude distributions of Elterman (Elterman, 1964). Our analytic approach contrasts with the ab initio inversion approach used by some other groups who assume no prior information. I think our approach will prove quite useful for some time since it is much less sensitive to noisy data. Perhaps at a later time we will reach a level of measurement sophistication when ab initio inversion techniques might be as practical.

The type of size distribution we have used is shown in Figure 6. This is a cumulative size distribution which is a generalization of the Junge power law but

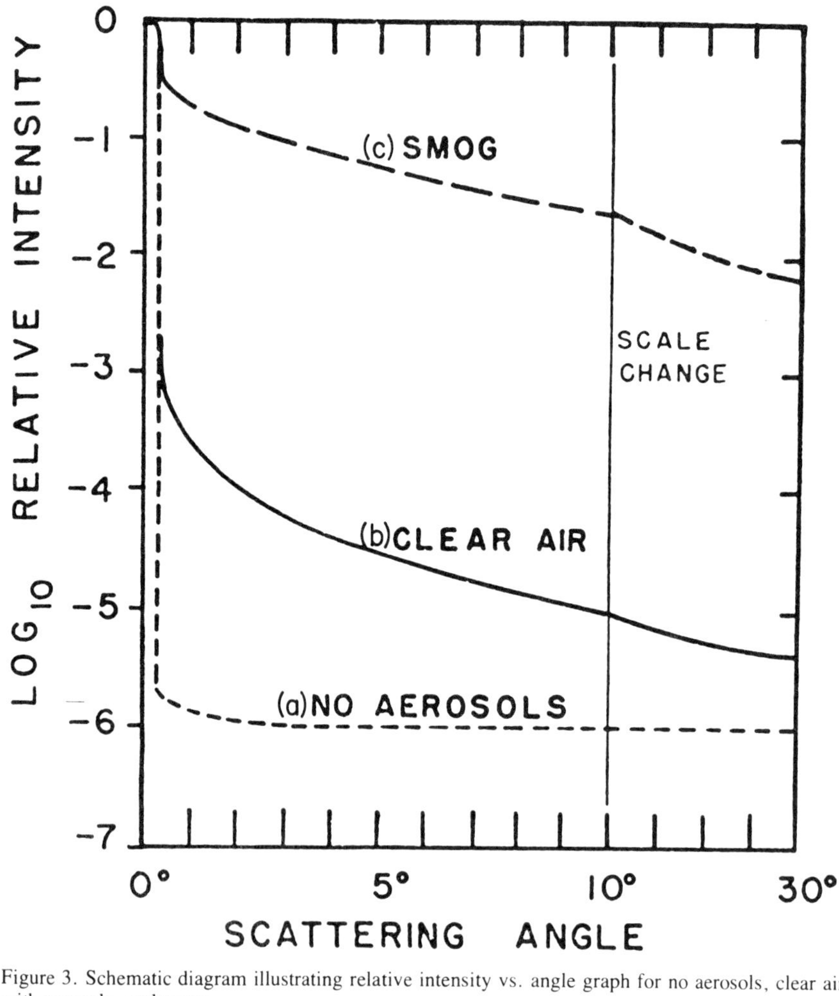

Figure 3. Schematic diagram illustrating relative intensity vs. angle graph for no aerosols, clear air with aerosols, and smog.

is much better behaved at small sizes. The derivative of our oversize or cumulative distribution has a peak at small sizes dictated by the parameter a. At large sizes it has the property of falling off as the power $\nu+1$, just as the Junge power law.

We have incorporated distributions of this sort in our Mie scattering calculations. Figures 7a and 7b show the phase functions for $\nu = 4$ and 6 for small angles for various scale parameters. Figure 8 shows the phase function for a = 0.1 and a = 1.0 for various ν. In effect, the family of angular distributions functions maps the family of size distribution functions. Thus an experimental determination of the single scattering phase function can be used in conjunction with our library of phase function to arrive at a relatively sharp determination of the parameters ν and a.

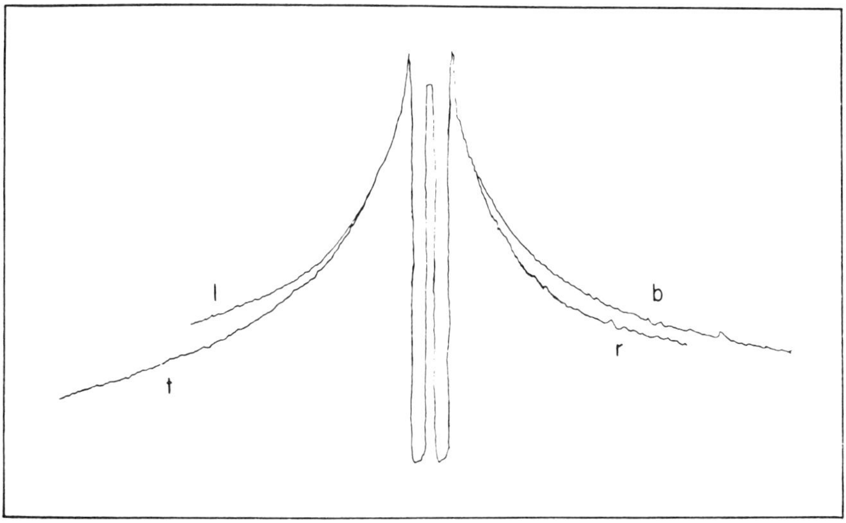

Figure 4. Densitometric trace of aureole photograph with attenuating disc for horizontal scan r (right) l (left) for vertical scans t (top) b (bottom).

Figure 9 is an example showing measurements of an angular distribution at 640 nm and shows how sharply the ν is determined. Generally, at 640 nm, multiple scattering which could confuse our single scattering analysis is relatively small.

We have been worried for a long time about multiple scattering corrections (McPeters and Green, 1976). When these are large we must go to some multiple scattering analyses such as Monte Carlo code (Furman et al., 1976) or a solution of the radiative transfer equation (Green and Wyatt, 1965). Unfortunately, the best solution for the radiative transfer equation for highly anisotropic phase functions has not yet been determined. In a recent work McPeters and Green (1976) find that single aerosol scattering in conjunction with the multiple Rayleigh scattering (Coulson et al., 1960) gives a fairly good characterization of observed angular distribution function for a reasonable array of optical depths.

I might say that in working in the ultraviolet at 309 nm we do not find the problems that Deirmendjian (1957) found in 1955 with negative turbidities. We have been able to account for our own ultraviolet aureole measurements in a straightforward way in terms of reasonable atmospheric aerosol distributions.

Now on some occasions we have run into a problem in which we have to utilize a third size distribution parameter. In particular, a large radius cut-off is needed if $\nu \leq 3$ since then, for example, the aerosol mass loading will blow up. To deal with this problem we have used another generalization of the

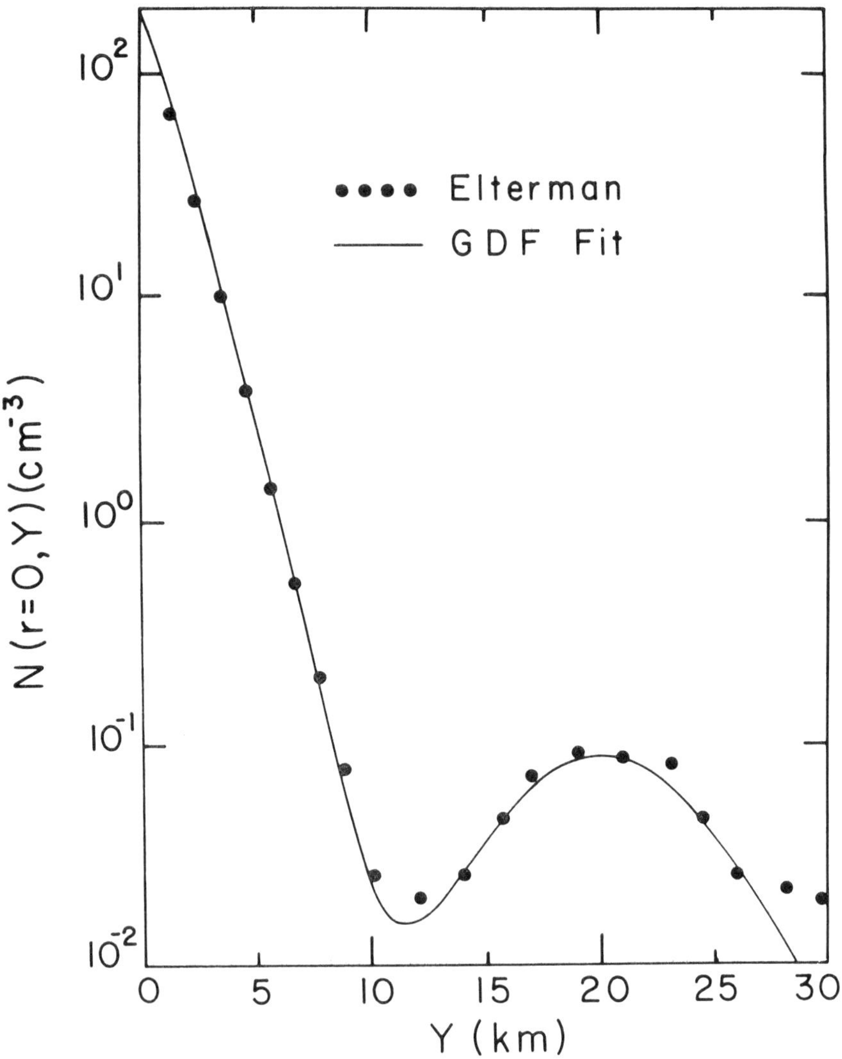

Figure 5. Representative altitude distribution function for aerosols and equations which characterize it. (See Green and Wyatt, 1965, p. 449.)

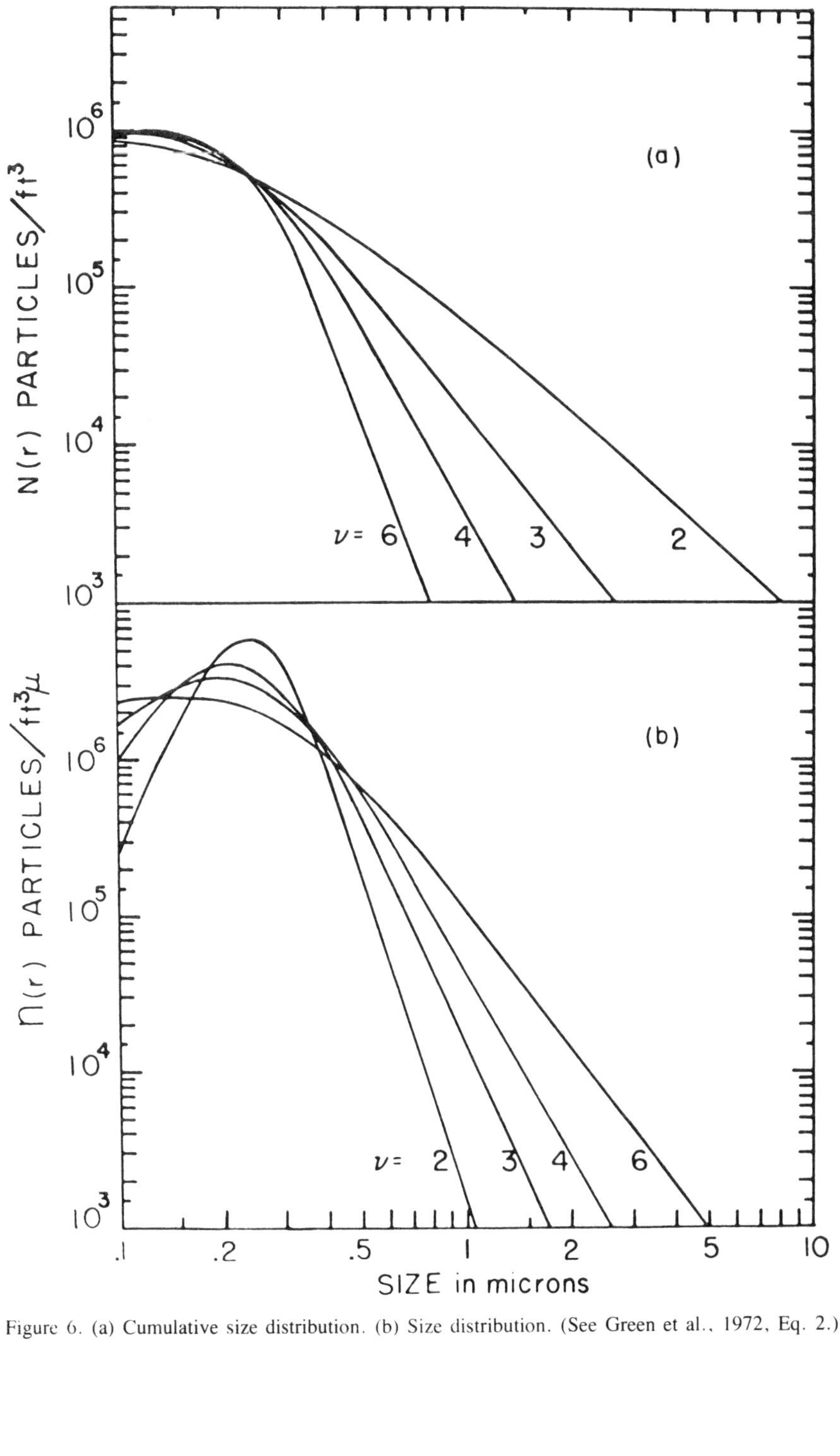

Figure 6. (a) Cumulative size distribution. (b) Size distribution. (See Green et al., 1972, Eq. 2.)

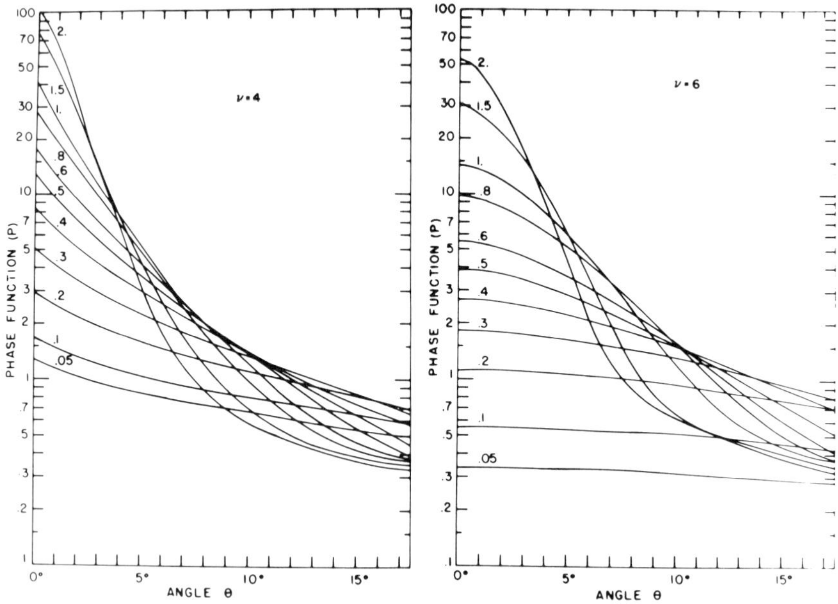

Figure 7. Phase functions for $\nu = 4$ and $\nu = 6$ for various values of size parameters.

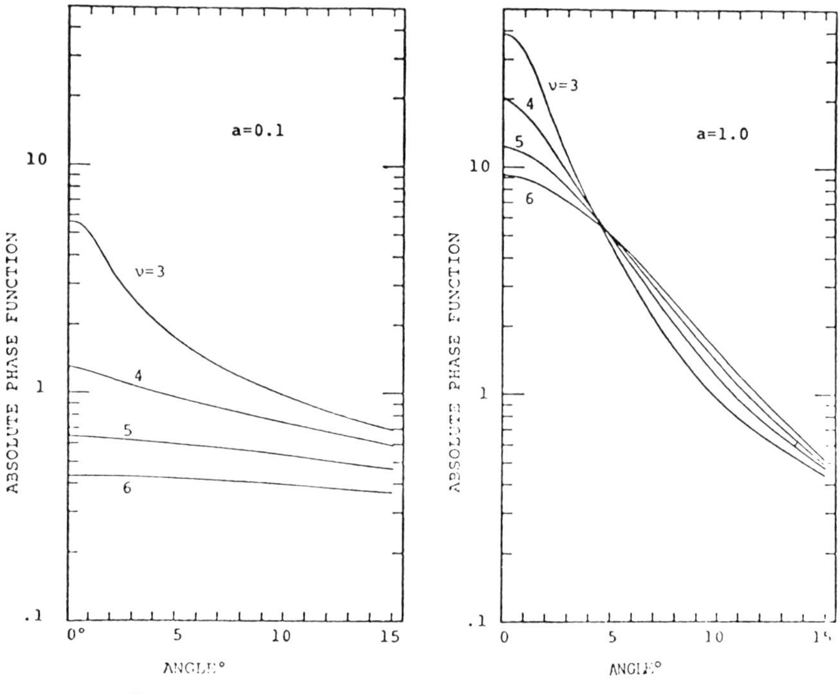

Figure 8. Phase functions for various ν at a = 0.1 and at a = 1.0.

power law size distribution which goes over to our generalized power law at medium r and still preserves the good behavior at small r. This is illustrated in Figure 10. We think this function is more realistic as a description of ordinary atmospheric aerosols than a Gaussian or the Deirmendjian distribution. Figure 10 also shows a size distribution measured with a Millipore system several years ago. This large cutoff parameter influences phase functions at very small angles (e.g., 1°~2°) so that one might hope in time to pull 3 parameters out of the phase function: a general scale parameter a, a large size cutoff parameter b as well as the power for the middle range of radii. I think we can do this in aureole studies in the relatively near future.

DIFFUSE SKYLIGHT

I would like next to describe briefly some of our work on the diffuse solar ultraviolet spectral irradiance (Chai and Green, 1976). Calculations of direct solar flux which allow for attenuation by Rayleigh and aerosol extinction are

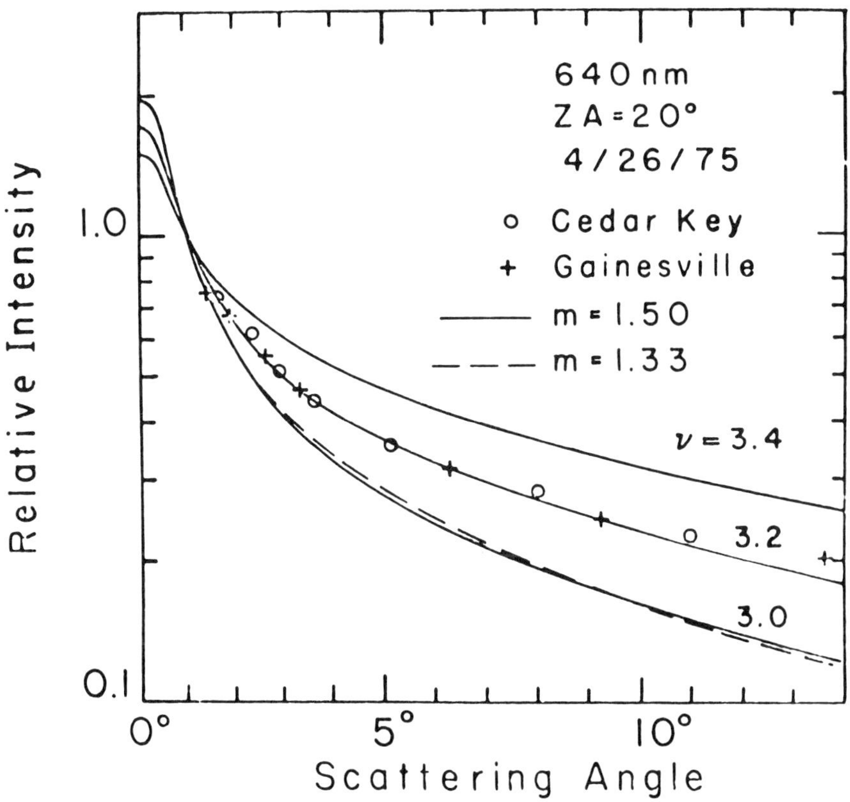

Figure 9. Angle distribution experimental vs. theoretical at 640 nm. (See McPeters and Green, 1976.)

relatively straightforward. However, the calculation of the diffuse flux is quite complicated and it is only recently that reasonable attempts have been made to allow for the presence of atmospheric aerosols. Indeed there are many papers in the literature in which skylight is ignored completely. This is rather bad science since in the ultraviolet, skylight usually contributes more to the radiation impinging upon a horizontal surface than does the direct sunlight itself. Figure 11 illustrates the direct and diffuse spectral irradiance in the ultraviolet for a representative model atmosphere. This may be of additional interest in view of the importance of UV in smog formation. The main point I wish to bring out is that the relative magnitudes of the direct and diffuse UV flux are quite sensitive to aerosol loading so that one might envisage some aerosol assessment technique which takes advantage of this ratio. Some work directed towards taking advantage of these differences is already underway.

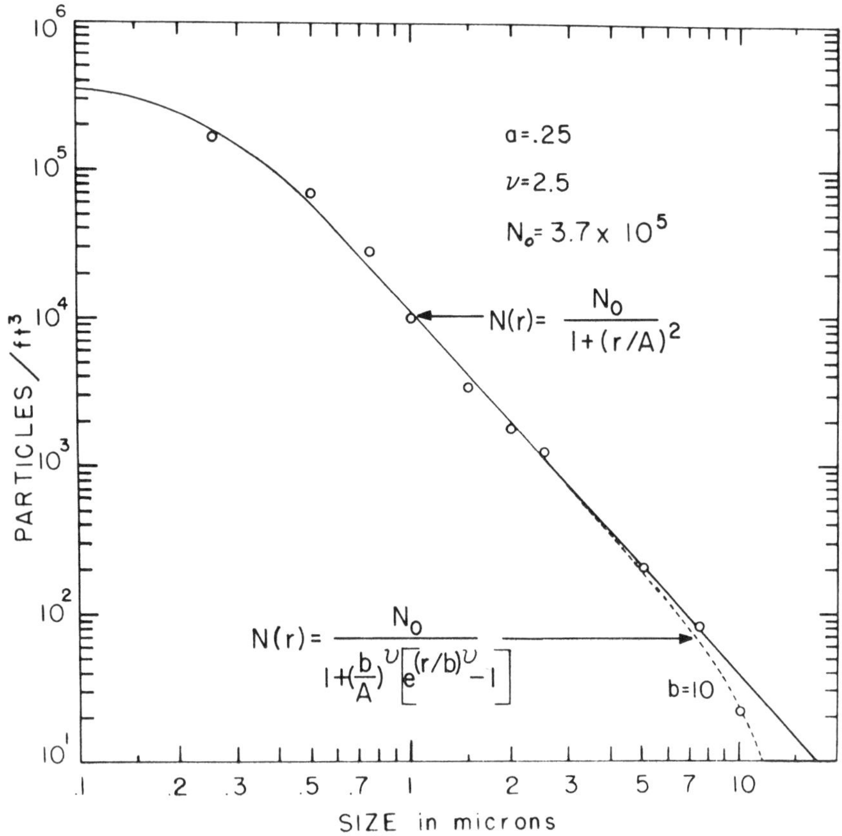

Figure 10. Power law size distribution vs. generalized power law size distribution from Millipore system.

REMOTE SENSING FROM SATELLITES

Remote sensing of atmospheric aerosols from satellites has gotten its greatest impetus from the scientific space satellites as well as military spy satellites. Of course it would be difficult to summarize everything that has been done or is underway but perhaps we can illustrate the state of the art with Figure 12 (or table) which summarizes the planned atmospheric remote sensing experiments to be carried out on NIMBUS G which will be launched in 1978.

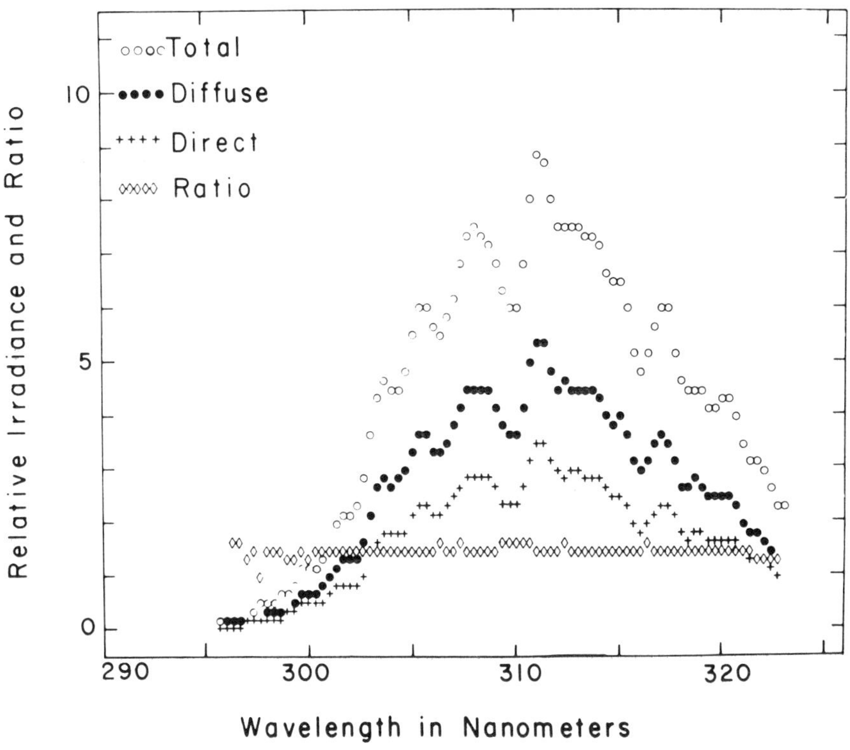

Figure 11. Direct and diffuse spectral irradiance in ultraviolet region. (See Chai and Green, 1976.)

The Nimbus Experiment team members and their affiliations are given in columns 1 and 2 of Figure 12. Columns 3 and 4 give the type of device used and the parameters of the earth's atmosphere which will be measured. For example, the first experiment is a combination study of solar UV backscatter experiment and total ozone mapping (Heath, et al., 1975). The object here is to arrive at a vertical ozone concentration profile. Particulates are involved in this context as a complicating aspect of the inversion problem.

The fifth experiment, the stratospheric aerosol measurement experiment, is of particular interest. Here the team members have proposed limb viewing to mea-

NIMBUS—G NET MEMBERS —(NIMBUS EXPERIMENTAL TEAM)

		Type of Device	Parameter Determinations
SBUV/TOMS—Solar and Backscattered Ultraviolet and Total Ozone Mapping System			
*Heath, D.	GSFC		
Mateer, C. L.	Atm, Envir. Service (Canada)	Sun and Earth Viewing	Solar UV Irradiance,
Belmont, A. D.	Control Data Corp.	Monochromators, Nadir	Ozone Profiles, Global
Miller, A. J.	NOAA, Nat'l Meteorological	Viewing and Nadir	Maps of Total Ozone
Green, A. E. S.	U. of Florida	Scanning	
Cunnold, D. M.	MIT		
Imhof, W. L.	Lockheed Space Sc. Lab.		
ERB—Earth Radiation Budget			
*Jacobowitz, H.	NOAA		
Vonder Haar, T. H.	Colorado State University		
House, F. B.	Drexel University	Sun and Earth Viewing	Solar Irradiance, Flux
Coulson, K. L.	U of California	Spectroradiometer,	and Radiance of Earth
Hickey, J. R.	Eppley Lab. Inc.	Nadir Viewing and	Reflected Short-wave
Stowe, L.	NOAA/NESS	Nadir Scanning	and Emitted Long-
Ingersoll, A. P.	Calif. Inst. of Technology		wave Radiation
Smith, G. L.	LaRC		
CZCS—Coastal Zone Color Scanner			
*Hovis, W.	GSFC		
Yentsch, C.S.	N. E. Research Foundation		
Clark, D.	NOAA/NESS	Earth Viewing	Chlorophyll, Sediment
Apel, J.R.	NOAA/ERL	Radiometer Nadir	Distribution,
El-Sayed, S. Z.	Texas A & M	Scanning	Gelbstoffe
Gordon, H. R.	U. of Miami		Concentration
Wrigley, R. C.	AMES		(Salinity),
Harris, T. F. W.	CSIR (South Africa)		Coastal Water
Newth, F. H.	EURASEP-CEC		Temperature
			Currents
SMMR—Scanning Multichannel Microwave Radiometer			
*Gloersen, P.	GSFC		
Ramseier, R. O.	Dept. of Environment. (Canada)	Earth Viewing Microwave Radiometer,	Sea Ice, Sea Surface Temperatures and
Staelin, D. H.	MIT	Nadir	Winds, Cloud Liquid
Campbell, W. J.	Dept. of Interior/Geo. Sur	Scanning	Water Content Precipitation, Water
Gudmandsen, P.	Tech. Inst. of Denmark		Vapor
Ross, D. B.	NOAA/ERL		Soil Moisture, Snow
Windsor, E. P. L.	British Aircraft Corp.		Cover
Kunzi, K. F.	Univ. of Berne, Switzerland		
SAM II—Stratospheric Aerosol Measurement			
*McCormick, M. P.	LaRC	Solar Extinction	Aerosols—Vertical
Pepin, T. J.	U. of Wyoming	Photometer, Limb	Profiles
Grams, G. W.	NCAR	Viewing	
Herman, B. M.	U. of Arizona		
Russell, P. B.	Stanford Research Institute		
LIMS—Limb Infrared Monitor of the Stratosphere			
*Russell, J.	LaRC	Limb Scanning	H_2O, HNO_3, NO_2, O_3,
*Gille, J. C.	NCAR	Infrared	Temperature—
House, F. B.	Drexel University	Radiometer	Vertical Profiles
Remsberg, E. E.	LaRC	Limb Scanning	CH_4, CO, H_2O, NO,

NIMBUS—G NET MEMBERS—(NIMBUS EXPERIMENTAL TEAM)

		Type of Device	Parameter Determinations
Loevy, C. B.	U. of Washington	Pressure	N_2O Temperature—
Drayson, S. R.	U. of Michigan	Modulated Infrared	Vertical Profiles
Fischer, H.	U. of Munchen, W. Germany	Radiometer	
Planet, W. G.	NOAA/NESS		
Girard, A.	ONERA, France		
Harries, J. E.	National Physical Labs., U.K.		
	THIR-Temperature Humidity Infrared Radiometer	Earth Viewing Infrared Radiometer, Nadir Scanning	Correlative Imagery for Other Experiments, Sea Surface Temperature

*Sensor scientists.

Figure 12. Summary of Nimbus G experiments and team members.

sure solar extinction with photometers to arrive at a vertical aerosol profile. While these satellites are costly, their tremendous information gathering capability make them quite cost effective. Undoubtedly you will hear much in the future about remote sensing from satellites.

In concluding the work I would like to acknowledge the efforts of my past and present colleagues in our aerosol remote sensing program. Their names can be found as co-authors in the list of references originating in our group.

REFERENCES

Chai, A. T. and A. E. S. Green, 1976: Measurement of the ratio of diffusion to direct solar irradiances in the middle ultraviolet. *Appl. Optics, 15,* 1182–1187.

Coulson, K. L., J. V. Dave and Z. Sekera, 1960: *Tables Related to Radiation Emerging From a Planetary Atmosphere with Rayleigh Scattering.* University of California Press, 548 pp.

Deirmendjian, D., 1957: Theory of the solar aureole I. *Ann. Geophys., 13,* 286–306.

Elterman, L., 1964: Parameters for attenuation in the atmospheric windows for fifteen wavelengths. *Appl. Optics, 6,* 745–749.

Flowers, E. C., R. A. McCormick and K. R. Kurfis, 1969: Atmospheric turbidity over the United States 1961–1966. *J. Appl. Meteor., 8,* 955–962.

Furman, D. R., T. Mo and A. E. S. Green, 1976: Multistream and Monte Carlo calculations of the sun's aureole. *J. Atmos. Sci., 33,* 537–543.

Green, A. E. S. and P. J. Wyatt, 1965: *Atomic and Space Physics.* Addison-Wesley, 619 pp.

———, A. Deepak and B. J. Lipofsky, 1971: Interpretation of the sun's aureole based on atmospheric aerosol models. *Appl. Optics, 10,* 1263–1279.

———, G. D. Ward, T. Sawada, R. S. Sholtes, J. M. Schwartz, A. Deepak, D. Eisenhart, R. D. McPeters and B. D. Reller, 1972: Light scattering and the size-altitude distribution of atmospheric aerosols. *J. Colloid Interface Sci., 39,* 520–535.

Heath, D. F., A. J. Krueger, H. A. Roeder and B. D. Henderson, 1975: The Solar Backscatter Ultraviolet and Total Ozone Mapping Spectrometer (SBUV/TOMS) for NIMBUS G. *Optical Engr., 14,* 323–331.

McPeters, R. D. and A. E. S. Green, 1976: Photographic aureole measurements and the validity of aerosol single scattering. *Appl. Optics, 5,* 2457–2463.

Volz, F., 1954: Photometer mit Selen-Photoelement zur spektrolung und zur Bestimmung der Wellenbangen abhangigkilt der Dunsttribung. *Arch. Met. Geoph., 10,* 1 (with translation).

———, 1973: Personal communication.

Ward, G. D., K. M. Cushing, R. D. McPeters and A. E. S. Green, 1973: Atmospheric aerosol index of refraction and size-altitude distribution from bistatic laser scattering and solar aureole measurements. *Appl. Optics, 12,* 2585–2592.

AEROSOL PARTICLE SIZE MEASUREMENTS USING THE ELECTRICAL RESISTANCE PRINCIPLE

S. Kinsman

Coulter Electronics, Inc., Hialeah, Fla.

Sensing of one by one particle volumes at preselected size levels using an electric current through a liquid conductor in a small orifice as the detector is described. Advantages and limitations of this particle size analysis method are discussed. Compatible aerosol collection methods and devices are outlined. Several specific applications of this device are presented citing appropriate scientific literature references.

The Coulter principle instruments count and size particles individually at rates up to 3000 per second. A dilute sample suspension in a conductive liquid is stirred in a beaker. A test tube shaped glass part with a small orifice in the bottom projects downward into the sample suspension as shown in Figure 1. A vacuum causes the sample to pass through the orifice. Two submerged platinum electrodes cause electric current to flow through the orifice also. Each particle passing through the orifice momentarily changes the electrical resistance of the liquid in the orifice causing a pulse. The pulse amplitude is proportional to the volumetric size of the particle. The series of pulses thus formed are counted at preselected size levels. This data represents size distribution of the number count at several particle size levels. Figure 2 shows the complete equipment.

Standard methods of particle sampling are used. These include Greenburg Smith or midget impingers, membrane surfaces, alumdum thimbles, impactors and fabrics. Removal of particles from sampling surfaces is done by wash off, back wash, scraping or membrane dissolving. In some instances blank surfaces are run if the collections are anticipated to be numerically small. The efficiency of particle removal from surfaces often is increased by use of ultrasonic energy in the washing process. Also, Coulter comparisons of a split test dust, one split direct and one split through a membrane and then washed, will give an indication of the membrane washing efficiency for the particular test dust being used. Less than milligram size samples can be analyzed. The size range is from $< 0.4\ \mu m$ to $> 200\ \mu m$ with a 1 to 20 size range limit for any given orifice. Count accumulation time is usually less than 30 seconds.

This method requires that careful consideration be given to the solubility of particles and to the state of agglomeration or deagglomeration. Either water or

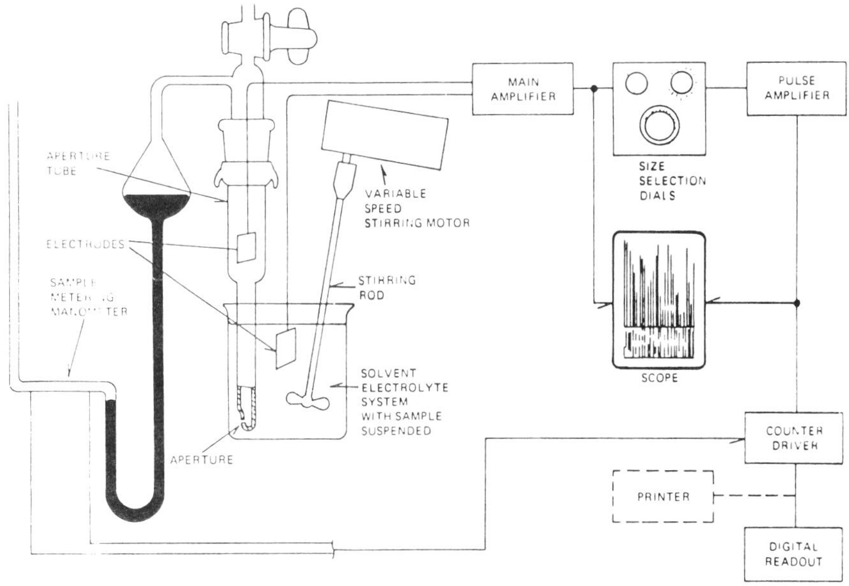

Figure 1. Typical electronic counter block diagram.

Figure 2. Coulter counter.

alcohol electrolyte systems can be used to suspend the particles. This gives some flexibility where particle solubility is a problem.

The method does not measure the so called aerodynamic diameter but Tomb and Corn (1972) have developed size ratios for converting from particle volume

DISPERSANT: IPA, 15 SEC. ULTRASONICS
EQUIPMENT: TAII
OPERATOR: M.B.

APERTURE	CAL. PART.	DIAM.	W	IA	A1
140	MICROSPHERES	9.69	8	3.2	120.8
50	LATEX	2.02	6	1.6	176.8

CH. DIAM.	W	IA	A2
10.1	8	3.2	107.3
2.0	6	1.6	182.2

DIAM.	CUM. %	DIF. %
1	100	.5
1.26	99.5	.6
1.587	98.9	1.3
2	97.6	2.3
2.52	95.4	3.5
3.175	91.9	5.4
4	86.5	7.3
5.04	79.2	9.8
6.35	69.4	12
8	57.5	12.7
10.1	44.7	12.9
12.7	31.8	12.9
16	18.8	10.4
20.2	8.4	4.3
25.4	4.1	2.6
32	1.5	1.1
40.3	.4	.4
50.8	0	0
64	0	0

MEAN: 8.685611
MODE: 12.69921
1 S.D.: (4.402379, 17.13615)

```
X                  1   .5%
X                  1.26   .6%
X                  1.587   1.3%
XXX                   2   2.3%
XXXXX                   2.52   3.5%
XXXXXXXX                   3.175   5.4%
XXXXXXXXXX                   4   7.3%
XXXXXXXXXXXXXX                   5.04   9.8%
XXXXXXXXXXXXXXXXXX                   6.35   12%
XXXXXXXXXXXXXXXXXXX                   8   12.7%
XXXXXXXXXXXXXXXXXXX                   10.1   12.9%
XXXXXXXXXXXXXXXXXXX                   12.7   12.9%
XXXXXXXXXXXXXXX                   16   10.4%
XXXXXX                   20.2   4.3%
XXX                   25.4   2.6%
X                  32   1.1%
X                  40.3   .4%
  50.8   0%
```

Figure 3. Computer printout of aerosol distribution data.

to aerodynamic diameter, and Randall (1970) has shown conversion factors for quartz, rutile and gold mine dust. In most cases, by use of mechanical dispersing such as ultrasonics and/or surface active agents, individual particles are presented to the sensing orifice for measurement.

Calibration of the particle size scale is made by passing near monosized particles of known diameter through the orifice thus generating a series of pulses of similar heights. An adjustable size control threshold on the instrument is adjusted to coincide with these known pulses. This gives a direct relationship for the threshold to a known standard. The calibration procedure requires less than 10 minutes.

When sample or test dust densities are known or can be approximated, the Coulter response of particle volume can easily be related to mass values. Add to this the accuracy and repeatability of the calibration for the multiple size classes, then data on mass efficiencies of separation and collection equipment at the selected size classes are quickly available. Both particle mass curves and particle count printout can be executed by the equipment from each analysis. Our lab transfers the data to a mini computer which in turn prints out requested data on a teletype per Figure 3. This printout usually includes calibration data, size distribution curve, mean, mode, and standard deviation.

There have been many applications for this device in health problems and in stack and environment control. This paper limits itself to presented or published information and a bibliography is available. For instance, Sundberg (1974) suggests that efficiencies of scrubbers and cyclones follow a log normal distribution using size distributions of fed and collected silica dust. Stenhouse et al. (1975) investigated the effect of humidity on the behavior of model filters, using a 2 to 20 μm feldspar and Durham et al. (1975) studied bag fabric retention changes with humidity using fly ash, limestone, cement dust, and silica. Miller et al. (1974) carefully determined fly ash retention of nonwoven fabric using several fiber cross sectional shapes, surface roughness, and degree of crimp. Dunson (1970) used talc and fly ash below 10 μm diameter for efficiencies of prototype dust collection equipment. He sampled the dust with midget impingers for Coulter analysis. Very recently Kaldor and Phillips (1976) used fly ash below 10 μm diameter to study aerosol scrubbing by a column of foam. Efficiencies for particle size, residence time, foam bubble size and concentrations were determined. Likewise Tomb et al. (1971) evaluated 0.68 μm–10 μm coal dust suppression with water spray and steam using midget impinger samples obtained in the ducting to a test chamber. They determined influent and effluent particle counts at several size levels using the Coulter device. Drasche et al. (1970) studied inhaled and exhaled limestone in a pit with humans and a simulated lung. He used membrane samplers and dissolved these before particle size analysis.

Comparisons with other particle size methods are numerous. Rendall's work (1970) previously mentioned, compared gas permeability analysis, sedimentation analysis and Coulter analysis by applying a shape factor for the particles. Svarovsky (1974) designed a three stage cyclone stack sampler and successfully compared his data with both the Coulter and sedimentation analyses using garnet test dust. Davison et al. (1974), when studying trace elements in fly ash, verified the particle size on two stages of an Anderson impactor by washing collected particles off the stages into a methanol and water mixture and using ultrasonics to aid particle removal. Some elements increased with decreasing particle size.

Spiga and Cherchi (1970) made a thorough study of mine dust collections on membranes using ultrasonic vibration to free the dust from the membrane for Coulter counting. Coal, limestone, silicate, and quartz were involved. Flux et al. (1968), wishing to determine design of stack collection devices, removed stack dust from an alumdum thimble and analyzed it for particle size distribution between 1 and 50 μm diameter. Tomb and Corn (1972) correlated Stokes diameters and Coulter diameter with shape factors for coal, limestone and mica. Correlation coefficient for coal was 0.997, for limestone 0.999 and for mica 0.995.

When using standard methods for aerosol sampling, additional particulate characterization is obtainable in the form of particle size distributions using the Coulter technique. The limitations must be understood and the data used accordingly. The speed of analysis, reliability of calibration and versatility of data forms favor its use. There are many printed examples of the use of Coulter principle devices for aerosol particle size measurements.

REFERENCES

Davison, R. L., D. Natusch, J. R. Wallace and C. A. Evans, Jr., 1974: Trace elements in fly ash, dependence of concentration on particle size. *Environ. Sci. Technol., 8,* 1107–1113.

Drasche, H. and R. Reiter, 1970: On the deposition characteristic of solid particles in the human respiratory tract. *Staub-Reinhalt. Luft, 30,* 12–30.

Dunson, J. B., Jr., 1970: Dust collector performance evaluation by midget impinger and Coulter counter. Presented at the 63rd Annual Meeting of AIChE.

Durham, J. F. and R. F. Harrington, 1970: Influence of relative humidity on filtration resistance and efficiency. Presented at 63rd Annual Meeting of AIChE.

Flux, J. H., D. J. Smithson and R. N. Smithson, 1968: Simplified dust sampling apparatus for use in iron and steel works. *J. Iron and Steel Inst.,* 1188–1193.

Kaldor, T. G. and C. R. Phillips, 1976: Aerosol scrubbing by foam. *Ind. Eng. Chem. Proc. Dev., 15,* 199–206.

Miller, B., G. E. R. Lamb and P. Costanza, 1974: Nonwoven fabrics as air filters: A study of critical fiber variables. *Tappi, 57,* 102–104.

Rendall, R. E. G. and G. C. H. van Sittert, 1970: The determination of the size distribution of mine dusts using the Coulter counter. Pneumoconiosis Research Unit Report, No. 10/70.

Spiga, G. and P. Cherchi, 1970: Translation: A method for analyzing airborne dust with an electronic particle counter. *La Medicina del Lavoro, 61,* 273–282.

Stenhouse, J. I. T., P. J. Lloyd and R. E. Buxton, 1975: The retention of large particles (> 2 μm) in fibrous filters, paper No. 8, presented at the Amer. Industrial Hygiene Assoc. Conference.

Sundberg, R. E., 1974: The prediction of overall collection efficiency of air pollution control devices from fractional efficiency curves. *J. Air Pollution Control Assoc., 24,* 758–765.

Svarovsky, L., 1974: Particle size analysis of chimney emissions *In Situ*. Presented at Soc. for Anal. Chem., Colloquium, University of Nottingham.

Tomb, T. F., L. Cheng and R. L. Stein, 1971: Suppression and collection of respirable coal dust using water and steam. Presented at the Internat'l. Conf. of Coal Worker's Pneumoconiosis.

——— and M. Corn, 1972: Comparison of equivalent spherical volume and aerodynamic diameters for irregular shaped particles. Presented at Amer. Industrial Hygiene Assoc. Conference.

AEROSOL MONITORS BASED ON THE CONTACT ELECTRICITY PRINCIPLE

WALTER JOHN, GEORG P. REISCHL, AND WILLIAM DEVOR

Air and Industrial Hygiene Laboratory, California State Department of Health, Berkeley

The contact electrification monitor offers some important advantages such as measurements in real time with a short response time. It is a mass monitor in that the electrical charge correlates accurately with the mass determined gravimetrically. The instrument is relatively simple and the electrical signal is convenient for monitoring and data processing. The probe can be inserted directly into a stack to avoid the complications associated with extraction and external analysis.

Although it has undergone over a decade of development, the contact electrification monitor is not widely used. This may be due to uncertainty arising from a lack of definitive data on the performance of the instrument and poor understanding of the charge transfer mechanism. However, because of the attractive advantages of this type of instrument, there is continuing interest in its development.

The influence of important variables on the response of the contact electrification monitor will be reviewed briefly here. (The reader is also referred to an earlier, more comprehensive review by John, 1976.) In addition new data obtained recently by the authors will be presented. Finally, the potential applications and limitations of the technique will be discussed.

REVIEW OF TECHNIQUE DEVELOPMENT

Contact Electricity Monitoring Instruments

In monitoring instruments based on contact electrification, a flow of the aerosol is directed at a probe. The transfer of charge from particle-probe collisions results in a current which is continuously monitored with a sensitive meter. Contact electrification, which refers only to charge transferred between two bodies as a

Work supported by U.S. Environmental Protection Agency Grant R 803719.

result of pure contact, will be used here to describe the charging process for the monitor, although sliding or rubbing may be present. Little is known about the latter charging mechanisms. We emphasize that no electrical potential is applied to the probe. The charge is transferred as a result of the contact potential between the two materials involved. The contact potential between two metals is illustrated in Figure 1.

Schutz (1966) described one of the first contact electrification monitors, an instrument having a cone-shaped probe in a nozzle as shown in Figure 2. The in-line filter allows the accumulation of a sample for weighing. Calibration of the instrument is accomplished by dividing the integrated current (charge) by the simultaneously collected mass. Similar instruments were constructed by Soo and his collaborators (Cheng and Soo, 1970; Soo et al., 1969; Cheng et al., 1970). Kisler (1972) described a monitor using a wire probe. The IKOR Air Quality Monitor[1] is also similar to the instrument in Figure 2 except for the absence of the taper in the pipe. The IKOR instrument (Gruber and Bastress, 1973) utilizes a bullet-shaped probe of Inconel. A more recent IKOR probe, designated a high turbulence probe, consists of metal twisted in the form of a helix. Mounted with the helical axis parallel to the pipe, the probe is designed to have a higher sensitivity (charge per mass) than streamlined probes. IKOR also manufactures a version of the instrument featuring an in-stack probe.

Prochazka (1964, 1966) developed the Konitest, an instrument produced commercially for a time in Germany. Illustrated in Figure 3, the Konitest has a tubular electrode of a semiconductor, steatite (magnesium hydrosilicate). The version in Figure 3a is designed to increase the probability that the particles impinge on the walls as a result of the helical path followed by the air. A new commercial instrument, the Kony Test,[2] has been developed in Japan, based on the Konitest concept.

Operating Characteristics of the Contact Electrification Monitor

Real Time Response

John (1975) compared the responses of the IKOR AQM and an optical particle counter to polydisperse aluminum oxide dust. The 1/e time constant of the IKOR AQM is approximately 10 seconds. An RC network was used to slow the response of the Climet optical particle counter to 4 seconds. The recorder traces are shown in Figure 4. The more extreme highs and lows of the Climet trace are attributed to the faster time constant. Over the first 250 seconds the average response of the IKOR is lower than that of the Climet. This growing sensitivity over the first few minutes of operation is typical. Otherwise, the correlation to the optical counter is good.

1. Manufactured by IKOR, Inc., Burlington, Mass. The mention of commercial products, their source or their use in connection with material reported herein is not to be construed as either an actual or implied endorsement of such products.
2. Manufactured by Kony Co., Ltd., Tokyo, Japan.

Correlation Between Charge and Mass

It is important to establish the correlation between the integrated current (charge) and the mass measured gravimetrically. Most investigators have found good correlations. Prochazka (1966) found the Konitest to be accurately linear, as shown in Figure 5. Ito et al. (1972) and Schnitzler (1970) have also reported good correlations for the Konitest. On the other hand, Schutz (1966) found the current to depend on the mass concentration raised to a power of 1.26 to 1.30. Cheng and Soo (1970) and Cheng et al. (1970) obtained plots showing small deviations from linearity. John (1975) found the IKOR AQM charge to be accurately proportional to mass, as shown in Figure 6.

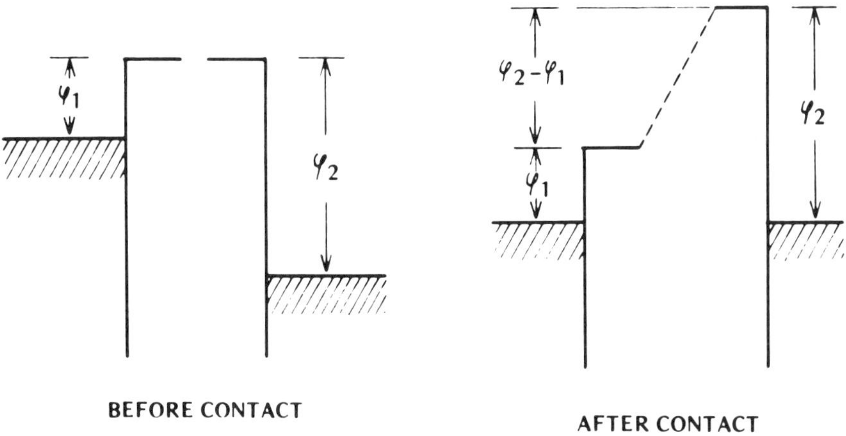

Figure 1. Energy level diagrams for two metals. Upon contact, charge flows until the Fermi levels are equalized; the contact potential is given by the difference in the work functions φ_1 and φ_2.

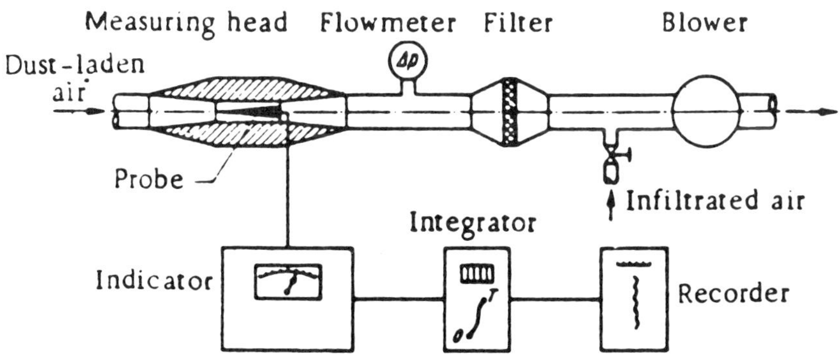

Figure 2. A dust monitor based on the contact electricity principle using a probe in a nozzle according to Schutz (1966).

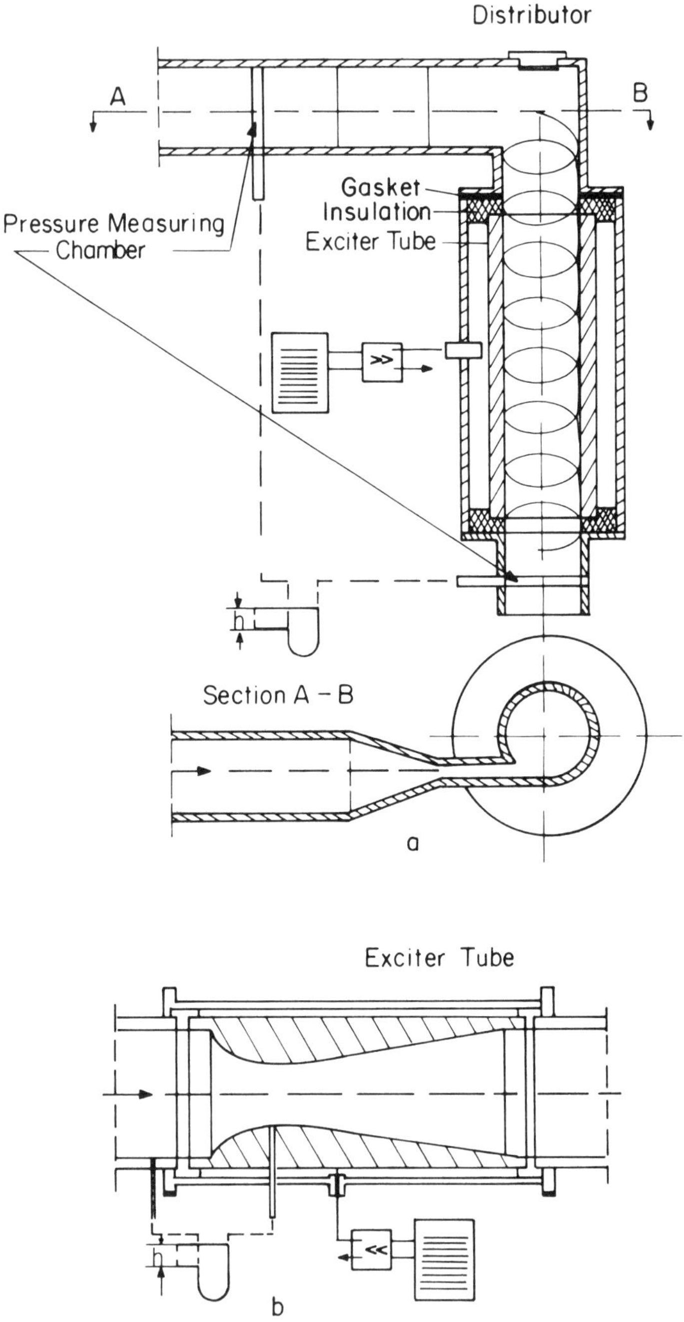

Figure 3. Two versions of the Konitest dust sampler. a: The electrode is a cylindrical turbulence chamber. b: A Venturi nozzle is employed. (Prochazka, 1966.)

Dependence of the Sensitivity on the Electrical Resistivity of the Particulate Matter

Using laboratory-generated aerosols, John (1975) measured the sensitivity (charge/mass) of the IKOR AQM for a number of substances. In Figure 7 the sensitivities are plotted vs. the logarithm of the electrical resistivity. It is apparent

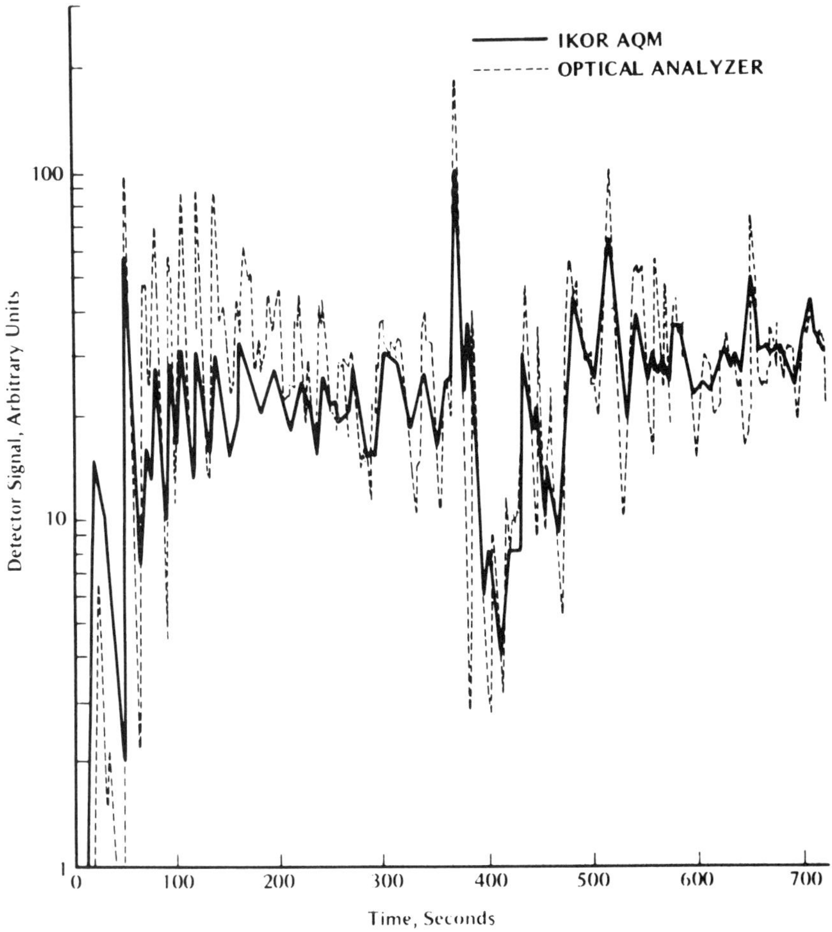

Figure 4. Dynamic response of the IKOR AQM compared to that of an optical particle counter for aluminum oxide particles (John, 1975).

that the sensitivity is correlated with the resistivity, the metallic conductors having the highest sensitivities. This result is not unexpected since within the limited time a particle is in contact with the probe, more charge can be transported in a good conductor than in an insulator.

Influence of Particle Size

Schutz (1966) reported higher sensitivity for small particles. Ito et al. (1972) reported the Konitest to be little affected by particle size when sampling cement dust. John (1975) found the IKOR AQM to have no strong dependence on particle size except for a cutoff in sensitivity for very small particles from cigarette smoke. This was explained by the tendency of small particles to follow streamlines around the probe. New results on the particle-size dependence of the sensitivity are presented in the next section.

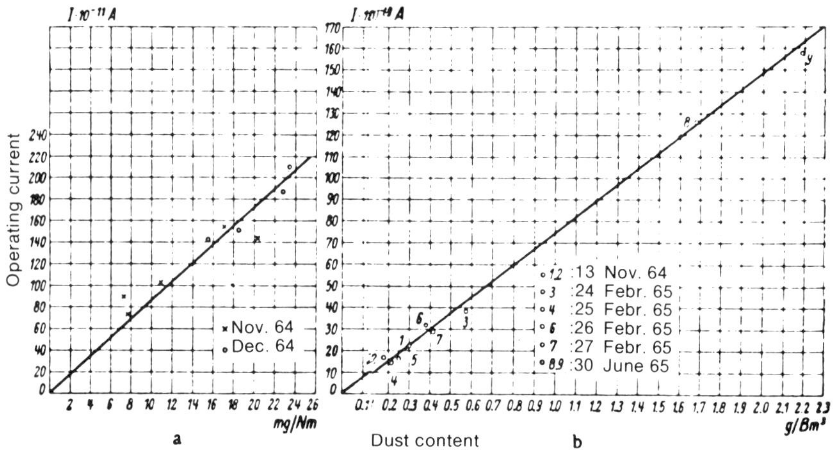

Figure 5. Calibration curves showing the Konitest current vs. the mass concentration. a: For blast furnace gas. b: For flue gas. (Prochazka, 1966.)

Effects of Humidity

Schutz (1966) reported no effect on his instrument for relative humidity up to 99%. However, droplets of water caused erroneous indications. Ito et al. (1972) stated that humidity caused no problems with the Konitest. Kisler (1972) stated that the humidity should be below the dew point. John (1975) found that humidity up to 69% did not affect the IKOR AQM, but droplets caused the instrument to break down.

Surface Condition of the Probe

Since contact electrification is a surface phenomenon, the physical and chemical condition of the probe's surface is a determining factor in the sensitivity of the instrument. In measurements with an IKOR AQM having an Inconel probe, John (1975) observed an increase in sensitivity by a factor of 5 to 10 over a period of a month. When the probe was scrubbed with abrasive cleanser and detergent, the sensitivity dropped to nearly the initial value. A new observation on the surface effect is presented in the next section.

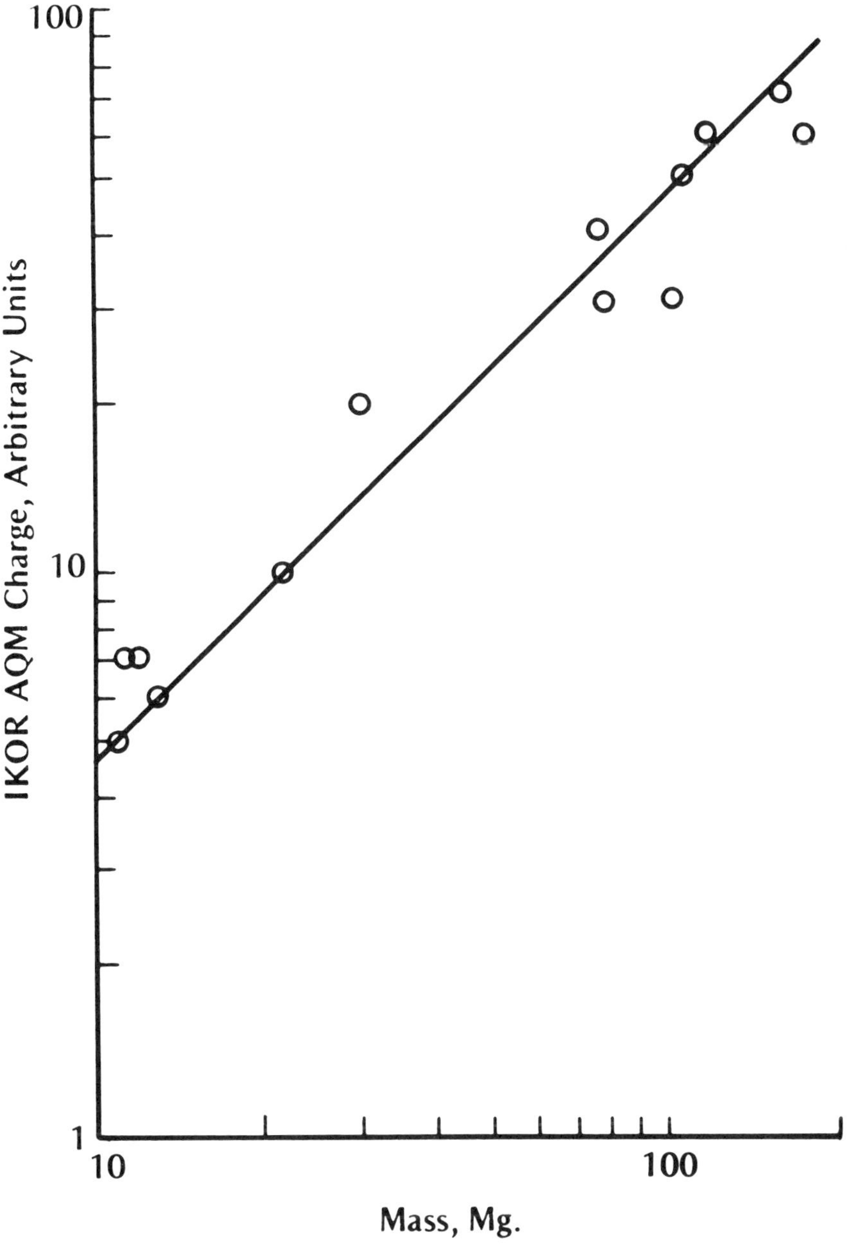

Figure 6. Charge vs. mass of aluminum oxide particles measured with an IKOR monitoring instrument by John (1975). The line is drawn at 45°, corresponding to a linear dependence of charge on mass.

NEW DATA ON THE CHARACTERISTICS OF THE CONTACT ELECTRIFICATION MONITOR

Experimental Methods

This section presents new experimental data which we have obtained on the response of the monitor to particles which are charged prior to entering the detector. Data on the dependence of monitor sensitivity on particle diameter and probe surface condition are also presented. The data were taken with the ar-

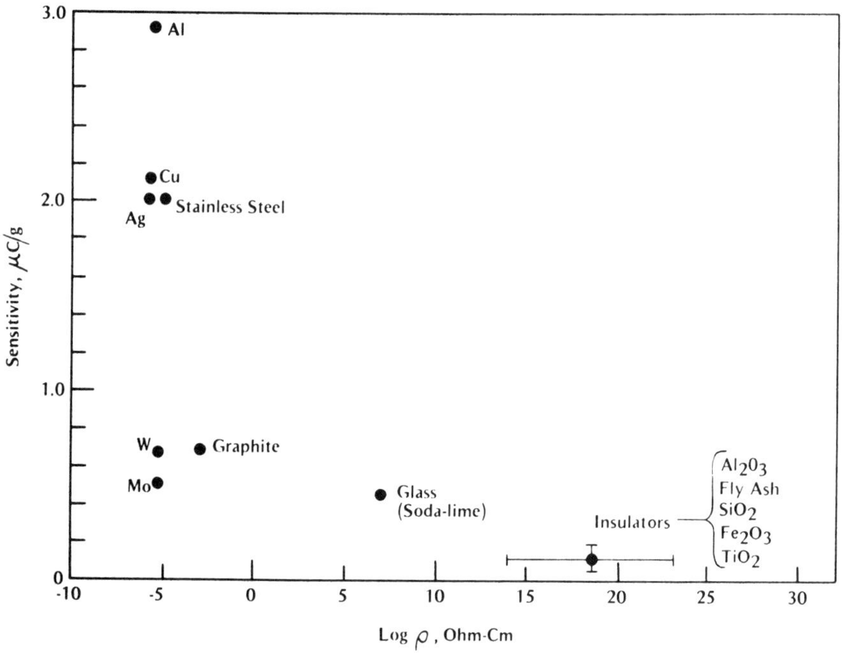

Figure 7. Sensitivity (charge/mass) of the IKOR AQM vs. the log of the electrical resistivity of the material sampled (John, 1975).

rangement shown in Figure 8. Monodisperse sodium chloride and methylene blue particles were generated by a Berglund-Liu aerosol generator. A particle charger of our own design induced charges on the droplets as they were formed in the vibrating orifice generator. This allowed selection of a highly uniform particle charge by control of the induction voltage as shown in Figure 9. The figure also shows that the particles could be neutralized by application of the appropriate induction voltage.

In monitoring instruments of this type the charge per particle depends on the product of the particle-probe impaction probability and the charge transfer efficiency. A new impaction probe was designed for these experiments in order to ensure that each particle had nearly 100% probability of striking the probe, thus allowing the measurement of the charge transferred per particle. Shown in Figure 10, the impactor consists of a rectangular slit 0.25 mm wide and an insulated stainless steel impaction plate. The beveled surface of the plate ensures that particles will not be reflected back to the plate again, but will strike the plate only once. At a nominal flow rate the cutoff particle diameter of the stage is approximately 0.6 μm. To ensure a high impaction probability, particles 1.5 μm and larger were used.

Transfer of Precharge

Results obtained for the transfer of charge to the probe as a function of the precharge on the particle are graphed in Figure 11. One set of data labelled "untreated probe surface" was taken with the probe which had been aged for a period of time. The other set was taken after the probe was wiped with paper tissue wetted with alcohol and water. Comparison of the two data sets shows that the "untreated" probe surface behaved as a rectifier with low resistance to the flow of positive current to the probe and high resistance to negative current. Presumably, this effect results from the formation of an oxide layer on the probe surface. After cleaning, the curve is symmetric with respect to the sign of the charge. A large fraction of the precharge is transferred to the probe, although the transfer is not complete; the dashed line represents complete charge transfer.

Dependence of the Contact Charge on Particle Diameter

The intercepts on the vertical axis of the curves in Figure 11 give the contact charge transferred to the probe by initially neutral particles. Similar data taken for several particle sizes are shown in Figure 12. The data for the untreated and

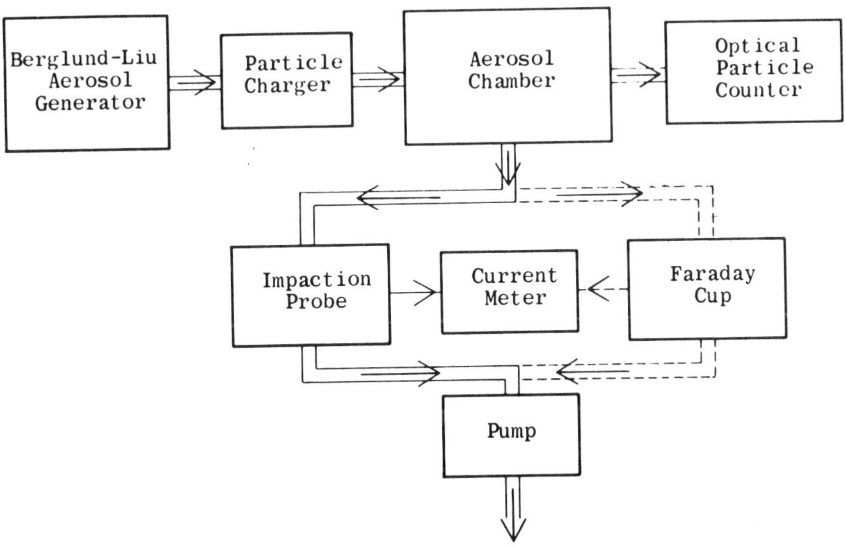

Figure 8. Experimental arrangement for charge transfer measurements.

the cleaned probe surface show no definite difference. It should be noted however, that the contact charge for sodium chloride on stainless steel happens to be positive, i.e., in the "forward" direction of the rectifying layer on the probe surface. A pronounced effect would be anticipated for particles having a negative contact charge. The line is drawn with a slope of 2.9 on the log-log plot.

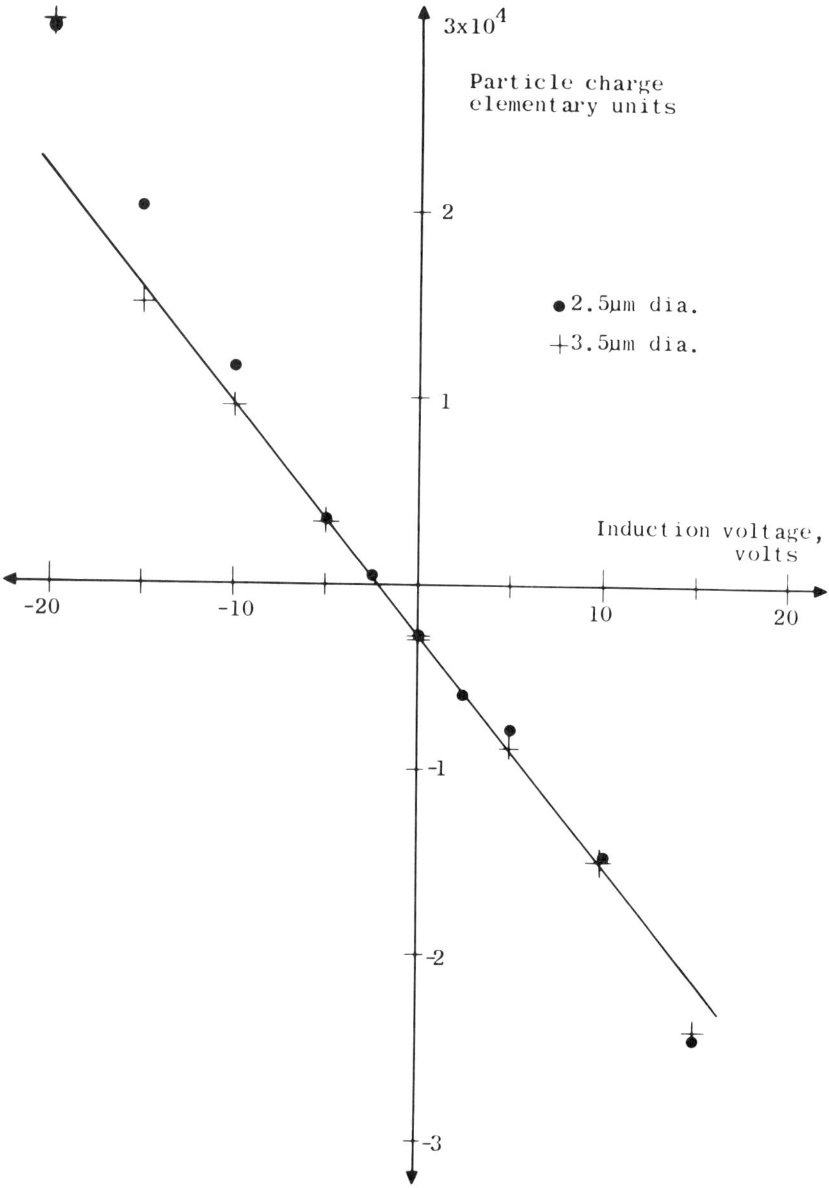

Figure 9. Calibration curve for the particle charger used to charge or neutralize particles.

Additional data, for methylene blue particles, are shown in Figure 13 for two particle velocities. The methylene blue particles were neutralized by passing the aerosol through the radiation field of a Kr-85 source. However, it was found necessary to remove the resulting charged gaseous ions by applying an electrical field transverse to the aerosol stream.

The lines in Figure 13 were drawn with a slope of 3. It is concluded that all the

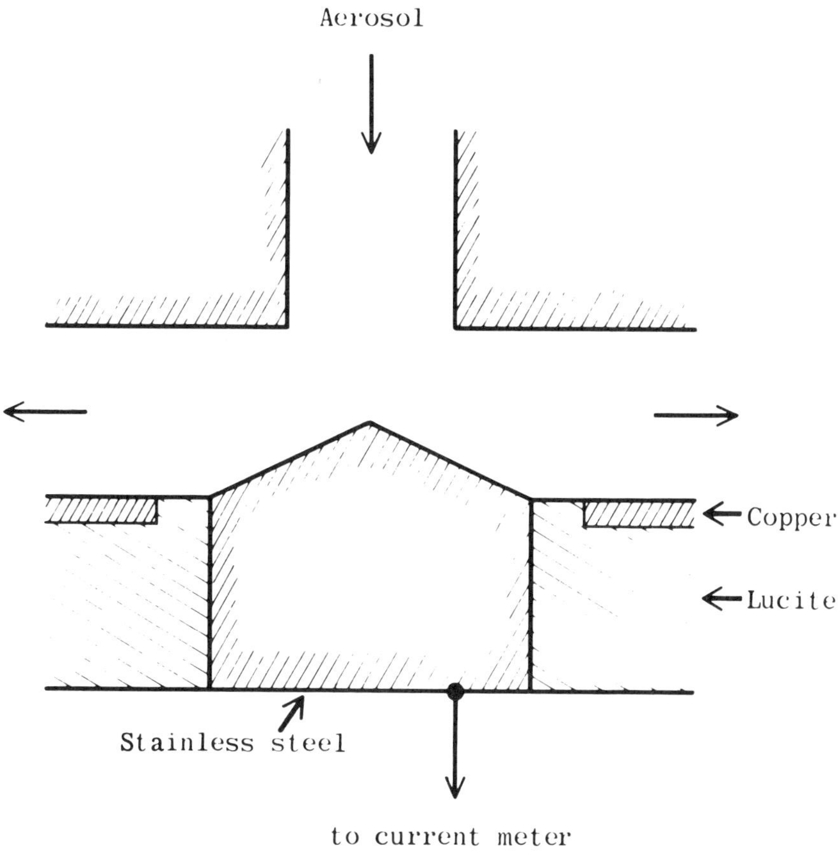

Figure 10. Impaction probe used for the charge transfer experiments.

data are consistent with a contact charge proportional to the particle diameter cubed. This implies that the total charge collected for a given mass of material will be independent of particle size. We emphasize that this result has been shown only for this ideal probe design. Other probe designs usually allow a particle to make multiple impacts. For multiple impacts we expect deviations from the cube law since the size dependence for precharge transfer is different. In some designs the impaction probability is small and dependent on particle size.

Dependence of Contact Charge on Particle Velocity

Data are shown in Figure 14 for the contact charge from methylene blue particles as a function of particle velocity. The straight lines drawn through the data for three particle sizes show that the contact charge is directly proportional to the particle velocity.

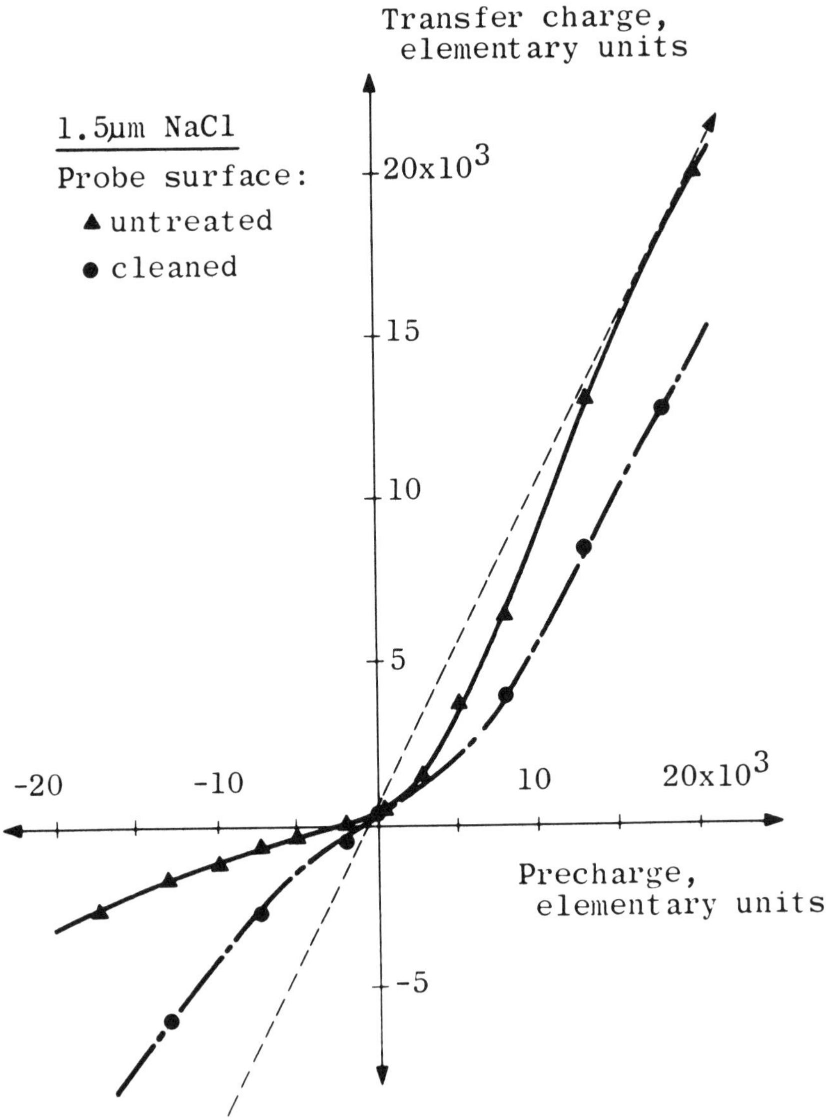

Figure 11. Charge transferred to the probe as a function of the particle precharge. The dashed line corresponds to complete charge transfer.

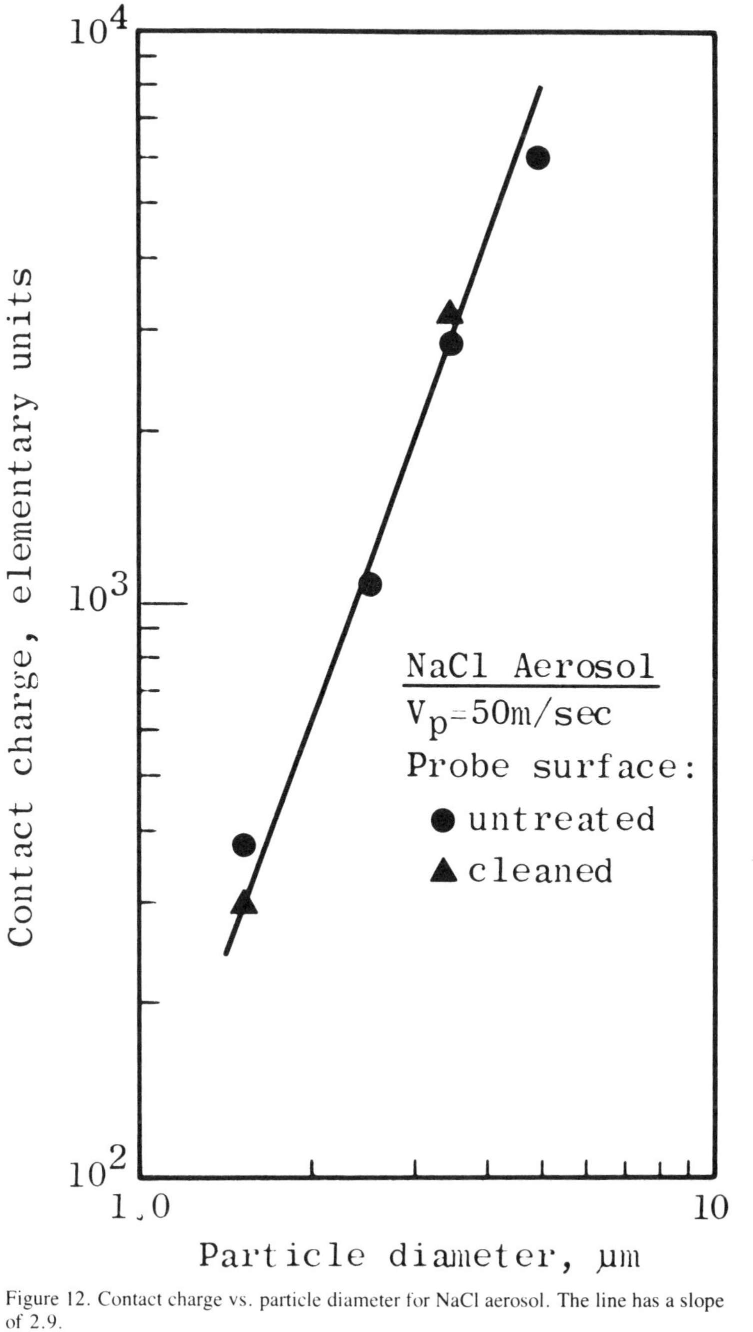

Figure 12. Contact charge vs. particle diameter for NaCl aerosol. The line has a slope of 2.9.

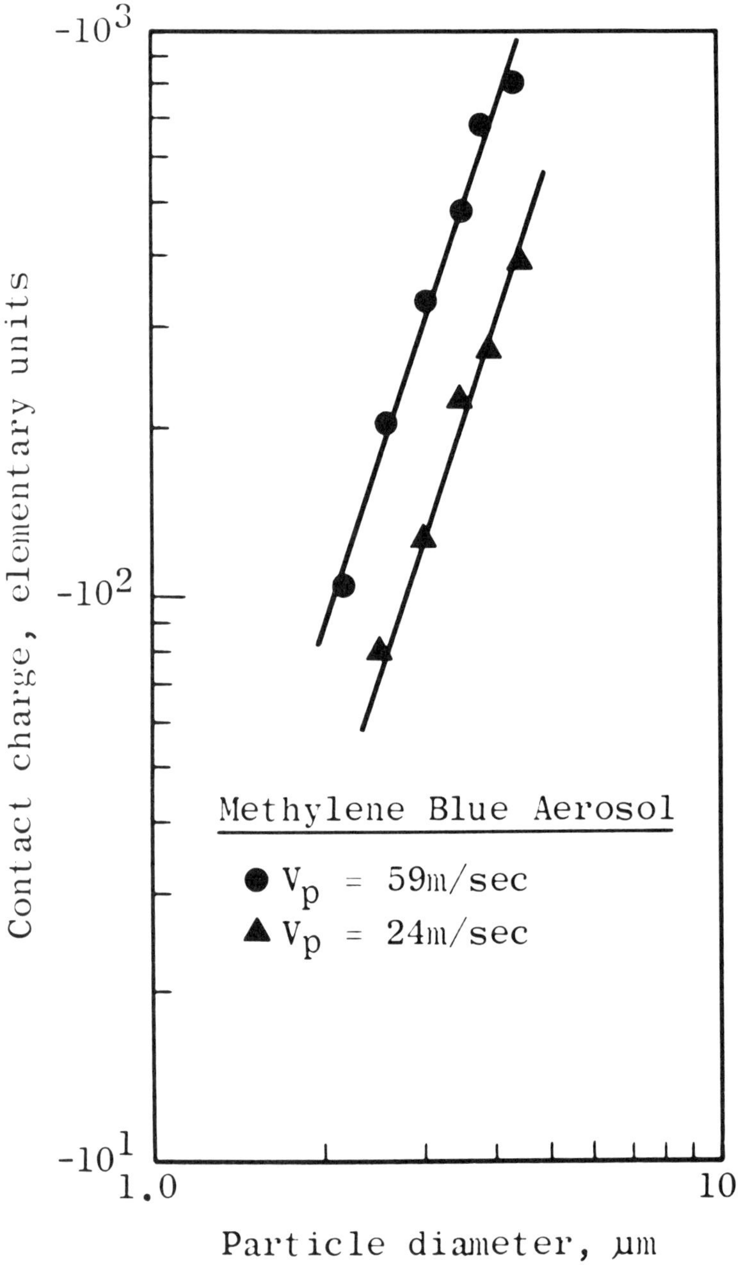

Figure 13. Contact charge vs. particle diameter for methylene blue aerosol. The lines have a slope of 3.0.

APPLICATION OF THE TECHNIQUE TO PRACTICAL MONITORING

Stationary Source Monitoring

The IKOR AQM and Kony Test instruments have been installed on a number of sources. Gruber and Bastress (1973) have described their field experience. The instruments appear to have had some success. However, there are few data

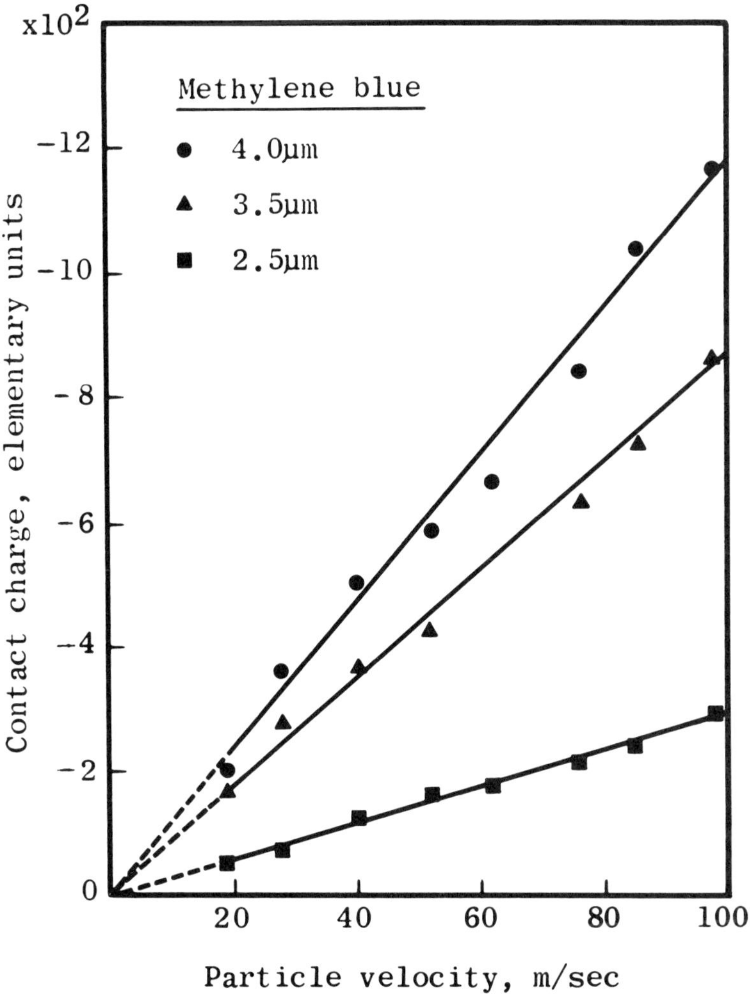

Figure 14. Contact charge vs. particle velocity for methylene blue aerosol.

available to indicate the limitations to the technique which have been encountered in practice.

On the basis of the characteristics of the monitor which have been presented above it is possible to list some precautions which should be taken to ensure proper performance of the monitor:

1. Sticky material should not be monitored since it will coat the probe and affect the charge transfer.
2. The monitor should not be used to sample material whose composition varies with time since the sensitivity and calibration of the instrument depends on the electrical resistivity of the material.
3. The aerosol should be neutralized because the monitor responds to pre-charge on the particles. The aerosol is likely to become charged as a result of mechanical and electrical processes existing within the source such as electrostatic precipitation.
4. Gaseous ions should be removed from the aerosol to avoid collecting spurious charge.
5. All parts of the monitoring system contacted by the aerosol should be constructed of metal and grounded to avoid charging effects on particles and gas exposed to surfaces of insulators.

Ambient Air Monitoring

The technique does not appear to be suitable for the monitoring of ambient air for several reasons:

1. Particulate matter in ambient air has a composition which varies with time.
2. Urban aerosol tends to be sticky.
3. The sensitivity of the field monitor is marginal for the low mass concentrations encountered. Such concentrations can be measured with a laboratory instrument. However, the measurement of such minute currents in the field does not seem to be practical.

CONCLUSION

While the monitor has shown some promise, there is need for more data obtained from operating experience and particularly, for information delineating the types of sources for which the monitor is suitable. Additional data are needed on the probe surface effect so that the choice of probe surface can be optimized. Further development of the contact charging theory is desirable to provide a sound basis for instrument design.

Acknowledgment.—We appreciate the interest and support of John Nader and Jerome J. Wesolowski.

REFERENCES

Cheng, L. and S. L. Soo, 1970: Charging of dust particles by impact. *J. Appl. Phys.*, *41*, 585–591.

———, S. K. Tung, and S. L. Soo, 1970: Electrical measurement of flow rate of pulverized coal suspension. *J. of Eng. for Power, Trans. of the ASME*, *92A*, 135–145.

Gruber, A. H. and E. K. Bastress, 1973: Application of the triboelectric effect to the measurement of airborne particles. 2nd Joint Conf. on Sensing of Environ. Pollutants, Washington, DC, Instrument Soc. of Amer., 161–168.

Ito, T., H. Saito, and N. Furuya, 1972: Continuous dust content measurement Konitest (In Japanese). *Proc. Japan Soc. of Air Poll.,* 247.

John, W., 1975: Investigation of particulate matter monitoring using contact electrification. U.S. Environmental Protection Agency Publication No. EPA-650/2-75-043. (Available from NTIS, Springfield, VA).

———, 1976: Contact electrification applied to particulate matter-monitoring. *Fine Particles. Aerosol Generation, Measurement, Sampling and Analysis,* B. Y. H. Liu, Ed., Academic Press, 649–667.

Kisler, S. Y. A., 1972: Monitoring of concentration of disperse phase of aerosol flows by electric contact method (In Russian). *Mekh. Avtomat. Proizvod., 26,* 27–28.

Prochazka, R., 1964: Neueste entwicklung des auf kontaktelektrischer basis beruhenden staubgehaltsmessgerates Konitest. *Staub-Reinhalt. Luft, 24,* 353–359.

———, 1966: Recording dust measurement with the Konitest. *Staub-Reinhalt. Luft, 26,* 22–28.

Schnitzler, H., 1970: Messtand fur die prufund und kalibrierung von registrierenden staub-und gasmessgeraten in einem steinkohlengefeuerten kraftwerk. SchrReihe Ver. Was-Boden Lufthyg, Berlin-Dahlem, Stuttgart, *33.*

Schutz, A., 1966: A recording dust-measuring instrument based on electric contact, with logarithmic indication. *Staub-Reinhalt. Luft, 26,* 18–22.

Soo, S. L., J. J. Stukel, and J. M. Hughes, 1969: Measurement of mass flow and density of aerosols in transport. *Environ. Sci. Technol., 3,* 386–393.

PERFORMANCE EVALUATION OF A NEW PIEZOELECTRIC AEROSOL SENSOR

GILMORE J. SEM

TSI Incorporated, St. Paul, Minn.

AND

MAJOR PETER S. DALEY

Air Force Civil Engineering Center, Tyndall AFB, Fla.

Instruments for the measurement of aerosol mass concentration using piezoelectric sensors have been commercially available since about 1970 (Olin and Sem, 1971; Olin, Sem and Christenson, 1971; Chuan, 1970). In 1974, Daley and Lundgren (Daley, 1974; Daley and Lundgren, 1975; Lundgren, Carter and Daley, 1976) at the University of Florida, completed an extensive evaluation of the piezoelectric aerosol measurement technique using 2 commercially available piezoelectric instruments. Sem and Tsurubayashi (1975) introduced a new instrument, based on a piezoelectric sensor, in 1975. Sem, Tsurubayashi, and Homma (1977) recently reported design improvements and additional performance data on the new instrument, called the TSI Model 3500 Piezobalance.[1] This paper discusses piezobalance performance characteristics and limitations in the light of Daley and Lundgren's evaluation of the piezoelectric aerosol measurement technique and Sem, Tsurubayashi, and Homma's piezobalance performance data.

PIEZOELECTRIC MICROBALANCE PRINCIPLES

A typical piezoelectric microbalance sensor is a disk of AT-cut quartz, about 13 mm diameter and 0.2 mm thick. The crystal oscillates at a stable resonant frequency when placed in an appropriate electrical circuit. When particles are deposited onto and adhere to the sensor, the resonant frequency decreases in direct proportion to the mass of the adhering particles. The sensitivity, S, of the piezobalance sensor is about 200 Hz/μg. The minimum detectible mass is about 5×10^{-9} g corresponding to a frequency change of 1 Hz.

A decrease in resonant frequency Δf (Hz) during sampling time Δt (sec) with a

1. PIEZOBALANCE is a registered trademark of TSI Incorporated.

constant sampling rate Q (m^3/sec) is related to aerosol concentration C ($\mu g/m^3$) by Eq. (1):

$$C = \frac{1}{SQ}\frac{\Delta f}{\Delta t}. \qquad (1)$$

The sensitivity S is calibrated for each piezobalance and, ideally, for each aerosol. Calibrated sensitivities can automatically account for the basic crystal sensitivity, aerosol collection efficiency, and particle sensing efficiency. Typically, the piezoelectric technique is applicable for particles below approximately 5 μm diameter. The size limit for any given aerosol depends upon its physical adhesive properties. Particles larger than those efficiently sensed by the crystal should be removed prior to collection because they introduce an unpredictable parameter into the measurement and may interfere with the proper operation of the electrostatic precipitator.

The nominal frequency of the quartz crystal sensor in the piezobalance is 5 MHz. This frequency is effectively reduced to a more convenient range (1–3 kHz) by electronically "beating" the sensor frequency signal against a reference crystal which has a frequency 1–3 kHz higher than the sensor frequency. The resulting "mixed" frequency is equal to the difference between the reference and sensor frequencies. The Δf in Eq. (1) then becomes the increase in mixed frequency during sampling time Δt.

The piezobalance operates with $S \simeq 200$ Hz/μg and $Q = 1$ ℓpm $= 16.7 \times 10^{-6}$ m^3/sec. In the direct concentration display mode, the sample time is either $\Delta t = 24$ sec or $\Delta t = 120$ sec. The piezobalance displays mass concentration in an xx.xx mg/m^3 format at the end of each 120 sec sample time. In the 24 sec mode, the operator must multiply the displayed concentration by 5 to obtain measured concentration. Because the piezobalance incorporates a 3.5 μm diameter cutoff impactor in the inlet, the device is sensitive primarily to particles in the "respirable" size range.

For certain applications, especially low level concentration measurements, the operator may calculate mass concentration from the frequency display (xxxx. Hz format) as described in detail by Sem, Tsurubayashi, and Homma (1977). The smallest increment of measurable frequency is 1 Hz. From Eq. (1), the minimum detectable concentration for a 2-minute sample time is 2.5μg/m^3 and for 20 minutes is 0.25 μg/m^3.

DESCRIPTION

Figure 1 is a photograph and Figure 2 is a schematic diagram of the TSI Model 3500 Piezobalance. A small pump sucks particle-containing air into the instrument at 1 ℓpm. Particles greater than 3.5 μm, the nonrespirable fraction, are removed by an impactor. Particles smaller than 3.5 μm are carried by the air to an electrostatic precipitator which deposits them onto a piezoelectric microbalance sensor which is oscillating at its natural frequency.

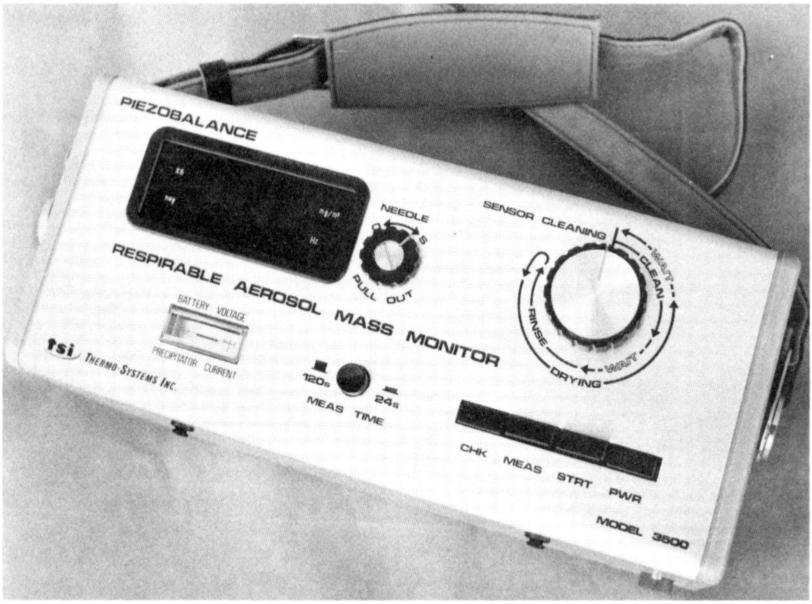

Figure 1. Photograph of the TSI model 3500 piezobalance.

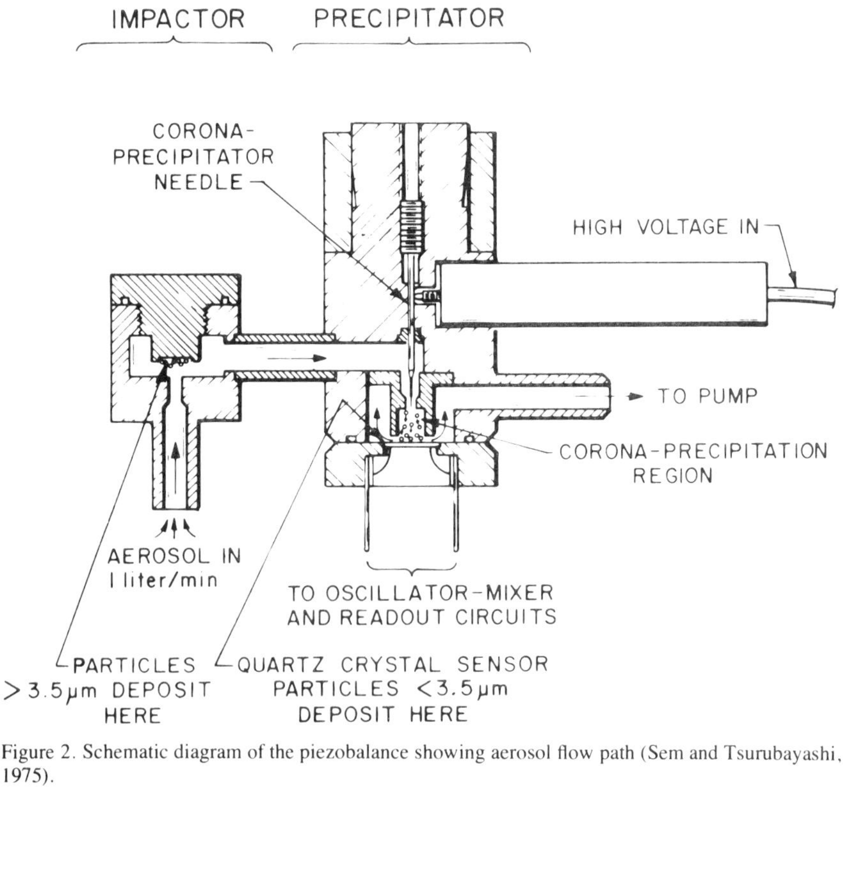

Figure 2. Schematic diagram of the piezobalance showing aerosol flow path (Sem and Tsurubayashi, 1975).

The mass of particles on the quartz crystal sensor causes the natural oscillation frequency to decrease by an amount proportional to the mass of particles. Each 10-sec frequency deviation is detected by a counter and displayed by the digital readout. In addition, each 24-second or 2-minute frequency deviation is converted to units of mg/m^3 by a conversion circuit and displayed on the digital readout.

The piezobalance is equipped with a mechanism for wiping the dust particles off the sensor when the accumulated weight exceeds an overload limit of the instrument. The cleaner operates by a manual turn of the cleaning knob.

The instrument is battery-operated; a full charge will operate the instrument for 4 hours. A carrying strap and light-weight, compact case are also included. The total weight of the device is 4.5 kg (10 pounds).

TEMPERATURE EFFECTS

Daley and Lundgren (1975) found that a temperature change rate of 1.0°C/min resulted in an error of about 5 μg/m^3 with a sensor nearly identical to the one used in the piezobalance. They suggest that temperature change rates higher than this are usually of short duration and can be eliminated by a heat sink, such as a short length of thermally conductive tubing, on the inlet aerosol stream. The piezobalance combines about 6 cm of brass tubing with a brass impactor on the inlet aerosol stream to effectively damp out short term temperature fluctuations. This reduces thermal noise and significantly increases precision.

Previous piezoelectric aerosol sensor designs have placed the reference crystal in the air stream behind the sensing crystal. It was thought that the reference crystal could then compensate for changes in air temperature, humidity, and other effects. Daley and Lundgren (1975) suggested that the reference crystal did not "see" the same air conditions as the sensing crystal on both commercial designs studied. Also, effective use of a reference crystal to compensate for temperature changes requires carefully matched crystal pairs, an expensive requirement. The reference crystal on the piezobalance has, therefore, been located in a sealed container on the oscillator circuit board where it experiences relatively constant temperature. It is used only to reduce the frequency of the signal which is measured so that a 4-digit counter and display is sufficient. The effects of temperature changes are nonetheless reduced to acceptably low levels by the aforementioned heat sink.

HUMIDITY EFFECTS

Daley and Lundgren (1975) found two errors resulting from changes in relative humidity during a measurement: (1) errors associated with the crystal and its electrode in the absence of aerosol, and (2) errors resulting from the hygroscopic nature of the aerosol deposit. The two errors are additive. The former is a function of the design of the device and the latter reflects the hygroscopic influence on aerosol mass in all mass measurement procedures.

Figure 3A shows that the error associated with the crystal and its electrodes in the absence of aerosol is not very large. The maximum specified relative humidity change rate for the piezobalance is 0.5%/measurement or 0.25%/min for the 2 min sample time, with a maximum associated error of less than 2 $\mu g/m^3$. This error would only be likely to be exceeded during outdoor measurements in unstable atmospheres.

Figure 3B shows that the errors associated with the hygroscopic nature of the aerosol deposit are potentially greater. With a relative humidity change rate of 0.25%/min, 2-min sample time, and a sensor fully loaded ($5\mu g \simeq 1000$ Hz) with moderately hygroscopic particles, the maximum expected error is about 10 $\mu g/m^3$. If the concentration of the aerosol being measured is 1.0 mg/m^3, this

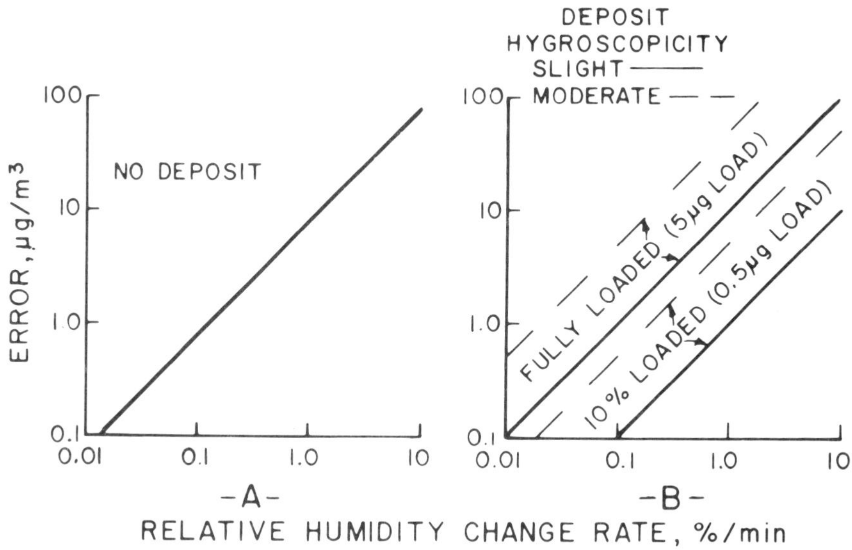

Figure 3. Errors resulting from inlet air relative humidity changes as they affect (A) the clean crystal in the absence of an aerosol deposit and (B) the mass of particles previously deposited on the crystal surface (Daley and Lundgren, 1975).

error is only 1%. However, if the aerosol concentration is 100 $\mu g/m^3$, the error is a significant 10%. The operator may reduce this error by reducing the relative humidity change rate or by reducing the sensor loading (cleaning more often). For example, if the sensor is cleaned after a loading of 0.5 μg ($\simeq$ 100 Hz), the error will be reduced by 10 times.

PARTICLE COLLECTION CHARACTERISTICS

Daley and Lundgren (1975) found that at least 20% of 0.2 μm diameter particles were collected in the inlet region of the TSI Model 3200A piezoelectric instrument with a slightly different electrostatic precipitator design. The inlet region was shortened considerably in the Model 3500 design to reduce inlet losses. Sem

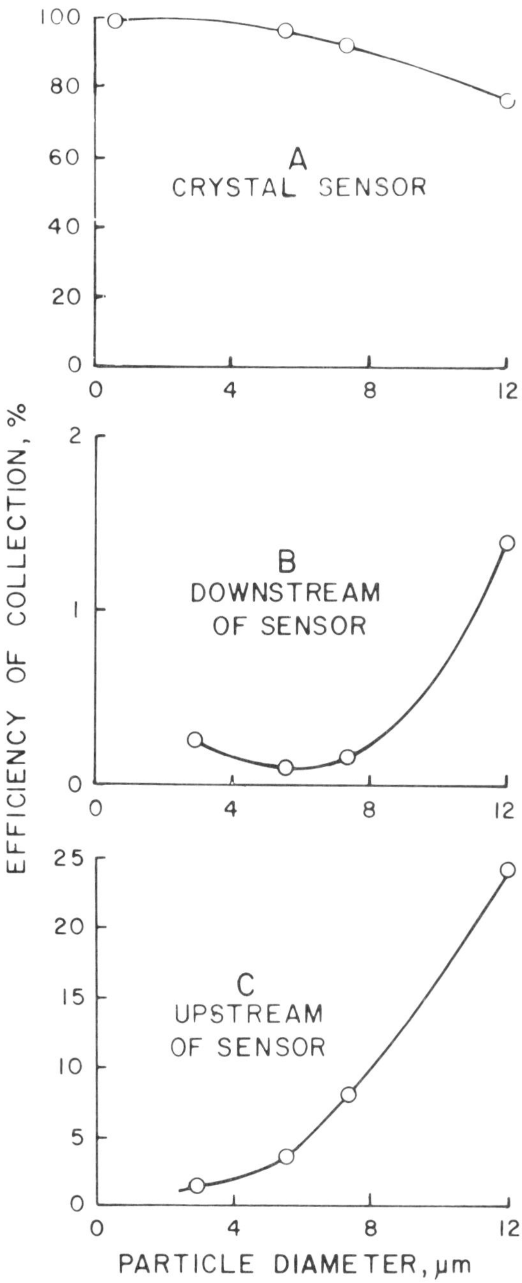

Figure 4. Collection efficiency onto the sensor, onto the precipitator body downstream of the sensor, and onto the precipitator upstream of the sensor for several particle diameters with no impactor upstream of the precipitator (Sem and Tsurubayashi, 1975).

and Tsurubayashi (1975), using the uranine tracer technique suggested by Daley and Lundgren, found >95% of the aerosol particles between 0.2 and 5.5 μm were collected on the sensor of the piezobalance, as shown in Figure 4.

To use the piezobalance as a respirable aerosol mass sensor, only respirable particles should reach the crystal. The American Conference of Governmental Industrial Hygienists (ACGIH) has defined respirable aerosols in the U.S. as any

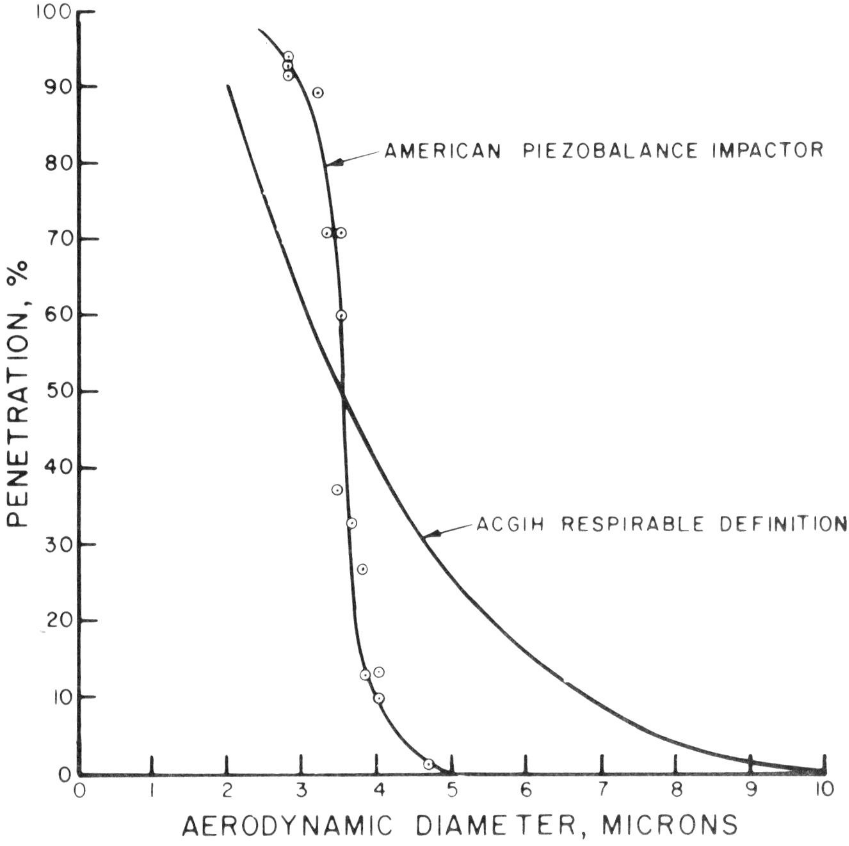

Figure 5. Penetration of unit density particles as a function of diameter for the American piezobalance compared to the ACGIH respirable definition (Sem, Tsurubayashi and Homma, 1977).

particles passing a size classifier matching the curve shown in Figure 5. Because of the unavailability of a cyclone or other particle classifier capable of matching the ACGIH curve at a flow rate of 1 ℓpm, the piezobalance was designed with an inlet impactor to remove large particles. The 50% cut diameter of the impactor is 3.5 μm, matching the 50% cut diameter of the ACGIH curve. However, Figure 5 shows that the impactor cutoff characteristics for particles larger and smaller than 3.5 μm do not match the ACGIH curve. For aerosols containing primarily small particles, such as smokes, fumes, and other combustion aerosols, the use of the impactor will cause very little error in the respirable measurement. The error may be significant if the aerosol primarily contains particles in the 2 to 5 μm diameter range, such as dusts produced by grinding and cutting processes. The piezobalance should currently be used, not as a legal compliance measurement tool, but rather, to locate problem areas where compliance measurements and/or dust control are needed. As such, the impactor upper size classifier appears adequate.

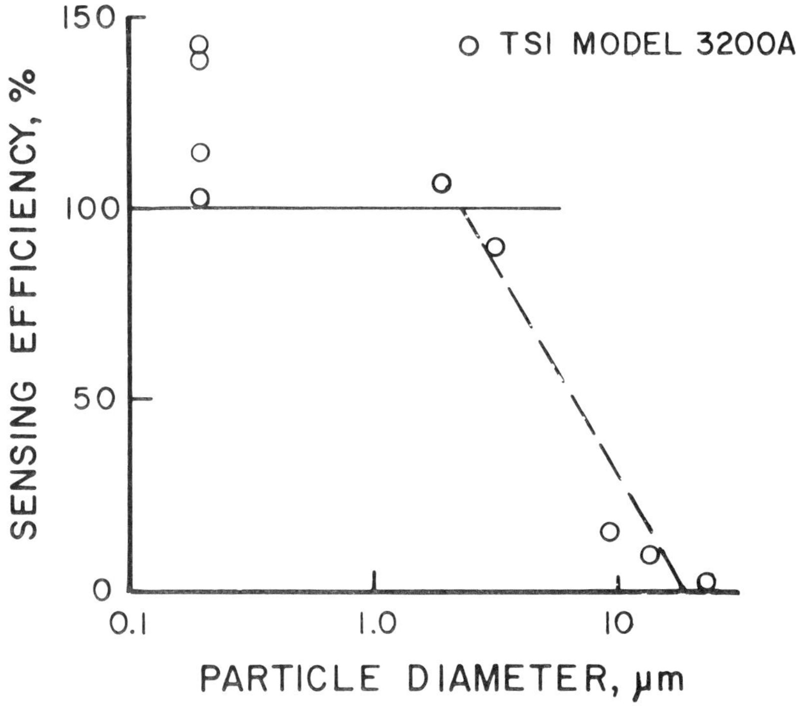

Figure 6. Ability of piezoelectric crystal to sense particles of various sizes (Daley and Lundgren, 1975).

SENSOR CHARACTERISTICS

Daley and Lundgren (1975) found that piezoelectric detectors do not sense large particles. Figure 6 shows their results for the TSI Model 3200A. The piezobalance sensor probably has similar characteristics, particularly in the size range above 3 μm. Figure 6 shows good sensing efficiency for particles below 2 μm diameter, the primary region of respirable particles.

A study reported by Sem, Tsurubayashi, and Homma (1977) using a TSI Model 3200A found 4 characteristic responses, illustrated on Figure 7, as particulate mass accumulates on the sensor. Daley and Lundgren found similar responses with a TSI Model 3200A. As aerosol particles deposit on the sensor, some types of particles cause an increase in mass sensitivity (Hz/μg) as illustrated by curve A, while other types of particles cause a decrease (curve C).

Curve B represents the ideal constant sensitivity while Curve D represents one type of artificially generated dry carbon black aerosol.

Table I lists the aerosols tested by Sem, Tsurubayashi, and Homma (1977), showing the class of response curve and the linear response limits for each aerosol. The linear response limit, illustrated on Figure 7, is defined as the sensor loading at which the mass sensitivity deviates more than ± 10% from the initial sensitivity with a clean sensor. TSI recommends cleaning the sensor before the

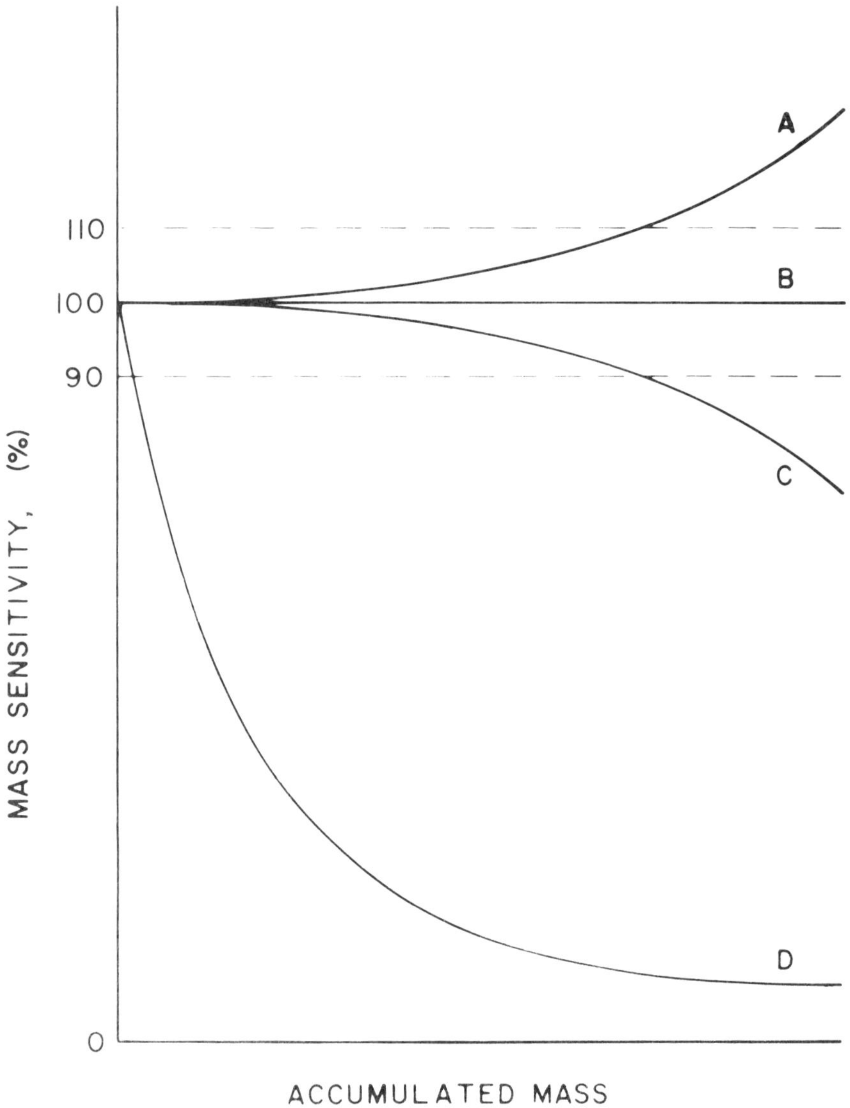

Figure 7. Mass sensitivity of a piezoelectric sensor as it becomes loaded with particulate mass (Sem, Tsurubayashi and Homma, 1977).

loading reaches 1000 Hz, well within the linear response range of all tested aerosols except a dry carbon black aerosol and a Japanese Industry Standard earth powder. Daley and Lundgren (1975) found similar linear response limits in their tests.

Table I represents a range of particle characteristics considerably greater than normally expected in U.S. industrial environments. First, the dry dispersed aerosols were tested at relative humidities $< 5\%$, causing poor adhesion of particles to the vibrating sensor. Higher relative humidity will result in better adhesion. Second, all tests reported in Table I were performed with the Japanese version of the instrument which uses an inlet impactor with a 50% cut diameter

TABLE I. Linear Response for Several Kinds of Aerosol

Aerosol	Aerosol Generation	Class of Curve	Linear Response (Hz)
Stearic acid	Evaporation and condensation[a]	A	2000
KCl		A	3000
Airborne dust		C	2000
Calcium carbonate (precipitated)	Dry dispersed	A	3000
Welding fumes	Arc welding	A	2300
Ferric oxide (hematite)	Dry dispersed	C	1400
Quartz dust	Dry dispersed	A	2500
Carbon black (Jis.-no. 12)	Dry dispersed	C	1600
Carbon black (Phil-O)	Dry dispersed	D	150
Jis. no. 8 (powder of kanto loam)	Dry dispersed	A	440
D.O.P.	Evaporation and condensation[a]	C	1200
Lead fumes	Evaporation and condensation[b]	A	3500

SOURCE: Sem, Tsurubayashi and Homma (1977).
a. LA-MER.
b. Microwave heating.

of about 8 μm. Figure 6 shows that particles in the 3–10 μm diameter range are sensed with less than 100% efficiency. Third, all tests in Table I were performed with homogeneous particles of one composition. A mixture of particulate compositions and sizes enhances adhesion to the sensor, particularly if submicron smokes and/or partially liquid particles are present. With the American 3.5 μm impactor and realistic particle compositions and relative humidity, the linear response limits are expected to be greater than shown on Table I.

The initial mass sensitivity of the aerosols in the tests reported in Table I were all repeatable within $\pm$ 10%, excluding the 2 carbon black aerosols. Thus, for relative measurements of similar aerosols, absolute field calibration may not be necessary. However, the mass sensitivities from one type of aerosol to another did vary significantly, from 115 Hz/μg for ferric oxide to 230 Hz/μg for welding smoke. This wide range probably resulted from the wide particle size range and the dryness of the particles. Thus, for absolute measurement of mass concentrations, Sem, Tsurubayashi, and Homma (1977) recommend field calibration with filter samples and describe one field calibration technique.

The reason for the differing sensitivity characteristics illustrated by Figure 7 and Table I appear to be quite complex. Sem, Tsurubayashi, and Homma have proposed a reasonable theory. They have found that, as loading on the sensor increases, the approaching particles deposit nearer the center of the electrode where the mass sensitivity is greater. Figure 8 shows results of a measurement of the deposit mass profile across a single sensor with 3 different accumulated mass loadings for $CaCO_3$. Notice that the loading at the outer edge increases much more slowly than the loading at the center. Figure 9 shows the measured mass sensitivity distribution across a sensor. The sensitivity near the center is greatest.

Sem, Tsurubayashi, and Homma (1977) propose the following theory. As mass loading on the sensor increases, particles deposit nearer the center of the electrode, possibly because increased surface resistivity of the particulate deposit deflects the corona current. In the case of characteristic Curve A (Figure 7), the later particles adhere well to the sensor, causing an increase of mass sensitivity with loading. In the case of Curve C, the buildup of particles near the center of the electrode causes progressively poorer adhesion of each particulate layer resulting in a decrease in mass sensitivity. Curve B represents an ideal situation where increased sensitivity and decreased adhesion cancel each other. Curve D occurs with carbon black, a long-chain agglomerate aerosol. The carbon black particles probably contact the sensor at 2 or 3 small surfaces with the center of mass of the particle a considerable distance above the sensor. Thus, adhesive forces are very low, causing the agglomerate to vibrate independently of the sensor. It is believed that this test is representative of dry carbon black such as found in a low humidity carbon black handling facility, but is probably not representative of carbon particles emitted from normal atmospheric combustion processes.

FILTER COMPARISON

Each piezobalance is factory calibrated with arc welding smoke using a filter comparison method described by Sem, Tsurubayashi, and Homma (1977). Figure 10 shows results of some tests comparing 8 piezobalances with filter samples while sampling arc welding smoke after being calibrated with similar smoke. Nearly all piezobalance measurements are within ± 10% of the filter measurements.

Figure 11 shows results of some filter comparison tests using other combustion-generated and oil mist aerosols. The piezobalances were first calibrated with arc welding smoke. Nearly all piezobalance measurements are within ± 15% of the filter measurements over a concentration range of 0.05 to 5.5 mg/m^3.

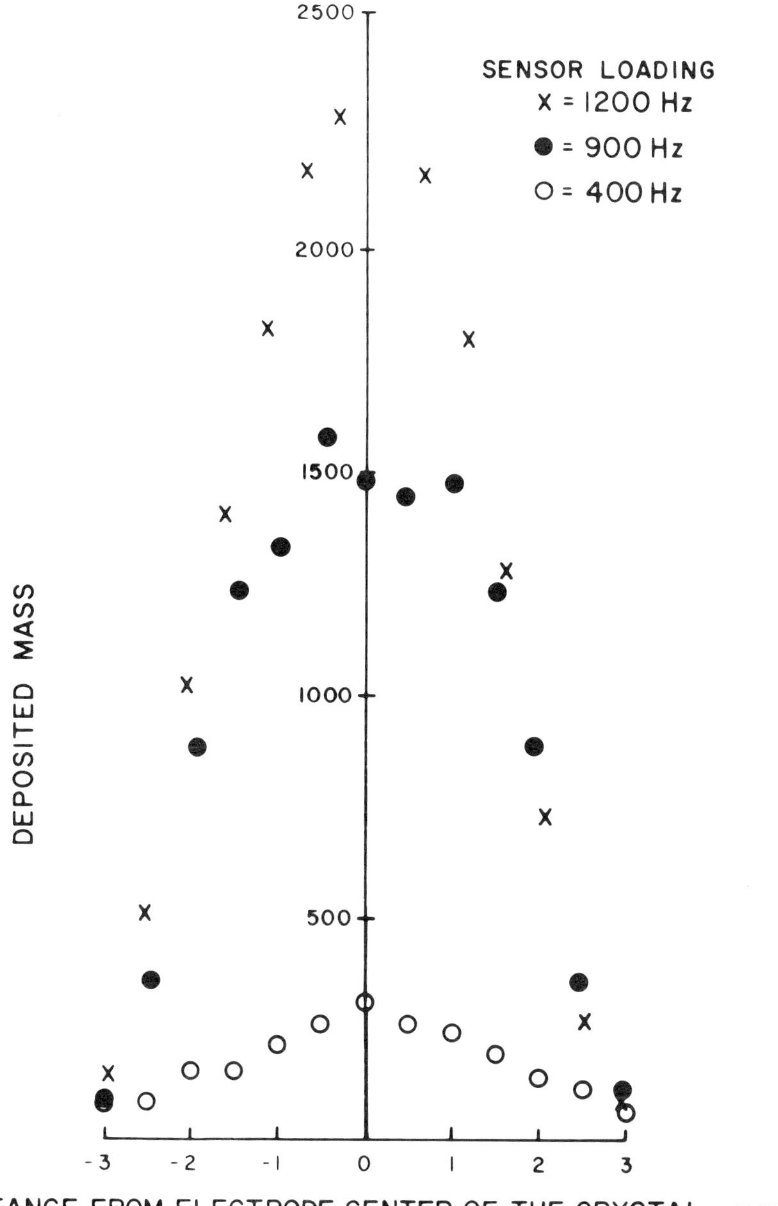

Figure 8. Distribution of $CaCO_3$ particle deposition on the piezobalance sensor. Vertical scale is in units of measured frequency (Hz) by "Kevex-Ray" technique (Sem, Tsurubayashi and Homma, 1977).

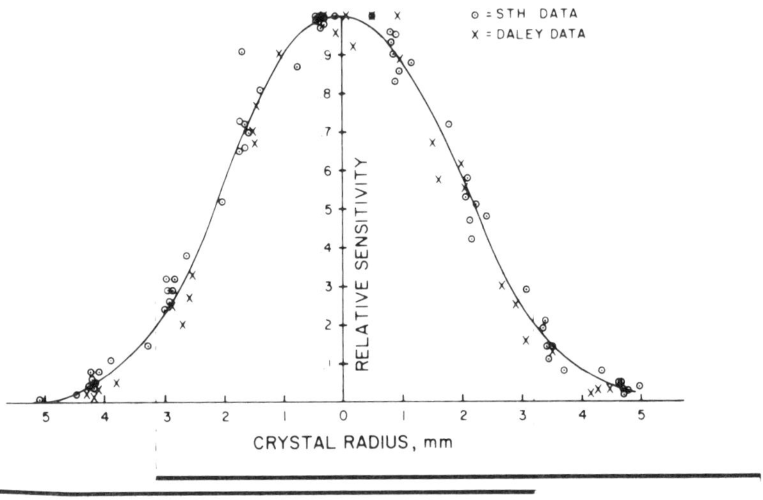

Figure 9. Mass sensitivity distribution of the piezobalance sensor (Sem, Tsurubayashi and Homma, 1977).

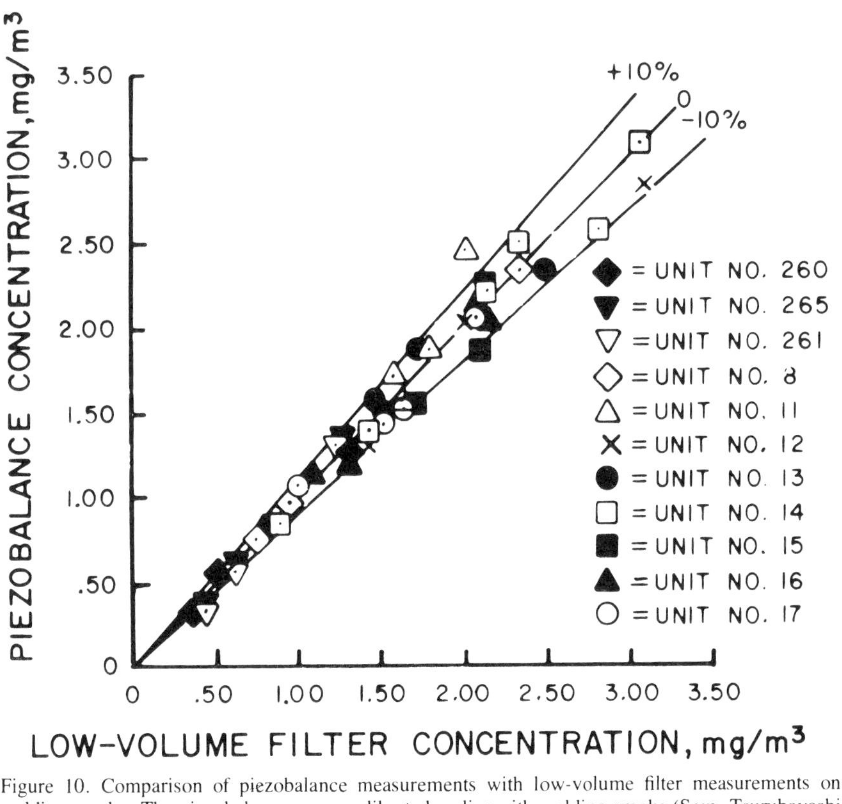

Figure 10. Comparison of piezobalance measurements with low-volume filter measurements on welding smoke. The piezobalances were calibrated earlier with welding smoke (Sem, Tsurubayashi and Homma, 1977).

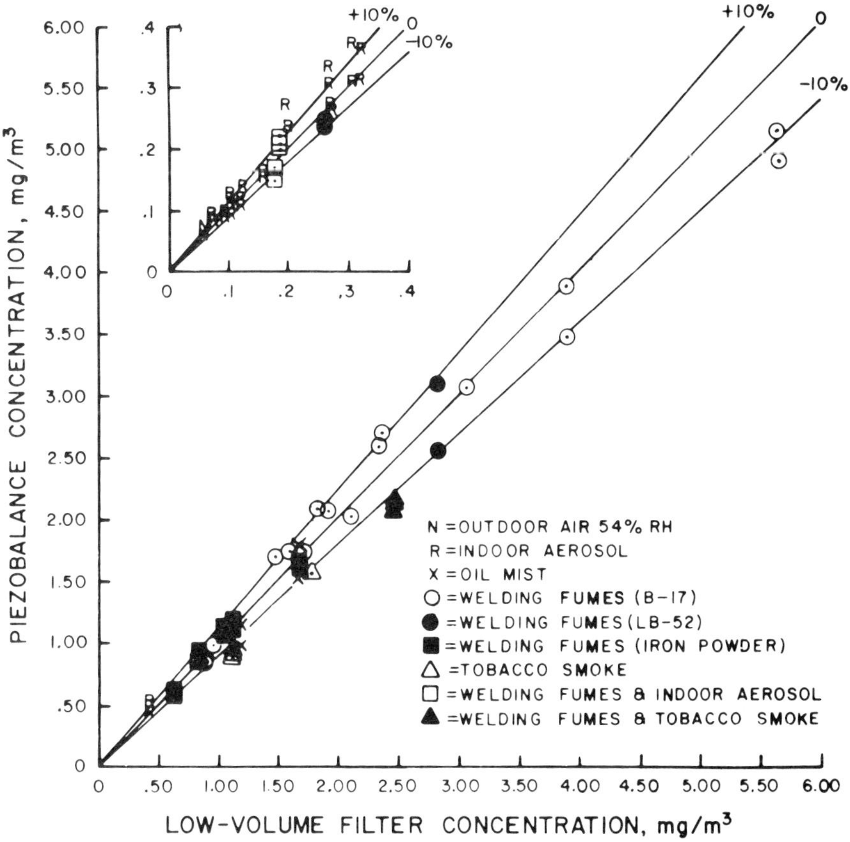

Figure 11. Comparison of piezobalance measurements with low-volume filter measurements on several aerosols. The piezobalances were calibrated earlier with welding smoke (Sem, Tsurubayashi and Homma, 1977).

CONCLUSION

The piezobalance appears to be a useful tool for respirable aerosol mass measurements in the 0.05–10 mg/m³ range when particle size is less than 3.5 μm diameter. It is especially suited for combustion smokes and fumes, but, with proper calibration, can also accurately measure other types of respirable dust. The operator should carefully avoid sensor over-loading by use of the built-in, manually-operated sensor cleaner. The built-in battery allows industrial hygiene, office smoke, and outdoor street level measurements in locations previously inaccessible to aerosol instruments.

REFERENCES

Chuan, R. L., 1970: An instrument for the direct measurement of particle mass. *Aerosol Sci.*, *1*, 111–114.

Daley, P. S., 1974: The use of piezoelectric crystals in the determination of particulate mass, Ph.D. Dissertation, University of Florida, 189 pp.

——— and D. A. Lundgren, 1975: The performance of piezoelectric crystal sensors used to determine aerosol mass concentration. *J. Amer. Industrial Hygiene Assoc.*, *36*, 518–532.

Lundgren, D. A., L. D. Carter, and P. S. Daley, 1976: Aerosol mass measurement using piezoelectric crystal sensors. *Fine Particles, Aerosol Generation, Measurement, Sampling, and Analysis*, B. Y. H. Liu, Ed., Academic Press, 485–510.

Olin, J. G. and G. J. Sem, 1971: Piezoelectric microbalance for monitoring the mass concentration of suspended particles. *Atmos. Environ.*, *5*, 653–668.

———, ———, and D. L. Christenson, 1971: Piezoelectric-electrostatic aerosol mass concentration monitor. *J. Amer. Industrial Hygiene Assoc.*, *32*, 209–220.

Sem, G. J. and K. Tsurubayashi, 1975: A new mass sensor for respirable dust measurement. *J. Amer. Industrial Hygiene Assoc.*, *36*, 791–800.

———, ———, and K. Homma, 1977: Performance of the piezoelectric microbalance respirable aerosol sensor. *J. Amer. Industrial Hygiene Assoc.*, *38*, 580–588.

NEW INSTRUMENTS IN LIGHT SCATTERING TECHNIQUES

FRANKLIN S. HARRIS, JR.*

Old Dominion University, Norfolk, Va.

Elsewhere in this volume, commercially available, widely used light scattering instruments have been discussed by Lieberman (Royco), Lepper (Climet), Petralli (TechEcology), Knollenberg (Particle Measurement Systems), and Whitby and Willeke (also on Bausch and Lomb). There have been also a large number of other systems either made available commercially, developed for special research or laboratory applications, or just proposed; the later cases are usually one of a kind. At the Workshop on which this volume is based there were ten men who had designed light scattering instruments, only parts of which were discussed. The purpose of this summary is to indicate literature references where more details can be found on other types of light scattering instruments which measure one particle at a time. The Russian (except for Bakhanova and Ivanchenko, 1973 and Filippov et al., 1974) and the Japanese instruments are not covered here because either they are not significantly different from those referenced, or detailed information is not readily available. There are other types of instruments which measure a large number of particles at the same time in a volume from which by various, often complicated techniques, a mean size or size distribution is derived. These include nephelometers, laser Doppler velocimeters, holographic, aureole, transmission and extinction and other photometers, and are not covered here.

There is one type of light scattering instrument in which detailed information on an individual particle is obtained by measuring the angular distribution of radiation scattered. Hence, information not only on size, but also on shape and refractive index is obtained. One instrument is commercially available (Phillips and Wyatt, 1972) in which the particle is suspended while the scattering as a function of angle is scanned. At least three other similar instruments have been

*Present address: Rockville, UT. 84763.

designed as a part of theses at universities. Since this type is slow, requiring several minutes per particle, the more difficult technique has been attempted in which the angular scattering distribution is scanned rapidly as the particle moves through a light beam. Such systems have been built at the Universities of Indiana (Marshall, Parmenter and Seaver) and Karlsruhe (Moser).

Some systems, such as the one constructed in the laboratory of the author (Harris et al.) in which the particle moved down the light beam to obtain uniform illumination, have not been given sufficient development to bring them to operational status. It should be kept in mind that an instrument which works well under special care in the laboratory might not be immediately useful for field operations. Even commercial instruments with years of development and application have problems as brought out in this volume. Many of the recently designed systems have used lasers for a light source for many appealing reasons, but when cost is paramount an incandescent lamp has been used with its limitations. There are many interesting and clever ideas in the systems referenced and they are representative if not all inclusive.

REFERENCES

Bakhanova, R. A. and L. V. Ivanchenko, 1973: The calibration curve of photoelectric counters and computation of particle size distribution when the relationship between particle size and electrical pulse amplitude is ambiguous. *J. Aerosol Sci., 4,* 485–490.

Bricard, J. and P. Y. Turpin, 1971: Un nouveau spectrophomètre d' aérosols. *J. Aerosol Sci., 2,* 207–217,

Cornillault, J., 1972: Particle size analyzer. *Appl. Opt., 11,* 265–268.

Filippov, V. L., V. N. Kazakov, S. O. Mirumyants, V. A. Galyashkin and L. D. Tsigler, 1974: An opto-electronic analyzer of the dispersed composition of aerosols. *Opt. Techn., 41,* 30–33.

Gebhart, J., J. Heyder, C. Roth and W. Stahlhofen, 1976: Optical aerosol size spectrometry below and above the wavelength of light—a comparison. *Fine Particles, Aerosol Generation, Measurement, Sampling and Analysis,* Benjamin Y. H. Liu, Ed., Academic Press, 793–815.

Grafendorfer, W., 1970: A classifying optical particle counter with variable adjustment of scattering angle. *Aerosol Research at the First Physics Institute of the University of Vienna.* First Physics Institute, University of Vienna.

Gravatt, C. C., 1974: Light scattering methods for the characterization of particulate matter in real time. *Aerosol Measurements,* Cassatt and Maddocks, Eds., National Bureau of Standards NBS SP 412., 21–32.

Harris, F. S., Jr., F. L. Morse, Jr. and W. C. Taliaferro, 1969: Laser photometer for particle size measurement. *J. Opt. Soc. Am., 59,* 1518.

Hoffmann, D. J., J. M. Rosen and T. J. Pepin, 1974: Global measurements of the time variations and morphology of the stratospheric aerosol. *Climatic Impact Assessment Program, Proceedings of the Third Conference, February 26–March 1, 1974.* Broderick and Hard, Eds., U.S. Department of Transportation, DOT-TSC-OST-74-15, 284–297.

Jacobi, W., 1968: Particle size spectrometry of aerosols by light scattering in a laser beam. *Staub-Reinhalt. Luft, 28,* 15–22.

Krebs, K. D., 1976: Scattering measurements of atmospheric aerosol particles in several size ranges. *Symposium on Radiation in the Atmosphere, Garmisch-Partenkirchen, 19–28 August 1976, Proceedings.* Science Press, Princeton, 71–73.

Kreikebaum, G. and F. M. Shofner, 1975: Design considerations and field performance for an *in situ* continuous fine particulate monitor based on ratio-type laser light scattering. Presented at the International Conference of Environmental Sensing and Assessment, Las Vegas, Nevada, September 14–19, 1975, and published in the *Proceedings.*

Lloyd, P. J., D. Rimberg and F. J. Haughey, 1975: A laser light scattering particle size spectrometer sensitive in the submicron diameter range. *J. Aerosol Sci., 6,* 183–189.

Marshall, T. R., C. S. Parmenter, and Mark Seaver, 1974: 360° scattering diagrams from individual aerosols in a flowing stream. *Aerosol Measurements,* Cassatt and Maddocks, Eds., National Bureau of Standards NBS SP 412, 41–56.

———, ——— and ———, 1975: Precision measurement of particulates by light scattering at optical resonance. *Science, 190,* 375–377.

McSweeney, A. and W. Rivers, 1972: Optical-fiber array for measuring radial distributions of light intensity for particle size analysis. *Appl. Opt., 11,* 2101–2102.

Moser, H. O., 1974: Instrument for measuring angular intensity distributions of light sources within a few milliseconds. *Appl. Opt., 13,* 173–176.

Neitzel, W. E., 1969: A high-volume, real-time aerosol monitor. Sandia Laboratories, Albuquerque, N.M., Report SC-DR-69-56, June (STAR N69-38396).

Phillips, D. T. and P. J. Wyatt, 1972: Single-particle light-scattering measurement: photochemical aerosols and atmospheric particulates. *Appl. Opt., 11,* 2082–2087.

Ryan, R. T., H. H. Blau, Jr., P. C. von Thüna, M. L. Cohen and G. D. Roberts, 1972: Cloud microstructure as determined by an optical cloud particle spectrometer. *J. Appl. Meteor., 11,* 149–156.

Schehl, R., S. Ergun and A. Headrick, 1973: Size spectrometry of aerosols using light scattering from the cavity of a gas laser. *Rev. Sci. Instr., 44,* 1193–1196.

She, C. Y. and P. W. Chan, 1975: Real-time particle sizing: increasing the capability of the instantaneous scattered intensity-ratio technique. *Appl. Opt., 14,* 1767–1768.

Schmitt, R. J., B. A. Boyd, R. M. F. Linford and R. G. Richmond, 1976: A laser particulate spectrometer for space simulation. *Space Simulation, Eighth Conference, Silver Spring, Maryland, November 3–5, 1975,* NASA SP 379, 647–661.

Sinclair, D. 1967: A new photometer for aerosol particle size analysis. *J. Air Poll. Control Assoc., 17,* 105–108.

PARTICULATE MASS MONITORING TECHNIQUES APPLIED TO EMISSION SOURCES

JOHN S. NADER

Environmental Sciences Research Laboratory, U.S. Environmental Protection Agency, Research Triangle Park, N.C.

Measurement of particulate mass emissions is prescribed in Federal New Source Performance Standards (NSPS) and State Implementation Plans (SIP) governing the control of emissions from stationary sources (U.S.E.P.A., 1971a, 1974a; U.S.E.P.A., 1971b). The reference measurement method given in these NSPS and SIP regulations is a manual technique that determines whether or not the source is in compliance with the required emission standards. Continuous monitoring instrumentation is also required for certain pollutant-industry combinations identified by the regulations. Continuous monitoring data provide information on the adequacy with which control equipment is operated and maintained and on the emission levels that exist at start up, shut down, and malfunction of the source and control operations (U.S.E.P.A., 1973).

One of the essential prerequisites for a monitoring requirement is the availability of a cost effective monitoring system that has been demonstrated to be reliable and accurate in its application to stationary source emissions. For example, NSPS identifies a monitoring requirement for emissions of sulfur dioxide (SO_2), nitrogen oxides (NO_x), and visible particulate matter (opacity) from coal-fired steam generating power plants. Performance specifications on the prescribed monitoring systems have been proposed (U.S.E.P.A., 1974). These specifications delineate the performance parameters which include accuracy and reliability, place performance limits on these parameters, and provide test procedures necessary to verify conformance of a candidate monitor with the specifications. Monitoring techniques and available monitoring systems were systematically evaluated to provide technical data in support of these performance specifications (Nader, Jaye and Conner, 1974).

Particulate mass monitoring has not been promulgated or proposed in NSPS or SIP as of this time. However, mass monitoring techniques are being reviewed and evaluated (Sem et al. 1971, 1973; Nader, 1976) and prototype mass monitoring systems are being developed (Lilienfeld and Cooper, 1973; Ensor and Bevan, 1973). Several mass monitoring systems are commercially available (Coles and Shak, 1973) and are being evaluated for their applicability to various source emissions. Although the status of particulate mass monitoring systems includes techniques in various stages of development (Table I), discussion will be limited

TABLE I. Status of Mass Monitoring Techniques for Particulate Matter

Analytical Scheme	Sampling Approach[a]	Development Status[b]
Electromechanical		
Gravimetric	E, electrostatic precipitation	P
Frequency Change (piezoelectric)	E, electrostatic precipitation or impaction	P
Impact Momentum	I	F
Optical		
Light attenuation	I	F
Light scatter	I	P, C
Electrical		
Ion attenuation	E	P
Charge transfer	E	C
Charge transfer	E, I	P
Nuclear		
Beta radiation attenuation	E, Filter	C
Beta radiation attenuation	E, Filter	P

a. E = extractive; I = in-situ; R = remote.
b. C = commercial; P = prototype; F = feasibility.

to source monitoring systems that are commercially available and that have been applied to source emissions (Table II).

These activities raise several questions:

What are the available monitoring techniques?

What are the source-monitor combinations that have been or are being evaluated?

What are the problem areas?

What are the promising candidate source-monitor combinations?

This paper will provide information and discussion to help answer these questions.

MASS MONITORING TECHNIQUES AND APPLICATIONS

Mass monitoring techniques are either direct measurements or indirect measurements. A direct measurement involves an analysis in which the parameter of interest is sensed directly. For example, a gravimetric measurement of mass

TABLE II. Applications of Particulate Mass Monitoring Systems to Stationary Sources

	Monitoring Techniques			
Source Industry	Piezoelectric	Beta Gauge	Transmissometry	Charge Transfer
Power Plants				
Coal	(Chuan, 1974)	(Dresia and Spohr, 1971) (Davis, 1974)	(Beautner, 1974) (Conner, 1974) (Conner, 1975)	(Gruber and Bastress, 1973) (Schnitzler, 1970)
Oil		(Jackson, 1975)	(Reisman, Gerger and Potter, 1975)	
Lignite		(Dresia and Spohr, 1971)	(Beautner, 1974) (Duwel, 1968)	
Cement plants		(Dresia and Spohr, 1971) (Jackson, 1975)	(Beautner, 1974) (Conner, 1974) (Buhne and Duwel, 1972)	(Gruber)
Incinerators		(Sem et al., 1971, 1973)	(Beautner, 1974) (Reisman, Gerger and Potter, 1975)	(Gruber)
Asphalt-concrete			(Reisman, Gerger and Potter, 1975)	
Petroleum refineries			(Reisman, Gerger and Potter, 1975)	
Iron and steel mills	(Chuan, 1974)	(Sem et al., 1971, 1973)		(Gruber)
Brass/lead smelters			(Reisman, Gerger and Potter, 1975)	
Ferro-alloy plant		(Dresia and Spohr, 1971) (Jackson, 1975)		
Kraft pulp mill			(Beautner, 1974) (Conner, 1974) (Larssen, Ensor and Pilat, 1972) (Hood and Caron, 1974)	(Gruber and Bastress, 1973)

would be a direct measurement. An indirect measurement involves an analysis of a parameter that is more or less related to the parameter of interest. For example, the optical attenuation of light by particulate matter is a factor that is relatable to the mass of particulate matter present, but additional factors enter into the relationship between them. Accordingly, an opacity measurement is potentially an indirect measurement of mass. The validity of an indirect measurement is predicated upon the uniqueness of the relationship to the measured parameter of interest. This unique relationship must be established by the particular circumstances under which the measurement is made, or by a verification that the other factors are constant or vary in a known and predictable manner.

Verification is particularly necessary for indirect methods of measurement applied to various industrial emissions. Demonstration of correlation, for example, in one industry between opacity and mass does not necessarily translate to another industry, nor to another operation within the same industry. In each instance, correlation must be established on a case by case basis.

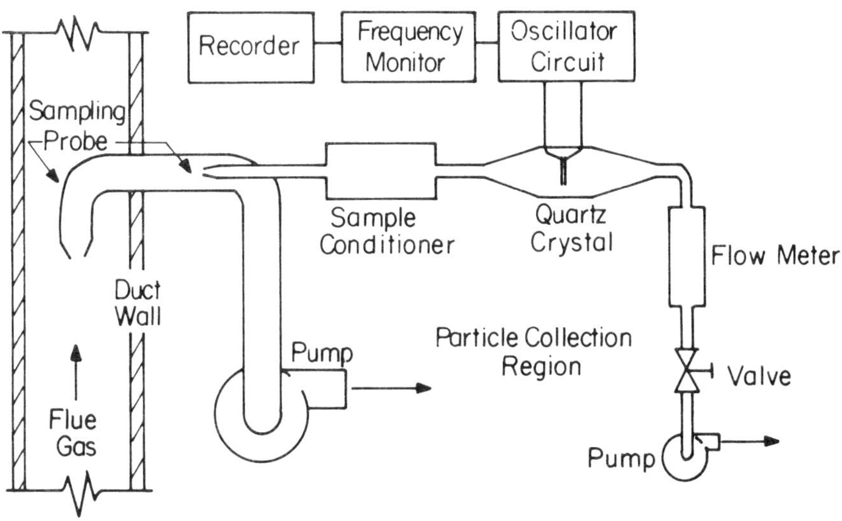

Figure 1. Schematic diagram of the piezo-electric (microbalance) system for monitoring particulate mass concentration.

Direct Methods

Electromechanical.—When a piezo-electric crystal is made part of an electronic oscillating circuit, the oscillating frequency of the circuit is determined by the natural vibrational frequency of the crystal (Figure 1). The deposition and adhesion of particulate matter on the crystal changes its vibrational frequency. The magnitude of this frequency change is directly proportional to the mass of the adhering material. Frequency changes can be measured electronically to one part in 10^7.

Electrostatic precipitation and impaction have been used to deposit particulate matter onto the crystal surface. Application of this method to a stack source involves extractive sampling and is therefore subject to particulate matter entrainment in the sampling probe. Although electrostatic precipitation has the potential for collecting particles of all sizes, in practice there is loss of efficiency for particles which are inadequately charged for the sampling conditions. When particle collection is by impaction, collection efficiency drops off significantly for particle sizes much below the impactor cut-off size.

An error is introduced in the measurement by the particles which fall off the vibrating crystal surface. Adhesion of particles to the crystal surface involves an interaction of molecular attraction forces. Particle size becomes a critical parameter in this interplay of forces such that particles above a critical size do not adhere

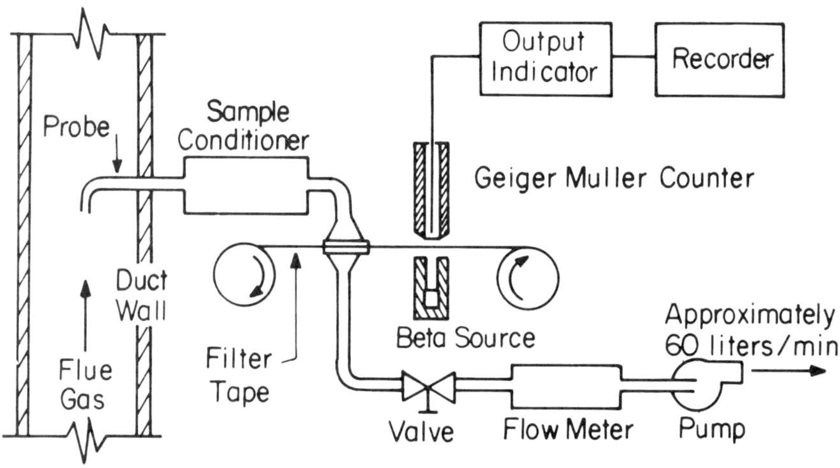

Figure 2. Schematic diagram of the beta radiation attenuation (radiometric) system for monitoring particulate mass concentration.

to the crystal. These large particles may affect the vibrational frequency of the sensor, but the effect is no longer proportional to mass.

Commercial devices of this type have been developed for ambient air monitoring, but their application to emissions from stacks has been with prototype sampling interface components and has been limited to a few special studies (Sem et al., 1971, 1973; Chuan, 1974).

Nuclear.—When beta particles impinge on matter, some are absorbed, some are scattered, and some are transmitted. The reduction in the incident beam intensity is known as beta radiation attenuation (Figure 2). The attenuation is primarily a function of the beta particle energy, the mass concentration of the attenuating matter in the radiation path, and the electron density of the matter. The electron density is the ratio of the atomic number (number of electrons per molecule) to the atomic weight (mass of the molecule). This ratio is essentially the same (between 0.45 and 0.50) for most of the elements below the very heavy ones. (Hydrogen is an exception but its presence does not normally cause a significant error.) Consequently, beta radiation attenuation is considered a direct method of mass measurement.

Both gases and particulate matter attenuate beta radiation. Therefore, in-situ application of beta attenuation as a particulate matter measuring technique is not feasible because the mass of gas molecules in a gas stream far outweighs the mass of the particulate matter, and beta attenuation methods cannot discriminate between the two. Consequently, implementation of this technique requires an extractive system in which a particulate sample is collected by one of the several available particulate matter sampling methods, and is subsequently analyzed.

The feasibility of beta radiation attenuation as an analytical technique has been demonstrated for stationary source particulate matter. Commercial beta radiation attenuation instruments (beta-gauge monitors) are available. All use a filter media for particle collection. Potential sources of sampling and analytical errors are loss of sample in the probe, varying filter thicknesses, uncertain flow measurement, losses in filter efficiency, statistics of radiation counting, and deviation from the Beer-Lambert law for attenuation.

Results reported on applications of commercially available beta-gauge monitors to various source emissions present conflicting conclusions on the accuracy of measurement.

Two first generation commercial beta-gauge instruments have been evaluated in a coal-fired power plant in comparison with manual gravimetric measurements (Sem et al., 1971, 1973). Results showed correlation coefficients of 0.94 and 0.81 for 29 and 43 measurements, respectively. The deviations at the 95% confidence level were ± 22% and ± 31% for a mass concentration value of 150 mg/m^3.

Data have been reported on experience with the beta-gauge monitor on a coal-fired and a lignite-fired power plant, a cement plant, and a ferro-alloy plant (Dresia and Spohr, 1971). Accuracy on coal-fired emissions was stated to be within 30%. The report stated that no dust deposits could be detected in the probes in the measurement of the emissions from the lignite-fired power plant. In this case the accuracy was within 16%. No information was given on how the gravimetric measurements were made. It is conceivable that the gravimetric measurements did not include cleaning but were limited to a filter catch. Under such circumstances, a favorable comparison is not surprising.

Other data on application of beta-gauge monitors to a coal-fired power plant (Davis, 1974), a cement plant, and a ferro-alloy plant, and an oil-fired power plant (Jackson, 1975), have indicated beta-gauge probe losses ranging from 30% to 86% of the total catch determined by the EPA reference method 5 for particulate matter emissions (U.S.E.P.A., 1971, 1974). The data showed an unpredictable and variable loss that tended to invalidate the assumption that the build up of particulate in the probe came to an equilibrium condition or that the probes remained clean. Correlation between the EPA reference method and the beta-gauge monitor data varied from 0.14 to 0.80.

The poor accuracy as a result of probe losses and the impact on correlation indicate the need for further development of an improved sampling interface to overcome the problems associated with transport of particulate matter in extractive monitors. A boundary layer interface system for extractive sampling of particulate emissions has been developed and will be evaluated with the beta-gauge monitor applied to various source emissions (Ranade, 1973).

Indirect Methods

Optical.—The light attenuation or transmittance measurement instruments commercially available as monitors of visible emissions (transmissometers) can be used to monitor indirectly the mass concentration of particulate emissions (Figure 3). The mass concentration of the particulate matter must be shown to correlate with the measured optical property and the calibration of the instrument system must be shown to hold constant for the emissions of a source under all operating conditions of the source. The physical characteristics of the particulate

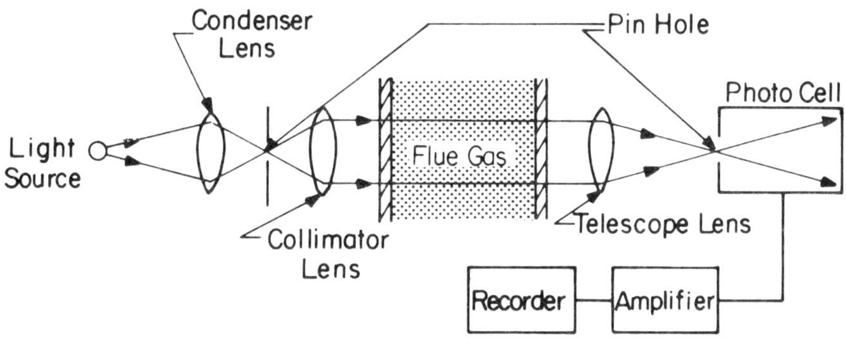

Figure 3. Schematic diagram of the light attenuation (opacity or transmittance) system for monitoring particulate mass concentration.

emissions affecting the calibration include particle shape, size, index of refraction, and density. These physical properties must not change if the calibration is to remain constant. In many source emissions the physical properties may not change significantly under normal operating conditions.

If, for a specific application, a valid correlation can be made between light attenuation and mass concentration, a light attenuation instrument can be calibrated in terms of mass concentration. Studies have been conducted to evaluate this correlation for emissions of particulate matter from a variety of industries (Table II). Generally, in such evaluations particulate mass concentration is determined gravimetrically by Method 5. These gravimetric measurements are made at a time interval concurrent with light attenuation measurements.

Data reported on emissions from a coal-fired power plant showed correlation coefficients for two different instruments of 0.63 and 0.87 for about 300 measurements; deviations at a concentration of 150 mg/m^3 were ± 68% and ± 39%, respectively, at the 95% confidence interval (Sem et al., 1971, 1973). The poorer correlation was attributed to contamination of the optical windows by the particulate matter. Additional measurements were made under minimum load operation, and operation with and without soot blowing. The results showed different correlation curves for each operating condition; one instrument performed relatively better than the other.

Correlation and performance evaluations were conducted on a coal-fired power plant over 1 three-month interval and 2 two-month intervals (January–March, July–August, October–November, 1973) representing different seasons of power plant operation (Conner, 1973). Emissions were increased at various times by cutting off one or more electrostatic precipitator stages. Correlation curves were essentially the same for the three different time periods with coefficients of 0.93, 0.98 and 0.99. Size distribution measurements were made with full electrostatic precipitator controls on, and at times when controls were partially cut back. No significant size differences were observed. Mass concentration ranged from 55 to 360 mg/m^3, corresponding to an opacity range from 3% to about 25%, respectively. The composite correlation curve for the data for all three time intervals has little scatter of points and a correlation coefficient of 0.97.

Comparative measurements at lignite-fired power plants (Duwel, 1968) gave results initially similar to those at coal-fired plants. Extended operation, however, showed that calibration at the beginning and end of a 3-month test period yielded regression lines of differing slope and parallel off-set. The sampling approach had an effect on the scatter of points about the regression line. Very little scatter was shown when gravimetric mass concentration was measured along the axis of the optical beam. Scatter for particulate mass concentration measured on a grid pattern to represent the entire duct cross-section was very great. The conclusion on reliability of performance was that monitors on lignite-fired plants were subject to dirtying of the optical surfaces within a short period of time; weekly maintenance was deemed necessary if calibration was expected to hold.

Comparative measurements were conducted on emissions from the gas duct of a rotary cement kiln with a suspension gas preheater and a subsequent rotary dryer (Buhne and Duwel, 1972). Correlation between light attenuation and mass concentration was studied for operating conditions that included normal operations, shut down and start up, and other operations with deviations. A total of 57 comparative values for all the operations gave a linear correlation with a coefficient of 0.98 for a dust loading that ranged from 10 to 450 mg/m^3 and a deviation of correlation of $\pm$ 11% at a particulate mass concentration of 150 mg/m^3. The correlations for the individual modes of operation were not significantly different from the correlation for the data from the combined modes, ranging from 99% to 100%. After 3 months of continuous operation with servicing according to operating instructions, the monitors were removed and inspected. Cleaning of the main filter (for clean purge air to keep the optical windows free of contamination) was not yet necessary. The optical boundary surfaces and the purging air attachments for the light source, reflecting mirror, and detector were also practically clean and free of dust. Performance results indicated that the zero point and the sensitivity remained constant during the 3-month interval.

An interesting study (Larssen, Ensor and Pilat, 1972) was conducted on the relationship between plume opacity and particle mass concentration measured downstream from electrostatic precipitators on a kraft pulp mill recovery furnace. The results of 18 tests of simultaneous measurements of particle mass concentration, size distribution, and light transmittance were analyzed. In the 10-day period of the tests random sampling of various operating conditions included conditions with both electrostatic precipitators on, with one precipitator on, and with both precipitators off. The range of light transmittance was from 83% to 41% corresponding to mass concentrations of about 700 to 5000 mg/m^3, respectively. The correlation coefficient was 0.99. The experimental correlation was also found to be in good agreement with the theoretical prediction based on Mie light scattering theory and measured particle size distribution. A linear multiple regression analysis of the measured data showed that variation in particle size distribution from 0.34 to 9.6 μm mass mean diameter had little effect on the correlation. This result is in agreement with predictions based on theoretical calculations. It is important to note, however, that the transmittance was measured using a device with a light source having an average wavelength of 1.2 μm as compared to 0.55 μm in the center of the visible range. Theoretical calculations show that the operating wavelength has an important role, relative to the geometric mass mean particle size and the geometric standard deviation of the size distribution, in whether or not a good correlation can be expected. Thus, it is possible, assuming other conditions are favorable, that the proper choice of operating wavelength can optimize the correlation for a given industry having particulate emissions with adequately characterized physical properties.

A continuous mass monitor has been developed to measure in situ the back-scattered radiation of a laser beam incident on the aerosol in a gas stream and to give a direct readout in terms of mass concentration. The monitor is reported to be independent of particle size distribution in the particle size range from approximately 0.05μm to 10μm diameter (Shofner et al., 1975).

Electrical.—The electrical properties of particles provide another parameter for indirect measurement of particulate mass. The ability of particles to transfer accumulated charges is a physical characteristic that is measurable with relatively simple electronic instrumentation. The electrical characteristics of particles depend on their size, shape, resistivity, dielectric constant and composition. Instruments utilizing this characteristic (Sem et al., 1971, 1973) of charge transfer must be calibrated for each specific application, and the constancy of this calibration must be validated periodically. The theory of operation of these instruments has not been adequately developed, but limited empirical data are available to judge instrument performance.

The charge transfer technique consists of the transfer of electrical charge between the particles in the sample stream and a sensing element when they come into physical contact with each other (Figure 4). The transfer of charge by physical contact of dissimilar materials is identified with the "triboelectric effect." Charge transfer is measured as flow of electrical current from the sensor element by appropriate electronic circuiting. The composition and surface properties of the particles and the sensor, and the flow characteristics of the sample stream, determine the probability and nature of physical contact between the particles and the sensor element, hence the electrical signal that is generated and the correlation of this signal with particulate mass flow. These interdependencies are poorly understood but are being investigated (John, 1975).

A German instrument (Konitest, no longer manufactured) samples the particles from a stack and introduces swirling motion to the air stream in the sensor by means of a tangential inlet duct (Schutz, 1966). Centrifugal force throws the particles against the sensor tube wall where the electrical charge is generated. A later Konitest design used a venturi nozzle instead of a cyclone as the sensing element. This instrument gave good correlation with mass concentration when applied to a power plant stack. Over a range of 0.1 to 2.2 gm/m^3, its response was linear, with a signal output of 7.4×10^{-9} amperes per 100 mg/m^3 of dust

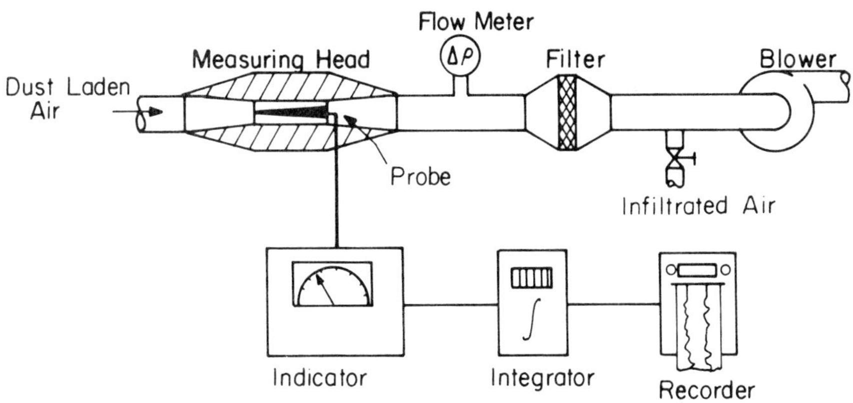

Figure 4. Schematic diagram of the charge transfer system for monitoring particulate mass concentration.

concentration. A test program evaluating the Konitest on the emission of a coal-fired power plant (Schnitzler, 1970) consisting of 312 measurements, gave a correlation coefficient of 92% and a deviation, at the 95% confidence level, of ± 33% at a mass concentration of 150 mg/m^3.

Problems potential to the application of the Konitest to emissions from coal- or oil-fired combustion facilities are: long-term contamination of the sensing tube; sensitivity to submicron particles which do not contribute proportionately to mass; change in calibration due to changes in particle composition.

A contact charge transfer instrument which senses particles passing through a venturi nozzle with a conical metal probe in its throat has been developed. A transfer of charge takes place when particles hit the probe at velocities above 100 m/sec. A permanent installation is available as well as a portable unit (Gruber and Bastress, 1973). The portable system has been utilized in the laboratory and in industrial facilities including coal, oil, and incinerator flyash, metal oxides, auto and jet engine exhausts. The monitor has a range of five decades (10^5 to 1). In a coal-burning combustion source, this range corresponds to a mass concentration of 1 to 10^5 mg/m^3. Its lower limit on sensitivity has not been fully explored. Recalibration of the instrument is necessary if substantial changes in particle size distribution, particularly in the submicron size, occurs. Its response with changes in particulate matter composition can vary sufficiently to require recalibration.

DISCUSSION AND CONCLUSIONS

Two direct and two indirect methods of continuous monitoring of particulate emissions are reviewed as commercially available systems that have been applied to some source emissions. The beta-gauge monitor (direct) and the opacity monitor (indirect) have had the greater utilization to date compared to the piezo-electric monitor (direct) and the charge transfer monitor (indirect). A significant consideration is the relative cost of these systems. The beta-gauge system is in the range of \$14,000 to \$20,000. The other three systems range from \$4,000 to \$8,000.

At present the beta-gauge is available only as an extractive system and appears to have a sampling interface problem that will require further investigation. An in situ beta-gauge system is presently in the early stages of development as a research prototype for size distribution studies. The attendant problems are the hostile source environment of temperature and some form of particulate collection within the gas stream.

The opacity system is an in-situ technique and has the advantage of avoiding the problems of extractive sampling. However, it suffers an interference from any water vapor that is in the form of condensed liquid particles. This situation emerges as wet scrubbers are introduced to control emissions of pollutant gases. For such applications, effort is underway to develop an opacity system that incorporates an extractive system for sample conditioning prior to analysis.

The piezo-electric monitor is available commercially but not as a complete system with a sampling interface. The technique as presently executed is not equally sensitive in its detection of particles over the size range of interest. The evaluation of its applicability to source emissions has been very limited, both in types of sources and in the configuration of a complete commercially available system.

The charge transfer system appears to have promise. Some in situ prototype developments are in the early stages as commercially available systems. Experience with application to various source emissions and evaluation of performance will require some time for appropriate assessment.

Transmissometry (opacity system) has had the maximum exposure and experience to date. Empirical data indicate it as a viable technique applicable to a number of source emissions. The technique as a commercial system has been optimized in its design and execution for measurement of opacity as an optical property of particulate matter. There is the option to optimize the design and execution of the technique specifically as an indirect measure of particulate mass.

The data reported in the literature have been obtained under a variety of test conditions such that comparison of data is difficult and conclusive assessment of performance would be premature. In any event, particulate mass monitoring systems need to be evaluated in the context of a prescribed set of performance requirements for application to specified sources. Experience to date should provide the technical base for the development of performance specifications and test procedures that can be used to properly evaluate performance capabilities of these systems at a future date.

REFERENCES

Beautner, H. P., 1974: Measurement of opacity and particulate emissions with the Lear Siegler on-stack transmissometer. *J. Air Pollution Control Assoc., 24,* 865.

Buhne, K. W. and L. Duwel, 1972: Recording dust emission measurements in the cement industry with the RM4 smoke density meter made by Messrs. Sick. *Staub-Reinhalt. Luft, 32,* 19.

Chuan, R. L., 1974: Application of an oscillating quartz crystal to measure the mass of suspended particulate matter. *Analytical Methods Applied to Air Pollution Measurements,* R. K. Stevens and W. F. Herget, Eds., Ann Arbor Science Publications, Inc., Ann Arbor, Michigan, 163–193.

Coles, R. R. and B. J. J. Shak, 1973: The measurement of particulate immissions and emissions. Report No. CEL/CH/29/73, Corporate Engineering Laboratory, British Steel Corporation, London England, 41 pp.

Conner, W. D., 1974: Measurement of the opacity and mass concentration of particulate emissions by transmissometry. EPA-650/2-74-128, United States Environmental Protection Agency, Research Triangle Park, North Carolina, 31 pp.

———, 1975: Personal communication. United States Environmental Protection Agency, Research Triangle Park, North Carolina.

Davis, J. W., 1974: in *Analytical Methods Applied to Air Pollution Measurement,* R. K. Stevens and W. F. Herget, eds., Ann Arbor Science Publications, Inc., Ann Arbor, Michigan, 285–297.

Dresia, H. and F. Spohr, 1971: Experience with the radiometric dust measuring unit 'beta staubmeter'. *Staub-Reinhalt. Luft, 31,* 19.

Duwel, L., 1968: Latest state of development of control instruments for the continuous monitoring of dust emissions. *Staub-Reinhalt. Luft* (English translation), *28,* 42.

Ensor, D. S. and L. D. Bevan, 1973: Application of nephelometry to the monitoring of air pollution sources. Paper No. 73-AP-14 presented at annual meeting of Air Pollution Control Assoc., Pacific Northwest Section, Seattle, Washington.

Gruber, A. H. and E. K. Bastress, 1973: A comparison of a new continuous electric charge particulate monitor with the EPA sampling train. Proceedings of the Technical Association of Pulp and Paper Industries Engineering Conference, Boston, Massachusetts.

———, Personal communication. Ikor, Inc., Burlington, Massachusetts.

Hood, K. T. and A. L. Caron, 1974: The relationship between particulate mass emission rate and observed plume appearance from kraft recovery furnaces. Paper No. 74-AP-08, Pacific Northwest Section Meeting of the Air Pollution Control Assoc., Boise, Idaho.

Jackson, M. R., 1975: Evaluation of field prototype beta gauges. Final Report on Contract No. 68-02-1216, in press, United States Environmental Protection Agency, Research Triangle Park, North Carolina.

John, W., 1975: Investigation of particulate matter monitoring using contact electrification. EPA-650/2-75-043, United States Environmental Protection Agency, Research Triangle Park, North Carolina.

Larssen, S., D. S. Ensor and M. J. Pilat, 1972: Relationship of plume opacity to the properties of particulates emitted from kraft recovery furnaces. *J. Tech. Assoc. of Pulp and Paper Industry, 55,* 88.

Lilienfeld, P. and D. Cooper, 1973: Literature review and selection of design principle for the development of a fine particulate source testing instrument. Interim report on EPA Contract No. 68-02-1314 with GCA Corporation, United States Environmental Protection Agency, Research Triangle Park, North Carolina.

Nader, J. S., F. Jaye and W. Conner, 1974: Performance specifications for stationary-source monitoring systems for gas and visible emissions. EPA-650/2-74-013, United States Environmental Protection Agency, Research Triangle Park, North Carolina, 82 pp.

———, 1976: Source monitoring. *Air Pollution,* A. C. Stern, Ed., 3rd ed., Volume III, Academic Press, Inc., New York, 589–645.

Ranade, M. B., 1973: Sampling interface for quantitative transport of aerosols. EPA-650–2-74-016, United States Environmental Protection Agency, Washington, D.C.

Reisman, E., W. D. Gerger, and N. D. Potter, 1975: In-stack transmissometer measurement of particulate opacity and mass concentration. EPA-650/2-74-120, United States Environmental Protection Agency, Research Triangle Park, North Carolina.

Schnitzler, H., 1970: *Schreihe ver Wass.-Boden Lufthyg. Berlin-Dahlem, 33,* Stuttgart.

Schutz, A., 1966: Possibilities for recorded dust measurement and control. *Staub-Reinhalt. Luft,* (English translation), *26,* 1.

Sem, G. J., J. A. Borgos, J. G. Olin, J. P. Pilney, B. Y. H. Liu, N. Barsic, K. T. Whitby and F. D. Dorman, 1971: State of the art: 1971 instrumentation for measurement of particulate emissions from combustion sources: Vol. I, Particulate mass-summary report, NTIS PB-202-655, 194 pp.; Vol. II, Particulate mass-detail report, NTIS PB-202-666, 225 pp.; 1973: Vol. IV, Experiments and final report, NTIS PB-231-919/AS, 121 pp., National Technical Information Service, Springfield, Virginia.

Shofner, F. M., G. Kreikebaum and H. W. Schmidtt, 1975: In situ continuous measurement of particulate mass concentration. Preprint 75-41.1, 68th Annual Meeting of the Air Pollution Control Association, Boston, MA.

United States Environmental Protection Agency, 1971a: Standards of performance for new sources, Federal Register *36,* 24876; 1974: *39,* 9308; 1974a: *39,* 37040; 1974: *39,* 37470.

———, 1971b: Requirements for preparation, adoption, and submittal of implementation plans. Federal Register, *36,* 15486.

———, 1973: Emissions during startup, shutdown, and malfunction. Federal Register, *38,* 28564.

———, 1974b: Proposed emission monitoring and performance testing requirements—stationary sources. Federal Register, *39,* 32852.

SUBJECT INDEX

AUTHOR INDEX